Basic Exploration Geophysics

Basic Exploration Geophysics

EDWIN S. ROBINSON
CAHIT ÇORUH

Virginia Polytechnic Institute and State University

JOHN WILEY & SONS
New York Chichester Brisbane
Toronto Singapore

Cover photograph by *Peter Neumann*

Original drawings prepared by *Kathryn A. Hawkins*

Library of Congress Cataloging in Publication Data:

Robinson, Edwin S.
Basic exploration geophysics.

Bibliography: p.
1. Prospecting—Geophysical methods. I. Çoruh,
Cahit. II. Title.
TN269.R54 1988 622'.15 88-10610
ISBN 0-471-87941-X

Printed in the United States of America

10 9 8 7 6 5 4 3

Printed and bound by Malloy Lithographing, Inc.

To Valarie and Dilek

PREFACE

This book presents a thorough introduction to the geophysical methods used to explore for natural resources and to survey earth structure for purposes of geology and engineering knowledge. These methods include seismic refraction and reflection surveying, gravity and magnetic field surveying, electrical resistivity and electromagnetic field surveying, and geophysical well logging. These topics are treated at a level suitable for a one-semester course offered to second- and third-year college students majoring in geology or engineering, and for professional geologists, engineers, and other scientists without prior formal instruction in geophysics. To understand the discussions, the reader should be competent in geometry, algebra, and trigonometry. A background in first-year college geology and physics is recommended.

Topics in the book make up a balanced discussion of modern field procedures and instruments, data processing methods, and the important aspects of interpretation. All basic surveying operations are described step by step and are further illustrated by practical examples. The important computer-based methods of processing and interpretation, as well as graphical methods, are introduced.

Although the topics in this book are basically the same as those found in other introductions to exploration geophysics, several important perspectives that depart from the conventional treatments are developed. We have made more extensive use of geometrical features of seismic wave paths and wave fronts to explain the relations between refracted and reflected waves and earth structure. By means of these geometrical features, the mathematical developments can be more fully coordinated with diagrams that illustrate the important physical processes. Expressions can be verified by inspection of these diagrams without extensive algebraic or trigonometric manipulation.

Modern seismic reflection surveying requires extensive computer processing to obtain meaningful data displays. In this book, each of the basic steps of data processing is presented in some detail, and the individual computer operations are explained by graphs and simple equations. No prior knowledge of the process of convolution or basic filtering operations is required to understand this presentation.

Considerable effort is devoted to explaining the physical meaning of gravity anomalies. A measurement of gravity is viewed as a value to be explained by comparing it with theoretical values calculated for idealized earth models. The conventional adjustments for effects of latitude, elevation, and mass excesses or deficiencies are applied to the theoretical value of gravity rather than to the measured value when calculating gravity anomalies. Although the result is numerically the same as that obtained by the traditional geodetic approach of adjusting the measured value, it is placed in a different perspective, one more appropriate for geologic and exploration purposes.

In the discussion of magnetic anomaly interpretation, the analysis of total-intensity fields is emphasized rather than the vertical-intensity component that is stressed in most introductory discussions. Much attention is given to how different field components are calculated and then combined to obtain total-intensity anomalies. The reason for this shift in emphasis from vertical- to total-intensity anomalies is that most modern magnetic field surveying is done with airborne total-intensity magnetometers.

Two aspects of electrical surveying have not

previously been stressed in introductions to exploration geophysics. The first is the geometrical analysis of the electrical resistance of a hemisphere surrounding an electrode, which is used to develop the conventional resistivity expressions without reference to integral calculus. The second aspect is the geometrical analysis of current density, which is used to estimate the relative effects of different zones in the earth on electrical measurements.

Modern geophysical instruments are described in some detail. Scientific principles underlying the instrument design and practical aspects of field operations are both stressed.

A number of persons have helped us in many ways to complete this book. Our friend and colleague John Costain enthusiastically participated in many discussions about effective ways of presenting various topics of exploration geophysics, and he has shared his ideas freely with us. He assembled much of the information used in Chapter 5 and assisted with the writing of that chapter. The book contains nearly 300 original drawings that were done by Kathryn Hawkins. The original manuscript was typed entirely by Marjorie Sentelle, who also helped in many ways with later editing and proofreading. Llyn Sharp prepared several original photographs. The expert editorial advice of Robert McConnin and the late Donald Deneck saw us through the earlier stages of manuscript preparation. We very much appreciate the guidance of our editor Clifford Mills, who helped with the revision and improvement of the manuscript and arranged for its publication. The careful copyediting of Priscilla Todd was very important in improving the clarity and accuracy of the manuscript. We gratefully acknowledge the efforts of Katharine Rubin and Joseph Ford for their day by day supervision and management of the assembly and production of this book. We thank our wives, Valarie Robinson and Dilek Çoruh, for their patience and encouragement.

Blacksburg, Virginia
April 1988

Edwin S. Robinson
Cahit Çoruh

CONTENTS

Chapter 4 REFLECTED SEISMIC WAVES AND EARTH STRUCTURE 81

Chapter 5 SEISMIC SURVEYING 117

Chapter 6 SEISMIC REFLECTION DATA PROCESSING AND INTERPRETATION 163

Chapter 10 EARTH MAGNETISM *333*

Chapter 11 SURVEYING THE ANOMALOUS MAGNETIC FIELD *375*

Chapter 12 MAGNETIC ANOMALIES AND THEIR GEOLOGIC SOURCES *401*

Chapter 13 GEOELECTRICAL SURVEYING *445*

ONE

The Search

A little more than a century ago, the search for oil began. At first, no one knew much about where to look for it, and so landowners and speculators eagerly sought advice. In these early days, a colorful assortment of individuals who claimed to have special oil-finding powers offered a helping hand. The first to come forward used the forked branches that for centuries had been the tools for locating water wells, but within a few years their clients demanded more impressive equipment. Wondrous devices were invented that supposedly pointed to the hidden treasure of oil with crackling discharges of electricity, mysterious liquids that changed color, or various peculiar sounds. Most of these devices proved to be useless. Skeptics started calling them doodlebugs, and their inventors became known as "doodlebuggers."

From these beginnings emerged a new breed of doodlebugger whose prospecting methods were based on scientific principle rather than witchcraft. These individuals began piecing together information about geologic environments favoring the occurrence of oil. In some places, the rock exposures and landscape features guided the search for anticlines and other structures in which oil might have become trapped. But what about places like the Gulf coast of Texas and Louisiana where the landscape of low plains and swamps gave scant information of the underlying geology? Could instruments be designed to detect the structures of interest?

Early in the twentieth century, seismologists recognized the value of earthquake wave vibrations for probing the deep interior of the earth. Could these kinds of vibrations be used in the search for oil-bearing structures? At the same time, geodesists studying the shape of the earth realized that small variations in the strength of the earth's gravity were related to differences in the densities of underlying masses of rock. *Density* is a physical property found by dividing the mass of a rock specimen by its volume. Perhaps instruments sensitive to these variations in gravity could be designed for oil prospecting.

Other kinds of instruments had been used to prospect for ores long before the birth of the oil industry. For more than two centuries magnetic devices, basically compasses, were used in the search for iron ore. Natural electrical fields produced by buried sulfide ores were first detected early in the nineteenth century.

Seismic waves, gravity, magnetism, and electrical fields in the earth are the foundation of modern *exploration geophysics.* The aim of exploration geophysics is the discovery of hidden geologic features by indirect methods. These methods involve measurements made some distance away from the feature of interest, which may be an oil trap, an ore body, or a structure fundamental to an understanding of the geology of some region. Modern electronics and computer technology have greatly enhanced the quality of instruments now used in exploration geophysics. Let us look more closely at the different kinds of modern doodlebugging.

EXPLORATION SEISMOLOGY

Our knowledge of the earth's deep interior has been largely conveyed by earthquakes, most of which are caused by the sudden movement of rock masses along a fault. As these rocks grind together, energy is released and produces vibrations which we call *seismic waves.* These waves spread throughout the earth like the ripples made by a pebble tossed into a quiet pond. Eventually, these seismic waves reach the earth's surface where they can be detected by instruments sensitive to ground vibrations. These instruments are called *seismometers.*

Earthquakes release the large amounts of energy needed to probe the deep mantle and core of the earth. But there are other ways to produce seismic waves that can be focused on geologic features closer to the earth's surface. These waves can be generated by explosions and then recorded on small seismometers, about the size of the human fist, which are placed nearby. This work is done by exploration seismologists who know how to control the paths of the seismic waves by the locations of explosives and seismometers. Their aim is to measure the *speed* of a seismic wave along different parts of its path in the earth. The speed changes as the wave moves from one kind of rock into another, depending on the physical properties of these materials.

Seismic wave speeds cover a large range of values in different kinds of rock and loose sediment. These values are most commonly given in the Système International (SI) units of meters per second (m/s), or in the non-SI units of kilometers per second (km/s) or feet per second (ft/s). For example, a wave moving 5000 m/s (16,400 ft/s) in sandstone may increase its speed to 6000 m/s (19,680 ft/s) in limestone. But these values are not the same for all layers of sandstone and limestone. The ranges of wave speed in different kinds of rock and the dependence of wave speed on the physical properties of rocks are discussed more fully in Chapter 2. It is important to point out that seismologists often interchange the terms *speed* and *velocity*. By strict definition, we should state the speed and direction of movement to describe velocity completely, but in common usage the direction is frequently omitted.

A seismic survey is usually conducted by placing seismometers along a straight line and then detonating an explosive close to one end. If the rock layers are horizontal or gently dipping, seismic waves follow uncomplicated paths. The two kinds of paths illustrated in Figure 1–1 are followed by *reflected seismic waves* and *refracted seismic waves*. The reflected waves have traveled downward to borders between rock layers where they bounce or echo back to the surface. In contrast, the refracted waves follow paths that bend at each border.

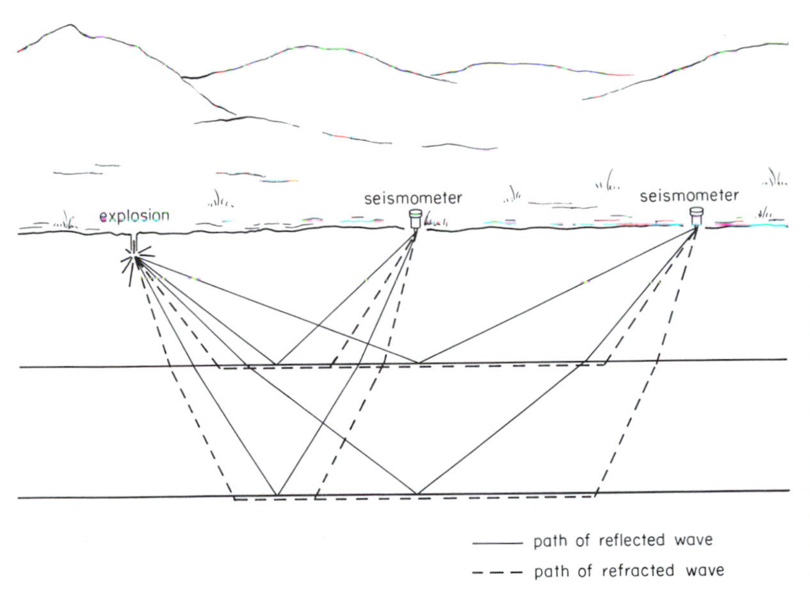

Figure 1–1
Seismic waves produced by a small explosion are reflected and refracted through different rock layers before reaching the seismometers. Paths show how these waves echo from a boundary or bend into a different direction when crossing a boundary from one layer into another.

Have you ever poked a stick into a pond? Recall how it appeared to bend at the water surface because of the bending, or refraction, of the light rays. This also happens to seismic waves.

Be sure to observe in Figure 1–1 that more than one seismic wave reaches each seismometer. Every wave reaching the seismometer produces a momentary impulse on a record of ground vibration. Such a record is called a *seismogram.* It indicates the times when different refracted and reflected waves reach the seismometer. These values of time are analyzed by an exploration seismologist to find out the wave speed in the different rock layers and thicknesses of these layers. The methods of analysis are described in Chapters 3 and 4.

Some seismic surveys are easily done by two or three people. Suppose that a highway department or construction company asks about the thickness of loose soil and gravel covering bedrock along a proposed roadway or building site. Such questions can perhaps be answered by placing a dozen seismometers in a line 5 meters apart (Figure 1–2). By detonating a one-half kilogram explosive charge, we

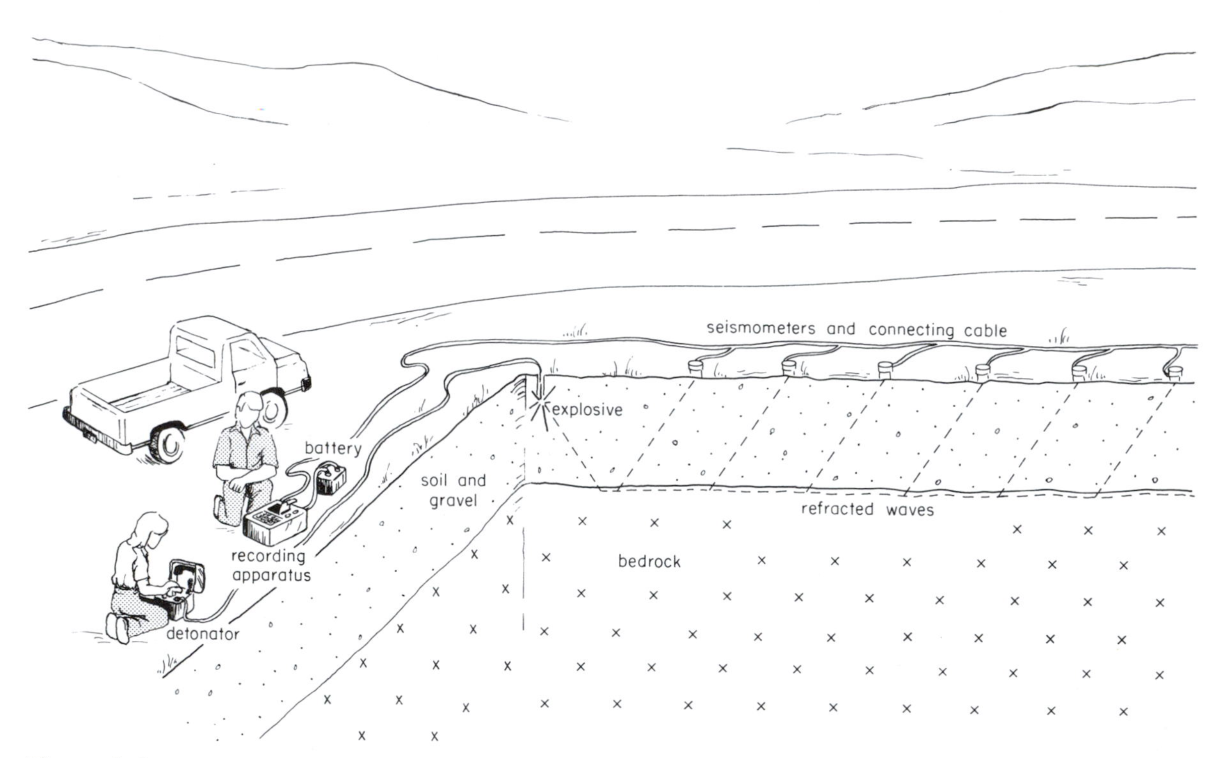

Figure 1–2
Seismic refraction experiment to measure the thickness of soil and gravel that covers solid bedrock. The times required for refracted waves to reach seismometers at different distances are measured. These times are analyzed to find the speed of the waves and the thickness of the top layer.

can produce refracted waves that will reach depths of about 15 meters. Instead of an explosive charge, we might produce the seismic waves by dropping a weight or pounding the ground with a sledge hammer. Ordinarily, two or three people would take about one-half hour to complete the measurements and calculate the depth to bedrock. In such a survey, the line of seismometers should extend a distance about four times greater than the maximum depth of interest. They would analyze only refracted waves, because the reflected waves would be too weak to detect.

In the search for oil, seismic surveys commonly probe as deep as 10 km (32,800 ft). Attention is focused on reflected waves that are recorded on hundreds or even thousands of seismometers in lines sometimes more than 5 km long. A crew of between 10 and 20 people is needed to move seismometers, cables, and other equipment. Rather than explosives, many doodlebugging crews are equipped with heavy vibrator trucks that press large vibrating pads on the ground to generate seismic waves. The most common survey procedure is to arrange the seismometers in a line along a road (Figure 1–3). An explosive or vibrator produces seismic waves that are recorded after reflecting, or echoing, from the buried rock layers. The equipment is then moved a short

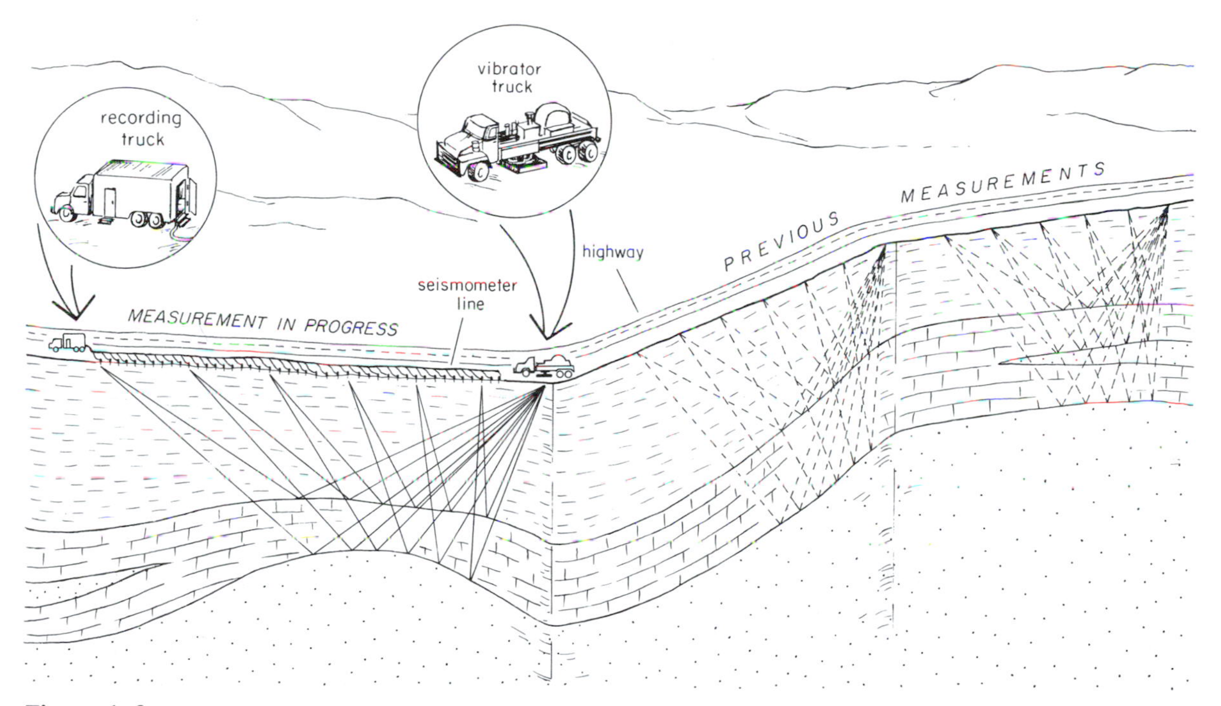

Figure 1–3
Seismic reflection experiment for detecting geologic structures in which oil and natural gas may be trapped, the times of echoes from different layer boundaries are measured, and this information is used to calculate layer thicknesses.

distance along the road, and the experiment is repeated. By examining seismograms recorded in this way, we can observe changes in the times that different reflected waves appear. Depths to different rock layers change as the survey passes over buried structures such as the anticline illustrated in Figure 1–3.

Professional seismic crews are an important part of the oil industry. The number of active crews fluctuates with the market for oil and natural gas. During times of high demand, more than 500 crews can be found operating in different parts of the world. Field procedures of these crews are described in Chapter 5.

A serious difficulty in exploration seismology is the often confusing pattern of unwanted vibrations received by the seismometer. Wind blowing on trees, traffic along a road, and waves reflected or refracted from geologic features other than those of interest all produce these unwanted vibrations. Sometimes they are so strong that they obscure the waves that the exploration seismologist hopes to record. The methods described in Chapter 6 for reducing or eliminating the undesirable vibrations rely on special arrangements of seismometers and computer processing of seismograms.

GRAVITY AND GEOLOGY

The attraction of gravity is not exactly the same everywhere on the earth's surface. There are small variations from place to place because of irregularities in rock density. Recall that the density of a rock specimen is obtained by dividing its mass by its volume. This physical property is described either in the SI units of kilograms per cubic meter (kg/m^3) or in units of grams per cubic centimeter (g/cm^3).

The exploration geophysicist hopes to distinguish different kinds of rock by detecting density irregularities from measurements of the attraction of gravity. For example, a buried salt dome that penetrates layers of shale (Figure 1–4) would produce a small but measurable decrease in the attraction of gravity, because the salt density of 2.0 g/cm^3 is smaller than the shale density of about 2.6 g/cm^3. Exploration geophysicists are particularly interested in locating salt domes because accumulations of petroleum and natural gas have been discovered above and on the flanks of many of these structures.

We can measure the earth's gravitational attraction with a small portable instrument called a *gravimeter.* Basically, it consists of a small object supported by a very sensitive spring which is stretched by the weight of the object. Because of variations in the attraction of gravity, however, this weight changes as the gravimeter is moved from place to place. Therefore, the stretch of the spring changes. It is possible to detect minute changes in the attraction of gravity by carefully measuring the stretch of the spring in different locations. The design and operation of gravimeters is discussed in Chapters 7 and 8.

A gravity survey is done by reading a gravimeter at many different locations in an area (Figure 1–4). These locations can be less than 1 km apart, or perhaps several kilometers apart depending on the size and depth of the geologic features of interest. It usually takes less than five minutes to operate the gravimeter at each location. This work can be done by one person, but additional people may be needed to make elevation and position measurements and to transport equipment in remote areas. Gravity survey crews of one or more persons have worked in most parts of the world. In addition to land-based surveys,

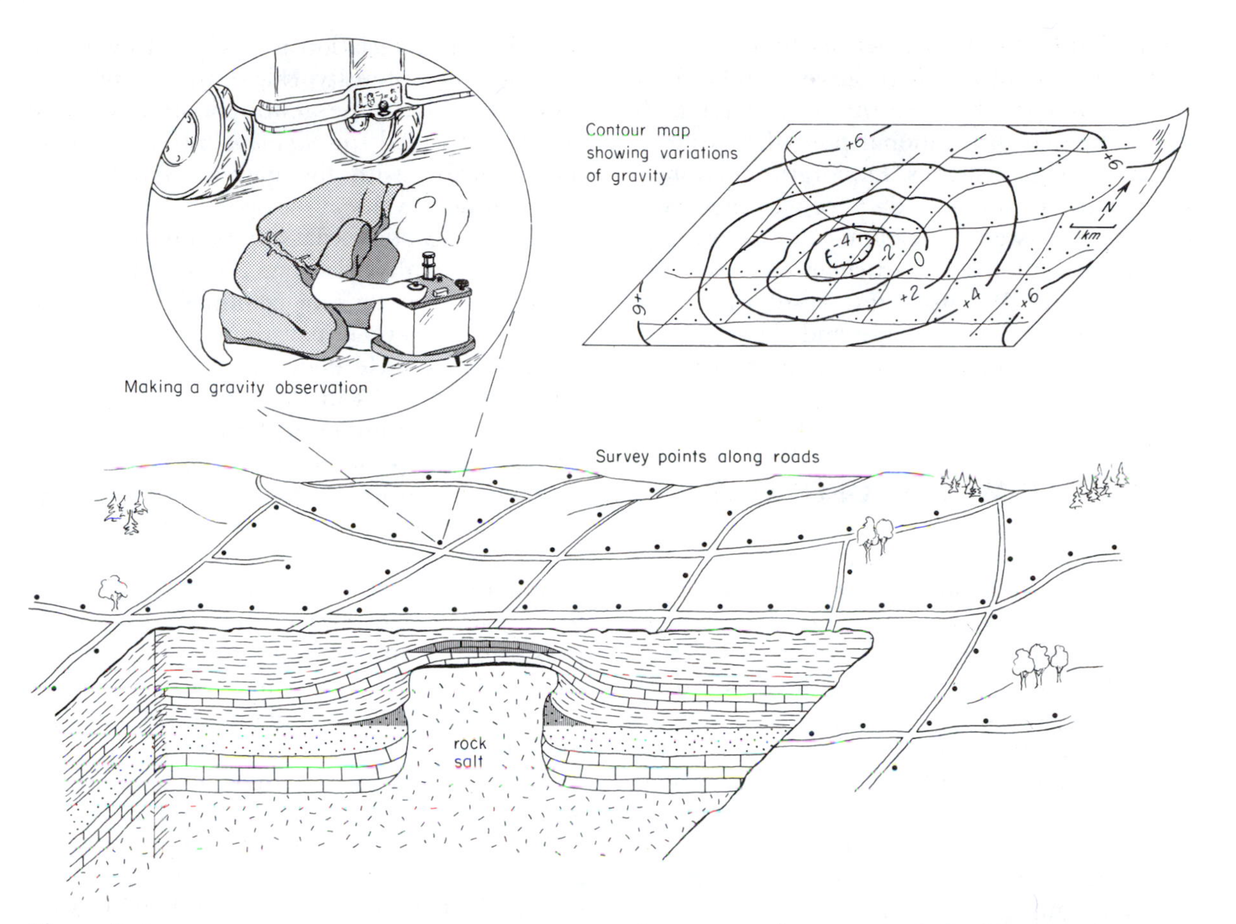

Figure 1–4
A gravity survey for detecting salt structures. Survey results are contoured on a map so that patterns of gravity variation indicative of these features can be recognized. In this example, a pattern showing a small decrease in gravitational attraction suggests the presence of a low-density salt dome.

gravimeters can be operated on ships, and recently they have been installed in helicopters for airborne surveys.

We cannot immediately make geologic interpretations from gravimeter measurements. First, adjustments are made to show how latitude and elevation also influence the attraction of gravity. After making the appropriate adjustments, discussed in Chapter 8, we obtain values that indicate irregularities in rock density. We can use these values to prepare profiles and contour maps that indicate gravity variations related to geologic features. Referring again to Figure 1–4, we can see how the particularly low values over a small area of such a map might indicate a hidden salt dome that had intruded heavier beds of shale.

The results of a gravity survey can be difficult to interpret because the gravimeter feels the combined gravitational pull of many different geologic features. Exploration geophysicists have developed data processing techniques, discussed in Chapter 9, to bring a gravity variation of particular interest into clearer focus. We use these results to make judgments about the shape and depth of the feature that produces that particular variation in gravity.

MAGNETISM AND GEOLOGY

The magnetic compass is one of our greatest inventions, but in some places it is not a reliable guide. Compass direction can be strongly deflected near concentrations of a few kinds of magnetizable minerals. Over three centuries ago, prospectors began using this feature of the compass to advantage in their search for ores associated with magnetic minerals such as magnetite.

Early instruments were designed to measure the direction of a delicately balanced magnet. Emphasis then shifted to measuring the strength of the earth's magnetic force on ingeniously designed test magnets. These measurements are made with instruments called *magnetometers.*

Earth magnetism has two principal sources, described in Chapter 10. By far the strongest part is produced in the molten core by flow of ionized fluids. The other part, which is much weaker, arises from contrasts in the concentration of magnetite and a very few other minerals in the rocks of the earth's crust. This second part is of greater interest to exploration geophysicists. Here the numerous local variations indicate different geologic features. The size of such a variation is described by a unit of magnetic intensity called the *gamma* (γ). More recently, some geophysicists have used an SI unit called the *nannotesla* (nT). It is interchangeable with the gamma and has the same numeric value.

Most magnetic variations of interest to exploration geophysicists are a few hundred, or perhaps a few tens of gammas, in size, although some are larger than 2000 gammas. They indicate differences in the magnetic susceptibility of rocks in the crust. *Magnetic susceptibility* is a number without units that describes the capacity of the rock to acquire magnetism. Suppose that a dike of gabbro with a susceptibility of 0.005 intrudes granite where the susceptibility is less than 0.001. Depending on the size of the dike, we might detect a variation of 100 or 200 gammas as we pass over it with a magnetometer. If the dike consisted mostly of magnetite and associated ore minerals, the susceptibility could be larger than 0.1. A variation of more than 1000 gammas might be measured, depending on its size and distance from the magnetometer.

Most magnetometers are compact devices small enough to carry in a suitcase. Some are operated on tripods or are hand held. Most magnetometers, however, are mounted in airplanes and record variations in magnetism more or less continuously along closely spaced flight paths. The variations obtained in this way indicate how an ore-bearing structure might be discovered (Figure 1–5). Airborne magnetic surveying has been done over large areas of the world. Surveys have also been done at sea using magnetometers trailed from ships.

Geophysicists compile profiles and contour maps to display the variations measured over an area. The data processing techniques described in Chapter 11 are similar to those ap-

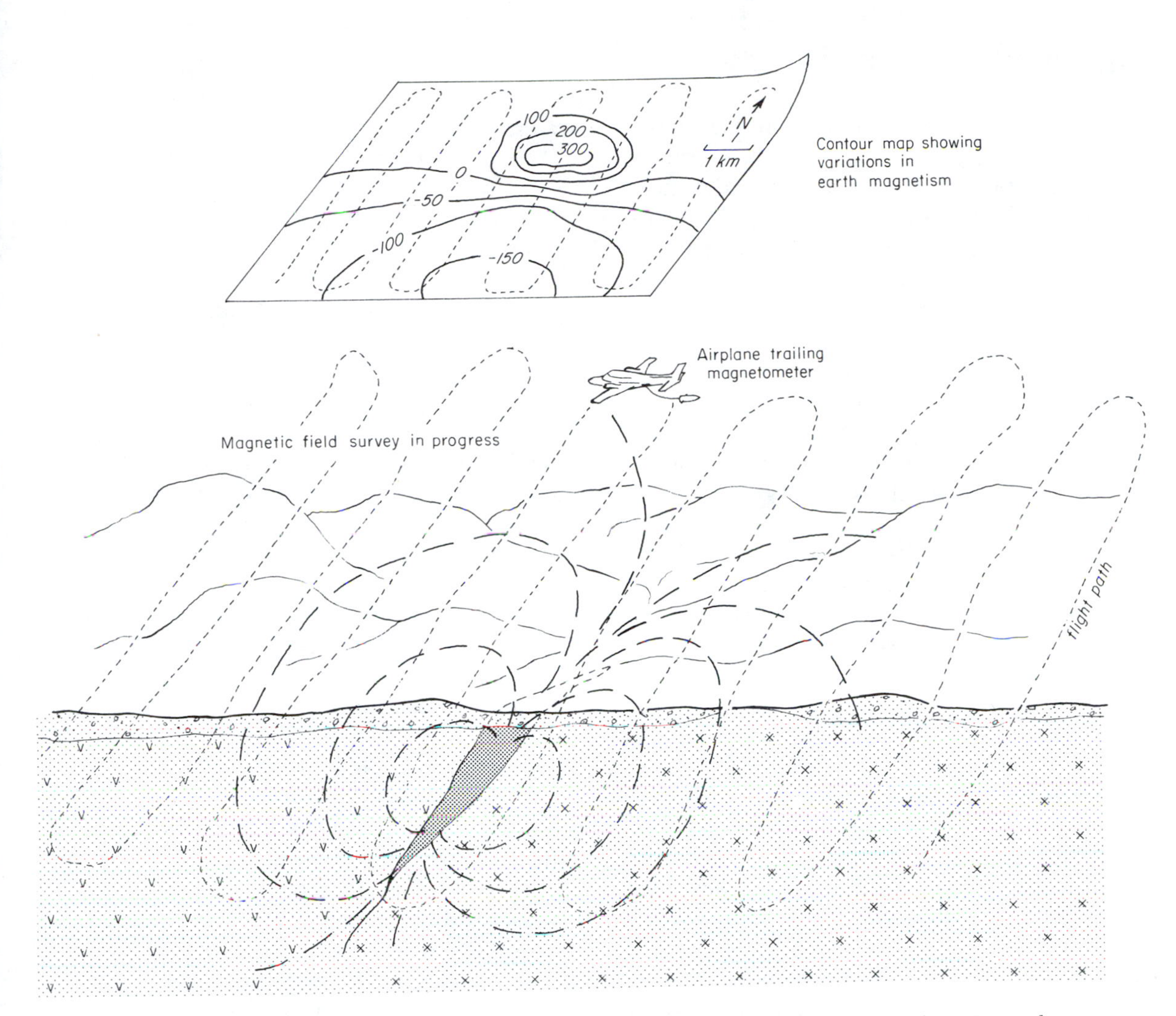

Figure 1–5
Airborne magnetometer survey for locating magnetized ores. A contour map prepared from the survey data indicates patterns of variation in the strength of earth magnetism. Irregular patterns on such a map are caused by concentrations of magnetized rock.

plied to gravity data and are used to bring certain variations into clearer perspective. Theoretical magnetic variations are then calculated for different geologic features to find the one that would produce a pattern that compares most closely to the measured pattern. These procedures are introduced in Chapter 12.

ELECTRICAL PROSPECTING

Geophysicists have devised several techniques for measuring electrical properties of rocks in the earth. One of these properties describes the resistance of rock to the flow of electric current. This property of *resistivity* can be used to distinguish different kinds of rock. Another property of interest is the capacity of a rock mass to become polarized in an electric field. Polarization results from a concentration of positive electric charge on one side of the mass and negative charge on the opposite side. We might think of this polarized rock as a natural battery buried in the earth.

The first electrical prospecting instruments were designed to detect naturally polarized ore bodies. For example, some massive sulfide bodies appear to be permanently polarized because of the way different ions became concentrated by weathering processes and the flow of ground water. By connecting a voltmeter between two electrodes placed in the ground, we can measure a voltage difference produced by such a polarized mass. Its location is indicated by a pattern of voltage variation obtained from measurements repeated at regular intervals over an area.

It is not necessary to rely exclusively on natural electric fields. By connecting a battery between two electrodes, we can compel current to flow through the earth. Then we can connect a voltmeter between two other electrodes to measure voltage differences produced by this current (Figure 1–6). The information is used to calculate rock resistivity. Measurements are made with the electrodes arranged in the different spacings described in Chapter 13 to determine how resistivity changes with depth. Surveys designed to reach depths of several hundred meters can be done with compact equipment that fits in a suitcase-sized container. Metal rods driven into the ground are satisfactory electrodes.

When we compel an electric current to flow in the ground, rocks containing metallic ions tend to become temporarily polarized. This process is called *induced polarization* (IP). It persists for a brief time after the current is shut off, depending on the kinds of metallic elements present and their concentrations. Measurements of the time required to dissipate this polarization have proved very useful in the search for sulfide ores. An IP survey is done by making these measurements with the same electrode spacing at uniformly spaced locations in the area of interest.

Other kinds of electromagnetic devices using test coils have been designed to study the flow of electric current in the earth. Basically, the equipment consists of a large primary input coil and a secondary receiver coil. An alternating current in the primary coil produces magnetic effects that induce the electric currents to flow in the ground in patterns that depend on rock resistivity. These ground currents, in turn, produce magnetism that compels current to flow in the receiver coil. The strength and frequency of this secondary current provides information about rock resistivity. Various coil configurations sensitive to different features of interest can be operated on the ground or in low-flying airplanes.

GEOPHYSICAL WELL LOGGING

The traditional way to find out what lies underground has been to drill a hole and see. But what kind of information does the drill provide? The rock cuttings that are flushed from the hole may have been mixed so that

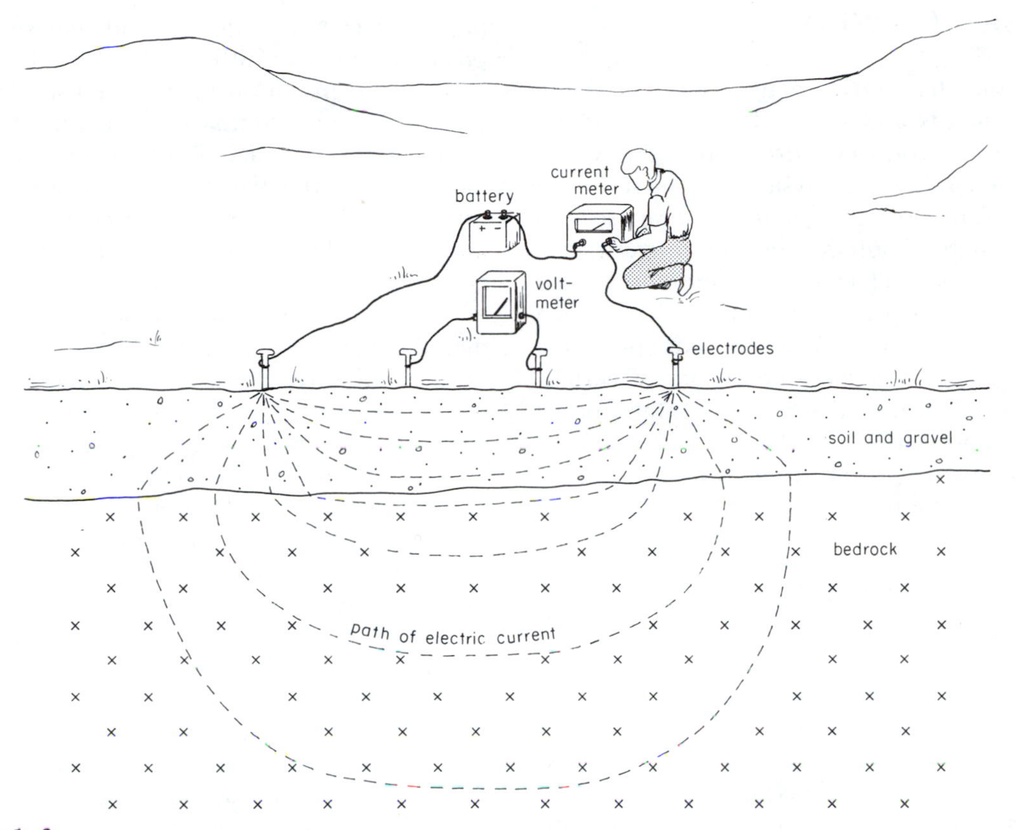

Figure 1–6
A survey of electrical resistivity can be used to find the thickness of soil and gravel that covers solid bedrock. Changes in the resistivity measured with different electrode spacings are related to the thickness of this layer.

we cannot tell exactly where they came from. A tool is needed that can be lowered into the hole to detect properties of the rock through which it passes. This is called a well logging tool.

The most widely used well-logging tools measure rock resistivity and natural voltages, seismic wave speeds, and radioactive properties. Less frequently used are special tools for measuring gravity variations, magnetism, and temperature. The well logging tools described in Chapter 14 produce continuous charts that indicate how these physical properties change while the various detectors move in the borehole.

Well logging is done in most oil wells, mining exploration wells, and in many water wells. Several companies offer a large selection of services. They maintain facilities in many parts of the world from which trucks or ships can be dispatched to make measurements in any well in a particular district.

PHYSICAL UNITS

The units that must be used for describing measurements and physical properties of the earth are introduced throughout this book. The literature of geophysics includes different combinations of the meter.kilogram.second (m.k.s) metric system, the centimeter.gram. second (c.g.s) metric system, and the British system. In recent years, efforts have been made to encourage worldwide adoption of a system based on m.k.s units. As noted earlier, this system is called the Système International and is abbreviated SI. Although SI units are becoming widely used, several non-SI units remain in common use.

BLENDING GEOLOGY AND PHYSICS

Geologists began learning about salt domes almost a century ago. The first information came from chance discoveries. Physics played no role at this point because no one knew enough about salt structures to figure out which principles of physics might be applicable. Using bits of information from drill cuttings, geologists eventually pieced together ideas about the shape and dimensions of these salt structures. With this knowledge they could turn to physics for ideas about how to search for them. Here was the source of knowledge about how to calculate the gravitational effects of objects of different shapes and how to measure these effects.

The search for salt domes is only one of many examples of how geophysicists look to geology for knowledge about what kinds of rocks and structure exist in nature and then turn to physics for ideas about how to find them. This book is concerned with the techniques used to calculate and measure the physical effects of different geologic features.

The interpretation procedure used by geophysicists is to compare a measured effect with values calculated for some standard model of earth structure. For example, the times of seismic waves measured from a seismogram could be compared with times calculated for travel through a sequence of flat layers. If the calculated times closely match the measured times, we conclude that the thicknesses and wave speeds of the flat layers in our model are a good representation of earth structure at that location. A poor comparison would indicate that the flat layer model is inappropriate. We might then use other formulas to calculate the times of seismic waves along paths through a more complicated model of sloping layers that dip at different angles. A better comparison with measured times would tell us that this interpretation is more realistic.

We can compare measured variations in gravity with those calculated for structures of different shape and density. A favorable comparison with values calculated for a vertical cylindrical form might tell us that the measured gravity variation could be explained by a salt dome. However, we might reject this interpretation if our measurements compared more closely with results calculated for a prism of different shapes. From these examples we can understand how knowledge of geology determines the standard models of earth structure that should be considered. Principles of physics can then be used to derive equations for predicting the seismic, gravimetric, magnetic, and electrical effects of these models. An understanding of geology again becomes important in our judgments about the actual rocks that are represented by the physical properties of a model.

STUDY EXERCISES

1. In each of the four principal branches of exploration geophysics, what particular property is measured?
2. What kind of measurements were made in the earliest exploration geophysical surveys?
3. Name two factors that can cause the path followed by a seismic wave to change direction.

SELECTED READING

Bathes, Charles C., Thomas Gaskell, and Robert B. Price, *Geophysics in the Affairs of Man.* Elmsford, N.Y., Pergamon Press, 1982.

Dobrin, Milton B., *Introduction to Geophysical Prospecting,* 3rd edition. New York, McGraw-Hill, New York, 1976.

Sheriff, Robert E., *Encyclopedic Dictionary of Exploration Geophysics,* 2nd edition. Tulsa, Okla., Society of Exploration Geophysicists, 1986.

Sheriff, Robert E., History of geophysical technology through advertisement in *Geophysics,* Geophysics, v. 50, n. 12, pp. 2299–2410, December 1985.

TWO

Seismic Waves

Seismic waves are messengers that convey information about the earth's interior. Basically, these waves test the extent to which earth materials can be stretched or squeezed somewhat as you can squeeze a sponge. They cause the particles of material to vibrate, which means that these particles are temporarily stretched out of position as they move back and forth. The capacity of a material to be temporarily deformed by passing seismic waves can be described by its properties of *elasticity.* These physical properties can be used to distinguish different materials. They influence the speeds of seismic waves through those materials. In this chapter, we will discuss the elastic properties of earth materials and the different kinds of seismic waves that can travel in these materials. Then we will examine how elastic properties influence the paths of seismic waves. Finally, we will look at the methods for recording seismic waves and the typical patterns of vibration displayed on seismograms.

ELASTICITY

Stress and Strain

We can deform an object in different ways by stretching, squeezing, and twisting it. To accomplish this distortion, we must apply force on the sides of the object. Such application of force on the surface of an object is called *stress,* and it is expressed in units of force divided by area. The SI unit of stress is called a *pascal* and is equivalent to one newton (N) of force applied over a surface one meter square: $1 \text{ Pa} = 1 \text{ N/m}^2$.

When an object is placed at some depth in a pond or lake, the water presses against its sides from all directions. This is a *nondirected* stress which we call hydrostatic pressure. It is the pressure felt in the ears of a swimmer who dives deeply into the water. The fact that the swimmer feels the same pressure on both ears is evidence of equal stress in all directions. Other kinds of stress can be applied in a particular direction. An example of such a *directed* stress is the force of a hammer striking the head of a nail.

What is the effect of stress on an object? It deforms that object by changing its shape and size. The deformation is called *strain.* The different kinds of strain that can be produced depend on the strength and direction of stress and the nature of the substance being deformed. First, let us consider *elastic* strain. This kind of strain is proportional to the applied stress, and it disappears when that stress ceases. For example, we can apply stress by pulling a rubber band, and the length to which it stretches is a measure of the strain. The harder we pull, the farther it stretches. But when we release one end, the stress vanishes, and the rubber band snaps back to its unstretched size. This is the nature of elastic strain.

Something different happens when we stretch a lump of soft clay. It remains stretched after we release it. This permanent deformation is evidence of *plastic* strain. Solid substances ordinarily respond elastically to relatively weak, short-term stresses, and then exhibit plastic strain if stronger or more prolonged stress is applied. Finally, if the stress exceeds the strength of the material, *rupture* occurs. By pulling very hard, we can break the rubber band or the lump of clay.

Exploration seismologists are concerned principally with elastic strain. To be sure, plastic deformation and rupture are caused by explosions or movements along faults that generate seismic waves. But this severe deformation is confined to a small zone close to the point of energy release. As the seismic waves travel away from this small source zone, they produce only elastic strain. Therefore, it is important for us to examine the formal ways of describing elastic strain.

Bulk Modulus

Suppose we placed a specimen of some substance in a fluid-filled container designed so that hydrostatic pressure could be adjusted

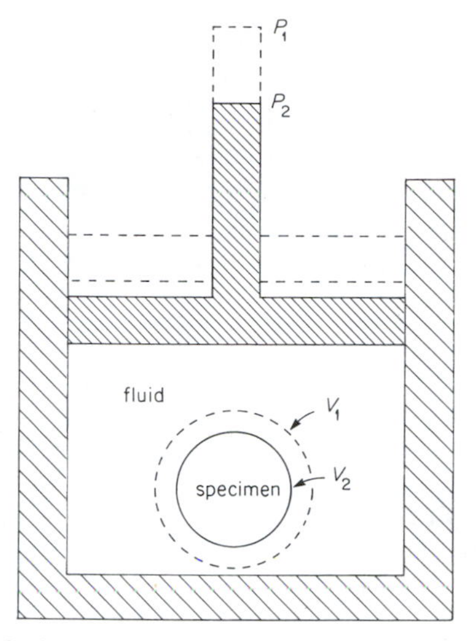

Figure 2–1
Testing the bulk modulus in a pressure cell. An increase in pressure from P_1 to P_2 causes the volume of the specimen to decrease from V_1 to V_2.

(Figure 2–1). If the pressure is maintained at a value of P_1, the volume of the specimen is observed to be V_1. Now, if the pressure is increased by a small amount to P_2, the specimen becomes compressed into a slightly smaller volume V_2. This indicates the strain produced by the change in stress. The mass and shape of the specimen remain unchanged, however. Therefore, if the density of the specimen is ρ_1 under pressure of P_1, an increase of pressure to P_2 causes an increase in density to ρ_2. Tests on elastic substances show that the changes in volume and density are proportional to the change in pressure. The constant of proportionality *(k)* is called the *bulk modulus,* and it is defined as

$$k = \frac{-\Delta P}{\Delta V/V_1} = \frac{\Delta P}{\Delta\rho/\rho_1} \qquad (2\text{–}1)$$

where $\Delta P = P_1 - P_2$, $\Delta V = V_1 - V_2$, and $\Delta\rho = \rho_1 - \rho_2$. The bulk modulus is a measure of the capacity of a substance to be compressed. Different values of this physical property can be used to distinguish one substance from another (Table 2–1). From Equation 2–1 we see that it is expressed in the same units as stress. The value of bulk modulus can be measured for any kind of solid, liquid, or gas.

Shear Modulus

Let us next consider a test of elasticity in which the shape of a specimen is distorted.

TABLE 2–1 Elastic Properties of Selected Rock Specimens

SPECIMEN	BULK MODULUS (N/m^2)	SHEAR MODULUS (N/m^2)	YOUNG'S MODULUS (N/m^2)	POISSON'S RATIO
Sandstone, quartzitic	4.17×10^{10}	4.28×10^{10}	9.6×10^{10}	0.118
Limestone, Solenhofen, Bavaria, West Germany	4.67×10^{10}	2.47×10^{10}	6.3×10^{10}	0.276
Granite, Quincy, Mass.	5.21×10^{10}	3.45×10^{10}	8.49×10^{10}	0.229
Gabbro, French Creek, Penn.	8.85×10^{10}	4.80×10^{10}	10.43×10^{10}	0.270
Marble, Vermont	7.19×10^{10}	3.33×10^{10}	8.7×10^{10}	0.299

From Francis Birch, J. F. Schairer, and H. Cecil Spicer (editors), *Handbook of Physical Constants,* Geological Society of America, Special Papers, No. 36, Table 5–8, p. 80, January 31, 1942, reprinted 1961.

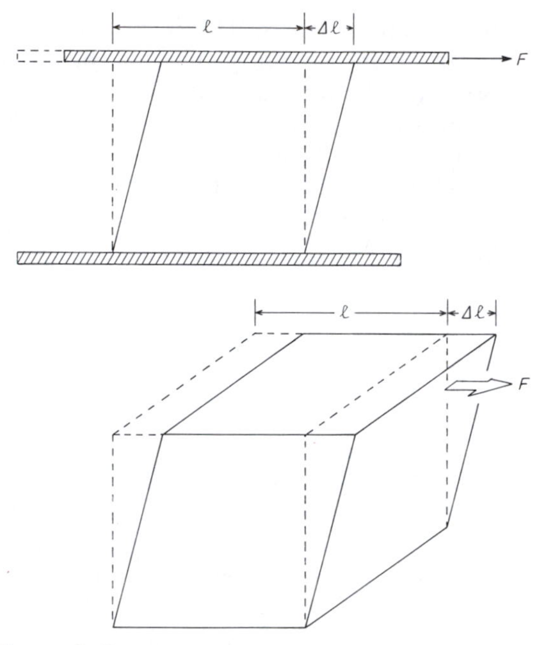

Figure 2–2
Testing shear modulus by distortion of a cube. One face of the cube is fixed, and the opposite parallel face is displaced by the small distance $\Delta \ell$ in response to the force of traction acting on that face.

We begin by attaching a cube of some substance between two parallel plates. Then force *(F)* is applied to displace one plate in a direction parallel to the other. Because one surface of the cube is attached to the plate, the force is applied over the entire area *(A)* of that surface. This creates a directed stress, $\tau = F/A$, which we call a *traction* or a *shear stress.* The cube becomes distorted as one side is shifted by the small distance $\Delta \ell$ relative to the opposite side (Figure 2–2). This distortion is an indication of shear strain. In an elastic cube of length ℓ, the small displacement will be proportional to the strength of the shear stress. The constant of proportionality (μ), called the *shear modulus,* is defined as

$$\mu = \frac{\tau}{\Delta \ell / \ell} \qquad (2\text{–}2)$$

Shear stress cannot be applied to ideal liquids and gases. For these substances, $\mu = 0$. Only solids possess the physical property described by the shear modulus. This property can be used to distinguish different solid substances (Table 2–1).

Young's Modulus and Poisson's Ratio

What happens to a specimen that is stretched or compressed by a directed stress? This effect can be tested by placing a cylindrical specimen in a press (Figure 2–3). Here, force *(F)* can be applied against the area *(A)* of the end of the specimen in the direction of the cylinder axis. This force produces a directed compressional stress: $\eta = F/A$. Suppose that when the stress has a value η_1, the length of the specimen is ℓ_1, and its diameter is d_1. Now, increase the stress to a slightly higher value η_2. Observe that the length of the specimen decreases to ℓ_2 and its diameter increases to d_2. The change in length is proportional to the change in stress for an elastic substance, and we call the constant of proportionality *Young's modulus (E).* It is defined as

$$E = \frac{-\Delta \eta}{\Delta \ell / \ell_1} \qquad (2\text{–}3)$$

where $\Delta \eta = \eta_1 - \eta_2$ and $\Delta \ell = \ell_1 - \ell_2$. Another constant, which we call *Poisson's ratio* (σ), is used to show that the change in diameter $\Delta d = d_1 - d_2$ is proportional to the change in length:

$$\sigma = \frac{\Delta d / d_1}{\Delta \ell / \ell_1} \qquad (2\text{–}4)$$

Young's modulus and Poisson's ratio, like the shear modulus, are constants that are use-

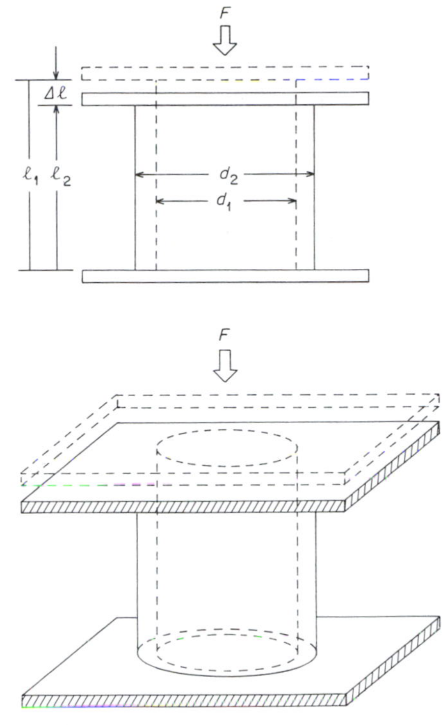

Figure 2–3
Testing Young's modulus and Poisson's ratio in a press. Young's modulus is determined from the change in length $\Delta\ell$, and Poisson's ratio is determined from the change in diameter d_2–d_1.

ful for describing the elastic properties of solids (Table 2–1). These constants cannot be used to describe liquids and gases for which elasticity is expressed solely by the bulk modulus.

SEISMIC BODY WAVES

We are more familiar with waves on ponds, lakes, and the ocean than with seismic waves. Tossing a pebble into a quiet pond produces ripples with certain features that are common to all waves. We will look at these features first and then describe the particular kinds of seismic waves. The ripples that spread over a pond are illustrated by the profile in Figure 2–4, which extends outward from the point of impact of the pebble. The shape of the pond surface changes from time t_1 to time t_2 because the ripples are advancing outward. The distance between two successive wave crests at some particular time is called the *wavelength* (λ). Observe the advance of one of these wave crests from position r_1 at time t_1 to position r_2 at time t_2. This shows that the speed of the wave is

$$V = \frac{r_2 - r_1}{t_2 - t_1} \qquad (2\text{–}5)$$

The wave amplitude *(H)* is the displacement of a water particle above or below the undisturbed surface of the pond. The accompanying graph shows how the amplitude at position r_1 changes with time. The length of time required for one cycle of oscillation is called the wave period *(T)*. This is the time needed for a wave crest to advance a distance of one wavelength (λ). Therefore, we can relate wavelength, period, and speed:

$$V = \lambda/T \qquad (2\text{–}6)$$

The *frequency* *(f)* of the waves is the number of oscillations that occur during a standard interval of time. It is the reciprocal of the period:

$$f = 1/T \qquad (2\text{–}7)$$

Waves on lakes and oceans commonly have periods of several seconds or tens of seconds, and wavelengths of several meters or more. Some earthquake waves have similar periods and wavelengths. But exploration seismologists usually work with wave periods that are fractions of a second. These waves have frequencies of tens or hundreds of cycles per second. The unit of frequency commonly used is the *hertz,* which is one cycle of oscillation per

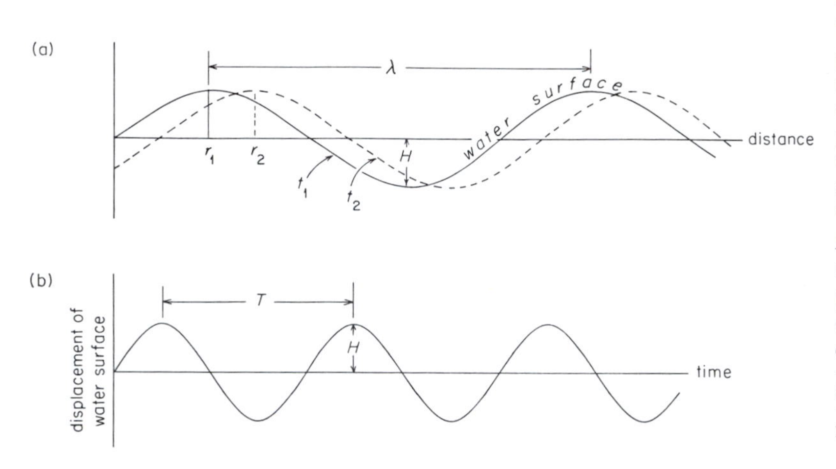

Figure 2–4
Waves produced by tossing a pebble into a pond. (a) Shape of the water surface along a profile extending outward from the point of impact first at time t_1 (solid line), then a moment later at time t_2 (dashed line). Wavelength λ and wave height H describe the shape of the surface. (b) Displacement of a point on the water surface above or below the undisturbed level changes with time because of passing waves. The wave period T and wave height describe the movement of this point.

second. The particular kinds of waves of most interest to exploration seismologists can now be described.

Compressional Waves

There are different ways to make vibrations in a specimen of rock. Let us consider what vibrations will be produced by directly striking one side with a hammer. Suppose that we have some way to detect the movement of rock particles inside the specimen. Actually, tiny pressure-sensitive devices called transducers can be used for this purpose. At the instant of impact, the particles on that side will be displaced in the direction that the hammer was moving. For a brief moment, these particles will move back and forth in this direction. A little later they will cease moving, but other particles farther inside will move back and forth in the same direction. In this way, a pulse of vibration moves through the specimen, causing particles farther and farther from the point of impact to vibrate momentarily. This pulse of vibration is the seismic wave. It moves through the specimen first compressing, and then stretching, the rock from place to place, as we can see in Figure 2–5. Observe that the vibrating particles move back and forth in the same direction as the path of the pulse through the specimen. A pulse causing this kind of vibration is called a *compressional wave,* a *longitudinal wave,* or a *P-wave,* all of which mean the same thing.

By placing transducers on both sides of the specimen to measure the instant of impact and the instant when the pulse reached the far side, we can find the time (t_p) required for the P-wave to travel the distance (x) through the specimen. From this information, the P-wave speed (V_p) can be calculated:

$$V_p = x/t_p \tag{2–8}$$

Next, suppose that experiments are done to measure the density and elastic constants of the specimen. We could then verify that the P-wave speed depends on these properties in the following way,

$$V_p = \sqrt{\frac{k + \frac{4}{3}\mu}{\rho}} \tag{2–9a}$$

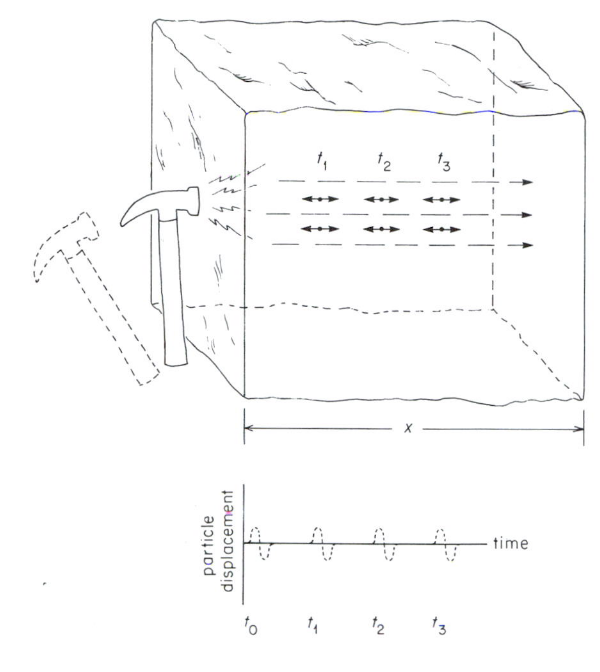

Figure 2–5

P-wave (also called a compressional or longitudinal wave) pulse of vibration traveling through a rock specimen. A simple pulse of vibration is produced on one side at time t_0 by the hammer. At later times, this wave pulse travels farther into the specimen, causing particles to vibrate back and forth in line with the direction of the advancing wave.

or

$$V_p = \sqrt{\frac{E}{\rho}\left[\frac{1-\sigma}{(1-2\sigma)(1+\sigma)}\right]} \qquad (2\text{–}9\text{b})$$

We can tell from these equations that *P*-waves will travel through any kind of substance: solid, liquid, or gas. They can do so because values of density and bulk modulus exist for all substances, even fluids for which the shear modulus is zero. Sound waves traveling through the air are *P*-waves.

Shear Waves

Can another kind of vibration be produced in a rock specimen? Rather than hammering directly on one side, suppose that we strike a glancing blow (Figure 2–6). At the instant of impact, the rock particles hit by the hammer will vibrate back and forth briefly in a direction parallel to this side. This pulse of vibration will then move through the specimen causing interior particles to vibrate in the same direction. After a length of time (t_s), the pulse will have traveled the distance (x) to the other side of the specimen. In Figure 2–6 the particles vibrate in a direction that is perpendicular, or transverse, to the path of the pulse. We call this kind of vibration a *transverse wave,* a *shear wave,* or an *S-wave.* These names are interchangeable. The speed V_s of the *S*-wave through the specimen would be

$$V_s = x/t_s \qquad (2\text{–}10)$$

If we had already measured the elastic properties of the specimen, we could verify the following relationships between elasticity, density, and *S*-wave speed,

$$V_s = \sqrt{\frac{\mu}{\rho}} \qquad (2\text{–}11\text{a})$$

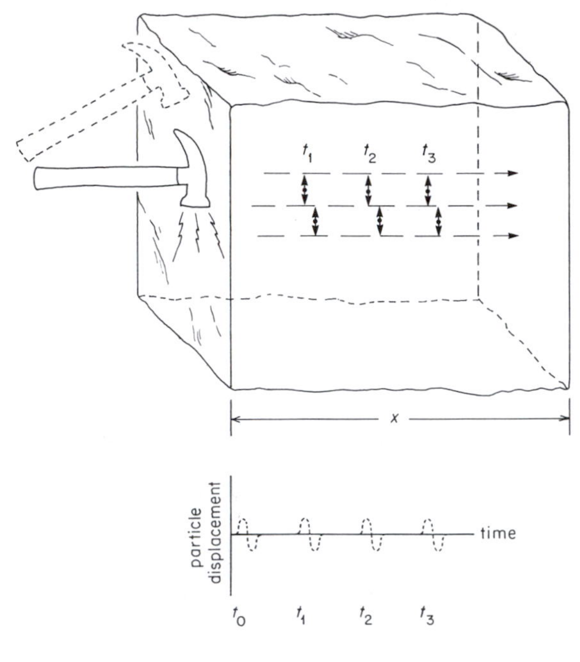

Figure 2–6
S-wave (also called a shear wave or a transverse wave) pulse of vibration traveling through a rock specimen. This wave pulse causes particles to vibrate back and forth in a line perpendicular to the direction of the advancing wave.

or

$$V_s = \sqrt{\frac{E}{2\rho(1 + \sigma)}} \qquad (2\text{–}11b)$$

We know that $\mu = 0$ for ideal liquids and gases. Therefore, Equation 2-11 clearly shows that *S*-waves do not travel in fluids. These transverse vibrations travel only in solids.

Body Waves

Both *P*- and *S*-waves are called *body waves* because they can travel directly through a mass of some substance in any direction. Other kinds of seismic waves called *surface waves* can travel only near the surface of such a mass, or close to the border between two different substances. The *P*- and *S*-waves are the most basic kinds of seismic waves. Surface waves can be explained in terms of combinations of these body waves that interfere with one another in distinctive ways when traveling near the surface of a mass. Surface waves will be described more fully later in this chapter.

We can learn some things about elasticity and body wave speeds by combining Equations 2–9 and 2–11 to get the ratio

$$\frac{V_p}{V_s} = \sqrt{\frac{k}{\mu} + \frac{4}{3}} \qquad (2\text{–}12a)$$

or

$$\frac{V_p}{V_s} = \sqrt{\frac{1 - \sigma}{\frac{1}{2} - \sigma}} \qquad (2\text{–}12b)$$

The fact that both k and μ are positive numbers indicates that $V_p/V_s > 1$. For this to be true, the *P*-wave must always travel faster than the *S*-wave through the same material, which is to say that $V_p > V_s$. It also restricts the positive range of values for Poisson's ratio: $0 < \sigma < 1/2$.

REFRACTION AND REFLECTION OF SEISMIC BODY WAVES

Rays and Wave Fronts

Think again about the ripples made by tossing a pebble into a quiet pond. Each wave moves away from the point of impact in an ever-expanding circle. The circle at the outer edge of an advancing wave marks the *wave front.* Lines drawn outward are *rays* that show directions along which the wave is advancing. Now, rather than the surface of a pond, think about waves produced at a point inside a mass of rock (Figure 2–7), perhaps by a small explosion. The movement of these waves is illustrated by spherical wave fronts and rays that radiate outward in all directions from the source. Observe that the rays are everywhere perpendicular to the wave fronts.

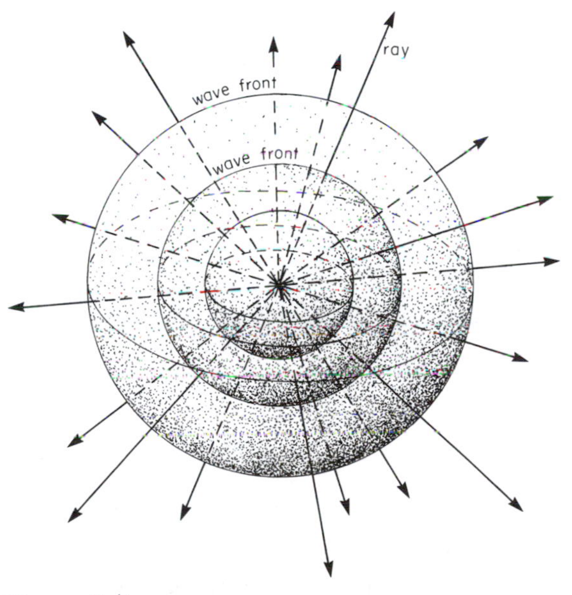

Figure 2–7
Spherical wave fronts and radiating rays illustrate the advance of waves outward in all directions from a source. The wave fronts show the position of a wave at successively later times t_1, t_2, t_3, and t_4.

Wave Conversions

What happens when a wave reaches a boundary between two substances in which the wave speeds are different? It divides into waves that bounce, or *reflect,* from the boundary and other waves that pass, or *refract,* across the boundary. The rays in Figure 2–8 illustrate the movement of these different waves. The original wave that travels to the boundary is called the *incident* wave. At the boundary, it is converted into *reflected* and *refracted* P-waves and S-waves that travel away from the boundary in different directions.

Observe that an incident P-wave converts into S-waves as well as P-waves (Figure 2–8a). An important feature of these S-waves is that the transverse particle motion occurs only in a plane that is perpendicular to the boundary and that contains the wave path. For a horizontal boundary, this will be a vertical plane, and the shear waves causing vibrations in this plane are called SV-waves. An SV-wave that is incident on a boundary converts into reflected and refracted SV-waves and P-waves (Figure 2–8b).

A shear wave that produces horizontal transverse particle motion is called an SH-wave. If this kind of horizontally polarized shear wave is incident on a boundary, it converts into reflected and refracted SH-waves (Figure 2–8c) with no associated P-waves.

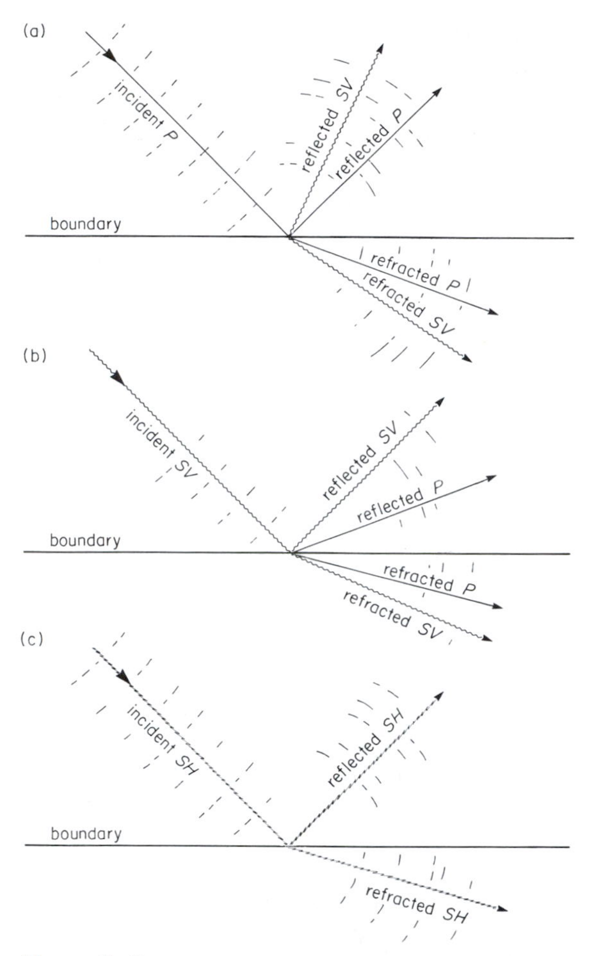

Figure 2–8
Wave conversions at the boundary between two rock layers. Each incident *P*-wave ray or *SV*-wave ray divides into four reflected and refracted *P*- and *SV*-wave rays. An incident *SH*-wave ray produces only reflected and refracted *SH*-wave rays.

Snell's Law

The directions of refracted and reflected waves traveling away from a boundary depend on the direction of the incident wave and the speeds of the waves. To explain this phenomenon, let us suppose that the source of incident *P*-waves is situated far from the boundary. One implication is that two rays very close together are almost parallel at the place where they reach the boundary. Another implication is that the portion of a wave front reaching between these two rays is so slightly curved that it is nearly a straight line. To simplify things further, we will look only at incident and refracted *P*-waves, ignoring for the moment other reflected and refracted *P*- and *SV*-waves.

Now examine in Figure 2–9 the parallel rays and straight wave fronts close to a boundary. The incident wave moves at speed V_1. It takes the same time for one point on the wave front to move from A to C as is needed for another point to move from B to D, because

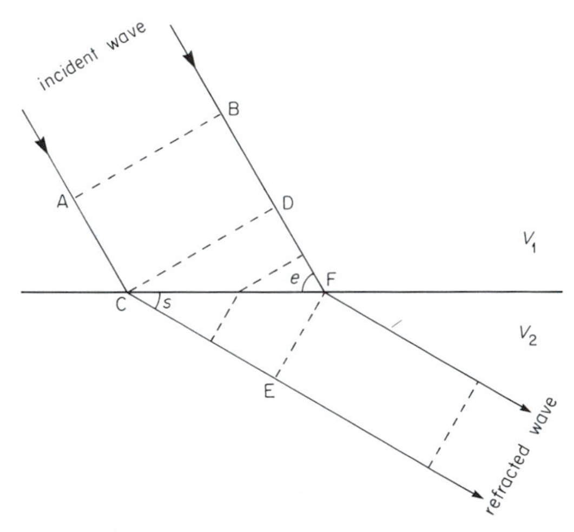

Figure 2–9
Wave refraction at the boundary between two rock layers. Notice that one point on the wave front advances from D to F at speed V_1 during the same time that another point on the same wave front advances from C to E at speed V_2. The wave front bends at the boundary as it moves along.

both parts of the wave front are moving at speed V_1. After C is reached, that point on the wave front refracts into a new direction and travels at speed V_2 toward E. Let t be the interval of time needed for the wave at that point to travel from C to E. Then, by rearranging Equation 2–8, we can show that the distance CE = V_2t. During the same interval of time, the other point continues from D to F at speed V_1. This requires that the distance DF = V_1t. Observe that the wave front bends where it crosses the boundary.

The incident rays are inclined from the boundary at the angle e. Because rays are perpendicular to wave fronts, the figure CDF is a right triangle. Therefore, from trigonometry we recognize that $\cos e = DF/CF = V_1t/CF$. This expression can be rearranged so that

$$CF = V_1t/\cos e \qquad (2\text{–}13a)$$

The refracted rays are inclined at the angle s from the boundary. Since the figure CEF is a right triangle, we see that $\cos s = CE/CF = V_2t/CF$. We can rearrange this expression to get

$$CF = V_2t/\cos s \qquad (2\text{–}13b)$$

Combining and rearranging Equations 2–13a and 2–13b shows how wave speeds and directions are related:

$$\frac{\cos e}{\cos s} = \frac{V_1}{V_2} \qquad (2\text{–}14)$$

Next, let us consider the relationship between a reflected wave and the incident wave that produced it. The case of an incident *P*-wave and the reflected *SV*-wave is illustrated in Figure 2–10, where the same letters that appear in Figure 2–9 are used. Now reread the previous two paragraphs, making the following changes. Replace the word "refract" with the word "reflect" and change V_1 and V_2

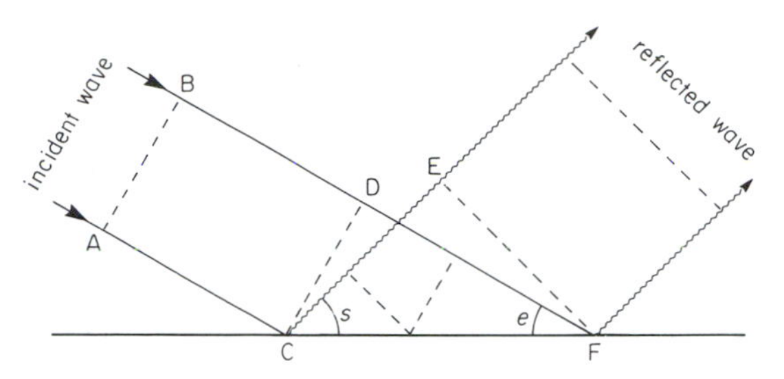

Figure 2–10
Example of wave reflection at the boundary between two rock layers in which a reflected *SV*-wave is produced by an incident *P*-wave. A reflected *P*-wave (not shown) would also be produced. Notice that a point on the incident *P*-wave front travels from D to F during the same time that a point on the reflecting *SV*-wave front travels from C to E.

to V_p and V_s. We will see that Equation 2–14 also shows how incident and reflected wave speeds and directions are related. Be sure to understand that an incident *P*-wave also will produce a reflected P-wave (Figure 2–8) that travels at the same speed. For this case, Equation 2–11 shows that the incident and reflected rays are inclined from the boundary at the same angle.

It is a more common practice among geophysicists to draw a line perpendicular to the boundary and to measure the angle i of the incident ray and the angle r of the refracted or reflected ray from this line (Figure 2–11). We will follow this practice and rewrite Equation 2–14 in the form

$$\frac{\sin i}{\sin r} = \frac{V_i}{V_r} \qquad (2\text{–}15)$$

where V_i is the speed of the incident wave and V_r is the speed of the refracted or reflected wave. This equation is called *Snell's law.* It governs the directions of all refracted and re-

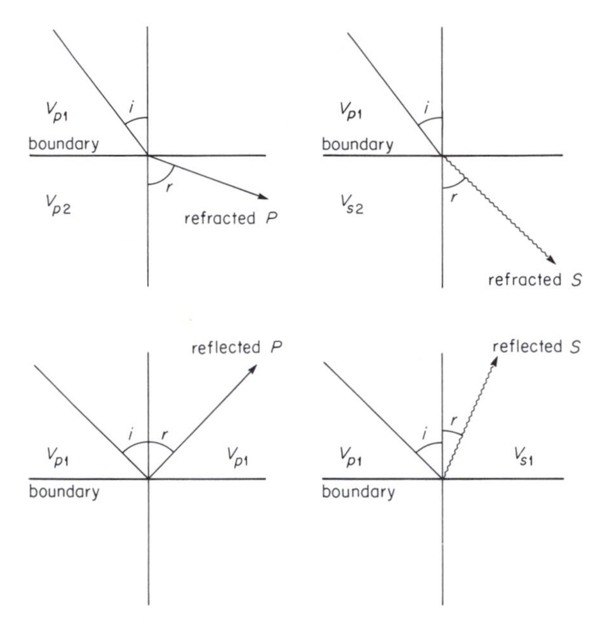

Figure 2–11
Angles of incidence *(i)* and reflection and refraction *(r)* measured from a line perpendicular to the boundary between two rock layers are shown separately for the four *P*- and *SV*-waves that are produced by an incident *P*-wave.

flected *P*- and *S*-waves produced by an incident *P*-wave or an incident *S*-wave.

Critical Refraction

Let us look further at how seismic waves cross a boundary between two different kinds of rock. To keep things as simple as possible, we will first consider only incident and refracted *P*-waves. In Figure 2–12a, rays from a source in the upper layer reach the boundary at different angles of incidence and then continue in the lower layer at different angles of refraction in accordance with Snell's law. Observe that the angle refraction is 90 degrees for one particular ray. It is the *critically refracted* ray that shows how the wave travels at speed V_2 right along the top of the lower layer.

The critically refracted wave is produced by an incident wave traveling along the ray that reaches the boundary at the *critical angle of incidence* i_c. For this case of critical refraction, note that $\sin r = \sin 90° = 1$, so that according to Snell's law we can write

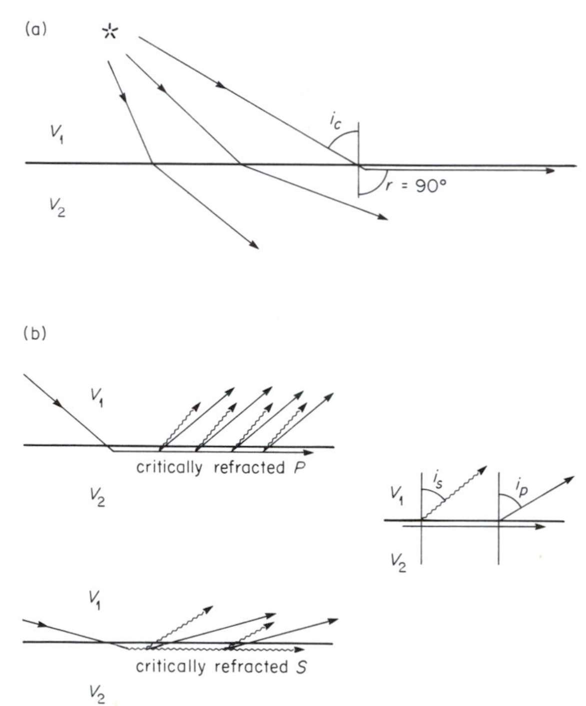

Figure 2–12
Refraction of rays at the boundary between two layers includes (a) a critically refracted ray for which the angle of refraction r = 90 degrees. This ray is produced from an incident ray reaching the boundary at the critical angle of incidence i_c. (b) Critically refracted *P*- and *SV*-waves at each point along the ray produce *P*- and *SV*-waves that refract across the boundary at angles i_p and i_s.

$$\sin i_c = V_1/V_2 \qquad (2\text{–}16)$$

which shows how the critical angle depends on the wave speeds. It is obvious that the angle of refraction (r = 90 degrees) is larger than the critical angle of incidence. We can tell from Snell's law that this condition is possible only if the refracted wave travels faster than the incident wave. Recall from Figure 2–8 that refracted *P*- and *SV*-waves are both produced by either an incident *P*-wave or an incident *SV*-wave. Critical refraction can result from any of these combinations if the refracted wave travels faster than the incident wave.

The critically refracted wave does an interesting thing as it travels close to a boundary between two layers. Figure 2–12b shows how a critically refracted *P*-wave traveling at speed V_{p2} along the top of the lower layer continually produces *P*- and *SV*-waves that refract across the boundary. They travel at speeds V_{p1} and V_{s1} in the upper layer. Let i_p and i_s be the angles of the upward directed rays. They are related to the wave speeds according to Snell's law in the same way as Equation 2–16 relates angles of incident and critically refracted rays: $\sin i_p = V_{p1}/V_{p2}$ and $\sin i_s = V_{s1}/V_{p2}$.

A critically refracted *SV*-wave traveling along the top of the lower layer at speed V_{s2} can also produce *P*- and *SV*-waves that refract into the upper layer (Fig. 2–12b). For this case, the angles are related to wave speeds in the following way: $\sin i_p = V_{p1}/V_{s2}$ and $\sin i_s = V_{s1}/V_{s2}$. The *P*-wave is generated only if $V_{p1} < V_{s2}$.

How does a critically refracted wave continually produce the upward traveling waves illustrated in Figure 2–12? A complete answer would be very complicated. We can glean some understanding from *Huygen's principle,* a statement which asserts that every point on a wave front is a source of new waves that travel away from it in all directions. This idea is illustrated in Figure 2–13, which shows new wave fronts spreading out from a point on an old wave front situated close to a boundary between two layers.

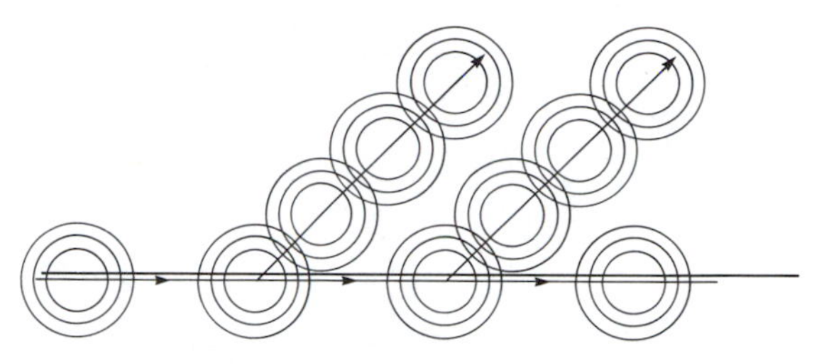

Figure 2–13
Huygen's principle asserts that each point on a wave front is a source of spherically radiating waves. This is shown by spherical wave fronts spreading from different points along paths of advancing waves.

Paths of Seismic Body Waves

The purpose of exploration seismology is to discover what lies underground based on measurements made on the earth's surface. Seismologists detect seismic waves produced at a source close to the surface that follow different paths into the earth, and then return to the surface. Rays from the source radiate in all directions and refract and reflect at each boundary between different substances in accordance with Snell's law. Which of these rays will reach a detector located some distance away from the source?

First, consider a very simple structure consisting of two horizontal layers (Figure 2–14) in which the wave speeds are $V_{p2} > V_{p1} > V_{s2} > V_{s1}$. Suppose that only *P*-waves are produced at the source. We can see that one ray extends along the land surface marking the

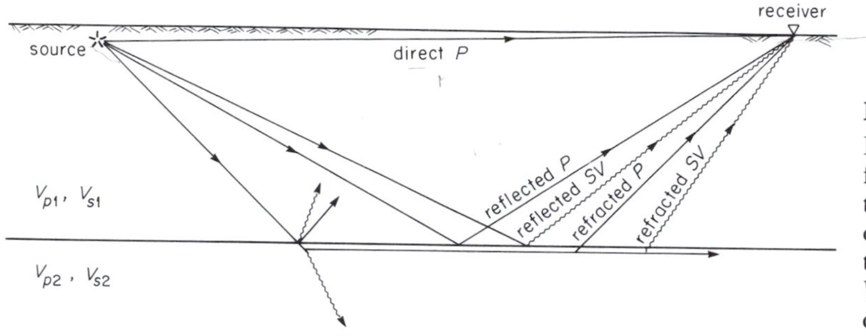

Figure 2–14
Paths followed by seismic waves from a source to a receiver for the idealized structure consisting of only two rock layers in which the wave speeds $V_{p2} > V_{p1} > V_{s2} > V_{s1}$. Each path is consistent with Snell's law.

path of the *direct wave*. Other rays mark paths of the reflected P- and SV-waves produced by different incident P-waves. Still another incident P-wave produces the critically refracted ray from which P- and SV-waves refract to the surface. Altogether, there are five different paths leading to the detector.

Next, look at a structure with several horizontal layers (Figure 2–15). We know that each ray divides into four refracted and reflected P- and SV-wave rays when it reaches a boundary. Clearly, there will be a large number of paths leading to a detector. By omitting all SV-wave rays from the diagram and showing only refracted and reflected P-waves, we

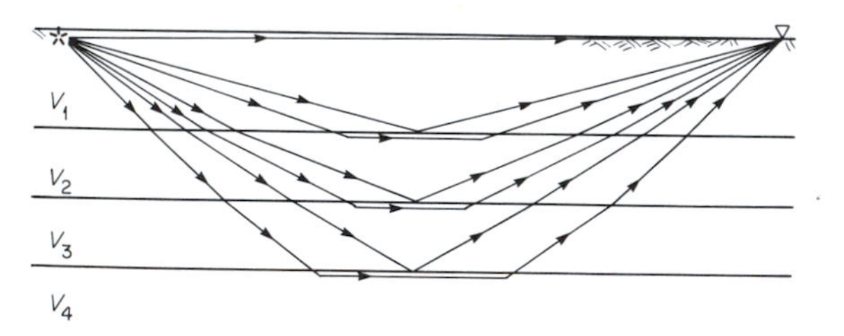

Figure 2–15
Paths followed by P-waves from a source to a receiver in a structure consisting of four parallel layers in which the wave speeds $V_{p1} < V_{p2} < V_{p3} < V_{p4}$. The paths of SV-waves that also would be produced are not shown in this example.

still have many rays converging on the detector.

How much time is required for a seismic wave to follow one of these paths? This time interval can be calculated from Equation 2–8 if we know the length of the path and the wave speed in each layer. For example, the total time t needed for a wave to travel the path ABCD in Figure 2–16 would be

$$t = t_{\mathrm{AB}} + t_{\mathrm{BC}} + t_{\mathrm{CD}}$$
$$= \frac{\mathrm{AB}}{V_{p1}} + \frac{\mathrm{BC}}{V_{p2}} + \frac{\mathrm{CD}}{V_{p1}} \qquad (2\text{–}17)$$

Suppose for the moment that seismic waves were not constrained by Snell's law but could follow other paths as well. What if waves could be refracted along various paths such as the

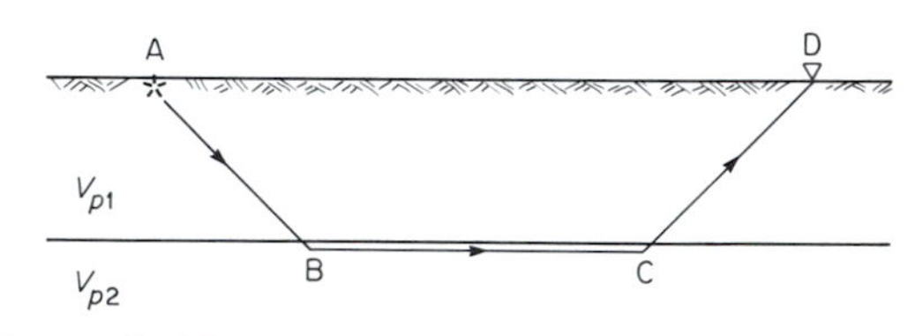

Figure 2–16
Segments of the path of the refracted wave traveling from a source at A to a receiver at D. Wave speed is V_{p1} along segments AB and CD. Along the segment BC, the wave travels at speed V_{p2}.

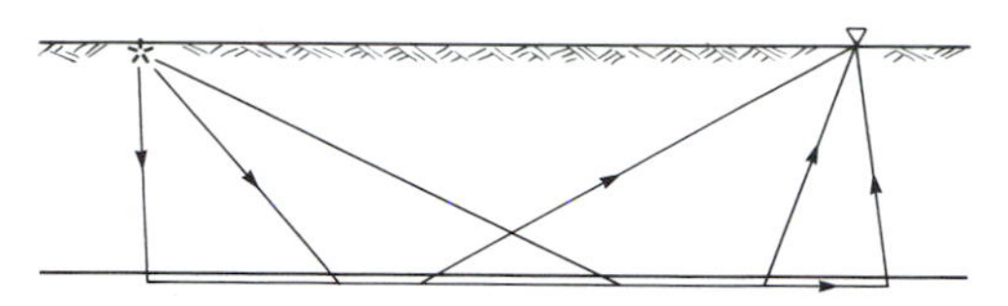

Figure 2–17
Alternate wave paths for testing Fermat's principle. By comparing the travel times along these different paths, we can find the smallest value for the path that is consistent with Snell's law.

ones illustrated in Figure 2–17? Calculating the travel times for these different routes reveals that the particular path requiring the least time is the path predicted from Snell's law. But we know that seismic waves are governed by Snell's law, which means that they do follow paths of minimum travel time. A statement of this fact is called *Fermat's principle,* which asserts that elastic waves travel between two points along paths requiring the least time.

SEISMIC SURFACE WAVES

Recall our discussion of the nature of vibrations produced by passing seismic body waves. As a *P*-wave pulse moves through a rock layer, particles vibrate back and forth in the direction of the ray that marks the path of the wave. The vibrations produced by an *S*-wave are perpendicular to the ray pointing out its path. Material close to the earth's surface experiences these kinds of *P*- and *S*-wave vibrations and other more complicated patterns of vibration as well. These additional kinds of vibration can be measured only at locations close to the surface. Instruments placed away from the surface in boreholes will not detect them. Such vibrations must result from waves that follow paths close to the earth's surface. We call them seismic surface waves.

Rayleigh Waves

Vibrations produced by seismic surface waves are separated into two types. One involves ground movement in a vertical plane aligned with the path of the wave (Figure 2–18). This type of vibration is produced by the particular surface wave bearing the name of the physicist J. W. S. Rayleigh (1842–1919), who made important contributions to our understanding of elasticity. Unlike body waves that produce simple movements back and forth along a straight line, passing Rayleigh waves cause a point on the ground to move along a path shaped like an ellipse.

Another important difference is the dura-

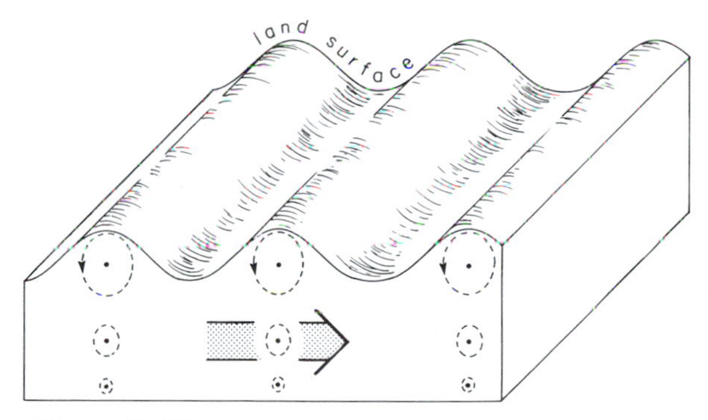

Figure 2–18
Ground vibration produced by Rayleigh waves. Notice that the amplitude of vibration diminishes with depth. As the waves pass, a point on the land surface moves through an elliptical orbit oriented in a vertical plane.

tion of the Rayleigh wave vibration. Ordinarily, a passing body wave pulse will cause a point on the ground to move back and forth a few times before coming to rest. But the Rayleigh wave vibration tends to persist much longer. The point on the ground will cycle around its elliptical path many times.

Rayleigh wave vibrations change with time in an interesting way. As the wave travels past, the ground vibrates slowly at first and then more and more rapidly. The wave frequency increases with time. As we have noted previously, this kind of vibration can be detected close to the earth's surface but becomes weaker with depth below the surface, rapidly diminishing to a level that cannot be measured.

The Rayleigh wave travels at a speed that is slower than the direct *S*-wave. It follows a path along the earth's surface.

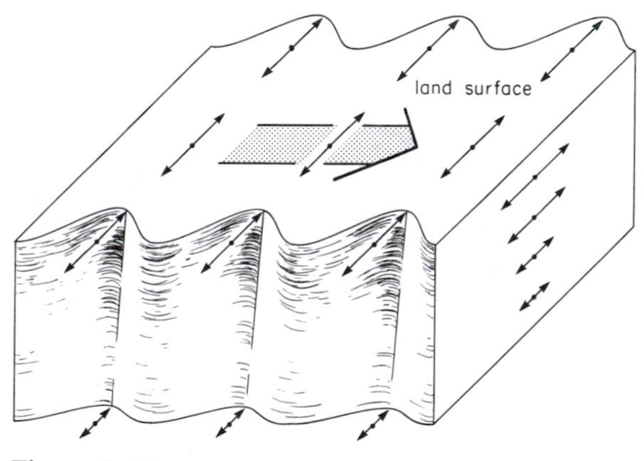

Figure 2–19
Ground vibration produced by Love waves. A point on the land surface cycles back and forth on a horizontal line perpendicular to the direction in which the wave is moving. Amplitude of this horizontal ground movement diminishes with depth.

Love Waves

The other type of seismic surface wave is named after A. E. H. Love (1863–1940), a pioneer geophysicist. Like the Rayleigh wave, the Love Wave travels at a slower speed than the direct *S*-wave. But the direction of the vibration is different. A passing Love wave causes horizontal ground movement perpendicular to the path of the wave (Figure 2–19). And, unlike an *S*-wave, which produces ground movement in the same direction, the Love wave vibration persists much longer, following many cycles of oscillation. One feature of this persistent Love wave vibration is similar to Rayleigh wave ground movement. After beginning slowly, the oscillation continually grows faster as the wave passes. In other words, frequency increases with time.

Wave Guides

In this discussion, we will not give a thorough theoretical explanation of seismic surface waves. But we can gain some understanding by looking at *P*- and *S*-waves that become trapped in layers close to the earth's surface.

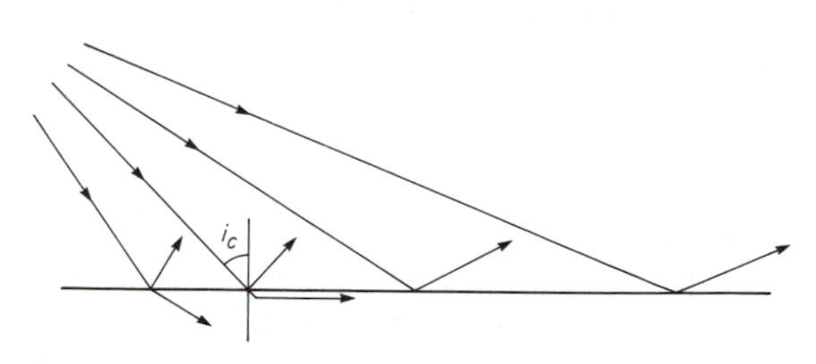

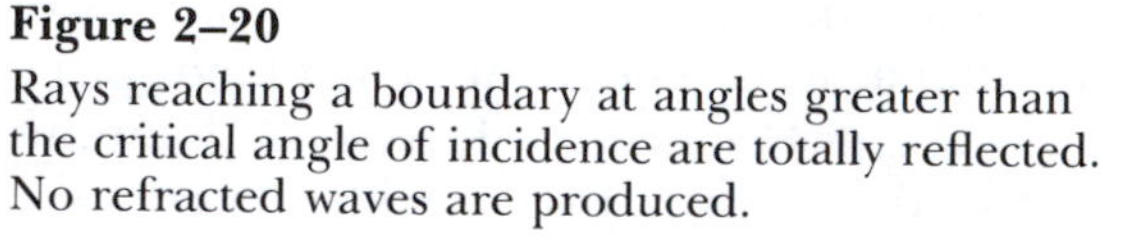

Figure 2–20
Rays reaching a boundary at angles greater than the critical angle of incidence are totally reflected. No refracted waves are produced.

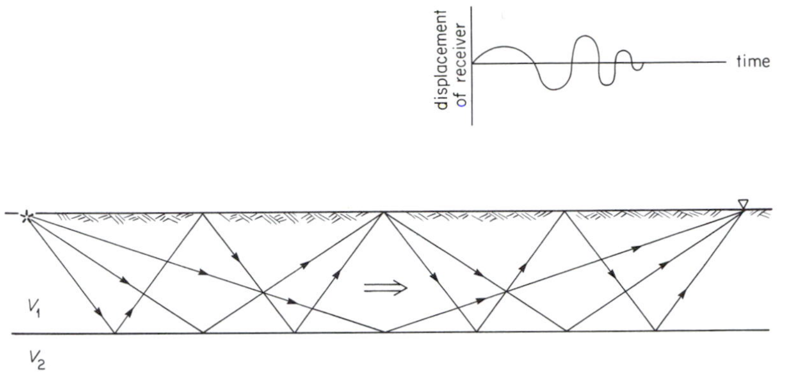

Figure 2–21
Body waves that are totally reflected at the upper and lower boundaries of a "waveguide" layer can interact to produce surface waves.

We have already seen what happens to body wave rays that are incident on a boundary at angles less than or equal to the critical angle i_c (Figure 2–11). These divide into refracted and reflected *P*- and *SV*-waves. But what about a ray with an angle of incidence larger than i_c? It produces only a reflected wave! No wave will refract across the boundary (Figure 2–20). What does this have to do with surface waves?

Typically, layers very close to the earth's surface are "low-speed" layers. Body waves travel more slowly in these layers of soil and weathered rock than in the underlying harder rock. Figure 2–21 shows what happens to *P*- and *S*-waves that are incident on the base of a single "low-speed" layer at angles larger than the critical angle i_c. These waves remain trapped in the low-speed layer, repeatedly reflecting from its base and its upper surface. Because the waves are guided along paths completely within the layer and cannot escape from it, the low-speed layer is called a *waveguide.*

Look at the individual paths in Figure 2–21 of waves traveling in the waveguide from a source to a seismometer. The length of a path depends on the number of times it is reflected. Shorter paths have fewer reflections. Because more time is needed to travel a longer path, a succession of waves will reach the seismometer. Each successively later wave will have followed a path with one more reflection from the base of the waveguide.

Now suppose that each wave adds one cycle of oscillation to the ground movement at the seismometer. The succession of repeatedly reflected waves will then produce a long sequence of oscillations. *P*-waves reflected in this fashion contribute to the oscillations that we recognize as Rayleigh waves. Repeatedly reflected *SH*-waves produce the persistent sequence of vibrations that we call Love waves.

SEISMOGRAMS

Seismic waves follow many different paths between a source and a receiver located some distance away. At the receiver, the wave vibrations are recorded on a chart that we call a *seismogram.* The pattern of vibrations displayed on a seismogram depends on the nature of the waves produced at the source and on how the size of a wave changes along its path.

The Source Wavelet

Suppose that an explosive is detonated at some point in the earth. This is a particularly effective way to produce *P*-waves. Severe deformation occurs in the zone very close to the charge. Rock is shattered and partly melted in this small *source zone.* Outside the source zone, however, there is only momentary elastic deformation indicated by vibrations.

Let us examine the vibration just outside the source zone. In Figure 2–22, *P*-wave rays extend outward in all directions. An individual particle vibrates back and forth in the direction of a ray. This motion is described on the accompanying graph, which shows how the displacement of the particle from its original position changes with time. The wave causing this brief pulse of vibration is called the *source wavelet.* It advances outward from the source zone in all directions.

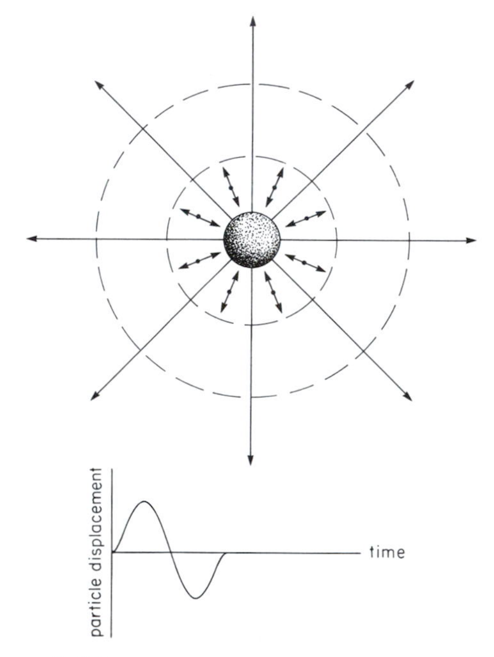

Figure 2–22
Wavelet produced in the source zone is the pulse of vibration that travels outward in all directions. In this idealized example, the source wavelet consists of a single cycle of oscillation.

We can think of the graph in Figure 2–22 as the seismogram that would be recorded at a location very close to the source. We say that it shows the "shape" of the source wavelet. In the example, it has a simple form showing only one cycle of oscillation. Source wavelets of more complicated shape are discussed in Chapter 5.

Geometrical Spreading and Absorption

As a source wavelet travels farther and farther from the source, its amplitude of vibration grows smaller. The ground vibration is caused by the energy of the wave, which comes from the source. After the wavelet leaves the source zone, it receives no more energy. But an advancing wavelet has a continually expanding spherical wave front (Figure 2–7). As the wave front expands, the same quantity of energy must be distributed over a larger spherical surface. This means that less energy is available for each particle on that surface than was available to other particles closer to the source where the wave front was smaller. Therefore, particles farther from the source vibrate at smaller amplitudes than particles closer to the source. This is the result of the effect we call *geometrical spreading.*

How much change in wave amplitude can we expect from geometrical spreading? The surface area of a sphere with radius x is $4\pi x^2$. The energy available to a particle is found by dividing the total energy Ω by the area: $\Omega/4\pi x^2$. It is possible to prove that the amplitude of the particle vibration is proportional to

$\sqrt{\Omega}$. This means that amplitude decreases directly with distance from the source,

$$H = H_0/x \qquad (2\text{–}18)$$

where H_0 is the amplitude of the wavelet leaving the source zone.

When something begins to vibrate, it usually begins to grow warmer at the same time. Heat is produced by the internal friction of particles rubbing together during the process of vibration. In this way, some of the energy of a seismic wave is converted into heat. We say that this wave energy has been absorbed by the material through which the wavelet is traveling. Energy used in this way is not available to cause vibration. So the effect we call *absorption* causes the amplitude of vibration to decrease with distance from the source.

Although we do not understand the process of absorption very well, we have learned from experiments that its effect on wave amplitude can be predicted from the formula

$$H = H_0 e^{-\alpha x} \qquad (2\text{–}19)$$

where α is called the *absorption coefficient,* and $e = 2.71828$ is the Napierian constant. The value of α is different for different materials.

We can combine the equations for geometrical spreading and absorption to predict how the amplitude of a seismic wave should change as it travels away from the source:

$$H = H_0 e^{-\alpha x}/x \qquad (2\text{–}20)$$

Transmission and Reflection Coefficients

The amplitude of a seismic wave is changed by reflection and refraction. Recall from Figure 2–8 that a wave incident on a boundary divides into reflected and refracted *P*- and *SV*-waves. Because the energy must be shared between these reflected and refracted waves, they will have smaller amplitudes than the incident wave.

Ratios that compare the amplitudes of reflected and refracted waves to the amplitude of the incident wave can be calculated. Suppose that an incident wave of amplitude H_0 produces reflected *P*- and *SV*-waves with amplitudes of H_{1p} and H_{1s}. The ratios

$$R_{1p} = H_{1p}/H_0 \quad \text{and} \quad R_{1s} = H_{1s}/H_0 \qquad (2\text{–}21a)$$

are called the *reflection coefficients.* The same incident wave also produces refracted *P*- and *SV*-waves with amplitudes H_{2p} and H_{2s}. The ratios

$$T_{2p} = H_{2p}/H_0 \quad \text{and} \quad T_{2s} = H_{2s}/H_0 \qquad (2\text{–}21b)$$

are called either *transmission coefficients* or *refraction coefficients.*

Values of transmission and reflection coefficients depend on the angle of incidence, densities of the two layers, and the wave speeds in these layers. For most angles of incidence, the values of R_{1p} and T_{2p} will be larger than R_{1s} and T_{2s} for an incident *P*-wave. The opposite is true for an incident *SV*-wave where R_{1s} and T_{2s} tend to be larger than R_{1p} and T_{2p}. If the angle of incidence $i = 0$ degrees for a *P*-wave, then $R_{1s} = R_{2s} = 0$, and

$$R_{1p} = \frac{\rho_2 V_{2p} - \rho_1 V_{1p}}{\rho_2 V_{2p} + \rho_1 V_{1p}}, \qquad T_{2p} = \frac{2\rho_1 V_{1p}}{\rho_2 V_{2p} + \rho_1 V_{1p}}$$

More complicated formulas are needed to compute the transmission and reflection coefficients when the angle of incidence is not zero.

Vibrations at a Receiver

What pattern of vibration do seismic waves from a source some distance away produce at a receiver? Suppose that a wavelet of simple form (Figure 2–22) is produced at the source. Look again at Figures 2–14 and 2–21, which show that many different paths lead this simple pulse of vibration from the source to a receiver. Because of the different distances and wave speeds along these various paths, pulses of vibration reach the receiver at different times. The effects of geometrical spreading, absorption, refraction, and reflection are different along each of these paths. Therefore, the pulses of vibration will have different amplitudes when they reach the receiver.

Now suppose that we plot a graph of the ground vibration at the receiver that would be produced by waves following the paths in Figures 2–14 and 2–21. The result is the *seismogram* seen in Figure 2–23. It consists of several discrete pulses followed by a continuous series of oscillations. Each pulse has the same shape as the source wavelet, but a different amplitude. These pulses indicate waves that followed the paths shown in Figure 2–14. The continuous oscillations that follow are surface waves related to paths seen in Figure 2–21.

The example in Figure 2–23 shows several pulses that are clearly separated from one another. The exploration seismologist is usually presented with a somewhat more complicated pattern of vibrations. In nature, source wavelets are more irregular, and they follow paths through many more layers. Like the example, however, these more complicated seismograms consist of discrete pulses of vibration, one for each different path through the earth, together with surface wave oscillations.

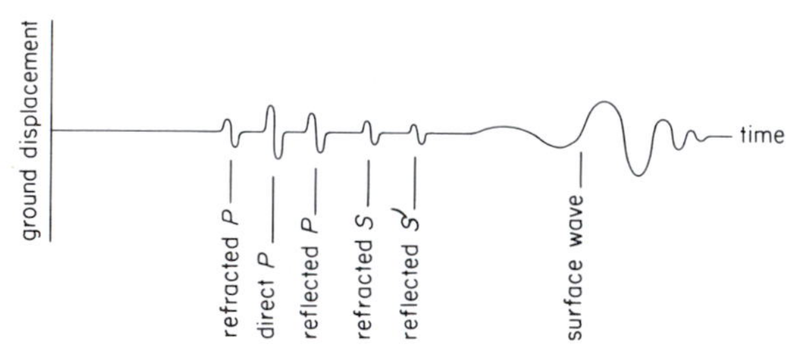

Figure 2–23
Seismogram consisting of body wave pulses, each having the same shape as the source wavelet, and surface wave oscillations with frequency that increases with time. The pulses represent the ground vibration produced by body waves that followed the paths in Figure 2–14.

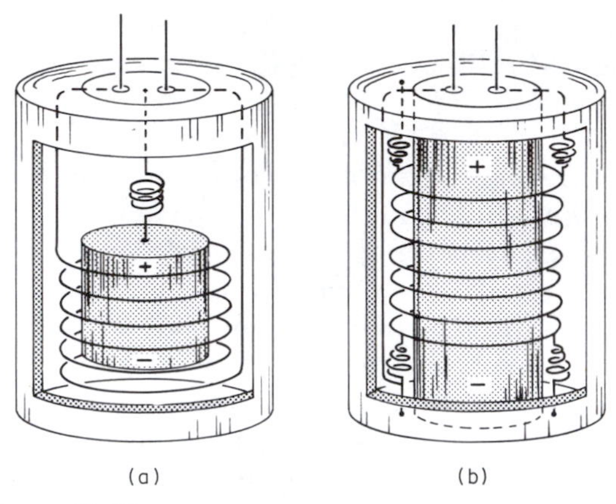

Figure 2–24
A typical geophone (a) has a magnet attached to a spring that can move inside a coil when the ground vibrates. Movement of a magnet in a coil causes an electric current to flow in the coil. (b) An alternate design consists of a coil suspended on springs that encloses a magnet fixed to the container.

Recording Seismic Waves

The exploration seismologist detects ground vibrations with a device that is called either a *seismometer* or a *geophone,* an instrument that is usually about the size of a fist. Two basic designs are illustrated in Figure 2–24. One design consists of a coil of wire mounted in a

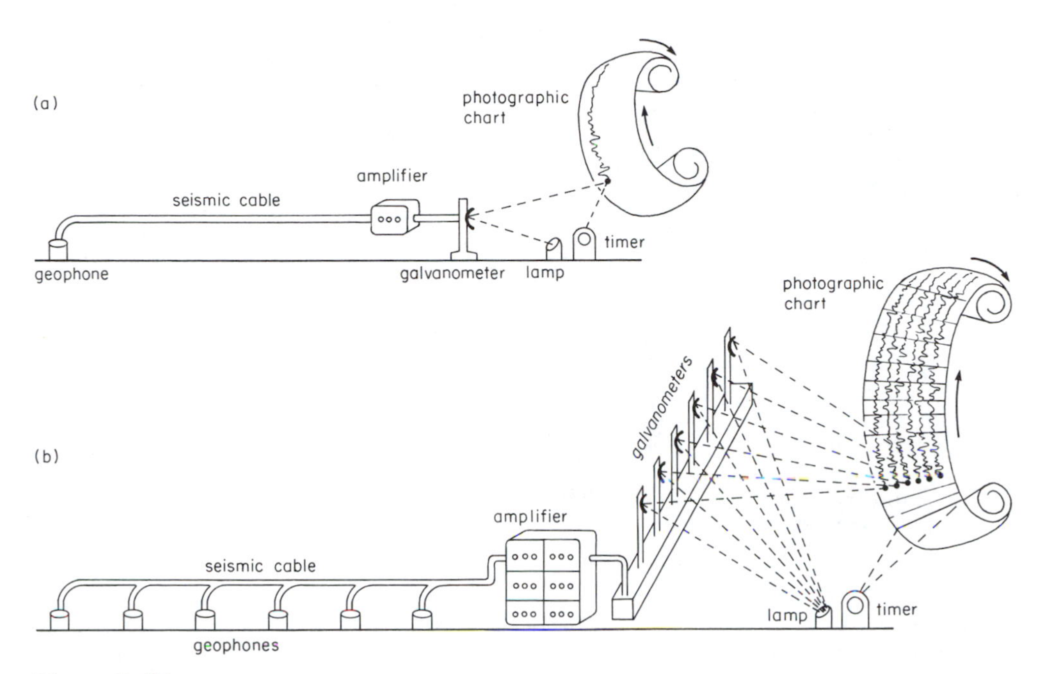

Figure 2–25

Optical–electronic recording of seismic waves. (a) Electric current from the geophone is transmitted through the seismic cable to an amplifier. Then the amplified current activates the galvanometer mirror, which reflects a beam of light onto an advancing photographic chart. (b) A six-channel exploration seismic system with a separate amplifier and galvanometer circuit for recording the electric signal from each geophone.

cylindrical container. A magnetized mass attached to a spring is suspended inside the coil. Ground vibration causes movement of the mass within the coil. Because the mass is a magnet, an electric current that is related to the amplitude of ground vibration is generated in the coil. In the second basic design, the magnet can be fixed to the container and the coil mounted on the spring.

The electric current from a geophone is carried through a line called the *seismic cable* to a recorder. In older seismic recorders, this current was first amplified electronically and then transmitted to a galvanometer (Figure 2–25). This device contains a coil that rotates on a suspension in response to an electric current. A concave mirror attached to this coil focuses a point of light from a nearby lamp onto a chart of photographic paper. As the coil and mirror rotate back and forth and the chart paper advances, usually at a speed of several centimeters per second, an irregular line is exposed on the chart. This irregular line showing all the pulses of ground vibration is the seismogram. Marks are also placed on the chart by a timing device.

The most common practice in exploration seismology is to record simultaneously on the same chart the vibrations detected by 6, 12, 24, 48, or 96 separate geophones (Figure 2–

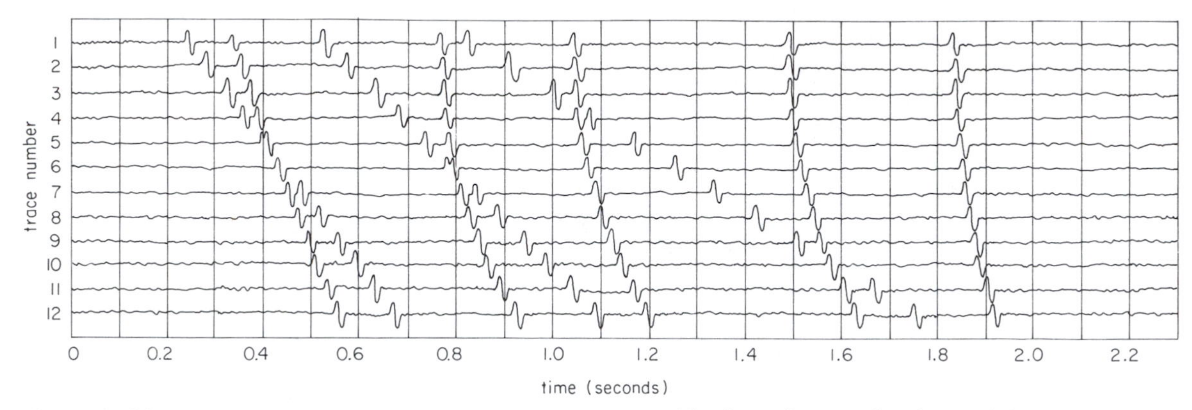

Figure 2–26
A 12-trace exploration seismogram showing pulses of seismic waves refracted and reflected from several rock layers. Vertical time lines on the chart provide the reference for determining the times at which these various pulses arrived at the different geophones.

25). The galvanometers in the recording system are mounted in a row. Lines crossing the chart are exposed every 0.01 second by the timing device. The result is an exploration seismogram (Figure 2–26).

Each geophone–amplifier–galvanometer unit of the system functions independently of the others. Such a unit is one *channel* of an exploration seismic system. The ground vibrations detected by this channel are displayed by one of the irregular lines on the exploration seismogram. Each irregular line is called a *trace.* A 12-channel seismic system, which is larger but otherwise similar to the 6-channel system in Figure 2–25, would produce the 12-trace seismogram shown in Figure 2–26. Modern seismic systems with 6, 12, 24, 48, and 96 channels are described more fully in Chapter 5. These systems are equipped with digital magnetic tape recorders as well as the optical recorders we have described. Records on digital magnetic tape are more convenient for data processing by computer, which is discussed in Chapter 6.

How does a geophysicist interpret the pulses of ground vibration that you see on an exploration seismogram? By displaying traces from a line of geophones on the same chart, we can recognize different alignments of pulses. Observe in Figure 2–26 that some pulses lie along straight lines drawn across the record at different angles. These are pulses from refracted waves. Methods for analyzing them are presented in Chapter 3, whereas analysis of reflected waves is the subject of Chapter 4. In Figure 2–26, reflected waves are indicated by the pulses that lie along curved lines.

STUDY EXERCISES

1. Suppose that a *P*-wave ray produced by an explosion detonated 5 m beneath the ocean surface travels down through the water and reaches solid rock on the ocean bottom at an angle of incidence $i = 30$ degrees. *P*-wave speed is 1500 m/s in the water and 5000 m/s in the rock.

   a. How many reflected and refracted rays are generated by this incident ray?

   b. What is the angle of refraction of the refracted *P*-wave ray?

2. An *SH*-wave originating in a solid layer where the shear modulus is μ_1 and the density is ρ_1 refracts across a boundary into another solid layer where the shear modulus μ_2 is equal to μ_1, but where the density ρ_2 is greater than ρ_1. This implies that the angle of incidence is (a) larger than or (b) smaller than the angle of incidence. Explain!

3. Consider two solid layers, one resting on the other. In the top layer, the *P*-wave speed $V_{p1} = 4000$ m/s and the *S*-wave speed $V_{s1} = 2200$ m/s, and in the deeper layer the *P*-wave speed $V_{p2} = 4000$ m/s and Poisson's ratio $\sigma = 0.25$. Suppose that an *SH*-wave originating in the top layer refracts into the deeper layer.

   a. Is the angle of refraction larger or smaller than the angle of incidence? Explain!

   b. If the density in the deeper layer is $\rho_2 = 2.7$ g/cm^3, what are the values of shear modulus and bulk modulus in that layer?

4. A *P*-wave traveling at a speed of 3000 m/s in one layer refracts across a boundary into another layer where its speed increases to 4000 m/s. If the frequency of vibration is 30 Hz, is the wavelength of the refracted *P*-wave longer or shorter than the wavelength of the incident *P*-wave? Explain!

5. An incident *P*-wave traveling at a speed of 3000 m/s is critically refracted at a boundary. If the critical angle is 30 degrees, what is the speed of the refracted *P*-wave?

6. At a distance of 100 m from a source, the amplitude of a *P*-wave is 0.1000 mm, and at a distance of 150 m the amplitude diminishes to 0.0665 mm. What is the absorption coefficient of the rock through which the wave is traveling?

SELECTED READING

Clark, Sidney P. (editor), *Handbook of Physical Constants,* revised edition. Boulder, Colo., Memoir 97, Geological Society of America, 1966.

Dobrin, Milton B., *Introduction to Geophysical Prospecting,* 3rd edition. New York, McGraw-Hill, 1976.

Garland, George D., *Introduction to Geophysics,* 2nd edition. Philadelphia, W. B. Saunders Co., 1979.

Jaeger, J. C., *Elasticity, Fracture, and Flow.* London, Methuen and Co., Ltd., and New York, Wiley, 1962.

Telford, W. M., L. P. Geldart, R. E. Sheriff, and D. A. Keys, *Applied Geophysics.* London and New York, Cambridge University Press, 1976.

Timoshenko, S., and J. N. Goodier, *Theory of Elasticity.* New York, McGraw-Hill, 1951.

White, J. E., *Seismic Waves—Radiation, Transmission, and Attenuation.* New York, McGraw-Hill, 1965.

THREE

Refracted Seismic Waves and Earth Structure

Seismic refraction surveying is one of our most powerful methods for detecting subsurface structure. This method had its beginnings in the second half of the nineteenth century when the first measurements were made of the speed, or velocity, of seismic waves through various earth materials. Early in the twentieth century, seismologists were able to discern the principal interior zones of the earth from analysis of refracted earthquake waves. During World War I German scientists used some of their methods to detect enemy artillery positions by means of refracted seismic waves produced by cannon recoil. In the decade following the war, these ideas were modified and expanded for application to geologic problems. During that time seismic refraction surveying emerged as one of our most useful tools in the exploration for natural resources.

Our aim in this chapter is to describe how refracted seismic waves can be used to detect the thicknesses and physical properties of buried rock layers. We have already introduced basic procedures for detecting seismic waves with geophones placed in a line extending away from an energy source (Figure 1–2), and for displaying the results on a seismogram (Figure 2–26). How do we identify refracted seismic waves on such a seismogram, and how do we analyze them to measure seismic wave velocities and layer thicknesses?

We learn to identify refracted waves from characteristic alignments of pulses on a seismogram. From the positions of these pulses we find the times required for refracted waves to travel from a common source to geophones at various distances. These travel times are then plotted at the appropriate geophone distances on a graph that we call a *travel time curve,* or a *time–distance curve,* often abbreviated *t–x curve.*

Alignments of points on a *t–x* curve indicate the velocities of seismic waves through different rock layers and provide the information needed to calculate layer thicknesses. We will consider some examples that illustrate how the information on a *t–x* curve is related to earth structure by application of Snell's law, the principles of Huygens and Fermat, and the methods of trigonometry and geometry.

THE SINGLE-LAYER REFRACTION PROBLEM

Critical Refraction

We begin by describing a simple structure consisting of a single horizontal layer of thickness h_1 that lies on another deeper material (Figure 3–1). To keep things as uncomplicated as possible, let us consider only the direct and refracted *P*-waves that are produced at the source S. The *P*-wave velocities will be V_1 in the top layer and V_2 in the underlying material. In addition, let the velocity be faster in the deeper material, that is, $V_2 > V_1$.

Now, suppose that *P*-waves produced at S travel outward in all directions such as those indicated by the rays in Figure 3–1. These rays refract at the boundary according to Snell's law (Equation 2–15), which tells us that larger angles of refraction correspond to larger angles of incidence. For example, because the incident angles are $i'_1 > i_1$, it follows that the angles of refraction will be $i'_2 > i_2$, and so on. Recall from Chapter 2 that we have a limit where the angle of refraction $i_2 = 90$ degrees corresponds to the critical angle of incidence i_c. For that particular case Snell's law takes the form

$$\sin i_c = V_1/V_2 \qquad (3\text{–}1)$$

which is the same as Equation 2–16. Remember also the principles of Huygens and Fermat which tell us that the critically refracted wave travels at velocity V_2 along the boundary of the deeper material, at the same time producing waves that refract back into the upper layer at the critical angle i_c (Figure 2–13). In this way it is possible for the critically re-

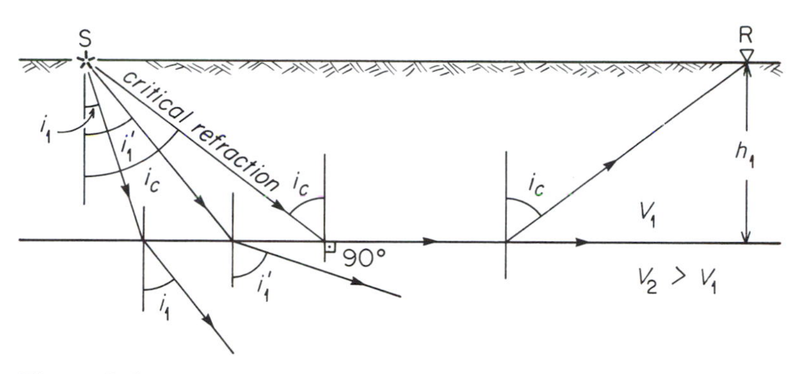

Figure 3–1
Refraction of seismic waves in a structure consisting of an upper layer in which wave velocity is V_1 separated by a plane horizontal boundary from underlying material in which wave velocity is V_2. The first layer has thickness h_1, and the velocity V_2 is greater than the velocity V_1. Three rays departing from the energy source (S) illustrate refraction at the boundary. The ray corresponding to a refraction angle of 90 degrees is called the critically refracted wave and can be observed at a receiver (R) on the surface.

fracted wave in Figure 3–1 to travel from the source S to a receiver at R along a path that reaches the boundary of the deeper material. This boundary is also called the *refractor* because it is here that the paths of seismic waves are refracted.

Now let us draw attention to an important fact. In order to detect a refracted wave, we must place a receiver far enough from the source for critical refraction to be possible. The minimum distance, indicated by X_{crit} in Figure 3–2, is called the *critical distance*. Observe that a refracted wave reaching a receiver at R′ would have traveled only an infinitesimally short distance along the refractor at point A before refracting upward to R′. A similarly refracted wave reaching a receiver at R would have traveled the distance AB along the refractor. The triangle SAO tells us that the critical distance is related to the critical angle of incidence and the layer thickness. We see that

$$\tan i_c = \frac{X_{\text{crit}}/2}{h_1} \tag{3–2}$$

which can be rearranged to give

$$X_{\text{crit}} = 2h_1 \tan i_c \tag{3–3}$$

Because we are concerned with critical refraction, we can use Equation 3–1 and some well-known trigonometric identities to show that

$$\begin{aligned}\cos i_c &= (1 - \sin^2 i_c)^{1/2} \\ &= \left[1 - \left(\frac{V_1}{V_2}\right)^2\right]^{1/2} = \left(\frac{V_2^2 - V_1^2}{V_2^2}\right)^{1/2}\end{aligned} \tag{3–4}$$

and therefore,

$$\tan i_c = \frac{\sin i_c}{\cos i_c} = \frac{V_1}{(V_2^2 - V_1^2)^{1/2}} \tag{3–5}$$

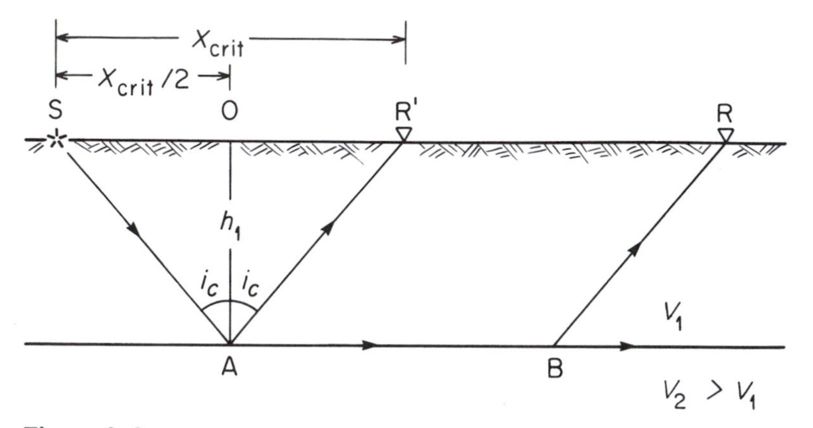

Figure 3–2
Minimum observation distance for waves refracted from a horizontal plane boundary. Because velocity V_2 is greater than V_1, the critical ray from the source (S) is refracted along the interface and back to the surface to be recorded at receivers R′ and R. The wave refracted from the point A to the surface receiver R′ is the first observable refracted wave, and the minimum distance for detecting the refracted wave is called the critical distance. Any point on the refractor beyond the point A refracts rays to the surface. The critical distance (X_{crit}) is controlled by the thickness of the layer and the velocity contrast between V_1 and V_2.

This equation can be rearranged to obtain

$$X_{crit} = \frac{2h_1}{\left[\left(\frac{V_2}{V_1}\right)^2 - 1\right]^{1/2}} \tag{3–6}$$

This equation tells us that the critical distance is directly proportional to the layer thickness and indirectly proportional to the velocity ratio V_2/V_1. For a given layer thickness, the larger the velocity contrast, the shorter the critical distance along which refracted waves cannot be detected. Equation 3–6 can be useful in planning the spacing of geophones for a seismic refraction survey if we can make a preliminary estimate of the velocity contrast we expect to encounter. For example, suppose that $V_1 = 1000$ m/s and $V_2 = 1414$ m/s, so that the ratio $V_2/V_1 = 1.414 = \sqrt{2}$. For this case, Equation 3–6 tells us that

$$X_{crit} = 2h_1 \tag{3–7}$$

Therefore, we can expect to detect a refracted wave if a boundary exists at a depth that is less than one-half the distance of a geophone from the source.

Preparing a Travel Time Curve

So far, we have discussed how the path of a refracted seismic wave is related to the wave velocities above and below a refractor, and the layer thickness. However, before we do a seismic refraction survey, we do not know the values of V_1, V_2, and h_1. The survey will measure these values. The first step, then, is to place several geophones in a line, and produce the seismic waves, perhaps by detonating an explosive and recording a seismogram. The next step is to prepare a travel time curve using information obtained from the seismogram.

Look again at the seismogram in Figure 2–26. Notice that several pulses appear on each of the traces. For the purposes of many seismic refraction surveys, only the first pulse on each trace is used in the analysis. This pulse is called the *first arrival,* because it indicates the first of several waves that reach the geophone.

Now let us examine Figure 3–3, which helps explain how a travel time curve is prepared. Here we see a source S and a line of geophones $R_1, R_2, \ldots, R_{12}$ on the surface of a simple one-layer structure. Above this structure is an idealized seismogram that shows only the first arrivals. After such a seismogram is recorded, the seismologist can write the geophone distances $x_1, x_2, \ldots, x_{12}$ on the seismogram traces. Then the time (t) of onset of the first arrival is measured for each trace. These times, $t_1, t_2, \ldots, t_{12}$, are marked on the seismogram. Points corresponding to these distance and time values are plotted on the graph at the top of Figure 3–3. This graph is the travel time curve.

Measuring Seismic Wave Velocities

What does this travel time curve tell us about the structure? First, look at the travel time points corresponding to the nearest geophones, R_1, R_2, and R_3. Observe that a straight line can be drawn through these points that also passes through the origin of the graph. This fact indicates that the first arrivals to reach these geophones must be the direct waves that have traveled straight along the surface from the source. The straight alignment shows that the additional time required for the wave to travel to a farther geophone is directly proportional to the additional distance to that geophone. Now let us find the slope of the straight line. This can be done by using the time and distance values for any two geophones, for example, R_1 and R_3, in the following way:

$$\text{slope} = \frac{x_3 - x_1}{t_3 - t_1} = \frac{\Delta_x}{\Delta_t} = \frac{1}{V_1} \qquad (3\text{–}8)$$

This relationship is very important! It tells us that we can calculate the velocity V_1 of the seismic waves in the top layer from the slope of a straight line drawn through the travel time points.

Next, examine the travel time points corresponding to the more distant geophones R_5, $R_6, \ldots, R_{12}$. These points plot along a different straight line. Rather than intersect the origin of the graph, the straight line intersects the vertical axis at the time value of T_1, which we call the *intercept time.* The importance of T_1 will be discussed later in this chapter.

Because the travel time points corresponding to the more distant geophones plot on a different straight line, they cannot be representing direct waves. Therefore, they must be representing refracted waves. But how can a refracted wave be a first arrival, reaching a geophone ahead of the direct wave? The ray paths in Figure 3–3 make it obvious that the refracted wave must always travel a longer distance than the direct wave. However, the refracted wave has the advantage of traveling at higher velocity through the deeper material. Since $V_2 > V_1$, there will be a point beyond which this advantage of speed overcomes the disadvantage of distance so that the refracted wave arrives before the direct wave.

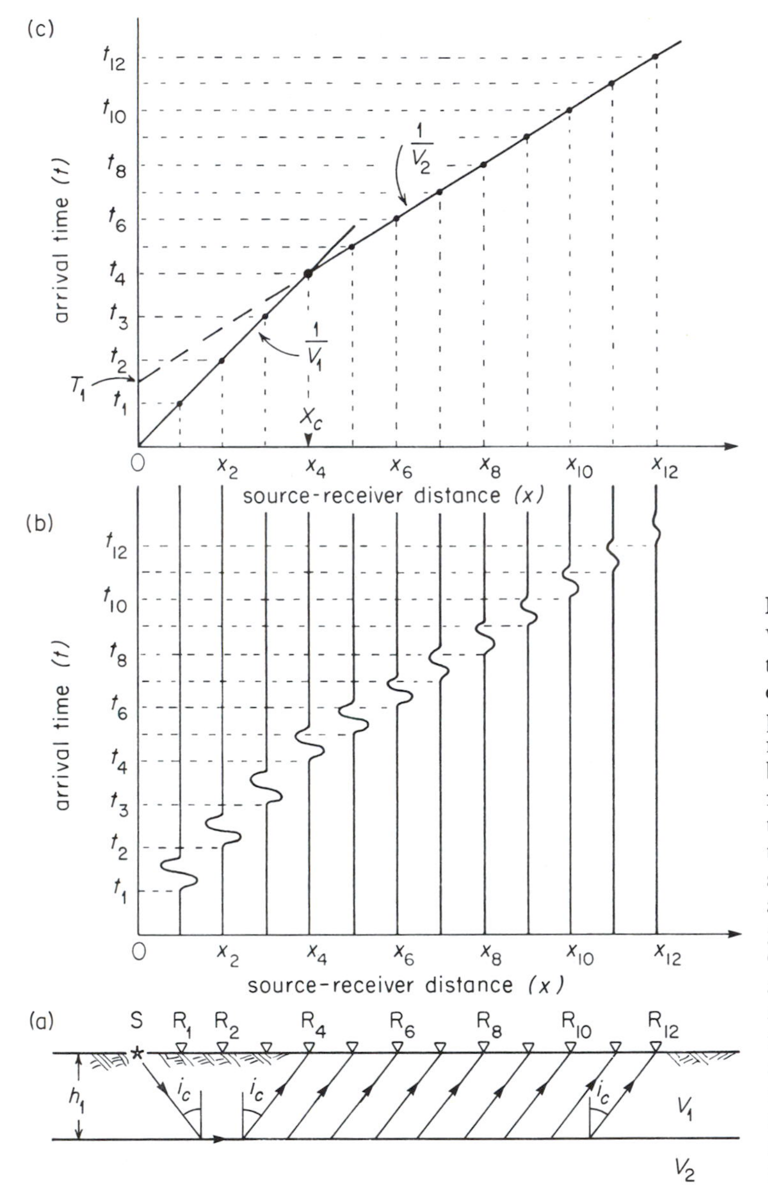

Figure 3–3
Wave paths, seismogram, and travel time curve for direct waves and waves critically refracted from a horizontal plane boundary. The figure shows recording of direct and refracted waves by a 12-channel system. (a) The critical ray from the source (S) refracts along the interface and back to the surface at the receivers from R_3 to R_{12}. (b) The seismogram consists of traces observed at receivers. Each trace represents ground vibration as a function of time. Onset times of the first arriving waves are marked on the vertical time axis, and the distances from the source to receivers are marked on the horizontal axis. (c) Time–distance curves are constructed by drawing lines through alignments of points that show arrival times at different distances. Slopes of the lines indicate the velocities.

The fact that travel time points for the refracted waves plot on a straight line indicates that the additional time required for the wave to reach a farther geophone is directly proportional to the additional distance to that geophone. We see in Figure 3–3 that all the refracted waves follow the same path from the source down to the refractor, and they all follow similar paths from the refractor upward to the geophones. The only difference between their paths is the distance traveled along the refractor. For any two geophones, this difference equals the distance between those geophones, because the upward paths through the layer are parallel. Therefore, the slope of the straight line, found from time and distance values for any pair of geophones, for example, R_5 and R_{10}, can be used to calculate the seismic wave velocity V_2 in the deeper material:

$$\text{slope} = \frac{x_{10} - x_5}{t_{10} - t_5} = \frac{\Delta x}{\Delta t} = \frac{1}{V_2} \qquad (3\text{–}9)$$

Let us summarize the basic procedure for measuring V_1 and V_2 by means of a seismic refraction survey.

1. Place an energy source, such as an explosive, and a line of geophones along the surface.
2. Record a seismogram.
3. Measure the times of first arrivals from the seismogram traces.
4. Plot these times at the corresponding geophone distances on a graph.
5. Draw lines through straight alignments of points to get the t–x curve.
6. Measure slopes of the straight lines and calculate velocities from the reciprocals of these slopes.

Another feature of the travel time curve in Figure 3–3 should be pointed out. The distance at which the two straight lines intersect is called the *crossing distance* X_c. A geophone placed at this distance would receive both the direct wave and the refracted wave at exactly the same time. For all distances beyond X_c, the refracted wave will be the first arrival, and the direct wave will become a later arrival.

What is the relationship between the crossing distance X_c and the critical distance X_{crit} that we introduced earlier? Look again at Figure 3–2. We see that a direct wave reaching R′ travels the distance SR′ $= X_{crit}$ at velocity V_1. But the refracted wave must follow the longer path SAR′, traveling at the same speed. Because it is immediately refracted back to R′ at the point of incidence A, it does not travel along the refractor at velocity V_2. Therefore, the refracted wave must arrive later than the direct wave at R′, implying that $X_{crit} < X_c$. At distances between X_{crit} and X_c, the direct wave will be the first arrival, and the refracted wave will be a later arrival on the seismogram.

Calculating Layer Thickness

Next, let us find out how to calculate the layer thickness h_1 defined by a horizontal interface from information on the travel time curve. Two approaches can be used to analyze this problem. One makes use of the crossing distance X_c, and the other utilizes the intercept time T_1. We will begin with the analysis based on X_c.

Suppose that a source S and a receiver R are separated by the distance x, as in Figure 3–4. Then the travel time t_D for a direct wave will be simply

$$t_D = x/V_1 \qquad (3\text{–}10)$$

The travel time for the refracted wave will be

$$t_R = \frac{\text{SA}}{V_1} + \frac{\text{AB}}{V_2} + \frac{\text{BR}}{V_1} \qquad (3\text{–}11)$$

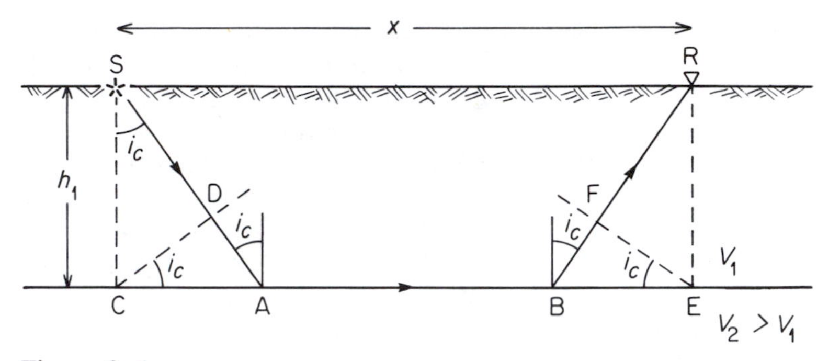

Figure 3–4
Geometrical features of the travel path and the wave front of a seismic wave critically refracted along a horizontal plane boundary. The critically refracted wave between the source (S) and receiver (R) can be studied by the geometry of wave fronts. If SA and BR are the paths of the critically refracted ray in the first layer, then CD and EF are the critically refracted wave fronts at the times when the critical wave reaches the interface and departs from the interface, respectively. The travel time for the ray path defined by SABR is equivalent to the travel time for the wave front moving between S and D, C and E, and F and R.

Now examine the right triangles SCA and BER that show us that

$$\mathrm{SA} = \mathrm{BR} = h_1/\cos i_c \qquad (3\text{–}12)$$

$$\mathrm{CA} = \mathrm{BE} = h_1 \tan i_c \qquad (3\text{–}13)$$

and finally,

$$\mathrm{AB} = x - \mathrm{CA} - \mathrm{BE} \qquad (3\text{–}14)$$

Substituting from Equations 3–12, 3–13, and 3–14 into Equation 3–11 gives

$$t_R = \frac{2h_1}{V_1 \cos i_c} + \frac{x - 2h_1 \tan i_c}{V_2} \qquad (3\text{–}15)$$

Then, using the identity $\tan i_c = \sin i_c/\cos i_c$, and rearranging the terms, we obtain

$$t_R = \frac{x}{V_2} + \frac{2h_1}{V_1 \cos i_c}\left(1 - \frac{V_1}{V_2}\sin i_c\right) \qquad (3\text{–}16)$$

According to Equation 3–1, however, we can write

$$1 - \frac{V_1}{V_2}\sin i_c = 1 - \sin^2 i_c = \cos^2 i_c$$

and therefore,

$$t_R = \frac{x}{V_2} + \frac{2h_1}{V_1}\cos i_c \qquad (3\text{–}17)$$

We know that at the crossing distance X_c the direct and refracted waves arrive at the same time, that is, $t_D = t_R$. Therefore, if $x = X_c$, we can set Equation 3–10 to equal Equation 3–17 to obtain

$$\frac{X_c}{V_1} = \frac{X_c}{V_2} + \frac{2h_1}{V_1}\cos i_c$$

Substituting from Equation 3–4 for $\cos i_c$, we get

$$X_c\left(\frac{1}{V_1} - \frac{1}{V_2}\right) = X_c\left(\frac{V_2 - V_1}{V_1V_2}\right) = \frac{2h_1}{V_1}\left(\frac{V_2^2 - V_1^2}{V_2^2}\right)^{1/2} \qquad (3\text{–}18)$$

Rearranging to solve for h_1, we have

$$h_1 = \frac{X_c}{2}\left(\frac{V_2 - V_1}{V_1V_2}\right)\left[\frac{V_1V_2}{(V_2^2 - V_1^2)^{1/2}}\right] \qquad (3\text{–}19)$$

and finally,

$$h_1 = \frac{X_c}{2}\left(\frac{V_2 - V_1}{V_2 + V_1}\right)^{1/2} \qquad (3\text{–}20)$$

This very important result can be used for interpreting a seismogram such as the one shown in Figure 3–3. After the travel time curve has been prepared, the velocities V_1 and V_2 can be found from the slopes of the two straight lines. Then the crossing distance X_c can be observed from the intersection of these lines. These three values are used in Equation 3–20 to calculate the layer thickness.

Now let us look at the other method for finding layer thickness that is based on the intercept time T_1. Recall from Figure 3–3 that T_1 is the point at which the straight line representing the refracted waves intersects the vertical axis of the travel time graph. This method involves analysis of wave fronts that were introduced in Chapter 2. It would be helpful to review the discussion of Figure 2–9 which introduces some important relationships between wave fronts and refracted rays.

In Figure 3–4 the line CD lies on the wave front of a wave traveling downward from the source S to the refractor. Similarly, the line EF lies on the wave front of a wave moving upward from the refractor. Note that these wave fronts are perpendicular to the lines SA and BR, which indicate the directions of the advancing waves. Now, observe carefully that whereas one point on the wave front is at D, another part of that wave front has already encountered the refractor at C. Then, as one part advances from D to A at velocity V_1, another part of the wave front is sweeping along the refractor from C to A at velocity V_2. Similarly, part of the upward traveling wave front, moves from B to F at velocity V_1 while another part sweeps along the refractor from B to E at velocity V_2. Therefore, we have the relationship

$$\frac{\text{DA}}{V_1} = \frac{\text{CA}}{V_2} = \frac{\text{BF}}{V_1} = \frac{\text{BE}}{V_2} \qquad (3\text{–}21)$$

The travel time of the refracted wave reaching a geophone at R in Figure 3–4 can be expressed as

$$t_R = \frac{\text{SD}}{V_1} + \frac{\text{DA}}{V_1} + \frac{\text{AB}}{V_2} + \frac{\text{BF}}{V_1} + \frac{\text{FR}}{V_1} \qquad (3\text{–}22)$$

or by substituting from Equation 3–21 we can obtain

$$t_R = \frac{\text{SD} + \text{FR}}{V_1} + \frac{\text{CE}}{V_2} \qquad (3\text{–}23)$$

But the triangles SDC and RFE tell us that

$$\text{SD} = \text{FR} = h_1 \cos i_c \qquad (3\text{–}24)$$

and we know that CE $= x$. Therefore, the travel time can be expressed as

$$t_R = \frac{x}{V_2} + \frac{2h_1}{V_1}\cos i_c \qquad (3\text{–}25)$$

which is the same as Equation 3–17. For the structure we have been studying, the velocities V_1 and V_2, the layer thickness h_1, and the critical angle are fixed. Therefore, the second term must be equal to a constant k,

$$\frac{2h_1}{V_1}\cos i_c = k \qquad (3\text{–}26)$$

so that

$$t_R = \left(\frac{1}{V_2}\right)x + k \qquad (3\text{–}27)$$

We recognize that this formula is used for a straight line on a graph of t and x, which has a slope $= 1/V_2$ and an intercept k. Equation 3–27, then, is the expression for the straight line in Figure 3–3 that represents the refracted waves. The constant k turns out to be the intercept time, so that

$$k = T_1 = \frac{2h_1}{V_1} \cos i_c \qquad (3\text{–}28)$$

Solving for the layer thickness, we get

$$h_1 = \frac{T_1}{2} \frac{V_1}{\cos i_c} \qquad (3\text{–}29)$$

and substituting from Equation 3–4 for $\cos i_c$ gives

$$h_1 = \frac{T_1 V_1 V_2}{2(V_2^2 - V_1^2)^{1/2}} \qquad (3\text{–}30)$$

This result is just as important as the one in Equation 3–20. From the travel time curve (Figure 3–3) velocities V_1 and V_2 are found, as before, from the slopes of the lines. Then the intercept time T_1 can be observed from the intersection of the line representing the refracted waves and the vertical axis of the graph. These three values can be used in Equation 3–30 to calculate the layer thickness.

Relationships Between Intercept Time and Crossing Distance

We know that layer thickness can be calculated by means of the velocities V_1 and V_2, and either the crossing distance X_c or the intercept time T_1. If we set Equations 3–20 and 3–30 equal to each other, we obtain

$$h_1 = \frac{X_c}{2}\left(\frac{V_2 - V_1}{V_2 + V_1}\right)^{1/2} = \frac{T_1 V_1 V_2}{2(V_2^2 - V_1^2)^{1/2}} \qquad (3\text{–}31)$$

The equation can be rearranged to express the crossing distance in terms of the intercept time and the velocities:

$$X_c = T_1 \left(\frac{V_1 V_2}{V_2 - V_1}\right) \qquad (3\text{–}32)$$

Next, by rearranging Equation 3–6 to solve for h_1 and setting the result equal to Equation 3–20, we get

$$h_1 = \frac{X_{crit}}{2}\left(\frac{V_2^2 - V_1^2}{V_1^2}\right)^{1/2} = \frac{X_c}{2}\left(\frac{V_2 - V_1}{V_2 + V_1}\right)^{1/2} \qquad (3\text{–}33)$$

which can be rearranged to relate the critical distance and the crossing distance

$$\frac{X_{crit}}{X_c} = \frac{V_1}{V_1 + V_2} \qquad (3\text{–}34)$$

This proves our earlier conclusions that $X_{crit} < X_c$, so that in the interval between X_{crit} and X_c the first arrivals on a seismogram must be the direct waves and the refracted waves will be later arrivals.

Let us look once again at Figure 3–3. We see that for geophones $R_5, R_6, \ldots, R_{12}$, the first arrivals are refracted waves. For each receiver let us calculate

$$\Delta t_x = t_x - \frac{x}{V_2} \qquad (3\text{–}35)$$

where t_x is the travel time of the refracted wave and x is the distance of that receiver.

The term Δt_x is called the *delay time*. We might think of it in this way: If the wave could travel directly to the receiver at velocity V_2, it would arrive sooner than the actual travel time t_x. Because the wave must travel at the slower velocity V_1 through the upper layer, it is "delayed." We know from Equations 3–25 and 3–28 that the intercept time

$$T_1 = t_R - \frac{x}{V_2} \tag{3-36}$$

where the travel time t_R of a refracted wave is the same as t_x in Equation 3–35. Thus, we know that the delay times calculated for all the receivers should have the same value, which is identical to the intercept time. This would be true for an ideal structure for which the values of h_1, V_1, and V_2 are everywhere constant. But in nature there are no perfectly uniform rock layers. Small changes from place to place can produce small differences in the delay times calculated for different receivers. Where such irregularities exist, we can calculate an average layer thickness by using an intercept time found from the average of the delay times for the different receivers.

Application

A practical application of these methods is a short refraction survey of the layer of soil and alluvium that overlies bedrock at a construction site. The procedure and results of such a survey are presented in Figure 3–5. For this example the field procedure began by arranging 12 geophones in a line at 2-meter intervals. The seismic cable connects them to the portable amplifier–recorder unit. *P*-waves were then produced at the source by pounding a hammer on a steel plate. A switch mounted on the hammer turns on the amplifier–recorder unit at the moment of impact, and a seismogram is recorded. First arrival times, read from this seismogram, were entered in the table of values at the corresponding geophone distances. These data were plotted on the travel time curve, and straight lines were drawn through the alignments of points.

From the slopes of the lines we obtain

$$V_1 = 415 \text{ m/s} \quad \text{and} \quad V_2 = 2055 \text{ m/s}$$

and we observe that the intercept time is 0.025 second and the crossing distance is close to 12.8 meters. Using Equation 3–20, we calculate

$$h = 5.2 \text{ m}$$

and from Equation 3–30 we get

$$h = 5.3 \text{ m}$$

There is only a small difference in these values because we cannot read the crossing distance and the intercept time more accurately from the travel time curve.

This survey indicates that the thickness of soil and alluvium overlying bedrock is about 5¼ meters. Field operations took approximately 15 minutes, and the analysis required about the same amount of time. Altogether, then, this sounding was done in about half an hour by means of refraction surveying.

This small-scale survey demonstrates a procedure that is followed in larger surveys that are used to probe deeper into the earth. Here a geophone line 24 meters long was used to detect a shallow refractor. The same basic procedure can be used with much longer geophone lines. On a very large scale, geophones have been placed in lines reaching several hundred kilometers to measure the thickness of the earth's crust and to detect structure in the upper mantle. For such a long line, shortwave radios rather than a seismic cable are

(a)

(b)

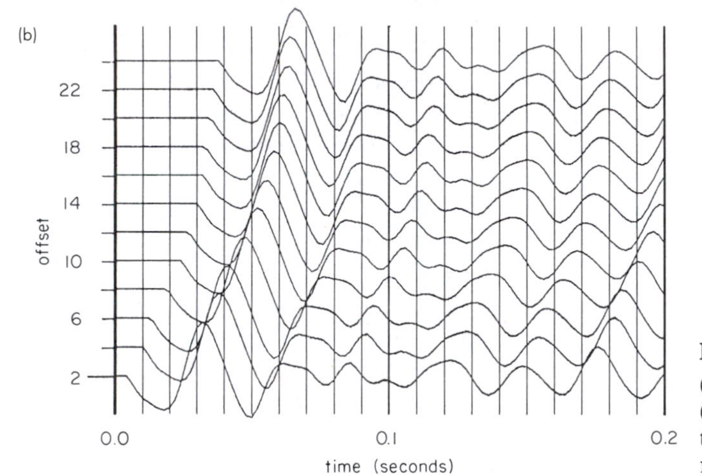

(c)

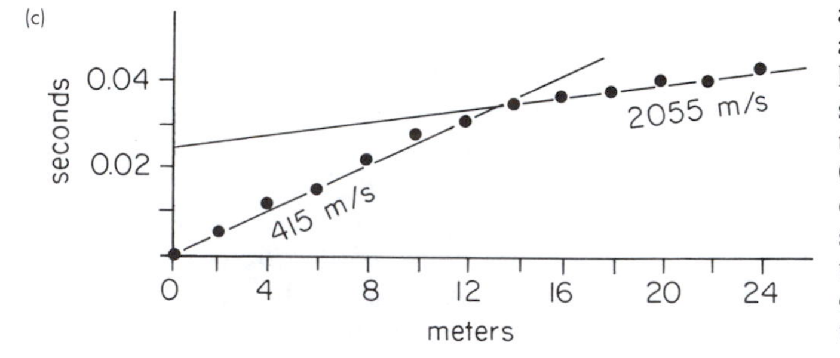

(c)

x (meters)	t (seconds)
2	0.005
4	0.011
6	0.014
8	0.020
10	0.025
12	0.028
14	0.031
16	0.033
18	0.034
20	0.036
22	0.037
24	0.039

Figure 3–5

(a) Portable seismic recording equipment, (b) refraction seismogram, and (c) travel time curve of first arrival times for a short refraction survey in southwestern Virginia. (a) Twelve receivers with 2-meter intervals are used to record the direct and refracted arrivals. Waves were generated by hammering a metal plate. (b) The seismogram horizontal axis is the recording time that is marked by the vertical lines at 0.01-second intervals. (c) The time–distance curve prepared from the seismogram. Straight lines through the travel time points show that the crossing distance is at 12.8 meters and intercept time is 0.025 second.

used to connect geophone positions. Otherwise, the procedure is quite similar to that illustrated in Figure 3–5.

REFRACTED WAVES IN MULTILAYERED STRUCTURES

We can use the same basic method to analyze refracted seismic waves in a multilayered structure that we have introduced for a single layer. Snell's law, critical refraction, and the geometry of rays and wave fronts are used in the same way.

The Ray Parameter

The multilayered structure in Figure 3–6 consists of four horizontal layers with thicknesses

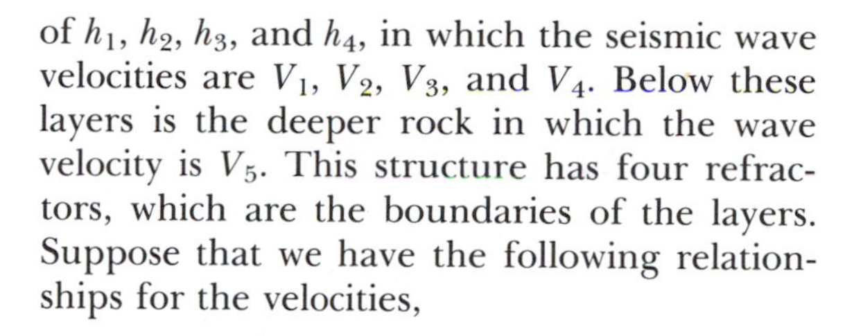

of h_1, h_2, h_3, and h_4, in which the seismic wave velocities are V_1, V_2, V_3, and V_4. Below these layers is the deeper rock in which the wave velocity is V_5. This structure has four refractors, which are the boundaries of the layers. Suppose that we have the following relationships for the velocities,

$$V_1 < V_2 > V_3 < V_4 < V_5$$

Let us consider a ray path originating at the source S that is critically refracted at the deepest refractor. At the other refractors noncritical refraction occurs. By applying Snell's law at each refractor, we obtain

$$\frac{\sin i_{12}}{\sin i_{23}} = \frac{V_1}{V_2}, \quad \frac{\sin i_{23}}{\sin i_{34}} = \frac{V_2}{V_3},$$

$$\frac{\sin i_{34}}{\sin i_{45}} = \frac{V_3}{V_4}, \quad \frac{\sin i_{45}}{\sin i_5} = \frac{V_4}{V_5}$$

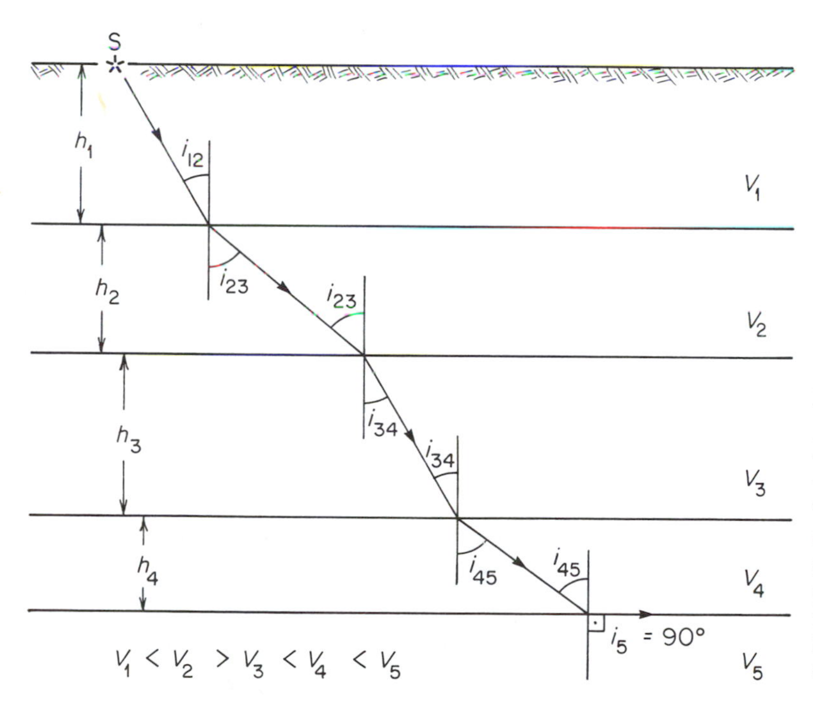

Figure 3–6
Path of a seismic wave refracted through a multi-layered structure with horizontal plane refractors. The ray path is defined by the departure angle (i_{12}), velocities, and layer thicknesses. The ray path represents the critically refracted wave at the interface between layers where the velocities are V_4 and V_5.

Because we can rearrange the terms in the following way,

$$\frac{\sin i_{12}}{V_1} = \frac{\sin i_{23}}{V_2} = \frac{\sin i_{34}}{V_3}, \text{ etc.}$$

and because $i_5 = 90$ degrees, we can write

$$\frac{\sin i_{12}}{V_1} = \frac{\sin i_{23}}{V_2} = \frac{\sin i_{34}}{V_3} = \frac{\sin i_{45}}{V_4} = \frac{\sin i_5}{V_5} = \frac{1}{V_5} = p \quad (3\text{–}37)$$

where p is a constant called the *ray parameter* or the *slowness* of the ray path. We see that p is the same for all parts of the path that a refracted wave follows through a multilayered structure.

Wave Fronts and Rays

Now let us consider how a wave front advances through a multilayered structure. This discussion will allow us to recognize some very important geometrical relationships that can be used later to find layer thicknesses. To keep the discussion as uncomplicated as possible, let us consider the structure in Figure 3–7 which consists of two horizontal layers resting on a deeper material in which the

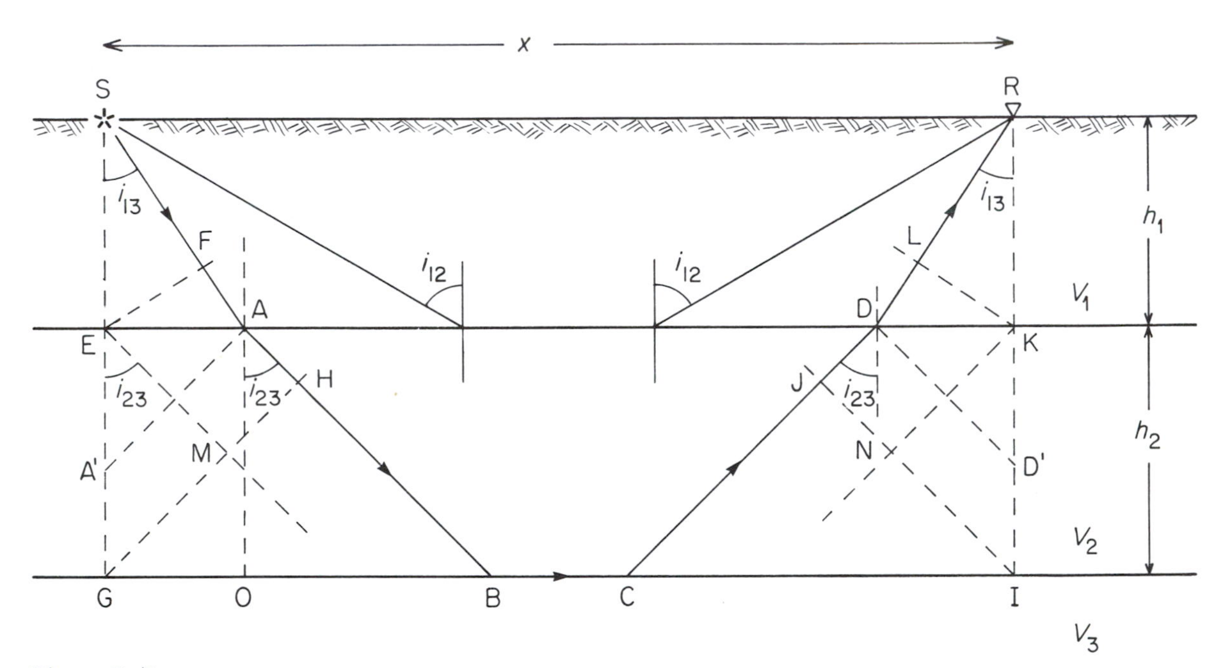

Figure 3–7
Geometrical features of the travel paths and the wave fronts of seismic waves critically refracted at two horizontal plane boundaries. The critically refracted wave from the second interface is defined by the ray path SABCDR. EF, AA′, GH, JI, DD′, and LK are the wave fronts. The wave represented by the path SABCDR can be viewed as a critically refracted wave front traveling from S to E with the velocity V_1, E to G with the velocity V_2, G to I with the velocity V_3, I to K with the velocity V_2, and K to R with the velocity V_1.

wave velocities are $V_1 < V_2 < V_3$. The receiver at distance x from the wave source S is far enough away to receive the direct wave, a wave critically refracted along the upper refractor, and a wave critically refracted along the deeper refractor. Critical refraction along the upper refractor is the same as we have already discussed for the single-layer structure. Therefore, let us concentrate on the wave that is critically refracted along the deeper refractor. Its ray parameter is

$$p = \frac{\sin i_{13}}{V_1} = \frac{\sin i_{23}}{V_2} = \frac{1}{V_3} \qquad (3\text{–}38)$$

Several lines perpendicular to its ray path in Figure 3–7 illustrate an advancing wave front. When we follow the ray path, we see that the travel time from S to R is

$$t = \frac{\text{SA} + \text{DR}}{V_1} + \frac{\text{AB} + \text{CD}}{V_2} + \frac{\text{BC}}{V_3} \qquad (3\text{–}39)$$

To understand more clearly how the wave is traveling, we must look closely at different parts of the wave front. First, let us examine how the wave front EF advances downward to A′A. As one part moves from F to A at velocity V_1, other parts are simultaneously sweeping from E to A and from E to A′ at velocity V_2. In addition, look at the geometrically similar upward advance of the wave front DD′ to LK. In the same time that one part moves from D to L at velocity V_1, other parts sweep from D to K and from D′ to K at velocity V_2. Therefore, we have the following relationships:

$$\frac{\text{FA}}{V_1} = \frac{\text{EA}}{V_2} = \frac{\text{EA}'}{V_2} = \frac{\text{DL}}{V_1} = \frac{\text{DK}}{V_2} = \frac{\text{D}'\text{K}}{V_2} \qquad (3\text{–}40)$$

Next, examine the advance from A′A to GMH at velocity V_2. We see that one part moves from A to H during the same time that another part sweeps from A′ to G. Similarly, the wave front JNI moves to DD′ at velocity V_2. Therefore, it moves from J to D at the same time that it sweeps from I to D′. This shows that

$$\frac{\text{AH}}{V_2} = \frac{\text{A}'\text{G}}{V_2} = \frac{\text{JD}}{V_2} = \frac{\text{ID}'}{V_2} \qquad (3\text{–}41)$$

From Equations 3–40 and 3–41 and the geometry in Figure 3–7, we can see that

$$\frac{\text{EA} + \text{AH}}{V_2} = \frac{\text{EM}}{V_2} = \frac{\text{JD} + \text{DK}}{V_2} = \frac{\text{NK}}{V_2} \qquad (3\text{–}42)$$

What we have shown so far is that while one part of the wave front moves from S to H, another part sweeps from S to G. Moreover, while one part of a wave front moves from J to R, another part of that same wave front sweeps from I to R.

Finally, we observe that as the wave front advances from H to B at velocity V_2, it is also sweeping from G to B at velocity V_3. It also sweeps from C to I at velocity V_3. Therefore,

$$\frac{\text{HB}}{V_2} = \frac{\text{GB}}{V_3} = \frac{\text{CJ}}{V_2} = \frac{\text{CI}}{V_3} \qquad (3\text{–}43)$$

The positions of wave fronts in Figure 3–7 show us that the travel time of the refracted wave from S to R is

$$t = \frac{\text{SF} + \text{LR}}{V_1} + \frac{\text{EM} + \text{NK}}{V_2} + \frac{\text{GI}}{V_3} \qquad (3\text{–}44)$$

Travel Time and Layer Thicknesses

Continuing with our study of Figure 3–7, we find from the triangles SEF and RKL that

$$SF = LR = h_1 \cos i_{13} \qquad (3\text{–}45)$$

and from the triangles EMG and NIK, it is evident that

$$EM = NK = h_2 \cos i_{23}$$

In addition, we observe that

$$GI = x$$

We can insert these terms into Equation 3–44 to obtain

$$t = \frac{x}{V_3} + \frac{2h_1}{V_1} \cos i_{13} + \frac{2h_2}{V_2} \cos i_{23} \qquad (3\text{–}46)$$

Because of the similarity in the form of the second and third terms, we can express them as a summation:

$$t = \frac{x}{V_3} + 2 \sum_{k=1}^{2} \frac{h_k}{V_k} \cos i_{k3} \qquad (3\text{–}47)$$

We can rearrange Equation 3–38 to get

$$\sin i_{13} = \frac{V_1}{V_3} \text{ and } \sin i_{23} = \frac{V_2}{V_3}$$

or, in general,

$$\sin i_{k3} = \frac{V_k}{V_3} \qquad (3\text{–}48)$$

and since

$$\cos i_{k3} = (1 - \sin^2 i_{k3})^{1/2} = \left[1 - \left(\frac{V_k}{V_3}\right)^2\right]^{1/2} \qquad (3\text{–}49)$$

we can obtain

$$t = \frac{x}{V_3} + 2 \sum_{k=1}^{2} \frac{h_k}{V_k V_3} (V_3^2 - V_k^2)^{1/2} \qquad (3\text{–}50)$$

Now we have an equation that relates layer thicknesses and velocities with the travel time of a refracted wave traveling between a source and a receiver separated by the distance x.

Next, we will examine the travel time curve that corresponds to a structure with two horizontal refractors. The example in Figure 3–8 illustrates the paths of refracted waves reaching 12 receivers spaced at intervals of Δx. The travel time curve above this structure shows the arrival times of these waves. We already know that the critical distance for the upper refractor can be calculated from Equation 3–6. Receiver R_4 is situated at that distance. How can we determine the critical distance for the deeper refractor? Ray paths in Figure 3–8 show that receiver R_5 is at this distance, which turns out to be twice the distance CA + DB. The triangles SCA and ADB show that

$$X_{crit} = 2(CA + DB) = 2h_1 \tan i_{13} + 2h_2 \tan i_{23}$$

which can be rewritten as the summation

$$X_{crit} = 2 \sum_{k=1}^{2} h_k \tan i_{k3} \qquad (3\text{–}51)$$

Because $\tan i_{k3} = \sin i_{k3}/\cos i_{k3}$, we can use Equations 3–48 and 3–49 to get

$$\tan i_{k3} = \left[\left(\frac{V_3}{V_k}\right)^2 - 1\right]^{-1/2}$$

Inserting this into Equation 3–51 gives us

$$X_{crit} = 2 \sum_{k=1}^{2} h_k \left[\left(\frac{V_3}{V_k}\right)^2 - 1\right]^{-1/2} \qquad (3\text{–}52)$$

Now let us consider the arrival time difference between the receivers R_8 and R_{10} where the interval is $2\Delta x$. If we follow the ray paths from S to R_8 and R_{10}, we see that SABE is a common ray path and the ray paths ER_8 and FR_{10} are equal. Therefore, the arrival time difference between the receivers R_8 and R_{10} is due to the ray path EF. This time difference may be given twice Δt where

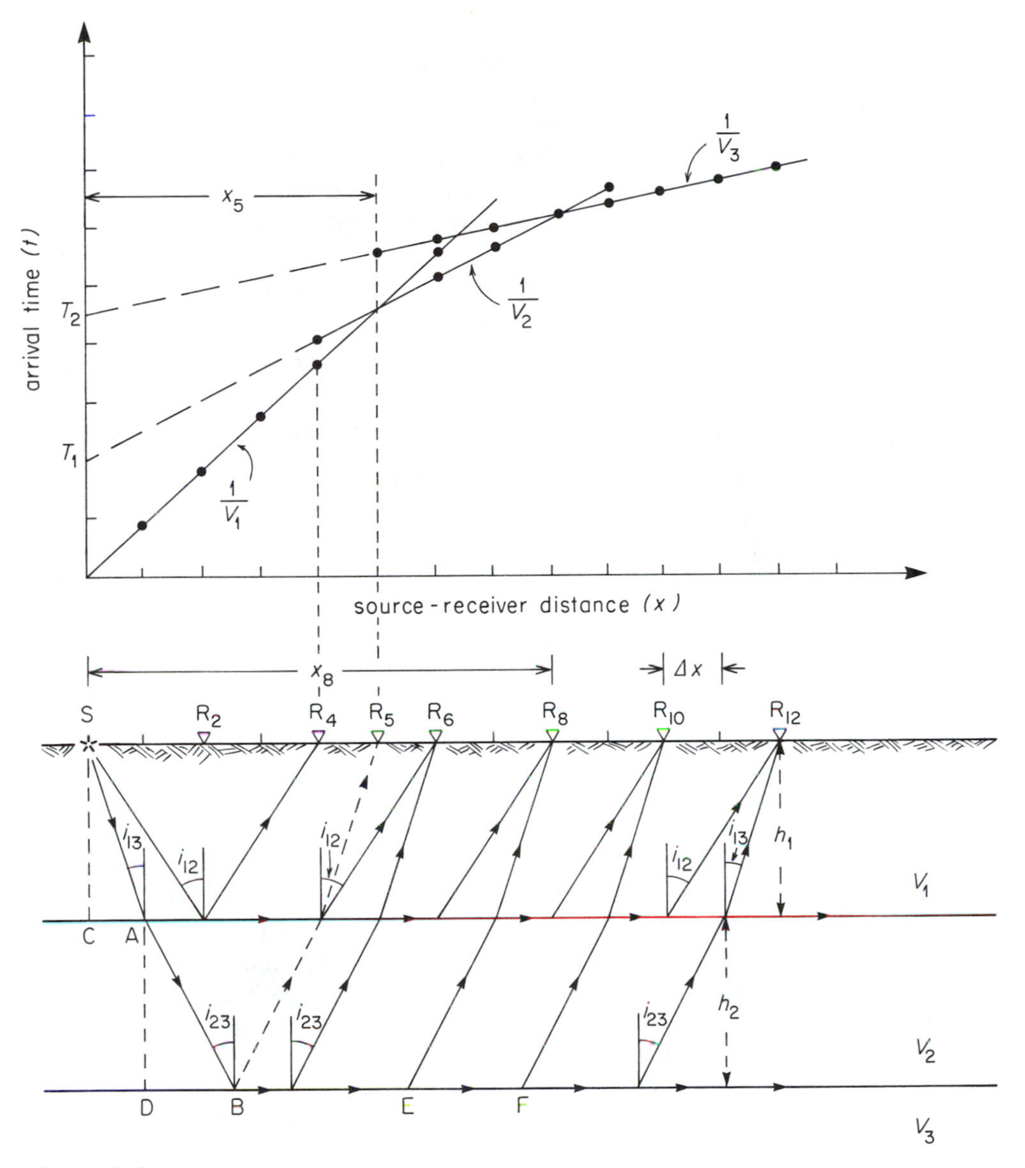

Figure 3–8
Travel paths and corresponding travel time curve for seismic waves critically refracted in a structure consisting of horizontal layers in which the wave velocities are V_1 and V_2 overlying deeper material in which the wave velocity is V_3. The first critically refracted arrivals from the top and bottom refractors can be observed at receivers R_4 and R_5, respectively. The first arrivals at the receiver groups R_1 to R_5, R_5 to R_8, and R_8 to R_{12} represent the direct wave from the top interface and refracted waves from the middle interface and the bottom interface, respectively. T_1 and T_2 are the intercept times for the first and second interfaces.

$$\Delta t = \Delta x / V_3$$

We can now see that the arrival times of waves critically refracted along the deeper refractor will plot on the travel time graph in a straight line with the slope of $1/V_3$. Equation 3–47 tells us that if $x = 0$, we have

$$t = T_2 = 2 \sum_{k=1}^{2} \frac{h_k}{V_k} \cos i_{k3} \qquad (3\text{–}53)$$

where T_2 is the intercept time found by extending that straight line to the vertical axis of the travel time graph.

Let us review the important features of the travel time graph in Figure 3–8.

1. There are three straight-line segments with slopes of $1/V_1$, $1/V_2$, and $1/V_3$.
2. There are intercept times T_1 and T_2 that correspond to waves critically refracted along the upper refractor and the deeper refractor, respectively.
3. There are crossing distances corresponding to each refractor.

At the nearest receivers, the direct waves traveling at velocity V_1 are the first arrivals. Between the first critical distance and the first crossing distance, the direct waves are first arrivals, and the waves traveling along the upper refractor at velocity V_2 are later arrivals. Beyond the first crossing distance, these refracted waves become the first arrivals, and the direct waves become later arrivals. Then at the second critical distance, waves traveling at velocity V_3 along the deeper refractor also begin to reach the receivers as later arrivals. Beyond the second crossing distances, these deeper refracted waves become first arrivals.

In a multilayered structure, either crossing distances or intercept times can be used for calculating layer thicknesses. For simplicity, we will rearrange Equation 3–53 and use the intercept times. To illustrate the procedure, let us suppose that the slopes of the straight lines on the travel time graph in Figure 3–8 indicate that $V_1 = 1500$ m/s, $V_2 = 2500$ m/s, and $V_3 = 3200$ m/s. Furthermore, suppose that we extend the upper two lines to the vertical axis and find intercept times of $T_1 = 0.2$ second and $T_2 = 0.45$ second. Now we can use Equation 3–29 to obtain

$$h_1 = \frac{T_1 V_1 V_2}{2(V_2^2 - V_1^2)^{1/2}} = 187.5 \text{ m}$$

Then we rearrange Equation 3–53 to get

$$h_2 = \left(\frac{T_2}{2} - \frac{h_1}{V_1} \cos i_{13} \right) \frac{V_2}{\cos i_{23}} \qquad (3\text{–}54)$$

We use Equation 3–49 to solve for $\cos i_{13}$ and $\cos i_{23}$, which are used with our previously calculated value of h_1 and the measured values of V_1, V_2, and T_2 to obtain

$$h_2 = 635.8 \text{ m}$$

This procedure illustrates how a travel time curve prepared from seismic refraction survey data can be interpreted to detect three different materials by means of seismic wave velocities and to determine the depths at which these materials will be encountered.

The same reasoning that we have used to analyze one-layer and two-layer structures can be used to find thicknesses and wave velocities in structures possessing any number of horizontal layers. Whatever the number of layers, the intercept time is expressed by an equation having the same general form as Equation 3–53. For example, in a structure with three refractors, the intercept time for waves refracted along the deepest one is

$$T_3 = 2 \sum_{k=1}^{3} \frac{h_k}{V_k} \cos i_{k4} \qquad (3\text{–}55)$$

This equation can be rearranged to obtain the thickness of the third layer:

$$h_3 = \left(\frac{T_3}{2} - \frac{h_1}{V_1}\cos i_{14} - \frac{h_2}{V_2}\cos i_{24}\right)\frac{V_3}{\cos i_{34}}$$

$$= \left(\frac{T_3}{2} - \sum_{k=1}^{2}\frac{h_k}{V_k}\cos i_{k4}\right)\frac{V_3}{\cos i_{34}} \qquad (3\text{–}56)$$

Finally, if a structure has n horizontal layers with thicknesses $h_1, h_2, \ldots, h_n$, and wave velocities $V_1, V_2, \ldots, V_n$ resting on a deeper material in which the wave velocity is V_{n+1}, we would expect a travel time curve with $n + 1$ straight-line segments. The intercept time determined from waves reaching the deepest refractor will be

$$T_n = 2\sum_{k=1}^{n}\frac{h_k}{V_k}\cos i_{k(n+1)} \qquad (3\text{–}57)$$

and the thickness of the deepest layer will be

$$h_n = \left[\frac{T_n}{2} - \sum_{k=1}^{n-1}\frac{h_k}{V_k}\cos i_{k(n+1)}\right]\frac{V_n}{\cos i_{n(n+1)}} \qquad (3\text{–}58)$$

REFRACTION IN STRUCTURES WITH DIPPING LAYERS

To be able to account for more of the variety found in nature, we need to analyze how refracted seismic waves travel in a structure with inclined refractors. Recall that in horizontal layers the downgoing and upgoing parts of a refracted raypath are similar. This is not true for waves that are critically refracted in a structure possessing dipping layers. Therefore, the calculation of layer thickness becomes more complicated.

Reciprocity

One feature of refracted wave travel time is the same for both dipping layer structures and horizontal layer structures. If the positions of a source and a receiver are interchanged, the travel time remains the same. This fact is called the *condition of reciprocity* and is illustrated in Figure 3–9a for a horizontal refractor. Here we have two opposite travel time curves. Observe that the travel time t_D for a refracted wave traveling from the source S_1 to the farthest detector R_{12} is exactly the same as the travel time t_R for the refracted wave traveling from the source S_2 to the receiver most distant from it. That $t_D = t_R$ should be obvious, because both waves follow exactly the same path. This is an example of reciprocity and is a necessary condition for geometrical manipulations of seismic waves.

For the other receivers such as R_6 in Figure 3–9a, the waves from S_1 travel through different parts of the layer than those from S_2, but the paths are geometrically similar. Therefore, the opposite travel time curves indicate identical velocities and intercept times.

Now look at Figure 3–9b, which shows an inclined refractor. Here we again see that waves from S_1 and S_2 reach the most distant receivers R_{12} in identical travel times $t_D = t_R$. That is obvious because these waves follow exactly the same path, but in opposite directions. Here we have the condition of reciprocity in a structure with a dipping refractor.

Figure 3–9b also shows that critically refracted waves from S_1 follow paths to intermediate receivers that are not geometrically similar to those of waves from S_2. This is evident from the paths reaching R_6 which is midway between S_1 and S_2. We see that the oppositely traveling waves arrive at different times. The effect of the inclination of the re-

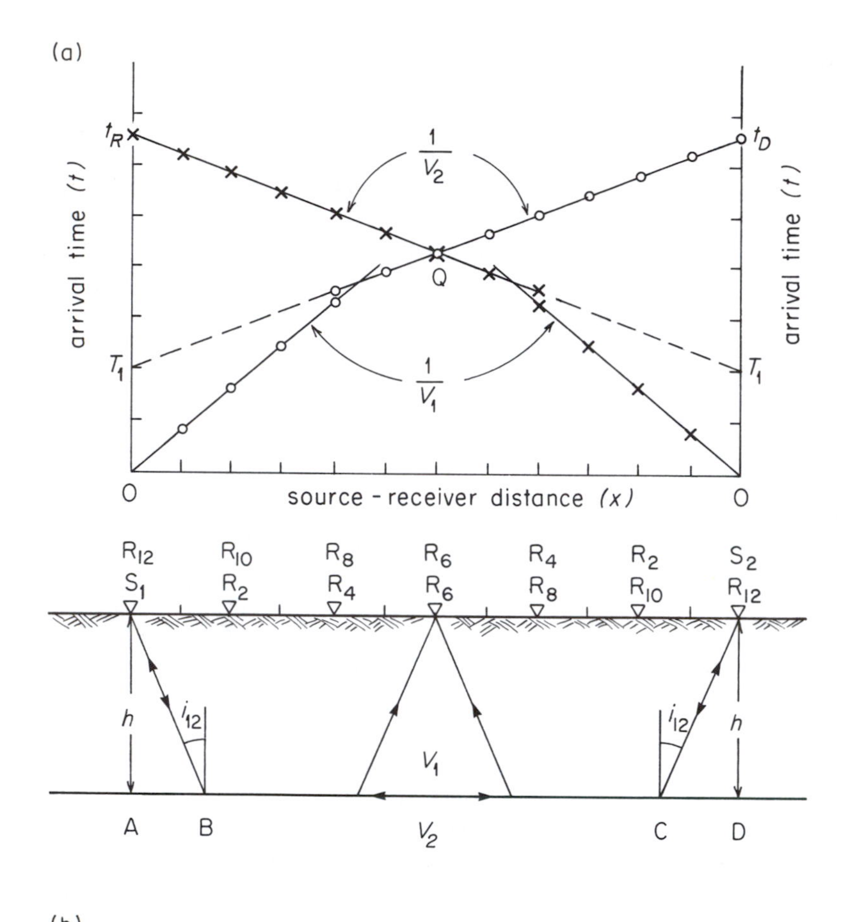

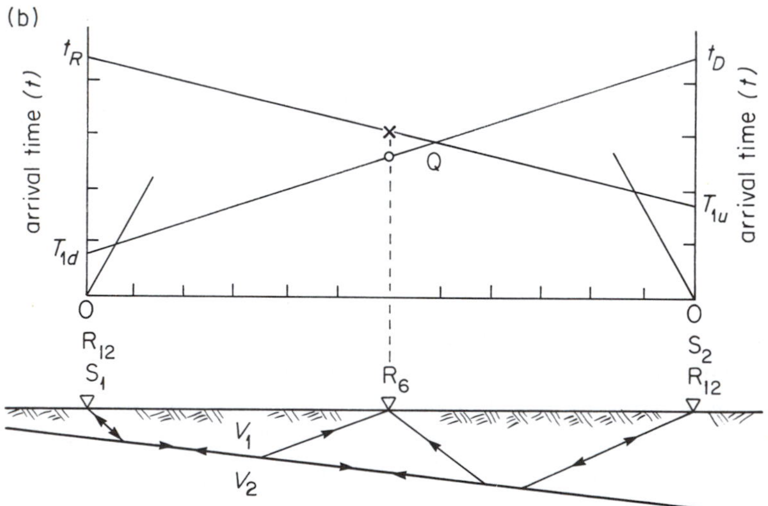

Figure 3–9
Travel paths and corresponding travel time curves for reversed refraction profiles (a) in a structure with a horizontal plane refractor and (b) in a structure with an inclined plane refractor. Intercept time T_1 is the same from the direct and reversed recordings, and the crossing point (Q) of the time–distance curves is at the center of the profile over (a) the horizontal interface. The intercept times (T_{1d}, T_{1u}) are different in the case of (b) a dipping interface. The crossing point (Q) is shifted away from the center of the profile in the down-dip direction.

fractor is to produce these differences, and similarly to produce different intercept times T_{1d} and T_{1u} on the opposite travel time curves. Notice that waves from S_1 travel along the refractor in the down-dip direction. Consequently, d is the subscript of the intercept time T_{1d}, corresponding to the waves that are refracted down dip. For the waves from S_2 that are refracted up dip, we use u in the subscript of the corresponding intercept T_{1u}.

We must have information from opposite travel time curves to analyze the depth and inclination of a dipping refractor. The seismic refraction survey field procedure is to set out the geophones in a line. A seismogram is recorded with the source at one end of this line. Then, another seismogram is recorded using a source at the opposite end of the line. This procedure is called "reversing the line." A survey done in this manner is called a *reversed refraction survey*.

Travel Time and Layer Thickness

For structures with dipping refractors, the geometry of refracted waves follows the same rules that we have already introduced in our analysis of horizontal layers. But where layers are inclined we must use more terms in the time–distance equations. In this book we will limit the discussion to the structure in Figure 3–10, which has a single refractor inclined at the angle α.

Let us begin with the source at S′ and a receiver at R′. With this arrangement we see that the critically refracted waves travel in the down-dip direction. As before, for critical refraction we have $\sin i_{12} = V_1/V_2$. The ray path in Figure 3–10 is S′BCR′, and the advance of wave fronts shows that

$$\frac{\text{EB}}{V_1} = \frac{\text{AB}}{V_2} \quad \text{and} \quad \frac{\text{CF}}{V_1} = \frac{\text{CD}}{V_2}$$

Therefore, the travel time will be

$$t = \frac{\text{S}'\text{B}}{V_1} + \frac{\text{BC}}{V_2} + \frac{\text{CR}'}{V_1} = \frac{\text{S}'\text{E} + \text{FR}'}{V_1} + \frac{\text{AD}}{V_2} \qquad (3\text{–}59)$$

But the triangles S′AE and R′FD show that

$$\text{S}'\text{E} = h_{1d} \cos i_{12} \quad \text{and} \quad \text{FR}' = h_{1u} \cos i_{12}$$

Furthermore, because S′G is parallel to the refractor, we have

$$\text{S}'\text{G} = \text{AD} = x \cos \alpha$$

where x is the distance between S′ and R′. Therefore, the travel time can be expressed as

$$t = \frac{x \cos \alpha}{V_2} + \frac{h_{1d} + h_{2d}}{V_1} \cos i_{12} \qquad (3\text{–}60)$$

Observe next that

$$\text{GR}' = h_{1u} - h_{1d} = x \sin \alpha$$

so that

$$h_{1u} = h_{1d} + x \sin \alpha \text{ and } h_{1d} = h_{1u} - x \sin \alpha$$

Now let us call the travel time for waves refracted in the down-dip direction the down-dip travel time t_d. Substituting the identities into Equation 3–60 gives

$$t_d = \frac{x \cos \alpha}{V_2} + \frac{2h_{1d} + x \sin \alpha}{V_1} \cos i_{12} \qquad (3\text{–}61)$$

Now, if we set $x = 0$, we obtain an expression for the intercept time,

$$T_{1d} = \frac{2h_{1d}}{V_1} \cos i_{12} \qquad (3\text{–}62)$$

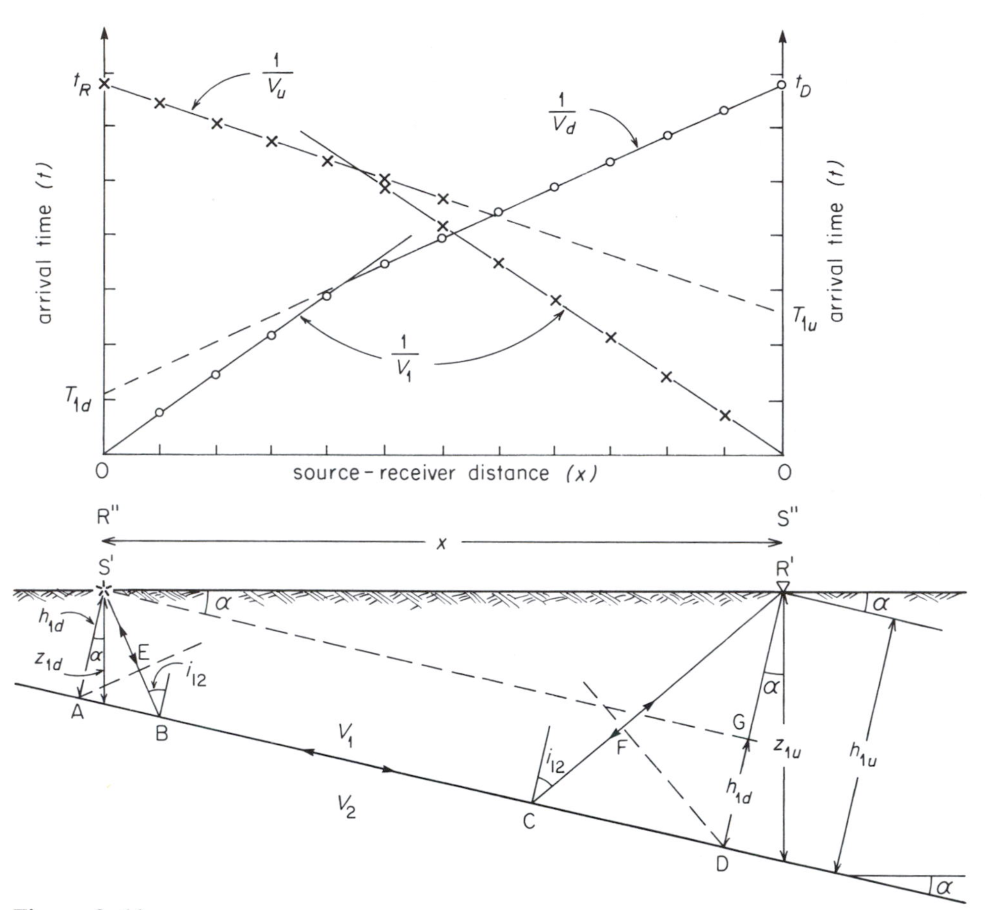

Figure 3–10
Geometrical features of the travel paths, the wave fronts, and the corresponding travel time curves for seismic waves critically refracted along an inclined plane refractor. Reversed time–distance curves over a dipping interface indicate different intercept times (T_{1d}, T_{1u}) and apparent velocities (V_d, V_u). Because of the reciprocity, $t_D = t_R$. AE and FD represent the wave fronts of critically refracted waves reaching and departing from the sloping interface. The depths to interface at S′ and S″, z_{1d} and z_{1u}, are respectively.

Then by rearranging Equation 3–60 into the form

$$t_d = \frac{x}{V_1}\left(\frac{V_1}{V_2}\cos\alpha + \sin\alpha\cos i_{12}\right) + \frac{2h_{1d}}{V_1}\cos i_{12}$$

and substituting from Equation 3–62, the relationship $\sin i_{12} = V_1/V_2$, and the identity

$$\sin i_{12}\cos\alpha + \sin\alpha\cos i_{12} = \sin(i_{12} + \alpha)$$

we obtain

$$t_d = \frac{x}{V_1}\sin(i_{12} + \alpha) + T_{1d} \qquad (3\text{–}63)$$

Now suppose that we reverse the positions of the source and receiver. For this arrangement we see in Figure 3–10 that critically refracted waves produced at S″ travel to the receiver at R″ following a path along the refractor in the up-dip direction. To obtain an expression for the up-dip travel time t_u, we now substitute using $h_{1d} = h_{1u} - x \sin \alpha$ in Equation 3–60, which gives

$$t_u = \frac{x \cos \alpha}{V_2} + \frac{2h_{1u} - x \sin \alpha}{V_1} \cos i_{12} \quad (3\text{–}64)$$

By setting $x = 0$, we find the up-dip intercept time

$$T_{1u} = \frac{2h_{1u}}{V_1} \cos i_{12} \quad (3\text{–}65)$$

Rearranging and making substitutions similar to those used to get Equation 3–63, we obtain the expression

$$t_u = \frac{x}{V_1} \sin(i_{12} - \alpha) + T_{1u} \quad (3\text{–}66)$$

Now let us compare Equations 3–63 and 3–66 with the result presented earlier in Equation 3–27 for a horizontal refractor. By setting the dip angle $\alpha = 0$, we see that because $\cos 0° = 1$, both Equations 3–63 and 3–66 reduce to

$$t_u = t_d = \frac{x}{V_2} + T_1$$

which is Equation 3–27 where $t_R = t_u = t_d$ and $k = T_1$.

Features of Reversed Travel Time Curves

Look again at the opposite travel time curves in Figure 3–10. We know from the condition of reciprocity that $t_R = t_D$. But we can also see from Equations 3–62 and 3–65 that the intercept times $T_{1u} > T_{1d}$ because $h_{1u} > h_{1d}$. We know that the line on the travel time curve extending from T_{1d} to t_D shows the alignment of travel times for waves refracted in the down-dip direction. The slope of this line is $1/V_d$ where V_d is the apparent speed of these waves. Similarly, the apparent speed V_u indicated by waves refracted in the up-dip direction is found from the reciprocal of the slope of the line extending from T_{1u} to t_R. It is clear from the facts that $t_D = t_R$ but $T_{1u} > T_{1d}$, that $V_u > V_d$. We can verify this result by means of Equations 3–63 and 3–66. Arranged in the following way,

$$t_d = x\left(\frac{1}{V_d}\right) + T_{1d} \quad \text{and} \quad t_u = x\left(\frac{1}{V_u}\right) + T_{1u}$$

we can recognize that both expressions have the familiar form of equations for straight lines that have slopes:

$$\frac{1}{V_d} = \frac{\sin(i_{12} + \alpha)}{V_1} \quad \text{and} \quad \frac{1}{V_u} = \frac{\sin(i_{12} - \alpha)}{V_1}$$

Inverting these expressions gives

$$V_d = \frac{V_1}{\sin(i_{12} + \alpha)} \quad (3\text{–}67)$$

and

$$V_u = \frac{V_1}{\sin(i_{12} - \alpha)} \quad (3\text{–}68)$$

Again, it is clear that $V_u > V_d$.

These relationships are also evident from inspection of Figure 3–9b. Compare the paths of down-dip and up-dip refracted waves reaching receiver R_6 which is midway between S_1 and S_2. The down-dip path is altogether shorter but has a longer portion along the refractor where velocity is higher compared with the up-dip path. Therefore, it is obvious that at R_6 we should find $t_u > t_d$. A line extending through t_R and t_u clearly has a larger intercept and a smaller slope than a line extending through t_D and t_d. This implies that $T_{1u} > T_{1d}$ and $V_u > V_d$.

We should also point out that the direct waves are not affected by the dipping refractor. Therefore, the straight lines on opposite travel time curves that represent direct waves should have identical slopes of $1/V_1$.

Suppose that we have just completed a reversed refraction survey and that we have plotted travel times read from the seismograms and drawn straight lines through the alignments of points. Simply by inspecting the newly prepared travel time curves, we can reach one or another of the following conclusions.

1. If corresponding straight lines on opposite travel time curves have identical slopes, we have a horizontal refractor.
2. If corresponding straight lines on opposite travel time curves have different slopes, we have a dipping refractor.
3. The refractor dips downward toward the source at which the intercept time is largest.
4. The refractor dips downward toward the source for which the crossing distance is greatest.

Calculating Velocity, Thickness, and Dip

By means of a reversed refraction survey, we can find V_1 from the slope of the direct wave line. But how do we find the velocity V_2 in the material below the refractor? We begin with a trigonometric inversion of Equations 3–67 and 3–68, which gives us

$$i_{12} + \alpha = \arc \sin \frac{V_1}{V_d} \quad \text{and}$$

$$i_{12} - \alpha = \arc \sin \frac{V_1}{V_u} \qquad (3\text{–}69)$$

We can add these expressions to get the critical angle

$$i_{12} = \frac{1}{2}\left(\arc \sin \frac{V_1}{V_d} + \arc \sin \frac{V_1}{V_u}\right) \qquad (3\text{–}70)$$

and then from Equation 3–1 we get

$$V_2 = \frac{V_1}{\sin i_{12}}$$

By trigonometric inversion of this result, we have

$$i_{12} = \arc \sin \frac{V_1}{V_2}$$

which can be used in Equation 3–70 to get

$$\arc \sin \frac{V_1}{V_2} = \frac{1}{2}\left(\arc \sin \frac{V_1}{V_d} + \arc \sin \frac{V_1}{V_u}\right)$$

For small angles we know that $\sin \alpha \cong \alpha$. For this case we can obtain the following approximation,

$$\frac{V_1}{V_2} \cong \frac{1}{2}\left(\frac{V_1}{V_d} + \frac{V_1}{V_u}\right)$$

or

$$V_2 \cong 2\left(\frac{V_d V_u}{V_d + V_u}\right) \qquad (3\text{–}71)$$

Now, by rearranging Equations 3–62 and 3–65, we find

$$h_{1d} = \frac{V_1 T_{1d}}{2 \cos i_{12}} \qquad (3\text{–}72)$$

and

$$h_{1u} = \frac{V_1 T_{1u}}{2 \cos i_{12}} \qquad (3\text{–}73)$$

where the $\cos i_{12}$ can be found from Equation 3–4.

By subtracting the expressions (3–69), we obtain the dip of the refractor

$$\alpha = \frac{1}{2}\left(\text{arc sin}\,\frac{V_1}{V_d} - \text{arc sin}\,\frac{V_1}{V_u}\right) \qquad (3\text{–}74)$$

Finally, we observe in Figure 3–10 that h_{1d} and h_{1u} are distances to the nearest points on the refractor beneath S′ and R′, respectively. The vertical depths to the refractor are

$$z_{1d} = h_{1d}/\cos\alpha \qquad (3\text{–}75)$$

and

$$z_{1u} = h_{1u}/\cos\alpha \qquad (3\text{–}76)$$

This completes the analysis of how to measure the depth and inclination of a single sloping refractor. The reasoning is similar to that introduced in our study of a horizontal refractor, but the procedure is necessarily more complicated. It should be obvious that this kind of analysis can be extended to a structure with multiple dipping refractors, but we will not discuss that topic in this book. For an analysis of multiple dipping layers, see *Exploration Seismology,* Volume 1, by Sheriff and Geldart (1982).

Application

The method for analyzing a single dipping refractor was applied in a reversed refraction survey in central Virginia. The purpose of the survey was to find the depth and inclination of the bedrock surface beneath the weathered zone. The geophones were placed in a line 140 meters long, and two 2-kg explosive charges were detonated at both ends of this line. First arrival times read from the direct and reverse seismograms were plotted, and straight lines were drawn through the aligned points. The results are shown in Figure 3–11.

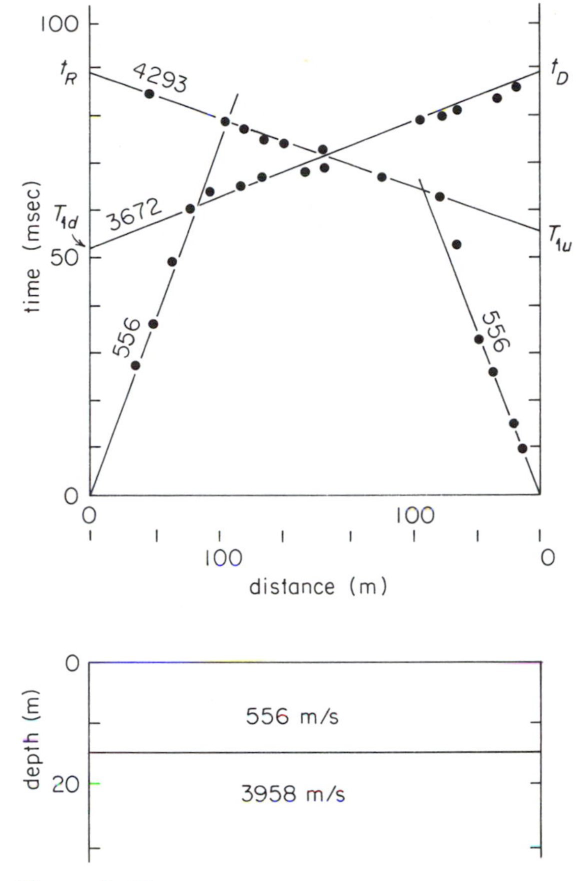

Figure 3–11
Results of a reversed seismic refraction survey in central Virginia. (From M. S. Bahorich, C. Coruh, E. S. Robinson, and J. K. Costain, Geophysics, v. 47, p. 1543, 1982.) The intercept times are T_{1d} = 0.052 and T_{1u} = 0.056 and $t_D = t_R$. Using the apparent up-dip and down-dip velocities, we can determine a true velocity of 3958 m/s for the second layer, which requires a critical angle of 8.075 degrees and a dip angle of 0.6 degree. The refractor depths z_{1d} = 14.6 m and z_{1u} = 15.7 m.

The following values can be measured from these travel time curves:

$$V_1 = 556 \text{ m/s} \quad V_d = 3672 \text{ m/s}$$
$$V_u = 4293 \text{ m/s}$$
$$T_{1d} = 0.052 \text{ s} \quad T_{1u} = 0.056 \text{ s}$$

Then Equation 3–71 was used to calculate

$$V_2 = 3958 \text{ m/s}$$

making it possible to find the critical angle

$$i_{12} = \text{arc} \sin(V_1/V_2) = 8.075 \text{ degrees}$$

Then by means of Equations 3–72, 3–73, and 3–74, we obtain

$$h_{1d} = 14.6 \text{ m} \quad h_{1u} = 15.7 \text{ m} \quad \alpha = 0.6 \text{ degree}$$

and finally from Equations 3–75 and 3–76, the depths are

$$z_{1d} = 14.6 \text{ m} \quad \text{and} \quad z_{1u} = 15.7 \text{ m}$$

These results are presented in the structure drawn beneath the travel time curves.

REFRACTION ALONG A DISCONTINUOUS BOUNDARY

So far, we have discussed seismic wave refraction in layers with plane boundaries, but we know that discontinuous boundaries also exist in nature. For instance, suppose that a refractor is offset by a fault or an erosional scarp. A simple example of this kind of discontinuous refractor is illustrated in Figure 3–12.

As in the previous structures we have studied, the direct waves are unaffected by the refractor. Their arrival times plot along a straight line having the slope of $1/V_1$, which is shown on the travel time curve. In this example they are the first arrivals at receivers R_1 through R_6.

Waves originating at S that are critically refracted along the upthrown part of the layer boundary between A and B are recorded by receivers R_4 through R_{13}, and they are first refracted arrivals beginning at R_8. Their arrival times plot along a straight line with the slope $1/V_2$. We already know how to calculate the layer thickness from information on this part of the travel time curve by means of Equation 3–20 or Equation 3–30.

Now look closely at the short interval between receiver R_{13} and the point F. No refracted waves can arrive in this interval. Instead, we have *diffracted* waves such as the one following the path from B to R_{14}. How is such a wave produced? According to Huygen's principle, we can view the point B as a source from which waves diffract outwardly along paths in all directions. Along a continuous plane boundary, the particular path at the critical angle is the one that satisfies Fermat's principle. In our example this is the path from B to R_{13}. But no path at the critical angle can reach R_{14}. Therefore, Fermat's principle can be satisfied only by the diffracted wave path that leads directly from B to R_{14}. Observe that this path through the top is longer than the paths, such as the one from B to R_{13}, that lead to the receivers closer to S. Therefore, the arrival time at R_{14} does not plot on the straight line drawn through the refracted arrival times at receivers R_4 through R_{13}. Rather, it plots above an extension of this line.

At distances beyond the point F, we find that Fermat's principle is satisfied by paths from S to A to C, and then along the deeper boundary from which they are critically refracted up to the receivers. Because the paths to receivers R_{15} through R_{24} are identical from S to C and similar from the refractor up to the receivers, the arrivals will plot along a straight line with the slope of $1/V_2$, where V_2 is the wave velocity along the refractor between C

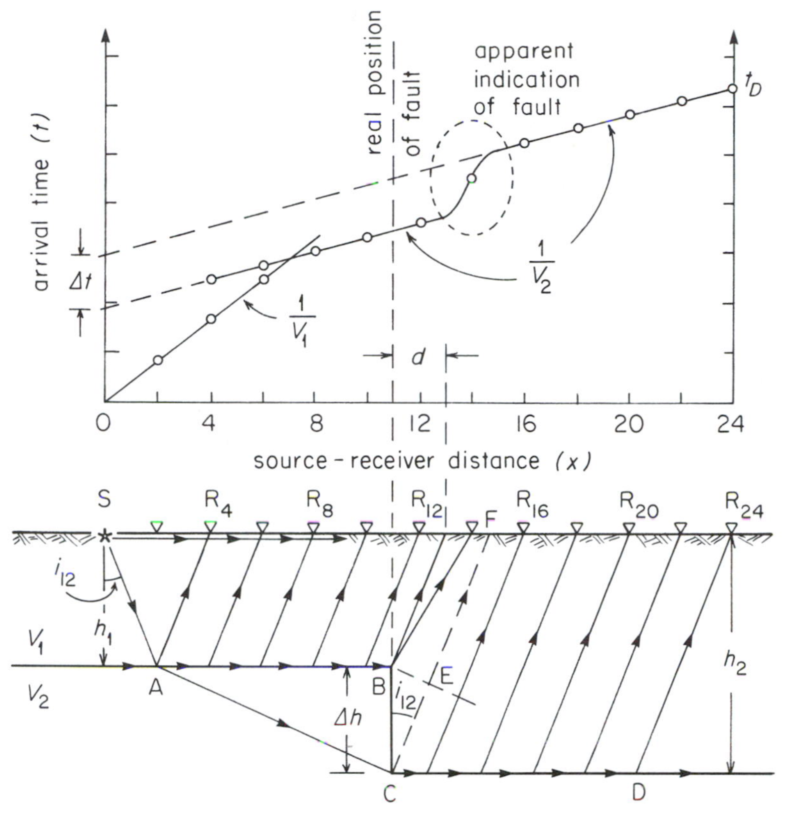

Figure 3–12
Travel paths and corresponding travel time curve for seismic waves that are refracted and diffracted in a structure with a horizontal plane refractor that is offset along a vertical surface such as a fault. There are no arrivals of refracted waves between the receiver R_{13} and point F. Instead, the diffraction arrivals from point B appear. The time–distance curve of the refracted wave is separated in two pieces by this gap, and the time difference (Δt) is caused by the offset (Δh). Notice that the offset in the travel time curve is not directly above the vertical offset of the refractor.

and D. Observe how this line is parallel to the line of arrival times for nearer receivers, but is offset from it by the time interval Δt. This additional time is necessary because the travel paths from CD up to the receivers are longer than the paths from AB up to the nearer receivers.

We see that the interruption in the travel time curve is displaced to the right of the actual offset of the layer boundary. This displacement is related to the inclination of the paths reaching the receivers. Since the interruption begins at the distance where the refracted wave from B reaches the surface, which is at receiver R_{13} in our example, the distance of the actual offset to the left of this point is given by

$$d = h_1 \tan i_{12} \tag{3–77}$$

The vertical offset Δh of the refractor can be calculated from the time shift Δt between the two segments of the travel time curve for which the slope is $1/V_2$. Consider the wave front BE that is perpendicular to the ray path BR_{13} and the deeper ray path CF. The difference in travel times along these two paths is the time Δt required for a wave to travel from C to E at velocity V_1. Therefore, the vertical displacement of the boundary can be found from the triangle BCE to be

$$\Delta h = \frac{\Delta t V_1}{\cos i_{12}} = \frac{\Delta t V_1 V_2}{(V_2^2 - V_1^2)^{1/2}} \qquad (3\text{–}78)$$

If we do a reversed refraction survey over an offset refractor, we can obtain opposite travel time curves with the features illustrated in Figure 3–13. The travel time curve for waves originating at S′ is the same as the one in Figure 3–12. For waves produced at S″, we see that the refracted arrival times plot along two straight-line segments with the same slope of $1/V_2$. But here the more distant segment is offset down by the time interval of Δt. The nearer segment terminates at a point to the left of the actual offset of the layer boundary. The ray paths in Figure 3–13 tell us that the distance of the actual offset to the right of the termination of this segment is also given by Equation 3–77. We see that the offset in the layer boundary is situated midway between the interruptions in the opposite travel time curves.

Because the straight-line segments corresponding to refracted waves in Figure 3–13 all have the same slope of $1/V_2$, we know that the offset refractor is horizontal. If the opposite

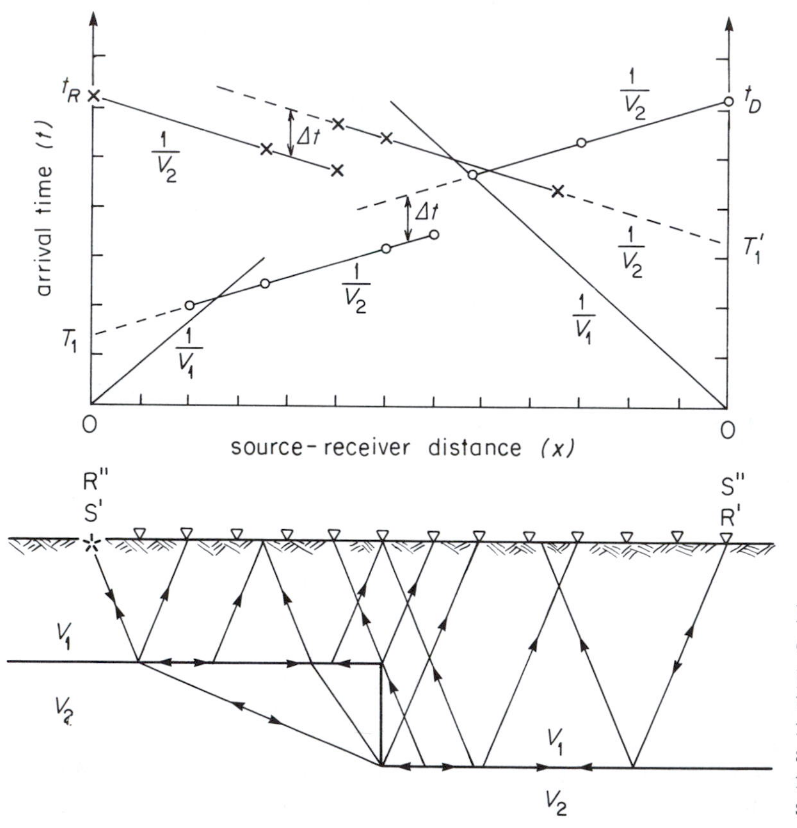

Figure 3–13
Geometrical features of the travel paths and corresponding travel time curves for reversed seismic refraction profiles over a structure with a horizontal plane refractor offset along a vertical surface.

travel time curves were to reveal different slopes, we would recognize the more complicated case of an offset dipping refractor. We discovered such a refractor from a reversed refraction survey that was done in Goochland County, Virginia, the results of which are shown in Figure 3–14. Offset travel time curves indicated the existence of a fault, and different slopes for opposite travel time curves pointed to dipping layers. Additional seismograms were then recorded with the source close to the fault. By combining these with the earlier measurements, we also obtained two sets of shorter reversed travel time curves, one to the southwest and the other to the northeast of the fault. On the southwest the travel time curves could be interpreted in terms of plane-dipping refractors. But on the northeast, the structure turned out to be more complicated, so that yet another offset in the deeper refractor was discovered. The dip of this refractor was found to be different from that along the southwest line.

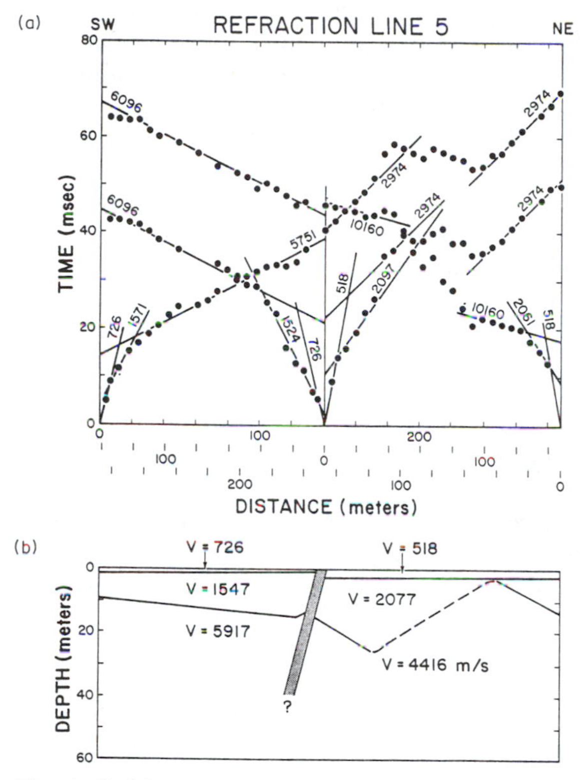

Figure 3–14

Results of a reversed seismic refraction survey over a structure with inclined refractors offset by faults. (From M. S. Bahorich, C. Coruh, E. S. Robinson, and J. K. Costain, *Geophysics*, v. 47, p. 1544, 1982.)

SOME LIMITATIONS OF SEISMIC REFRACTION SURVEYS

All the methods we have presented for interpreting earth structure by means of refracted seismic waves assume two conditions.

1. Wave velocities increase in successively deeper layers, that is, $V_1 < V_2 < V_3 \ldots < V_{n+1}$.
2. Waves critically refracted from each of the boundaries produce first arrivals along some interval of distance from the source.

Are there circumstances in which one or both of these conditions are not realized?

Consider what happens if the wave velocity in some layer is lower than the velocities in the layers above and below it. Waves must be critically refracted along a refractor in order to detect the refractor. But according to Snell's law, critical refraction can occur only if the incident wave velocity is slower than the refracted wave velocity. Otherwise, we see from Figure 3–6 and Equation 3–37 that a refracted wave that travels more slowly than the incident wave must have a smaller angle of refraction. Thus, the refracted wave is more steeply inclined than the incident wave.

Look at the example in Figure 3–15 where $V_1 > V_2 < V_3 < V_4$. We see that critical refraction occurs along the top refractor and plots along a line with the slope $1/V_3$. Similarly, we have critical refraction along the deepest refractor that is indicated by the line with the slope of $1/V_4$. Together with the direct wave arrivals, we have a travel time curve with three straight-line segments. But critical refraction cannot occur along the first boundary. There is no suggestion of its existence on the travel time curve. Yet, the waves reaching the deepest refractor must pass through the middle layer. This distorts the intercept time T_3. From the form of the travel time curve we would analyze for only two refractors. Not only would the middle one go undetected, but its presence would cause an incorrect depth for the deeper refractor to be calculated.

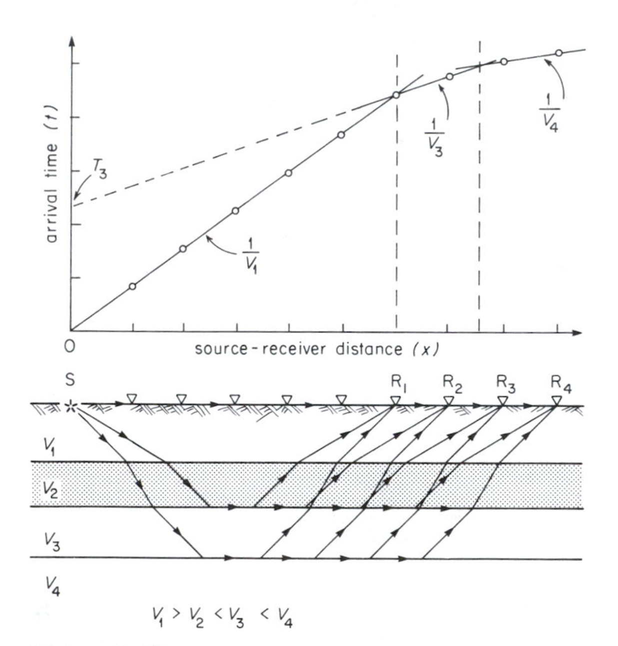

Figure 3–15
Travel paths and corresponding travel time curves for seismic waves refracted in a multi-layered structure in which the velocity V_2 in an intermediate layer is slower than the velocities V_1 and V_3 in the layers above and below it. Because of this pattern of velocity variation, only two refractors are evident on the travel time curve. The existence of an undetected middle refractor leads to error in the depth calculated for the deeper refractor. The undetected low-velocity layer is sometimes called a "blind zone."

This shortcoming cannot be corrected by means of refraction data alone. It remains an undetectable demon, unless we have some independent information from drilling or reflected seismic waves. For this reason, a "low-velocity" zone is sometimes called a *blind zone.*

Critically refracted waves that cannot be recognized as first arrivals also pose problems. We already know that such waves can appear as first arrivals only in the interval between the crossing distances related to refractors above and below. In Figure 3–16 this interval includes receivers R_5 through R_8. Elsewhere at R_4 and R_9 through R_{12} in this example, waves from the upper refractor are later arrivals. Now, in some instances, the interval between crossing distances is too small to be observed. Such may be the case for waves critically refracted at the upper and lower boundaries of a thin layer. This is illustrated in Figure 3–16 where the three lines with slopes $1/V_2$, $1/V_3$, and $1/V_4$ all intersect at the same crossing distance X_c. Except at this point, the waves refracted along the middle refractor at velocity V_3 appear as later arrivals.

Under ideal circumstances we can recognize later arrivals on a seismogram. They are clearly seen on the seismogram in Figure 2–26. In such a case, the arrival times can be plotted on the travel time curve, as in Figure 3–16, and used to calculate the layer thickness without difficulty. But nature does not always provide ideal circumstances. Later arriving

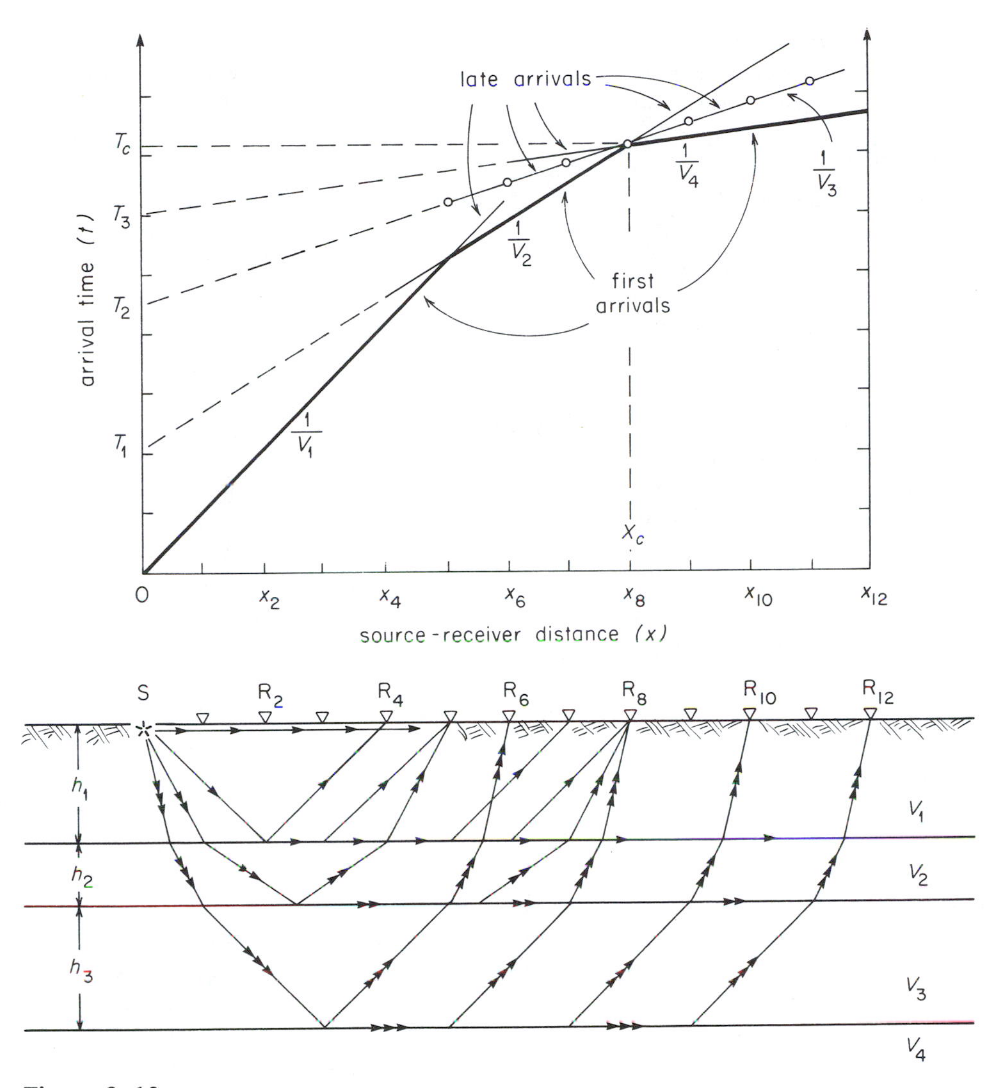

Figure 3–16

Travel paths and corresponding travel time curves for seismic waves refracted in a multi-layered structure in such a way that critically refracted waves from V_2/V_3 boundary are always later arrivals. When the crossing distance X_c for the deeper refractor is shorter than that for the overlying refractor, the waves from the upper refractor do not appear as first arrivals. A "hidden" layer causes error in the depth calculated for the deeper refractor. In this case, the second refractor is not seen in the first arrivals.

pulses often overlap the first arrival or one another so that their onset cannot be recognized. In such instances, the refractor producing them remains undetected. If only the first arrivals in Figure 3–16 are used to analyze the structure, the layer in which wave velocity is V_3 will be a *hidden layer,* and the depth calculated for the deeper refractor will be incorrect.

The hidden-layer problem can result from an especially large velocity contrast across the deeper refractor as well as from a thin layer. This circumstance will produce an especially small critical angle at the deeper refractor, resulting in very steep ray paths and a shorter crossing distance. In extreme cases, the crossing distance for the deeper refractor can be shorter than that for the overlying refractor. Then the waves from the upper refractor will never appear as first arrivals.

Because of the possibility of errors related to hidden layers and blind layers, insofar as possible refraction survey results should be verified by analysis of reflected waves. Methods for doing this analysis are presented in Chapter 6.

INTERPRETING A SEISMIC REFRACTION SURVEY

When we are planning a seismic refraction survey, we must first decide on the maximum depth of investigation. Suppose that we want to study the earth structure down to a depth h. The line of geophones must extend farther than the critical distances and crossing distances for any refractors that might be present. If we have any basis for knowing, say from previous drilling or geologic mapping, about what wave velocities might be encountered, we can use expressions such as Equations 3–6 and 3–20 to estimate the necessary length of the line of geophones, which is usually called the *spread.* Because we lack such information, the "rule of thumb" is to make the spread length about four times the maximum depth of interest. This usually insures that velocity contrasts commonly encountered in nature will be detected. Then, if we have no preliminary information about possible inclinations of the refractors that may exist, it is wise to plan a reversed refraction survey.

Static Corrections

Seismic refraction surveys are usually interpreted in terms of the idealized layered structure that we have already discussed. For all these cases, we have assumed that the geophone spread was situated on a plane observation surface. In nature, however, the land surface is somewhat irregular, so that the geophones and source points are actually at different elevations. Therefore, the observed arrival times must be adjusted to obtain values that would be expected were the geophones and source points on a plane surface. These time adjustments are called *static corrections.*

Although there is no practical way to determine perfect static corrections, approximate procedures in common use give values that are accurate enough for most purposes. Figure 3–17 shows the features to be considered. We can begin by choosing the average elevation of the land surface along the spread as the plane observation surface, otherwise called the *datum surface.* Then we find the vertical height ΔZ of each geophone above or below that surface. We see in Figure 3–17 that the refracted wave travels to the geophone by the path BR at velocity V_1. If we were to project the geophone down to the datum at D, the wave would follow the path BC at velocity V_2

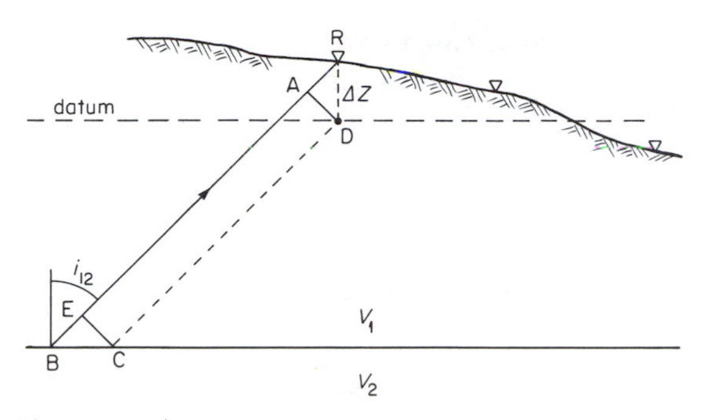

Figure 3–17
Actual travel path to a receiver at R on the land surface, and the hypothetical travel path to the point D on a horizontal datum surface that is situated directly beneath R. Using the wave front (AD) makes it obvious that reducing R to D is equivalent to reducing R to A. The static correction is determined by AR/V_1, where AR = $\Delta Z \cos i_{12}$ and ΔZ is the elevation difference between the receiver (R) and datum (D).

and the path CD at velocity V_1. Therefore, the static correction Δt_s would be the difference in travel times along these different paths,

$$\Delta t_s = \frac{BR}{V_1} - \left(\frac{BC}{V_2} + \frac{CD}{V_1}\right)$$

$$= \frac{BR - CD}{V_1} - \frac{BC}{V_2} \qquad (3\text{–}79)$$

Observe here that AR = $\Delta Z \cos i_{12}$, that BE = BC $\sin i_{12}$, and

$$BR = CD + \Delta Z \cos i_{12} + BC \sin i_{12}$$

Since AD = CE, we have

$$BC \cos i_{12} = \Delta Z \sin i_{12}$$

so that

$$BC \sin i_{12} = \frac{\Delta Z}{\cos i_{12}} \sin^2 i_{12}$$

$$= \frac{\Delta Z}{\cos i_{12}}(1 - \cos^2 i_{12}) = \frac{\Delta Z}{\cos i_{12}} - \Delta Z \cos i_{12}$$

which can be used to get

$$BR - CD = \frac{\Delta Z}{\cos i_{12}}$$

Substitution into Equation 3–79 gives

$$\Delta t_s = \frac{\Delta Z}{V_1 \cos i_{12}} - \frac{\Delta Z \tan i_{12}}{V_2}$$

Then, by means of Equation 3–1, which tells us that $V_2 = V_1/\sin i_{12}$ and the identity $\cos^2 i_{12} = 1 - \sin^2 i_{12}$, we obtain the expression

$$\Delta t_s = \frac{\Delta Z}{V_1} \cos i_{12} \qquad (3\text{–}80)$$

Let us examine the refraction static correction using wave fronts. It is clear that reducing R to D is equivalent to reducing R to A. So, from the triangle ARD, it is obvious that Δt_s is the travel time from A to R.

If we can estimate V_1 and V_2 from preliminary travel time curves prepared from arrival times without static corrections, we can find $\cos i_{12}$ from Equation 3–4 and proceed to calculate the static correction for each source point and geophone. For some circumstances there is yet a simpler method. We know that the wave velocity V_1 is usually very slow in the weathered zone close to the earth's surface. This insures that the arriving waves follow steeply inclined paths for which i_{12} is a small angle. If we assume that $i_{12} \cong 0$ so that $\cos i_{12} \cong 1$, Equation 3–80 reduces to the approximate form

$$\Delta t_s \cong \frac{\Delta Z}{V_1} \qquad (3\text{–}81)$$

As we can see from Figure 3–17, the distance between the datum and the land surface along the direction of BR does not differ very much from ΔZ. Therefore, the static correction estimated from Equation 3–81 turns out to be accurate enough for the requirements of many surveys.

Inspection of Travel Time Curves

After static corrections have been calculated for each source point and geophone, they are added to or subtracted from the arrival times of waves reaching those geophones, depending on whether the source and geophone positions is below or above the datum. Arrival times adjusted in this way are plotted, and lines are drawn through straight alignments of points that appear on the graph.

Now we can begin with the interpretation of earth structure. First, we can examine how closely the lines can be fitted to the arrival time points. If the points plot close to these lines, we can conclude that the refractors are nearly plane surfaces. A larger scatter of points may suggest a more irregular boundary. We can judge by the differences in the slopes of the lines on opposite travel time curves, whether the refractors are horizontal or inclined. We know that an inclined refractor dips downward toward the source that produces the largest intercept time and crossing distance. If we observe offset line segments with similar slopes, we recognize that the corresponding refractor must be offset, perhaps by a fault.

Qualitative interpretations such as these can be made by direct inspection of the travel time curves. They suggest which of the idealized models of earth structure we should use for calculating layer thicknesses, angles of dip, and positions of faults. Still, we must realize that the numbers we obtain give only an approximate indication of the positions of a refractor, which can prove to be much more irregular than those in our idealized structures. How can we learn more about these irregularities?

The Plus–Minus Method

Suppose that the refractor is a somewhat irregular undulating surface rather than a plane, as in Figure 3–18. A method called the *plus–minus method* or the *ABC method* makes use of reversed refraction survey data to obtain depths to the refractor at intermediate geophone positions.

Let us consider the points A, B, and C in Figure 3–18. For the direct refraction spread the source is at A, and B and C are receivers. Then, for the reversed spread the source is at C, and A and B are receivers. Because of the condition of reciprocity, the refracted wave travel times from A to C and from C to A are

$$t_{AC} = t_{CA}$$

Now consider the travel times t_{AB} and t_{CB} for refracted waves from the two sources to the receiver at B. We can use all these travel times to calculate

$$\Delta t = (t_{AB} + t_{CB}) - t_{AC}$$

What does Δt tell us? In terms of the wave fronts IF and JF, we see that Δt is the travel time of waves from I to B and from J to B. If the angles of dip of the refractor are quite small, angle θ is approximately the same as the angle of refraction i_{12}. Therefore, the depth of the refractor beneath B is

$$h_B \cong \frac{BI + BF}{2 \cos i_{12}}$$

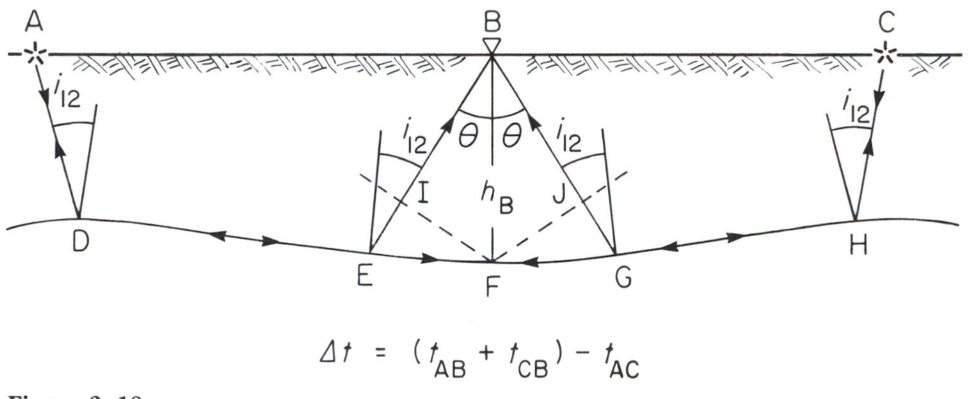

Figure 3–18
Seismic waves refracted from an undulating surface, and geometrical features used to determine refractor depth. The direct or reverse time (t_{AC} or t_{CA}) from A to C is equal to the sum of times from A to I and C to J. So, Δt is the time for the refracted wave to travel from I to B and J to B.

Since $\Delta t = (BI + BJ)/V_1$, we get

$$h_B = \frac{\Delta t}{2} \frac{V_1}{\cos i_{12}}$$

and substituting from Equation 3–4, we obtain the expression

$$h_B = \frac{\Delta t}{2} \frac{V_1 V_2}{(V_2^2 - V_1^2)^{1/2}} \qquad (3\text{–}82)$$

In this way, the depth to the refractor can be calculated at every intermediate geophone position where direct and reversed refracted waves are recorded. This approach gives us much more detailed information about an irregular refractor than can be found by assuming a structure with plane refractors. The method can be applied to deeper refractors after making travel time adjustments similar to static corrections to account for undulations of the overlying refractors.

The Wave Front Method

One of the earliest methods used to interpret reversed refraction surveys involves graphical construction of wave fronts for critically refracted waves. Examine the structure in Figure 3–19 where A and D are source points and B and C are receivers. These receiver positions are especially chosen to give the following relationship of refracted wave travel times:

$$t_{AD} = t_{AB} + t_{DC} \qquad (3\text{–}83)$$

For this condition we find that wave fronts that have reached B and C also intersect at the point E on the refractor. Knowing this fact, we can proceed to interpret a reversed refraction survey by selecting appropriate points from the direct and reversed travel time curves.

Consider first the travel time curves in Figure 3–20 which indicate a plane dipping refractor. Only the lines for refracted waves are

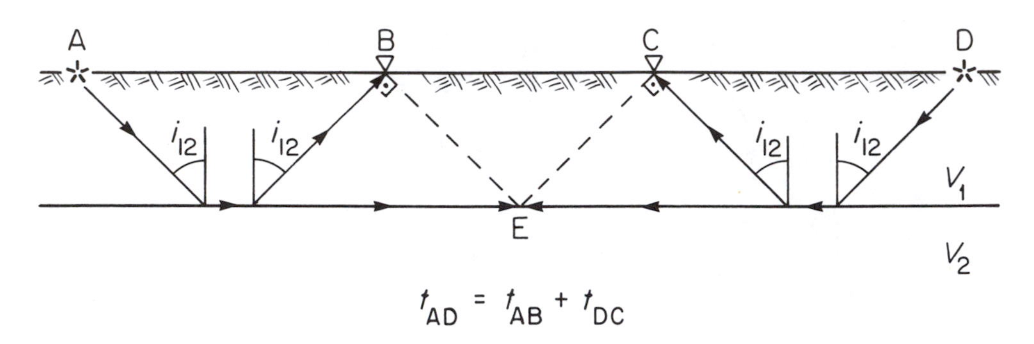

Figure 3–19
Paths of refracted waves for which the wave fronts arriving at points B and C on the observation surface also intersect at point E on the refractor. BE and CE are the wave fronts at times t_{AB} and t_{DC}. $t_{AD} = t_{AB} + t_{DC}$ and intersection of the wave fronts with this condition represents a point on the refractor.

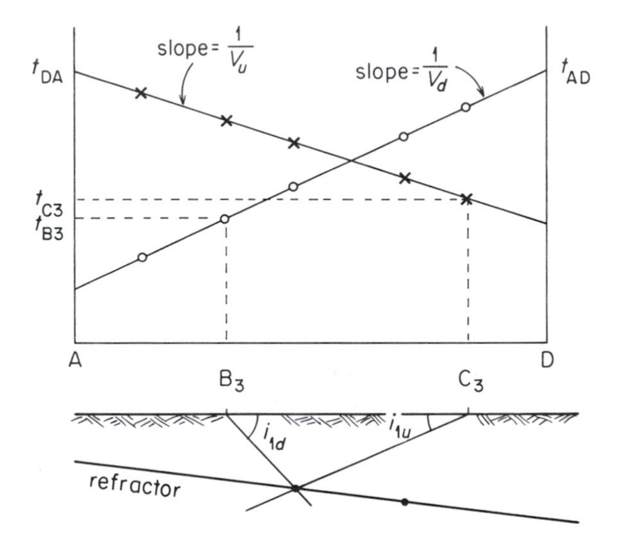

Figure 3–20
Features on reversed refraction travel time curves used to locate receiver positions for which the wave fronts also intersect on the refractor.

plotted here. To keep the diagram as simple as possible, we have left off the lines for direct waves, from which we would find the velocity in the top layer. To locate a point on the refractor, we arbitrarily choose a time, say t_{B3}, which as we see from the direct travel time curve is the arrival time at distance B_3. Then, using Equation 3–83, we calculate the corresponding time t_{C3} for the reverse spread, which the reverse travel time curve shows to be the arrival time at position C_3.

Next, we use the apparent velocities V_d and V_u, obtained from the slopes of the travel time curves, to find the angles of the wave fronts that reach B_3 and C_3. At position B_3 we have

$$\sin i_{1d} = \frac{V_1}{V_d} \quad \text{or} \quad i_{1d} = \text{arc sin} \frac{V_1}{V_d}$$

and at position C_3 we have

$$\sin i_{1u} = \frac{V_1}{V_u} \quad \text{or} \quad i_{1u} = \text{arc sin} \frac{V_1}{V_u}$$

On the diagram below the travel time curves, we draw wave fronts making these angles at

B_3 and C_3. The intersection of these wave fronts locates a point on the refractor.

If we repeat this procedure using another set of times, say t_{B4} and t_{C4} and their corresponding distances B_4 and C_4, we can locate another point on the refractor. By drawing a straight line through the two points, we have located the refractor.

Now let us look at a more complicated problem. If the refractor is not a plane surface, the travel time points will not align perfectly along straight lines as seen in Figure 3–21. Rather than attempting to draw straight lines through these groups of points, we sketch curved lines through them. Let us use the procedure just described to analyze these curved lines so that we may locate points on an undulating refractor. Again we choose a time, say t_{B3}, that is the actual arrival time at position B_3. Then applying Equation 3–83 we locate the corresponding time t_{C3} and distance C_3 on the reversed travel time curve. To find the angle of the wave front at position B_3, we draw a line tangent to the travel time curve at that distance. The slope of that line, $1/V_{B3}$, gives us the apparent velocity that is detected exactly at that location. Then we have

$$\sin i_{B3} = \frac{V_1}{V_{B3}} \quad \text{or} \quad i_{B3} = \arcsin \frac{V_1}{V_{B3}}$$

where i_{B3} is the angle of the wave front. In the same way we can solve for the angle i_{C3} of the wave front at C_3 from the slope of the line tangent to the reversed travel time curve at position C_3,

$$\sin i_{C3} = \frac{V_1}{V_{C3}} \quad \text{or} \quad i_{C3} = \arcsin \frac{V_1}{V_{C3}}$$

We can draw wave fronts at these angles and find a point on the refractor from their intersection, as seen in Figure 3–21. By repeating this procedure using other travel time values, we find that the intersections of several pairs of wave fronts describe an undulating refrac-

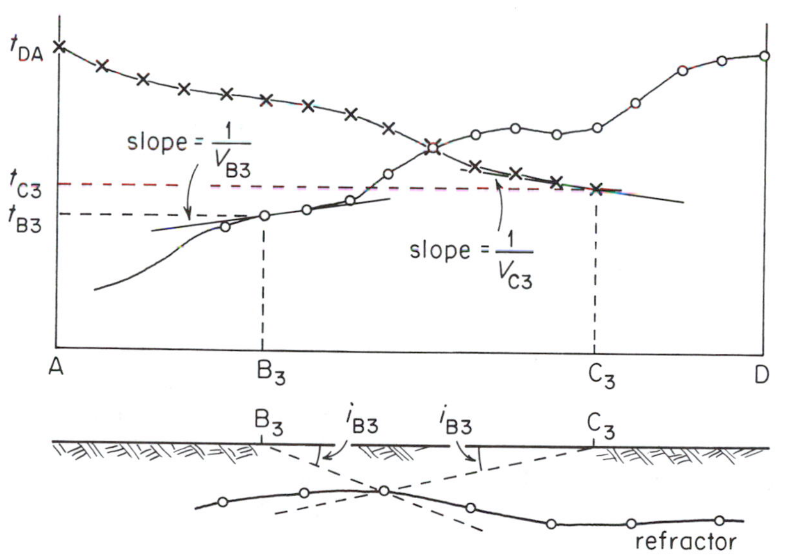

Figure 3–21
Features on undulating reversed refraction travel time curves used to locate receiver position for constructing wave fronts that intersect at points on an undulating refractor.

tor. We see that this method, like the ABC method, provides a way to interpret refractors that are not plane surfaces. But we need to remember that this is an approximate solution because undulating refractors will produce curved wave fronts.

APPLICATIONS OF SEISMIC REFRACTION SURVEYING

In the early days of geophysical exploration for oil and natural gas, seismic refraction surveying played a key role in locating salt domes and other shallow structures associated with reservoirs. As the resolution of seismic reflection surveying improved, however, the refraction methods were relegated to a supplementary capacity. In modern oil and gas exploration programs, refraction surveys are used to obtain important velocity information needed to interpret reflection data. In this role, refraction measurements are made with sources or receivers placed at different depths in deep wells, as well as on the land surface. Refraction surveys are also done to determine static corrections for reflection data.

Large-scale refraction surveys have been very important in exploring the deep structure of the earth's crust and upper mantle. For this work, the geophone lines extend for several hundred kilometers, and multi-ton explosives are used as sources. Most such surveys are conducted by university and government research groups rather than by industrial exploration groups.

Perhaps the most important applications of seismic refraction surveying are for geotechnical purposes. Shallow structures at construction sites and along proposed highway routes can be investigated by this method. Most consulting engineering firms concerned with these kinds of site surveys are equipped to do shallow refraction surveying.

STUDY EXERCISES

1. Show that the refracted ray path between the points A and B, as shown in Figure 3–22 and defined by Fermat's principle, satisfies Snell's law.

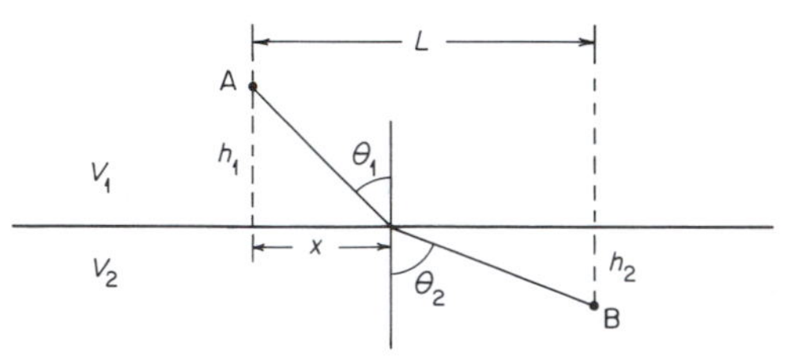

Figure 3–22

2. Suppose that a layer with a velocity of V_1 = 1500 m/sec and a thickness of 100 m lies above another layer with a velocity of V_2 = 3000 m/s. Compute the expected crossover distance and intercept time for the critically refracted waves.

3. A reversed seismic refraction survey indicates that a layer with velocity V_1 lies above another layer with velocity V_2 and that $V_2 > V_1$. We examine the travel times at a point C located midway between the shot points at sites A and B. The travel time of the refracted wave from A to C is less than the travel time of the refracted wave from B to C. Show that the apparent ve-

locity determined from the slope of the travel time curve for refracted waves produced from the source at A is less than the apparent velocity for refracted waves produced from the source at B. Does the boundary between the V_1 and V_2 layers dip downward toward A? Explain!

4. Suppose that based on well information a structure consisting of four layers is recognized. Layer thicknesses and velocities are shown in Figure 3–23. What is the critical distance for a refracted wave from the interface between layers defined by the velocities of 3500 and 4000 m/s? If you change V_2 to 3000 m/s, what will be the new value for the critical distance?

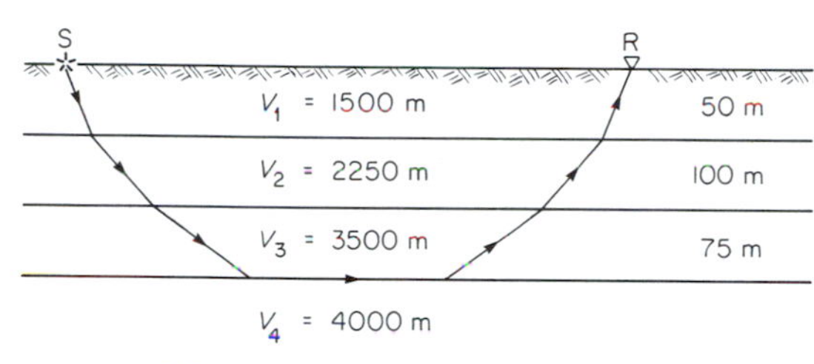

Figure 3–23

5. Suppose that a reversed refraction survey (using shots A and B) indicated velocities V_1 = 1500 m/s and V_2 = 2500 m/s from shot A and velocities V_1 = 1500 m/s and V_2 = 3250 m/s from shot B. Find the dip of the refractor. What would be the changes in velocities if the refractor had a slope 10 degrees larger than the one you computed?

6. From a seismic refraction survey, the travel time data given in the accompanying table were obtained for first arriving waves.

   a. Plot the data and determine velocities.

   b. How many layers are indicated by these data? What are the velocities for the direct and refracted waves? Determine the thicknesses of the layers.

DISTANCE (meters)	TIME (second)
10	0.010
20	0.020
30	0.030
40	0.040
50	0.045
75	0.055
100	0.065
125	0.075
150	0.080
175	0.085
200	0.090
250	0.100
300	0.110

7. Using the data given in the table, compute the delay times related to each refracting interface. Show the wave front pattern in each layer for the ray path critically refracted from the last interface.

SELECTED READING

Adachi, R., On a proof of fundamental formula concerning refraction method of geophysical prospecting and some remarks, *Kumanoto Journal of Science,* Series A, v. 2, pp. 18–23, 1954.

Barry, K. M., Delay time and its application to refraction pro file interpretation. In *Seismic Refraction Prospecting,* pp. 348–361 (ed. A. W. Musgrave). Tulsa, Okla., Society of Exploration Geophysics, 1967.

Barthelmes, A. J., Application of continuous profiling to refraction shooting, *Geophysics,* v. 11, n. 1, pp. 24–42, 1946.

Dix, C. H., *Seismic Prospecting for Oil.* New York, Harper, 1952.

Dix, C. H., Seismic velocities from surface measurements. *Geophysics,* v. 20, n. 1, pp. 68–86, 1955.

Dobrin, M. B., *Introduction to Geophysical Prospecting,* 3rd edition. New York, McGraw-Hill, 1976.

Gardner, L. W., Refraction seismograph profile interpretation. In *Seismic Refraction Prospecting,* pp. 338–347 (ed. A. W. Musgrave). Tulsa, Okla., Society of Exploration Geophysicists, 1967.

Griffiths, D. H., and R. F. King, *Applied Geophysics for Geologists and Engineers.* Oxford, England, Pergamon Press, 1981.

Hales, F. W., An accurate graphical method for interpreting seismic refraction lines, *Geophysical Prospecting,* v. 6, n. 3, pp. 285–294, 1958.

Johnson, S. H., Interpretation of split-spread refraction data in terms of plane dipping layers, *Geophysics,* v. 41, n. 3, pp. 418–424, 1976.

Kearey, P., and M. Brooks, *An Introduction to Geophysical Exploration.* Oxford, England, Blackwell Scientific Publication, 1984.

Laski, J. D., Computation of the time-distance curve for a dipping refractor and velocity increasing with depth in the overburden, *Geophysical Prospecting,* v. 21, n. 2, pp. 366–378, 1973.

McGee, J. L., and R. L. Palmer, Early refraction practices. In *Seismic Refraction Prospecting,* pp. 3–11 (ed. A. W. Musgrave). Tulsa, Okla., Society of Exploration Geophysicists, 1967.

Mooney, H. M., *Handbook of Engineering Geophysics.* Minneapolis, Bison Instruments, 1977.

Musgrave, A. W. (editor), *Seismic Refraction Prospecting.* Tulsa, Okla., Society of Exploration Geophysicists, 1967.

Parasnis, D. S., *Principles of Applied Geophysics,* 4th edition. London, Chapman and Hall, 1986.

Rockwell, D. W., A general wavefront method. In *Seismic Refraction Prospecting,* pp. 363–415 (ed. A. W. Musgrave). Tulsa, Okla., Society of Exploration Geophysicists, 1967.

Sharma, P., *Geophysical Methods in Geology.* Amsterdam, Elsevier, 1986.

Sheriff, R. E., *Encyclopedic Dictionary of Exploration Geophysics.* Tulsa, Okla., Society of Exploration Geophysicists, 1973.

Sheriff, R. E., and L. P. Geldart, *Exploration Seismology*. Volume 1: *History, Theory, and Data Acquisition*. Cambridge, England, Cambridge University Press, 1982.

Slotnick, M. M., A graphical method for the interpretation of refraction profile data, *Geophysics*, v. 15, n. 2, pp. 163–180, 1950.

Slotnick, M. M., *Lessons in Seismic Computing*. Tulsa, Okla., Society of Exploration Geophysicists, 1959.

Telford, W. M., L. P. Geldart, R. E. Sheriff, and D. A. Keys, *Applied Geophysics*. Cambridge, England, Cambridge University Press, 1976.

Wyrobek, S. M., Application of delay and intercept times in the interpretation of multilayer refraction time-distance curves, *Geophysical Prospecting*, v. 4, n. 2, pp. 112–130, 1956.

FOUR

Reflected Seismic Waves and Earth Structure

The first convincing demonstration of a procedure for detecting seismic waves reflected from a sedimentary bed was done in Oklahoma. After his successful tests in 1921, J. C. Karcher introduced reflection seismology as a practical method for petroleum exploration. During the 1920s this method together with the refraction method proved their worth in the discovery of several important oil fields.

At first, the refraction method was favored in the search for shallow oil-bearing structures.

Many shallow salt domes as well as other structures were found by refraction surveying. Because reflected waves never occur as first arrivals, they were more difficult to detect than refracted waves.

As the search moved deeper, the advantages of the reflection method soon became evident. The critical distance for detecting refracted waves from deeper layers turned out to be impractically large for routine surveying. From the 4:1 "rule of thumb," we see that a refraction spread would have to reach 30 km to probe depths of 7 to 8 km. In contrast, waves reflected from these and greater depths can be detected with much shorter geophone spreads and smaller energy sources.

Another advantage is that the reflection method is not "blind" to low-velocity layers that cannot be detected by the refraction method. A property of every substance is its *acoustical impedance,* which is the product of density and seismic wave velocity, ρV. Seismic waves will reflect from any boundary where the acoustical impedance changes. Therefore, reflections from deeper layers are produced regardless of whether the velocity is faster or slower in those layers.

The principal disadvantage of the seismic reflection method is the difficulty in recognizing the arrivals of reflected waves on a seismogram. The reflections are always later arrivals. In some instances they are strong enough to be clearly identified. Perhaps more often they are weak arrivals which are obscured by overlapping refracted waves and surface waves. A tremendous effort has gone into developing ways to enhance reflections so that they can be clearly identified. Modern methods involve both field operations and computer processing of seismograms recorded on magnetic tape.

The product of a seismic reflection survey is called a *seismic section.* It is prepared by combining seismogram traces from an appropriate selection of source–receiver positions. The example in Figure 4–1 illustrates the amazing detail that can be revealed. In some ways this seismic section has the appearance of a geologic cross section. The reflections are enhanced by shading to show more clearly the arching rock layers and the positions of faults. There is a very important difference. This seismic section indicates the arrival times of reflections rather than the depths of the reflecting layers. Qualitatively, it reveals some important features of the structure beneath the survey route. But to convert this picture into a more correct geologic cross section, we must determine seismic wave velocities and layer thicknesses from the arrival times of the reflected waves.

Our discussion of reflection seismology is in three parts. First, in this chapter we give the procedure for calculating wave velocities and the thickness and inclination of layers from reflected wave arrival times. For the purposes of this discussion, we will assume that we can properly identify the reflections on a seismogram. The second part deals with field operations and is the subject of Chapter 5. There we describe the variety of equipment and the survey procedures that are used to obtain high-quality reflection seismograms. The third part of our discussion, presented in Chapter 6, concerns data processing. We introduce the techniques for enhancing reflection quality and for combining the information on different seismogram traces in ways that reveal the most about the structure.

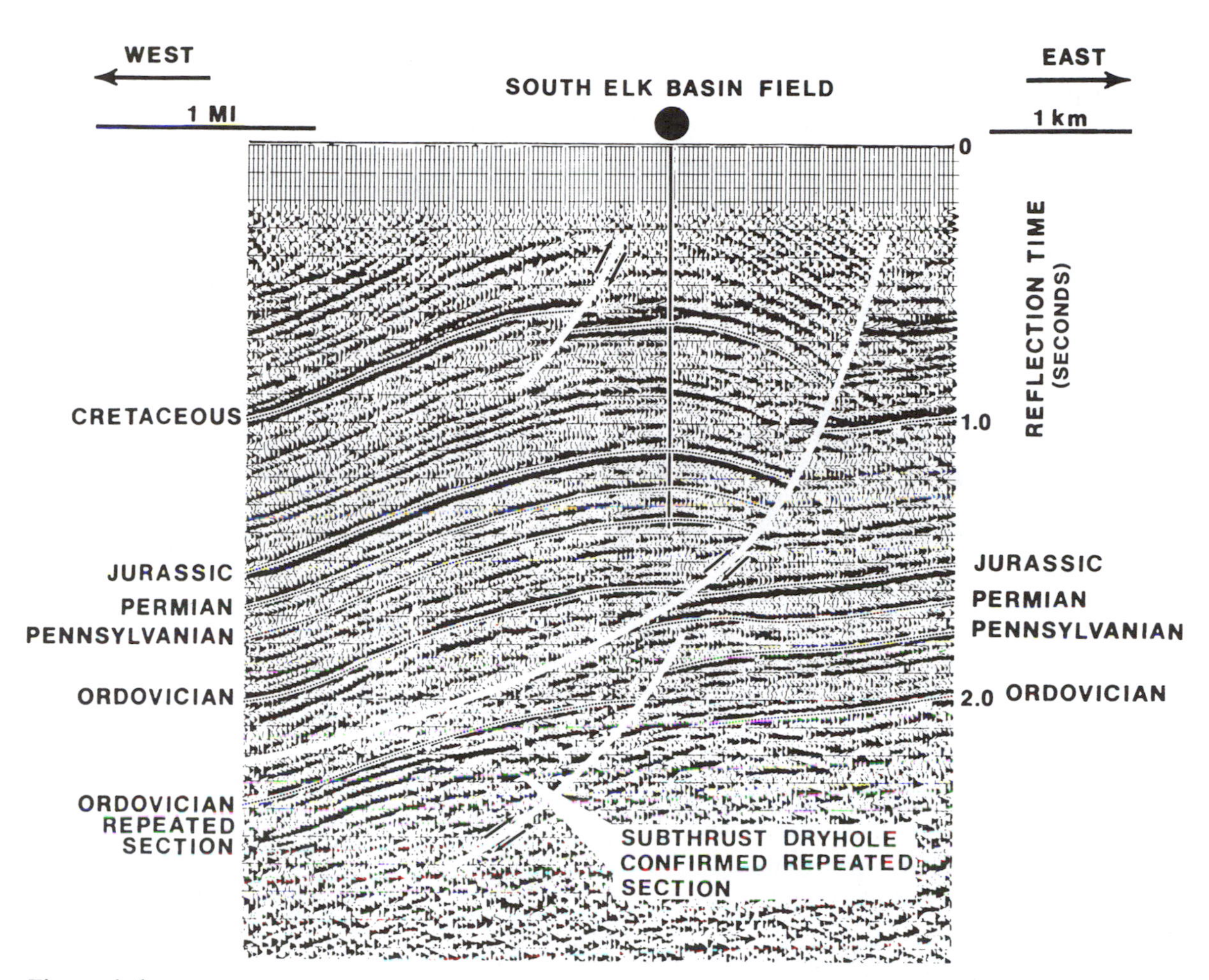

Figure 4–1
A seismic section that indicates folded and thrust-faulted structure. This is the conventional representation of seismic reflection data after lengthy processing steps. The horizontal axis at the top corresponds to the profile, and the vertical axis is the two-way vertical reflection time. The data are obtained by the powerful method called stacking. This type of data may be viewed as a subsurface cross section in reflection time. (Courtesy of Conoco, Inc.)

REFLECTION FROM A SINGLE HORIZONTAL SURFACE

Let us begin with the same simple structure that we used to introduce the refraction method. It consists of a single horizontal layer in which the seismic wave velocity is V_1 and the thickness is h_1, which lies above a deeper material in which the velocity is V_2. In earlier chapters we referred to the boundary between them as the refractor. For purposes of the present discussion we call this same boundary the *reflector*. To keep the analysis as uncomplicated as possible, we will look only at reflected *P*-waves. The results of this analysis apply equally to *S*-waves.

The Reflection Travel Time Curve

Basically, we should be able to detect reflected waves by applying the same surveying procedure we used to detect refracted waves. Several geophones are arranged in a line to receive the waves produced by a source such as an explosion. The paths of reflected waves are illustrated in Figure 4–2, together with a seismogram and travel time curve. Although the paths of direct waves and refracted waves are not shown, the arrivals of these waves appear on the seismogram together with the later-arriving reflected waves pulses. For this example, the critical distance for refraction is at receiver R_4. The following features should be noted.

1. The reflected wave pulses describe an arc across the seismogram, and their arrival times plot along a curve on the travel time graph. This pattern differs from the straight alignments of direct and refracted waves.
2. Reflected waves arrive at all receivers including R_1, which is at the same location as the source.
3. The reflection travel time curve is tangent to the refracted-wave travel time curve at the critical distance.
4. The reflection travel time curve lies above the direct-wave travel time curve but approaches it asymptotically with increasing distance.

The first of these four features will be analyzed in some detail. Before beginning our analysis, let us look at Figure 4–3, which helps to explain the other three features. It is obvious that there will be reflected paths leading to all receivers, including one placed close to the source. At the critical distance, indicated by R_4 in this example, the refracted wave and the reflected wave follow exactly the same path. In fact, they are one and the same wave. When we are discussing a refraction survey, we refer to it as a critically refracted wave that follows a path of zero distance along the refractor. But when we are concerned with a reflection survey, we call it the *critically reflected wave*. Because it carries the energy allocated for both reflection and refraction, it produces a pulse of higher amplitude on the seismogram than the nearby reflected and refracted pulses. And since this wave corresponds to both reflection and refraction, its arrival time must plot on both the reflection and refraction travel time curves. At all other distances the reflection arrival times plot above the refraction travel time line. Hence, the two travel time curves must be tangent at the critical distance.

The reflected waves travel to all the receivers at velocity V_1, which is the same as the velocity of the direct wave. The path of the direct wave is always shorter, however, and so it must always arrive before the reflection. This is why the reflection travel time curve lies above the direct-wave travel time curve. As receiver distance increases, the difference between the lengths of the direct and reflected paths diminishes, so that the two travel time curves approach asymptotically. This can be seen by comparing the paths to R_1 and R_7.

Reflection Arrival Time

The path of a reflected wave from a source at S to a receiver at R is shown in Figure 4–4. Because the angle of incidence and the angle of reflection are equal, the downgoing and

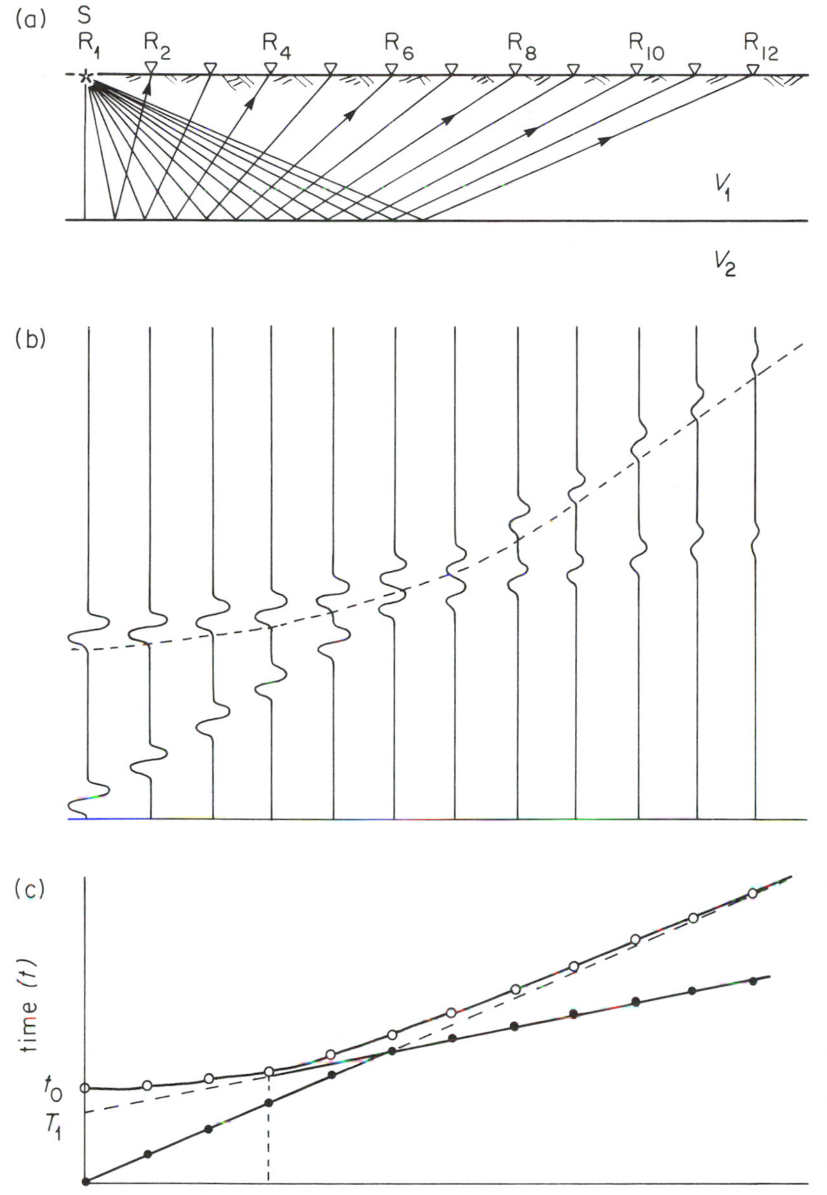

Figure 4–2

(a) Paths of waves reflected from a horizontal boundary, (b) the corresponding seismogram with reflected and refracted pulses, and (c) the reflection–refraction travel time graph (c). The travel time curves show that the curve of reflected waves has hyperbolic geometry, whereas the curves of direct and refracted waves are straight line segments.

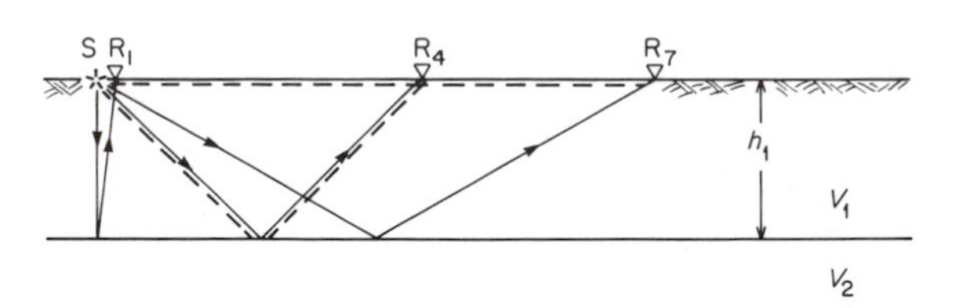

Figure 4–3
Direct and reflected wave paths from a source to receivers at small distance, critical distance, and beyond critical distance. The waves produced at the source (S) are reflected from the interface and recorded at receivers at near or short offset (R_1), the critical offset (R_4) where angle of incidence is defined by $\sin i_e = V_1/V_2$ and postcritical offset (R_7).

upgoing parts of the path will also be equal. In as much as the wave travels at velocity V_1, the travel time along the path distance of $2r$ is

$$t_x = \frac{2r}{V_1} \tag{4–1}$$

and since

$$r = \sqrt{\left(\frac{x}{2}\right)^2 + h_1^2} \tag{4–2}$$

where x is the distance between the source and the receiver, and h_1 is the depth to the reflector, we have

$$t_x = 2\sqrt{\frac{x^2/4 + h_1^2}{V_1}} \tag{4–3}$$

By squaring both sides, we get

$$t_x^2 = \frac{x^2}{V_1^2} + \frac{4h_1^2}{V_1^2}$$

and dividing by $4h_1^2$ and rearranging gives

$$\frac{t_x^2}{4h_1^2/V_1^2} - \frac{x^2}{4h_1^2} = 1 \tag{4–4}$$

Because h_1 and V_1 are constant properties of the structure, we see that Equation 4–4 expresses a hyperbola that is symmetric about $x = 0$. This important result explains why the reflection arrival times plot along a curve in Figure 4–2. We see that time t_x varies with receiver distance x according to an hyperbolic curve, as shown by Figure 4–5.

If the source and receiver are placed at the same location so that $x = 0$, Equation 4–3 reduces to

$$t_0 = \frac{2h_1}{V_1} \tag{4–5}$$

where t_0 is called the *zero-offset time.* It is the travel time of the reflected wave along a vertical path. If we now express h_1 in terms of t_0

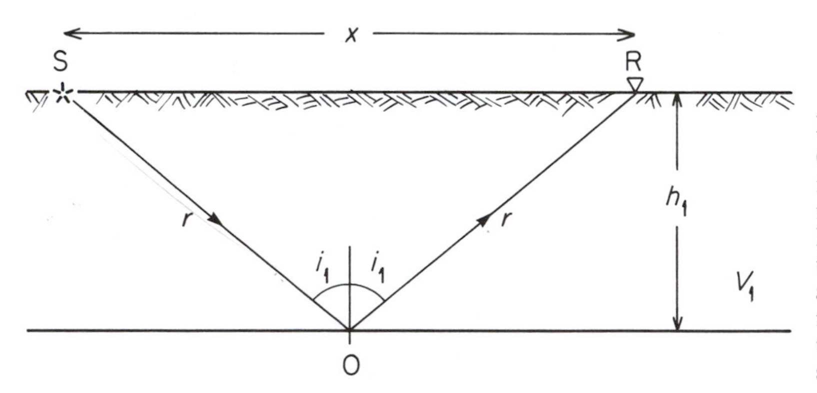

Figure 4–4
Geometry for a wave reflected from a single boundary. The SOR ray path is determined by Snell's law, which states that the angle of the incident ray is equal to the angle of the reflected ray. The reflection point (O) corresponds to the midpoint between the source (S) and the receiver (R)

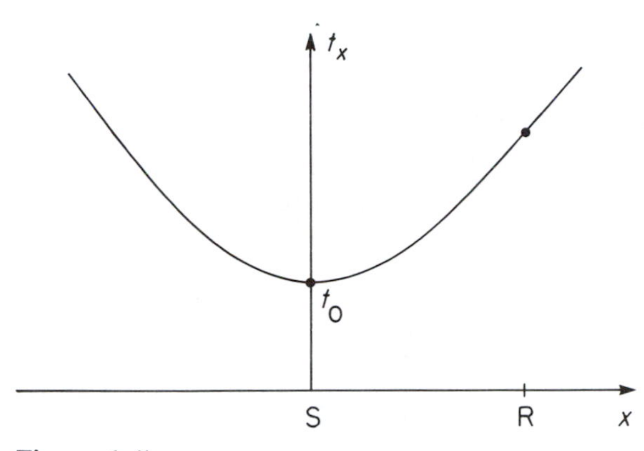

Figure 4–5
Hyperbola showing the form of a reflection travel time curve. The arrival times of reflected waves from a horizontal boundary recorded at different offsets on both sides of the source (S) plot along a hyperbolic time–distance curve. The minimum time (t_0) is recorded at the zero offset ($x = 0$).

and V_1 and substitute the result into Equation 4–4, we obtain

$$\frac{t_x^2}{t_0^2} - \frac{x^2}{t_0^2 V_1^2} = 1 \qquad (4\text{–}6)$$

which, again, shows that the reflection arrival time varies hyperbolically with distance.

An actual seismogram consisting of 120 traces is presented in Figure 4–6 to illustrate the hyperbolic pattern of reflected curves. In this illustration the source is placed at the center of the geophone spread. The straight alignments of pulses indicate refracted waves. Several hyperbolic arcs are marked by wave pulses reflected from a succession of layer boundaries.

Normal Move-out

A common practice in reflection seismology is to express the reflected-wave travel time t_x as the sum of the zero-offset time t_0 and the additional increment of time Δt that is needed because the receiver is offset a distance x from the source:

$$t_x = t_0 + \Delta t$$

The time increment Δt is called the *normal move-out (NMO) time.* It provides us with a way to express travel time that is more convenient for some methods of analyzing reflection data.

Now let us substitute from Equation 4–5 into Equation 4–3 to obtain

$$t_x = \sqrt{t_0^2 + x^2/V_1^2} = t_0\sqrt{1 + x^2/t_0^2V_1^2} \qquad (4\text{–}7)$$

and the normal move-out time,

$$\Delta t = \sqrt{t_0^2 + x^2/V_1^2} - t_0 \qquad (4\text{–}8)$$

If we let

$$a = \frac{x}{t_0 V_1}$$

then

$$t_x = t_0\sqrt{1 + a^2}$$

which can be expressed as the binomial expansion series

$$t_x = t_0\,(1 + \frac{1}{2}a^2 + \ldots)$$

or

$$t_x = t_0\left(1 + \frac{x^2}{2t_0^2V_1^2} + \ldots\right) \qquad (4\text{–}9)$$

For purposes of reflection seismology, we retain only these first two terms of the series and neglect the other higher-order terms. If we compare the expression

$$t_x = t_0\left(1 + \frac{x^2}{2t_0^2V_1^2}\right) = t_0 + \frac{x^2}{2t_0V_1^2}$$

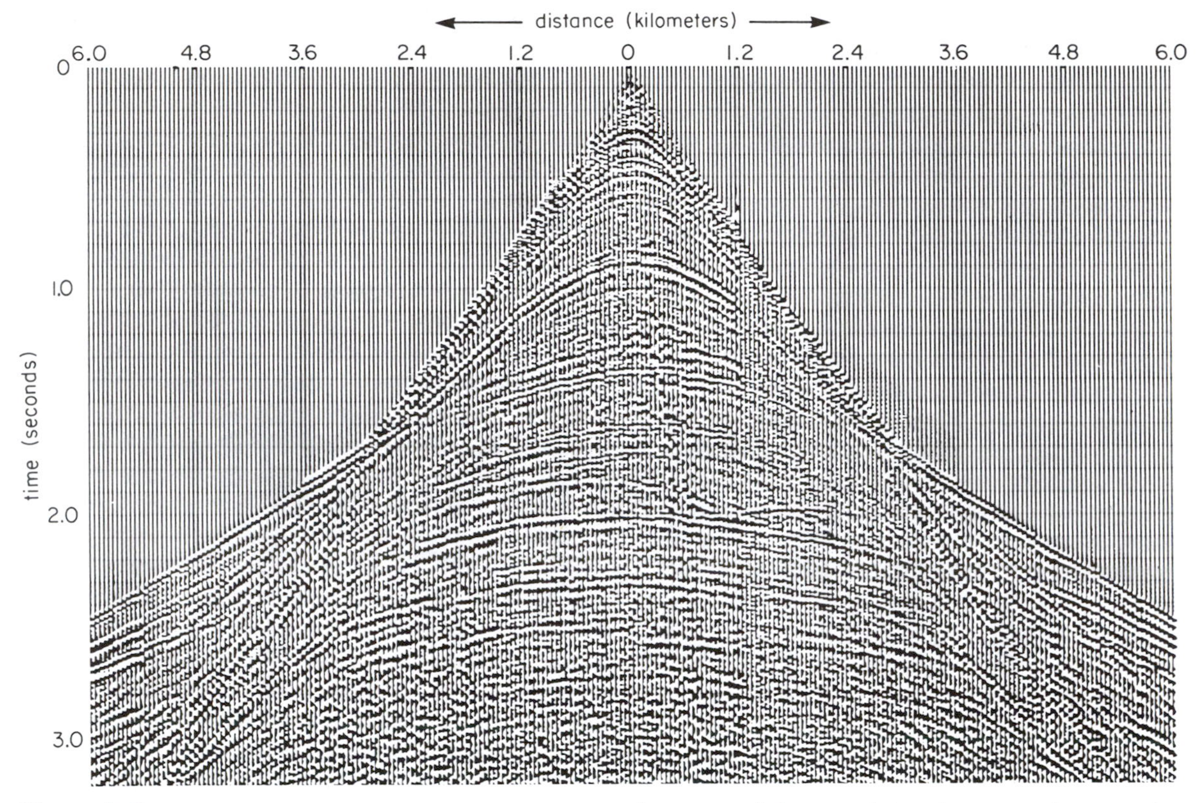

Figure 4–6
An actual seismic reflection record. Receiver distances for the different traces are given across the top, and the travel time increases downward from the top. (Courtesy of Prakla–Seismos AG.)

with Equation 4–7, we see that the NMO time is

$$\Delta t = \frac{x^2}{2t_0V_1^2} \qquad (4\text{–}10)$$

This expression allows us to examine how travel time increases as the receiver distance increases. For reflected waves Δt is directly proportional to x^2 and inversely proportional to t_0 and V_1^2. These patterns are shown graphically in Figure 4–7. Here we see that an increase in the depth to the reflector, which increases t_0, decreases NMO time. This means that the hyperbolic arc is less sharply curved for deeper reflectors. This curvature is evident in Figure 4–6 which shows that the hyperbolic arcs of successively deeper reflections are not parallel. Curvature along these arcs decreases as reflector depth increases.

Measuring Velocity and Reflector Depth

The depth h_1 to the reflector is easily calculated if we know the zero-offset time t_0 and

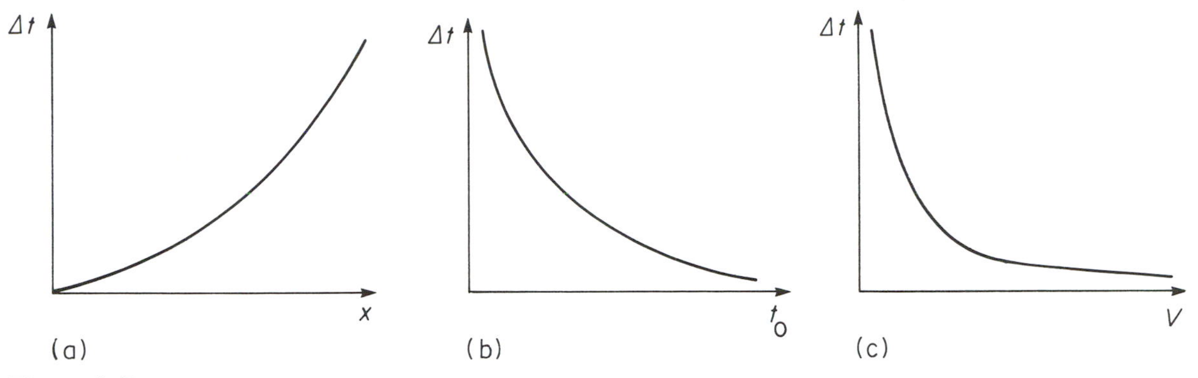

Figure 4–7
Variation of normal move-out: (a) variation of Δt with receiver distance; (b) variation of Δt with depth; and (c) variation of Δt with velocity. The normal move-out is directly proportional to the source–receiver distance while being inversely proportional to reflection time and velocity.

the wave velocity V_1. By rearranging Equation 4–5, we get

$$h_1 = \frac{t_0 V_1}{2} \qquad (4\text{–}11)$$

We can measure t_0 from the arrival time of the reflection on the seismogram trace corresponding to the zero-offset geophone. But suppose we cannot clearly identify the reflected pulse on that particular trace. Then we can use the arrival time t_x from another trace on which the reflection is more clearly distinguishable. To determine t_0 from t_x, we can multiply both sides of Equation 4–6 by t_0^2 and take the square root to obtain

$$t_0 = \sqrt{t_x^2 - x^2/V_1^2} \qquad (4\text{–}12)$$

Now we must find a way to calculate V_1 from the reflected-wave arrival times. One way this calculation can be done requires reflection arrival times t_{x1} and t_{x2} from any two receivers at offset distances of x_1 and x_2, as in Figure 4–8. Again, we multiply both sides of Equation 4–6 by t_0^2 and rearrange to obtain

$$t_0^2 = t_{x1}^2 - \frac{x_1^2}{V_1^2} = t_{x2}^2 - \frac{x_2^2}{V_1^2}$$

which allows us to write

$$t_{x2}^2 - t_{x1}^2 = \frac{1}{V_1^2}(x_2^2 - x_1^2)$$

so that we have finally

$$V_1 = \sqrt{(x_2^2 - x_1^2)/(t_{x2}^2 - t_{x1}^2)} \qquad (4\text{–}13)$$

Now we can find the values of t_0 and V_1 needed to solve Equation 4–11.

There is another interesting way to determine t_0 and V_1 which utilizes all the reflection arrival times that can be read from the seismogram. We see that Equation 4–4 can be rearranged to express

$$t_x^2 = \left(\frac{1}{V_1}\right)^2 x^2 + \frac{4h_1^2}{V_1^2} \qquad (4\text{–}14)$$

Now, if we let $t_x^2 = \tau$, $1/V_2 = M$, $x^2 = \chi$, and $4h_1^2/V_1^2 = B$, we can write

$$\tau = M\chi + B$$

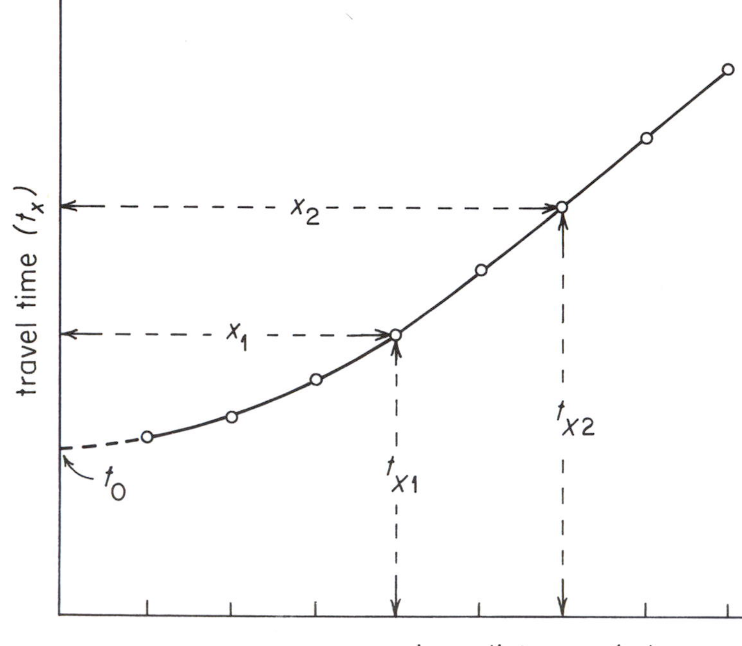

Figure 4–8
Information from a reflection travel time curve that can be used to calculate velocity and reflector depth. Here x_1 and x_2 are the offsets where corresponding reflection times t_{x1} and t_{x2} are determined. Using two offsets and related reflection times, we can compute the velocity V_1. Then, the zero-offset reflection time (t_0) and this velocity V_1 can be used to determine the layer thickness. If the reflector is horizontal, t_0 is the minimum reflection time.

which is the formula for a straight line. It tells us that if we plot squared values of travel time (t_x^2) and distance (x^2) corresponding to many receivers, the points should lie along a straight line. This kind of x^2–t^2 graph is illustrated in Figure 4–9. Observe that the slope of this alignment of points is $1/V_1^2$, and the intercept of this line at the t^2 axis is

$$\frac{4h_1^2}{V_1^2} = t_0^2$$

The following procedure, then, can be used to interpret the reflection seismogram in Figure 4–2.

1. Obtain reflection arrival times $t_1, t_2, t_3, \ldots$, at the receiver distances $x_1, x_2, x_3, \ldots$, from the seismogram.
2. Square these values to obtain $t_1^2, t_2^2, t_3^2, \ldots$, and $x_1^2, x_2^2, x_3^2, \ldots$, and plot these results on an x^2–t^2 graph.
3. Draw a straight line through the alignment of points in the x^2–t^2 graph. Calculate the velocity from the slope of this line, which is $1/V_1^2$, and determine the zero-offset time from its intercept, which is t_0^2.
4. Calculate the reflector depth h_1 using Equation 4–11 and these values of t_0 and V_1.

Reflected Waves and Direct Waves

Earlier, in Figure 4–2, we observed that the reflection travel time curve asymptotically approaches the direct-wave travel time curve. Now that we know how reflection travel time is related to velocity and depth of the reflector, we can examine this feature more closely.

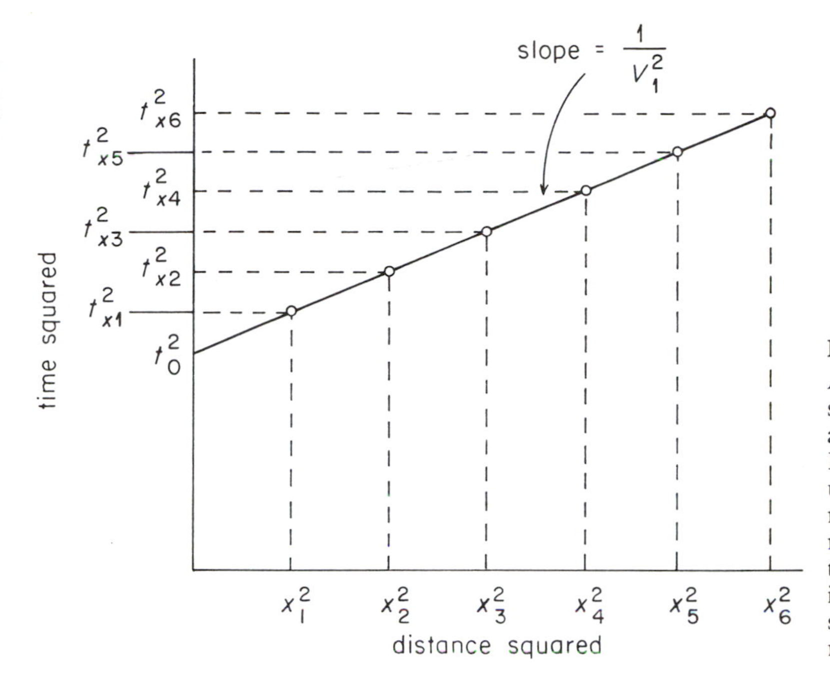

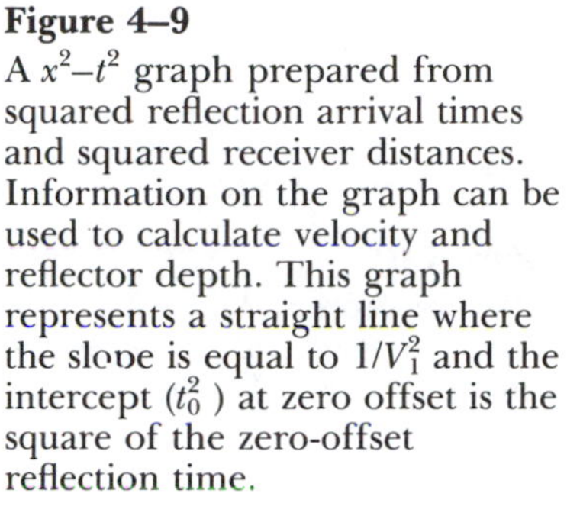

Figure 4–9
A x^2–t^2 graph prepared from squared reflection arrival times and squared receiver distances. Information on the graph can be used to calculate velocity and reflector depth. This graph represents a straight line where the slope is equal to $1/V_1^2$ and the intercept (t_0^2) at zero offset is the square of the zero-offset reflection time.

Let us write Equation 4–3 in the form

$$t_x = \sqrt{\frac{x^2}{V_1^2} + \frac{4h_1^2}{V_1^2}}$$

We can divide both sides by x to obtain

$$\frac{t_x}{x} = \sqrt{\frac{1}{V_1^2} + \frac{4}{V_1^2}\frac{h_1^2}{x^2}}$$

As distance $x \rightarrow \infty$, the term $h_1/x \rightarrow 0$, so that

$$\lim_{x\rightarrow\infty} \sqrt{\frac{1}{V_1^2} + \frac{4}{V_1^2}\frac{h_1^2}{x^2}} = \frac{1}{V_1}$$

Therefore, as $x \rightarrow \infty$, we obtain

$$\frac{t_x}{x} = \frac{1}{V_1} \quad \text{or} \quad t_x = \frac{x}{V_1}$$

We know from Equation 3–10 that this result expresses the direct-wave travel time. It confirms our earlier assertion that the direct-wave travel time curve is the asymptote of the reflection travel time curve.

REFLECTION FROM A SLOPING SURFACE

For the purpose of analyzing waves reflected from a sloping surface, let us assume that the receivers are placed in lines that extend in opposite directions from the source. Our aim is to determine the depth and inclination of the reflector.

Paths of Reflected Waves

The structure in Figure 4–10 consists of a single reflector inclined at the angle α. Waves re-

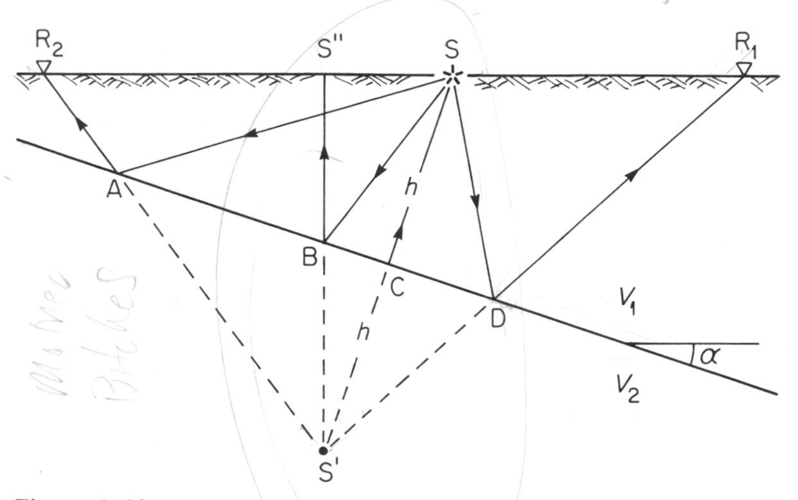

Figure 4–10
Reflection from a sloping boundary. Geometrical relationships of source–receiver paths and distances from the image of the source. S is the source, and R_1 and R_2 are receivers. The reflected waves recorded at R_1 and R_2 can be considered as waves from S′, the mirror image of S with respect to the reflector, which is the boundary between layers characterized by velocities V_1 and V_2. S″ is the vertical projection of the image onto the surface. This point is the location for the minimum reflection time (see Figure 4–11).

flected from this surface travel at velocity V_1, and the depth to the nearest point on the reflector beneath the source at S is h. To simplify our analysis of reflection paths, we introduce the point S′ which is called the *image* of the source at S. Imagine that the reflector is a mirror. The image that we would see in this mirror would appear to be at S′. The reflection point C lies exactly halfway between S and S′, and line SS′ is perpendicular to the reflector.

The image point is very useful for constructing the path of a reflected wave. It can be constructed in the following way. First, draw a line from S′ to the receiver. Then, at its intersection with the reflector, extend another line to the source. The lines above the reflector indicate the wave path. For example, in Figure 4–10 a line was drawn from S′ to the receiver R_1. It intersects the reflector at D. A second line was then drawn from D to the source at S. The path of the reflected wave is SDR_1. In the same way, the path SAR_2 was constructed for the reflected wave reaching the receiver at R_2. For all paths constructed by this method, the angle of incidence is equal to the angle of reflection as required by Snell's law.

Now consider another important feature. The total length of a reflected wave path is equal to the distance of the receiver from the image. To confirm that this statement is true, look again at the path reaching receiver R_1 in Figure 4–10. Because the point C is equidistant between S and S′, we see that SDS′ is an isosceles triangle, which means that it has two equal sides,

$$\mathrm{SD} = \mathrm{S'D}$$

so that the reflection path length

$$\mathrm{SD} + \mathrm{DR_1} = \mathrm{S'R_1}$$

We have this relationship for any travel path, regardless of receiver position, because the line SS′ will always form one side of an isosceles triangle whose third corner is the reflection point.

Reflection Travel Time

We already know that, in general, reflection travel time is found by dividing the path length by the wave velocity. For example, in Figure 4–10 the travel time to receiver R_1 will be

$$t_1 = \frac{\mathrm{SD}}{V_1} + \frac{\mathrm{DR_1}}{V_1} = \frac{\mathrm{S'R_1}}{V_1}$$

Because the velocity is the same along all reflection paths, it is evident that travel time is directly proportional to path length. This finding brings an interesting fact to our attention. Lines drawn from the image S′ in Figure 4–10 show clearly that the shortest reflection path is the one reaching the point S″ on the surface. We see that S″ is the vertical projection of the image onto the surface. It is obvious that a line from the image to any other surface point must be longer than this vertical line. Therefore, the minimum reflection travel time t_{min} will be along the path SBS″. We can see that the zero-offset time t_0 must be greater,

$$t_0 > t_{min}$$

because the path lengths

$$\mathrm{SS'} > \mathrm{S''S'}$$

We can conclude from Figure 4–10 that the minimum travel time will be measured by a receiver placed at the vertical projection of the image.

Now let us obtain a general expression for the travel time of a wave reflected from a dipping reflector. Look at Figure 4–11 where the receiver is offset from the source by the distance

$$\mathrm{SR} = x$$

and the path of the reflected wave is

$$\mathrm{SD} + \mathrm{DR} = \mathrm{S'R}$$

so that the travel time will be

$$t_x = \frac{\mathrm{SD} + \mathrm{DR}}{V_1} = \frac{\mathrm{S'R}}{V_1} \qquad (4\text{–}15)$$

Because we have the right triangle S′RS″, we can express the path length S′R in terms of the other two sides of that triangle:

$$(\mathrm{S'R})^2 = (\mathrm{S'S''})^2 + (\mathrm{S''R})^2 \qquad (4\text{–}16)$$

Observe from the geometry in Figure 4–11 that

$$\angle \mathrm{SS'S''} = \alpha$$

and that the line

$$\mathrm{SS'} = 2h \qquad (4\text{–}17)$$

Therefore, the line

$$\mathrm{SS''} = 2h \sin \alpha \qquad (4\text{–}18)$$

and the line

$$\mathrm{S'S''} = 2h \cos \alpha \qquad (4\text{–}19)$$

and, finally, the line

$$\mathrm{S''R} = \mathrm{SS''} + \mathrm{SR} = x + 2h \sin \alpha \qquad (4\text{–}20)$$

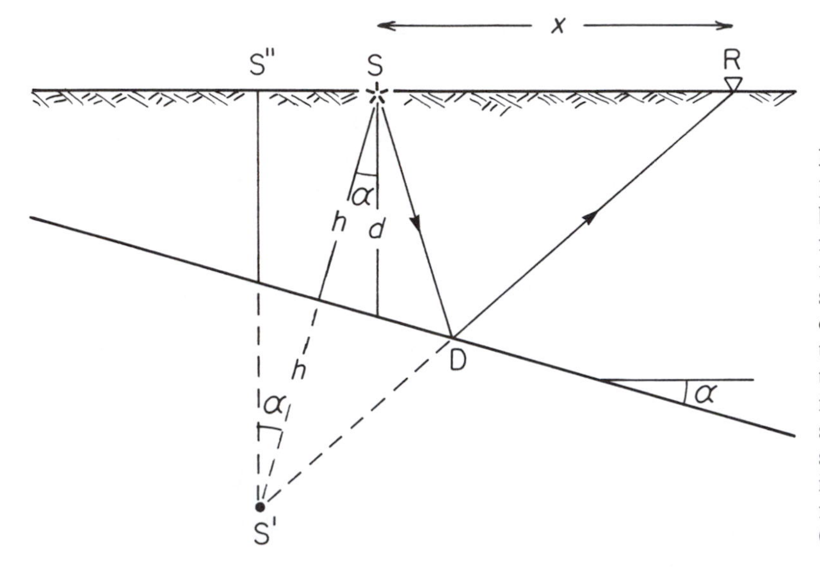

Figure 4–11
Reflection from a sloping boundary. Geometrical relationships involving a source–receiver path, the image of the source, and the vertical projection of the image on the surface. S′ is the mirror image of S, and S″ is the vertical projection of the image onto the surface. The shortest distance from S′ to the surface is the line S′S″. This indicates that the minimum reflection time is at S″. When $\alpha = 0$, S″ is at S.

Then, by substituting from Equations 4–19 and 4–20 into Equation 4–16, we obtain the travel time expression

$$t_x^2 = \left(\frac{2h\cos\alpha}{V_1}\right)^2 + \left(\frac{x + 2h\sin\alpha}{V_1}\right)^2 \qquad (4\text{–}21)$$

where x is the distance of the receiver from the source.

This is the equation of an hyperbola, which tells us that the reflection travel time curve for a dipping reflector has the form of an hyperbola. If we graph this equation, as in Figure 4–12, we see that the hyperbolic travel time curve is offset from the source point S. The minimum travel time $t_{\min}$ is observed at S″, which, as we know from Figure 4–11, is the vertical projection of the image.

Reflector Depth and Dip

Two time values are easily read from the travel time curve in Figure 4–12. First is the zero-offset time t_0, which corresponds to the path distance of $2h$, as shown in Figure 4–11 and Equation 4–17. Since $x = 0$ for this case, Equation 4–21 reduces to

$$t_0^2 = \frac{4h^2}{V_1^2} \qquad (4\text{–}22)$$

Second is the minimum travel time $t_{\min}$, which corresponds to the path length $2h\cos\alpha$, as shown by Equation 4–19 and Figure 4–11. For this case, we have

$$t_{\min} = \frac{2h\cos\alpha}{V_1} \qquad (4\text{–}23)$$

By squaring both sides, we obtain

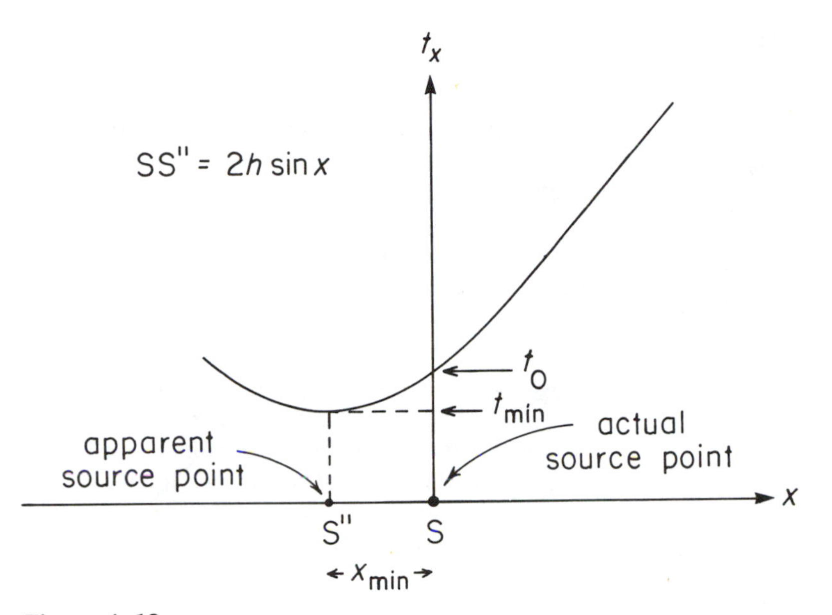

Figure 4–12
An offset hyperbola that shows the form of the reflection travel time curve for a sloping boundary has information that can be used to determine the position of the reflector. Indication of a sloping reflector is a hyperbolic time–distance curve that is not symmetric about the vertical axis at the source (S). The reflection time (t_0) at the zero offset is not the minimum reflection time. The minimum reflection time is at the point S″ on the up-dip side. From the travel time curve, S″ is the apparent source where the minimum reflection time (t_{min}) is observed. The distance between the actual source (S) and the apparent source (S″) is determined by $2h \sin \alpha$, where α is the slope of the reflector.

$$t_{min}^2 = \frac{4h^2}{V_1^2} \cos^2 \alpha \tag{4–24}$$

Substituting from Equation 4–22 gives

$$t_{min}^2 = t_0^2 \cos^2 \alpha$$

or

$$\cos \alpha = \frac{t_{min}}{t_0} \tag{4–25}$$

Now let us define the offset distance x_{min}, where travel time is minimum as the distance between the source S and the vertical projection of the image S″, that is,

$$x_{min} = SS''$$

From Equation 4–18 we know that

$$x_{min} = 2h \sin \alpha$$

This equation can be rearranged to find the reflector depth,

$$h = x_{\min}/2\sin a \qquad (4\text{–}26)$$

where from Equation 4–25 we know that

$$\alpha = \text{arc}\cos\left(\frac{t_{\min}}{t_0}\right) \qquad (4\text{–}27)$$

Looking again at Figure 4–11 and Equation 4–25, we see that the depth d of the reflector beneath the source is

$$d = \frac{h}{\cos\alpha} = h\,\frac{t_0}{t_{\min}} \qquad (4\text{–}28)$$

and by means of Equation 4–26 we get

$$d = \frac{x_{\min}t_0}{2t_{\min}\sin\alpha} \qquad (4\text{–}29)$$

Now we have a very simple way to locate the dipping reflector by means of a seismic reflection survey.

1. Plot the arrival times obtained from the reflected pulses on the seismogram and draw the hyperbolic travel time curve through these points.
2. Read the values of $x_{\min}$, $t_{\min}$, and t_0 from this graph, as in Figure 4–12.
3. Calculate the dip angle α by means of Equation 4–27 and the depth d by means of Equation 4–29.

Here we see that it is not even necessary to calculate the velocity V_1 to determine the reflector position. Of course, the velocity can now be calculated quite easily. By rearranging Equation 4–23 to get

$$V_1 = \frac{2h}{t_{\min}}\cos a$$

and then by substituting from Equations 4–25 and 4–28, we obtain

$$V_1 = \frac{2t_{\min}}{t_0^2}\,d \qquad (4\text{–}30)$$

Suppose that reflections are difficult to identify on some seismogram traces. Then it may also be difficult to draw the hyperbola through the travel time points so that $x_{\min}$ and $t_{\min}$ can be accurately determined. To bring these features into clearer focus, we can prepare an x^2–t^2 graph. As in the case of the horizontal reflector, points on this graph plot in straight alignments, as shown in Figure 4–13. Because the receivers are placed in lines extending in opposite directions from the source, we obtain two straight alignments on the x^2–t^2 graph. The intersection of the two lines drawn through these points is at the

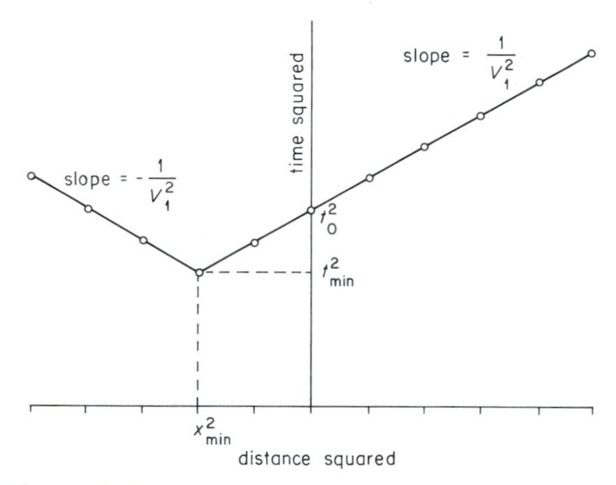

Figure 4–13
The x^2–t^2 graph, prepared from squared arrival times and squared receiver distances for reflections from a sloping boundary, has information for determining the reflector position. This x^2–t^2 graph displays two line segments with opposite slopes. Absolute value of the slope is equal to $1/V_1^2$, and the crossing point of the curves defines $t_{\min}^2$ and $x_{\min}^2$. Here t_0^2 is determined from the intercept time at the zero-offset vertical axis.

point $x^2_{\min}$, $t^2_{\min}$, and the slopes of these lines are $1/V_1^2$ and $-1/V_1^2$.

Finally, let us compare our results for a dipping reflector with those we obtained earlier for a horizontal reflector. As shown in Figure 4–11, if the reflector were horizontal, the line SS′ would be vertical. For this case, the dip angle $\alpha = 0$, and it is obvious that $d = h$, $x_{\min} = 0$, and $t_{\min} = t_0$. Then Equation 4–21 reduces to Equation 4–4, and Equation 4–23 reduces to Equation 4–5. So our analysis of a dipping reflector also applies to the case of $\alpha = 0$. We see that tilting the reflector causes the line to rotate from the source to the image in the up-dip direction, as seen in Figure 4–14. Therefore, all the geometrical relations seen in Figure 4–10 also exist for a horizontal reflector.

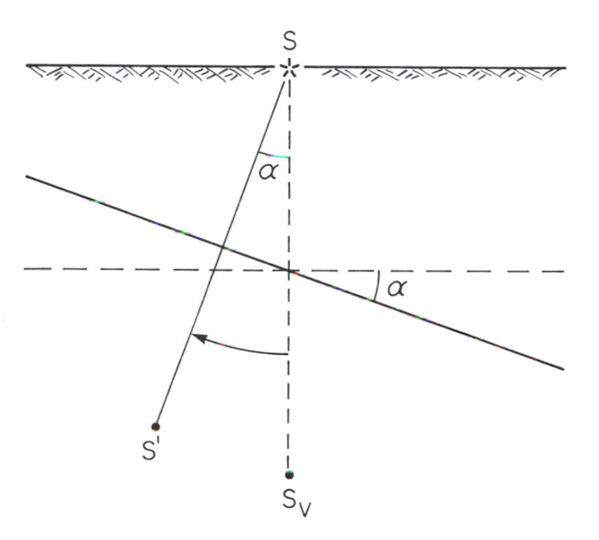

Figure 4–14
Comparison of image points for inclined and horizontal reflectors. If the sloping reflector is rotated to be horizontal, S′ will move to S_V, the mirror image of S with respect to the horizontal reflector. In other words, the mirror image of the source moves to a point in the up-dip direction as an originally horizontal reflector is tilted.

Alternate Analysis

There is another way to find the dip of a reflector that is more familiar to many exploration seismologists. This method requires the reflection arrival times at two receivers situated at equal distances in opposite directions from the source, as illustrated in Figure 4–15.

First, consider the travel time t_x to the down-dip receiver R_x. According to Equation 4–21, we have

$$t_x^2 = \left(\frac{2h \cos \alpha}{V_1}\right)^2 + \left(\frac{x + 2h \sin \alpha}{V_1}\right)^2$$

or

$$t_x^2 = \frac{4h^2}{V_1^2} \cos^2 \alpha + \frac{x^2}{V_1^2} + \frac{4hx \sin \alpha}{V_1^2} + \frac{4h^2}{V_1^2} \sin^2 \alpha$$

Then, since $4h^2 \cos^2 \alpha + 4h^2 \sin^2 \alpha = 4h^2$, we obtain

$$t_x^2 = \frac{4h^2}{V_1^2} + \frac{x^2 + 4hx \sin \alpha}{V_1^2}$$

and substituting from Equation 4–22 gives us

$$t_x^2 = t_0^2 + \frac{x^2 + 4hx \sin \alpha}{V_1^2}$$

for which the square root is

$$t_x = t_0 \sqrt{1 + \frac{x^2 + 4hx \sin \alpha}{4h^2}}$$

If we expand this expression in a binomial series and use only the first two terms of that series, we have

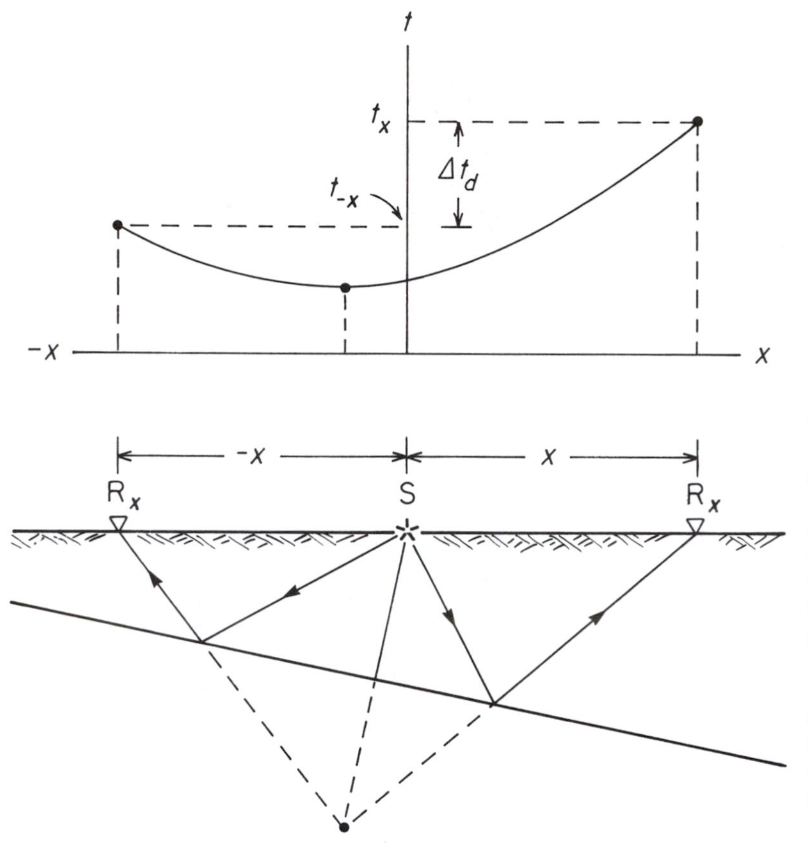

Figure 4–15
Information for calculating the dip angle of a reflector by means of arrival times at receivers along opposite lines at equal distance from the source. The reflection times t_x and t_{-x} are observed at receivers R_x and R_{-x}, respectively, and $\Delta t_d = t_x - t_{-x}$ is called dip move-out time. For a horizontal reflector $\Delta t_d = 0$. So, as a qualitative interpretation, $\Delta t_d \neq 0$ indicates a sloping reflector that dips toward the receiver where a greater travel time reading was measured.

$$t_x = t_0\left(1 + \frac{x^2 + 4hx \sin\alpha}{8h^2}\right)$$

or

$$t_x = t_0 + \frac{x^2 + 4hx \sin\alpha}{4hV_1} \qquad (4\text{–}31)$$

Using the same procedure, we can obtain the travel time t_{-x} to the up-dip receiver R_{-x}, which is

$$t_{-x} = t_0 + \frac{x^2 - 4hx \sin\alpha}{4hV_1} \qquad (4\text{–}32)$$

The difference between these two travel times is the *dip move-out time* Δt_d which can be expressed as

$$\Delta t_d = t_x - t_{-x}$$

or, by substituting from Equations 4–31 and 4–32, we get

$$\Delta t_d = \frac{2x \sin\alpha}{V_1} \qquad (4\text{–}33)$$

Now we can solve for the dip angle

$$\alpha = \text{arc}\sin\left(\frac{\Delta t_d V_1}{2x}\right) \qquad (4\text{–}34)$$

To obtain the dip by this method, we calculate the dip move-out time from the travel time curve as shown in Figure 4–15, and we determine the velocity from the slope of the lines on the x^2–t^2 graph as in Figure 4–13.

Three-Dimensional Dip Calculations

The methods of dip calculation that we have just presented assume that seismic reflection travel times have been measured along a profile aligned in the dip direction. These are sometimes called two-dimensional methods because we are concerned only with (1) distance along the profile and (2) depth of the reflector.

Suppose that the reflection measurements are made in some other direction. Then the computed angle will not represent true dip. Rather, it is a value called *apparent dip*. To determine true dip, we need to have apparent dip measurements in two different directions. For this, we require a three-dimensional method of dip calculation that involves (1) distance along a profile in one direction, (2) distance along another profile in a different direction, and (3) depth of the reflector.

Developing an exact three-dimensional method of dip calculation is a complicated problem in analytic geometry,[1] for the reflection travel paths do not lie within a vertical plane. Instead, they are in an inclined plane, and the alignment of reflection points differs from the direction of the source–receiver profile. Figure 4–16 shows that the subsurface reflection points (A, B) not only are inclined relative to the source–receiver profile on the surface (S–R_1–R_2) but also point in a different direction.

Where the reflector dip angle is quite small, say, less than 10 degrees, this difference can be neglected for the purpose of making an approximate dip calculation. By assuming that the reflection travel paths lie in a vertical plane, we can use Equation 4–27 or Equation 4–34 to estimate the apparent dip angle in the direction of a source–receiver profile.

Apparent dip angles α_x and α_y along two intersecting source–receiver profiles can be used to calculate the direction and angle of the true dip of the reflector. Figure 4–17 illustrates how this is done. Here two vectors are drawn

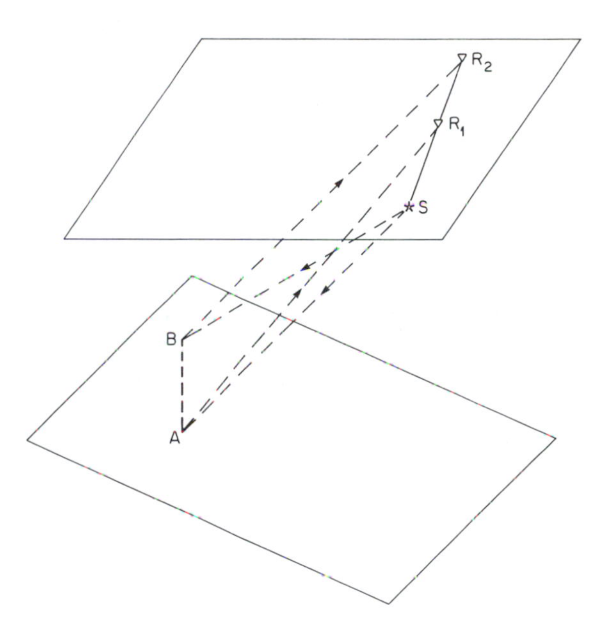

Figure 4–16
The subsurface reflection profile of the points A and B is inclined relative to the surface profile of source (S) and receivers (R_1 and R_2). Notice that the reflection paths are in an inclined plane, and that the alignment of reflection points is not parallel to the source–receiver profile.

[1]A thorough discussion of three-dimensional dip problems is presented by Morris Miller Slotnick in *Lessons in Seismic Computing*, Tulsa, Okla., Society of Exploration Geophysicists, 1959.

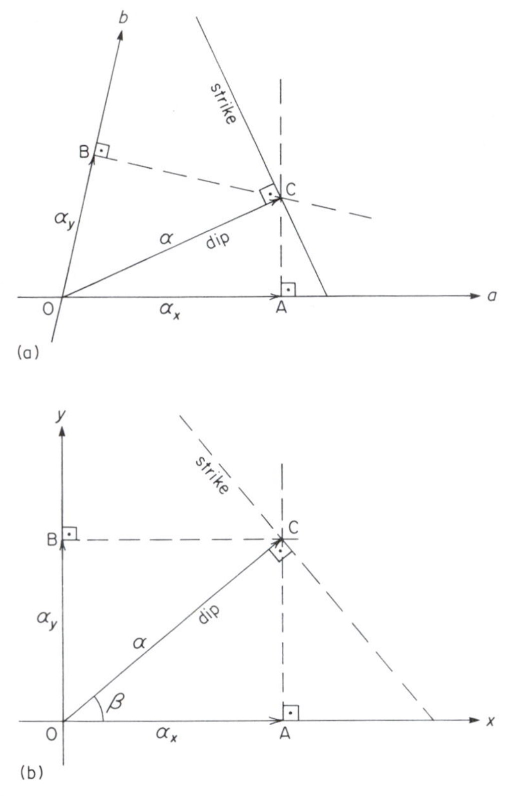

Figure 4–17
(a) Vector diagram for estimating the true reflector dip from apparent dip measurements in two directions *a* and *b*. Lengths of vectors OA and OB correspond to the apparent dip angles. The length of OC gives the true dip angle, and its position indicates the true dip direction. (b) A special case of two perpendicular source–receiver profiles.

in the directions of the source–receiver profiles. The lengths of these vectors, OA and OB, are proportional to the values of the dip angles. Lines extending perpendicularly from the ends of these vectors intersect at point C. The vector OC is then drawn to indicate the true dip direction. The length of OC is proportional to the true dip angle α.

For the particular case of two perpendicular source–receiver profiles (Figure 4–17b), the true dip direction relative to the profile in the x direction is specified by the angle β where

$$\tan \beta = \alpha_y/\alpha_x$$

The true dip angle for this case is simply

$$\alpha = (\alpha_x^2 + \alpha_y^2)^{1/2}$$

Procedures for acquiring and processing seismic reflection data suitable for three-dimensional analysis are introduced in Chapters 5 and 6.

REFLECTED WAVES IN A MULTILAYERED STRUCTURE

Seismic waves reflect from any boundary where the acoustical impedance ρV changes. On reflection seismograms it is common to see reflections from a succession of boundaries, such as those appearing in Figure 4–6. Our purpose in this section is to explain how to calculate depths to a succession of reflectors and the velocities in the layers between them by means of reflection arrival times. In our discussion we will consider only structures with horizontal layers.

Average Velocities

We begin with the structure in Figure 4–18, which has three horizontal reflectors and thicknesses h_1, h_2, and h_3 for layers in which velocities are $V_1 > V_2 < V_3$. Reflection paths to a receiver R that is offset from the source S by the distance x are shown. Observe that the paths reflected from deeper boundaries are

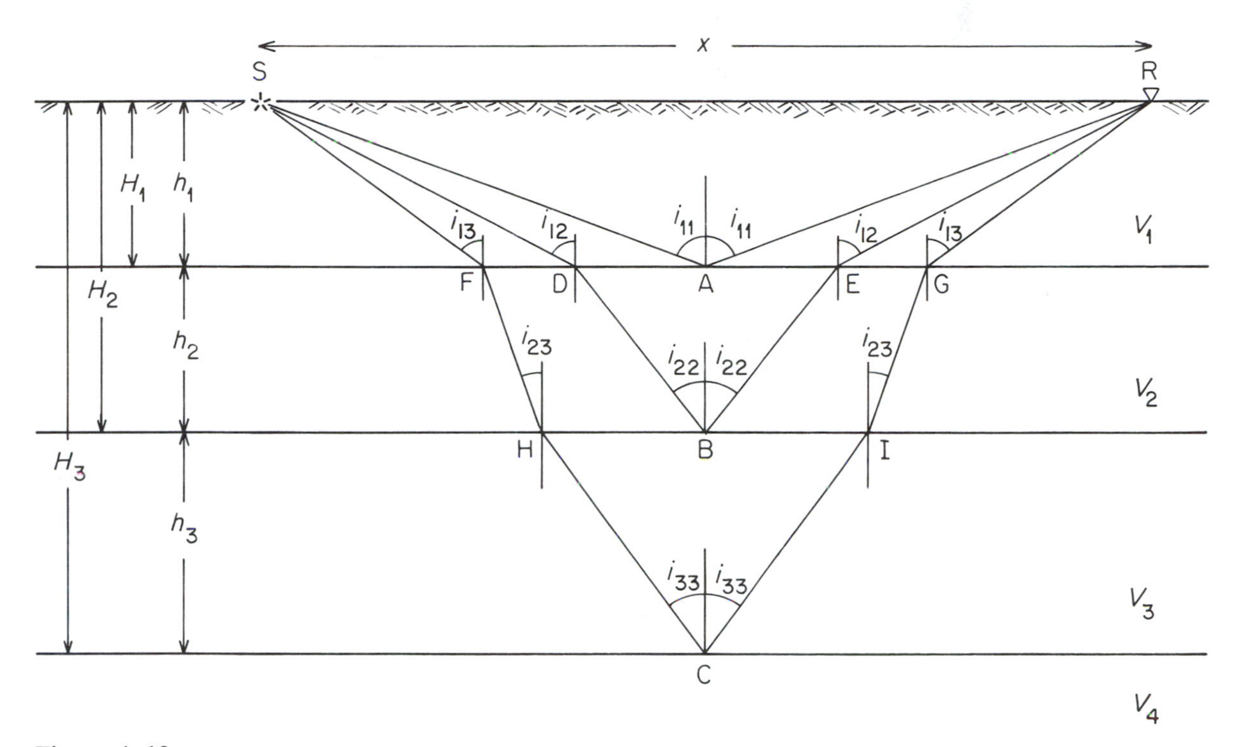

Figure 4–18
Reflection paths in a structure with three horizontal reflectors.

refracted where they cross overlying boundaries.

Let us consider the zero-offset times for reflections from these three boundaries. Applying Equation 4–5, we see that

$$t_{0,1} = 2h_1/V_1 = 2\Delta t_1 \tag{4–35a}$$

$$t_{0,2} = \frac{2h_1}{V_1} + \frac{2h_2}{V_2} = 2\Delta t_1 + 2\Delta t_2 \tag{4–35b}$$

$$t_{0,3} = \frac{2h_1}{V_1} + \frac{2h_2}{V_2} + \frac{2h_3}{V_3} = 2\Delta t_1 + 2\Delta t_2 + 2\Delta t_3 \tag{4–35c}$$

where Δt_1, Δt_2, and Δt_3 are the one-way travel time increments through the layers.

Next, consider the depths to the reflectors that are

$$H_1 = h_1 \tag{4–36a}$$

$$H_2 = h_1 + h_2 \tag{4–36b}$$

$$H_3 = h_1 + h_2 + h_3 \tag{4–36c}$$

By combining Equations 4–35 and 4–36, we obtain

$$V_1 = \frac{2H_1}{t_{0,1}} = \frac{h_1}{\Delta t_1} \tag{4–37a}$$

$$V_{2,\text{ave}} = \frac{2H_2}{t_{0,2}} = \frac{h_1 + h_2}{\Delta t_1 + \Delta t_2} \tag{4–37b}$$

$$V_{3,\text{ave}} = \frac{2H_3}{t_{0,3}} = \frac{h_1 + h_2 + h_3}{\Delta t_1 + \Delta t_2 + \Delta t_3} \tag{4–37c}$$

where we define $V_{2,\text{ave}}$ and $V_{3,\text{ave}}$ to be the *average velocities* along the zero-offset reflection paths to the two deeper reflectors. According to Equations 4–35, we can also express the average velocities in terms of the individual layer velocities,

$$V_{2,\text{ave}} = \frac{V_1\,\Delta t_1 + V_2\,\Delta t_2}{\Delta t_1 + \Delta t_2} = \frac{\sum_{i=1}^{2} V_i\,\Delta t_i}{\sum_{i=1}^{2} \Delta t_i} \qquad (4\text{–}38\text{a})$$

$$V_{3,\text{ave}} = \frac{V_1\,\Delta t_1 + V_2\,\Delta t_2 + V_3\,\Delta t_3}{\Delta t_1 + \Delta t_2 + \Delta t_3} = \frac{\sum_{i=1}^{3} V_i\,\Delta t_i}{\sum_{i=1}^{3} \Delta t_i} \qquad (4\text{–}38\text{b})$$

It is clear that we can generalize from the form of these expressions to obtain the average velocity along the zero-offset reflection path to any reflector in a multilayered structure. If there are n horizontal reflectors, the average velocity along the zero-offset path to the deepest one is

$$V_{n,\text{ave}} = \frac{\sum_{i=1}^{n} V_i\,\Delta t_i}{\sum_{i=1}^{n} \Delta t_i} \qquad (4\text{–}39)$$

The depth H_n to the deepest reflector is

$$H_n = \frac{V_{n,\text{ave}}t_{0,n}}{2} \qquad (4\text{–}40)$$

where $t_{0,n}$ is the zero-offset time of the reflection from that boundary. Later, we will describe how to determine the layer velocities that are needed to calculate the average velocity.

Root-Mean-Square (RMS) Velocities

The average velocity expressed by Equation 4–39 cannot be used for analyzing waves that follow paths to receivers offset from the source. The reason is that the path distance through any layer is not simply the layer thickness, as is the case for the zero-offset paths. Instead, the distance varies according to the angle at which it refracts into the layer together with the layer thickness. Snell's law tells us that the angle of refraction increases as velocity increases, which indicates that for layers of equal thickness the path distance will be longer in the higher-velocity layers. This is obvious in Figure 4–18 where $h_1 = h_2$.

In order to analyze the travel time along an offset path, we need to determine a weighted average velocity that accounts for changes in path direction as well as changes in layer thickness. To see how this calculation can be done, we will look first at the offset path for the wave reflected from the upper reflector in Figure 4–18. This is the path SAR. Observe that the directions of the lines SA and AR can be specified by the angle i_{11}. Suppose that δt_1 is the travel time along the line SA and, similarly, along the line AR. Then we can express the weighted average velocity along the path in the following way:

$$V_{1,\text{rms}} = \left(\frac{V_1^2\,\delta t_1}{\delta t_1}\right)^{1/2} \qquad (4\text{–}41)$$

This weighted average velocity is called the *root-mean-square velocity*, or the *RMS velocity* along the path of the reflected wave. Here it is obvious that $V_{1,\text{rms}} = V_1$ because the path is entirely in the top layer. Nevertheless, by expressing the velocity this way, we establish a form that can be used for deeper layers. From Equation 4–35 we know that the vertical travel time through the top layer is $\Delta t_1 = h_1/V_1$. Because both a vertically traveling wave and a wave following the path SA move at velocity V_1, we have

$$\cos i_{11} = \frac{h_1}{\text{SA}} = \frac{\Delta t_1}{\delta t_1}$$

or

$$\delta t_1 = \frac{\Delta t_1}{\cos i_{11}} \tag{4–42}$$

Substituting the result into Equation 4–41 gives us

$$V_{1,\text{rms}} = \left(\frac{\dfrac{V_1^2 \, \Delta t_1}{\cos i_{11}}}{\dfrac{\Delta t_1}{\cos i_{11}}} \right)^{1/2} \tag{4–43}$$

Here we can begin to see what root-mean-square velocity means. We are taking the square root of a sum of squared velocity terms.

In this form we take account of the path direction by the term $\cos i_{11}$, and we account for the layer thickness by the term Δt_1.

Now we can proceed to the deeper reflectors. The second reflection follows the path SDBER in Figure 4–18. To find the RMS velocity for this path, we need terms that account for path directions and thicknesses of the upper two layers. For the angles involved we see that

$$\cos i_{12} = \frac{h_1}{\text{SD}} = \frac{\Delta t_1}{\delta t_1} \quad \text{or} \quad \delta t_1 = \frac{\Delta t_1}{\cos i_{12}}$$

and

$$\cos i_{22} = \frac{h_2}{\text{DB}} = \frac{\Delta t_2}{\delta t_2} \quad \text{or} \quad \delta t_2 = \frac{\Delta t_2}{\cos i_{22}}$$

where δt_1 and δt_2 are travel times along SD and DB, and Δt_1 and Δt_2 are vertical travel times through the layers. Because these expressions have the same form as Equation 4–42, it appears that we can use the form of Equation 4–43 for each of the layers to find the RMS velocity along the offset reflection path from the second reflector:

$$V_{2,\text{rms}} = \left(\frac{\dfrac{V_1^2 \, \Delta t_1}{\cos i_{12}} + \dfrac{V_2^2 \, \Delta t_2}{\cos i_{22}}}{\dfrac{\Delta t_1}{\cos i_{12}} + \dfrac{\Delta t_2}{\cos i_{22}}} \right)^{1/2} \tag{4–44}$$

Similarly, for the deepest offset reflection that follows the path SFHCIGR in Figure 4–18, we use the path angles, velocities, and vertical travel time increments to obtain

$$V_{3,\text{rms}} = \left(\frac{\dfrac{V_1^2 \, \Delta t_1}{\cos i_{13}} + \dfrac{V_2^2 \, \Delta t_2}{\cos i_{23}} + \dfrac{V_3^2 \, \Delta t_3}{\cos i_{33}}}{\dfrac{\Delta t_1}{\cos i_{13}} + \dfrac{\Delta t_2}{\cos i_{23}} + \dfrac{\Delta t_3}{\cos i_{33}}} \right)^{1/2} \tag{4–45}$$

From the pattern you see in these expressions, it is clear that, in general, for a structure with many horizontal layers the RMS velocity along an offset path reflected at the nth boundary will be

$$V_{n,\text{rms}} = \left(\frac{\displaystyle\sum_{k=1}^{n} \frac{V_k^2 \, \Delta t_k}{\cos i_{kn}}}{\displaystyle\sum_{k=1}^{n} \frac{\Delta t_k}{\cos i_{kn}}} \right)^{1/2} \tag{4–46}$$

So far, we have described the RMS velocity as a weighted average velocity along the path of a reflected wave, and we have shown how it is related to velocities, thicknesses, and travel path directions through the layers above the reflector. But how do we actually measure the RMS velocity? To explain how this measurement can be taken, let us examine the expression for the travel time of a reflected wave following a path to a receiver that is offset a distance x from the source.

A simple way to express the travel time $t_{x,n}$ of a wave reflected from the nth reflector is by using the form of Equation 4–8. For the present case, however, we replace t_0 and V_1 by $t_{0,n}$ and the RMS velocity, which gives

$$t_{x,n} = \sqrt{t_{0,n} + x^2/V_{n,\text{rms}}^2} \qquad (4\text{–}47)$$

We see that this expression, like Equation 4–8, is the formula for an hyperbola. Therefore, let us proceed in the usual way to find $V_{n,\text{rms}}$. We can choose two arrival times from the reflection travel time curve and calculate $V_{n,\text{rms}}$ by means of Equation 4–13. In addition, we can prepare an x^2–t^2 graph using all the arrival times, draw a line through the alignment of points, and find the RMS velocity from the slope of this line which is $1/V_{n,\text{rms}}^2$.

In these ways an RMS velocity can be determined for each reflector in a multilayered structure. For example, three reflections are identified on the seismogram in Figure 4–19. Arrival times read from the seismogram traces are then plotted, and hyperbolic travel time curves can be drawn, as in Figure 4–20. Arrival times and offset distances are then squared and plotted on an x^2–t^2 graph, also shown in Figure 4–20. From the slopes of the lines on this x^2–t^2 graph, we calculate the RMS velocities for reflection paths to the three reflectors.

We should point out a fact that we have overlooked up to now. The RMS velocities along different offset paths from the same reflector are not exactly the same. This is easy to understand, because for different offset distances the path angles in each layer will be different. However, these differences are usually quite small from one receiver to the next. So, for practical purposes, they are ignored. The result is that, except for the uppermost reflector, points on an x^2–t^2 graph will not plot exactly along a line. Ordinarily, however, their alignment is so nearly straight that the exploration geophysicist gets satisfactory results by drawing a straight line as close as possible to these points.

Layer Thickness and Velocity

Reflections identified on seismograms indicate the existence of layer boundaries and provide the information for calculating RMS velocities along wave paths reflected from these boundaries. How can we use this information to find the thickness and velocity for individual layers in the structure? We accomplish this objective by using RMS velocities and zero-offset reflection times. We already know how to use Equation 4–11 to find the thickness h_1 of the top layer in the structure, and we know how to find V_1 from a travel time curve or an x^2–t^2 graph. What about h_2 and V_2 for the layer directly beneath it? First, look at Equations 4–35 which allow us to express the vertical travel times Δt_1 and Δt_2 through these layers in terms of the zero-offset reflection times $t_{0,1}$ and $t_{0,2}$,

$$\Delta t_1 = \frac{t_{0,1}}{2} \quad \text{and} \quad \Delta t_2 = \frac{t_{0,2} - t_{0,1}}{2} \qquad (4\text{–}48)$$

Now look at Equation 4–44. For the zero-offset path, which is vertical, the angles $i_{12} = i_{22} = 0$ degrees, so that $\cos i_{12} = \cos i_{22} = 1$. For this case, Equation 4–44 reduces to

$$V_{2,\text{rms}} = \left(\frac{V_1^2\Delta t_1 + V_2^2\Delta t_2}{\Delta t_1 + \Delta t_2}\right)^{1/2}$$

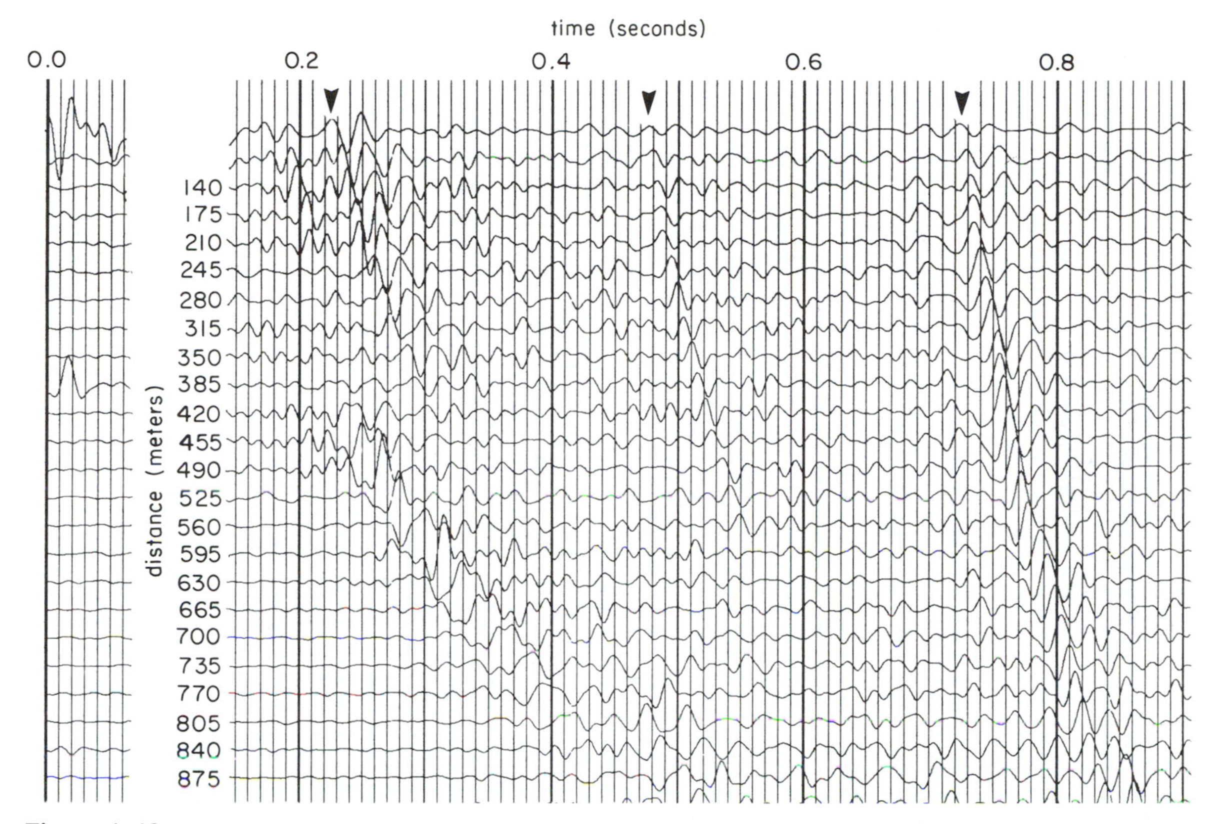

Figure 4–19
Reflection seismogram recorded in eastern Virginia. Receiver distances and the time scale are indicated, and three reflections are identified. The data were recorded with a 24-channel system. The receiver interval was 35 meters, and the source was 70 meters from the first detector, allowing a spread from 70 to 875 meters. The energy source was vibroseis.

and by substituting from Equations 4–48, we get

$$V_{2,\text{rms}} = \left[\frac{V_1^2 t_{0,1} + V_2^2(t_{0,2} - t_{0,1})}{t_{0,2}}\right]^{1/2} \qquad (4\text{–}49)$$

We can square both sides and rearrange to obtain

$$V_2^2 = \frac{V_{2,\text{rms}}^2 t_{0,2} - V_1^2 t_{0,1}}{t_{0,2} - t_{0,1}} \qquad (4\text{–}50)$$

All the values on the right-hand side of this expression can be determined from x^2–t^2 graphs of the two reflections, as seen in Figure 4–20b. Then, having calculated V_2 in this way, we can use Equations 4–35 and 4–48 to find

$$h_2 = \frac{V_2(t_{0,2} - t_{0,1})}{2} \qquad (4\text{–}51)$$

Proceeding to the next layer, we can rearrange Equation 4–45 according to the same reasoning we used to adapt Equation 4–44. Again, for the zero-offset path, which is vertical, we have $\cos i_{13} = \cos i_{23} = \cos i_{33} = 1$ and $\Delta t_3 = (t_{0,3} - t_{0,2})/2$. Therefore, Equation 4–45 reduces to

$$V_{3,\text{rms}} = \left[\frac{V_1^2 t_{0,1} + V_2^2(t_{0,2} - t_{0,1}) + V_3^2(t_{0,3} - t_{0,2})}{t_{0,3}}\right]^{1/2}$$

According to Equation 4–49, the first two terms in the numerator can be expressed as

$$V_1^2 t_{0,1} + V_2^2(t_{0,2} - t_{0,1}) = V_{2,\text{rms}}^2 t_{0,2}$$

which can be substituted above to get

$$V_{3,\text{rms}} = \left[\frac{V_{2,\text{rms}}^2 t_{0,2} + V_3^2(t_{0,3} - t_{0,2})}{t_{0,3}}\right]^{1/2}$$

This expression can be squared and rearranged to obtain

$$V_3^2 = \frac{V_{3,\text{rms}}^2 t_{0,3} - V_{2,\text{rms}}^2 t_{0,2}}{t_{0,3} - t_{0,2}} \qquad (4\text{–}52)$$

Again, all the terms on the right-hand side can be determined from an x^2–t^2 graph, such as the one in Figure 4–20. After calculating V_3 we can find the layer thickness

$$h_3 = \frac{V_3(t_{0,3} - t_{0,2})}{2} \qquad (4\text{–}53)$$

The pattern is now clear. In a multilayered structure we can find the velocity V_n and thickness h_n in any layer that is bounded by the nth reflector below and the $n-1$th reflector above,

$$V_n^2 = \frac{V_{n,\text{rms}}^2 t_{0,n} - V_{n-1,\text{rms}}^2 t_{0,n-1}}{t_{0,n} - t_{0,n-1}} \qquad (4\text{–}54)$$

and

$$h_n = \frac{V_n(t_{0,n} - t_{0,n-1})}{2} \qquad (4\text{–}55)$$

wave velocity in the interval between two reflectors. Therefore, we see that the Dix equation is used to calculate interval velocities.

Equation 4–54 was developed by the well-known exploration seismologist C. H. Dix and is commonly referred to as the *Dix equation.* The individual layer velocities are called *interval velocities* because they indicate the specific

Reflector Depth

There are two procedures for calculating reflector depth. One is obvious from Equations 4–36. For a multilayered structure, this procedure can be expressed as the summation

$$H_n = \sum_{i=1}^{n} h_i \qquad (4\text{–}56)$$

To find the depth in this way, we must first solve Equation 4–54 and then Equation 4–55 for each layer.

The second procedure makes use of the average velocity $V_{n,\text{ave}}$ along the zero-offset path to a reflector and the zero-offset time. Reflector depth H_n is given by Equation 4–40. Observe that to determine the average velocity according to Equation 4–39, we must know the interval velocities. Therefore, the Dix equation must be solved for each layer, but it is not necessary to solve Equation 4–55 for each layer, if the goal is to find the depth of one particular reflector.

Practical Example

The seismogram in Figure 4–19 was recorded in eastern Virginia to measure the thicknesses

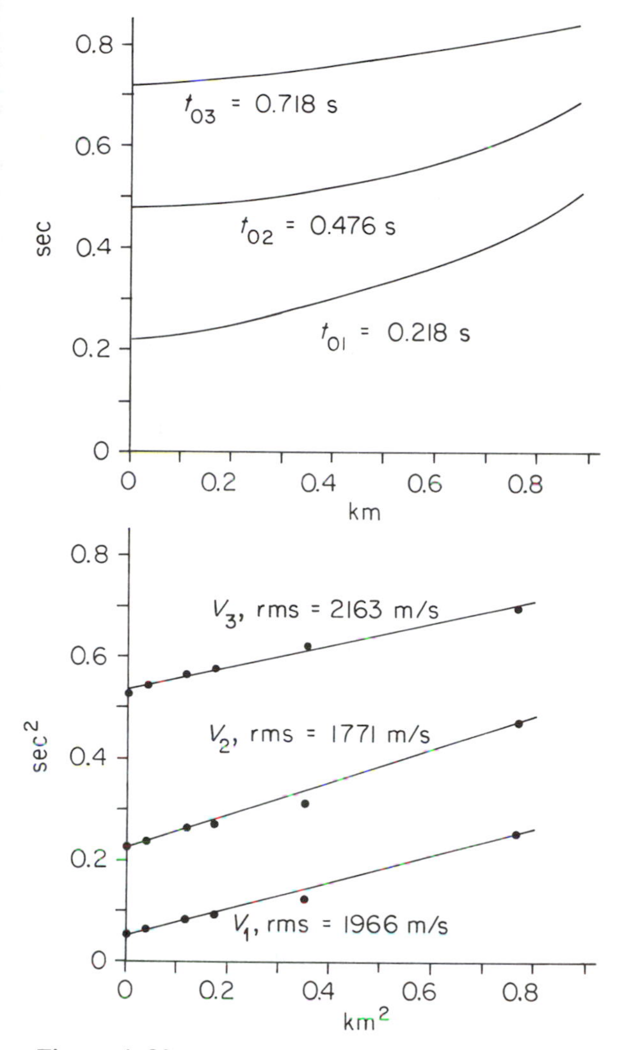

Figure 4–20
Travel time curves and x^2–t^2 graphs prepared from arrival times and receiver distances of the reflections identified on the seismogram in Figure 4–19. RMS velocities and zero-offset times can be used to calculate interval velocities and layer thicknesses. The t_0 times were determined assuming horizontal reflectors.

of sand and clay layers that make up the sequence of Atlantic Coastal Plain sediments. Geophones were placed in a line at 35-meter intervals, and the source was located 70 meters from the first geophone.

Arrival times for the three reflections were read from the seismogram. They were used to prepare the travel time curves and x^2–t^2 graphs in Figure 4–20. From the x^2–t^2 graphs we obtain the zero-offset time of

$$t_{0,1} = 0.218 \text{ s} \quad t_{0,2} = 0.476 \text{ s} \quad t_{0,3} = 0.718 \text{ s}$$

and from the slopes of the lines we calculate

$$V_1 = 1966 \text{ m/s} \quad V_{2,\text{rms}} = 1771 \text{ m/s} \quad V_{3,\text{rms}} = 2163 \text{ m/s}$$

Using these values in the Dix equation, we find

$$V_2 = 1588 \text{ m/s} \quad \text{and} \quad V_3 = 2777 \text{ m/s}$$

and then from Equation 4–55 we calculate

$$h_1 = 214 \text{ m} \quad h_2 = 205 \text{ m} \quad h_3 = 336 \text{ m}$$

MULTIPLE REFLECTED WAVES

From our discussion of reflected and refracted waves, it should now be clear that many paths lead from a source to a receiver. The paths of multiply reflected waves are one class of wave paths not yet described. These waves reflect more than one time from the same boundary. Examples of these paths are illustrated in Figure 4–21. We see that multiple reflections can occur at any boundary. In fact, the land surface itself plays an important role in the downward reflection of waves.

According to Figure 4–21, we could expect to record an infinite number of reflections even from a structure with only one reflector, but in practice only the first few multiples possess enough energy to produce a distinguish-

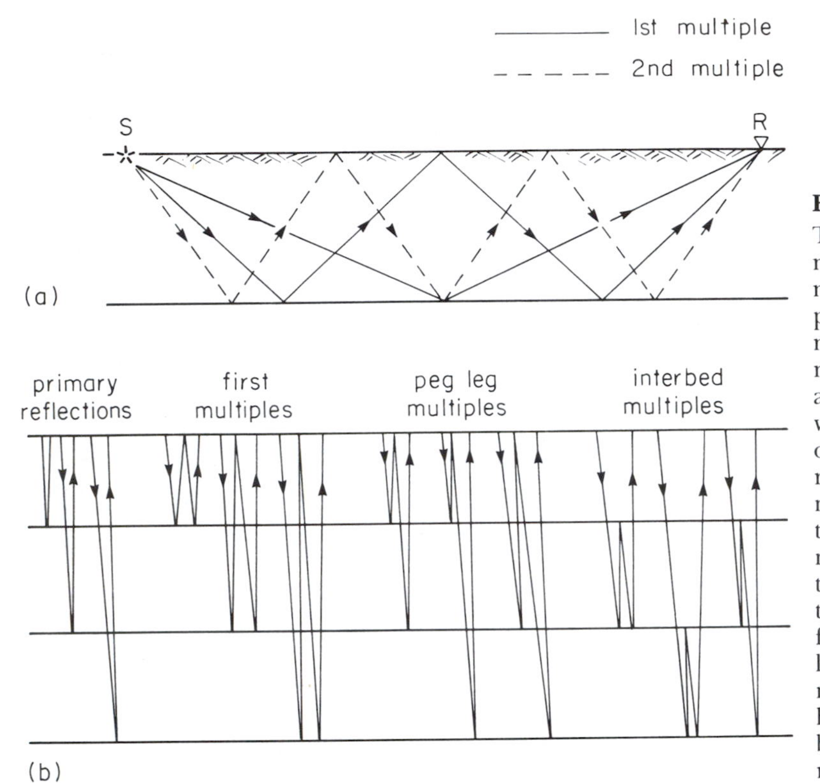

Figure 4–21
Typical paths for multiply reflected waves. (a) A single reflector model of paths for the primary reflection, first multiple reflection, and second multiple reflection. Because of the later arrival times of multiples, there will be three distinct reflections on the seismogram. (b) Primary reflection and different types of multiples from a model with three reflectors. First multiples may be recognized with arrival times that are twice the arrival times of the primary reflections from the different reflectors. Peg leg multiples and interbed multiples have path segments that have reflected from different boundaries than the primary reflector.

able pulse on a seismogram. To understand why, look again at our discussion of reflection coefficients in Chapter 2. According to Figure 2–8, at each point of incidence on a boundary the wave is partitioned into reflected and refracted waves that travel away from the boundary. It is clear that the energy of *P*-waves following the paths in Figure 4–21 is partitioned between a transmitted and reflected component at each boundary. For zero-offset paths, which are vertical, the resulting change of amplitude can be calculated from Equations 2–21 and 2–22. They show that this decrease in amplitude depends on the acoustical impedance contrast at each boundary.

Some numerical examples presented in Figure 4–22 indicate the decline in amplitude of a succession of multiples. In this illustration a wave amplitude of 1.0 at the source is assumed, and the corresponding amplitudes of the multiples are given by the numbers above the structures. These amplitudes were calculated from Equations 2–21 and 2–22 for the zero-offset receiver using velocities given on the structures. The paths, which are meant to represent zero-offset paths, are drawn at angles to illustrate the number of multiple reflections that occurred. These results demonstrate that only the first few multiples might possess amplitudes that can be detected on a seismogram. We know that geometrical

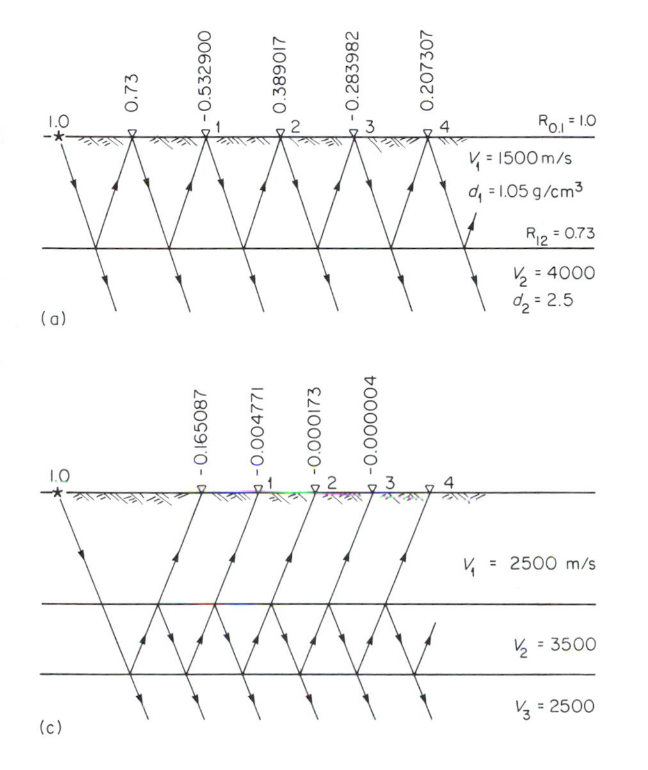

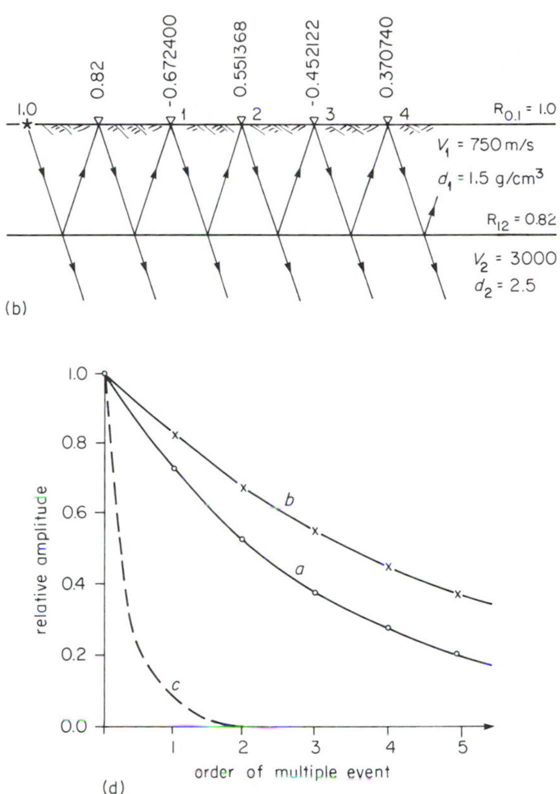

Figure 4–22
Relative amplitudes of multiply reflected waves in three different structures that represent (a) a layer of seawater over bedrock, (b) a low-velocity weathered layer over bedrock, and (c) interbed multiples in a relatively high-velocity layer. Numerical results for each model are plotted on the graph (d) to show how amplitude diminishes for a succession of multiples. A low-velocity weathered layer and a water layer cause multiples with relatively high amplitudes.

spreading, attenuation, and scattering act to diminish the amplitudes of multiples still further.

The few multiples that do possess amplitudes sufficient to be detected on a seismogram can cause interpretation problems. These problems can be especially acute for seismic surveying at sea where the first reflector is the ocean bottom. Because of the large acoustical impedance contrast between the water and the underlying sediment or rock, especially large multiples can be recorded.

How can we distinguish a multiple reflection from a primary reflection coming from a deeper boundary? Figure 4–23 shows that the path SABCR of a multiple has the same length as the path SDR to an imaginary reflector located at $2h_1$, the depth of the actual reflector. Suppose that we read the arrival times of a multiple reflection on a seismogram and

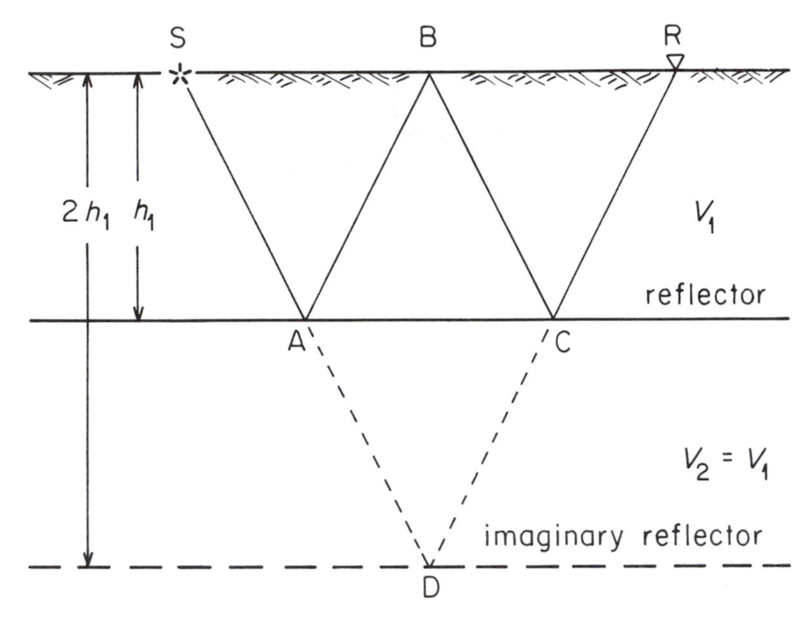

Figure 4–23
Geometrical relationships for a multiple reflection and for an apparent reflection from an imaginary boundary. The ray path SABCR is equivalent to SADCR, which gives an image of reflection from a point D on an imaginary reflector at a depth of $2h_1$ where h_1 is the depth of the primary reflection.

then prepare travel time and x^2–t^2 graphs. By analyzing the information in the usual way, we would calculate an identical velocity V_1 and $2h_1$, the reflector depth which we would obtain from analysis of the primary reflection. In other words, the multiple would indicate a second layer of equal thickness and velocity. When analysis of a seismogram produces such a result, the exploration seismologist knows that a multiple has been mistaken for a primary reflection. Then the travel time curves and x^2–t^2 alignments corresponding to the multiple can be deleted, and the analysis can be repeated using only the primary reflections.

DIFFRACTED WAVES

Some seismic wave paths do not follow from Snell's law, but they can be understood from Huygen's principle. These are the paths of diffracted waves which we introduced in Chapter 3 in our discussion of waves refracted from a discontinuous, or faulted, boundary. Some travel time characteristics of diffracted waves turn out to be similar to those of reflected waves. Diffracted waves are recorded on seismograms from reflection surveys over discontinuous or nonplanar boundaries.

A simple structure with a discontinuous, or faulted, boundary is shown in Figure 4–24. Consider the horizontal wave front AOB as it starts to travel vertically downward from the land surface at velocity V_1. By the time t_1 it has advanced downward a distance $s = V_1t_1$ to the position CDE. Then, the left-hand side CD reflects upward and reaches the land surface along AO at time $t_2 = 2t_1$. During this same time interval the right-hand side DE continues downward to the lower boundary along FG. While all this is happening, a circular diffracted wave front advances away from D, which by Huygen's principle is a wave source.

Therefore, at time t_2 we have the continuous wave front AOFG, where the part from O to F is the semicircle on the right-hand side of D. Observe that this circular part of the wave front is tangent to the linear parts at O and F, and that its radius $s = V_1 t_1$. Diffracted portions of a wave front, such as this one, are produced at discontinuities on a reflector.

Now let us consider the travel times of waves reaching the receivers R_1 and R_2, which are offset by the distance x from point O in Figure 4–24 lying directly above the discontinuity. At R_1 the reflection arrives at time t_2 as noted earlier. Later, the diffracted wave front, shown by the quarter circle on the left-hand side of O and D, will advance to R_1. Its travel time t_d will be

$$t_d = t_1 + \Delta t_d$$

where Δt_d is the travel time along the path R_1D. Since the triangle R_1DO is a right triangle, we see that

$$\Delta t_d = \sqrt{t_1^2 + x^2/V_1^2}$$

Therefore, the diffracted wave reaches R_1 at the time

$$t_d = t_1 + \sqrt{t_1^2 + x^2/V_1^2} \qquad (4\text{–}57)$$

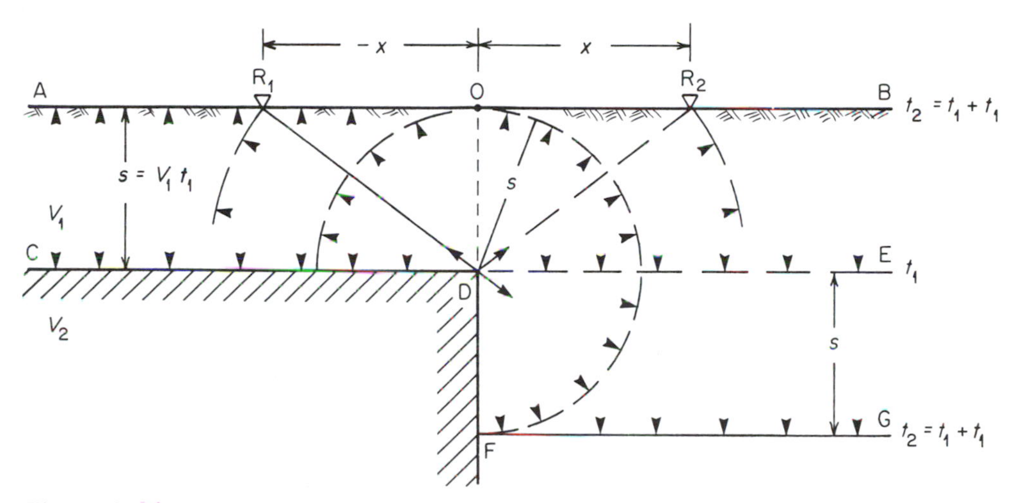

Figure 4–24
Geometrical features of wave fronts reflected and diffracted from a discontinuous or faulted boundary. AOB is the downgoing wave front at time $t = 0$. CDE is the downgoing wave front at time $t = t_1$ where $t_1 = s/V_1$. At time $t = t_1$, the part of the wave front (CD) is reflected and becomes an upgoing wave front while the other part (DE) continues to go down. At time $t = t_2$, the upgoing wave front is at AO, and the downgoing wave front is at FG. There is another wave front at time $t = t_2$ represented by the circular wave front from a Huygen's point source at point D. This wave front is needed to make a continuous full wave front (AOFG) at time $t = t_2$. The upgoing part of the circular wave front is observed at different times at different locations on the surface. The result of this circular wave front is an apparent reflecting surface that is due to point D.

This is also the travel time of the diffracted wave reaching R_2. However, we see that the diffracted wave makes up part of the first wave front to reach R_2.

If several receivers were located along lines in opposite directions from O, the corresponding arrival times for reflected and diffracted waves would plot in the pattern seen in Figure 4–25. The diffracted arrival times form a hyperbolic arc, which we would expect because Equation 4–57 expresses a hyperbolic relationship between time t_d and distance x. This indicates that on reflection travel time curves we could expect irregularities of hyperbolic form that are symmetrical about points lying above discontinuities on a reflector.

MULTIFOLD REFLECTIONS

As we pointed out earlier, one of the principal difficulties in reflection seismic surveying is recognizing weak reflected pulses on a seismogram. A very common modern practice is to enhance these weak pulses by *multifold reflection* surveying. We obtain a multifold reflection by combining many reflections from the same point on the reflector that were recorded separately with different source and receiver positions. Observe in Figure 4–26 how six different source–receiver combinations have paths reaching the same point on the reflector. This point of reflection is called the *common depth point.*

Because of the different source–receiver spacings, all the reflection arrival times will be

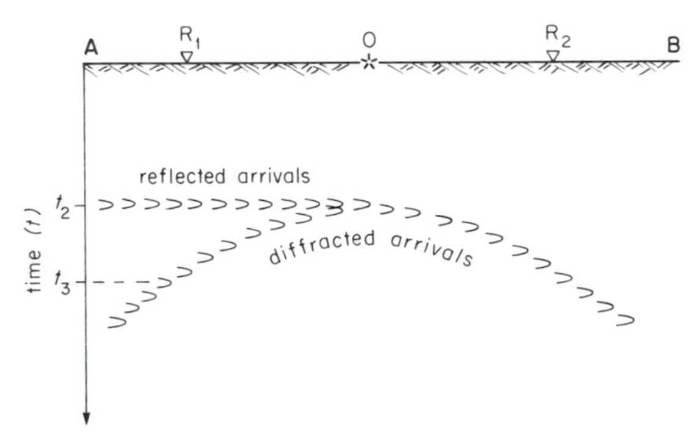

Figure 4–25
Pattern of arrival times of waves reflected and diffracted from a discontinuous boundary such as the one in Figure 4–24. The reflections forming the horizontal pattern are from the upthrown side of the fault, and arrivals with hyperbolic geometry are from the upper corner of the fault. Similar images may also be due to other subsurface structures, such as the stratigraphic termination of a layer from which reflections occur.

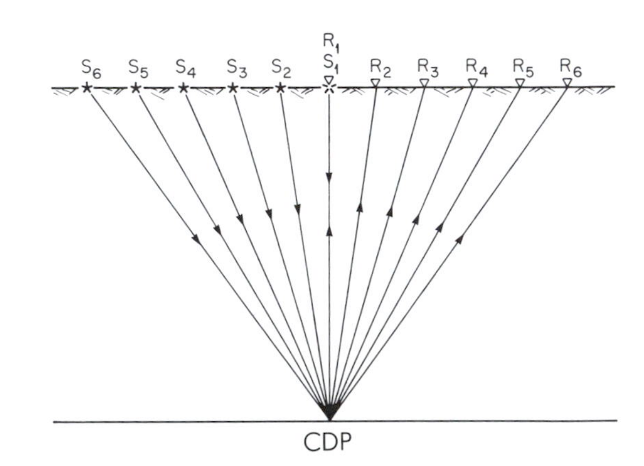

Figure 4–26
Source–receiver paths reflected from a common depth point on the reflector. Waves reflecting from the same subsurface point follow paths with different source (S) and receiver positions. The common reflecting point is below the common midpoint (CMP) of several source–receiver pairs. Therefore, it is referred to as the common depth point (CDP) on the reflector.

different. By applying an appropriate normal move-out time adjustment with the help of Equation 4–8, however, we can obtain corrected seismogram traces on which the reflections should appear at identical times as in Figure 4–27. But other pulses that tend to obscure the weak, individual reflections are not adjusted to identical times. Therefore, by adding all the traces together to form a single composite trace, we hope to obtain a large pulse from the sum of the reflections at identical times; the other pulses would act to cancel one another. Now we have obtained a multifold composite trace.

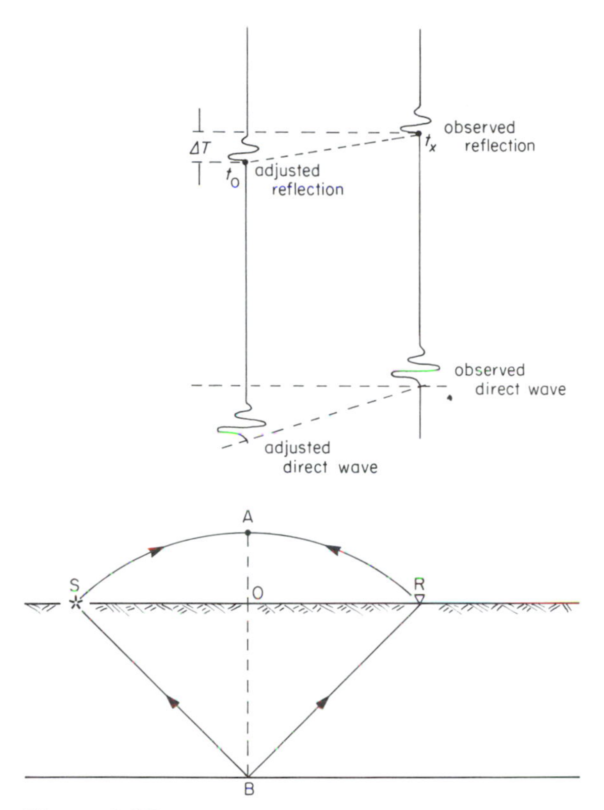

Figure 4–27
A normal move-out adjustment is made to shift a reflection recorded with a source–receiver offset of SR to the time expected for zero offset. The path segments SB and RB are "folded" to the vertical position AB, and reflection time is adjusted to the corresponding time along the path OB.

The name "multifold" comes from the fact that we have a multiple trace combination and that by making normal move-out adjustments we are, in a sense, "folding" the downgoing and upgoing parts of each path into a vertical path. We can assign a fold number that tells how many individual traces were used to obtain the composite. In Figure 4–26 we have sixfold coverage of the common depth point.

The process of adding together the adjusted traces is called *stacking*. In Chapter 6 we will discuss in some detail how normal move-out corrections are applied and how traces are stacked to obtain multifold reflection data. The seismic section in Figure 4–1 is a result of this kind of processing. Field procedures for efficiently obtaining the necessary seismograms are described in Chapter 5. The purpose of these surveying and processing techniques is to produce identifiable reflections that we can interpret by methods we have been developing in this present chapter.

STUDY EXERCISES

1. Using Fermat's principle, show that in Figure 4–28 the reflected ray between the source point S and receiver R has a reflection angle θ_2 equal to incidence angle θ_1.

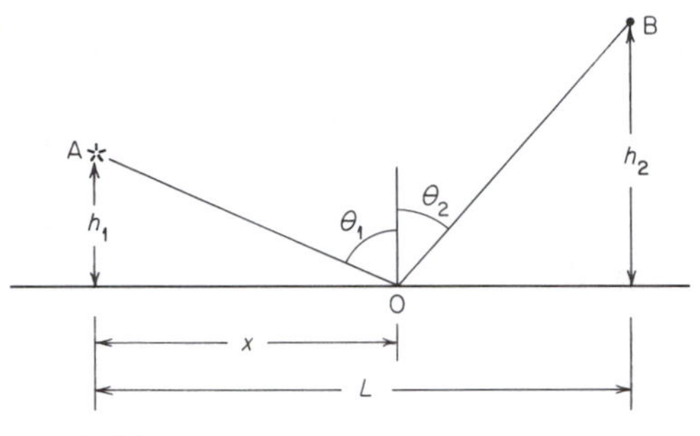

Figure 4–28

2. Suppose that a reflection survey indicates a depth of 750 m to the first reflector. The horizontal reflecting interface is the boundary between layers defined by the velocities $V_1 = 1500$ m/s and $V_2 = 2500$ m/s. What will be the two-way reflection time at offsets equal to zero and 1500 m?

3. Using the model described in Exercise 2, construct the time–distance curves for the direct, refracted and reflected arrivals. Then suppose that the reflecting interface has a slope equal to 20 degrees, up dip from source to receivers, and that its depth is 750 m below the source point. Reconstruct the time–distance curve for the sloping case.

4. Suppose that a flat layer with velocity $V_1 = 1500$ m/s lies above another layer with velocity $V_2 = 2500$ m/s. What will be the values of the zero-offset, two-way reflection time and the refraction intercept time? Show the relation between the refraction intercept time and the zero-offset, two-way reflection time.

5. From a two-sided reflection survey, an asymmetrical time–distance curve was obtained. From this curve, a minimum reflection time of 0.050 s was determined at an offset of 173 m (x_{min}). The zero-offset reflection time is 0.1 s. Determine the slope of the reflector and its depth. What is the velocity of the top layer?

6. Suppose that a seismic reflection survey was done over the layered sequence shown in the Figure 4-29 where interval velocities and layer thicknesses are given. Determine the average and root-mean-square velocities as functions of zero-offset reflection time.

S
15°
V_1 = 1000 m/s   50 m
V_2 = 2500 m/s   100 m
V_3 = 1500 m/s   75 m
V_4 = 3000 m/s   150 m
V_5 = 4000 m/s

Figure 4–29

7. Using the layered model given in Figure 4-29, determine the offset for a ray path with a departing angle of 15 degrees from the source point S.

8. Suppose that from a reflection time–distance curve we obtained the zero-offset times of $t_{01} = 0.2$ s, $t_{02} = 0.45$ s, $t_{03} = 0.70$ s, and $t_{04} = 0.95$ s, and the root-mean-square velocity values of

$V_{1,\text{rms}} = 1500$ m/s, $V_{2,\text{rms}} = 2250$ m/s, $V_{3,\text{rms}} = 2100$ m/s, and $V_{4,\text{rms}} = 3000$ m/s. Compute the interval velocities and layer thicknesses.

9. Consider a three-layer model with the interval velocities and densities shown in Figure 4-30. Suppose that the ray paths are vertical. What are the primary and secondary interbed multiple reflection times? What is the amplitude ratio of the primary reflection and the multiple reflection. Consider an output from sources equal to unity and use the reflection and transmission coefficients.

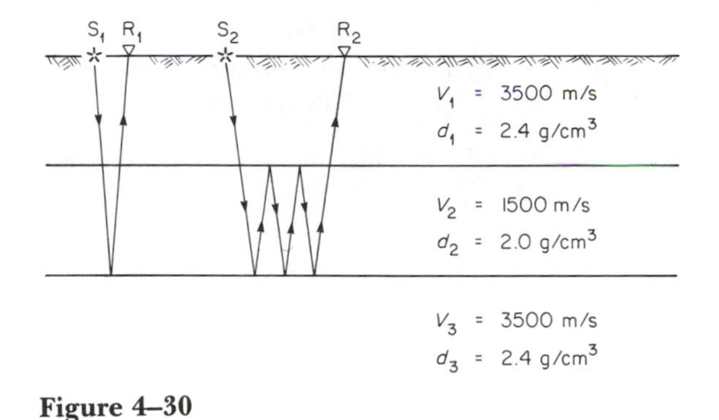

Figure 4–30

SELECTED READING

Braddick, H. J. J., *Vibrations, Waves and Diffractions.* New York, McGraw-Hill, 1965.

Coffeen, J. A., *Seismic Exploration Fundamentals.* Tulsa, Okla., Petroleum Publishing Co., 1978.

Dix, C. H., *Seismic Prospecting for Oil.* New York, Harper, 1952.

Dix, C. H., Seismic velocities from surface measurements, *Geophysics,* v. n. 1, 20, pp. 68–86, 1955.

Dobrin, M. B., *Introduction to Geophysical Prospecting,* 3rd edition. New York, McGraw-Hill, 1976.

Griffiths, D. H., and R. F. King, *Applied Geophysics for Geologists and Engineers.* Oxford, England, Pergamon Press, 1981.

Hagedoorn, J. G., A process of seismic reflection interpretation, *Geophysical Prospecting,* v. 2, n. 2, pp. 85–127, 1954.

Kearey, P., and M. Brooks, *An Introduction to Geophysical Exploration.* Oxford, England, Blackwell Scientific Publication, 1984.

Mayne, W. H., Common-reflection-point horizontal data-stacking techniques, *Geophysics,* v. 27, n. 6, pp. 927–938, 1962.

Meidav, T., Hammer reflection seismics in engineering geophysics, *Geophysics,* v. n. 3, 34, pp. 383–395, 1969.

Nettleton, L. L., *Geophysical Prospecting for Oil.* New York, McGraw-Hill, 1940.

Parasnis, D. S., *Principles of Applied Geophysics,* 4th edition. London, Chapman and Hall, 1986.

Sengbush, R. L., *Seismic Exploration Methods.* Boston, IHRDC, 1983.

Sharma, P., *Geophysical Methods in Geology.* Amsterdam, Elsevier, 1986.

Sheriff, R. E., *Encyclopedic Dictionary of Exploration Geophysics.* Tulsa, Okla., Society of Exploration Geophysicists, 1973.

Slotnick, M. M., A graphical method for the interpretation of refraction profile data, *Geophysics,* v. n. 2, 15, pp. 163–180, 1950.

Slotnick, M. M., *Lessons in Seismic Computing.* Tulsa, Okla., Society of Exploration Geophysicists, 1959.

Telford, W. M., L. P. Geldart, R. E. Sheriff, and D. A. Keys, *Applied Geophysics.* Cambridge, England, Cambridge University Press, 1976.

Waters, K. H., *Reflection Seismology*, 3rd edition. New York, Wiley, 1987.

FIVE

Seismic Surveying

Since its beginnings in the 1920s, seismic surveying has been used mainly to search for oil. By far the largest investment of money and resources for developing practical methods of acquiring seismic reflection data has come from the petroleum industry. As its value for detecting subsurface geologic features became recognized by those outside the industry, seismic surveying began to be applied to other objectives, especially engineering site evaluations, groundwater exploration, and, to a lesser extent, mining exploration. Government agencies and university research groups use seismic surveys to obtain more general information for regional planning and resource evaluation, and to gain a better basic understanding of earth structure.

Central to all seismic surveying, whatever the purpose, is the task of recording reflected and refracted waves. Our aim in this chapter is to describe the recording procedures and equipment that are in common use. Although they have been briefly introduced in earlier chapters, the common recording problems and ways of overcoming them to obtain satisfactory data have not been covered. Most attention is given to seismic reflection surveying. The field operations range from relatively simple soundings, which can be done by one or two people, to the complex activities of a modern petroleum survey, which can require more than 50 people to handle the geophones and cable, to attend to the explosives or other energy sources, and to operate and maintain the recording apparatus.

For older exploration surveys, the seismograms were recorded on photographic paper charts. Afterward there was no way to improve any particular seismogram. If the results were unsatisfactory, it was necessary to record another one after changing the instrument settings, geophone arrangement, and the position and strength of the source. After some experimentation, a procedure might be found that produced satisfactory reflections, but not always.

Improvements in the methods of seismic surveying have paralleled the growth of the modern electronics and computer industries. A new dimension was added in the 1950s with the introduction of magnetic tape recording which made it possible to store the ground vibration measurements in a form allowing later experimentation with amplification and frequency filtering. In the 1960s digital recording technology created opportunities for combining, or stacking, seismogram traces as well as for testing a larger range of amplification and filters. In addition, new ways of graphically displaying the results were devised. These latter aspects are part of the modern data processing operations, that follow field operations; these are presented in Chapter 6.

The availability of data processing techniques for the later improvement of seismograms influences the planning of field procedures. Obtaining recognizable reflections on the original field seismograms is not the only consideration. Some thought must also be given to recording data in a form that is suitable for efficient processing at a later time. Even if reflections can be identified without additional enhancement, the subsequent velocity and layer thickness calculations and machine-drawn seismic sections can be obtained directly by computer processing of digital magnetic tapes recorded in the field.

INSTRUMENTS FOR SEISMIC SURVEYING

Seismic surveying consists of placing some receivers at different locations and then using them to detect vibrations produced by an energy source. The receivers convert the mechanical vibrations into electric current that is transmitted to a recorder; the recorder is designed to preserve the information in a form that can be displayed and analyzed. Seismic surveying can be done on land or at sea.

We will begin by looking at the instruments that are available for seismic surveying. These include the receivers, the cables, or other transmitters that connect them, the recorders, and the energy sources. Later we will discuss the conventional arrangements of receivers and sources in lines or other patterns.

Geophones

The receiver used to detect ground vibrations is called a *geophone* or a *seismometer*. It is used for seismic surveying on land, and it can be operated on the ocean floor if mounted in a suitable container.

The basic design of a geophone was introduced in Chapter 2. A magnetized mass fixed to the container is surrounded by a wire coil suspended on a spring, as seen in Figure 2–24. When the ground vibrates, the coil moves back and forth around the magnetic mass. We know from the principles of electromagnetism that the motion of a coil around a magnet induces electric current to flow in the coil. The strength of that current depends on the speed of the motion and is greatest when the coil moves the most swiftly. Therefore, the electric current emanating from a geophone tends to be proportional to the velocity of the ground movement.

The geophones most widely used in exploration seismology are sensitive to vertical ground motion. However, instruments are also available for measuring horizontal movement. The design is illustrated schematically in Figure 5–1, together with pictures of typical vertical and horizontal geophones now in use. They are equipped with spikes that can be pushed into soft ground or base plates that set on rocky or frozen ground.

A geophone does not respond in the same way to different frequencies of vibration. At certain frequencies the electric current is stronger than at other frequencies, even though the amplitude of vibration is the same. This phenomenon is not difficult to understand when you think about the different springs and masses that could be used in the geophone. Suppose that the coil is attached to a stiff spring. Then a low-frequency oscillation of the ground may be too slow to produce any stretch in the spring, so that there is no motion of the coil with respect to the magnet. But a more rapid high-frequency vibration might produce a cyclic stretching of the same spring. A relatively heavy coil on a weak spring is more sensitive to lower-frequency vibrations than a smaller coil on a stiffer spring.

The response of a geophone to vibrations of different frequency can be tested with a device called a *shake table*. As the name implies, it consists of a small platform that can be caused to vibrate at different frequencies by means of an electronic oscillator control system. The amplitude of vibration is the same for all frequencies. A geophone is placed on this platform, and its electric current output is measured for different vibration frequencies. This information is plotted to obtain the *response curve* for the geophone, as seen in Figure 5–2.

The frequency of vibration that stimulates the strongest geophone response is recognized as the *natural frequency* of the geophone. It is found from the highest point on the response curve. Typically, the response curve diminishes quite abruptly for frequencies lower than the natural frequency. Above the initial frequency, the curve at first diminishes modestly and then tends to reach a more or less constant response level for higher frequencies.

Geophones commonly have natural periods in the range of 5 to 40 hertz. The smaller, more compact ones ordinarily have higher natural frequencies. The geophone is in a class of instruments that we call harmonic oscillators. The natural frequency f_0 of such a

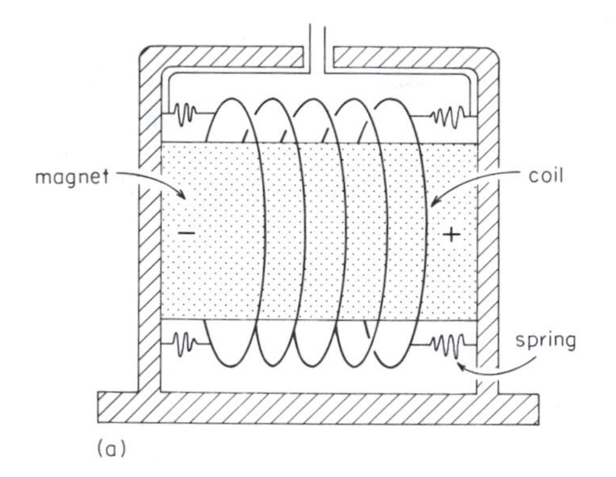

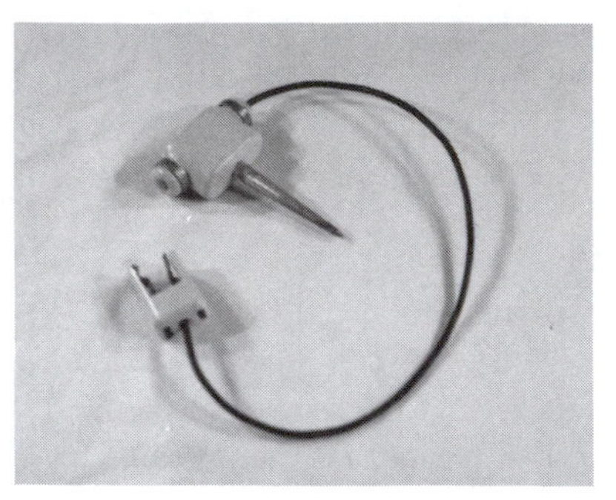

(b)

Figure 5–1
Vertical and horizontal geophones used in seismic surveying. (a) Geophones consist of a bar magnet fixed to the container and a surrounding coil suspended by springs. (b) Vertically suspended systems are sensitive to vertical ground motion, and (c) horizontally suspended systems are sensitive to the horizontal component of ground motion in the direction of the coil axis.

device can be expressed in terms of the mass m and the spring constant k of the system:

$$f_0 = \frac{1}{2\pi} \sqrt{k/m} \qquad (5\text{–}1)$$

Hydrophones

For most seismic surveying at sea, the receivers trail behind a recording vessel submerged just below the ocean surface. These receivers are sensitive to changes in water pressure produced by passing seismic waves. This kind of pressure-sensitive receiver is called a *hydrophone.*

The basic element of most hydrophones is a piezoelectric crystal. As the hydrophone case is deformed by changing water pressure, the interior mounting is also deformed in a way that changes the stress on the crystal. This causes the crystal to emit an electric signal.

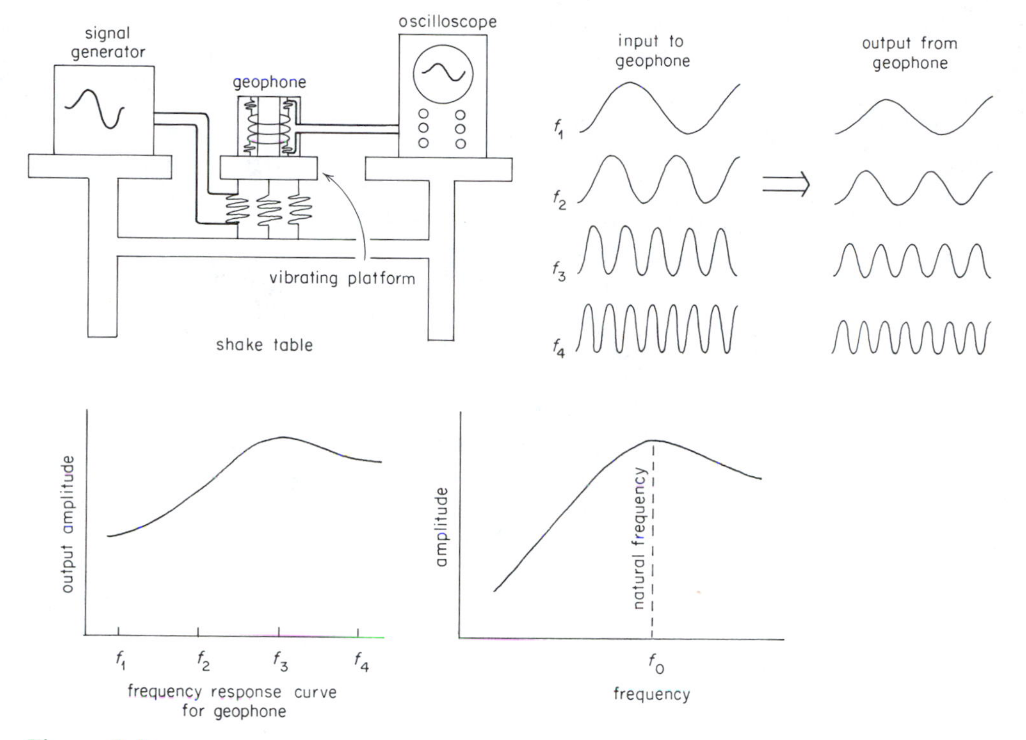

Figure 5–2
Testing the geophone response on a shake table. Typical response curve displays maximum amplitude at the natural frequency f_0 of the geophone. A geophone is an oscillating system that has a resonance frequency (natural frequency f_0) at which it shows its highest response. A resistor connected in parallel with the coil modifies the geophone to respond similarly to motions with different frequencies above f_0. This produces damping, which forces the oscillations of the coil to cease when ground vibration ceases.

The Seismic Cable

The geophone signal, which is the electric current produced by ground vibration, is transmitted to the recording system by means of the seismic cable. Each geophone requires two wire conductors. Therefore, the number of conductors in the seismic cable depends on the number of geophones being used in the survey. For small-scale engineering surveys, cables usually have 24 conductors which carry the signals of 12 geophones.

At regular intervals along the cable are "takeout" points where a geophone can be connected to its pair of conductors, as in Figure 5–3. Seismic cables come with takeout intervals as short as a few meters and as long as several hundred meters. Manufacturers can ordinarily supply cables with 6, 12, 24, 48, and even 96 takeouts.

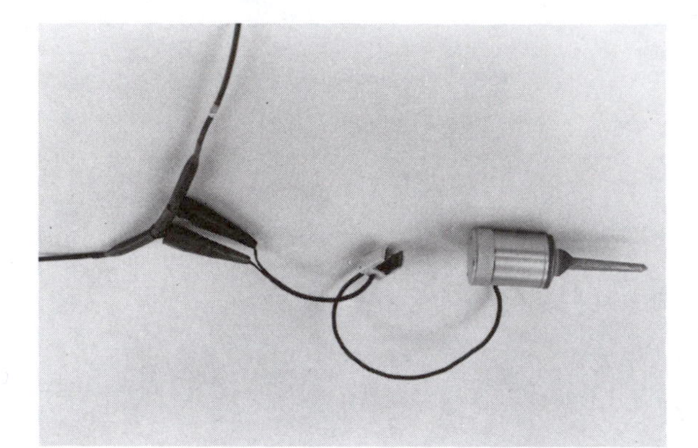

(a)

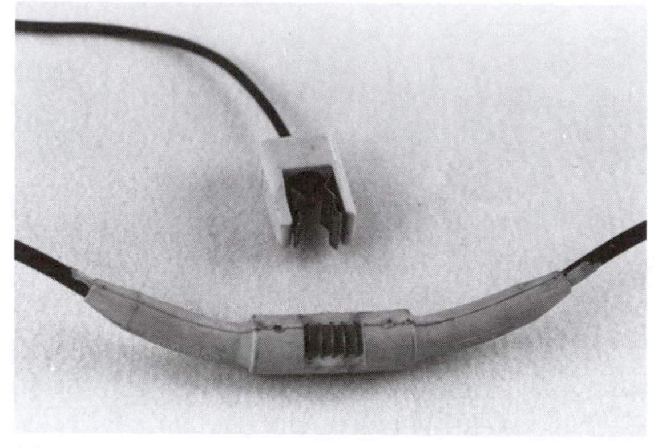

(b)

Figure 5–3
Takeouts on a seismic cable, and two different types of clips for connecting a geophone to the cable. (Photograph by Llyn Sharp, VA Tech.)

For certain survey procedures, it is more convenient to use a segmented seismic cable. Each segment, which is usually a few tens to a few hundreds of meters long, has one takeout point and multiprong plugs on both ends for connecting to adjacent segments. With this arrangement, a single cable segment and geophone can be moved from one end of a line to the other without disconnecting all the other geophones. This is an important advantage for the common depth point reflection surveying procedure, which we will discuss later.

Marine Streamer Cables

For marine seismic surveying the hydrophones are mounted in a flexible plastic tube about 3 inches in diameter. Several hydrophones spaced at equal intervals, together with their conducting wires, are included in a segment of tube, which is otherwise filled with oil to make it buoyant. Tube segments can be joined in a line to make up a *streamer* which is towed by the recording vessel. Weights and fins are attached to the streamer to hold it submerged just below the ocean surface while the vessel is in motion.

Analog Recording Systems

When the ground is vibrating, it is in continuous motion. A geophone responding to this motion produces a continuously varying electric signal. A seismogram is a graph that shows how the amplitude of this signal varies with time. It is a permanent record of how the ground was vibrating during that interval of time at the receiver location.

Different techniques may be used for obtaining a seismogram. One such technique is a recording device that draws a continuous graph at the same time that it is receiving the geophone signal. One of these systems is illustrated in Figure 2–25. Here the graph is drawn by the point of light that is focused on a photographic chart by a galvanometer. This kind of system has been in use since the beginning of exploration seismology.

In another recording technique a tape with a magnetizable coating is used. As the tape advances, the geophone signal continuously magnetizes the portion passing by the recorder head. The strength of the magnetism varies continuously along the tape in proportion to the strength of the geophone signal. Later, this magnetic tape can be used in place of a geophone to produce an electric signal for activating a galvanometer in a recorder such as the one shown in Figure 2–25. Frequency modulation techniques are utilized in the analog magnetic tape recording.

A third recording technique is basically different from the two just discussed. At discrete intervals of time, it measures the strength of the geophone signal and records it as a number. Rather than a continuous graph, it presents us with a series of numbers. Later, these numbers can be plotted and a seismogram prepared by connecting the points, as in Figure 5–4.

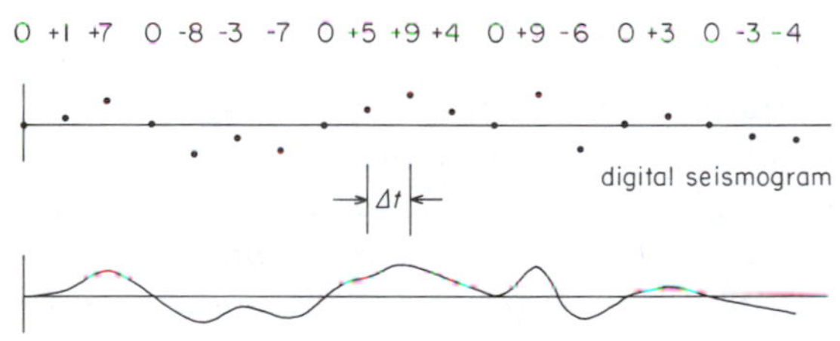

Figure 5–4
Comparison of analog and digital seismograms. An analog seismogram obtained directly from the output of a geophone is a continuous record of ground motion as a function of time. A digital seismogram is a series of numbers that indicate amplitudes of ground motion at discrete moments in time. By plotting these numbers and connecting the points, we can reconstruct a record that is similar to an analog seismogram.

The technique of recording the geophone output at discrete moments is called *digital recording* because the record consists of a series of digits or numbers. For purposes of discussion, the older techniques, which produce continuous graphs, were named *analog recording* techniques. Although seismic surveying for petroleum now utilizes digital recording systems almost exclusively, analog systems are still used for some nonpetroleum-related surveys. Furthermore, some analog electronics are required to precondition the geophone signal for digital recording.

Before the geophone signal is recorded by an analog system, it can be electronically amplified and filtered. The amplifier, like the volume control on a radio, is used to increase the strength of the weak geophone signal. The vacuum tubes used in older systems were replaced by transistors during the 1960s, which greatly reduced the battery power required for operation.

Amplification of a geophone signal is commonly given in *decibels* (db), a term borrowed from audio electronics. It is simplest to explain changes of amplification in the following way. Each 6-db increase in amplification doubles the signal strength. Magnifications resulting from different decibel amplifications are given in Table 5–1.

Oscillations covering a wide range of frequencies may be present in the geophone signal. Some of these oscillations may obscure reflections or may be undesirable for other reasons. It is possible to remove them by means of electronic filtering before recording the signal. These filters are like the treble and base controls on a radio. Electronic components arranged in different circuits transmit signals in one frequency range while blocking the transmission of signals in other frequency ranges. Simple circuits in Figure 5–5 illustrate

TABLE 5–1 Magnifications Resulting from Various Decibel Amplifications

AMPLIFICATION (db)	MAGNIFICATION
6	× 2
12	× 3.98
18	× 7.94
20	× 10
24	× 15.85
30	× 31.62
36	× 63.1
40	× 100
42	× 125.89
48	× 251.19
60	× 1,000
80	× 10,000
100	× 100,000

the basic designs of high-pass, low-pass, and band-pass filters.

An important feature of a seismic recording system is its *dynamic range,* which is the ratio of the strongest and weakest signals that can be usefully recorded. Older analog systems that recorded directly on photographic paper charts were limited by the galvanometers, which could not accept signals exceeding a specified voltage. If a geophone signal was amplified too much, a distorted unusable record was the result. The dynamic range of such a system was usually not more than about 24 db. Analog magnetic tape recorders increased the dynamic range to as much as 45 db. Here the limitation is the magnetic saturation level of the tape. If a signal is too strong, the tape cannot become magnetized strongly enough to represent it properly.

When we examine the range of amplitudes of ground vibration that we would like to record, we sometimes find differences of as much as 100 db between the strong refracted arrivals and surface waves on the early part of a seismogram and the weak reflections that may arrive later. How can all these arrivals be properly recorded by a system with a dynamic range of 24 or 45 db?

A partial solution to this problem is *automatic gain control* (AGC). Rather than constant amplification of a geophone signal, the amplifier is adjusted by processing the incoming signal through a feedback circuit that keeps the strength of the signal reaching the galvanometer or magnetic tape recorder within a specified range. The result is low amplification for the part of the seismogram dominated by high-amplitude waves; the amplification gradually increases as the amplitudes of later arriving waves diminish. AGC is very useful for strengthening weak signals that arrive at substantially different times than strong signals. But for strong and weak signals that arrive more or less simultaneously, nothing can be done to preserve the weak one if the amplitude difference exceeds the dynamic range of the system.

An analog seismic recording system is equipped with a separate amplifier–filter circuit and a separate galvanometer or track on a magnetic tape for each geophone. These components make up one *channel* of the recording system, which produces a trace on a seismogram that is independent of the other traces. Most of the older galvanometer systems used for petroleum exploration were 24-channel recorders. With the introduction of analog magnetic tape recording, 48-channel and even 96-channel systems could be used efficiently. For smaller-scale engineering surveys, 1-channel, 6-channel, and 12-channel systems are available.

Analog seismic recording systems with 24 or 48 channels that were used for petroleum exploration were compact enough to install in a small truck, such as the one in Figure 5–6.

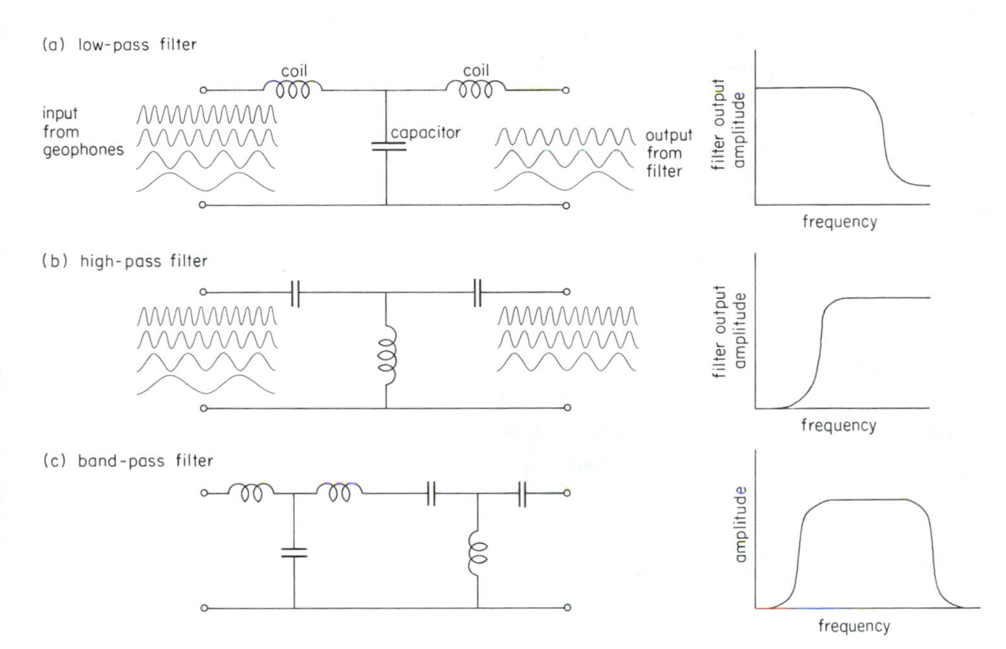

Figure 5–5
Electronic circuits for (a) low-pass, (b) high-pass, and (c) band-pass filtering of alternating electric signals. Seismic wave motions are transformed into electric signals by receivers (i.e., geophones, hydrophones). These electric signals can be filtered by the electronic filters to reduce or eliminate certain frequencies. A filter that removes higher frequencies (a) is called a high-cut or low-pass filter. Similarly, filters can be designed (b, c) to pass higher frequencies (low cut or high pass) or only a band of frequencies (band pass). The band of frequencies that is removed is called a band-reject filtering.

Portable 12- and 24-channel systems consisted of one or two amplifier cases, a recorder case, and a battery-operated power supply unit. These components could be hand-carried or transported by automobile.

Digital Recording Systems

The 1960s saw a fundamental change in seismic recording systems. Digital records, which were first available for field units in 1963, had virtually replaced the older analog systems in use for petroleum exploration by the end of the decade. Digital recording is an outgrowth of modern computer technology, and the capability of computer processing of digital data is essential for the purposes of exploration seismology. All the pieces of digital equipment needed for a typical engineering survey are shown in Figure 5–7 and include 24 geophones, a seismic cable on a reel, and a 24-channel amplifier–filter–recorder unit.

A digital recorder makes use of *binary numbers* to store the measurements of geophone signal strength. Let us compare the binary

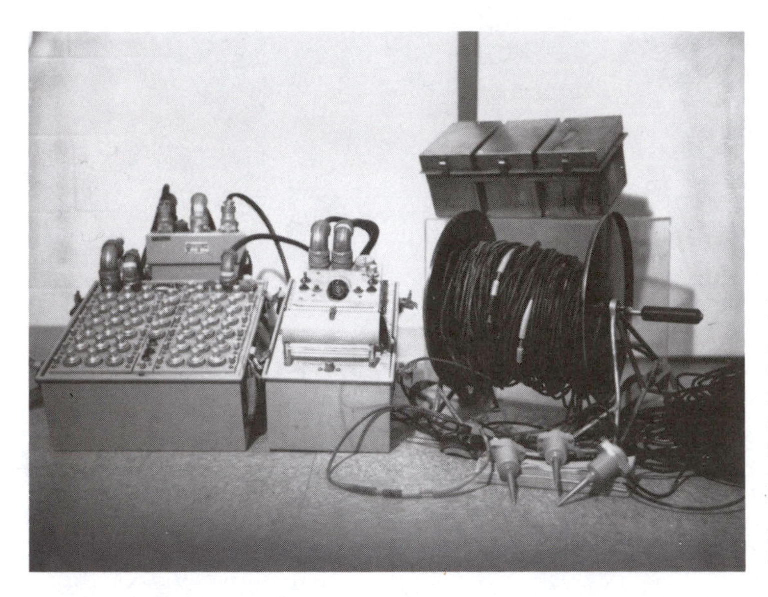

Figure 5–6
Analog seismic recording system. Containers can be hand-carried or mounted in a recording truck. An analog seismic recording system has independent channel circuits for producing a separate seismogram trace from the output of each geophone. (Photograph by Llyn Sharp, VA Tech.)

Figure 5–7
Portable 24-channel seismic recording system with cable and geophones. These systems are most commonly used in shallow surveys such as foundation surveys for engineering purposes. Depths to bedrock or the thickness of the weathered layer can easily be determined from the seismograms recorded with these units. Output from such modern portable systems can also be recorded on magnetic tape. (Courtesy of Oyo Corporation.)

number system and the decimal number system with which we are most familiar. The decimal number 245 is a short way to write

$$2 \times 10^2 + 4 \times 10^1 + 5 \times 10^0 = 245$$

Suppose we were to express this same value in a sequence of powers of two. Then we obtain

$$1 \times 2^7 + 1 \times 2^6 + 1 \times 2^5 + 1 \times 2^4 + 0 \times 2^3 + 1 \times 2^2 + 0 \times 2^1 + 1 \times 2^0 = 11110101$$

where

$$\text{binary } 11110101 = \text{decimal } 245$$

We see that the binary number with eight digits expresses the same value as the decimal number with three digits. This may seem less efficient, but it has a distinct advantage for purposes of computer processing. This and any binary number can be expressed with only two symbols, 0 and 1, rather than the ten symbols, 0, 1, 2, . . ., 9, required for decimal numbers. Therefore, a binary number can be stored on magnetic tape by means of a succession of small areas that are either magnetized (1) or left unmagnetized (0).

The usual procedure is to arrange the areas for storing binary numbers in rows across a magnetic tape. Generally, a system with nine heads is used to record on ½-inch tape, or 21 heads for 1-inch tape. Two or four heads, respectively, are assigned for other purposes, leaving the remainder to write the binary number. Suppose that seven heads are used to write on ½-inch tape, each one of which is in contact with a small area on the tape. In Figure 5–8, these contact areas are shown as squares. Each square is called a *bit,* and each row of squares crossing the tape is called a *byte.* One bit is used to store one digit of a binary number. If the head magnetizes this bit, the digit 1 is implied. If the head does not

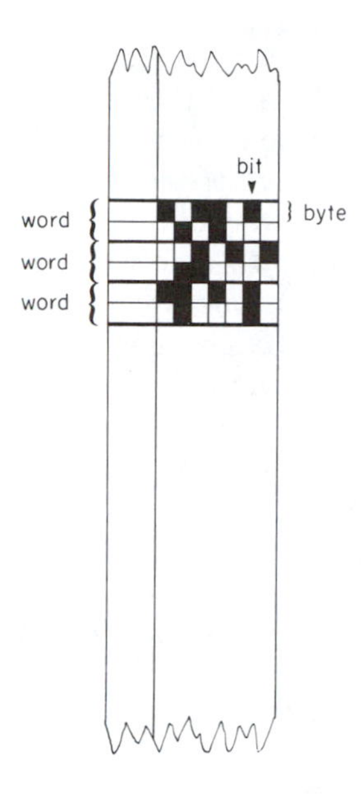

Figure 5–8

Schematic illustration of a digital magnetic tape. Magnetized areas, the shaded squares, represent a digit equal to 1, and unmagnetized areas, the unshaded squares, represent the digit 0.

magnetize the bit, the digit 0 is implied. The total number of bits assigned for one binary number makes up a word. In Figure 5–8, a word consists of 14 bits, which is two rows, or bytes, on the tape. By summing powers of two up to 2^{14}, we see that the largest decimal number that can be expressed by a 14-bit word is 16384. For purposes of exploration seismology, words as large as 32 bits are sometimes required.

Now let us see where the binary numbers come from that are stored by a digital seismic recorder. In a multichannel system, each geophone signal is first amplified and filtered by analog electronic components. Then the signal is changed to digital form by a unit called the *analog to digital converter,* or the A/D converter.

Only one A/D converter is needed to process the signals of all the geophones, a procedure accomplished by means of a high-speed switch called a *multiplexer*. The procedure is illustrated schematically in Figure 5–9. The multiplexer first connects to channel 1 for a period of about one microsecond, which is sufficient time to charge a capacitor to the voltage of the channel 1 amplifier output at that moment. The signal is then amplified and transmitted to the A/D converter.

In the A/D converter, different combinations of standard voltages are generated and tested to find the particular combination that exactly balances the signal. The standard voltages increase by powers of two, which corresponds to the binary numbering system. Voltages making up the combination that balances the signal are transmitted to the formatting unit. Here, the voltage pulses are converted to control signals that activate the recorder heads to magnetize the appropriate bits on the magnetic tape.

This entire sequence of steps requires less than 30 microseconds. Then the multiplexer switches to channel 2, and the sequence is repeated. After recording signals from all the geophones in this way, the multiplexer returns to channel 1 and again begins to sample the signals of all geophones a second time, and so on for the duration of the intended seismogram.

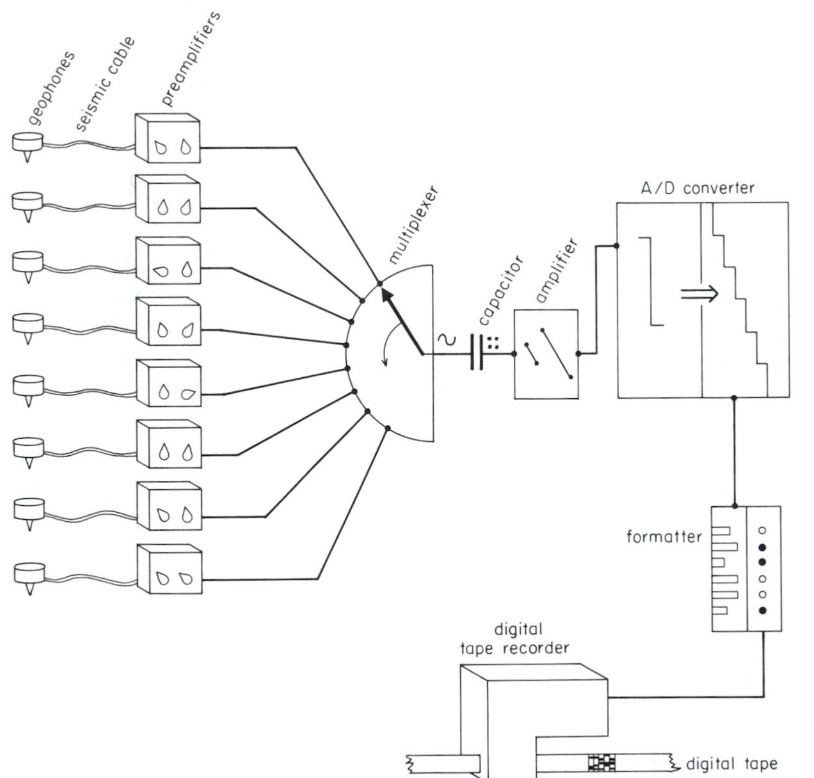

Figure 5–9
Schematic illustration of a digital seismic recording system. The main feature of a digital system is the multiplexer, which rapidly samples outputs of all geophones and then passes these samples in sequence through a single channel, which processes and records them on digital magnetic tape.

We see that one set of signals from a 24-channel system can be recorded in about 720 microseconds, or 0.72 millisecond. For a 48-channel system, the time would be 1.44 milliseconds. For purposes of petroleum exploration, satisfactory digital seismograms can sometimes be prepared from geophone signals that are sampled at 4-millisecond intervals. The multiplexer and A/D converter work fast enough for this to be possible for a 96-channel system. If samples at 2-millisecond intervals are needed, however, only 48 channels can be recorded. Some very detailed surveying requires intervals of 1 millisecond or less, which can be accomplished only by recording fewer channels. The digital recording system is controlled by an oscillating crystal timer that restarts the multiplexer on channel 1 at prescribed intervals such as 4, 2, or 1 millisecond.

Now let us look at the sequence of binary numbers that are recorded on the magnetic tape. To prepare a digital seismogram, as in Figure 5–4, we must sort out from this sequence the particular numbers representing channel 1, channel 2, and so on. Suppose we are using a 24-channel system. We see that the first word on the tape gives the first sample point for the channel 1 seismogram. But the next 23 words pertain to channels 2 to 24, and then the twenty-fifth word gives us the second sample point for channel 1. Similarly for channel 2, the first two sample points are obtained from the second and twenty-sixth words on the tape. For these and other channels we must sort out the following word sequences:

$$
\begin{array}{ll}
\text{channel 1} & \rightarrow 1,25,49,73,97,121,145,169 \ldots \\
\text{channel 2} & \rightarrow 2,26,50,74,97,122,146,170 \ldots \\
\text{channel 3} & \rightarrow 3,27,51,75,98,123,147,171 \ldots \\
\text{channel 4} & \rightarrow 4,28,52,76,99,124,148,172 \ldots \\
& \\
\text{channel 24} & \rightarrow 24,48,72,96,120,144,168,192 \ldots
\end{array}
\qquad (5\text{–}2)
$$

This process of sorting the data from the magnetic tape into individual channel sequences is called *demultiplexing*. Ordinarily, the recording systems are not equipped to perform this function in the field. Rather, the tape is sent to a data processing center where computers are available for demultiplexing and for preparing digital seismograms. This data processing procedure is described in Chapter 6.

Digital seismic recording systems are usually supplied with an auxiliary galvanometric recorder for producing an analog seismogram at the same time that the magnetic tape is being recorded. This seismogram is sometimes called a *field monitor record*. It indicates whether the system is functioning properly and may reveal some of the stronger reflections. But it is not of the high quality that can be produced later by appropriate processing of the digital magnetic tape.

One of the important advantages of digital seismic recording systems is the large increase in dynamic range over what was available in analog systems. Almost no limits are imposed by the A/D converter or the magnetic tape recorder. Because the tape is used to store binary digits, magnetic saturation is not a problem. The size of a number that can be stored is limited only by the number of bits in a word. Otherwise, the only restriction on dynamic range comes from the analog amplifier–filter system that preprocesses the geophone signal before it is transmitted to the

A/D converter. A modern digital seismic recording system has a dynamic range somewhat in excess of 100 db. Therefore, AGC is not essential as it is in analog systems. Weak signals occurring at the same time as strong signals can be preserved. By means of data processing techniques introduced in Chapter 6, these weak signals can be distinguished from very strong signals that override them.

Now that miniature electronic components are available and because only one A/D converter and one tape recorder are needed to service a multichannel system, digital seismic systems are as compact as the older analog systems. Figure 5–10 shows a truck-mounted, 96-channel system typical of those used for petroleum exploration.

One final aspect of modern seismic recording systems should be noted. Most systems are equipped with memory units suitable for storing individual geophone output before it is transmitted to the A/D converter. The geophone output produced by one source can be stored in a form that can be overprinted by the output produced at later times by other sources. This process of adding together the signals from more than one source is called *summing*. The summed output can then be processed through the A/D converter and recorded on the digital magnetic tape.

Seismogram Displays

The older galvanometric recording systems produced seismogram traces by focusing points of light on advancing photographic paper charts. At the same time, a light inside a rotating slotted cylinder exposed lines crossing the chart. The rotation speed was controlled by a tuning fork or crystal oscillator so that these so-called *timing lines* were exposed at 10-millisecond intervals. After the recorder was turned off, the operator treated the chart with conventional photographic chemicals in a developing box or darkroom to obtain the seismogram. The result, illustrated in Figure 5–11, is the traditional *wiggly trace* exploration seismogram, named for the appearance of the individual traces.

Modern analog recorders can produce the same kind of wiggly trace seismogram on

Figure 5–10
Truck-mounted digital seismic recording system. Modern seismic recording systems are computerized to display the data and to undertake some preliminary processing. (Courtesy of Prakla–Seismos AG.)

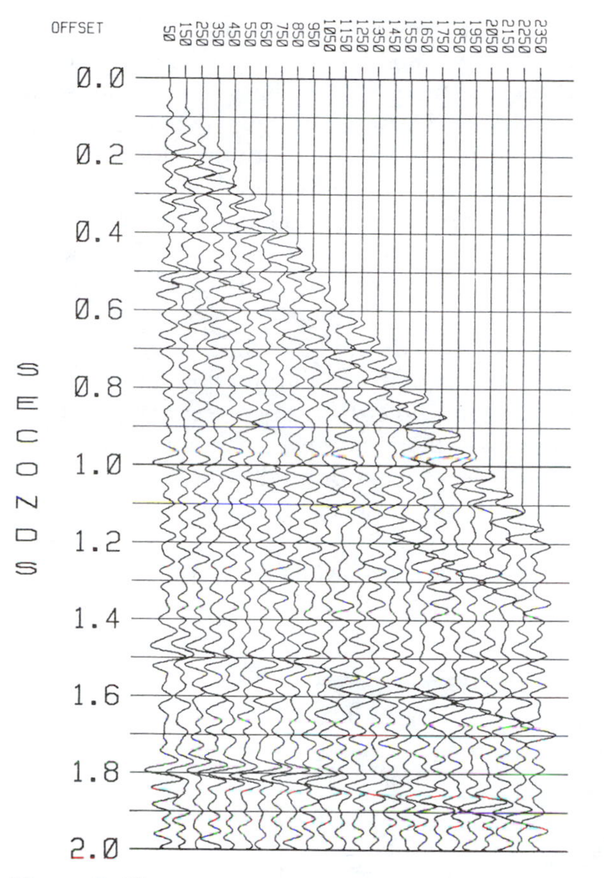

Figure 5–11

Wiggly trace reflection seismogram. Source–receiver distances (meters) are on the traces. The timing line interval is 0.1 second. A single vibrator was used as energy source, and 24-geophone groups were at 100-meter intervals along a spread extending 50 to 2350 meters from the source.

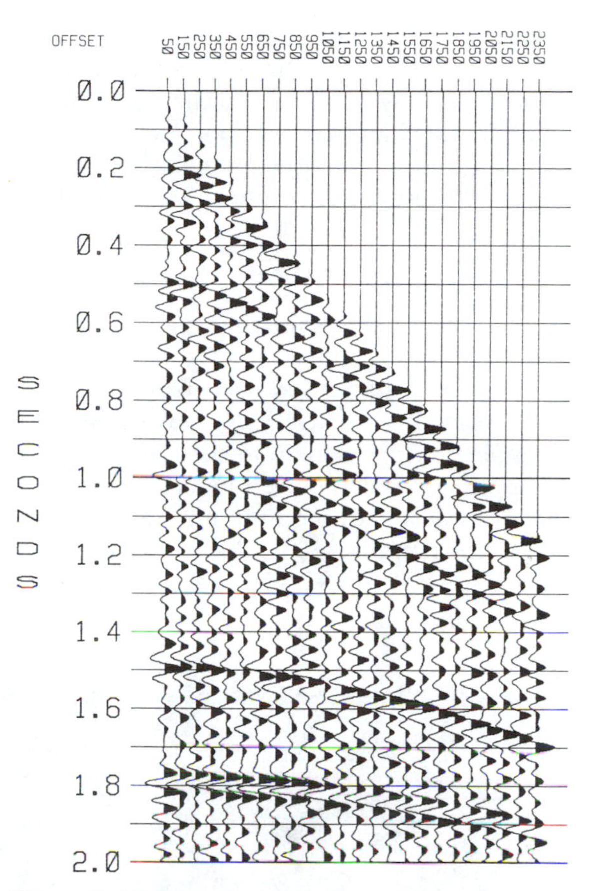

Figure 5–12

Wiggly trace reflection seismogram with variable-width shading added. Source–receiver distances at end traces are in meters, and timing lines are at 0.1-second intervals. This is same seismogram shown in Figure 5-11.

newer light-sensitive or electrostatic paper charts that do not need further chemical processing in the dark. In addition, it is conventional to prepare wiggly trace seismograms with plotting machines that connect the data points obtained from the digital magnetic tape.

Many modern recorders have the additional capability of emphasizing arrivals on the seismogram by shading alternate half cycles of oscillation on each trace, as in Figure 5–12. Variable-width shading focusses attention on variations of frequency. Variable-density shading emphasizes amplitude with gradational tones of gray. These kinds of shading produce a more visually dramatic seismogram than is given by the wiggly trace alone. Different colors have also been used in a limited way to reveal different seismogram features more vividly.

Impulsive Energy Sources

Since the first surveys in the 1920s, explosives have played a central role in exploration seismology. The dynamite used in the early surveys has been replaced by other, safer explosives, mostly ammonium nitrate. Manufacturers can supply it in canisters of different sizes. Figure 5–13 shows several one-pound cans of ammonium nitrate that are threaded at the ends so that they can be screwed together to form larger charges. Also shown is an electric blasting cap that produces a concussion needed to detonate the charge. It is a small capsule of fulminated mercury which can be exploded by an electric current.

For most seismic surveying, the explosive charge is detonated in a hole. The depth of this *shot hole* can range from a few feet to a few hundred feet, depending on various circumstances that we will mention later. Therefore, a shot hole drill is necessary for field operations where explosive energy sources are used. The truck-mounted drill in Figure 5–14 is typical of those used for petroleum exploration.

Explosives are available in other forms as

Figure 5–13
Cans of ammonium nitrate explosive connected to make a stick charge, and an electrical blasting cap containing fulminated mercury, which is used to detonate the charge. (Courtesy of DuPont.)

Figure 5–14
Truck-mounted drills preparing shot holes for a seismic reflection survey. Depending on the size of the crew and recording parameters, four to five heavy-duty drilling units are used. These units are ordinarily used to drill shot holes up to 50 meters deep. (Courtesy of Prakla-Seismos AG.)

well as canisters and tubes. Sacks of pellets or granules are convenient for pouring into shot holes in different amounts. Another useful explosive comes in the form of a cord wound around a reel. It is called primacord. Different lengths can be unwound and cut from this reel to make up charges of various sizes. It is possible to produce an explosion along a line rather than at a point location by stretching a length of primacord on the ground.

In addition to explosions, seismic waves can be produced by the impact of a sledge hammer or a falling weight. Except for small-scale engineering surveys, these energy sources have had relatively little success. They usually produce insufficient energy and proportionally larger surface waves that are undesirable. Some improvement has been realized by means of modern recording systems equipped for summing geophone signals. Then it is possible to sum the contributions of many impacts, any one of which would be too weak to produce satisfactory results. It is necessary to develop such techniques for use in places where explosives are prohibited or are otherwise unsuitable.

Explosives can be used for marine surveying, but other more convenient devices are satisfactory for many purposes. One such device is called a *sparker*. Two electrodes are mounted on a frame that can be trailed from a ship. An on-board electric circuit produces a buildup of opposite electric charge on the electrodes. When the charge difference becomes large enough, a spark jumps the gap separating the electrodes. Seismic waves are generated at the "crack" of the spark. Sparks can be produced at different uniform intervals of time by adjusting the gap distance.

Another marine device capable of producing a stronger impulsive burst of energy than the sparker is called an *air gun*. It is constructed with two air chambers in a cylinder, connected by a piston assembly, as shown in Figure 5–15. Pumping compressed air into these chambers causes a pressure difference to develop, which forces the piston to move. At a critical pressure the movement of the piston opens parts through which a burst of air is released from the lower chamber. When trailed

(a)

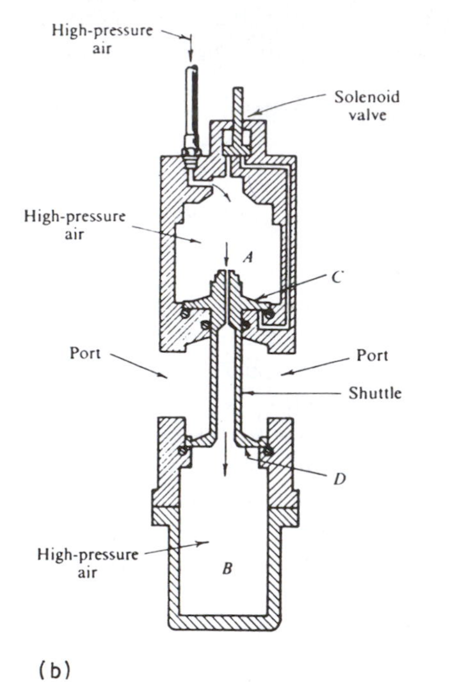

(b)

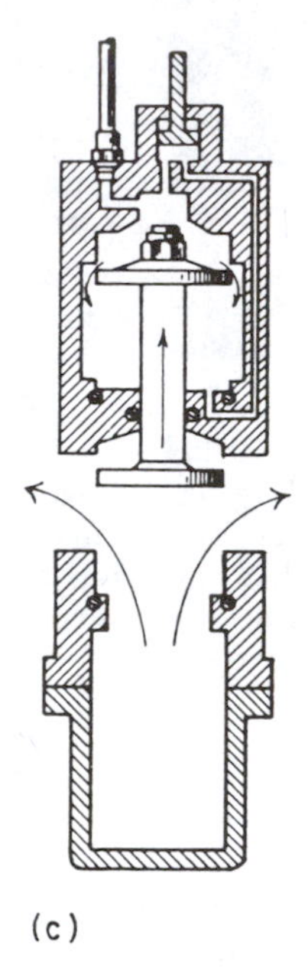

(c)

Figure 5–15

(a) Vessel for marine seismic surveying and (b, c) schematic illustrations of an air gun. The most commonly chosen energy source in marine surveys is the air gun. Pressures as high as 10,000 psi are used. Air compressed to high pressure is released from the chambers of air guns into the water to generate pressure pulses. (Courtesy of Prakla–Seismos AG.)

from a ship, this sudden burst of air produces seismic waves in the water by creating a pressure pulse. The most commonly used pressures are about 2000 psi (pounds per square inch), but pressures as high as 10^4 psi have been used for deeper penetration of waves. Air guns are used in array form to improve the resulting signature of the source system with higher energy output.

Other devices for marine operation produce explosions by igniting flammable gas mixtures emitted from an element trailed from a vessel. Like the sparker and the air gun, they can be adjusted to release bursts of energy at regular intervals of time.

Nonimpulsive Energy Source

Explosives are very effective for producing seismic waves, but the potential for destructive side effects and the time, cost, and inconvenience of drilling shot holes provide incentive for developing alternate energy sources. Much seismic surveying must be done along public roadways where permits for drilling and detonating explosives can be difficult or impossible to obtain. The best alternative is the *vibroseis* system.

In the vibroseis system energy is produced by a pad pressed firmly to the ground, which vibrates in a carefully controlled way. A typical truck-mounted vibrator is pictured in Figure 5–16. The pad, which is about one meter square, is attached beneath the truck by hydraulic jacks. Extending these jacks allows part or all of the weight of the truck to be used to press the pad to the ground. Then an oscillatory pressure fluctuation is transmitted through the jacks to the pad, causing it to vibrate at a fixed amplitude. Depending on the weight pressing on the pad, a thrust of as much as 15 tons can be directed onto the ground.

The pressure fluctuation driving the pad is carefully monitored by an electronic oscillator. It produces oscillations of continuously varying frequency for a specified interval of time. Typical patterns of vibration produced in this way are shown in Figure 5–17. During the duration of vibration, which can range from a few seconds to more than 30 seconds, the frequency can vary from low to high or from high to low. One sequence of vibration is called a *sweep*. It can be an upsweep or a downsweep, depending on whether the frequency increases or decreases. The duration of vibration is called the *sweep time*. The sweep time and starting and ending frequencies, usually in the range of 10 to 100 hertz, can be set by means of electronic controls.

The sweep is the input signal produced by a vibroseis source. Recall our discussion in Chapter 2 about source wavelets. Here we see that the vibroseis source wavelet persists for several seconds or tens of seconds. This is quite unlike impulsive source wavelets, which vibrate for a few tens of milliseconds or less.

What kind of seismogram results from the refraction and reflection of the vibroseis source wavelet along the many paths leading to a geophone? Because the wavelet lasts so long, all the reflected and refracted arrivals overlap one another. None of these arrivals can be distinguished on the seismogram recorded in the field. But because we know the precise form of the input signal, we can use a processing technique called *correlation* to reduce these prolonged arrivals to equivalent pulses of short duration. The result is a seismogram, sometimes called a *correlogram*, which is very similar to a seismogram recorded from an impulsive source.

Correlation is done by computer processing

Figure 5–16
Vibrator truck used for Vibroseis surveying. Large pads pressed on the ground are compelled to vibrate through a preset range of frequencies by means of an electronically controlled hydraulic system. Depending on the weight of the truck resting on the pad, more than 25 tons of thrust can be delivered to the ground. (Courtesy of Prakla—Seismos AG.)

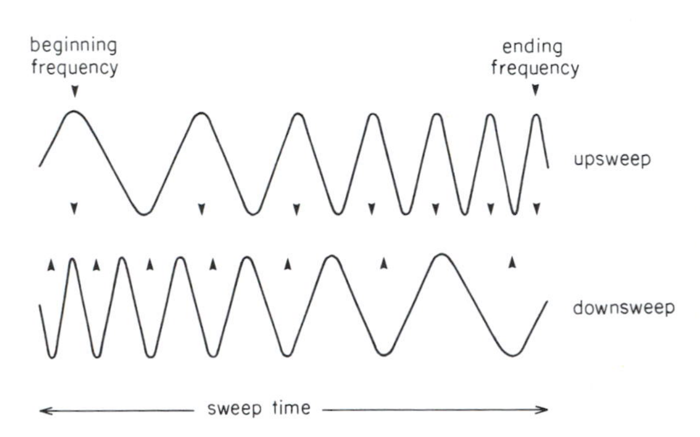

Figure 5–17
Vibroseis sweep signals indicate the pattern of pad vibrations. An upsweep signal begins at low frequency and continuously increases the frequency of vibration. A downsweep signal begins at high frequency and continuously decreases the frequency of vibration.

of the digital magnetic tape after it has been sent from the field site to a data processing center. The procedure is explained in Chapter 6. Most vibroseis digital recording systems have electronic field correlators, which make use of temporary memory units to produce analog monitor records that can be examined in the field.

Vibroseis has proved to be a highly successful alternative to explosive sources and is used for a larger proportion of the exploration surveys now being undertaken. The use of a prolonged low-energy source is much less destructive to property and is more rapidly and easily deployed than explosives that require deep shot holes. Vibroseis can be used along almost any public roadway and even within cities. Its success is based on being able to sum the results of several sweeps, which are usually required to accumulate enough energy to generate a satisfactory seismogram.

The Seismic Crew

In seismic exploration for petroleum, the equipment we have been describing is operated by the *seismic crew,* which, as mentioned in Chapter 1, is sometimes called a doodlebugging crew. Major oil companies maintain some company crews. There are in addition many contracting companies that will provide crews and equipment on a temporary contract basis. Larger contractors are prepared to undertake all aspects of a seismic survey from initial planning, field operations, data processing, to interpretation.

A seismic crew usually operates under the supervision of a *party chief,* who is responsible for fieldwork. Day-to-day field activities are directed by the *party manager,* who receives instructions from the party chief.

During the planning stages of a seismic survey, a *permit* person visits the area to contact landowners and local highway department officials and obtains the necessary permits to operate the equipment there. After the plan for a seismic survey has been worked up and permits are in hand, a transit survey must be conducted to locate source and receiver positions. Usually, this task is handled by two or more surveyors before the main crew arrives, although these surveyors can continue a short distance ahead of the main crew along the proposed route. Using transits and stadia rods, they establish position coordinates and elevations along the entire route which they mark by small flags, stakes, or marks on road pavement.

If explosive sources are to be used, *shot hole drillers* follow the surveyors. This work, too, is best done before the main crew arrives, not only because of the time required to drill the holes, but also because of the vibration of the drills that would interfere with seismic recording.

Now things are ready for the main crew. It consists of one or two observers who are responsible for operating and maintaining the recording equipment, which is usually mounted in one truck. If explosives are used, two or more *shooters* are employed to maintain the explosives cache, to load and tamp the shot holes, and to detonate each shot at a signal from the observer. After a shot hole is loaded, it is filled, or tamped, with water, sand, or soil to contain the explosion. If the source is vibroseis, from one to five or more *vibrator mechanics* are employed to operate and maintain the large vibrator trucks.

A *ground crew,* usually directed by a foreman, consists of *jug hustlers* who place the geophones at points marked by the surveyors, and *cable handlers* who lay out and maintain the seismic cable. Depending on the scale of the operation, the size of the ground crew can range from fewer than a half dozen individuals to more than two dozen.

Until the 1960s, most seismic crews were accompanied by two or more *interpreters,* who were trained seismologists. They examined the seismograms as they were recorded, identified reflections, and prepared preliminary interpretations of earth structure. Before the advent of computer processing, this activity was essential to ensure that usable seismograms were being obtained. If the interpreters could not recognize reflections, field operations could immediately be altered to improve record quality. Charge sizes, instrument settings, shot hole depth, and geophone arrangement could all be adjusted in the effort to obtain good reflections.

When digital recording and central data processing centers were introduced, this kind of shot-by-shot quality control moved away from the field site. The interpreters mostly remain at data processing centers, but maintain daily or more frequent contact with the field crew by telephone or radio. Insofar as possible, digital tapes are delivered daily or twice daily to the data processing center for immediate preliminary examination.

The basic core of a seismic crew, consisting of permanent company employees, usually includes the party chief and party manager, observers, surveyors, permit people, drillers, shooters or vibrator mechanics, ground crew foreman, and interpreters who may accompany the crew from time to time. This basic core of people moves from place to place for periods of a few days to a few months, depending on the scope of the survey.

Temporary employees hired at each survey site include jug hustlers and cable handlers. For off-road operations in open range country, jungles, swamps, and Arctic tundra regions, brush cutters, bulldozer operators, and other laborers may be employed temporarily to prepare trails for the recording truck, drill or vibrator trucks, and other service vehicles. In remote areas where food and housing must be provided, still more people will be included on the field crew. Considering both permanent and temporary personnel, a seismic crew can consist of fewer than 10 to more than 100 people, depending on the circumstances of the survey.

The makeup of a marine seismic crew differs in some respects from that of a land crew. Again, a party manager supervises the shipboard activities. Unlike a land crew that traditionally works from sunrise to sunset, the marine crew operates 24 hours a day. Therefore, shifts, or *watches* as they are called at sea, are established. This means that more individuals are needed for all job categories.

In addition to the party manager, the marine crew includes observers, a group of as many as six or more per shift to assemble and maintain the streamer containing the hydrophones and to operate the air gun, sparker, or other source. In addition, a navigation crew is needed to establish accurately the position of the vessel. This function is carried out by electronic navigation using shoran and loran, doppler positioning, and satellite positioning systems. These navigation systems, which are also used in airborne magnetic field surveying, are described in some detail in Chapter 11. Therefore, they will not be dealt with in this chapter.

Vessels used for marine seismic surveying are operated by licensed merchant marine officers and crew. By law, the captain is in command of the entire operation. Except for emergency situations, however, the captain is expected to operate the vessel according to directions from the party manager.

Marine surveying can require two or more vessels if large explosive charges that must be detonated at considerable distance from the receivers are needed. For some purposes, more than one streamer or other receiver units are deployed by means of additional vessels.

Some consulting engineering companies employ small seismic crews for site surveys. Although the field operations often involve a sequence of tasks similar to that used for petroleum exploration, the scale of activity is usually small enough to be carried out by a half dozen or fewer people.

FIELD OPERATIONS

Now that we have introduced the pieces of equipment and the personnel that are available for seismic surveying, we can examine the procedures for operating in the field. First, we will present some basic practices, some of which have been used since the beginning of exploration seismology. Then we will review some methods that have been developed to help overcome common problems of interference from surface waves, refracted arrivals, and vibrations from various unintended sources, all of which can obscure desired reflected arrivals. Unintended sources of vibration include wind motion in trees, passing traffic, and industrial activity. Desired arrivals are generally called a *signal* and undesired arrivals, whatever their source, *noise*. Field procedures as well as later data processing procedures have been developed to improve the signal-to-noise ratio, thereby producing more useful seismograms.

Basic Spreads

The arrangement of geophones that is used to record seismic waves is called a spread. For most purposes, interpretation of seismograms is simplest if the geophones are arranged in straight lines. The analyses that are worked out in Chapters 3 and 4 to relate earth structure and seismic wave arrival times assume straight spreads. The basic straight-line spreads are illustrated in Figure 5–18. End spreads reach away from the source in one direction. This pattern can be modified to an in-line offset spread by moving the source some distance away from the first geophone. This

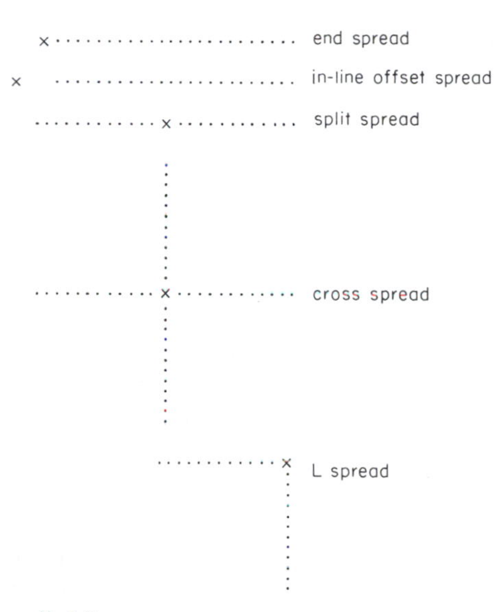

Figure 5–18
Conventional spreads for seismic reflection surveying include the end spread where the source is placed at one end of the line of geophones, the split spread where the source is at the center of the line, the cross spread, and the L spread where the source is at the corner of two lines.

spread is sometimes desirable because the energy from the source is too overwhelming at closer distances.

The most commonly used spread before the introduction of digital recording is the split spread. Geophones are arranged in two opposite lines with the source at the center. The lines may be offset to provide a shot gap. We know from Figure 4-15 that arrivals from a split spread are needed to detect dipping reflectors. However, unless the spread is aligned in the dip direction, only a value of apparent dip can be measured. Therefore, a cross spread is useful. It consists of two split spreads centered about the same source. When reflections are recorded in two directions, it is possible to calculate the true dip angle and direction. To save time, an L spread can be used in place of a cross spread. It consists of two lines extending in different directions from the same source.

Cross spreads and L spreads are not difficult to set out in wide-open range land, desert areas, and polar tundra, but they cannot easily be placed in more confined countryside. Although it is convenient to place a split spread along a roadside, fences, crops, or forest can make it impractical to set out a crossing spread or one leg of an L spread. It may only be feasible to do so where roads intersect.

Turning our attention to refraction surveying, we know from Figure 3–10 that an in-line spread with sources at both ends is needed to detect a dipping refractor. Again, refraction measurements along a crossing spread are needed to determine true dip angle and direction.

Another kind of refraction spread was introduced in the 1920s as a means of prospecting for shallow salt domes. The geophones were placed in an arc, fanning out in different directions from the source; this kind of survey is called fan shooting. The idea, illustrated in Figure 5–19, is that refracted waves should reach equidistant geophones in the same "normal" travel time unless they are refracted through salt. If they contact salt, their velocity becomes somewhat higher, shortening the travel time. Fan shooting proved successful for locating some shallow salt domes. Today it is occasionally used in engineering surveys to search for caverns or other localized features that would change refraction travel time.

Single-Coverage Reflection Profiling

The basic spreads that we have described are used for seismic sounding at a single location. That is, after recording the seismic wave arriv-

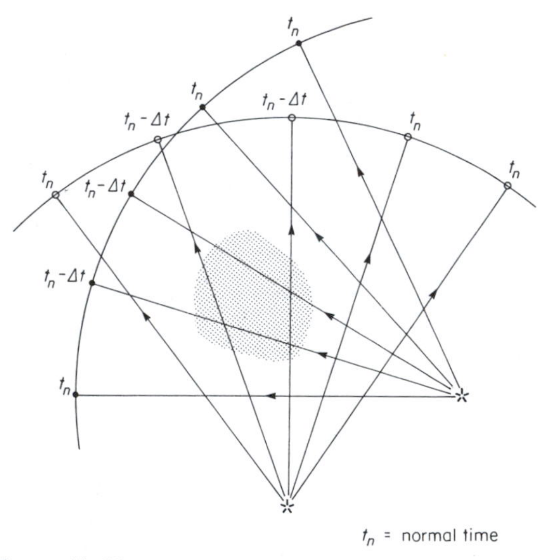

Figure 5–19
Refraction spreads for fan shooting designed to detect deviations from the "normal" travel time between the source and equally distance detectors in different directions.

als, we can calculate depths and dips of layer boundaries at the site of the spread. But to obtain a broader picture of earth structure, we need to make soundings in many places.

The usual procedure before digital recording became possible was to make split-spread soundings at equally spaced locations along lines many miles, tens of miles, or even hundreds of miles long. Insofar as possible, these lines extended along public roads, but off-road surveying was done as needed in open country, through swamps, and in jungles.

The individual soundings were ordinarily made using a 24-channel analog system. Geophones were spaced at 100- or 200-foot intervals, 12 on one side and 12 on the opposite side of the shot hole. The shot was detonated and the seismogram was recorded. Then the spread was picked up, set out in the same way farther down the line, and another seismogram was recorded. Depending on time and cost considerations, the spread was moved in one or another of the following ways (see Figure 5–20). The most detailed coverage was obtained by one-leg overlap spreads. Here, the line of 12 geophones reaching away in one direction from a shot point was left in place to become the line reaching in the opposite direction from the next shot point. Meanwhile, the seismic crew moved the other 12 geophones to the opposite side of the second shot point. In this way, reflections from a continuous sequence of equally spaced points on a layer boundary could be obtained. Analog seismograms from these overlapping spreads could be arranged side by side, as in Figure 5–21, in order to indicate how the reflector depths changed along the line.

Less detailed coverage was obtained from an end-to-end succession of spreads. Here, the geophone at one end of one spread remained in place as the geophone at the opposite end of the next spread, thereby producing gaps in the coverage of the reflectors. Still larger gaps resulted from soundings for a succession of offset spreads along the line.

The practice of making soundings at a succession of points along a line, or profile, is called *seismic* profiling. Observe in Figure 5–20 that each reflection point on a layer boundary is on the path of only one reflected wave. Because there is only one reflection from each point, the procedure is called *single-coverage profiling*. Later, we will contrast this procedure with multifold coverage, which we introduced in Chapter 4.

A seismic crew doing single-coverage profiling generally began work at sunrise. The observer tested the recording apparatus while the jug hustlers placed geophones at points previously marked by the surveyors along the road or trail. At the same time, the cable handlers unrolled seismic cables from reels so that the jug hustlers could clip the geophones to the takeouts. By the time this job was finished, the shooters had loaded and tamped the charge. Then the observer gave a call for all to stand still, signaled the shooter to detonate the shot, and recorded and developed an analog seismogram. All this work was done in about 30 to 45 minutes. The ground crew then picked up geophones and cable and moved them to the next spread position, while the shooters loaded and tamped the next shot hole. This procedure was repeated for the rest of the day until sunset. Except when there were delays for instrument repairs or for adjustments to improve reflection quality, a seismic crew could expect to record between 10 and 20 shots during a workday. Typical activities are seen in Figure 5–22.

When a seismic survey was conducted along parallel profiles and some crossing profiles, lit-

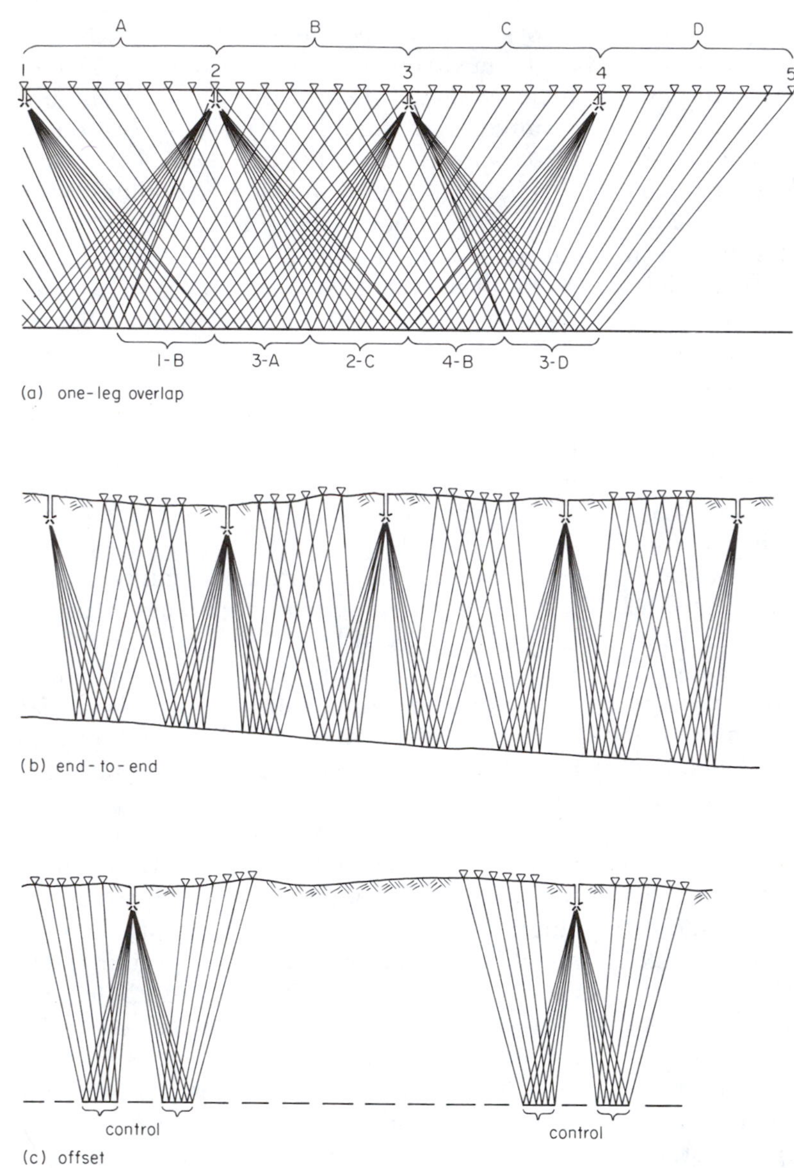

Figure 5–20

Single–coverage reflection-profiling procedures. (a) Continuous subsurface coverage obtainable by a one-leg overlap succession of spreads. (b) Discontinuous subsurface coverage obtainable by an end-to-end succession of spreads. (c) Sampled subsurface coverage obtainable by offset spreads.

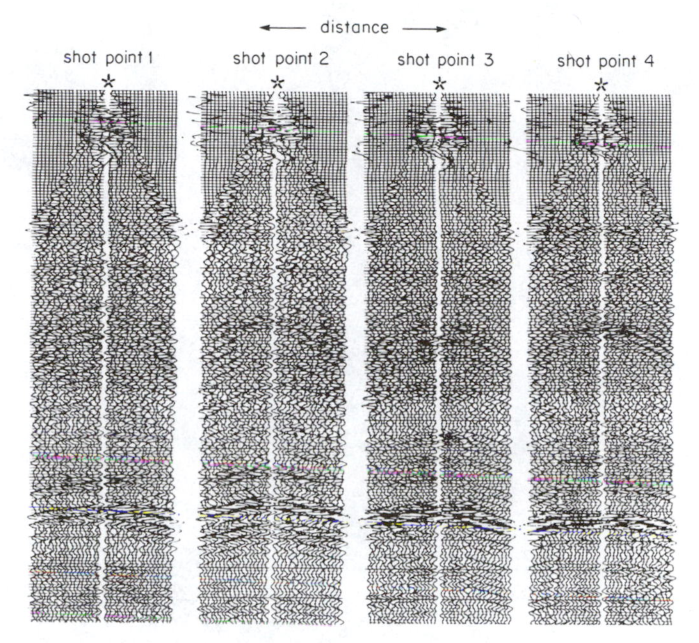

Figure 5–21
Sequence of split-spread reflection seismograms from four adjacent one-leg overlap spreads. The data are not corrected for normal move-out; therefore, the reflections on the traces of an individual seismogram outline hyperbolic curves across the seismogram. From the different minimum reflection times on different records, it is obvious that the reflector depths are changing along the profile. (Courtesy of Amoco Company.)

tle effort was made to set out cross spreads or L spreads for dip control. This information could be obtained from the changing reflector depths measured along the profiles.

Common Depth Point (CDP) Reflection Profiling

When digital recording and computer data processing became possible in the 1960s, a new method of reflection profiling was introduced. It is called *common depth point (CDP) reflection profiling*. This is the procedure for obtaining the multifold reflections that were discussed briefly at the end of Chapter 4. Why are multifold reflections important? For simple earth structures with constant layer velocities and plane boundaries, multifold reflections convey no more information than we can get from single-coverage data. But we know that real earth structures possess more irregular boundaries and some velocity differences from place to place in a layer. Multifold reflections help us to resolve these variations more accurately. More important, however, are noise problems which have always plagued seismologists. The ability to combine seismogram traces from different source–receiver offsets in order to obtain multifold reflections can vastly improve the signal-to-noise ratio.

To introduce the idea of CDP profiling, we will begin with a single source and a spread with single geophones connected to each takeout on the seismic cable. For this purpose, we will use a cable consisting of segments, one for each geophone, that can be connected by plugs to form the complete spread cable. This

Figure 5–22
Field activities on a single-coverage seismic reflection survey included laying out the geophone spread, drilling the shot hole, detonating the source charge, and recording the reflection seismogram.

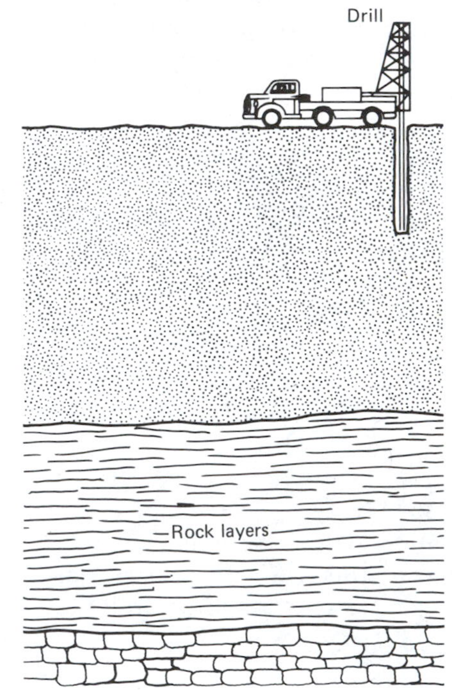

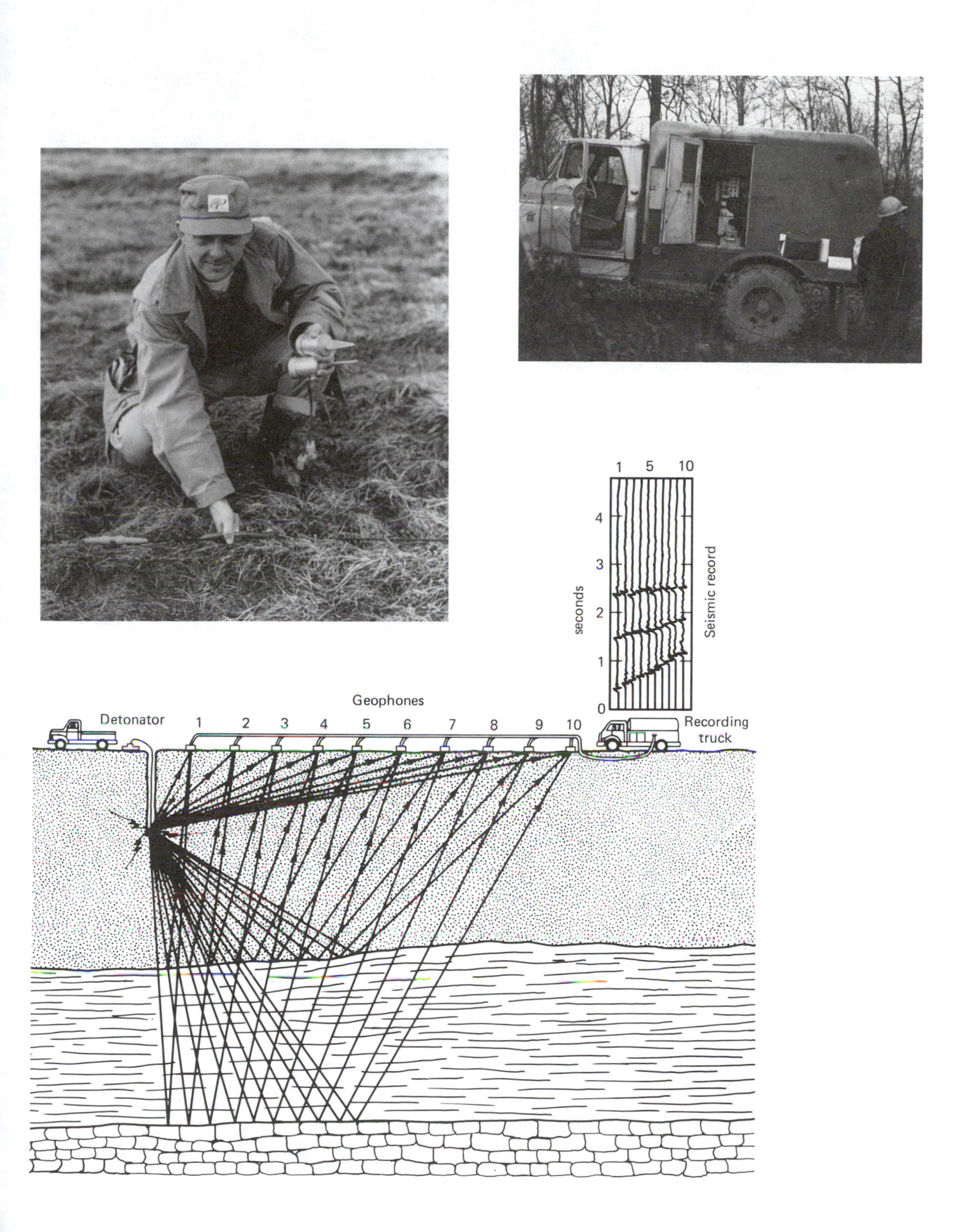
1
5
10
4
3
2
1
0
seconds
Seismic record
Geophones
Detonator
1
2
3
4
5
6
7
8
9
10
Recording
truck

setup is the same one we have been discussing all along. Later, when we look at field techniques for reducing noise, we will describe how more complicated combinations of sources and geophones are used in CDP profiling operations.

Basically, the procedure is simple. We use an end spread with an in-line offset source. For purposes of explanation, consider the spread of four geophones in Figure 5–23. We begin with the source at A and the geophones at positions 1, 2, 3, and 4; then we record a seismogram that has traces we can label A1, A2, A3, and A4.

Next, we disconnect the first geophone and its cable segment, move it to position 5, and connect it to the opposite end of the original spread. At the same time, we move the source to position B which is the same as geophone position 2. Again, we record a seismogram, this time labeling the traces B2, B3, B4, and B5.

Continuing in the same way, we move the geophone from 3 to 6, and the source from B

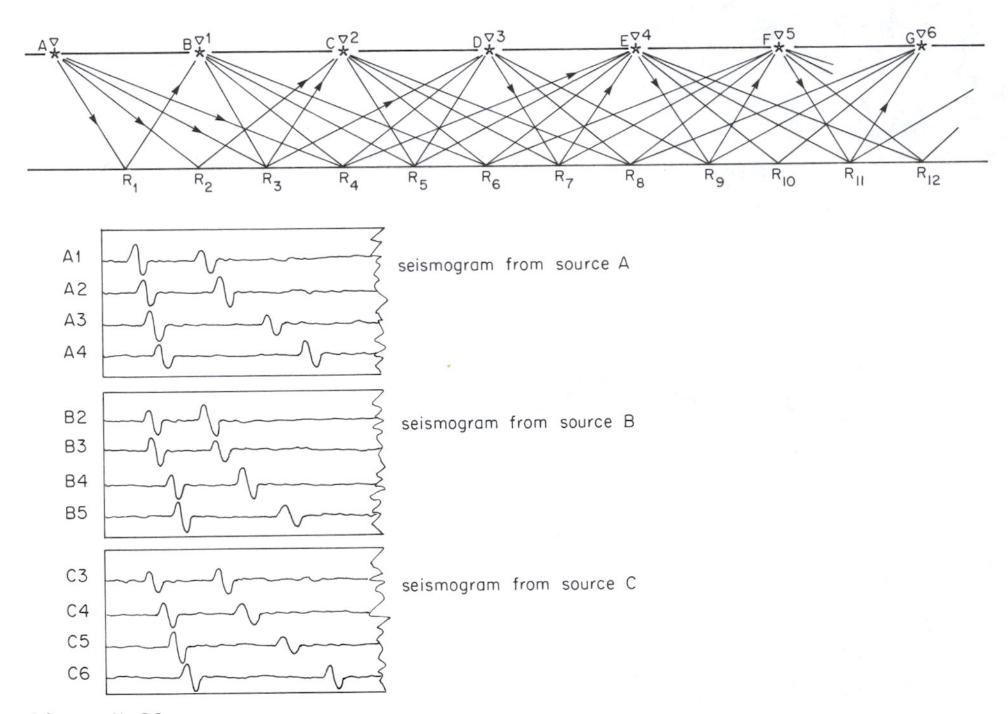

Figure 5–23

Common depth point recording procedure for obtaining multifold reflection data. The example illustrates twofold coverage. Reflections from the same common depth point but with different ray paths are recorded. Letters represent source points, and numbers identify receivers. This recording procedure is designed to give twofold coverage, as, for instance, at subsurface point R3 which is sampled by A3 (source A, receiver number 3) and B2 (source B, receiver number 2).

to C, which is the former geophone position 2. This time we record the seismogram with traces C3, C4, C5, and C6. Then we keep moving along the line by shifting one geophone at a time, extending the opposite end of the spread and advancing the source.

Looking carefully at Figure 5–23, we see that two reflection paths reach every reflection point from R3 and beyond. Therefore, we have *twofold coverage* of these reflection points. The traces with reflections from these common depth points come from different seismograms, for example, at

$$\begin{aligned} R_3 &\rightarrow \text{traces A3 and B2} \\ R_4 &\rightarrow \text{traces A4 and B3} \\ R_5 &\rightarrow \text{traces B4 and C3} \\ R_6 &\rightarrow \text{traces B5 and C4} \\ R_7 &\rightarrow \text{traces C5 and D4, etc.} \end{aligned} \tag{5–3}$$

Here we see that each combination includes traces from two different seismograms. We also see that by using four geophones at equal intervals, we obtain twofold coverage.

Suppose that we have a 48-channel system. By proceeding in the fashion we have just described, we would be able to obtain 24 separate reflections from a common depth point, as in Figure 5–24. To do so, we would advance the spread at one end and the source position at the other end by one geophone interval and then record a seismogram. The 24 traces with reflections from any common depth point would be on 24 different seismograms.

To save time, suppose that we decided to move two geophones from one end to the other of the spread, and correspondingly advance the source by two geophone intervals. Then we would obtain only 12 common depth point reflections, or 12-fold coverage. In general, we can calculate the multifold coverage in terms of the number of recording channels N, the geophone interval Δx, and the source interval Δs:

$$\text{fold number} = \frac{N}{2}\frac{\Delta x}{\Delta s} \tag{5–4}$$

For our example in Figure 5–23, we see that $N = 4$ and $\Delta x = \Delta s = 1$, so that we calculate twofold coverage.

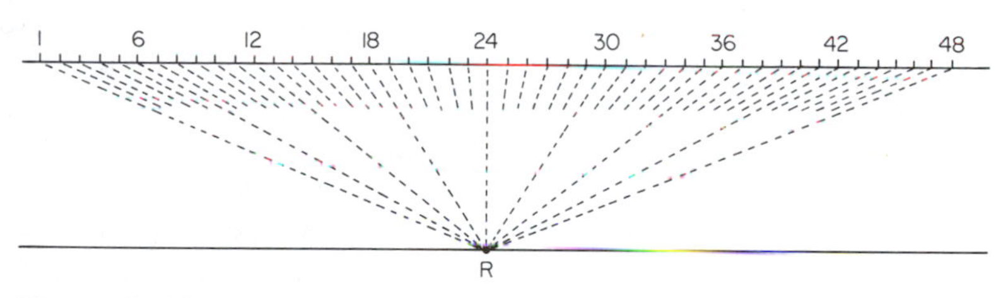

Figure 5–24

Twenty-four-fold reflection coverage obtained with a 48-channel recording system. Twenty-four source and receiver pairs, which are symmetric about a common midpoint, are used to obtain 24 reflections from different ray paths that all reflect from the same point. From the station numbers, the locations for source and receiver pairs can be identified, that is, the source at station 18 received at station 30 reflects from the same common depth point as the source at station 6 received at station 42.

Clearly, the only practical way to extract multifold reflection data from multichannel digital magnetic tapes in multiplexed form is by computer processing. First, the traces of each seismogram must be extracted from the tape by demultiplexing, as in expression 5–2. Then the *common midpoint* traces must be sorted from many seismograms, as in expression 5–3. The term *common midpoint* encompasses all traces for which the lines from source to receiver have the same center point. Common depth points on underlying reflectors should align vertically beneath this common midpoint if the reflectors are horizontal. These steps in data processing are discussed more fully in Chapter 6. The present chapter focuses on data acquisition. Before we discuss the activities of a digital seismic crew doing CDP profiling, we should introduce the common procedures for noise reduction, which have much to do with these activities.

Marine Seismic Profiling

Practically all modern marine seismic exploration for petroleum involves CDP profiling that can be done with a single vessel. The work can proceed much faster than land surveying, because segments of the spread do not have to be continually disconnected and then reconnected at the opposite end of the spread. Instead, the source and receivers are arranged at fixed intervals and towed by the vessel. A source, such as an air gun, remains about 300 meters astern followed by the streamer containing hydrophones at 100-meter intervals. Altogether, the line stretches 3 to 4 km behind the vessel.

To acquire CDP data, we must see that the source releases its impulse at intervals corresponding to some multiple of the hydrophone spacing. Suppose that a 48-channel system is being used, and the hydrophone interval is 100 meters. Then, by releasing an air gun impulse at 100-meter intervals, we obtain 24-fold coverage. For a vessel traveling at a speed of 5 knots, the distance is covered in about 38½ seconds, which means that the air gun must shoot every 38½ seconds. For 23-fold coverage, the impulse would have to be released every 200 meters, or at time intervals of 77 seconds.

Although seismic profiling can proceed swiftly at sea, there are stringent demands for accurate navigation. Great care must be exercised in keeping track of the ship position and the streamer position, which can shift with crosscurrents and waves. Accurate CDP profiling requires a reasonably calm sea. As mentioned earlier, the navigation procedures for accurate positioning at sea are described in Chapter 11.

Noise Control

The deeper reflections in Figure 5–21 are clearly recognizable and so would delight most seismologists. The shallower events are less distant, however, appearing on some but not all of the traces. On other parts of these seismograms there are hints of reflections, but arrival times certainly cannot be picked with confidence because the signals are overridden by noise. Are there ways to obtain more and clearer reflections? What can be done during a field survey if reflections are not evident on the seismograms?

Before the introduction of magnetic tape recording, all improvements in record quality had to be accomplished while the survey was underway. The methods developed to cope with noise problems involved electronic filtering and different groupings of geophones and sources. Most of these methods are still used. The newer ideas for signal-to-noise enhancement mostly require computer processing of digital data. Powerful as these new ideas appear to be, they do not completely solve noise problems. Furthermore, the task of computer-based improvement of reflections is greatly simplified when efforts in the field succeed in producing some recognizable reflections at the time of recording.

There are two basic kinds of seismic noise. *Coherent noise* displays some regular patterns on a seismogram. Often, it consists of recognizable waves such as surface waves, refracted waves, and multiples that are produced by the source. By examining the patterns of coherent noise, we can devise field procedures to reduce it. *Incoherent,* or *random,* noise displays no systematic pattern. It may arrive simultaneously from many sources such as wind blowing on trees and passing traffic. Other procedures are used to suppress this kind of noise.

The basic tools available for controlling noise in the field include

1. Source size.
2. Source depth.
3. Electronic filtering.
4. Receiver arrays.
5. Source arrays.
6. Electronic mixing.

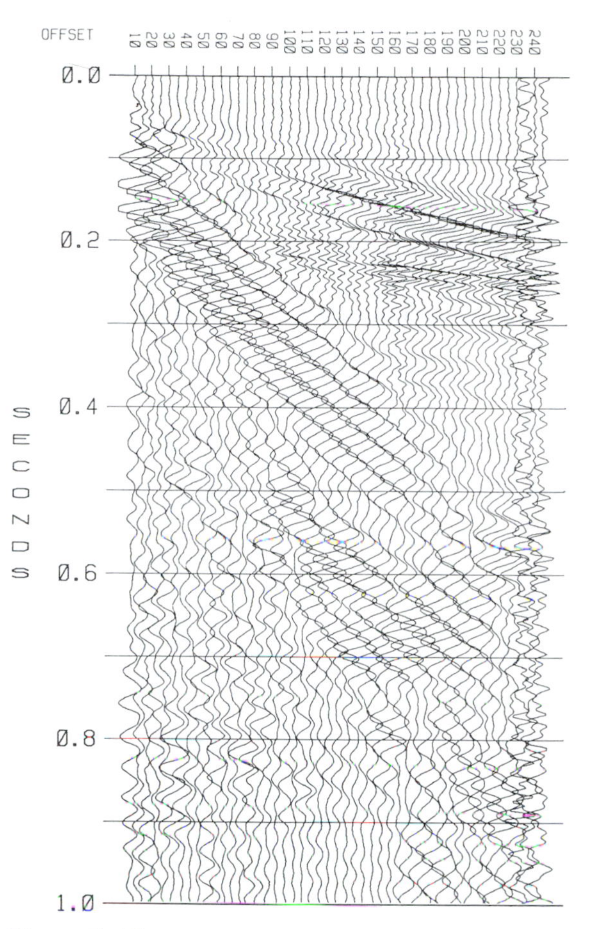

Figure 5–25
Seismograms displaying Rayleigh waves which are commonly referred to as ground roll. High-amplitude waves such as these can obscure weaker reflected and refracted arrivals.

Most of the energy that reaches a receiver is carried by surface waves (see Chapter 2). Like waves rolling over the ocean, Rayleigh waves move out over the land surface. Exploration seismologists often refer to them as *ground roll.* Figure 5–25 illustrates large-amplitude Rayleigh waves that persist for several seconds. For reasons we do not fully understand, ground roll this strong occurs in some survey areas, but in other areas, ground roll is much weaker.

Much thought has been given to the problem of suppressing ground roll. Reducing source size is usually not effective, because reflection amplitudes are diminished as well as the undesired surface waves. A more productive approach is to increase the depth of the source. Most Rayleigh wave energy propagates along the low-velocity weathered zone that extends down a few tens of feet from the earth's surface. A charge detonated at the surface or within this shallow weathered zone tends to stimulate much stronger Rayleigh waves than shots from deeper holes that reach below the weathered zone. Even though deeper shot holes are more costly and time-consuming, they may provide an answer to a ground roll problem. Of course, this solution is not appropriate for vibroseis surveying, which requires that the source be at the surface.

In some places ground roll frequency differs substantially from that of reflected waves. Where this is true, electronic filtering can be effective. If the ground roll is principally in the range of 10 to 20 hertz and the reflections have frequencies between 30 and 50 hertz, a high-pass filter may eliminate much of the problem. Also effective are geophones with natural frequencies higher than the ground roll frequency but lower than the frequencies of reflections. Such geophones also act as high-pass filters.

Unfortunately, ground roll often propagates with frequencies similar to those of reflections. Receiver and source arrays may then

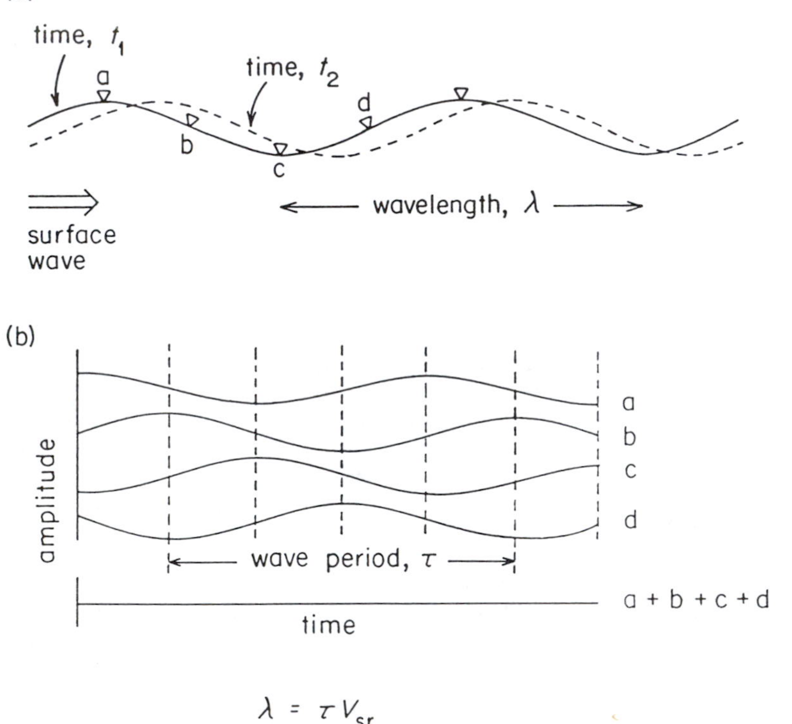

Figure 5–26
A linear geophone array for suppressing surface wave oscillations on a seismogram. Outputs from the geophones a, b, c, and d are summed by connecting them together. Depending on the period (τ) and velocity (V_{sr}) of the surface wave, each element is sensing a different part of the horizontally traveling wave train at the same time. Summing the outputs from the geophone elements results in attenuation.

help to overcome the problem. First, we will look at receiver arrays. Up to this point, we have described spreads in which one geophone is attached to each takeout on the cable. Suppose that several geophones, arranged in some pattern, were attached to the same takeout. To understand the effect, we must look at a fundamental difference in the way Rayleigh waves and reflected waves travel. Rayleigh waves spread out horizontally from the source, distorting the land surface much like the ripples in Figure 2–4 distort the surface of a pond. Now compare the motions of several geophones placed in a line, as shown in Figure 5–26. While some move up, others move down. Individually, each one would record high-amplitude ground motion. However, with all of them connected to the same takeout, the output from the rising geophones acts to cancel the output from the falling ones. The collective output reaching the recording channel is, therefore, reduced. Placing several geophones within a distance equal to the wavelength of the ground roll will ensure its effective suppression. We know that some of the phones so placed must be moving upward while others are moving downward.

How do we find out the wavelength of the ground roll? Where this is expected to be a problem, the seismic crew undertakes a *noise study*. By recording a surface shot on a long spread, the crew obtains a seismogram displaying well-developed ground roll. Then the periods, velocities, and wavelengths can be measured from this seismogram. The information can sometimes be estimated from an ordinary reflection seismogram. It usually turns out that the ground roll occurs with a range of wavelengths. So there is no ideal geophone spacing that will ensure complete suppression. Nevertheless, a group of geophones connected to one takeout can be quite effective in reducing the ground roll.

In the meantime, what about the reflected waves? How do they affect this line of geophones? Figure 5–27 shows a reflected wave front together with the ray paths to the different geophones. It is clear that the wave front will reach all the geophones at nearly the same time. This means that they will rise and fall in response to the reflection almost simultaneously. When connected to the same takeout, the individual geophone signals reinforce one another to produce a stronger reflected pulse signal.

Clearly, a line of geophones connected to the same takeout strengthens reflections at the same time as it is suppressing ground roll. This effect persists even if the ground roll frequency range and the reflection frequency range are similar.

A group of geophones connected to the

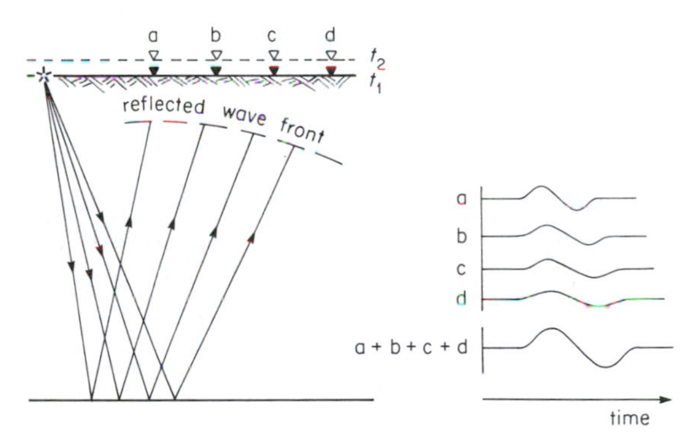

Figure 5–27

Reflected wave front approaching a geophone array. The reflection reaches all the individual geophones at about the same time. When geophones a, b, c, and d are connected in an array, the reflection arrivals at all of them constructively interfere.

same takeout is called a *receiver array* or a geophone array. We have been describing the simplest kind of receiver array, which consists of several identical geophones placed at equal intervals along a line parallel to the spread. Other array patterns are also used. In *weighted arrays,* clusters of one, two, three geophones are placed together at different points in the array. Feathered arrays consist of lines of geophones that angle away from the spread like feathers on an arrow. Although each array pattern was originally proposed for a specific purpose, their effectiveness will not be known until we have tested them during a survey. In petroleum surveys, it is common to use up to 48 geophones in a single array. With this number of geophones connected to each of 48 channels, the jug hustlers must set out 2304 geophones. If they are arranged in linear arrays at 5-meter intervals, the total spread length will be more than 11 km.

Ground roll can be suppressed and reflections enhanced by means of source arrays. Figure 5–28 shows how this can be done. Suppose that four charges at different positions along a line are detonated simultaneously. The horizontally traveling surface waves from these sources interfere destructively with one another. The result is suppression if the source spacing is chosen with the same regard to ground roll wavelength as was applied for receiver arrays. At the same time, the waves traveling downward along reflection paths have nearly coincident wave fronts. They interfere constructively to produce a stronger reflected signal.

Source arrays were introduced in the 1930s when different patterns of explosions were tested. Some arrays were arranged on the land surface or in shallow holes; others were placed on poles in the air. Line charges made up with primacord were also tried. Each of these patterns was effective in some circumstances. Major drawbacks are the time needed to set up the array and difficulty in securing permits to use explosives in these ways.

The availability of digital recorders with summing capability opened the way for vibroseis source arrays. The vibrator truck has proved to be well suited for this operation. A single truck can easily be used to create the entire array. As it moves from point to point through the array, the signal is recorded and stored in a temporary memory unit, and then summed with signals from other array points. To save time, however, most vibroseis crews are equipped with three, four, or five vibrator trucks that can be operated simultaneously at the source array points. In vibroseis work the pattern of vibrator points in the source array is called the drag pattern. The most widely used drag pattern consists of the set of trucks equally spaced in line with the spread.

Almost no effort is made in the field to sup-

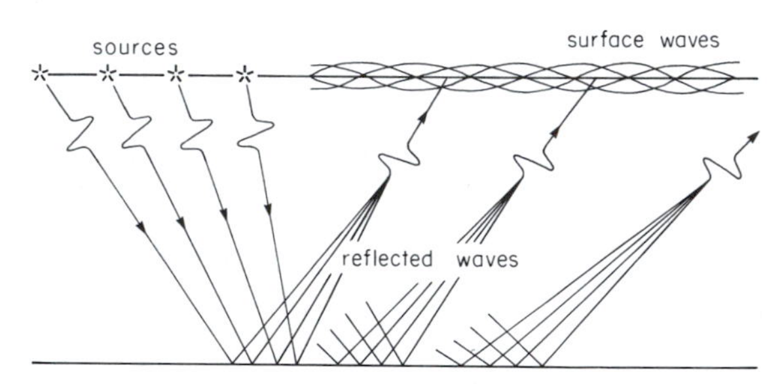

Figure 5–28
Constructive interference of reflections and destructive interference of surface waves from a source array. Nearly vertical reflections from the four separate sources reach an individual geophone at about the same time, resulting in constructive interference. Horizontally traveling surface waves from these sources tend to interfere destructively at the geophone.

press refracted arrivals or multiples. In marine seismic surveying, multiples can be a very serious noise problem; this problem can be addressed by data processing methods presented in Chapter 6.

Random noise tends to be suppressed by receiver arrays. Because these pulses tend to be different from one geophone to another, there is a tendency for destructive interference when they are combined by an array. Electronic filtering may prove useful if the noise frequency range is different from the reflection frequencies.

Many analog recording systems were equipped with electronic mixing circuits. Some proportion of a geophone signal was transmitted to the channels of adjacent geophones. In some ways this kind of mixing creates a modified receiver array. It can sometimes reduce random noise by means of destructive interference of the noise components of the signals being mixed.

Source size and depth can also be adjusted to help control random noise, since the sources of random noise are unlikely to change very much over a brief period of time. Seismogram quality may be improved by increasing the charge size or the thrust of the vibrator. This will increase reflection amplitude while the competing noise remains the same.

In these basic ways we can attempt to adjust our field recording procedures to control noise and improve reflections.

Noise Problems at Sea

Marine seismic crews face noise problems different from those encountered on land. Three kinds of marine seismic noise need to be mentioned. Operating noise is produced by the ship's vibrations and the movement of the streamer through the water. The usual procedures for coping with such noise include trailing the streamer far behind the ship and reducing the speed of the ship, both within practical limits.

Seismic sources of noise such as air guns and explosions release a large bubble in the water at the moment of initial impulse. At first, the bubble expands because the gas pressure exceeds the hydrostatic pressure of the water. However, a limit is reached at which the pressure difference is reversed, and the bubble collapses with an impact. But the gas still remains trapped in the water, so the bubble again expands and collapses as it rises to the surface. The succession of impacts resulting from the oscillating bubble size acts as a sequence of sources. Waves produced by each cycle of bubble expansion and collapse are called *bubble pulses.* Therefore, on a seismogram, all the reflected and refracted arrivals from the initial impulse are followed by a sequence of similar arrivals produced by the bubble pulses. The problem can be modified by locating the source close to the water surface so that the bubble can escape before collapsing, or by using later data processing procedures.

The marine equivalent of ground roll is produced by multiples that echo back and forth between the surface and the ocean floor. A sequence of multiples can produce such strong oscillations on a seismogram trace that other arrivals cannot be identified. Although source arrays are sometimes partially effective in suppressing the ringing of multiples, the most effective means of control requires later data processing to "de-reverberate" the seismogram. To accomplish this, we measure the water depth continuously by means of an echo sounder mounted in the ship.

Vibroseis CDP Profiling

Vibroseis crews perform much of the modern seismic reflection surveying. This work is routinely done along public roads, but off-road trails must be prepared for some operations. Use of source and receiver arrays is a regular part of the field procedure. Profiling is done with a digital recording systems with 120 channels or more. Ordinarily, three to five vibrator trucks are combined for the source arrays, and 10 to 48 geophones per takeout make up the receiver arrays, or geophone groups.

Beginning at sunrise, the jug hustlers and cable handlers lay out the initial spread. Geophones making up each group are already wired together. Attaching a single clip at the end of the conductor to a takeout on the seismic cable allows all geophone signals to feed into this takeout. Laying out a 48-channel spread may require two hours or more. With takeout spacings of 50 to 100 meters, the assembled cable will reach between 2.4 and 4.8 km. Several cable handlers are needed to stretch out and connect all the cable segments. At the same time, several jug hustlers must set out more than 2000 geophones.

Meanwhile, the observer tests the recording system, and the vibrator mechanics run tests to synchronize all the vibrator trucks so that the sweeps will be exactly together when all are in operation.

When all is ready, the daily recording operation begins. The vibrator trucks stand in line at fixed intervals at one end of the spread, as in Figure 5–29. At a signal from the observer, the pads are pressed to the ground, the sweeps are triggered simultaneously, and the recording system is switched on. Recording continues for 10 to 30 seconds depending on the sweep time, after which a field monitor seismogram is printed. It can be examined to determine whether any adjustments in the recording procedure are needed. If not, the profiling begins.

Jug hustlers then disconnect the geophone group and cable segment from one end of the spread, and drive to the far end of the spread where the geophones and cable segment are reconnected. Meanwhile, the trucks advance to the next source positions, and the recording process is repeated. Each recording sequence requires one to two minutes if there are no interruptions. It continues throughout the day until late afternoon when the spread must be picked up.

If there are no serious delays, a vibroseis

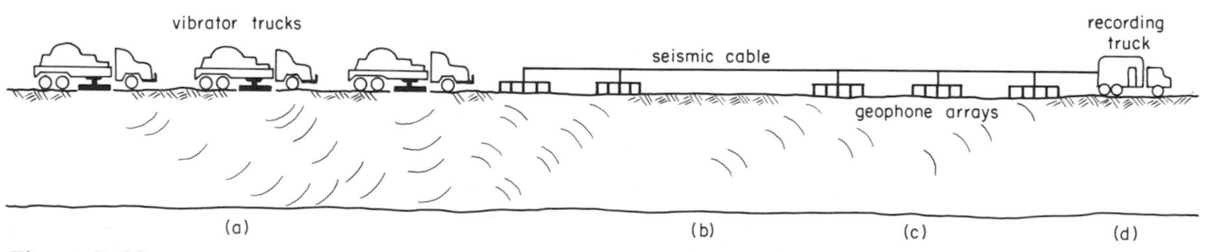

Figure 5–29
Field activities on a vibroseis CDP survey. (a) Vibrator trucks lined up to produce a source array, (b) a reel of seismic cable with clip points for geophone arrays, (c) a set of geophones connected together for use as a receiver array, and (d) a multichannel recording truck.

crew can make 50 to 100 recordings in a day. This allows time to set out and pick up the spread and to transport geophones and cable segments from one end of the spread to the other. The distance covered during the day depends on the interval between source arrays. If the trucks and the spread are advanced 50 meters after each recording, the distance normally adds up to between 2½ and 5 km. Greater distances, perhaps 7 or 8 km/day, can be covered by advancing, say, two receiver array spacings after each recording. But this option reduces the multifold coverage of the CDP profile.

Single-Receiver Marine Profiling

A relatively simple method of marine profiling involving one receiver can produce an image of a shallow subbottom structure while the survey is underway. A source, such as a sparker, and a single receiver are trailed from the ship. The receiver is connected to a special chart recorder. At the moment a spark is emitted, a stylus advances along a line crossing the electrosensitive chart paper. Each wave pulse that reaches the receiver produces an electric signal that is amplified and transmitted to the stylus, stimulating it to etch a mark on the chart. As the stylus moves across the chart, it leaves a trail of marks. The intensity of each mark is proportional to the signal strength.

After reaching the other side of the chart, the stylus snaps back to the beginning position. The chart paper is advanced a small distance so that the stylus moves along another line after the next spark is emitted. This process is repeated as sparks continue to be emitted at intervals of a few seconds or more, depending on water depth.

As the chart continues to advance, the points along the different stylus paths indicate patterns of alignment that image the reflectors, as seen in Figure 5–30. This is one kind of seismic section. It plots time variations rather than depth variations of the reflectors along the profile. Insofar as velocities in the layer are not too different from one another, the profile gives an approximate image of subocean structure.

Shear Wave Surveying

Shear waves reflect and refract according to the same physical laws as *P*-waves. Can they, too, be used for exploration purposes? The principal problem has been to develop a practical *S*-wave source. Explosions are effective for producing *P*-waves, but the *S*-waves are usually too weak to be of use. For small-scale surveys, it is possible to embed a vertical plate in the ground and strike it horizontally with a hammer, but too little energy is introduced to probe very deeply. Experiments have been completed using recoil of a bazooka to generate *S*-waves.

In recent years, interest in *S*-wave applications has increased. They are useful in studies of the porosity and fluid content of different rock layers. *P*- and *S*-wave velocities depend on the proportions of solid and fluid and the wave velocities in these separate parts. Because *S*-waves do not propagate in fluids, their velocity is more sensitive to porosity than the *P*-wave velocity.

The interest in *S*-wave exploration has led to the development of an *S*-wave vibrator for use in vibroseis profiling. This device looks quite like a *P*-wave vibrator except that large steel cones are mounted on the pad. The drive action is designed to cause the pad to vibrate horizontally. When the steel cones are pressed firmly into the ground, the horizontal vibration generates *S*-waves. Enough energy

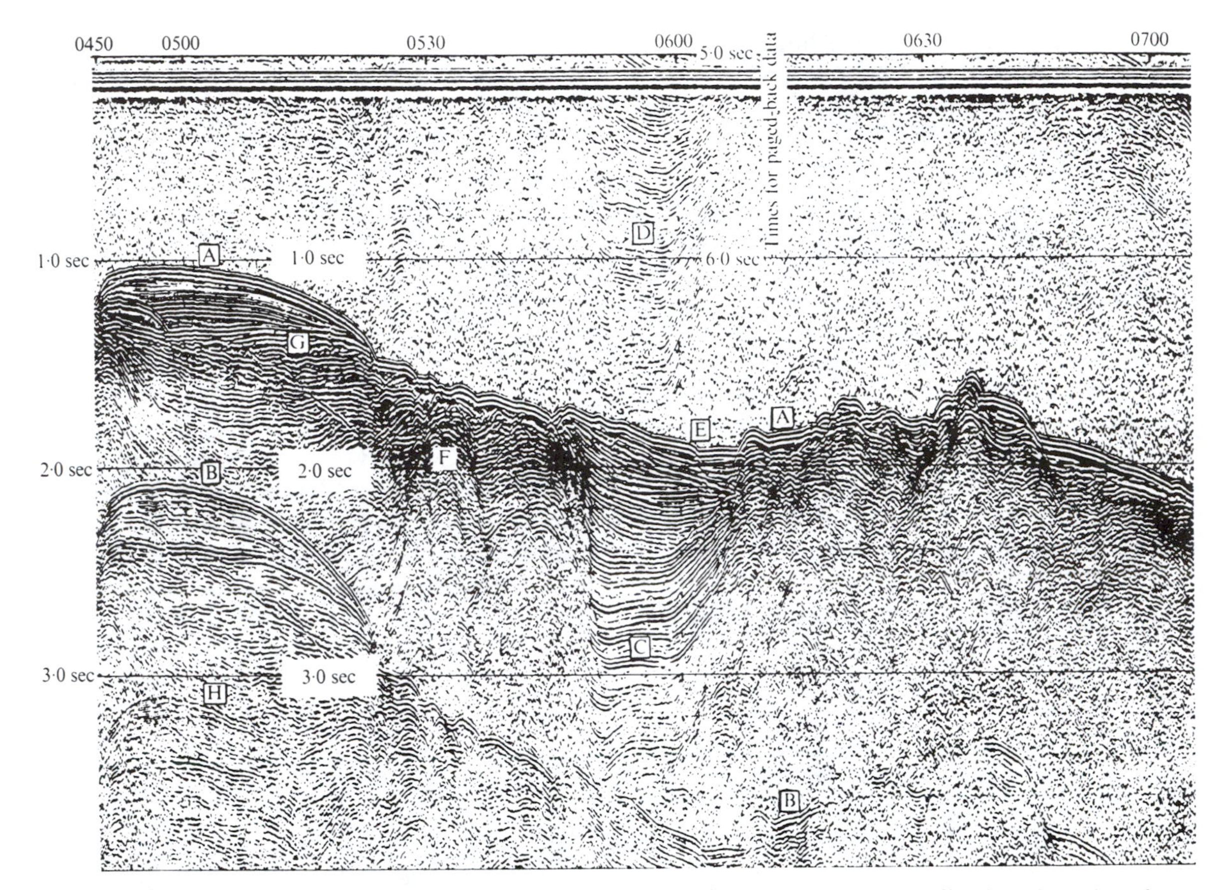

Figure 5–30
Seismic section from a marine sparker–single-receiver survey. Images on the chart actually represent the arrival times of reflections at different distances along the profile rather than the reflector depths. However, the patterns resemble a cross section indicating the subsurface structural features along the profile. It turns out that A is the primary reflection, and B and H are first and second multiples from the same reflector. (Courtesy of Teledyne Exploration Company.)

can be recorded to detect *S*-wave reflections by summing several sweeps. Some useful results have been obtained in areas where *P*-wave surveying has been of questionable value.

A difficulty with the *S*-wave vibrator is that it cannot be operated on a roadbed. It must be placed in an off-road position where the cones on the pad can be penetrated into the ground. For this reason, it is more difficult and slower to do *S*-wave reflection profiling than to do *P*-wave reflection profiling.

Velocity Surveys in Wells

Seismic velocities measured in wells are useful for interpreting seismic reflection data. This information provides an independent measure to compare with velocities found from arrival times of reflected and refracted waves on surface spreads.

Velocity surveys in wells can be done in two ways. One way, discussed in some detail in Chapter 14, is called *sonic logging*. It is done by means of a source and receivers mounted on a probe several feet long. The difference in arrival times at two receivers, usually one foot or three feet apart, indicates the P-wave velocity over a short interval of rock along the side of the well.

Another way to measure velocity in a well is through a special well geophone, which is waterproof and can be clamped to the side of the well. It is lowered to different positions where it is used to detect waves from small explosions on the land surface close to the well. The velocity V is found from the difference in arrival times, $\Delta t = t_2 - t_1$, over the interval of two geophone positions, $\Delta z = z_2 - z_1$:

$$V = \frac{\Delta z}{\Delta t} \tag{5–5}$$

If a well is available in the area of a seismic survey, it can be helpful to undertake this kind of vertical velocity survey.

Three-Dimensional Reflection Acquisition

Seismic reflection data recorded along a profile are not sufficient to determine the position of a reflector exactly, unless the profile is aligned in the dip direction of the reflector. If it is not, only a value of apparent dip can be obtained. Because the reflection travel paths do not lie in a vertical plane (Figure 4–16), a three-dimensional array of reflection travel times must be measured to locate a reflector as precisely as possible.

A common practice in three-dimensional seismic reflection surveying is to place the receivers at equal intervals along a square grid of profiles (Figure 5–31). Common midpoint reflection coverage of the area is then obtained by placing the source at each receiver point in turn and recording the reflections it produces at all the receivers. Through proper selection of the large number of source–receiver pairs, multifold reflection

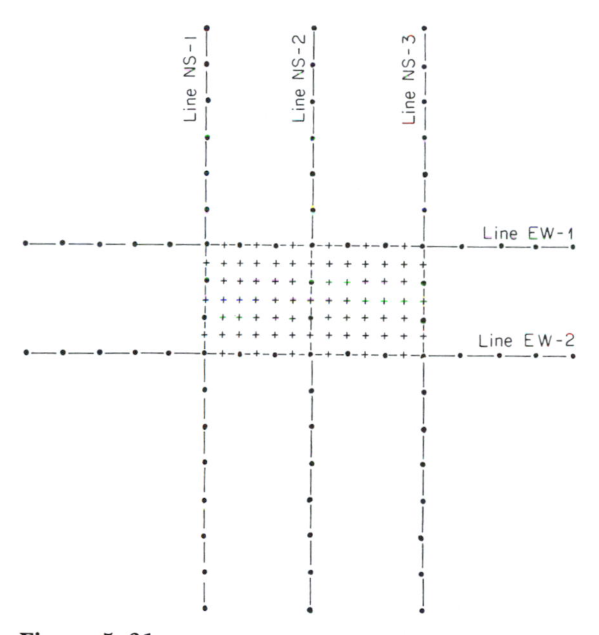

Figure 5–31
Diagram showing source and receiver locations (•) and common midpoints (+) for a three-dimensional seismic reflection survey.

coverage corresponding to all the common midpoints shown in Figure 5–31 becomes available.

Compared with seismic surveying along individual profiles, three-dimensional surveying obviously requires a tremendous increase in time, effort, and cost. For economical reasons, such surveys are confined to relatively small areas that have been identified by previous two-dimensional seismic reflection profiling as requiring special attention.

Three-dimensional reflection surveying can be most efficiently carried out at sea, for recording vessels can trail several parallel streamers and sources. Several vessels, each equipped like the one in Figure 5–32, can ad-

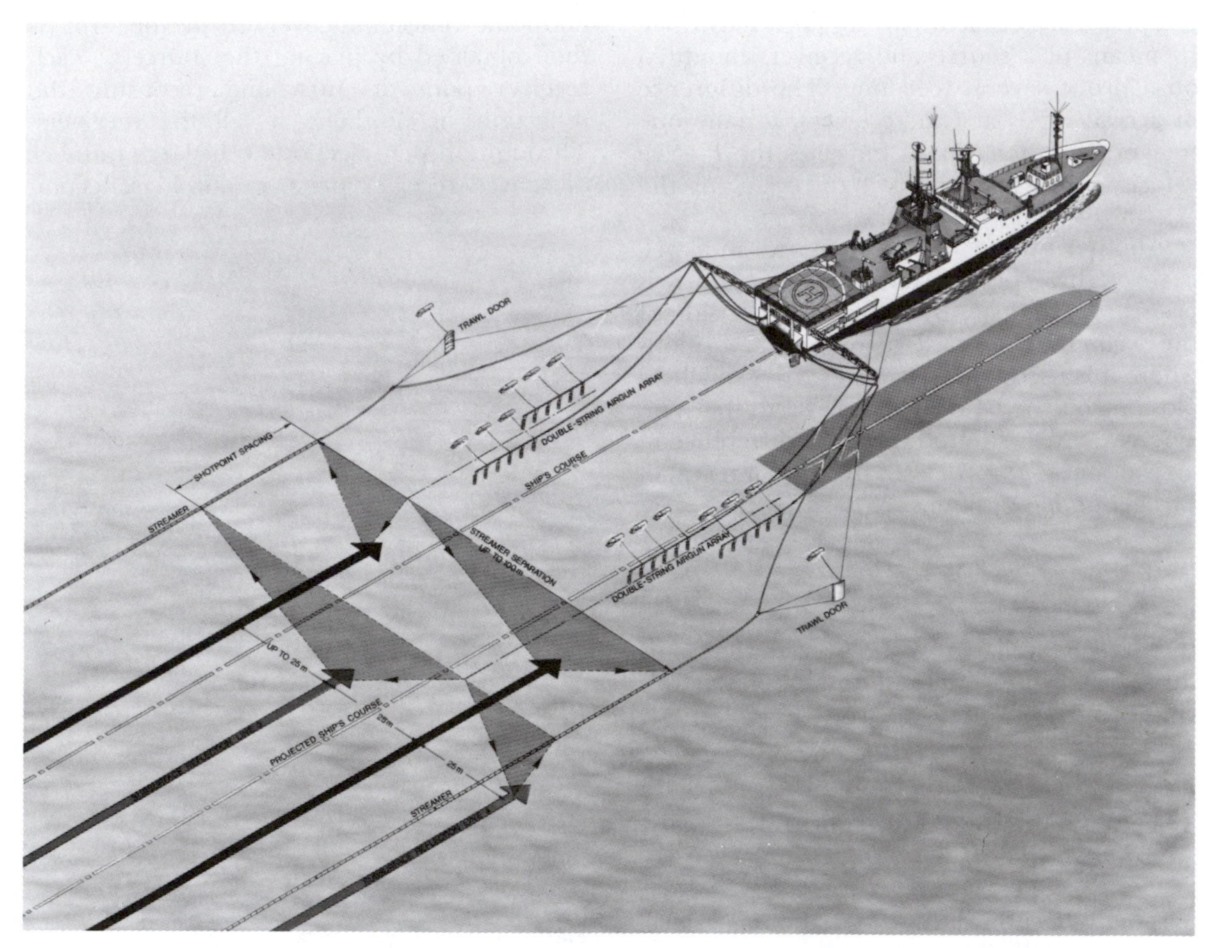

Figure 5–32
A vessel equipped with the multiple source arrays and steamers used for three-dimensional reflection surveying at sea. (Courtesy of Prakla–Seismos AG.)

vance together along parallel courses to survey rapidly a relatively large area. Much greater effort is necessary to obtain equivalent coverage on land, where each source and receiver must be handled individually.

Crooked-Line Reflection Surveying

Under ideal circumstances seismic reflection surveying would be done along straight profiles. For practical reasons most surveying is done along roads that wind through the countryside. These reflection surveys, then, are along profiles that follow all the bends and curves in the road.

It is not practical to obtain uniform common midpoint reflection data along a crooked profile. Figure 5–33 illustrates why this is true. Here we see that different source–receiver pairs do not have common midpoints along the profile. For a profile that curves only slightly, these individual midpoints may be grouped closely enough to neglect the small differences in position and to assume that the center of the group is the common midpoint. But for more irregular profiles, where individual midpoints are widely scat-

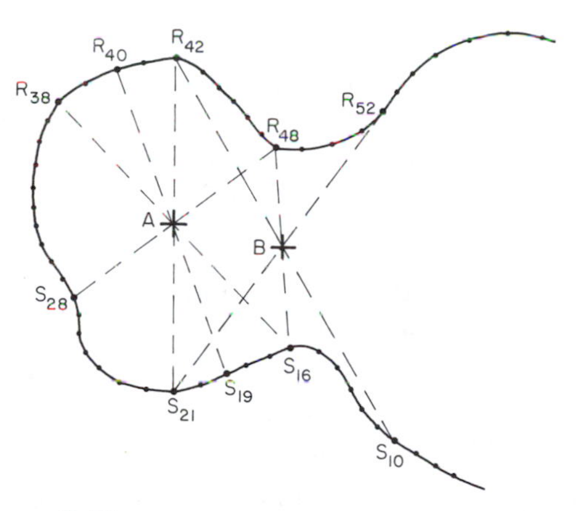

Figure 5–33
Source–receiver points and two common midpoints (A and B) for a crooked-line seismic reflection profile. Notice that the common midpoints obtained with source–receiver pairs do not lie along the source–reciever profile, and they do not have the same reflection coverage.

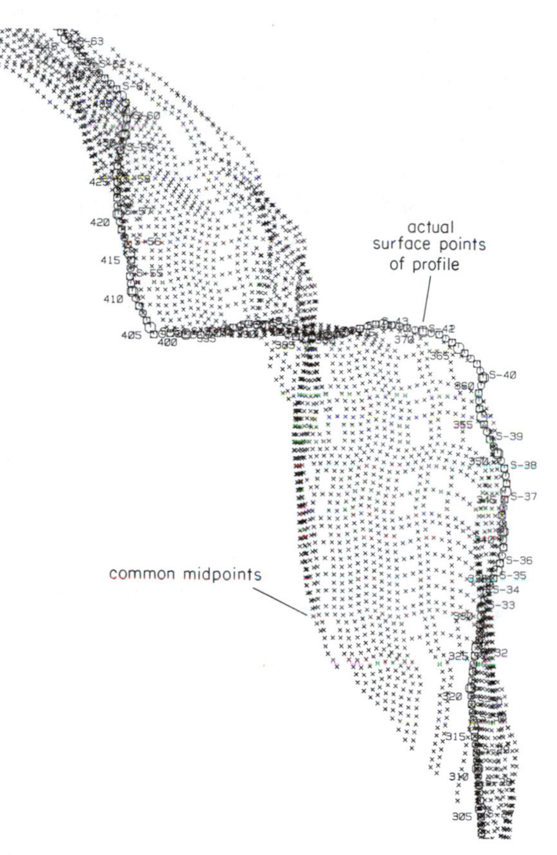

Figure 5–34
Source–receiver positions (□) and midpoints (x) for individual source–receiver pairs for a crooked-line seismic reflection survey done in South Carolina. Recording was done with a 120-channel split spread.

tered, special methods of crooked-line analysis must be used.

The example in Figure 5–33 illustrates some problems and some advantages of crooked-line seismic reflection profiling. Here we see that when all source–receiver combinations are tested, some are found to have common midpoints. The example shows that point A is the common midpoint for four different source–receiver pairs, but that point B is the common midpoint for only three pairs. This discrepancy illustrates the problem of nonuniform coverage along a crooked profile, that is, fourfold coverage at A, but only threefold coverage at B.

An advantage can also be recognized, however. Points A and B and other common midpoints are distributed over an area rather than being aligned in a straight profile. This shows that "accidental" three-dimensional coverage has been obtained. These crooked-line data may provide more information about the orientation of a dipping reflector than would have been obtained from a straight profile.

The coverage of an actual crooked-line seismic reflection survey done in South Carolina with a 120-channel recording system is shown in Figure 5–34. All source–receiver positions and all possible midpoints for individual source–receiver pairs are shown. Variations in multifold coverage are indicated by the grouping of the midpoints, which is denser where higher-fold coverage is obtained. Figure 5–34 illustrates the extent of accidental three-dimensional reflection coverage available from this crooked-line reflection survey.

STUDY EXERCISES

1. Design the field layout for a reflection survey to produce data with continuous subsurface coverage from refracted waves. If a receiver interval of 50 m is used with a 48-channel recording system, what distance between shot points will give continuous coverage?

2. Using the same system and receiver interval, design the field layout for a reflection survey that will produce continuous singlefold subsurface coverage. Define the distance between shot points.

3. Suppose that a 120-channel recording system is used in a splitspread fashion with a 50-m geophone interval. What is the maximum multiplicity (fold) obtainable? If 30-fold coverage is desired, what distance must the recording spread be moved after each shot? Determine the interval between common reflecting points. Prepare a subsurface stacking chart to define the shot and receiver locations related to the common reflection points.

4. Suppose that vibroseis will be used in a reflection survey. Maximum depth of investigation is 5000 m. The average velocity down to this depth is about 4000 m/s, and the instrument is capable of recording up to 10 s. What is the maximum sweep length we can use?

5. Suppose that we need high-resolution reflection data to recognize relatively thin layers. If the instrument has a minimum of 2 milliseconds sampling time, what is the highest frequency that can be recorded?

6. Assume that a vibroseis energy source generates

surface waves with dominant velocity V_{sr} = 800 m/s and dominant frequency f_{sr} = 20 Hz. What is the wavelength for the surface waves? If a receiver array is to be used, define a minimum length for the array.

7. Suppose that a spread of 48 geophones at 50-m intervals is used for a reversed reflection survey. Sources are offset 50 m from the end points. If a sloping reflector is at depths of 1000 m and 2000 m below the shot points (see figure), graphically determine subsurface coverage on the reflector. In other words, what is the length of the profile of reflection points and the interval of these reflection points?

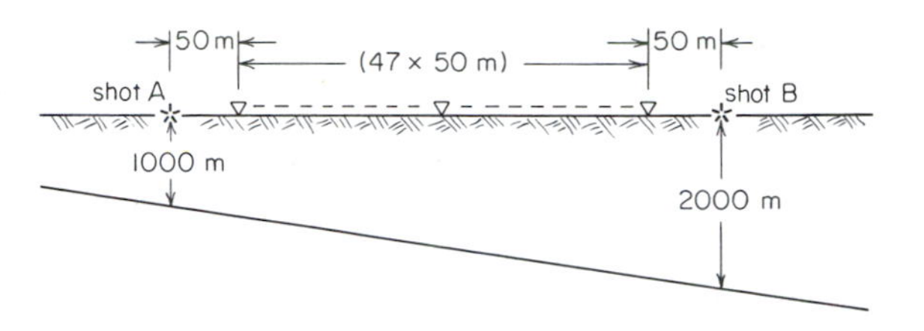

8. Suppose that the reflector configurations below a river have important implications for your survey objectives. You have a land crew to carry out the survey. Design a special procedure to record reflections below the river.

SELECTED READING

Anstey, N. A., *Signal Characteristics and Instrument Specifications.* Volume 1: *Seismic Prospecting Instruments.* Berlin, Gebruder Borntraeger, 1970.

Barbier, M. G., and J. R. Viallix, Sosie—A new tool for marine seismology, *Geophysics,* v. 38, n. 3, pp. 673–683, June 1973.

Barry, K. M., D. A. Cavers, and C. W. Kneale. Recommended standards for digital tape formats, *Geophysics,* v. 40, n. 2, pp. 344–352, April 1975.

Coffeen, J. A., *Seismic Exploration Fundamentals.* Tulsa, Okla., Petroleum Publishing Co., 1978.

Dennison, A. T., The design of electromagnetic geophones, *Geophysical Prospecting,* v. 1, n. 1, pp. 3–28, 1953.

Dix, C. H., *Seismic Prospecting for Oil.* New York, Harper, 1952.

Dix, C. H., Seismic velocities from surface measurements, *Geophysics,* v. 20, n. 1, pp. 68–86, February 1955.

Dobrin, M. B., *Introduction to Geophysical Prospecting,* 3rd edition. New York, McGraw-Hill, 1976.

Evenden, B. S., and D. R. Stone, *Instrument Performance and Testing.* Volume 2: *Seismic Prospecting Instruments.* Berlin, Gebruder Borntraeger, 1971.

Galperin, E. L., *Vertical Seismic Profiling* (translated by A. J. Hermont). Tulsa, Okla., Society of Exploration Geophysicists, 1974.

Goupillaud, P. L., Signal design in the Vibroseis technique, *Geophysics,* v. 41, n. 6, pp. 1291–1304, December 1976.

Green, C. H., John Clarence Karcher, 1984–1978, Father of the reflection seismograph, *Geophysics,* v. 44, n. 6, pp. 1018–1021, December 1979.

Kearey, P., and M. Brooks, *An Introduction to Geophysical Exploration.* Oxford, England, Blackwell Scientific Publication, 1984.

Mayne, W. H., Practical considerations in the use of common reflection point techniques, *Geophysics,* v. 32, n. 2, pp. 225–229, April 1967.

Mayne, W. H., and R. G. Quay, Seismic signatures of large air guns, *Geophysics,* v. 36, n. 6, pp. 1162–1173, December 1971.

McKay, A. E., Review of pattern shooting, *Geophysics,* v. 19, n. 3, pp. 420–437, June 1954.

Mossman, R. W, G. E. Heim, and F. E. Dalton, Vibroseis applications to engineering work in an urban area, *Geophysics,* v. 38, n. 3, pp. 489–499, June 1973.

Olhovich, V. A., The causes of noise in seismic reflection and refraction work, *Geophysics,* v. 29, n. 6, pp. 1015–1030, December 1964.

Parasnis, D. S., *Principles of Applied Geophysics,* 4th edition. London, Chapman and Hall, 1986.

Parr, J. O. Jr., and W. H. Mayne, A new method of pattern shooting, *Geophysics,* v. 20, n. 3, pp. 539–564, June 1955.

Sharma, P., *Geophysical Methods in Geology.* Amsterdam, Elsevier, 1986.

Sheriff, R. E., *Encyclopedic Dictionary of Exploration Geophysics.* Tulsa, Okla., Society of Exploration Geophysicists, 1973.

Sheriff, R. E., and L. P. Geldart, *Exploration Seismology.* Volume 1: *History, Theory, and Data Acquisition.* Cambridge, England, Cambridge University Press, 1982.

Slotnick, M. M., A graphical method for the interpretation of refraction profile data, *Geophysics,* v. 15, n. 2, pp. 163–180, April 1950.

Slotnick, M. M. *Lessons in Seismic Computing.* Tulsa, Okla., Society of Exploration Geophysicists, 1959.

Telford, W. M., L. P. Geldart, R. E. Sheriff, and D. A. Keys, *Applied Geophysics.* Cambridge, England, Cambridge University Press, 1976.

SIX

Seismic Reflection Data Processing and Interpretation

The success of modern seismic reflection surveying in imaging earth structure can be directly attributed to digital technology. We have already discussed one part of this technology, which is the acquisition of data by means of digital recording equipment. To bring meaning to the digital data recorded by seismic crews, we must make use of computer-based data processing.

The first step is to sort out data for individual seismograms from a multiplexed digital magnetic tape. This task of demultiplexing the field recordings would be impractical without the help of a computer. For vibroseis data, an additional processing procedure called *correlation* must be undertaken to obtain recognizable seismograms.

The seismograms obtained by these initial data processing operations appear to be the same as the seismograms obtained with analog equipment. But today much can be done with them. The computer is essential for accomplishing three important kinds of operations.

1. Stacking operations that combine groups of seismogram traces, and digital filtering operations that improve reflections and reduce noise.
2. Preparation of seismic sections that image many features of earth structure by means of computer-controlled plotting machines.
3. Calculations of seismic wave velocity, layer thicknesses, and other physical properties.

Modern data processing does not eliminate the need for judgments by seismologists. But it does provide the means to assemble far more information in forms that can be examined more systematically and efficiently than would be possible without digital technology. This means that more informed judgments can be made.

SEISMOGRAM PREPARATION

Demultiplexing

Most reflection seismic data are now being recorded on digital magnetic tape. Unlike analog seismograms, which are recorded in a form suitable for analysis, digital seismograms must be assembled from the digital tape by a sorting process. This sorting process, which was introduced briefly in Chapter 5, is called *demultiplexing*.

Let us look again at the form of the data on a digital magnetic tape. Suppose we have used a recording system with n channels and that each channel has been sampled m times. We can describe each value, A_{ij}, on the field tape by the subscript i, which refers to the channel, and the subscript j, which is the number of the point in the sequence of values obtained from that channel. For example, the sixth value recorded on channel 8 will be A_{86}. The digital recording system puts the values in the following order:

$$\begin{aligned}&A_{11}A_{21}A_{31}\ldots A_{i1}A_{i+1,1}\ldots A_{n1}A_{12}A_{22}A_{32}\ldots A_{i2}A_{i+1,2}A_{n2}A_{13}A_{23}\ldots\\&A_{i3}A_{i+1,3}A_{n3}\ldots A_{1j}A_{2j}A_{3j}\ldots A_{ij}A_{i+1,j}\ldots A_{nj}A_{1,j+1}A_{2,j+1}\\&A_{3,j+1}\ldots A_{i,j+1}A_{i+1,j+1}A_{n,j+1}\ldots A_{1m}A_{2m}A_{3m}\ldots A_{nm}\end{aligned} \tag{6–1}$$

Data in this form are demultiplexed by a computer programmed to sort these values into the sequences for separate channels. First, the computer reads the entire digital tape. Then, it repeatedly stores a value, counts past the next m values, stores another value, counts past another m value, and so on until the following sequences have been compiled separately:

$$
\begin{array}{l}
A_{11}A_{12}A_{13} \ldots A_{1m} \\
A_{21}A_{22}A_{23} \ldots A_{2m} \\
\\
A_{n1}A_{n2}A_{n3} \ldots A_{nm}
\end{array}
\qquad (6\text{–}2)
$$

Now the data have been demultiplexed, and we have a digital seismogram for each channel. Each digital seismogram is simply a list of numbers that are stored to await further processing. At this point, the seismologist usually wants to examine the data in a graphical form that looks like a conventional analog seismogram. The seismologist may obtain such a seismogram by programming a digital plotter to plot each value A_{ij}, which is the seismogram amplitude at a particular time, in sequence, and to connect the plotted points by lines. Other lines are plotted at appropriate spacing to represent timing lines. The result is a graphical display of the digital seismogram, as in Figure 6–1, prepared automatically by the computer and digital plotter.

Demultiplexing can be done with most digital computers that have sufficient memory to store all the data from the original digital magnetic tape and the individual channel sequences. In petroleum exploration for which impulsive sources are used, recording ordinarily continues for about 5 seconds. If a 96-channel system is used and each channel is sampled at 2-millisecond (0.002 second) intervals, there will be a total of 240,000 values per field record. Because these values must be stored in both multiplexed and demultiplexed form, there must be space for 480,000 values.

Considerably more computer memory is needed to store vibroseis data. Depending on the duration of the sweep time, it is not unusual to record for 20 seconds or more. Then about 2 million values would have to be stored, which would require a reasonably large computer.

Vibroseis Correlation

A seismogram from an impulsive source consists of a series of pulses indicating reflected and refracted waves that follow different paths from the source to the receiver. This concept was introduced in Chapter 2, where it is pointed out that each of these pulses has the same shape as the pulse originally released from the source. That original pulse is called the *source wavelet.*

If the differences in arrival times of the reflected and refracted waves are greater than the time duration of the source wavelet, the pulses on the seismogram will be clearly separate from one another, as in Figure 2–23. However, when the time intervals between successive arrivals are shorter than the duration of the wavelet, the pulses overlap one another. Then the onsets of the arrivals can be difficult to identify.

On an ordinary seismogram we might expect to see reflections arriving at time intervals of several tens to a few hundreds of milliseconds. Because the wavelet from an impulsive source usually has a duration of a few tens of milliseconds, most of these reflections can be distinguished from one another. But what happens if we have a vibroseis sweep rather than a brief impulsive source wavelet? Then, depending on the sweep time, the "wavelet" lasts several seconds, perhaps as long as 15 seconds. It is obvious that all the reflected and refracted signals in a vibroseis seismogram overlap one another extensively.

After vibroseis digital magnetic tapes have been demultiplexed and the results plotted, it

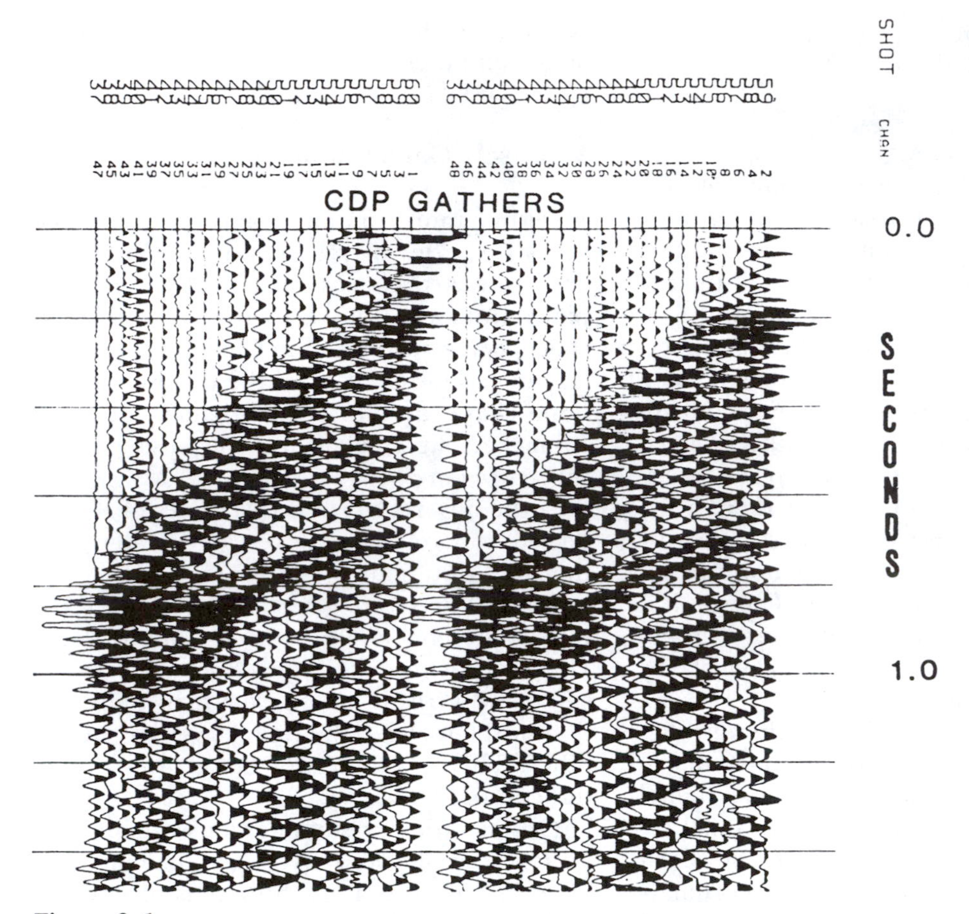

Figure 6–1
Digital seismogram prepared from demultiplexed magnetic tape data and plotted as wiggly lines with variable-width shading. The right half of each cycle of oscillation is shaded black. Such shading emphasizes the reflected arrivals on the seismogram.

is still impossible to recognize any reflections. Additional computer processing is needed to obtain a recognizable seismogram. This procedure, called *vibroseis correlation* enables us to extract from each of the long overlapping sweep signals a short wavelet much like those obtained with impulsive sources.

Let us proceed step by step with the vibroseis correlation of an idealized vibroseis seismogram trace. For this first simple example, we see in Figure 6–2 that the sweep signals corresponding to different arrivals do not overlap. In digital form, this seismogram trace is expressed by the sequence of values

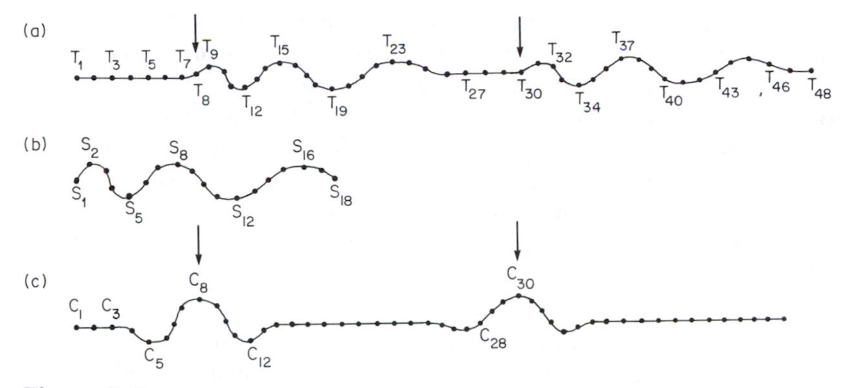

Figure 6–2

Idealized example of the vibroseis correlation procedure. (a) The seismogram recorded by a geophone can be represented digitally by the sequence of numbers $T_1, T_2, \ldots, T_{48}$. The two reflections on this trace begin at T_8 and T_{30}. For purposes of this simple example, these reflections were chosen far enough apart that they do not overlap. (b) The vibroseis input sweep signal can be represented digitally by the numbers $S_1, S_2, \ldots, S_{18}$. Notice that this is a downsweep signal. (c) The correlated seismogram trace, or correlogram, can be represented digitally by the numbers $C_1, C_2, \ldots, C_{40}$. It is obtained by the process of cross multiplication and summing the digital seismogram trace and the digital sweep signal. The correlogram therefore displays two reflections which are indicated by Klauder wavelets.

$T_1T_2T_3T_4, \ldots, T_{48}$, as indicated in the illustration. Below the trace the sweep input signal is shown. In digital form, it is represented by the sequence of values $S_1S_2S_3S_4, \ldots, S_{18}$.

We begin the correlation process by cross-multiplying, point by point, the first 18 values of the seismic trace with the values for the sweep. Then we sum these products to obtain the first value C_1 for the *correlated seismogram trace,* which is also shown in Figure 6–2:

$$C_1 = T_1 \times S_1 + T_2 \times S_2 + T_3 \times S_3 \ldots T_{18} \times S_{18}$$

Then we advance one point on the original trace and cross-multiply points T_2 through T_{19} with the sweep values. We sum these products to obtain the second point on the correlated seismogram trace,

$$C_2 = T_2 \times S_1 + T_3 \times S_2 + T_4 \times S_3 \ldots T_{19} \times S_{18}$$

We calculate each additional value for the correlated trace by advancing one point on the original trace, cross-multiplying, and summing:

$$C_3 = T_3 \times S_1 + T_4 \times S_2 + T_5 \times S_3 \ldots T_{20} \times S_{18}$$

$$C_4 = T_4 \times S_1 + T_5 \times S_2 + T_6 \times S_3 \ldots T_{21} \times S_{18}$$

$$C_i = T_i \times S_1 + T_{i+1} \times S_2 + T_{i+3} \times S_3 \ldots T_{i+17} \times S_{18} \qquad (6\text{–}3)$$

When we plot the values obtained in this way, we see an interesting result. Instead of a long sweep signal, we obtain a simple wavelet centered at the arrival time for each wave indicated on the original trace. What has happened is that during the correlation procedure, a large value is obtained by cross multiplication and summing where the sweep input signal is aligned with a signal having that same form on the original trace. But for positions where the sweep does not align with a similar form, the peaks and troughs of the input signal interfere destructively with those on the trace. Then the peaks on one overlap troughs on the other, causing cancellation of signal.

Vibroseis correlation produces the same result for overlapping signals on a seismogram trace, as illustrated by the example in Figure 6–3. The frequency of the resulting wavelet is equal to the middle frequency of the sweep signal. The particular form of wavelet produced by vibroseis correlation is called a *Klauder wavelet.* On a correlated seismogram trace, otherwise called a *correlogram,* the reflected and refracted arrivals appear as Klauder wavelets.

By plotting all the correlogram traces obtained from a multichannel recording, we obtain a seismogram with the appearance of one recorded from an impulsive source. As before, reflected waves align on hyperbolic arcs, and refracted arrivals have straight alignments. There is one important difference. Arrival times on a correlogram are indicated by the center peak of the Klauder wavelet, rather than by the onset of the wavelet as is the case for impulsive sources.

In practice, vibroseis correlation involves the processing of a much larger number of points than is shown in our idealized examples. If the sweep time is 10 seconds and samples are recorded at 2-millisecond intervals, each point on a correlogram trace requires 5000 cross multiplications. To obtain 96 cor-

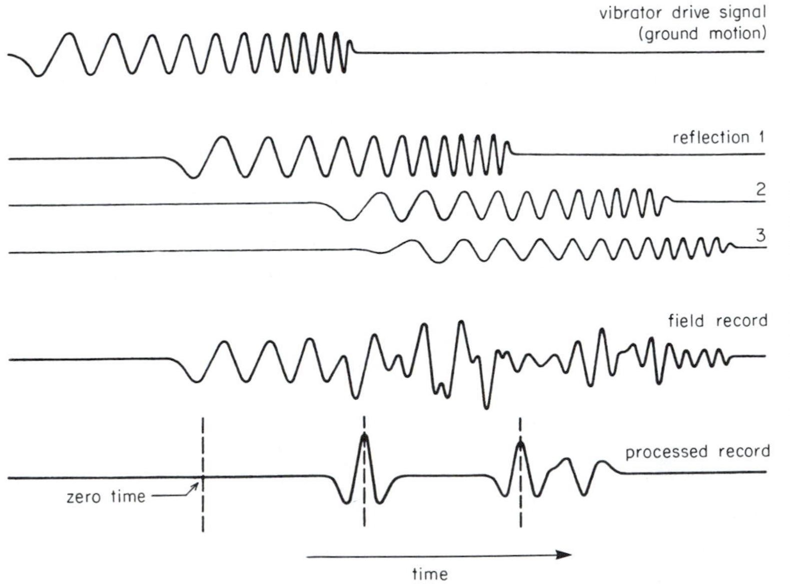

Figure 6–3
Correlation of a vibroseis seismogram trace that consists of three overlapping reflections. The sweep signal is shown at the top. Below it are three traces that show the three reflections separately. The seismogram trace is the sum of these three separate reflection traces. After correlation of this seismogram trace with the sweep signal, the correlogram trace is obtained, on which the three reflections are indicated by non-overlapping Klauder wavelets.

relogram traces, each consisting of 2500 values, we must perform a total of 1.2×10^9 multiplications and 1.2×10^9 additions. Obviously, this work can be done only by means of a large computer. After this processing, vibroseis data are in a form suitable for further processing and analysis.

Source Array Summing

Chapter 5 described how source arrays are commonly used in seismic reflection surveying to reduce noise and strengthen reflected signals. The usual practice is to add together the contributions of the sources in the array using temporary memory units attached to the field recording system. For special purposes, however, it is sometimes useful to record each source separately and to combine the results later at a data processing center. Then the parts of the array can be tested in different combinations to see whether one gives superior results.

After the data have been demultiplexed, computer summing of digital traces obtained for different sources is a simple procedure. Consider the digital traces recorded on one channel from three different sources that are expressed by the sequences

$$\begin{array}{c} E_1, E_2, E_3, \ldots, E_m \\ F_1, F_2, F_3, \ldots, F_m \\ G_1, G_2, G_3, \ldots, G_m \end{array} \qquad (6\text{–}4)$$

The trace produced by an array consisting of these three sources is simply

$$(E_1 + F_1 + G_1), (E_2 + F_2 + G_2), \ldots, (E_m + F_m + G_m) \qquad (6\text{–}5)$$

When vibroseis data are being processed, this kind of summing can be done before or after the correlation process.

STACKING PROCEDURES FOR DATA ENHANCEMENT

The process of stacking is our single most powerful tool for enhancing the quality of seismic reflections while at the same time reducing the level of noise. It is a simple process of adding together seismogram traces obtained from different sources and receivers. First, the traces must be selected and adjusted to ensure, insofar as possible, that reflections on all the traces, if they exist, have the same arrival times. Then, when the traces are combined, the individually reflected wavelets add together to produce a stronger signal. Noise, on the other hand, tends to be different from trace to trace, so that it is reduced by destructive interference when the traces are combined.

Stacking first became practical when analog magnetic tape recording was introduced in the 1950s. It was done electronically by playing back the taped signals from different traces into a recorder that combined them. Analog electronic stacking has been almost entirely replaced by digital stacking, which offers greater dynamic range and processing flexibility.

One or the other of two considerations is the basis for selecting the seismogram traces that will be stacked. *Common offset stacking* is done with traces that have the same source–receiver offset distance. *Common midpoint stacking* utilizes traces with different source–receiver offsets, all of which are centered on the same point. These two configurations are compared in Figure 6–4.

Traces for common offset stacking can be obtained either from single coverage or from the common depth point (CDP) reflection profiling procedures described in Chapter 5. The first efforts to employ stacking made use of common offset traces. In this connection,

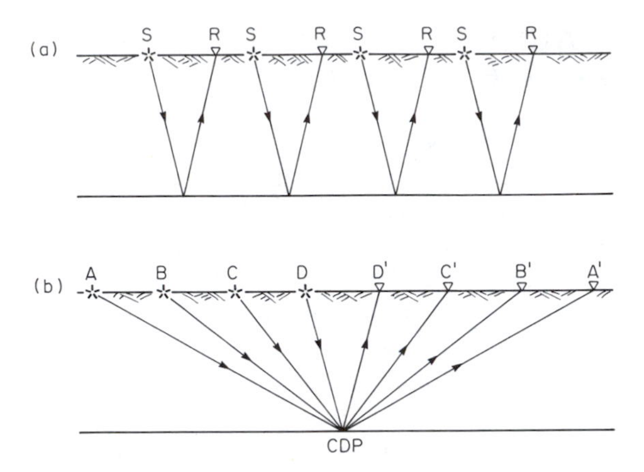

Figure 6–4
Reflection travel paths for (a) common offset and (b) common midpoint gathers. A common offset data gather consists of seismogram traces from source–receiver pairs that have the same offset but have recorded reflections from different points on the reflector. A common midpoint gather consists of traces from source–receiver pairs that have different offsets but are symmetrical about the same midpoint and have reflections from the same common reflecting point.

two problems must be considered. First, the reflections being stacked come from different points on a reflector. If depths to these points turn out to be different, the arrival times will also be different, so that the reflections are not well aligned for stacking. The second problem is that if reflection arrivals turn out to be identical, the arrivals of refracted waves and surface waves will also be identical. Therefore, this stacking procedure is effective only for reducing incoherent noise and tends to enhance coherent noise. Because of these problems, common offset stacking has been largely abandoned in favor of common midpoint stacking.

Common midpoint stacking is widely used today in seismic reflection data processing. This procedure consists of several steps. First, the traces to be stacked must be sorted out from the digital seismograms. Then static corrections must be applied that are similar to those described in Chapter 3 for refraction analysis. Because of the different offsets, reflection arrival times vary from trace to trace. These must be adjusted to common arrival times before stacking can commence. To do so, we must first determine root-mean-square (RMS) velocities and then apply normal moveout corrections. Finally, using the velocity information, amplitude adjustments that take into consideration geometrical spreading and absorption along the different reflection paths are made. After following these steps, which we will now describe in more detail, we can stack traces and prepare seismic sections.

Common Midpoint Sorting

Each trace to be included in a common depth point stack comes from a different digital seismogram. A large number of digital seismograms can be stored in computer memory after they have been prepared by demultiplexing. The process of selecting a set of traces to be stacked is called *sorting*. The set of traces sorted for a particular CDP stack is called the *CDP gather*. Most CDP reflection profiling is done with source and receiver arrays. For purposes of sorting, a source–receiver offset is measured from the center of the source array to the center of the receiver array.

Sorting a CDP gather can be done in a systematic way. The simple example in Figure 6–5 illustrates reflection paths for traces on a single seismogram and for traces from several seismograms that are included in fourfold

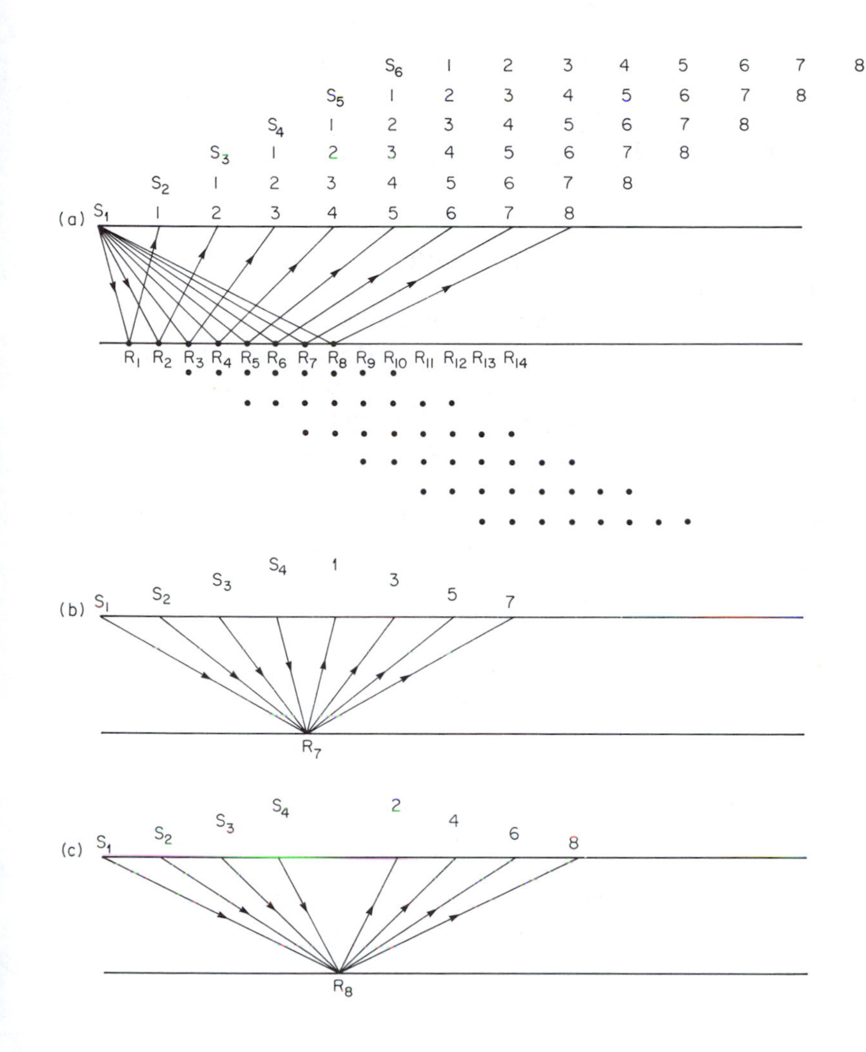

Figure 6–5
Reflection travel paths for seismogram traces to be included in a succession of CDP gathers. (a) Recording procedure with an eight-channel system to obtain four-field data. Six source–receiver spreads are shown from which six seismograms are recorded. Numbers identify the recording instrument channels connected to the geophones that make up these different spreads. Rows of points indicate the horizontal positions of reflection points for each spread. Such a diagram of points is called a stacking chart. It indicates the reflection coverage by the number of points appearing in a vertical column. For example, four points are arranged vertically at reflector positions R_7 and R_8 (b, c). These reflection paths indicate fourfold coverage for these points.

gathers. Observe that for a source at S_1, the interval between the reflection points R_1, R_2, . . ., R_8 is one-half the receiver spacing.

Reflection paths for sources S_2 . . . S_6 are not drawn in Figure 6–5, but these sources and receiver positions are indicated above the structure, and their reflection points are shown below the structure, set apart from one another for clarity. Looking carefully at the sequence of spread positions, we see that source–receiver paths

$$S_1–7, S_2–5, S_3–3, S_4–1 \qquad (6–6a)$$

all reflect from the common depth point R_7. These indicate the common midpoint traces that would be sorted to obtain the fourfold gather for the common depth point R_7, as well as underlying common depth points on any deeper reflectors. Similarly, another fourfold gather for the common depth point R_8 includes the traces

$$S_1\text{–}8,\ S_2\text{–}6,\ S_3\text{–}4,\ S_4\text{–}2 \qquad (6\text{–}6b)$$

Additional common depth points in Figure 6–5 are indicated by sets of points that are vertically set apart from one another in the drawing. They indicate the following trace gathers for common depth points R_9 through R_{14}:

for R9	S_2–7, S_3–5, S_4–3, S_5–1
for R10	S_2–8, S_3–6, S_4–4, S_5–2
for R11	S_3–7, S_4–5, S_5–3, S_6–1
for R12	S_3–8, S_4–6, S_5–4, S_6–2
for R13	S_4–7, S_5–5, S_6–3, S_7–1
for R14	S_4–8, S_5–6, S_6–4, S_7–2

(6–7)

Here we see that for alternate common depth points, the gathers consist of odd- and even-numbered traces.

The patterns indicated in Figure 6–5 and expression 6–7 indicate the kinds of sequences to be followed in sorting gathers with many more traces. If a 48-channel recording system is used, the 24-fold odd and even gathers for several common midpoints will be

S_1–47, S_2–45, ... S_{23}–3, S_{24}–1
S_1–48, S_2–46, ... S_{23}–4, S_{24}–2
S_2–47, S_3–45, ... S_{24}–3, S_{25}–1
S_2–48, S_3–46, ... S_{24}–4, S_{25}–2
S_3–47, S_4–45, ... S_{25}–3, S_{26}–1
S_3–48, S_4–46, ... S_{25}–4, S_{26}–2

(6–8)

and so on.

In a data processing center, the digital seismograms with the traces needed for stacking are read into the computer memory in sequence, and each digital trace is assigned a sequential code number. The computer is then programmed to scan the memory unit in order to sort the traces for each gather, which are then copied in a group onto another memory region. Each gather is then ready for further processing.

Static Corrections

Variations in elevation and the thickness of the low-velocity weathered zone introduce irregularities in the arrival times of seismic waves at the different receivers along the spread. This subject is discussed in Chapter 3 where a procedure for adjusting the arrival times of refracted wave is presented. Seismic reflection data must be adjusted in a similar way.

Factors to be considered for static corrections of reflection data are illustrated in Figure 6–6. The object is to adjust arrival times on all traces to values that would be measured if the sources and receivers were on the same horizontal datum surface. Because velocity V_0 in the weathered zone is usually much lower than the underlying bedrock velocity V_1, the travel paths through the weathered zones tend to be very steep. For practical purposes of making static corrections, the common practice is to assume that these paths are vertical. That assumption is discussed in more detail in Chapter 3.

Static corrections for both surface source and receiver points are calculated in the same way. Referring to Figure 6–6a, suppose that

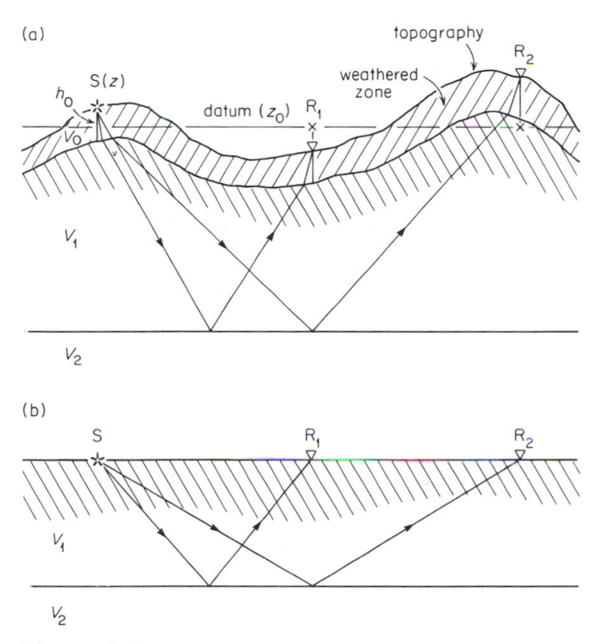

Figure 6–6

Travel paths and vertical projections involved in static corrections of seismic reflection data. Because of undulating topography and changes in the velocity and thickness of the weathered zone, reflection times are ordinarily adjusted to a horizontal reference surface. The static correction can be viewed as (a) replacing the weathered zone velocity (V_0) with bedrock velocity (V_1) and (b) then reducing source and receiver points to the same horizontal datum.

the elevation of such a point is z and that the elevation of the datum surface is z_0. Furthermore, let the thickness of the weathered zone be h_0. The static correction Δt is the travel time along a vertical path from the point to the datum surface:

$$\Delta t = \frac{h_0}{V_0} + \frac{z - z_0 - h_0}{V_1} \qquad (6\text{–}9)$$

This expression can be rewritten in the form

$$\Delta t = t_e + t_w \qquad (6\text{–}10)$$

where

$$t_e = \frac{z - z_0}{V_1} \qquad (6\text{–}11)$$

and

$$t_w = h_0\left(\frac{1}{V_0} - \frac{1}{V_1}\right) \qquad (6\text{–}12)$$

The terms t_e and t_w are called the *elevation correction* and the *weathering correction,* respectively.

Be sure to note that the elevation correction is made with the bedrock velocity. This implies that after static corrections have been made, the weathered zone is removed computationally, and the zone between the horizontal datum and the bottom of the weathered layer is filled with a material of velocity V_1, as shown in Figure 6–6b.

For each trace in a CDP gather, a static correction is applied for both the source and the receiver positions. As we see in expression 6–8, the same source and receiver positions are used repeatedly in different gathers. Because vertical travel paths are assumed, the same static correction is applied for a particular point whenever it is used. We know that as CDP reflection profiling is underway, each point is alternately a source point and a receiver point. Therefore, the static correction is calculated by means of Equation 6–10 only once for each point along the survey profile. That value can then be used repeatedly as the point appears in different gathers.

The choice of the datum surface must be given some thought. Because vertical travel paths are assumed, the static corrections are only approximations of the true travel time irregularities. Errors related to this assumption are reduced by choosing the datum surface so that the points lie more or less equally above and below. However, for long profiles along which there is a regional change in average elevation, it may be necessary to choose different datum surfaces for different parts of the profile. Later, these differences can be adjusted when a seismic section is being compiled for the entire profile.

Static corrections are applied to each digital trace by adding or deleting numbers at the beginning of the list of numbers that represents that trace, depending on the sign of the correction. This procedure is illustrated in Figure 6–7 for traces that correspond to the travel paths in Figure 6–6. Let us assume that these digital seismograms are represented by values at 2-millisecond intervals. Because the source lies above the datum surface, its static correction must be subtracted. Suppose that this static correction is 6 milliseconds. To account for it, we would delete the first three points, that is, the first three values from the list of numbers representing both traces.

Now look at receiver R_1 which lies below the datum surface. A static correction must be added to adjust the trace to the datum elevation. If the static correction happens to amount to 4 milliseconds, two numbers, both zeros, will be added to the beginning of the list. However, when we combine the source and receiver status, we see that the total correction is accomplished by deleting the first number from the list.

At receiver R_2, the static correction must be subtracted. Suppose that it turns out to be 8 milliseconds, which requires deletion of four values. These deleted values are combined with the three values removed for the source correction; we see that the combined static correction is made by deleting seven points from the beginning of the list. The adjusted digital seismograms are listed below the originals for both traces in Figure 6–7.

In hilly countryside, static corrections can differ by a few tens of milliseconds or more

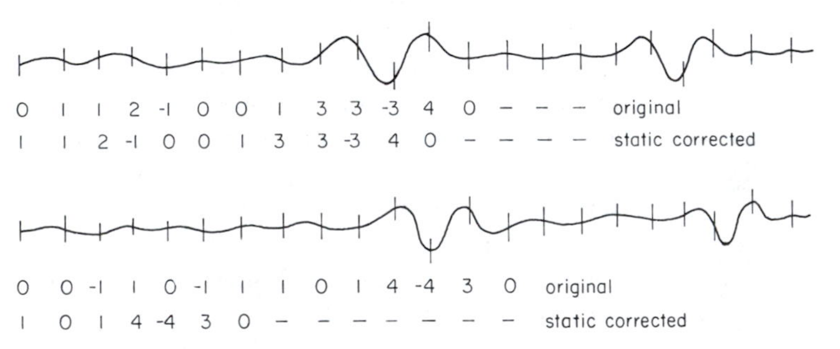

Figure 6–7
Application of the static correction to a digital seismogram trace is accomplished by adding or deleting values at the beginning of the sequence of numbers representing that trace. If the static correction is negative, numbers are deleted from the original sequence. In this example, one number (0) was deleted from the upper trace, and seven numbers (0, 0, −1, 1, 0, −1, 1) were deleted from the lower trace to make the static corrections. If the correction turns out to be positive, the appropriate number of zeros are inserted at the beginning of the sequence of numbers.

from trace to trace. It is essential to make these corrections for results of CDP stacking to be satisfactory. Having made these corrections, we can proceed to the next step of data processing.

Velocity Analysis

Because of the different source–receiver offsets of the traces in a CDP gather, the arrival time of the reflection from a common depth point will be different on each trace. Before the traces can be stacked, normal move-out corrections have to be applied to shift reflections on all traces to a common arrival time. However, we know from Equation 4–47 that arrival time differences depend on the RMS velocity as well as on the offset distance. How can we determine the RMS velocities needed for normal move-out corrections?

If reflections could be recognized on the traces, we could proceed as in Figure 4–20 to plot travel time curves and x^2–t^2 graphs from which RMS velocities can be obtained. Because of the very large number of seismograms and CDP gathers that must be processed and the difficulty in recognizing some reflections, it is impractical to expect a seismologist to measure arrival times and plot these graphs manually. Instead, a completely automated computer search is done to determine the appropriate RMS velocities. This is called *velocity analysis*.

A trial-and-error procedure is used for velocity analysis. The information needed for this analysis is illustrated in Figure 6–8, which shows a six-trace gather. The search is based on the idea that each point on each digital trace might be the onset of a reflection. Of course, most points are not, but the aim of the search is to discover the few that are.

To begin the search, we choose a starting point, say, t_{a0} on the zero-offset trace and a trial value V_k for the RMS velocity. Furthermore, we assume a length for the reflected

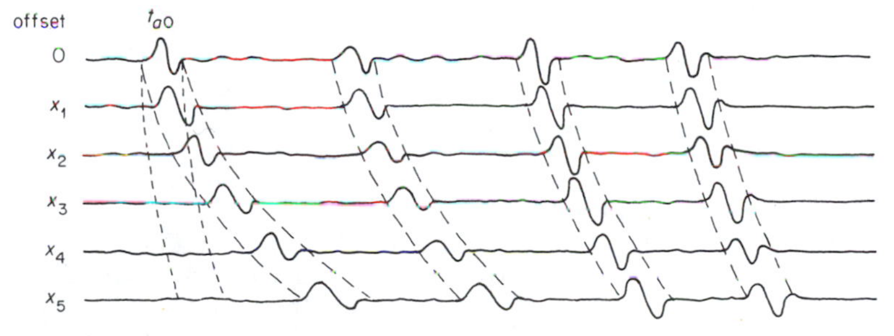

Figure 6–8

Arrangements of segments of digital seismogram traces that are tested by cross multiplication and summing during velocity analysis. The analysis is performed on traces in a common midpoint gather. Particularly high semblance coefficient values, obtained by trial calculations, indicate the sets of trace segments containing reflections from common depth points. The trial velocity corresponding to a high semblance coefficient is recognized to be the root mean square velocity in the zone above that reflector.

wavelet, for example, 20 milliseconds. Thus, if the sampling interval is 2 milliseconds, a wavelet beginning at t_{a0} would be represented by the following 10 points on the digital trace.

Next, we know that if t_{a0} is the zero-offset time for a reflection, the arrival times on the other traces are given by the expression

$$t_{ax} = \sqrt{t_{a0}^2 + x^2/V_k^2} \qquad (6\text{–}13)$$

which has the same form as Equation 4–47. These arrival times are indicated by a hyperbolic arc in Figure 6–8. A second parallel arc is then drawn 20 milliseconds to the right of the first one. The interval on each trace between these arcs is where the reflected wavelets should be found if, indeed, a reflection exists.

Now we cross-multiply the corresponding points of the trace segments between these arcs and sum the products to obtain a number C_{a1} which we call a *semblance coefficient.* If the trace segments have a similar form so that they closely resemble each other, the value of the semblance coefficient will be large. But if the segments are dissimilar, a small value will be found.

Next, we change the trial velocity by a small amount ΔV to get

$$V_{k+1} = V_k + \Delta V$$

Another set of arrival times is then calculated, and positions of another set of trace segments, shown by a second pair of hyperbolic arcs in Figure 6–8, are located. These are cross-multiplied, and the products summed to obtain another *semblance coefficient* C_{a2}. This process is repeated for a range of trial velocities to obtain more values in the series of semblance coefficients C_{a3}, C_{a4}, C_{a5}, . . ., C_{an}, all of which correspond to the zero-offset time of t_{a0}.

Now we can begin to prepare the *semblance coefficient map* in Figure 6–9. Observe how the

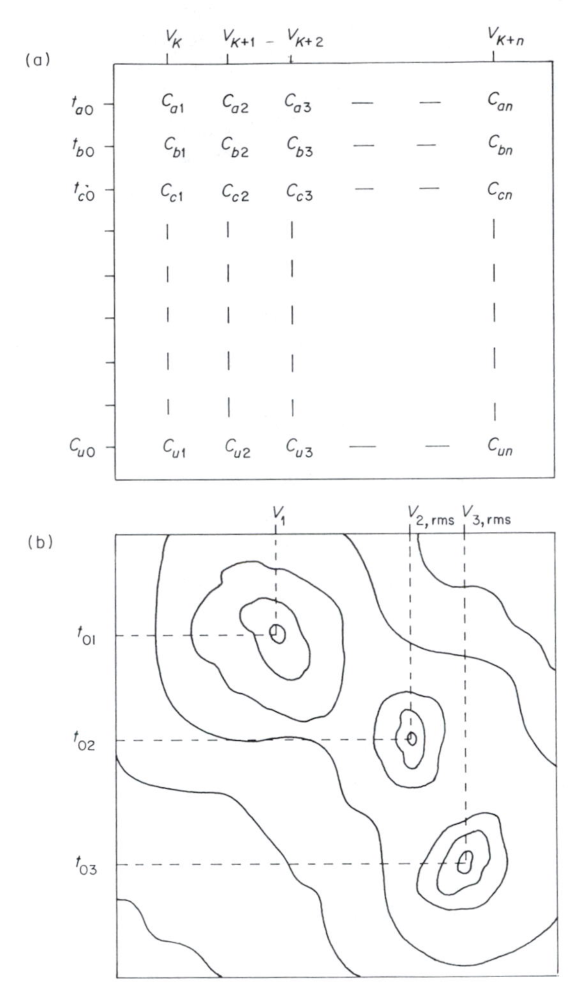

Figure 6–9
Semblance coefficient values (a) plotted on a graph at the corresponding time and velocity points and (b) illustrated by contours. Inner closed contours surround the highest values. These high values indicate the presence of reflections. From the positions of such points, root mean square velocity above the reflector and zero-offset reflection times can be read from the horizontal and vertical axes of the graph.

values C_{a1}, C_{a2}, . . ., C_{an} are plotted at their respective velocities along a line from the time t_{a0}.

To continue the search, we advance to the next point t_{b0} on the zero-offset trace, which we will now test as a possible zero-offset time for a reflection. Proceeding in the same way as before, we use the same trial velocities V_k, V_{k+1}, . . . in Equation 6–13 and calculate the series of semblance coefficients C_{b1}, C_{b2}, C_{b3}, . . ., C_{bn}. These are now plotted on the semblance coefficient map.

The search goes on, moving along point by point t_{c0}, t_{d0}, t_{e0}, . . . until we have tested the same range of trial velocities at all possible reflection arrival times. After each test calculation, the semblance coefficient is plotted on the map. When the search is completed, we have semblance coefficient values at an array of points equally spaced over the entire map.

The next step is to contour the semblance coefficients on the map. This process is handled in the same way that we would contour elevation values on a topographic map. Automatic contouring by means of a computer-controlled plotter is used for this purpose. The procedure is described in more detail in Chapter 11 where it is also applied in the preparation of magnetic field intensity contour maps.

What does a semblance coefficient contour map such as the one shown in Figure 6–9b tell us? Points marked at the centers of closed contours show the trial velocities and zero-offset times for which the semblance coefficient values are highest. For our example in Figure 6–8 where three reflections are obvious, the highest semblance coefficient values would be expected for the particular velocities and zero-offset times that align these wavelets most exactly. Elsewhere on these traces and at other trial velocities, the trace segments contain only noise. Cross multiplication and summing for such dissimilar trace segments yield low semblance coefficients that tell us there are no reflections corresponding to these combinations of velocity and zero-offset times.

By inspecting a semblance coefficient map, the seismologist can tell how many reflectors are present and the RMS velocities for waves reflected from them. The idealized example in Figure 6–9b shows three obvious patterns of closed contours that leave little doubt about the reflection times and velocities. More common are maps like the one in Figure 6–10, which was prepared from real seismic data. Here the patterns are more ambiguous, and the seismologist must use some judgment in selecting meaningful contour patterns. Closed contour patterns arranged on a line of constant velocity usually indicate multiple reflections that must be disregarded. Depending on how complicated the patterns turn out to be, some experience may be required to make sound judgments.

By comparing semblance coefficient maps from a series of adjacent CDP gathers, we can recognize which contour patterns are most consistent from one map to another. In this way, we can decide which ones are reliable indications of reflections. It is not unusual to detect the presence of weak reflections in this way, which are not obvious simply from inspection of the seismogram traces.

The velocity analysis method we have just described is sometimes called *hyperbolic scanning* or *velocity scanning*. Because of the very large number of trial calculations, a high-speed computer is essential for this kind of data processing. To a large extent, it lets the computer identify the reflections and calculate the RMS velocities. In this way a large volume of seismic reflection survey data can be analyzed and the results presented in a concise

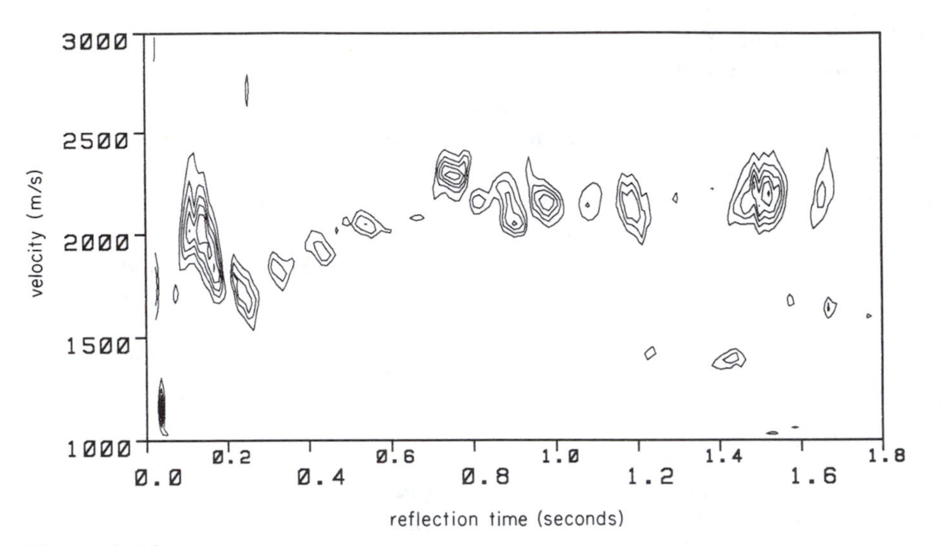

Figure 6–10
Semblance coefficient contour graph prepared from actual seismic data recorded on the Atlantic Coastal Plain in Virginia. Root-mean-square velocities for the Coastal Plain sedimentary layers vary from 1600 to 2200 m/s in this region. Closed contours surrounding a point at velocity of 2150 m/s and time 0.98 second indicate a primary reflection. Contours around another point at the same velocity but a time of about 1.18 second appear to indicate the first multiple from the same reflector.

form that leaves the seismologist relatively few interpretative judgments to make.

Amplitude Adjustment

The amplitudes of reflected waves diminish with offset distance because of geometrical spreading and absorption. These effects are discussed in Chapter 2. Before the traces of a CDP gather are used, amplitude adjustments as well as normal move-out adjustments must be made.

According to Equation 2-20, the effects of geometrical spreading and absorption can be combined to get

$$H_x = \frac{H_0}{r} e^{-\alpha r}$$

where H_x is the wavelet amplitude at a receiver offset a distance x from the source, H_0 is the amplitude of the source wavelet, α is the absorption coefficient, and r is the total distance along the travel path of the reflected wave. Because the wave travels the distance r in the time t_x at the RMS velocity, we can write the expression

$$r = V_{\text{rms}} t_x \qquad (6\text{–}14)$$

Substituting this expression into the amplitude equation above and rearranging, we obtain

$$H_x = \frac{H_0}{V_{\text{rms}} t_x} e^{-\alpha V_{\text{rms}} t_x} \qquad (6\text{–}15)$$

This tells us how wave amplitude diminishes as travel time increases.

Our purpose is to apply a correction that compensates for the effects of geometrical spreading and absorption. The effect will be to increase the amplitudes on the more distant traces. By rearranging Equation 6–15, we obtain the factor

$$K = \frac{H_0}{H_x} = V_{\text{rms}} t_x e^{\alpha V_{\text{rms}} t_x} \qquad (6\text{–}16)$$

We see that the value of K changes with arrival time and RMS velocity.

Each point on a digital seismogram trace indicates the amplitude of vibration at a particular time. Therefore, we can use this time and the corresponding RMS velocity to calculate the correction factor K. This factor is then multiplied with the value of the point to obtain an amplitude that is adjusted for the effects of geometrical spreading and absorption. Proceeding point by point in this way, each time calculating a new value of K, we obtain an amplitude-adjusted digital seismogram trace. The value of the absorption coefficient may be estimated from our knowledge of the general kinds of rocks encountered along the wave paths. This value may also be estimated by inspecting results after processing the same data set with a range of absorption coefficient values.

The question can be asked whether amplitude adjustment should precede or follow velocity analysis. The results of velocity analysis are usually clearer if prior amplitude adjustments have been made. But values of RMS velocity must be known to determine the values of the adjustment factor K.

The problem can be resolved through a preliminary velocity analysis of unadjusted traces to obtain preliminary RMS velocities. After these velocities have been used to make amplitude adjustments, the velocity analysis can be repeated to see whether the results are improved or otherwise changed. If necessary, this procedure can be repeated to further refine the results.

Normal Move-out Adjustment

The final step in preparing a CDP gather is the normal move-out adjustment. This adjustment is sometimes called the *dynamic time correction.* We know that reflected wavelets on unadjusted traces lie on an hyperbolic arc, as in Figure 6–11b. The idea of normal move-out or dynamic time corrections is to shift these wavelets to the common arrival time equal to that on the zero-offset trace, as in Figure 6–11c.

We know from Equation 4–47 that the difference Δt between zero-offset travel time t_0 and the travel time t_x of a wave arriving at a receiver offset a distance x from the source is

$$\Delta t = \sqrt{t_0^2 + x^2/V_{\text{rms}}^2} - t_0 \qquad (6\text{–}17)$$

This correction must be applied to shift reflected wavelets to common arrival times. It is based on the assumption that the waves have reflected from horizontal boundaries. An important fact to learn from this equation is that for the same offset distance and RMS velocity, the value of Δt decreases as the arrival time t_0 increases. This means that deep reflections show less normal move-out than shallower ones.

The procedure for making the normal move-out adjustment is to use points from each digital seismogram trace in the gather to prepare an equivalent zero-offset trace. Because of the difficulty in recognizing all the reflections that may be present, we treat each point on this equivalent trace as though it might be the zero-offset time of a reflection. To see how this can be done, look at the original digital trace in Figure 6–12 and the equivalent zero-offset trace that is being prepared

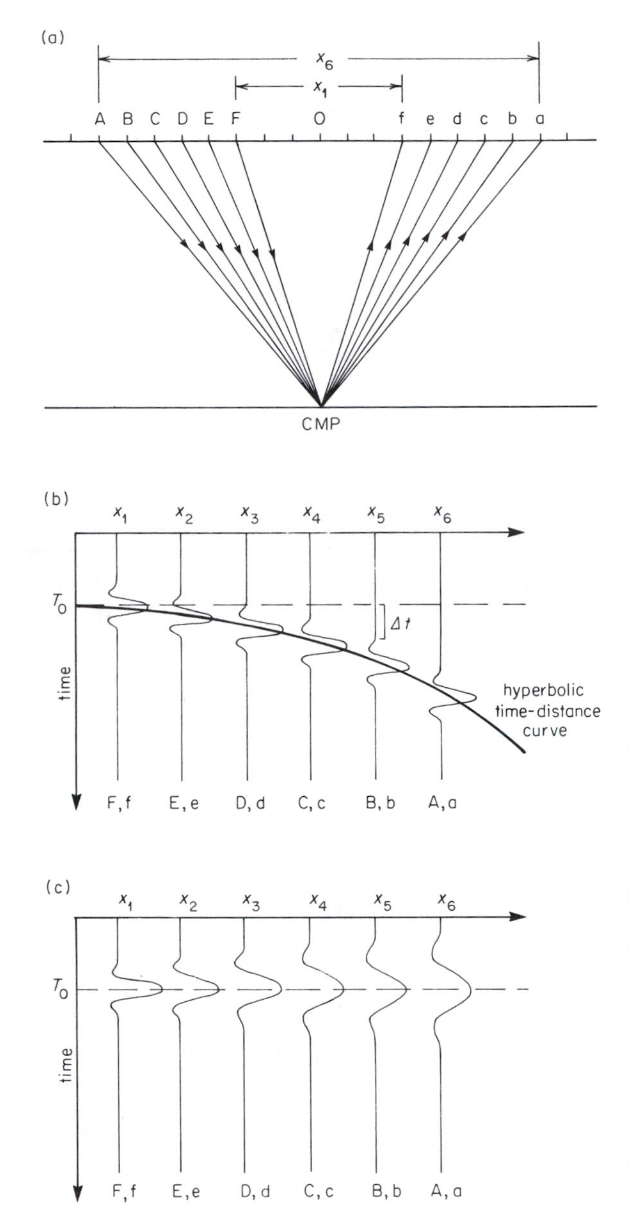

Figure 6–11

(a) Common depth point reflection travel paths, (b) seismogram traces not adjusted, and (c) seismogram traces adjusted for normal move-out. Observe that wavelet shapes are changed slightly in the process of normal move-out adjustment. Uppercase and lowercase letters represent sources and related receivers, respectively. Source–receiver pairs are symmetric about the common midpoint zero and have different offsets (x_i) and normal move-out times (Δt). Application of normal move-out corrections aligns the wavelets recorded from different paths by shifting them to the same zero-offset reflection time (t_0). The normal move-out correction process also deforms the wavelets by stretching them. The amount of stretching increases with increasing offset.

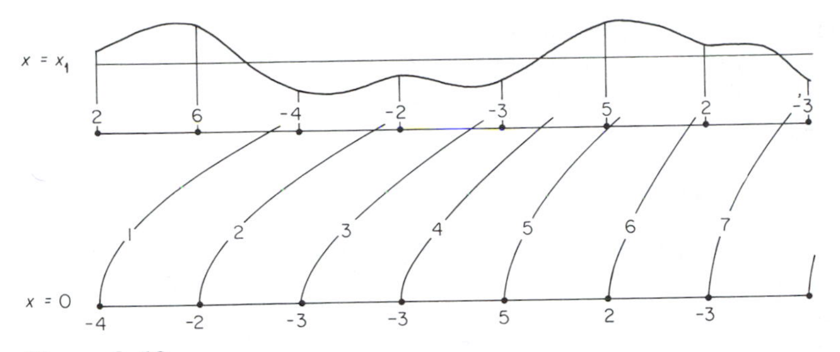

Figure 6–12
Procedure for selecting the points of an equivalent zero-offset trace from a digital seismogram trace. The numbers that make up a digital seismogram trace specify the amplitudes of vibration at a sequence of times. To determine the amplitude at some time t_0 on the equivalent zero-offset trace, we use Equation 6–17 to calculate the corresponding time, $t_x = t_0 + \Delta t$, on the trace with an offset of x. The amplitude value on this offset trace occurring at the time closest to t_x is used to represent the amplitude at t_0 on the equivalent zero-offset trace. The next time point on this zero-offset trace is then designated t_0. The procedure is repeated to find the corresponding amplitude value, and so on until amplitude values from the trace offset by x have been selected for all the points of the equivalent zero-offset trace.

below it. We begin by assuming that the first point on the equivalent trace is at the zero-offset time t_0. Then, using Equation 6–17, we calculate the corresponding arrival time t_x for such a reflection on the trace where the offset is x_1:

$$t_{x1} = t_0 + \Delta t \qquad (6\text{–}18)$$

It would be at the place where the first hyperbolic arc intersects the offset digital trace in Figure 6–12. Observe that this place turns out to be slightly to the left of the third point where the amplitude is -4. Since that point is closest to the intersection, we assign the amplitude of -4 to the first point on the equivalent zero-offset trace.

We then advance to the second point on the equivalent trace, setting that time equal to t_0. Again, the corresponding arrival time on the offset trace is calculated, which is indicated in Figure 6–12 by the place where the second hyperbolic arc crosses the offset trace. Because this is slightly to the left of the point where the amplitude is -2, this amplitude is assigned to the second point of the equivalent zero-offset trace.

As we continue in this way, observe that both intersections of the third and fourth hyperbolic arcs with the offset trace fall closer to the fifth point where the amplitude is -3. Therefore, this amplitude is assigned to both the third and fourth points on the equivalent trace. This is the standard practice in computer processing of CDP gathers.

This process is repeated point by point to prepare an equivalent zero-offset trace for each trace in the gather. For every point, an arrival time must be calculated by means of Equations 6–17 and 6–18. An RMS velocity value is required for each calculation. These values can be obtained from the preliminary velocity analysis we described earlier. From such an analysis, we obtain RMS velocities for a few reflections that produce easily recognized contour patterns on a semblance coefficient map, as in Figure 6–9. For purposes of preparing an equivalent zero-offset trace, the RMS velocity for a point situated between two known reflection times is determined by interpolation between the RMS velocities for these reflections.

Equivalent zero-offset traces that have been prepared from each of the offset traces in a CDP gather can now be stored in computer memory in the following form:

$$\begin{array}{l} A_1\ A_2\ A_3 \ldots A_m \\ B_1\ B_2\ B_3 \ldots B_m \\ C_1\ C_2\ C_3 \ldots C_m \end{array}$$

These values are the amplitudes at the sequence of times on traces, A, B, C, and so on. When data in this form have been obtained for all the gathers in a profile, CDP stacking can commence, followed by preparation of a seismic section.

Common Depth Point Stacking

The procedure for stacking the equivalent zero-offset traces of a CDP gather is simple. Amplitudes from corresponding points on these traces are added together, and the sum is taken as the amplitude of that point on the resulting stacked trace. Using the notation in expression 6–19, we can obtain the following stacked trace:

$$\begin{array}{l} (A_1 + B_1 + C_1 + \ldots), (A_2 + B_2 + C_2 + \ldots), \\ (A_3 + B_3 + C_3 + \ldots), \ldots, (A_m + B_m + C_m \ldots) \end{array} \qquad (6\text{–}20)$$

In this way, a single stacked trace, as in Figure 6–13, is obtained from each CDP gather.

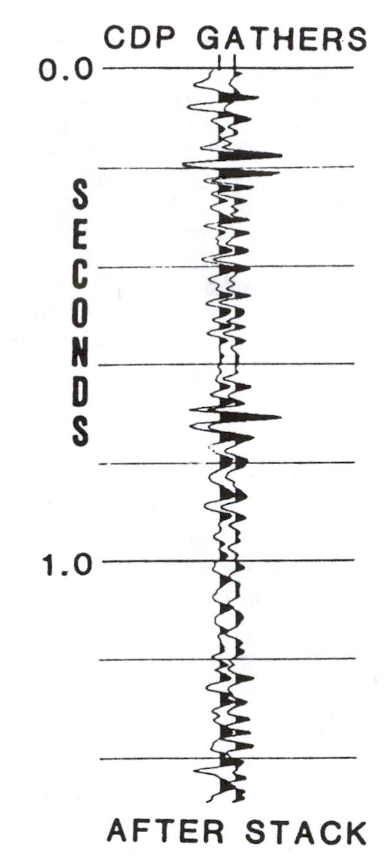

Figure 6–13
Common-midpoint-stacked traces. After normal move-out corrections are applied, the traces in a gather are summed point by point to obtain a stacked trace. Each trace in this figure was produced from a 24-fold CDP gather.

By means of our normal move-out adjustment procedure, we have done all that is practical to shift the reflected wavelets to the same arrival times on the equivalent zero-offset traces. If we have been successful, these aligned wavelets should combine to produce especially strong reflected signals on the stacked trace.

The same normal move-out adjustment acts to shift the coherent noise of refracted wavelets and surface waves into random positions. This happens because these waves, which show straight alignments on seismograms, have been shifted along hyperbolic arcs. Therefore, stacking should reduce coherent noise as well as other random noise.

Common depth point stacking has proved to be a remarkably effective data processing procedure. No other procedure has been as generally successful in strengthening reflections and reducing noise. In many seismic reflection surveys, this method is sufficient for this purpose.

Seismic Sections

Preparation of seismic sections involves plotting, side by side, all the stacked traces from a CDP reflection profile. Each trace is drawn as a vertical wiggly line, and the computer plotting machine is usually instructed to add variable-width shading, as in Figure 6–14. Here we see how effective this shading can be for illustrating the continuity of reflectors along the profile.

As we pointed out before, the seismic section has some of the features of a geologic cross section. Even though time rather than depth is the vertical dimension, the positions

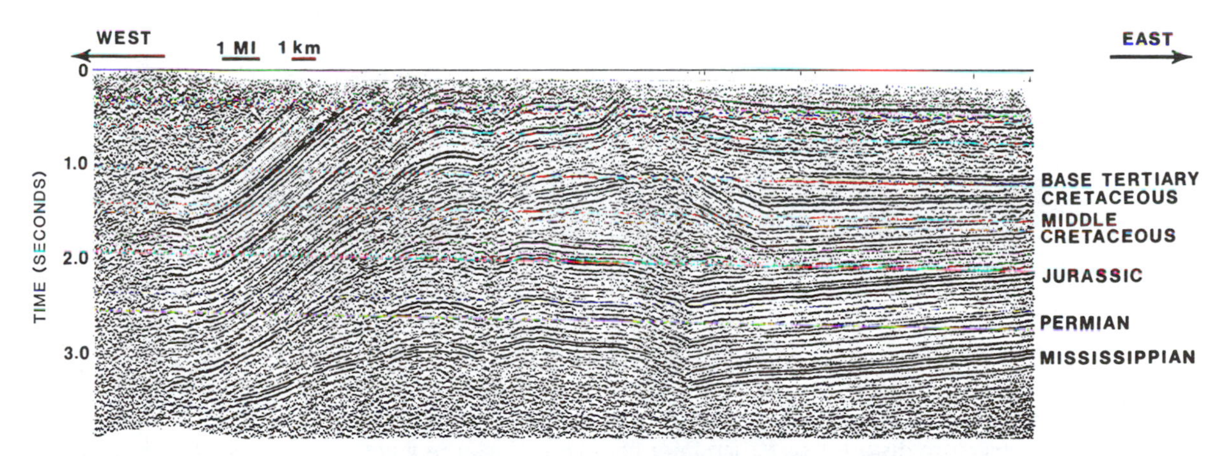

Figure 6–14
Seismic section prepared by side-by-side plotting of CDP-stacked traces. The data were recorded in Wyoming using three vibrators with a 14-second-long downsweep from 43 to 8 Hz. An off-end spread with 268-meter near offset and 1475-meter far offset was used. With a large number of stacked traces plotted in this way, the reflections, shown by variable-width shading, align to produce images of structural features along the survey profile. (Courtesy of Conoco, Inc.)

of folds and faults are not difficult to interpret. However, three important features can be misleading if they are not properly understood. (1) Velocity effects cause variations in the spacing between reflections as well as variations in layer thicknesses; (2) diffracted arrivals can indicate boundaries in positions where they do not exist; and (3) multiple reflections can indicate additional layers that do not exist. Let us examine these features in more detail.

First, consider the combined effects of velocity and layer thickness on the patterns displayed on a seismic section. We can illustrate these effects with the simple structures in Figure 6–15. Let us begin with two horizontal layers of equal thickness in which the velocities are V_1 and V_2. For this structure, the zero-offset time for reflections from the upper reflector will be

$$t_{01} = \frac{2h}{V_1}$$

and from the deeper reflector it will be

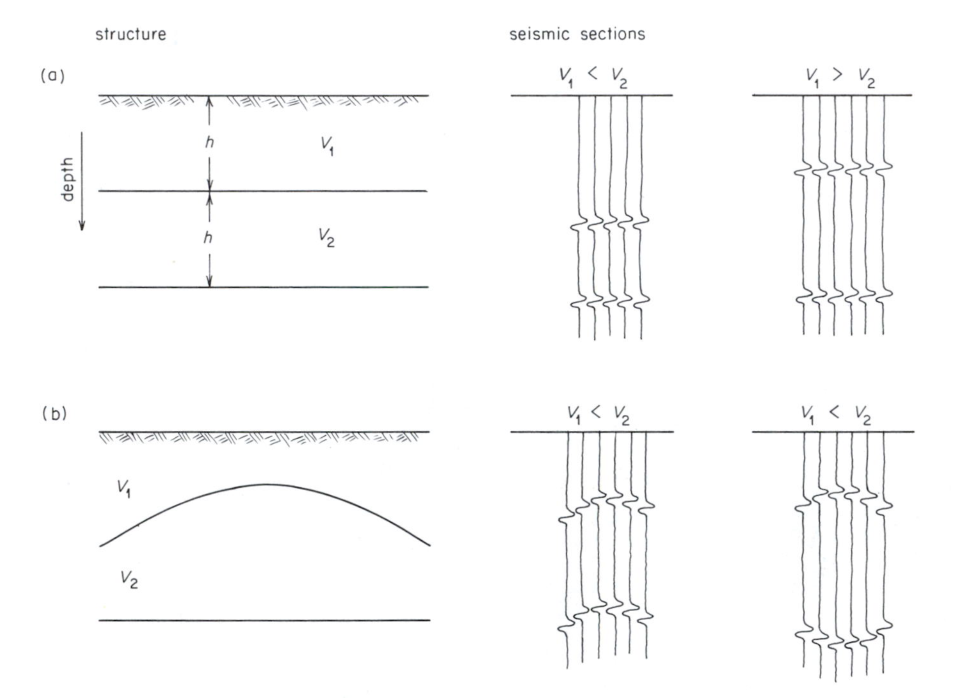

Figure 6–15
Images of earth structure distorted by variation of both layer thickness and velocity. Depending on velocity differences, two layers of equal thickness appear to have different thicknesses on the seismic sections. For the structure that actually has one curved boundary and one flat boundary, the corresponding seismic sections indicate two curved boundaries. Although the upper boundary is correctly imaged, the lower one appears to be distorted in different ways according to the difference of velocities in the layers.

$$t_{02} = \frac{2h}{V_1} + \frac{2h}{V_2}$$

We see that if $V_1 < V_2$, then $t_{02} < 2t_{01}$, but if $V_1 > V_2$, then $t_{02} > 2t_{01}$. Seismic sections corresponding to these two conditions are included in Figure 6–15a. It is clear that the spacing of reflections does not directly indicate that the layers are of equal thickness.

Next, look at the structure (Figure 6–15b) in which the first reflector arches upward and the deeper reflector is horizontal. We see that, although depth to the deeper reflector is constant, the reflection times will vary along the profile according to the proportions of the travel paths in the upper and lower layers. If $V_1 < V_2$, the reflection time from the deeper reflector will be shortest over the center of the arch because the travel path through V_2 is longest, and the path through V_1 is shortest. If $V_1 > V_2$, then for the same reason the reflection travel time for the deepest reflector will be greatest over the center of the arch. As a result, the deeper reflections on the seismic sections do not directly reveal a horizontal reflector. Rather, they present distorted images of the actual form of that reflector.

These examples illustrate how, through the combined variations in velocity and layer thickness, a seismic section may present a distorted image of the structure. The possibility of such distortion must be remembered in any attempt to interpret geologic features directly from inspection of a seismic section.

The second feature of a seismic section that can distort the image of earth structure is the presence of diffraction signals, which are discussed in Chapter 4. Look again at Figure 4–25, which indicates the pattern of wavelets that would be recorded over a faulted reflector. According to Equation 4–57, the diffracted arrivals align on an hyperbolic arc.

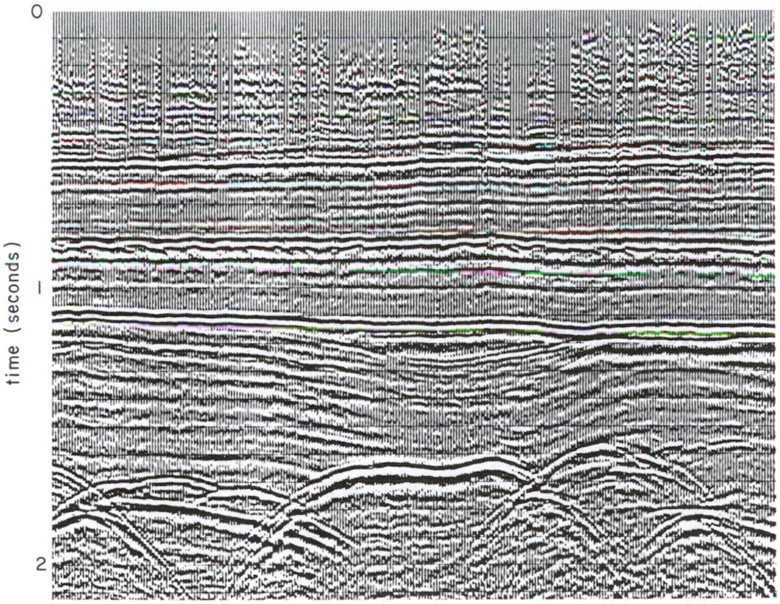

Figure 6–16
Seismic section that shows arc-shaped images produced by diffractions. Among the features helpful in distinguishing these spurious patterns from true reflections are the intersection or crossing of two images, forming a so-called "bow tie" pattern, and the fact that diffraction images tend to have hyperbolic forms. Data processing procedures referred to as migration procedures are used to delete these spurious images from a seismic section. (Courtesy of Prakla–Seismos AG.)

Therefore, the normal move-out adjustments and stacking are not completely effective in removing these arrivals from CDP-stacked traces.

Waves that were diffracted from irregularities along a layer boundary can produce arc-shaped images on a seismic section. Such diffraction patterns are clearly evident on the seismic section in Figure 6–16. These patterns can seriously interfere with efforts to interpret geologic structure. Therefore, additional data processing must be done to remove them. The method for accomplishing this objective is called *migration;* it is the subject of the next section.

The third source of distortion on a seismic section is multiple reflections. The nature of these signals is discussed in Chapter 4. Even though they exhibit hyperbolic normal move-out, the CDP stacking procedure can be partially effective in suppressing them. This happens because the RMS velocities used for normal move-out adjustments at times when multiples occur tend to be different from the RMS velocities along the paths of the multiples. Nevertheless, some multiples are likely to

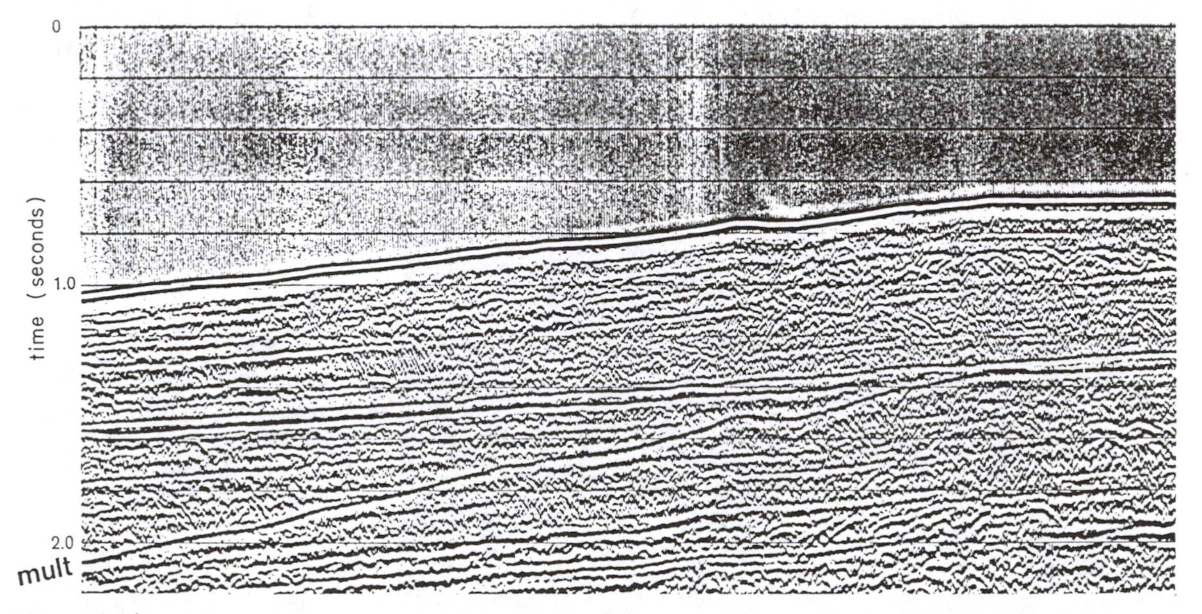

Figure 6–17
Marine seismic section displaying strong primary and multiple reflections from the ocean floor. The multiple reflection (mult) is inclined because of the slope of the ocean floor. Therefore, it crosses other reflections, which helps to distinguish it from these deeper primary reflections. But on the right-hand side of the section, the multiple event becomes coincident with another nearly horizontal reflection and becomes difficult to distinguish. Failure to identify multiples properly may cause incorrect interpretations. The data processing procedure called deconvolution is designed to delete multiples from seismic sections. (Courtesy of Prakla–Seismos AG.)

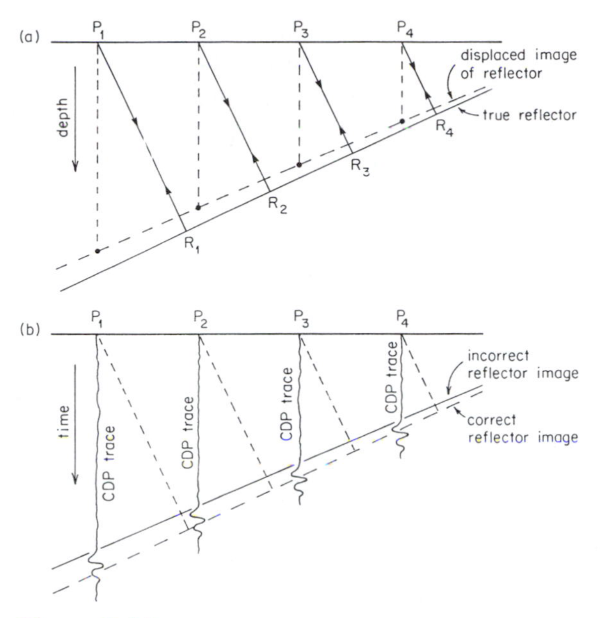

Figure 6–18
The image of a reflector displaced by plotting reflections along vertical lines. True reflection paths shown by inclined lines indicate the correct position of the reflector.

remain on the seismic section. Here they create images of boundaries that do not exist, or they mask primary reflections arriving at approximately the same time.

Especially strong multiples are common in marine seismic surveying. In Figure 6–17, most of the reflections are nearly horizontal. But one is different. It stands out because it is parallel to the image of the ocean floor. This is a typical feature of multiples on marine seismic sections. On many marine sections prepared directly after CDP stacking, very strong multiples mask most of the deeper reflections. Then, additional data processing is necessary to obtain reflections that can be interpreted. The procedure is called *deconvolution* or *inverse filtering;* some aspects of this procedure are introduced in a later section of this chapter.

SEISMIC MIGRATION

The standard procedure for common depth point stacking makes no allowance for reflector dip or irregularity. Each CDP-stacked trace in a seismic section is plotted to show reflec-

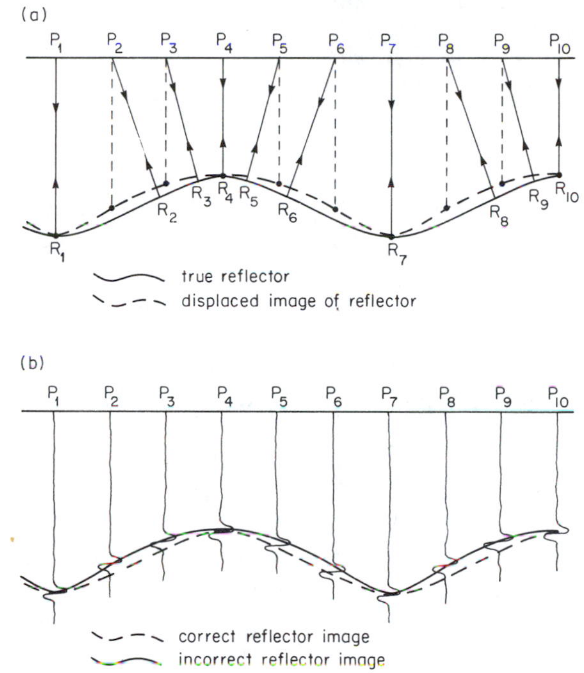

Figure 6–19
Distortion of the image of an undulating reflector produced by plotting reflections along vertical lines. (a) True reflection paths shown by solid lines indicate the correct position of the reflector. Vertical dashed lines lead to points that do not lie on the reflector and that give an incorrect indication of its position. (b) The reflector image observed in seismic section does not correspond to the true position of the reflector.

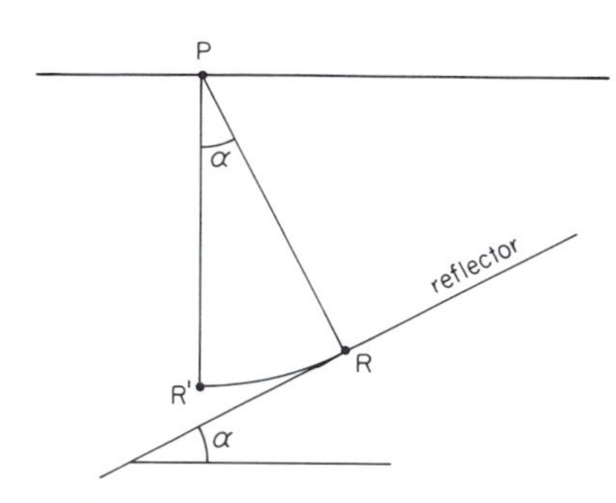

Figure 6–20
Adjustment of a reflection from a point on a vertical line to a point on the reflector by rotation along a circular arc through an angle equal to the dip angle α. This relation is the basis for a procedure of geometrical migration.

tions in positions that correspond to vertical travel paths. Yet, we can see on a section such as the one shown in Figure 6–14 that reflectors are inclined and irregular. What kinds of distortion result from analyzing such reflectors as if they were horizontal plane surfaces? Two simple examples illustrate what happens.

Displacement of Reflections

Let us consider the zero-offset reflections at several points along a profile over an inclined reflector. All these reflections have followed the inclined paths in Figure 6–18a, which are perpendicular to the reflector. Now suppose that distances to the reflector, calculated from the travel times along these paths, are plotted as vertical depths. By connecting the points at these depths with a line, we show the reflector incorrectly in a position that is different from its actual position.

Similarly, if we plot reflection travel times along vertical lines as they appear in a seismic section, we will image the reflector in a different position than we would obtain by showing these same times along inclined lines that correspond to the actual travel paths, as in Figure 6–18b. This distortion results from preparing

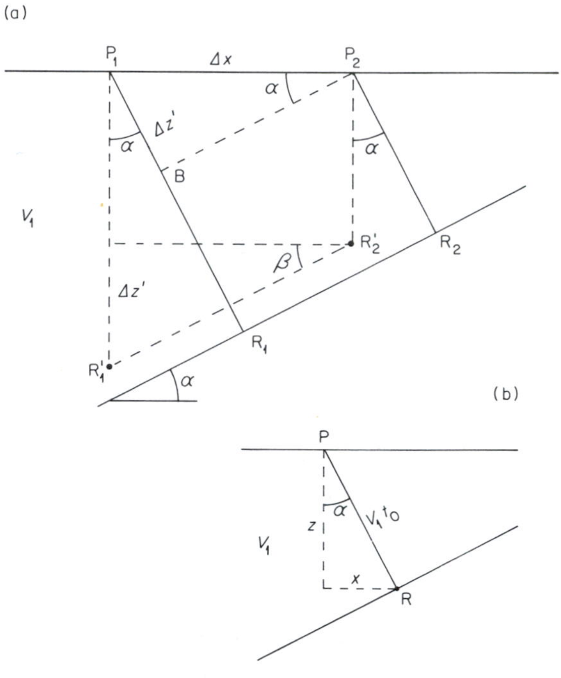

Figure 6–21
(a) Geometrical features used to determine the dip angle needed for migrating reflection points and (b) the components for correctly plotting the reflection point. R_1 and R_2 are reflector points on the refractor with a slope of α. If zero-offset source–receiver pairs are located at P_1 and P_2, the reflections from R_1 and R_2 can be recorded. Because the slope α is not initially known, these reflections are first considered to come from reflection points R_1' and R_2' along vertical lines. These reflection points define an incorrect reflector with a slope of β. However, the distance between P_1 and P_2 and the reflection times can be used in Equations 6–21, 6–22, and 6–23 to find the coordinates x and z of a point R on the true reflector relative to the source–receiver point P.

a seismic section in the standard way. Not only the position but also the dip angle is incorrectly imaged by vertical plotting.

We can extend this analysis to the undulating reflector in Figure 6–19. Here we see that the zero-offset paths reflect from points that are more concentrated on the high parts of the reflector. When plotted on vertical lines, the reflections produce an image with broader crests and narrower troughs than actually exist. Observe that the highest and lowest points are correctly plotted because in these places the reflection paths are vertical.

When we examine a schematic seismic section, as in Figure 6–19b, in which an undulating reflector is indicated, we should realize that the image reveals the proper spacing of undulations, but incorrect shapes. This distortion is produced by plotting along vertical lines the reflections that have followed inclined paths. The reflector surface mapped in Figure 6–19b does not exactly match with the true reflector shown in Figure 6–19a.

Migration of Displaced Reflections

We can correct the dip distortion by moving the reflection points away from their positions on vertical lines onto inclined lines that correspond to the travel paths. The process of shifting these points to positions that correctly image the reflector is called *migration*.

A simple geometrical procedure illustrates the basic idea of migration. The distortion we have been discussing does not result from incorrect arrival times. Rather, it results from plotting arrival times, or depths calculated from them, in incorrect positions. Because the arrival time itself is correct, we know that the position of the reflecting point must lie somewhere on a circular arc at the same distance from the zero-offset position as the point that was incorrectly plotted along a vertical line. This is seen in Figure 6–20, which shows that the actual travel path makes an angle α with the vertical line equal to the dip angle. If the dip angle can be determined, it will be a simple task to replot the point in its correct position.

If the reflector is an inclined plane, the dip angle can be found from two zero-offset reflections according to the geometry in Figure 6–21. Here the correct reflection points we wish to locate are indicated by R_1 and R_2. When plotted along vertical lines, these reflection points will be incorrectly placed at R_1' and R_2'. The difference in the depths of these incorrect reflections is $\Delta z'$, so that we can determine the inclination β of the line connecting R_1' and R_2' from

$$\tan \beta = \frac{\Delta z'}{\Delta x}$$

where Δx is the distance between the two zero-offset positions P_1 and P_2. But from the line BP_2 that is parallel to the reflector, we see that the correct dip α can be found from the expression

$$\sin \alpha = \frac{\Delta z'}{\Delta x}$$

or since

$$\Delta z' = V_1(t_{01} - t_{02}) = V_1 \, \Delta t$$

where t_{01} and t_{02} are the zero-offset reflection times, we can write

$$\sin \alpha = V_1 \frac{\Delta t}{\Delta x}$$

or

$$\alpha = \arc \sin\left(V_1 \frac{\Delta t}{\Delta x}\right) \tag{6–21}$$

Now we can find the correct position of the reflection point R relative to the zero-offset position P on the surface. According to Figure 6–21b, it will be at depth

$$z = V_1 t_0 \cos \alpha \tag{6–22}$$

and it will be displaced horizontally by

$$x = V_1 t_0 \sin \alpha \tag{6–23}$$

On a seismic section, the reflection time should be adjusted from time t_0 to

$$t_0' = t_0 \cos \alpha \tag{6–24}$$

Before computer data processing was possible, migration was done graphically. One method for accomplishing migration is illustrated in Figure 6–22. First, the reflection points are plotted on vertical lines to mark the radii of circular arcs. After these arcs are constructed, the reflector can be drawn tangent to them. The points of tangency will be at positions described by the equations just given.

This simple migration procedure could be programmed for computer processing. For an irregular undulating reflector, it would have to be assumed that the reflector between adjacent zero-offset positions could be represented by a plane surface.

The procedure we have described illustrates the basic idea of migrating reflections. Modern data processing schemes can be considerably more complicated because they attempt to reshape the forms of wavelets on different CDP-stacked traces. The aim is to create wavelets with appropriate amplitude and position on a vertically plotted trace to correctly image the reflector.

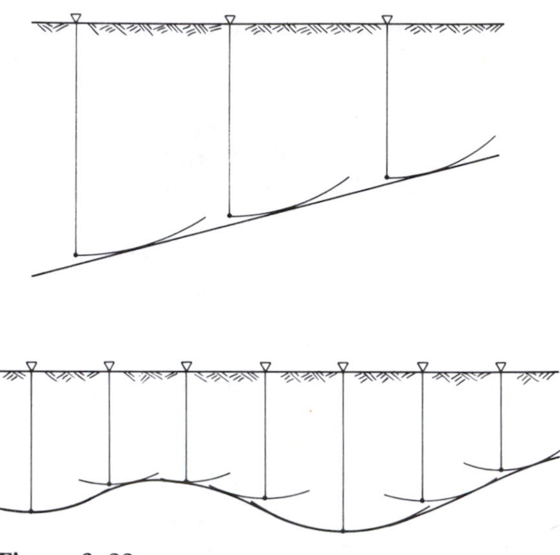

Figure 6–22
Procedure for graphical migration of reflection images. For each zero-offset reflection, a circular arc is drawn. Then the position of the reflector is sketched so that it is tangent to each of these circular arcs.

Migration of Diffracted Waves

Irregularities on a reflector produce diffracted wavelets which interfere with the reflections on a seismogram. On seismic sections, these diffractions produce arclike patterns like those pictured in Figure 6–16. The simple example in Figure 6–23 illustrates how this happens. The waves diffract from the point Q, which for example, could be the edge where the reflector is broken by a small fault. The travel paths radiate from Q to the zero-offset positions on the surface.

When CDP-stacked traces are plotted vertically to prepare a seismic section, diffracted wavelets create an image that could be mistaken for an arcuate reflector, as in Figure 6–24a. The diffraction image has the shape of

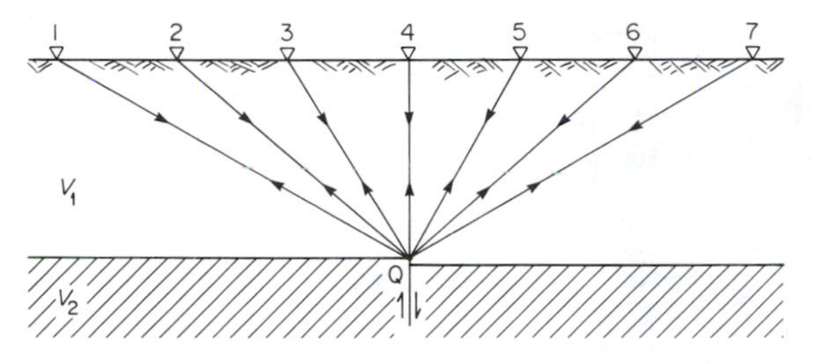

Figure 6–23
Travel paths of waves diffracted from point Q on a reflector with a vertical offset. Point Q behaves as a secondary source called a diffractor. Waves from the diffractor are received at every receiver.

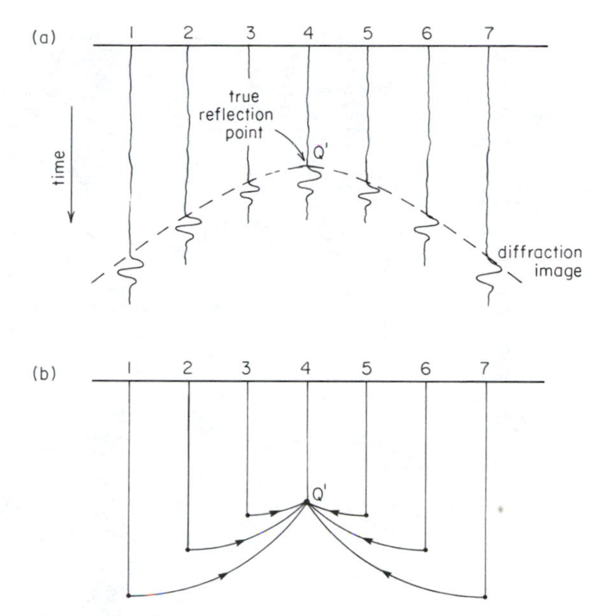

Figure 6–24
(a) Arc-shaped diffraction image produced by plotting diffracted arrivals along vertical lines and (b) subsequent migration of vertically plotted diffracted arrivals by displacement along circular arcs that intersect at the correct position of the diffraction point.

a hyperbola. Therefore, we have the interesting condition shown in Figure 6–24b. The circular arcs centered on the zero-offset positions with radii equal to zero-offset travel times all intersect at a common point Q′. This point marks the true zero-offset travel time for the point Q in Figure 6–23. This geometrical condition is the key to migrating diffracted wavelets.

One of the principal aims of migration is to remove diffraction images from seismic sections. The geometrical basis for doing so is shown in Figure 6–24, but the procedure is quite complicated. The computer must be programmed to test sets of CDP-stacked traces to locate wavelets displaying this geometrical arrangement. Because there can be many diffraction images overlapping one another, and one above another, it is difficult to test and sort out the particular wavelets making up one image. Furthermore, where diffracted wavelets are superposed on reflections, it is desirable to remove only the diffracted signal while retaining the reflection with relatively little wavelet distortion.

A general-purpose migration procedure, then, has two important tasks to perform. It must reposition reflections that were displaced because of dip, and it must remove diffraction images. Although the basic ideas for carrying out these tasks are quite simple, preparation of computer programs to implement these ideas has required much careful thought and attention. The need for migration can usually be judged by examining a seismic section. Where dipping images and diffraction images are seen, as in Figure 6–16, migration is recommended. The quality of such a seismic section can be considerably improved by modern migration processing. The migrated seismic section in Figure 6–25 provides evidence of this.

Figure 6–25
Seismic section along the profile in Figure 6–16 that has been adjusted by migration. Notice that the main effect of the migration process has been to adjust reflections with undulations, dip, and offsets. After removal of diffraction arrivals, "bow tie" patterns and hyperbolic arcs disappear. Compare this figure with Figure 6–16 and notice the clarity of deeper reflections arriving just before 2 seconds. (Courtesy of Prakla–Seismos AG.)

FILTERING SEISMIC DATA

One way to change something is to put it through a filter. Many kinds of filters are available, some of which are instruments and others mathematical procedures. A chemist may use a simple paper filter to remove suspended particles from a liquid. In this procedure the mixture of particles and liquid is the *input* that is poured into the filter. After the mixture passes through the filter, the *output* that is discharged from it is different from the input. The liquid without the particles has become trapped in the filter.

Similarly, a physicist may use a colored lens to extract a particular color of light from a sunbeam. Here the lens is the filter. The output is a beam of, for example, blue light that is different from the input.

Filters play an important role in processing seismic data. We have already discussed electronic frequency filtering (in Chapter 5), which is sometimes useful for separating low-frequency surface waves from higher-frequency reflections. Here the input is a seismic signal displaying high- and low-frequency vibrations. After filtering, we obtain an output signal that displays only, say, high-frequency vibrations.

Frequency filtering can also be done by mathematical processing of digital seismic data. Another important filtering process is performed to remove multiple reflections from seismograms. Still other techniques of filtering act to separate wavelets that have traveled at different velocities. Let us look more closely at some of these filters.

The Concept of Digital Filtering

Let us begin with a very simple method of filtering—the method of *running averages.* Consider the portion of a digital seismogram shown in Figure 6–26. Suppose that we filter this input signal by calculating average values using the amplitudes of sets of five adjacent points as follows:

$$C_3 = (A_1 + A_2 + A_3 + A_4 + A_5)/5$$
$$C_4 = (A_2 + A_3 + A_4 + A_5 + A_6)/5$$
$$C_5 = (A_3 + A_4 + A_5 + A_6 + A_7)/5$$
$$C_i = (A_{i-2} + A_{i-1} + A_i + A_{i+1} + A_{i+2})/5 \quad (6\text{-}25)$$

The values C_3, C_4, and so on, make up the output signal that is also shown in Figure 6–26. Unlike the input signal that shows both high- and low-frequency vibrations, the output signal shows only low-frequency vibrations. In the summing process, the high-frequency peaks and troughs act to cancel one another, leaving only the low-frequency part. By choosing only five consecutive points for averaging, we do not include long enough parts of the seismogram for the low-frequency peaks and troughs to cancel one another. Therefore, we have created a *low-pass* filter. That is, low-frequency signals pass through the filter to become part of the output, but

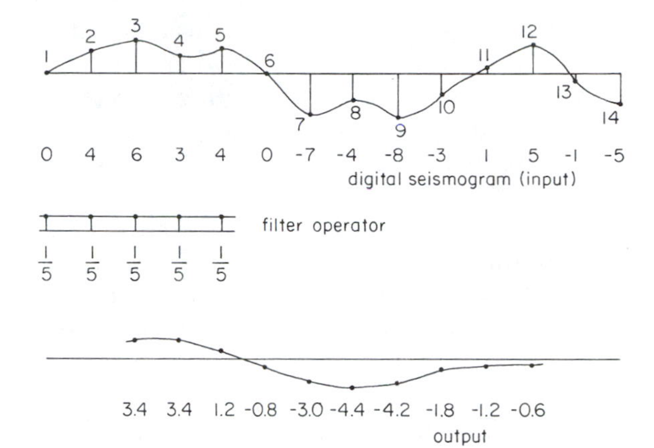

Figure 6–27
Filtering accomplished by cross correlation of a digital seismogram and a digital filter operator designed for five-point averaging. The process begins by multiplying each of the first five values of the digital seismogram by the corresponding filter coefficient, which is one of a sequence of five weighting factors that make up the filter operator. The products are summed to obtain the first value for the filtered seismogram. The procedure is then repeated using values at points 2 through 6, 3 through 7, and so on from the original digital seismogram to obtain the sequence of values that make up the filtered seismogram.

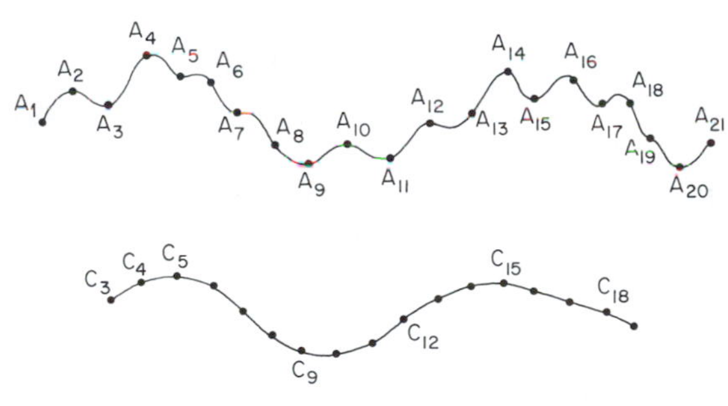

Figure 6–26
Low-pass filtering of a digital seismogram trace by five-point averaging. The method of running averages is a simple filtering process. It acts to suppress high frequencies and to bring low-frequency oscillations into clearer focus.

high frequencies become "trapped" by the filter.

Let us describe what we have just done in a slightly different way that may seem more complicated but has some advantages for explaining more complicated filters. Here we introduce the idea of a *filtering function,* which is otherwise called a *filter operator.* Like a digital seismogram, the filter operator is a list of numbers. Rather than vibration amplitudes, these numbers can be viewed as weighting factors. They are called *filter coefficients.*

For purposes of five-point averaging, we need a filter operator with five filter coefficients. Because we want to treat all the amplitude values equally and we will be averaging five values, each filter coefficient has the value of 1/5. Now look at Figure 6–27 where we have one row of numbers representing part of a digital seismogram, and below it is another row of numbers representing the filter operator.

We filter the seismogram first by cross-multiplying the filter coefficients with the first five amplitude values on the seismogram and summing the products. This gives us the first output amplitude value. To obtain the next output amplitude, we repeat the process using points 2 through 6 on the seismogram. Additional output values are calculated by advancing one interval on the seismogram, cross-multiplying, and summing

This process of cross multiplication and summing repeatedly after advancing the folded filter coefficients one interval along the seismogram trace each time is the process of *convolution.* Digital filtering is done, then, by convolving a filter operator with a digital seismogram trace. Observe that this procedure is similar to the one used for vibroseis correlation. There the folded sweep input was convolved with the seismogram trace. For that processing procedure, the "filter coefficients" are the amplitudes of the sweep signal. It is a filtering operation that changes the long input signal to an output of Klauder wavelets.

$$C_1 = (1/5 \times 0 + 1/5 \times 4 + 1/5 \times 6 + 1/5 \times 3 + 1/5 \times 4) = 3.4$$
$$C_2 = (1/5 \times 4 + 1/5 \times 6 + 1/5 \times 3 + 1/5 \times 4 + 1/5 \times 0) = 3.4$$
$$C_3 = (1/5 \times 6 + 1/5 \times 3 + 1/5 \times 4 + 1/5 \times 0 + 1/5 \times -7) = 1.2$$

and so on.

Our procedure has been to design a filter operator and to use it to filter the seismogram trace. This particular filter operator was designed specifically for five-point averaging. By choosing different filter coefficients, we can design filters that can do other specific kinds of weighted averaging, but the filtering procedure would be the same. In general, if the amplitudes for a digital seismogram are A_1, A_2, A_3, . . ., A_m, and a set of coefficients called folded filter coefficients are f_1, f_2, f_3, . . ., f_n, output amplitudes are

$$C_1 = (f_1A_1 + f_2A_2 + f_3A_3 \ldots f_nA_n)$$
$$C_2 = (f_1A_2 + f_2A_3 + f_3A_4 + \ldots f_nA_{n+1})$$
$$C_3 = (f_1A_3 + f_2A_4 + f_3A_5 + \ldots f_nA_{n+2}) \quad (6\text{–}26)$$
$$C_i = (f_1A_i + f_2A_{i+1} + f_3A_{i+2} - \ldots f_nA_{i+n-1})$$

Frequency Filtering

The basic idea of frequency filtering is that all the complex features on a seismogram can be duplicated by adding together the right combination of simple cosine waves. This concept is illustrated in Figure 6–28. By methods of *spectral analysis,* which we will not attempt to discuss, it is possible to calculate the frequencies and amplitudes of the particular waves in this combination from the amplitudes of a digital seismogram trace. Frequency filtering

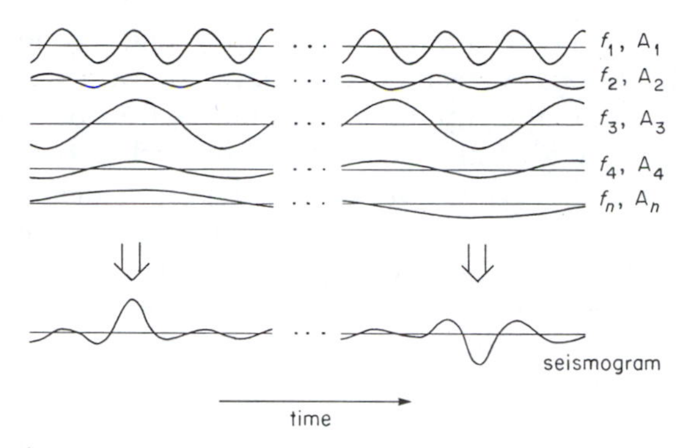

Figure 6–28
Cosine waves that add together to reproduce the form of a seismogram trace. The spectrum of the seismogram is described by specifying the frequencies and amplitudes of the sequence of such waves that is required to reproduce its form. Filtering is a process of spectrum modification which involves removing or suppressing certain frequencies. Removing frequency components from the lower end of the spectrum is called low-cut or high-pass filtering. Removing frequencies from the high end of the spectrum is called high-cut or low-pass filtering.

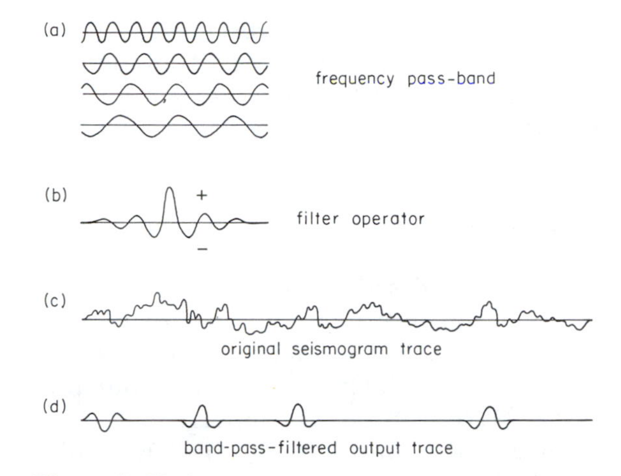

Figure 6–29
Band-pass filtering of a digital seismogram trace by convolution with an appropriate digital filter operator. The filter operator can be designed to pass or retain frequencies of a certain band while rejecting or suppressing the other higher or lower frequencies that may be present. It consists of a sequence of folded filter coefficients that are applied in the process of cross multiplication and summation that is described as convolution.

deletes certain of these waves from the combination.

Filter operators can be designed to pass certain frequencies and eliminate others as they are convolved with a digital seismogram trace. A band-pass filter eliminates both high and low frequencies and passes a specified band or range of intermediate frequencies. Figure 6–29a shows a band of frequencies that are to pass through a filter to make up the output signal. From an analysis of these frequencies, the filter operator in Figure 6–29b is determined. Observe that it is an oscillating function with both positive and negative filter coefficients. The form of this operator is typical of all band-pass filters. The result of convolving such a filter operator with the seismogram trace (Figure 6–29c) is an output trace with no very low frequency or very high frequency oscillations (Figure 6–29d).

In Chapter 5 we discussed the use of electronic frequency filtering as a means of separating signal from noise during field recording operations. With the development of digital filtering techniques such as we have been describing, the most common practice is to avoid frequency filtering during field recording operations unless it becomes absolutely necessary. The reason for avoiding frequency filtering is that weak reflections in the same frequency range as the noise may inadvertently be filtered out. A better practice is to attempt to retain these reflections while eliminating the noise by means of CDP stacking. In the event that further improvement might be possible by frequency filtering, an appropriate filter operator can then be designed for convolution with the traces.

The Earth Filter

Seismologists describe the earth itself as a filter that can convert an impulsive input signal into the series of vibrations we see on a seismogram. Earth filtering consists of four processes: (1) conversion of an inpulse into a wavelet within the source zone; (2) division of a wavelet into a succession of wavelets by reflection and refraction at boundaries; (3) geometrical spreading; and (4) absorption. These processes are shown schematically in Figure 6–30.

An explosive source is a sudden outward concussion that lasts less than 0.1 millisecond. It can be represented as a sharp narrow spike, which is the input signal entering the source zone. Within this small zone the rock is compressed, twisted, broken, and partly melted. This is the activity of the filter, which is reshaping the brief input spike into a wavelet that oscillates for a few tens of milliseconds.

The division of a wavelet by reflection and refraction is explained in Chapter 2. For purposes of describing how a zero-offset trace is produced, the filtering effect of reflection and refraction can be represented by the filter operator depicted in Figure 6–31a. It shows a vertical line at the time of each reflection. The length of each line is proportional to the reflection coefficient found from Equation 2-22, and its direction, up or down, depends on the sign of this reflection coefficient. By convolving the single wavelet in Figure 6–31b with

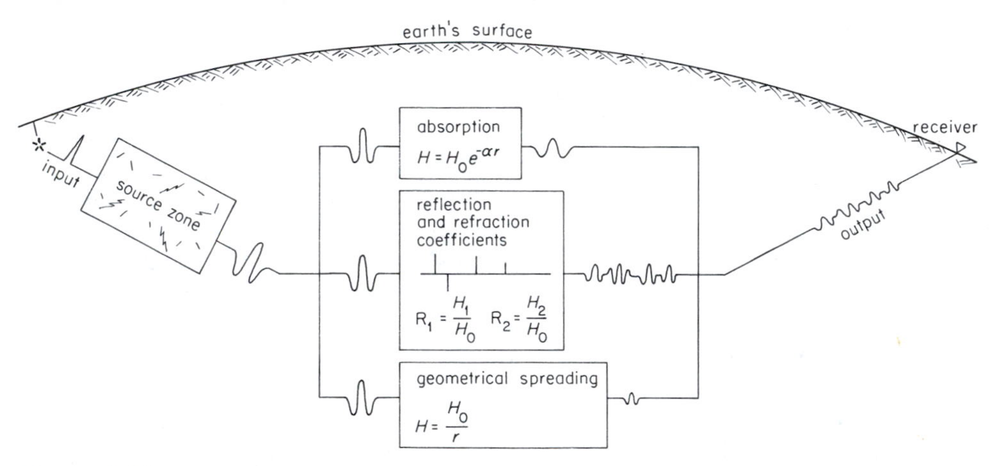

Figure 6–30
Components of the earth filter. These components illustrate the various effects that occur within the earth to change an input signal, which is introduced at one time and place, into the output signal, or seismogram, which is recorded at another time and place. First, the earth transforms an input pulse into a wavelet. Then, it divides that into several wavelets by refraction and reflection and alters these wavelets by absorption and geometrical spreading.

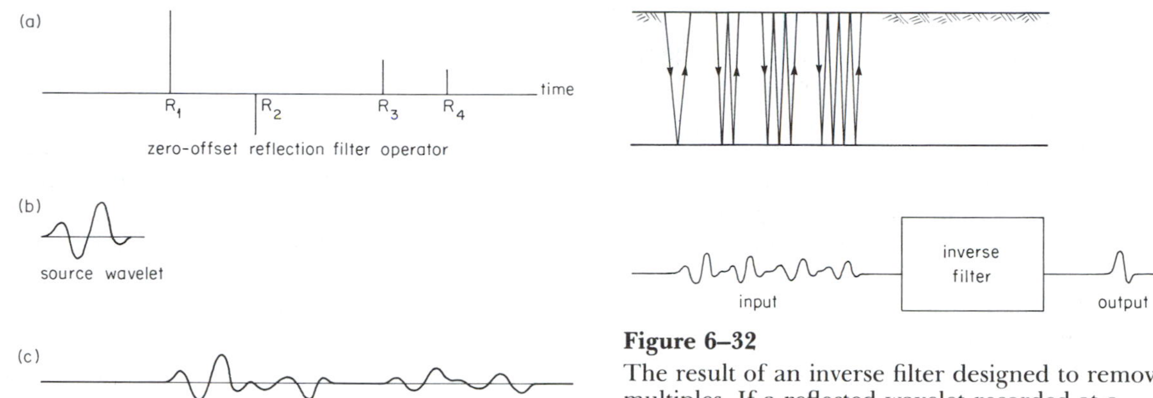

Figure 6–31
Convolution of a zero-offset reflection filter operator and a wavelet to produce a reflection seismogram trace. (a) Vertical lines are proportional to reflection coefficients. In digital form this filter operator would be expressed by a sequence of filter coefficients, all of which would be zeros except those at points R_1, R_2, R_3, and R_4. Convolution of such a sequence with (b) a digital source wavelet, which involves cross multiplication and summing, produces the sequence of values that could be plotted to obtain (c) the reflection seismogram trace.

Figure 6–32
The result of an inverse filter designed to remove multiples. If a reflected wavelet recorded at a receiver is considered to be the result of the earth's filtering effect, multiple reflections can also be viewed as another component of the filter. In general, multiples from one interface produce a series of wavelets. An inverse filter can be designed to suppress the multiples by removing wavelets that occur at times predicted for multiple arrival times. Inverse filtering is called deconvolution and is widely used for processing data when multiples are suspected.

the filter operator, we obtain the basic reflection seismogram, as in Figure 6–31c.

Both geometrical spreading and absorption act to reduce wavelet amplitude. Absorption also tends to alter the frequency spectrum of the wavelet by absorbing high-frequency energy more readily than lower-frequency energy. As such, it is a low-pass filter.

Inverse Filters

We know that the earth filter expands an impulse into a wavelet and then divides the wavelet into several wavelets by reflection and refraction. A class of operations called *inverse filtering* or *deconvolution* has been developed as a means of partially reversing the effect of the earth filter. The idea is to design filter operators whose effects are equivalent to moving backward through the earth filter. An inverse filter, then, has two objectives: (1) it seeks to compress the wavelet into a shorter impulse that is more like the initial explosive impulse; and (2) it seeks to shift some wavelets produced by reflection and refraction back into the incident wavelets that produced them. When relatively long reflected wavelets overlap one another so that their onsets cannot be clearly recognized, it is desirable to obtain shortened wavelets that are separate from one another. Where strong multiple reflections override important deeper reflections, it is desirable to shift them back into the primary wavelets from which they were derived. After the interfering multiples are removed, the deeper reflections can then be identified. These are the most important uses of inverse filters, otherwise called deconvolution filters.

Two kinds of information, which can be difficult to obtain, are needed to design an inverse filter when we use different methods. Wavelet shape must be known before a filter can be designed to shorten it. The position of a reflector must be known before a filter can be designed to shift wavelets produced by multiple reflections from it. Wavelet shape can be difficult to recognize because of distortions introduced by noise, absorption, and other superposed wavelets. One purpose of reflection seismology is to determine the position of reflectors. If we cannot recognize the reflections that give us this information, we cannot design a filter to shift some of the multiples.

Deconvolution has been most successfully applied in the processing of marine seismic data to remove strong signals multiply reflected from the ocean floor. Because the primary reflection from the ocean floor is strong and easily recognized, the arrival times of multiples can be accurately predicted. The inverse filter is then designed to suppress wavelets at those times on a seismogram trace.

Many studies have been conducted on the shapes of source wavelets produced by marine sources such as air guns. This information is

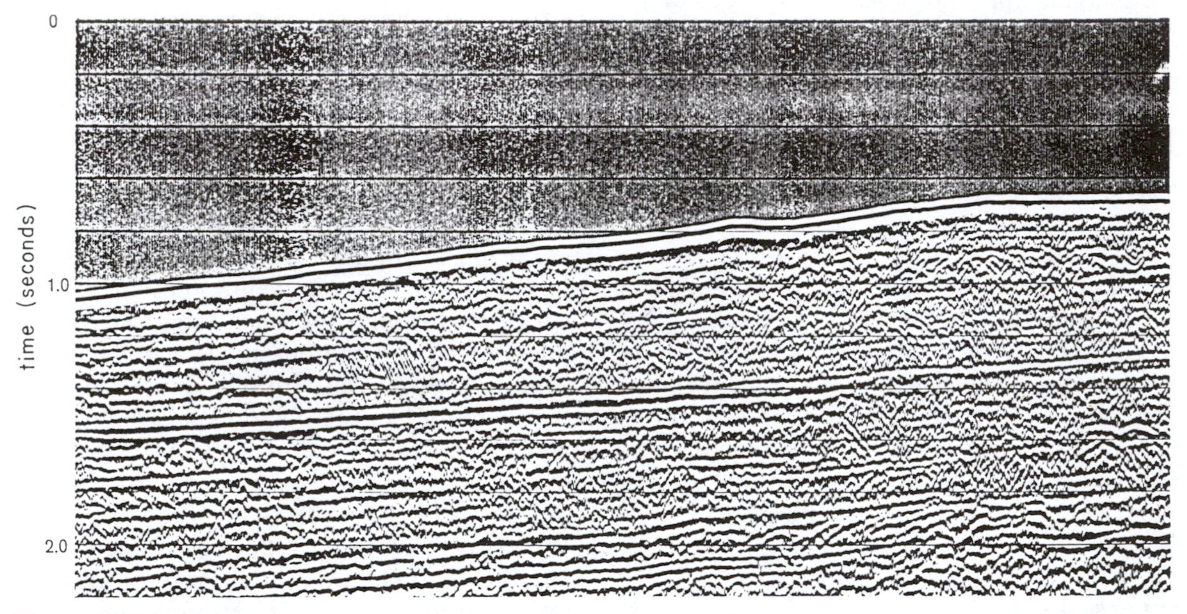

Figure 6–33
Seismic section along the profile shown in Figure 6–17, now processed by deconvolution to suppress the multiple reflections from the ocean floor. Notice how the diagonally crossing multiple event in Figure 6–17 has been eliminated by the deconvolution process, and overall signal-to-noise ratio has been improved. (Courtesy of Prakla–Seismos AG.)

being used for designing a filter for wavelet compression.

We will not attempt to describe the complicated procedure for designing inverse filters, but we can illustrate schematically what we expect it to produce. The input in Figure 6–32 consists of a primary reflected wavelet and several multiples. After deconvolution filtering, we obtain only the primary reflection that is indicated by a shortened wavelet.

Successful application of inverse filtering is seen in Figure 6–33. This seismic section was prepared by deconvolution of the data presented in Figure 6–17. The strong image paralleling the ocean floor in Figure 6–17 has been eliminated by deconvolution.

SEISMIC REFLECTIONS AND GEOLOGY

The aim of seismic reflection surveying is to reveal as clearly as possible the structure of the earth. In Chapter 4 we described how to calculate positions of layer boundaries and seismic wave velocities within these layers from reflection arrival times. In Chapter 5 and in the present chapter, we have been concerned with the procedures for obtaining recognizable reflections. For the most part, these calculations and data acquisition and processing procedures can be done objectively. The results, which can be displayed on profiles and maps, are the positions of reflections and the reflecting surfaces and the wave velocities that depend on the physical properties in the different layers. The seismologist's principal responsibility is to present this information as clearly as possible with a minimum of unsubstantiated assumptions and artifacts of processing.

What is the geological meaning of a seismic reflection, which is simply an indication of a boundary where acoustical impedance changes? Does this boundary mark a fault or the stratigraphic contact between two layers? How do we distinguish features that are not marked by sharp boundaries? Answers to these questions require the insight of geologists familiar with the patterns of folding, faulting, and stratigraphy that exist in different natural settings. Our principal aim in this book is to present the objective seismological procedures for properly identifying reflections, reflector positions, and wave velocities. However, as a test of the success of these procedures, it is important to show that the results can be explained in terms of geologically meaningful structural and stratigraphic features.

Depth Sections and Time Sections

So far, we have discussed seismic sections compiled from CDP-stacked traces that have been migrated and deconvolved as necessary to remove false images that are artifacts of processing. These seismic sections display reflection arrival time variations along a profile. They are called *time sections.*

It is possible to continue on to another procedure to prepare seismic sections that display reflector depths. At each common midpoint, the RMS velocities found earlier from velocity analysis are used in the Dix equation (Equation 4-54) to determine interval velocities. These values are then used in Equations 4-55 and 4-56 to calculate layer thicknesses and reflector depths below the common midpoints. A computer-controlled plotting machine can then be programmed to produce *depth traces* where the reflected wavelets are shifted to the appropriate reflector depths. A seismic section

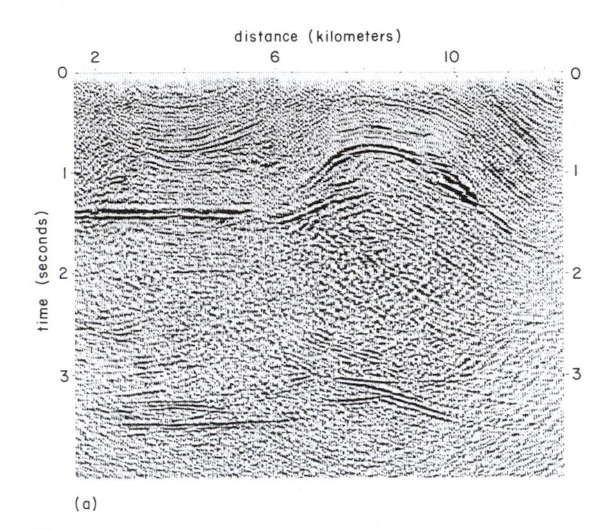

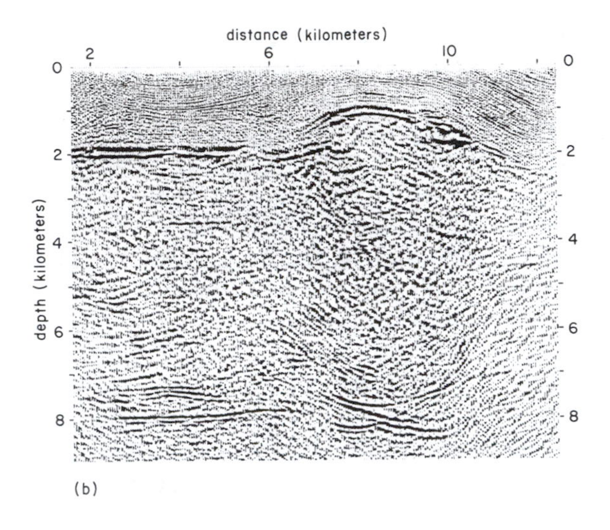

Figure 6–34
Comparison of (a) a land seismic time section and (b) a corresponding depth section. The seismic time section is an apparent cross section of subsurface structure expressed in terms of the variation in reflection arrival times along a profile. Because the time section has not been adjusted for velocity differences between rock units, the forms of structural images may be distorted. In a seismic depth section, interval velocities between reflections are used to calculate reflector depths along the profile. Insofar as correct interval velocities are used, the images seen along a profile indicate the undistorted form of structural features. (Courtesy of Western Geophysical.)

prepared from these traces is called a *depth section.* Here the wavelets image the reflector positions.

A time section and a depth section along the same profile are presented in Figure 6–34. Undulations and dips of the deeper reflectors, evident on the depth section, are not clearly displayed on the time section. Positions of reflections on the time section depend, of course, on velocity differences as well as on layer thicknesses. On the depth section, the wavelet positions depend only on layer thicknesses.

Seismologists have not reached a consensus on the extent to which preparation of a depth section is an interpretative rather than an objective data processing procedure. The basis for dispute is the velocity data used for this procedure. Because velocity analysis is performed by using some, but not all, of the reflections, there is some uncertainty about the accuracy of interval velocities that must be determined for depth calculations. The fact that standard velocity analysis procedures assume horizontal plane reflectors causes additional uncertainty about velocity accuracy. The errors introduced by these uncertainties are less noticeable on time sections than on depth sections.

For these reasons, some seismologists prefer to prepare only time sections. Later, if depth information is required for particular portions of the section, additional calculations can be made. For this purpose, velocity analysis can be repeated using more reflections to obtain more accurate interval velocities.

There are definite advantages to interpreting depth sections if reasonably good velocity data are available. Some seismologists prefer to add this extra step in data processing to determine additional structural and stratigraphic features. Then it is possible to identify places where additional refinement of velocity data will be desirable. Regardless of the advantages of depth sections, most interpretation of seismic reflection data begins with an examination of time sections.

Identification of Stratigraphic Boundaries

Geologists ordinarily group sequences of sedimentary rocks into units called *formations.* The rocks that make up a formation usually possess some features that remain more or less consistent over a large area. Formations include sequences thick enough that their occurrence can be clearly displayed on geologic maps and cross sections.

Formations can be described in terms of the age, thickness, and lithology of the constituent layers. For example, we can say that the Green Hill Formation consisting of Mississippian age rocks overlies the Parkdale Formation that is made up of Devonian age rocks. Moreover, we could describe the Green Hill Formation as ranging from 1000 to 1500 meters in thickness and containing three principal sandstone facies, each between 100 and 200 meters thick and interbedded with shale facies. The deeper Parkdale Formation might consist of alternating thin layers of shale and limestone.

Some formation boundaries mark distinct changes in lithology, such as an abrupt change from shale to limestone. But there are also boundaries that are recognized by a change in fossil assemblages rather than by lithologic changes. For example, the lower part of the Green Hill Formation may be a shale facies that is nearly identical to the upper shale facies of the Parkdale Formation. The boundary is recognized by a change from Devonian to Mississippian age fossils.

How do we distinguish different geologic formations by means of seismic reflections? This is an important question in interpreting seismic survey data. Some seismic surveys are planned so that a profile passes close to a well that penetrates the formations of interest. Then a travel time and velocity survey can be done with a special well geophone or by sonic logging.

By these methods, travel times to different depths are measured directly. These travel times can be marked on the stratigraphic column prepared by a geologist from examinations of rock cuttings and core. The stratigraphic column can then be replotted in terms of time rather than depth. In this form the reflection travel times to different boundaries and facies boundaries within a formation can be compared directly with the part of the seismic section that is close to the well. An example is given in Figure 6–35.

By comparing the positions of stratigraphic boundaries with the reflections in the seismic section, the interpreter can recognize directly which boundaries are producing reflections. Some formation boundaries may turn out to be reflectors, as may certain facies boundaries within a formation. What emerges are diagnostic patterns of reflections indicative of different formations. It is important to note that reflections are not necessarily produced at formation boundaries. This happens only if there is a significant change in acoustical impedance there.

The interpreter marks the positions of stratigraphic boundaries on the corresponding segment of the seismic section. Guided by the di-

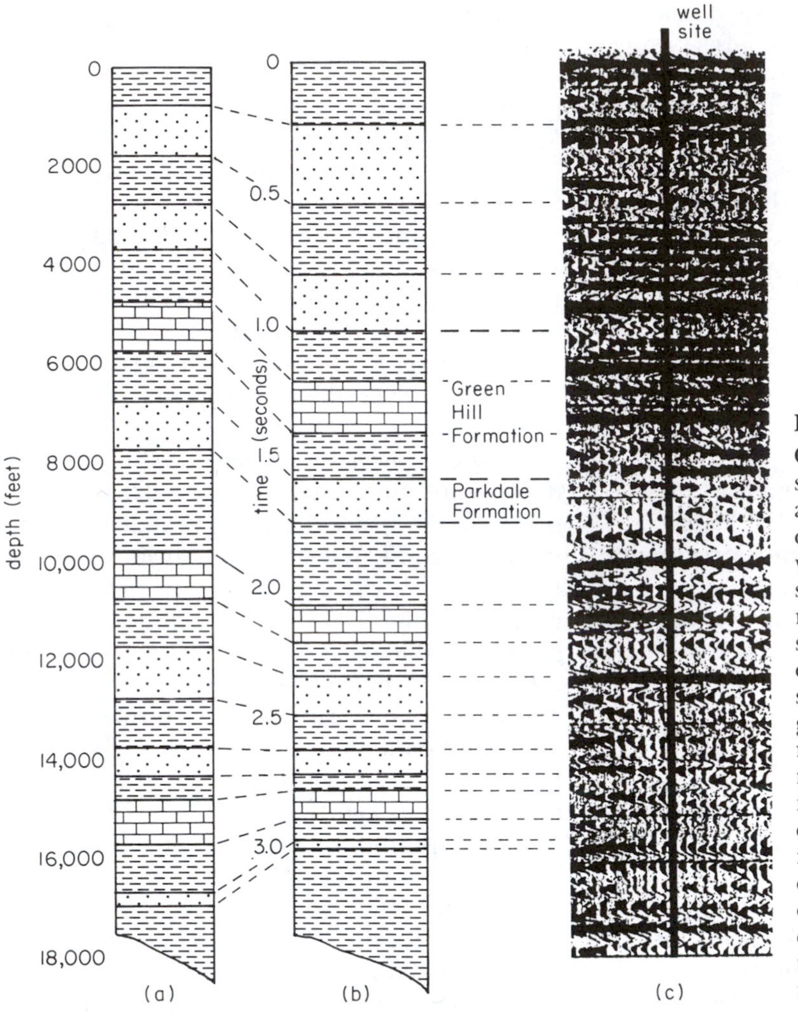

Figure 6–35
Correlation of boundaries on a stratigraphic column plotted according to (a) depth and (b) corresponding reflection times with (c) reflections on a seismic section. Identification of the reflectors imaged in seismic sections and calibration to actual depths are important steps in seismic data interpretation. In general, velocity–time–depth relationships obtained from well measurements form the basis for identification and calibration. These measurements furnish average velocities for conversion of reflection time, depth, and rock cuttings, and cores that are used to identify the lithologies between reflectors.

agnostic reflection patterns, the interpreter projects the positions of these boundaries away from the well by drawing them on the complete seismic section, as shown by the white lines in Figure 6–36. These lines delineate two formations.

Suppose that no wells are situated along the seismic profile. Then it is necessary to figure out what formations might be present from geologic maps and more distant wells. Using this information, the interpreter can prepare a stratigraphic column representative of the survey area. Then it may be possible to identify reflectors imaged on the seismic section. Synthetic seismograms can be prepared as aids in making such interpretations.

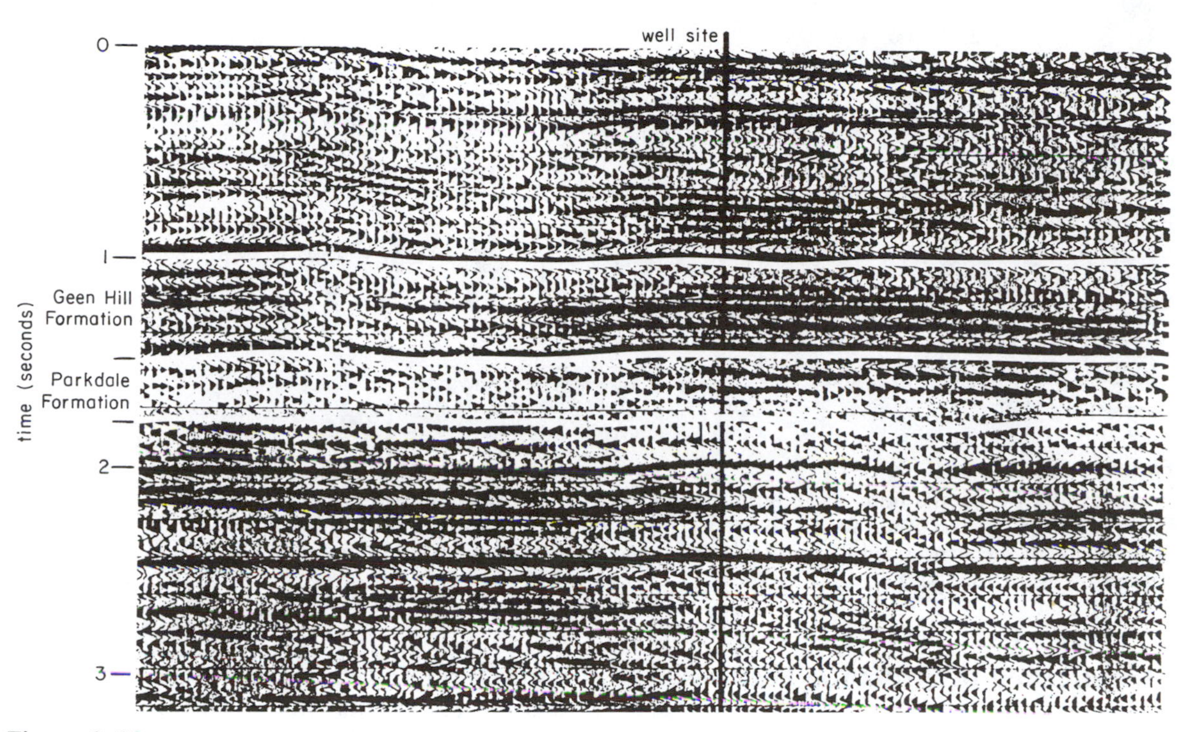

Figure 6–36
Formation boundaries on a seismic section are interpreted by correlating them with a stratigraphic column prepared from well survey data. Comparing seismic sections and lithologic data from wells is the first step in mapping reflecting lithological interfaces. Correlations in this example indicate the upper and lower boundaries of the Green Hill and Parkdale formations.

We have introduced all the analytical tools needed for preparing a synthetic seismogram. Rock samples of the different stratigraphic units in our representative stratigraphic column can be tested in the laboratory to determine wave velocity and density. Acoustical impedances determined from these values are used in Equations 2–22 to calculate reflection and transmission coefficients. Vertical lines representing these values can be drawn at the appropriate zero-offset times to compile a filter operator. We then choose a wavelet shape, convolve it with the filter operator, and apply absorption and geometrical spreading adjustments to obtain the synthetic seismogram. The procedure is illustrated in Figure 6–37.

A synthetic seismogram has diagnostic reflection patterns that would be expected for the representative stratigraphic column. Using it as a guide, as in Figure 6–38, the interpreter attempts to distinguish similar patterns on the seismic section. Although the reflections cannot be expected to occur in the same positions because layers vary in thickness along the profile, some of the patterns of successive reflections can be located in the seismic section.

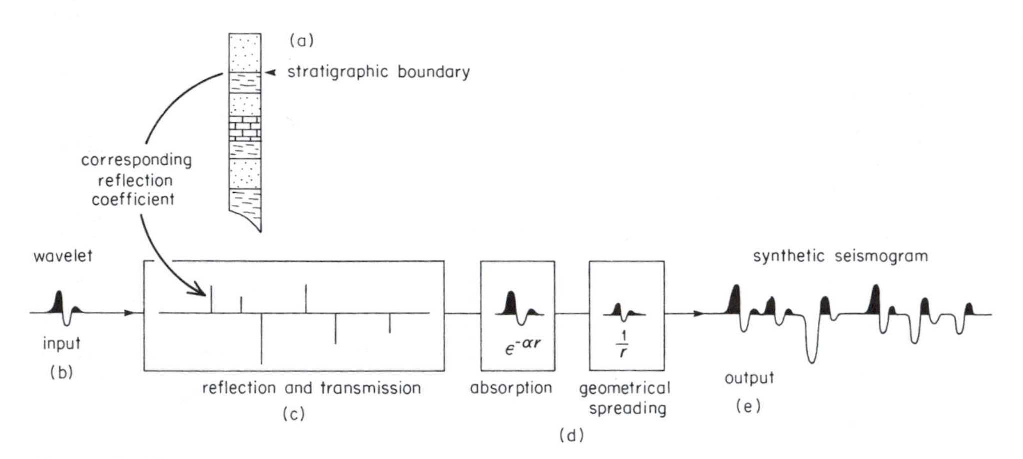

Figure 6–37
Preparation of (e) a synthetic seismogram by convolving (b) a wavelet with an earth filter operator that accounts for (c) the reflection and transmission seen at (a) the boundaries in the stratigraphic column, and by making (d) absorption and geometrical spreading adjustments.

Identification of Structural Features

Some seismic sections contain images that can be interpreted without difficulty. Discontinuous reflections clearly indicate faults, and undulating reflections reveal folded beds. Much more common, however, are ambiguous patterns that do not simply and clearly image a recognizable structure. For example, in Figure 6–39 are we looking at the image of a thrust fault, or do the reflections indicate the stratigraphic pinchout of a layer? Here the choice of interpretations should be made by an experienced earth scientist. On the basis of independent knowledge of other geologic features in the area, one interpretation may be judged more probable than another.

Most interpretations of structural features are marked directly on seismic time sections. In Figure 6–40, faults and folds that conform to discontinuities and curves in the pattern of reflections are drawn on the lower time section. Can an alternate interpretation be proposed? The faults on the left are indicated quite clearly by displaced reflections, but evidence for the fault on the far right is not as convincing. Here the pattern is more ambiguous because of diffraction arcs that have not been entirely removed by migration. The interpretation of a this fault in this position is consistent with the structural style of the region, however, and some of the strong reflectors appear to be interrupted by such a fault. The apparent continuation of other reflectors across the fault may be the result of diffraction.

The seismic time section in Figure 6–41 reveals a complicated sequence of folds and faults. In this geologic setting, diffractions and problems such as those illustrated in Figure 6–15 can seriously distort the reflection patterns. Although a depth section might be preferable for interpretation, it is unlikely that velocities could be determined accurately

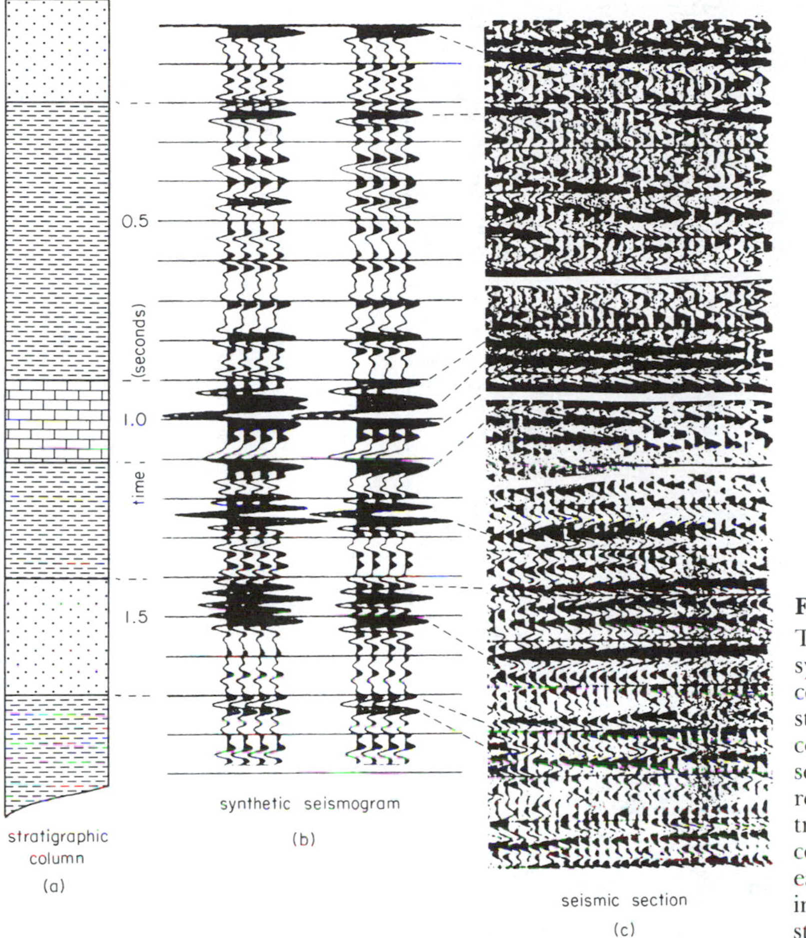

Figure 6–38
This example shows (b) a synthetic seismogram corresponding to (a) a stratigraphic column and compares it with (c) an actual seismogram recorded in the same region. The synthetic seismic traces were obtained by convolving a wavelet with an earth filter operator that incorporates the features of the stratigraphic column.

enough to obtain reliable results. For this reason, most interpreters work with time sections.

Preparation of synthetic seismic sections can be a useful way to verify interpretations marked on a time section. In preparing a synthetic section, interpreters must use all the factors in Figure 6–37 to construct individual traces. In addition, the computer must also be programmed to reproduce the displacement of reflections related to dip. This is a complicated procedure. In Figure 6–42, the structural interpretation of the time section was used to compute the synthetic section that reproduces the important features.

So far, we have discussed the interpretation of individual profiles. Where a survey consists of several profiles crossing an area, it is possible to compile structure contour maps that reveal the form of a reflector beneath this area, as in Figure 6–43. Interesting three-dimensional fault patterns can be displayed in this way.

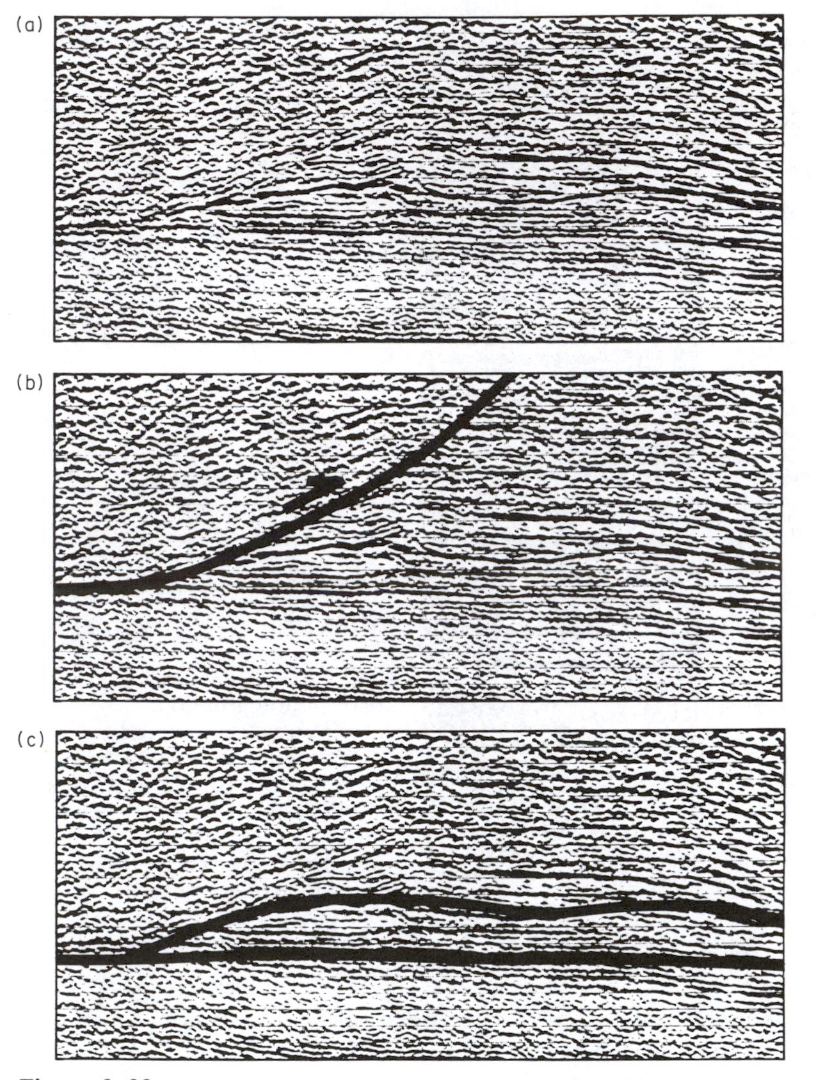

Figure 6–39
Alternate interpretations of images on a seismic section. For lack of other independent information about the geology along this profile, it can be difficult to tell whether (a) the reflection patterns appearing on the profile represent (b) a thrust faulted structure or (c) a stratigraphic pinchout. (Courtesy of Saga Petroleum.)

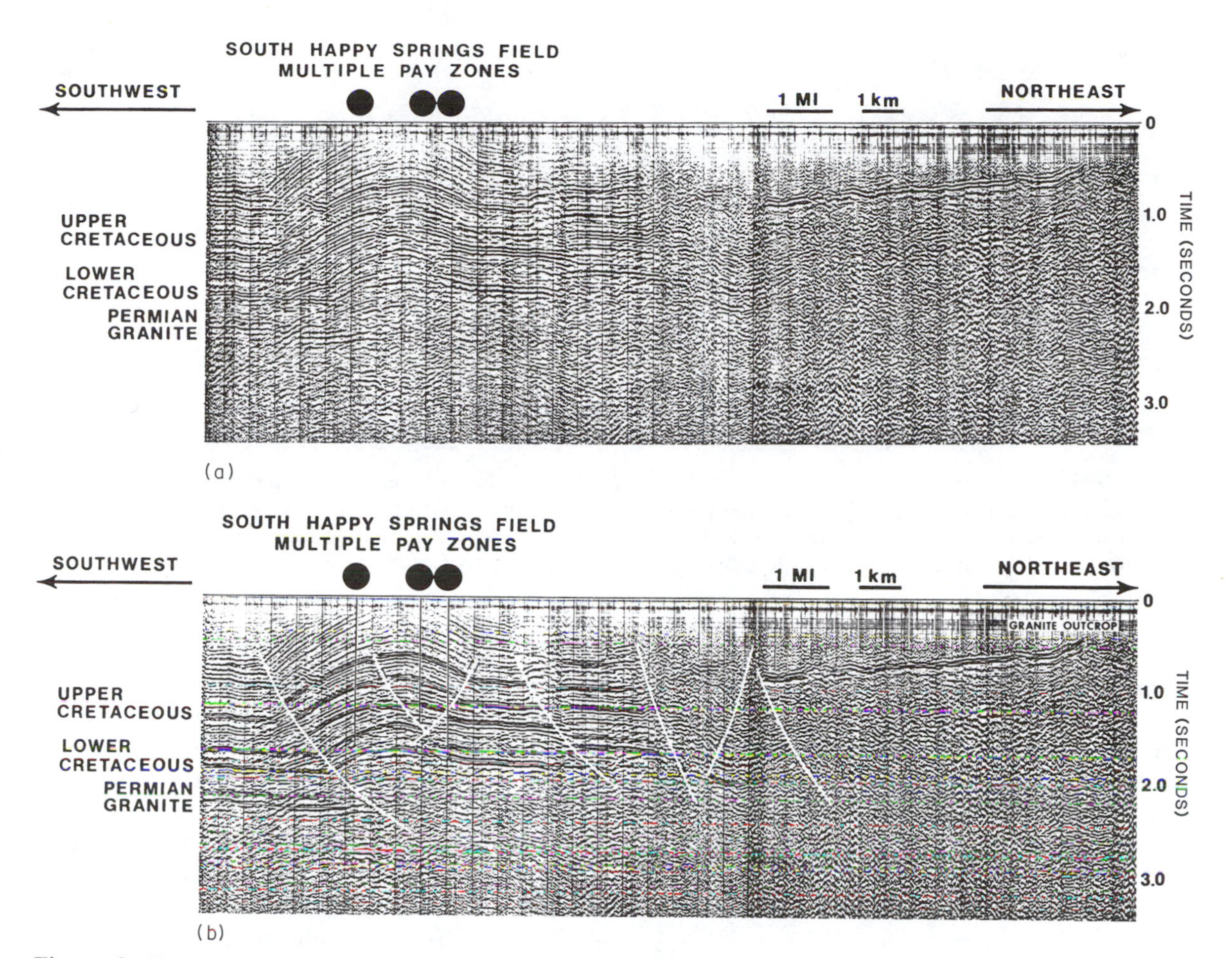

Figure 6–40
Seismic section (a) before interpretation and (b) after interpretation, with folds and faults drawn along reflection images. The seismic section shows complicated subsurface features defined by folded and faulted rocks. Well information was used to identify lithological boundaries. Black spots indicate well locations where multiple producing zones are present. The seismic time section is unmigrated, therefore, the presence of diffraction images had to be taken into account when making the interpolation. There can be some confusion about the curved patterns; some may be true reflections and others diffraction arcs. The interpretation includes folds and thrust faults that are consistent with structures expected in this region of Wyoming. (Courtesy of Conoco, Inc.)

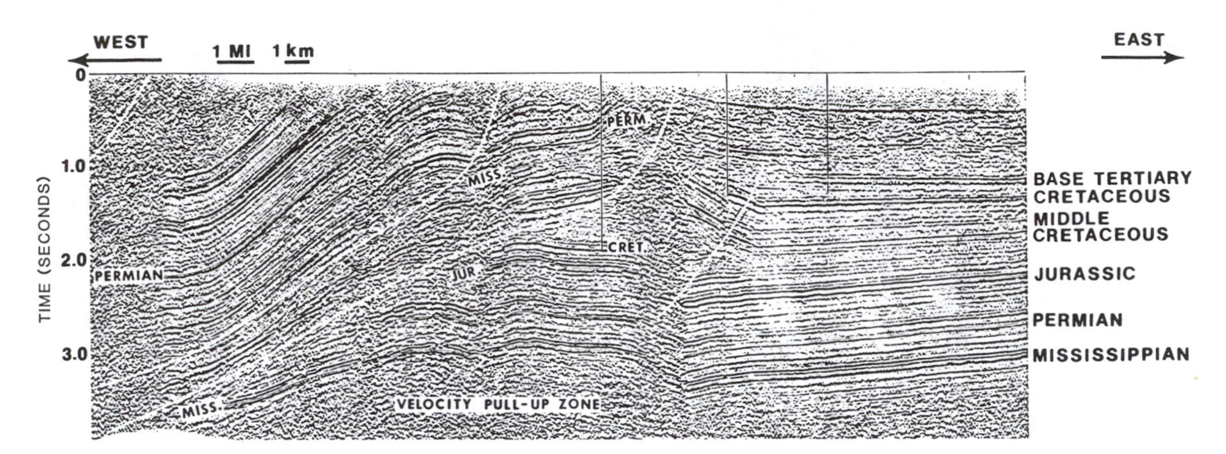

Figure 6–41
Complicated structural features interpreted from images on a seismic section in Figure 6–14. The data are from Wyoming along a profile crossing a thrust belt, and some of the structural features were confirmed by drilling. In the velocity pull-up zone there are images of false structures. (Courtesy of Conoco, Inc.)

Exploration Targets

The many structures in which oil and natural gas have been discovered are too numerous to describe in detail here. The methods we have discussed for identifying stratigraphic units and structural features are basic to the search for anticlines, fault traps, and stratigraphic traps.

Sometimes a target stands out clearly on a seismic section. This is the case of the salt dome recognized by the image in Figure 6–44 and the case of the reef structure imaged in Figure 6–45a. It is quite a success to obtain such a clear image of an exploration target. More often, the pattern is subtle and requires experience to identify it. Images over reef structures such as the one in Figure 6–45b would be unrecognizable unless the interpreter knows what to look for. This pattern was discovered from test surveys of previously discovered reef structures.

Efforts to detect the presence of oil have been largely unsuccessful. Because its acoustical impedance is too similar to that of water, seismic waves are insensitive to different proportions of oil and water. Procedures for detecting natural gas have been a little more successful, owing to its very low acoustical impedance. Because of this, low impedence, reflections from gas-saturated zones of reservoir rock tend to have much larger amplitudes than reflections from adjacent oil- or water-saturated zones. Parts of otherwise continuous reflections that show distinctively higher amplitude have been used with some success to identify gas-saturated reservoirs. These high-amplitude parts are called *bright spots*. An example is shown in Figure 6–46.

Recently, some effort has gone into seismic surveys of coal prospects. High-resolution CDP profiling techniques involving short spreads and digital sampling intervals have been successful for detecting shallow coal

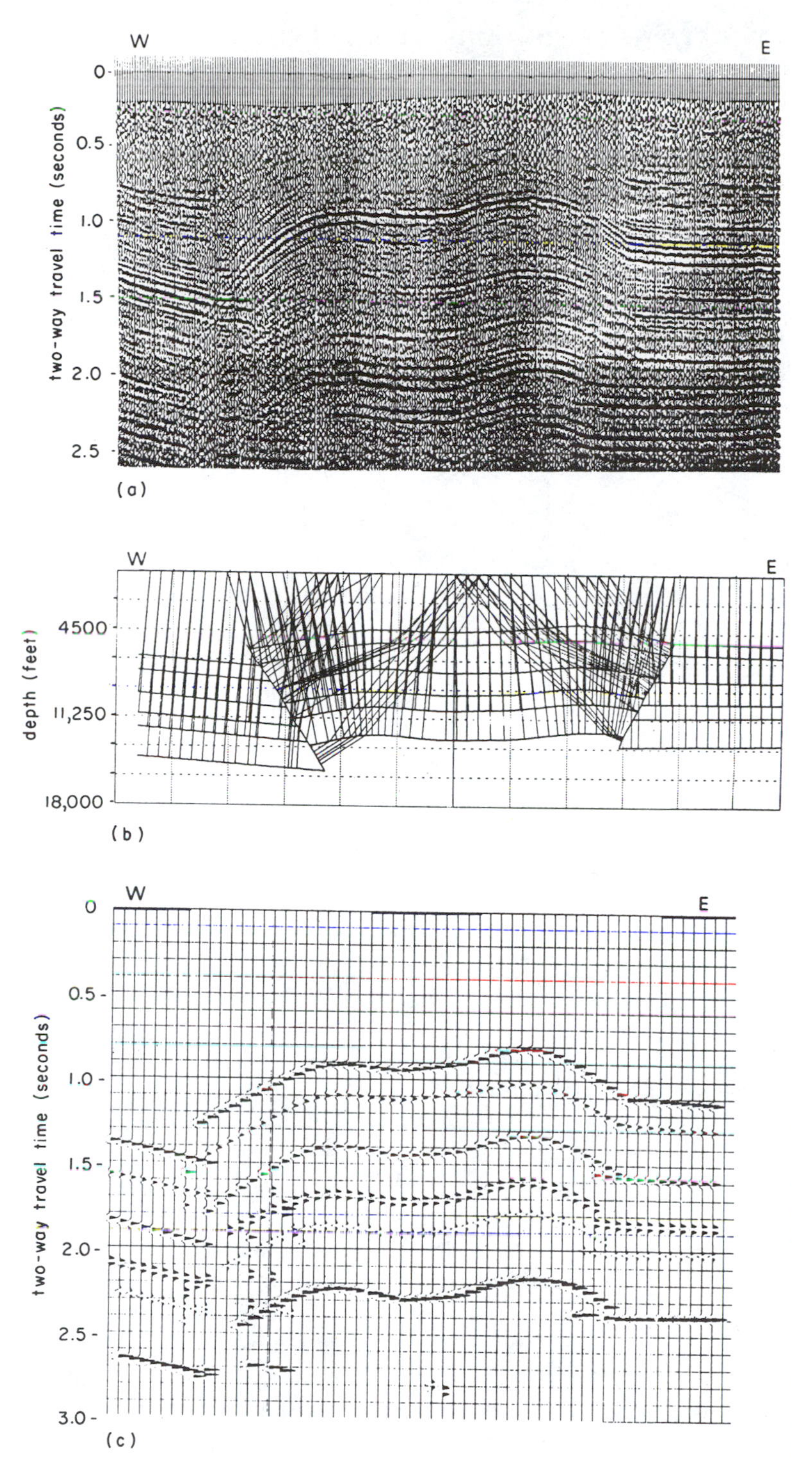

Figure 6–42

Comparison of (a) the images on a seismic section from Wyoming and (c) the images on a synthetic seismic section prepared from (b) the reflection paths on a cross section through a model that proposes a faulted structure. Reflection times along these paths were calculated from the reflector positions and layer velocities assumed for the model. (Courtesy of GeoQuest International and Seismograph Service Corporation.)

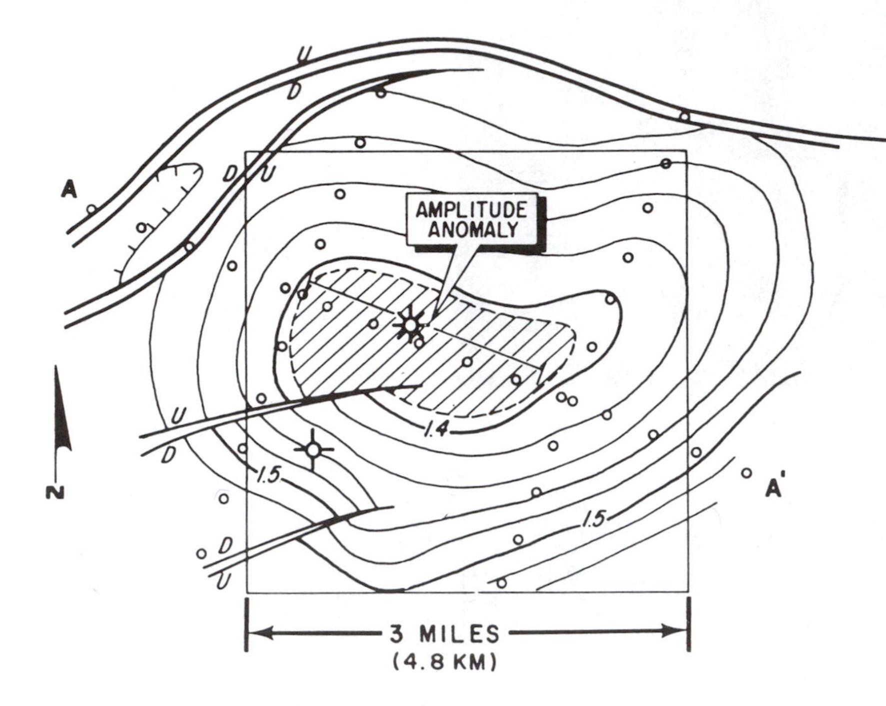

Figure 6–43
Structure contour map showing reflection times obtained from several seismic sections and the positions of faults indicated on these section. (From M. W. Schramm, Jr., E. V. Dedman, and J. P. Lindsey, Practical stratigraphic modeling and interpretation. *Seismic Stratigraphy—Applications to Hydrocarbon Exploration,* p. 489 (ed. C. E. Peyton). Tulsa, Okla., AAPG Memoir 26, 1977. Published by the American Association of Petroleum Geologists, reprinted with permission from AAPG.)

seams. The seismic section in Figure 6–47 clearly indicates coal at depths of a few hundred feet.

These high-resolution reflection profiling methods are being adapted for engineering and groundwater surveys. Common depth point processing techniques have been successful for enhancing shallow reflections that formerly were unrecognizable because of strong refracted waves and surface waves. These applications are evidence of the increasingly diverse uses of reflection seismology.

Three-Dimensional Seismic Reflection Surveys

For exploration targets of particular importance, three-dimensional reflection surveying may be required to image accurately the subsurface features of interest. Geometrical aspects of three-dimensional reflection measurements are introduced in Chapter 4, and field surveying procedures are described briefly in Chapter 5. Basic procedures for processing these measurements are an extension of the two-dimensional procedures we have already described.

Ideally, a three-dimensional survey records on receivers at a square grid of points reflections produced by sources located in turn at each of these same grid points. We see from Figure 6–48 that each grid point is the common midpoint for source–receiver pairs situated on a radial series of profiles. Therefore, by appropriate sorting it is possible to make up a three-dimensional CDP gather that includes source–receiver pairs from all these intersecting profiles.

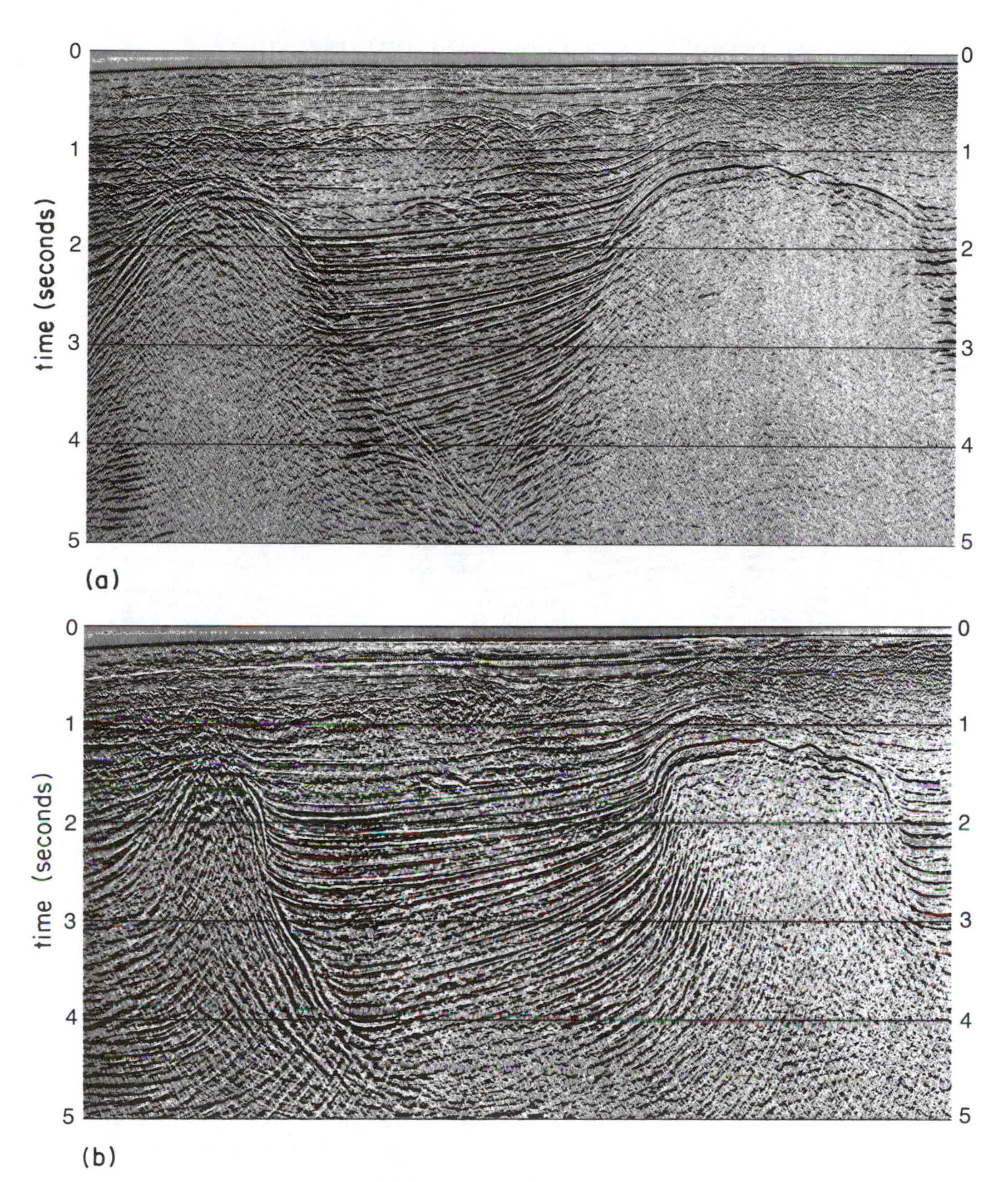

Figure 6–44

Clear image of a salt dome on a seismic section. (a) A 48-fold common midpoint stack from the Gulf of Mexico. (b) The data are shown after migration processing. Notice the change in size and clarity of the salt dome images on the migrated section. (Courtesy of Western Geophysical.)

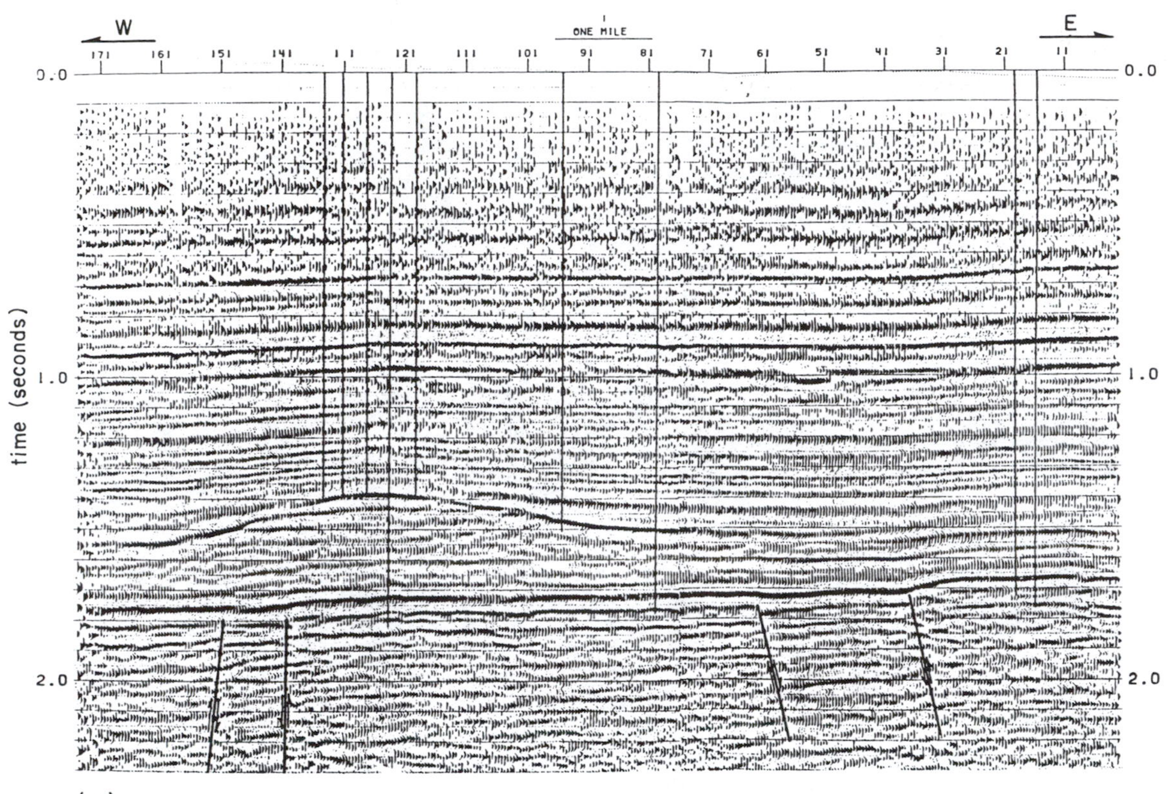

Figure 6–45
(a) Seismic section clearly displaying image of a reef structure in west Texas. (Courtesy of Conoco, Inc.).

Static corrections and normal move-out adjustments are done by the same procedures used in the processing profile data to obtain equivalent zero-offset traces for all source–receiver pairs in the three-dimensional gather. These traces are then stacked. This procedure is repeated to prepare CDP-stacked traces for all the points on the survey grid.

Results can be displayed along profiles extending in different directions and on diagrams showing a three-dimensional perspective. As with two-dimensional data, the three-dimensional stacked traces can be plotted side by side to prepare seismic time sections. Because data are available from a grid of common midpoints, these seismic sections can be compiled for profiles in many different directions.

Another way of displaying the results of a three-dimensional reflection survey is by horizontal sections. As shown in Figure 6–49, imagine that the five rectangles could be folded into the top and sides of a block. Each side is a vertical seismic section, where varying shades of gray indicate amplitude of oscillations on the stacked traces making up the sec-

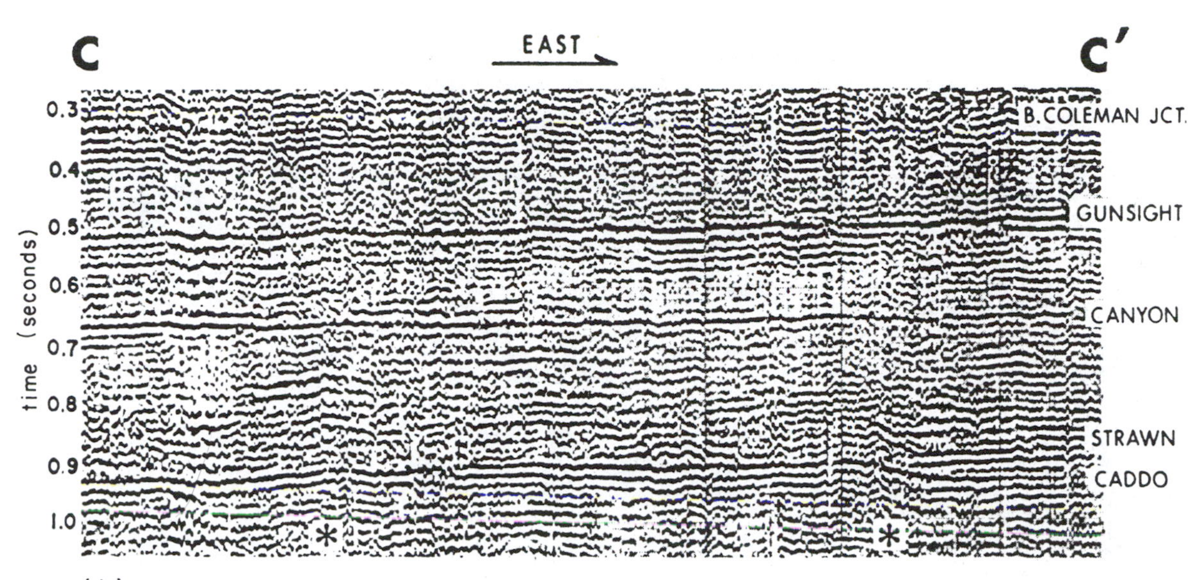

Figure 6–45 (cont.)
(b) Seismic section showing the subtle image of a reef structure. Identification of the reef structure was based on earlier knowledge of reflection patterns obtained over other similar features that had been confirmed by drilling. (From J. C. Harwell and W. R. Rector, North Knox City Field, Knox County, Texas. In *Stratigraphic Oil and Gas Fields—Classification, Exploration Methods, and Case Histories,* p. 457 (ed. R. E. King). Tulsa, Okla., AAPG Memoir 26, SEG Special Publication No. 10, 1972. Published jointly by the American Association of Petroleum Geologists and the Society of Exploration Geophysicists, reprinted with permission from AAPG and SEG.)

tion. Reflections appear as dark zones extending across parts of each section. The top of the block is a "horizontal time slice" which shows how oscillation amplitude at the same time on all stacked traces varies over the survey area. Here a continuous dark or white zone indicates the area of a reflector from which reflections have the same zero-offset arrival time.

Three-dimensional migration can be done to enhance the accuracy of a survey. To understand the nature of this procedure, look again at the two-dimensional graphical method illustrated in Figure 6–22. To carry out three-dimensional migration, we would replace the circular arcs with spherical surfaces. Then to locate the reflector, we would replace the line that is tangent to the arcs with a plane tangent to the spherical surfaces. Similarly, the principles used in more complicated two-dimensional migration procedures can also be applied in three dimensions. Figure 6–50 illustrates the improvement that can be obtained along a profile by means of three-dimensional migration.

Three-dimensional seismic surveying is our most advanced method for imaging an earth structure by means of measurements made on

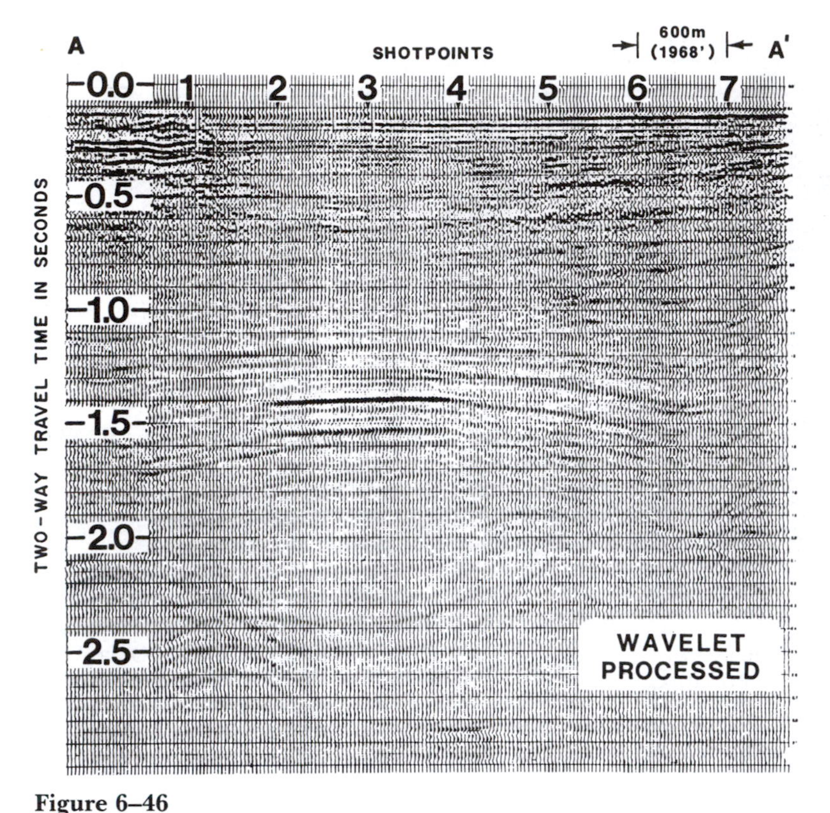

Figure 6–46

Seismic section with *a bright spot* along the reflection from gas-saturated reservoir rock in the Gulf of Mexico. The presence of gas rather than water or oil in the reservoir pore spaces produces a lower velocity in the reservoir and a greater contrast with the velocity in the overlying rock. The reflection coefficient is larger, which explains the large reflection amplitude over the gas-saturated part of the reservoir. (From M. W. Schramm, Jr., E. V. Dedman, and J. P. Lindsey, Practical stratigraphic modeling and interpretation. In *Seismic Stratigraphy—Applications to Hydrocarbon Exploration,* p. 483 (ed. C. E. Payton). Tulsa, Okla., AAPG Memoir 26, 1977. Published by the American Association of Petroleum Geologists, reprinted with permission from AAPG.)

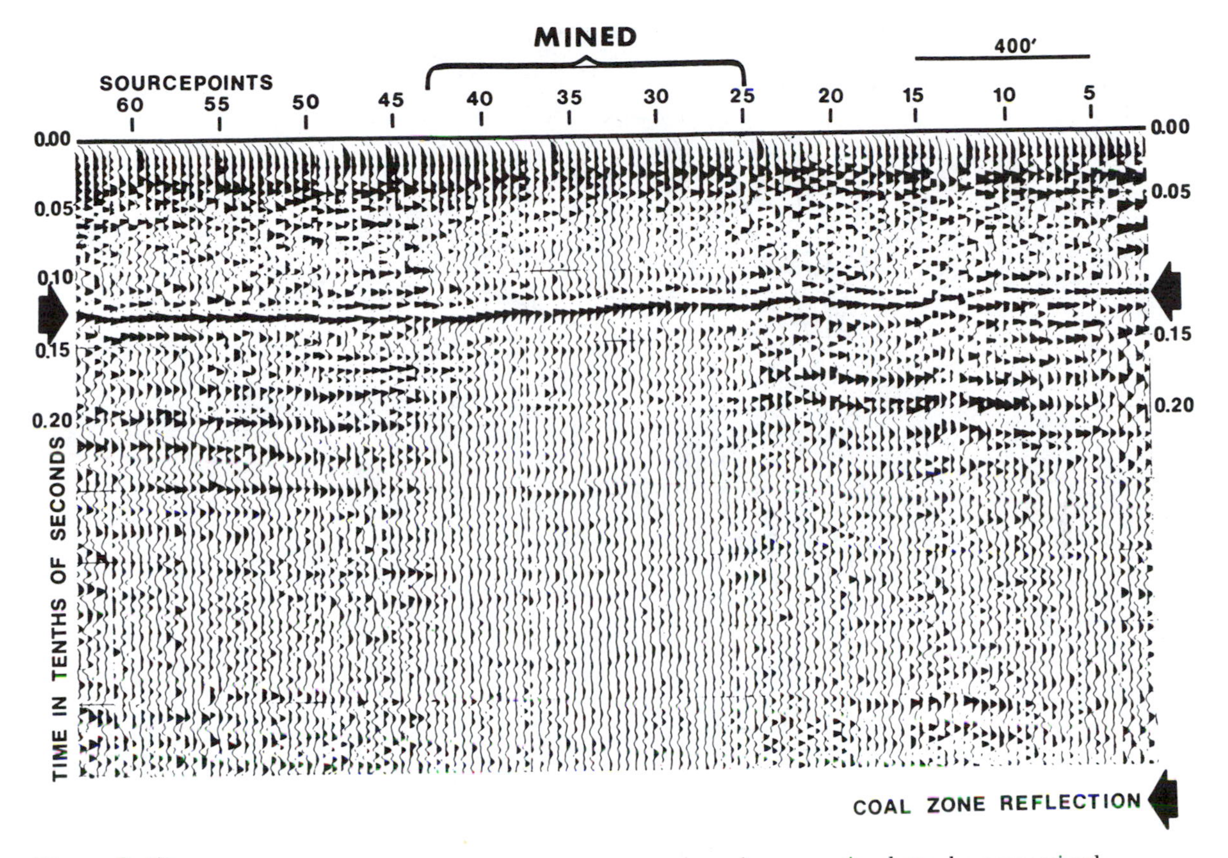

Figure 6–47
High-resolution seismic section indicating shallow coal seams. From the character change in the section, the area mined can be recognized. (Courtesy of Conoco, Inc.)

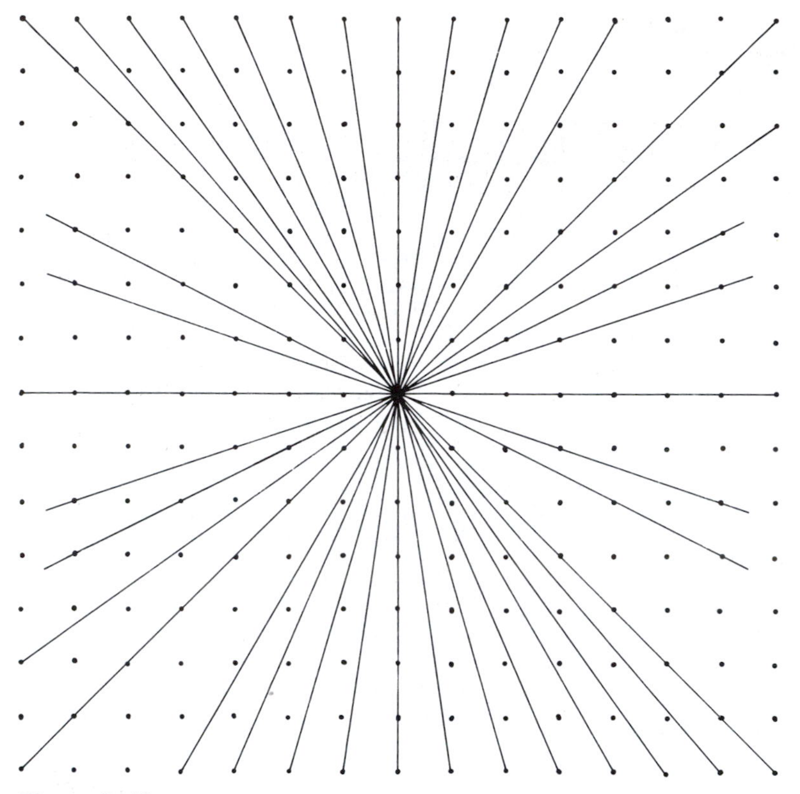

Figure 6–48

Grid showing source and receiver positions for a three-dimensional seismic reflection survey. Profiles connecting different alignments of grid points intersect at the common midpoint for all the source–receiver pairs that are included in a three-dimensional CDP gather.

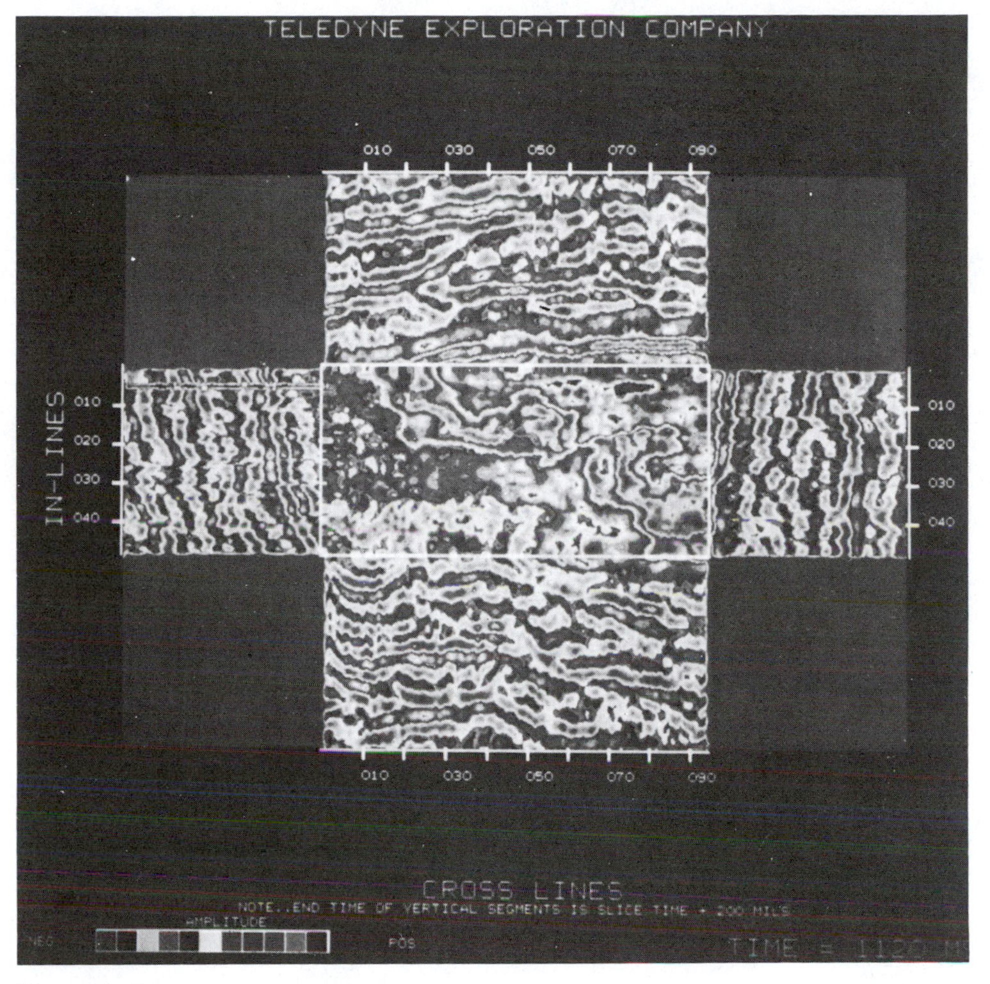

Figure 6–49

Three-dimensional display of seismic reflection data. Five rectangles could be folded to make the sides and top of a box diagram. The vertical sides correspond to different seismic time sections, and the top represents a horizontal time slice. Oscillation amplitude on zero-offset stacked traces is indicated by different shades of gray. Elongated dark zones on the vertical sides and continuous dark areas on the top tend to indicate reflections. (Courtesy of Teledyne Exploration Company.)

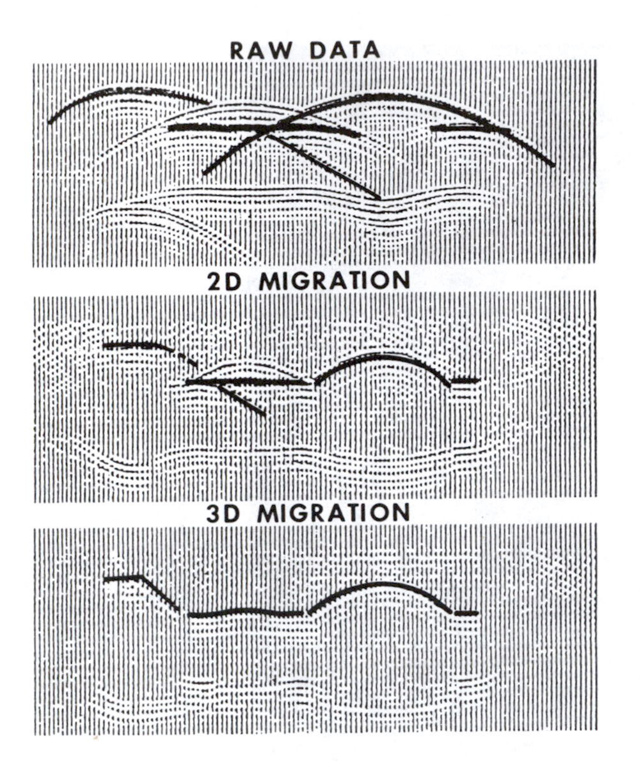

Figure 6–50
Seismic model sections consisting of unmigrated CDP-stacked traces, traces migrated along the two-dimensional profile, and traces migrated in three dimensions. (Courtesy of Geophysical Service, Inc., a subsidiary of Texas Instruments, Inc.)

the surface. The field operations and data processing procedures can be costly and time-consuming, and they require computing facilities only recently available. However, the additional effort should produce more accurate results that can be obtained by any other geophysical procedure.

STUDY EXERCISES

1. Suppose that a digital seismogram has the following form:

   0, 1, 2, 1, 0, −1, −3, −2, −1, −3, 0, 1, 0

   Perform a cross-correlation using the operator 1, 3, 5, 3, 1. Assign the computed values with respect to the middle value of the operator. Plot the original and correlated output. What is the difference between cross-correlation and weighted average? Is there a smoothing effect? Explain!

2. Suppose that you have planned a CDP reflection survey with 750 source–detector points at 50-m intervals. What will the distance between adjacent common reflecting points (or common mid-

points) be? What will the actual length of the profile on the reflectors be?

3. You have a seismic crew equipped with a 120-channel recording system. You are told that a common midpoint density of 40 reflection points per kilometer is necessary with 30-fold data. What will the distance between adjacent source points be?

4. A reflection survey is to be planned to map a seismic interface at a depth of 1000 m. The interval velocity for the top layer is 2000 m/s. Because of desired resolution, there is a limit of 0.08 s for the maximum normal move-out time. Define the maximum offset for a receiver in the recording spread.

5. Assume that a source point is at an elevation of 200 m and that a receiver is at an elevation of 125 m. A weathered layer with a velocity of 800 m/s lies over a layer with a velocity of 1500 m/s. A seismic refraction survey gives thicknesses of 20 and 30 m for the weathered layer at the points where the source and receiver are located, respectively. If a datum of 150 m is used, what total two-way static correction value in seconds will be applied to a reflection recorded at this receiver?

6. You have unmigrated and migrated seismic sections for a particular profile. A planar interface indicates slopes of 25 and 30 degrees before and after migration. Do these values fit with simple geometrical relations? If they do not fit, give an explanation.

7. On a seismic record section, a reflection is seen to extend between two points A and B, where average seismic velocity is 2500 m/s. Termination of the reflection is attributed to faults. Point A is at $x = 100$ m and $t_0 = 1.75$ s, and point B is at $x = 1500$ m and $t_0 = 1.25$ s. Determine the actual location of faults by computing coordinates for the points A and B after migration.

8. Compute the length of reflector segments in Exercise 7 before and after migration. Which one is shorter?

9. Compare the time and depth sections given in Figure 3–34 to point out differences in the apparent subsurface geometry.

SELECTED READING

Anstey, N. A., *Seismic Interpretation—The Physical Aspects*. Boston, International Human Resources Development Corp., 1977.

Cassand, J., B. Damotte, A. Fontanel, G. Grau, C. Hemon, and M. Lavergne, *Seismic Filtering*. Tulsa, Okla., Society of Exploration Geophysicists, 1971. (Translated by N. Rothenburg from *Le Filtrage en Sismique*, Paris, Editions Technip., 1966.)

Coffeen, J. A., *Seismic Exploration Fundamentals*. Tulsa, Okla., Petroleum Publishing Co., 1978.

Dix, C. H., *Seismic Prospecting for Oil*. New York, Harper, 1952.

Dix, C. H., Seismic velocities from surface measurements, *Geophysics*, v. 20, n. 1, pp. 68–86, February 1955.

Dobrin, M. B., *Introduction to Geophysical Prospecting*, 3rd edition. New York, McGraw-Hill, 1976.

Fitch, A. A., *Seismic Reflection Interpretation.* Berlin, Gebruder Borntraeger, 1976.

Grant, F. S., and G. F. West, *Interpretation Theory in Applied Geophysics.* New York, McGraw-Hill, 1965.

Kanasewich, E. R., *Time Sequence Analysis in Geophysics.* Edmonton, University of Alberta Press, 1973.

Kearey, P., and M. Brooks, *An Introduction to Geophysical Exploration.* Oxford, England, Blackwell Scientific Publication, 1984.

Kulhanek, O., *Introduction to Digital Filtering in Geophysics.* Amsterdam, Elsevier, 1976.

McQuillan, R., M. Bacon, and W. Barclay, *An Introduction to Seismic Interpretation.* Houston, Gulf Publishing Co, 1979.

Neidell, N. S., and F. Poggiagliolmi, Stratigraphic modeling and interpretation. In *Seismic Stratigraphy Applications Hydrocarbon Exploration,* pp. 389–416 (ed. C. E. Payton). Tulsa, Okla., AAPG Memoir 26, 1977.

Parasnis, D. S., *Principles of Applied Geophysics,* 4th edition. London, Chapman and Hall, 1986.

Payton, C. E. (editor), *Seismic Stratigraphy—Applications to Hydrocarbon Exploration.* Tulsa, Okla., AAPG Memoir 26, 1977.

Robinson, E. A., *Multichannel Time Series Analysis with Digital Computer Programs.* San Francisco, Holden-Day, 1967.

Robinson, E. A., and S. Treitel, *The Robinson-Treitel Reader.* Tulsa, Okla., Seismograph Service, 1973.

Robinson, E. A., and S. Treitel, *Geophysical Signal Analysis.* Englewood Cliffs, N.J., Prentice–Hall, 1980.

Sengbush, R. L., *Seismic Exploration Methods.* Boston, International Human Resources Development Corp., 1983.

Sharma, P., *Geophysical Methods in Geology.* Amsterdam, N, Elsevier, 1986.

Sheriff, R. E., *Encyclopedic Dictionary of Exploration Geophysics.* Tulsa, Okla., Society of Exploration Geophysicists, 1973.

Sheriff, R. E., Inferring stratigraphy from seismic data, *Bulletin of the American Association of Petroleum Geologists,* v. 60, n. 4, pp. 528–542, 1976.

Sheriff, R. E., Limitations on resolution of seismic reflections and geologic detail derivable from them. In *Seismic Stratigraphy—Applications to Hydrocarbon Exploration,* pp. 3–14 (ed. C. E. Payton). Tulsa, Okla., AAPG Memoir 26, 1977.

Sheriff, R. E., *A First Course on Geophysical Exploration and Interpretation.* Boston, International Human Resources Development Corp., 1978.

Sheriff, R. E., *Seismic Stratigraphy.* Boston, International Human Resources Development Corp., 1980.

Sheriff, R. E., and L. P. Geldart, *Exploration Seismology.* Volume 2: *Data Processing and Interpretation.* Cambridge, England, Cambridge University Press, 1982.

Society of Exploration Geophysicists. *Geophysical Case Histories,* Volumes 1 and 2. Tulsa, Okla., 1948, 1956.

Waters, K. H., *Reflection Seismology.* New York, Wiley, 1987.

Widess, M. B., How thin is a thin bed?, *Geophysics,* v. 38, n. 6, pp. 1176–1180, December 1973.

SEVEN

Gravity on the Earth

How can you weigh a geologic structure? It can be done by gravity surveying. The "scales" that you need to accomplish this is called a gravimeter. It is used to measure the attraction of gravity from place to place on the earth's surface. The aim is to detect effects of the differences in rock density related to subsurface geologic structure. But the attraction of gravity can be explained almost entirely in terms of the size and shape of our planet and its mass and speed of rotation. So the first step in correctly interpreting a gravity survey is to account for these very strong effects. After this is done, we may be able to recognize the extremely weak variations in gravitational attraction that are clues to subsurface irregularities in rock density.

This chapter is concerned with (1) some basic ideas about the nature of gravity, (2) the delicate instruments we use to measure gravity on the earth, and (3) the strong gravity variations related to the size and shape of this spinning globe. With this background, we can proceed in Chapters 8 and 9 to analyze the minute variations of gravity that can aid us in learning more about subsurface geology.

THE NATURE OF GRAVITY

Early in the seventeenth century, Johannes Kepler accurately described the motions of planets in the solar system by three simple statements. Elsewhere at about the same time, Galileo discovered that small objects close to the earth fall to its surface at uniformly accelerated speed. But neither man understood that the heavenly bodies follow their orbits, and that small objects fall to the earth because of the same kind of force.

A half century later, Isaac Newton (1643–1727) recognized that these seemingly different motions were governed by a force of gravitation that all objects exert on one another. He discovered the force governing these motions by combining his three laws of motion with Kepler's earlier statements, and he went on to formulate the *universal law of gravitation*, which relates a force F of attraction between two particles to their respective masses, m_1 and m_2, and the distance r between them:

$$F = G\frac{m_1 m_2}{r^2} \qquad (7\text{–}1)$$

The constant of proportionality G, which must be measured, is called the universal gravitational constant. More than a century passed before Henry Cavendish performed the experiments that yielded the first useful value of G. His apparatus (Figure 7–1) consisted of two large, identical spheres mounted on a horizontal rectangular frame. A thin horizontal rod with identical spherical masses at its ends was suspended on a vertical fiber. The position of the rod could be controlled by twisting the fiber. The idea was to hold the rod in the same position while shifting the large spherical masses from A and B to C and D. This was done by changing the twist of the fiber to counteract exactly the gravitational force F between adjacent large and small masses separated by a measured distance r. Because the shear modulus of the fiber had been determined in another experiment, it was possible to calculate F and then rearrange Equation 7–1 to find G. Maintaining all the masses in the same horizontal plane kept the vertical gravitational attraction of the earth from affecting the experiment. The Cavendish experiment has been repeated many times using modern apparatus to obtain the value

$$\begin{aligned} G &= 6.673 \times 10^{-8}\ (\text{g/cm}^3)/\text{s}^2 \\ &= 6.673 \times 10^{-11}\ (\text{kg/m}^3)/\text{s}^2 \end{aligned}$$

The universal law of gravitation expresses the mutual force of attraction between two particles. We should think of each of these particles as being infinitely small, so that its mass is concentrated at a single point. What can happen to these particles as a consequence of this gravitational force? To answer

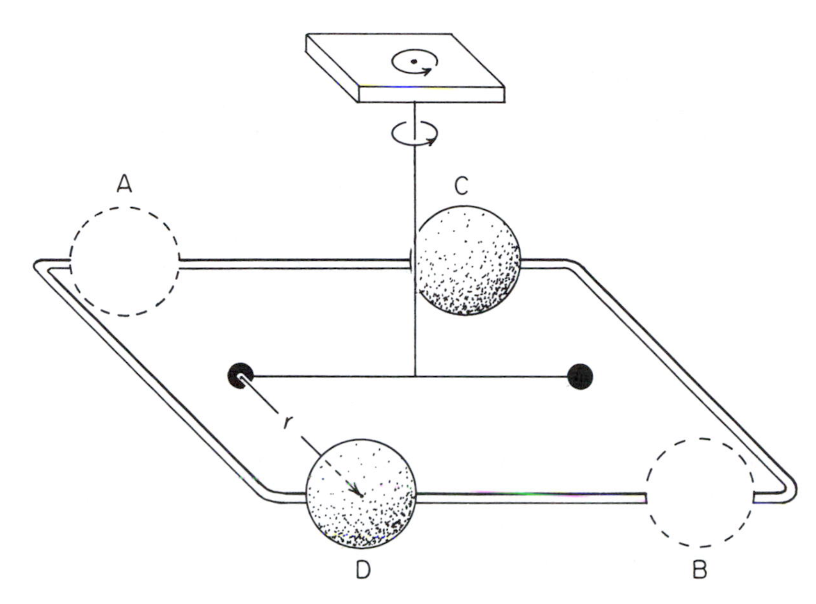

Figure 7–1
Arrangement of apparatus used in the Cavendish experiment for measuring the universal gravitational constant. Large masses can be moved along the frame from positions A to C and B to D. The rod with small masses at its ends can be rotated by twisting the fiber on which it is suspended.

this question, we can rearrange Equation 7–1 in the following ways:

$$F = m_1\left(\frac{Gm_2}{r^2}\right) = m_1 g_2 \qquad (7\text{–}2a)$$

$$F = m_2\left(\frac{Gm_1}{r^2}\right) = m_2 g_1 \qquad (7\text{–}2b)$$

Newton's second law of motion states that force is the product of mass and acceleration. This tells us that the terms g_1 and g_2 represent acceleration. If the particle m_1 is free to move, it will be drawn toward m_2 at a speed that constantly increases, or accelerates, at the rate g_2. We say that g_2 is the gravity of the mass m_2 at the distance r. Similarly, g_1 is the gravity of m_1 at this distance. It is the acceleration imparted on the particle m_2 by the presence of m_1.

Gravity g associated with the presence of any single particle is independent of all other particles. It depends on the mass m of this single particle and distance r away from it:

$$g = \frac{Gm}{r^2} \qquad (7\text{–}3)$$

Gravity is the capacity of the particle to accelerate other objects. A vector is useful for describing the acceleration of such an object. Because acceleration is toward the particle, the vector must point in that direction. The strength of acceleration, calculated from Equation 7–3, can be represented by the length of the vector.

Vectors in Figure 7–2 illustrate how the gravity of particle m_1 affects other particles at different locations. At the same time, the particle m_1 is affected by the gravity of each of these other particles. Vectors representing the gravity of these particles are combined in Figure 7–3 to determine their collective effect on the acceleration of m_1. Similar diagrams could be drawn for the other particles because all particles have a mutual gravitational attraction. This basic reasoning can be used to learn about the gravity of larger objects. The earth,

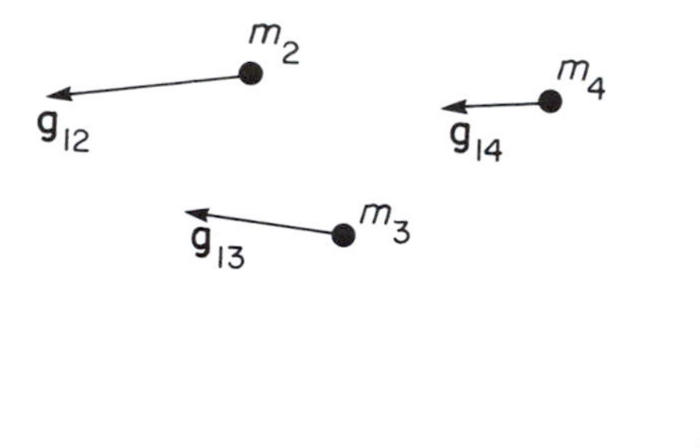

Figure 7–2
Vectors labeled $\mathbf{g}_{12}$, $\mathbf{g}_{13}$, $\mathbf{g}_{14}$, and $\mathbf{g}_{15}$ indicate the gravitational attraction of a single particle m_1 on other particles m_2, m_3, m_4, and m_5 at different distances. Notice that although increased vector length indicates greater gravitational attraction, the vectors are not drawn to scale.

after all, consists of a very large number of very small particles.

The basic cgs unit of acceleration used to describe gravity is the *gal,* named for Galileo. This unit and its subdivisions commonly used by many geophysicists are defined as

$$\begin{aligned} 1 \text{ gal} &= 1 \text{ cm/s}^2 \\ 1 \text{ gal} &= 1000 \text{ milligals} \\ 1 \text{ gal} &= 10{,}000 \text{ gravity units} \\ 1 \text{ gal} &= 1{,}000{,}000 \text{ microgals} \end{aligned}$$

The SI units of acceleration are

$$1 \text{ m/s}^2 = 10^5 \text{ milligals} = 10^6 \text{ gravity units}$$

GRAVITY ON A ROTATING ELLIPSOID

To describe the gravity of a large globe such as the earth, we must consider the combined effect of all the particles making up the globe. First, let us suppose that the globe is a stationary sphere of uniform density. We could use Equation 7–3 to calculate, step by step, the attractions of all its constituent particles on some object at a distance r from its center. We would discover that the total gravity is expressed by $g = GM/r^2$ and is directed toward the center of the globe which has a total mass M. If the radius of this globe is R, any location on its surface is at that distance from its center. Therefore, gravity anywhere on its surface would be $g = GM/R^2$. This is the attraction we would experience if we were standing on the surface of the stationary sphere.

A globe such as the earth is not a sphere, however, and it is not stationary. Because it rotates around an axis passing through the North and South Poles, each particle of its mass is subjected to an outward centrifugal force as well as the mutual gravitational attrac-

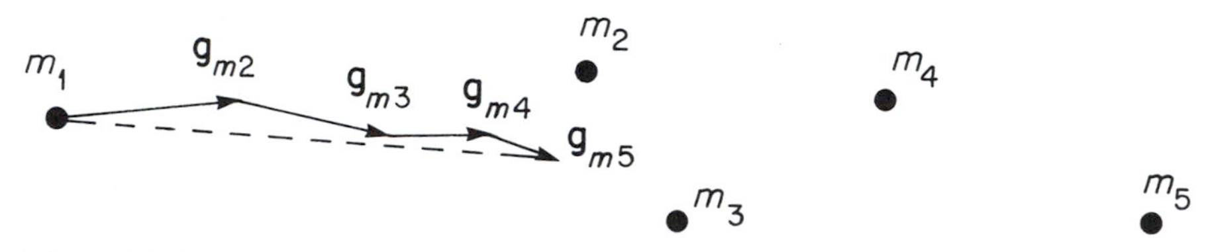

Figure 7–3
Vectors indicating the combined gravitational attraction of several particles on a single particle m_1. Notice that the vectors are not drawn to scale.

tion of all the other particles. The balance of gravitational and centrifugal effects shapes the globe into an ellipsoid, which is a flattened sphere (Figure 7–4). The extent to which a globe becomes flattened depends on the speed of rotation and the physical properties of the mass. The lengths of the equatorial radius R_e and the polar radius R_p are used to describe the ellipsoid. Another term for describing the shape of an ellipsoid is the flattening f, which is a combination of these radii:

$$f = \frac{R_e - R_p}{R_e} \tag{7–4}$$

An object resting on the surface of a rotating ellipsoid experiences the gravitational attraction of the total mass of the ellipsoid. This attraction is directed toward the center of the mass, but it is stronger at the poles of rotation than at the equator. An increase in the strength of the mass gravitational attraction on the surface of the ellipsoid is illustrated in Figure 7–5a by vectors that grow longer with increasing latitude. Equation 7–3 can help us to understand why this is true. Because the strength of gravity increases as the distance from the center of mass decreases, it is expected to be strongest at the poles where the ellipsoid radius is shortest. Gravity is also affected by the arrangement of mass in the ellipsoid, which is different from the arrangement of mass in a sphere.

The distance from the center is not the only reason for gravity being stronger on the surface of an ellipsoid close to its poles of rotation. The centrifugal effect, which acts on an object resting on the surface, must also be considered. Because the globe is turning, this object moves along a circular path centered about the axis of rotation. Look again at Figure 7–5. Observe that the centrifugal effect is not directed outward from the center of the globe; rather, it points outward from the center of the circular path of the moving object. The strength of the centrifugal acceleration a_c is

$$a_c = \omega^2 d \tag{7–5}$$

where ω is the angular speed of rotation and d is the distance of the object from the center of its circular path. We can see that the centrifugal effect acts the most strongly on an object situated at the equator. Here the value of d is largest, being equal to R_e. But d grows shorter with increasing latitude, reaching zero at the poles. Therefore, according to Equation 7–5, the centrifugal effect grows weaker away from the equator and vanishes at the poles. Vectors in Figure 7–5b illustrate the decrease in centrifugal acceleration with increasing lat-

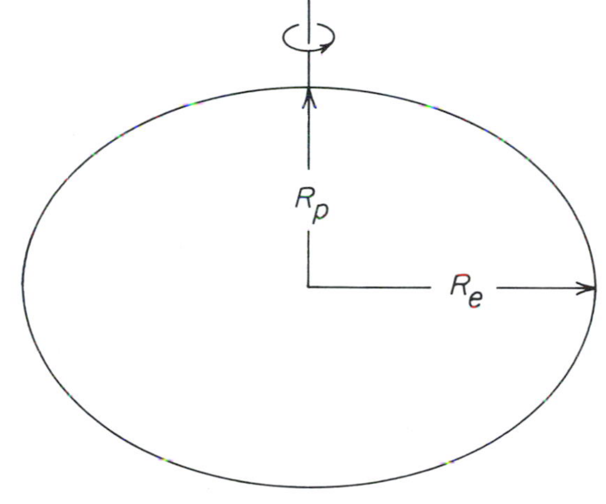

Figure 7–4
A rotating ellipsoid with an equatorial radius R_e and a shorter polar radius R_p. This is the equilibrium shape produced by the mutual gravitational attraction of the particles making up the mass plus the centrifugal force acting on these particles because the mass is rotating.

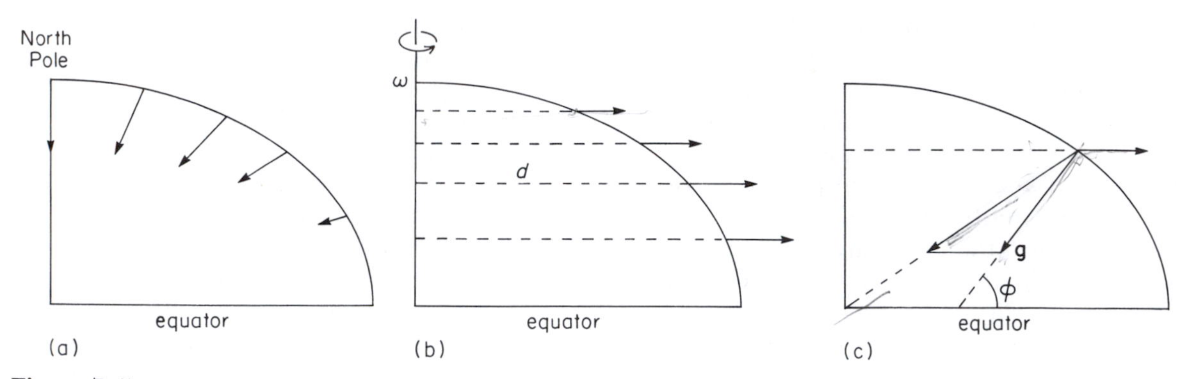

Figure 7–5
Vectors illustrating (a) mass gravitational attraction on the surface of an ellipsoid, (b) centrifugal acceleration on the surface of a rotating ellipsoid, and (c) complete gravitational attraction on a particle at the surface of a rotating ellipsoid, which is the vector sum of the mass gravitation and centrifugal accelerations. Notice that the vectors are not drawn to scale.

itude. Observe that all these vectors are parallel. They point outward from the axis of rotation rather than from the center of the globe.

Gravity on the surface of a rotating ellipsoid is described completely by combining the mass and centrifugal effects. Observe in Figure 7–5c how the complete gravitational attraction, represented by the vector **g,** is obtained from these two effects. But be sure to understand that the vectors in Figure 7–5 are not drawn to scale. Actually, the mass gravitational attraction is very much stronger than the centrifugal effects. For example, at the earth's equator the mass effect is almost 300 times stronger. Note that at the equator the directions of these two effects are opposite, so that **g** points directly toward the center of the ellipsoid. At the poles where there is no centrifugal effect, **g** also points at the center. This is not true at all other locations where **g** is not directed exactly at the center of the ellipsoid.

It should now be clear why gravity is not everywhere the same on a rotating ellipsoid. Recall that on a stationary sphere, $g = GM/R^2$. But on a rotating globe, this constant value must be modified to account for flattening and centrifugal effects, both of which change with latitude. Corrections are applied to the value of gravity g_e on the equator. The formula for calculating gravity $g(\phi)$ at any latitude ϕ on a rotating ellipsoid is

$$g(\phi) = g_e(1 + C_1 \sin^2 \phi + C_2 \sin^4 \phi) \quad (7\text{–}6)$$

The constants C_1 and C_2 depend on the flattening f and the rate of rotation ω. The values of these constants for the earth in particular are found from astronomical measurements and observations of orbiting satellites. For purposes of exploration geophysics, we need not be concerned with how this is done. But we should become familiar with the methods for measuring gravity on the earth's surface that are needed to find the value of g_e in Equation 7–6.

MEASURING GRAVITY

Gravity on the earth can be measured in at least four different ways: with (1) a free-falling test mass, (2) a swinging pendulum, (3) a test mass that stretches a spring, and (4) a test mass attached to a vibrating fiber. The physical principles that govern these measurements are quite simple and easy to understand. The real challenge is to design and construct instruments with sufficient sensitivity. In modern exploration geophysics, gravity values should have an accuracy within at least 0.1 mgal, and for some purposes, microgal precision is demanded. Furthermore, we must have portable instruments that can be transported to remote locations and that require only a few minutes for making a measurement. No single instrument meets all the requirements, but by combining them in different ways, we can obtain much useful information.

Measurements of a Free-Falling Mass

Suppose that a small object is released from a stationary position. The earth's gravitational attraction g will cause it to fall a distance x during the length of time t according to the well-known formula

$$x = \frac{1}{2}gt^2 \qquad (7\text{–}7a)$$

which can be rearranged to obtain

$$g = 2x/t^2 \qquad (7\text{–}7b)$$

This equation tells us that gravity can be determined simply by timing the fall of the object through a measured distance. It will fall 1 meter in slightly less than one-half second, indicating a gravitational acceleration of approximately 980 gals close to the earth's surface. But a few more simple calculations will show us that if the precision of g is to be within 0.1 mgal, the 1-meter distance must be accurate within 10^{-5} cm, and the time will have to be measured to closer than 10^{-8} second. Not until after 1960 did we have the laser and electronic technology to accomplish this precision.

One of the most accurate instruments combines two corner cube prisms and a laser. A *corner cube prism* is designed to produce by internal reflections a beam of reflected light that is parallel to, but offset from, the incident beam. A simplified sketch of the operation in Figure 7–6 shows one stationary prism and one falling prism. The sharply focused and extremely intense light beam produced by the *laser* is separated into two perpendicular parts

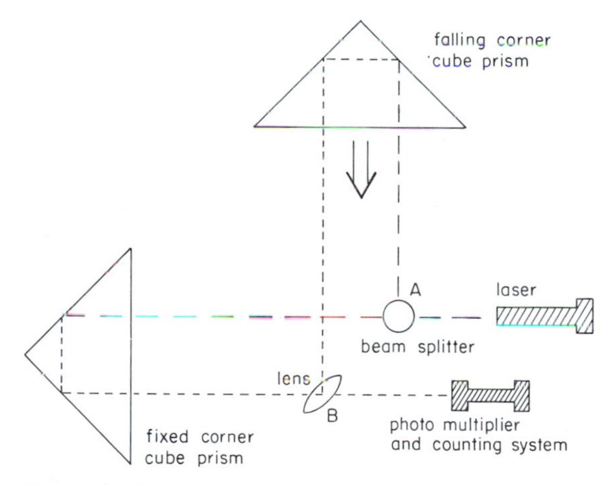

Figure 7–6
The free-falling-mass experiment for measuring the acceleration of gravity using corner cube prisms. The interference of light beams from the two prisms is used to measure the time required for the upper prism to fall a specific distance.

by the beam splitting device at position A. These beams travel to the two prisms where they are internally reflected and then rejoin at position B. There they are directed into a photomultiplier that is sensitive to light intensity.

These light beams can be considered as oscillating waves. If both are oscillating in the same way when they join at position B, a particularly bright composite beam is produced. But if they are oscillating in opposition, the composite beam will be dark. Because one of the prisms is falling, its beam oscillations are alternately aligned, and then opposed, to beam oscillations from the stationary prism. Therefore, the composite beam is alternately bright and dark. The cycles of brightness are counted in the photomultiplier apparatus. Because the speed of light and the wavelength of light produced by the laser are known to high precision, both time and distance can be determined by counting these cycles of brightness.

The experiment with a falling corner cube must be done in a vacuum chamber so that air resistance will not slow the motion of the prism. Because minute ground vibrations can introduce error into any single measurement, the test must be repeated many times. One standard procedure is to calculate gravity from an average of 50 drops. The equipment must be operated at a carefully prepared site. Although the instrument shown in Figure 7–7 can be taken apart for transport to other sites, reassembly and operation are time-consuming procedures. Therefore, the apparatus is not truly portable or easy to operate. Its use is largely restricted to a few well-established observatories. Nevertheless, the apparatus is of critical importance because it provides the only way we have yet devised of measuring g to a precision of better than within 0.1 mgal. Values determined at some well-known observatories are given in Table 7–1.

Other devices measure the motion of a free-falling object by means of laser interferometers. In one system a catapult projects the test mass vertically upward so that it passes through two horizontal light beams separated by a precisely measured distance. Motion of the mass is timed during its upward rise and again during its return fall as its passing interrupts the light beams. Still another device records the times that two markers on a vertically falling rod pass a sensor. These devices, like the falling corner cube device, must be operated in well-designed observatories. To find g at other places, we use a different kind of instrument that measures, quite precisely, the change in gravity Δg between one of these observatories and another more remote location. By adding or subtracting Δg to the value of g measured at the observatory, we can find the attraction of gravity at the remote location.

Pendulum Measurements

An object suspended from a stationary pivot so that it is free to swing back and forth is called a pendulum (Figure 7–8). The period τ of the pendulum is the time required for it to swing through one cycle of motion. The formula relating the pendulum period and the attraction of gravity is

$$\tau = 2\pi \sqrt{I/mgh} \qquad (7\text{–}8a)$$

where m is the mass of the pendulum, I is its moment of inertia about the point of suspension, and h is the distance between the point of suspension and the center of mass. If the pendulum consisted of a weightless string fixed at one end and attached to a point mass

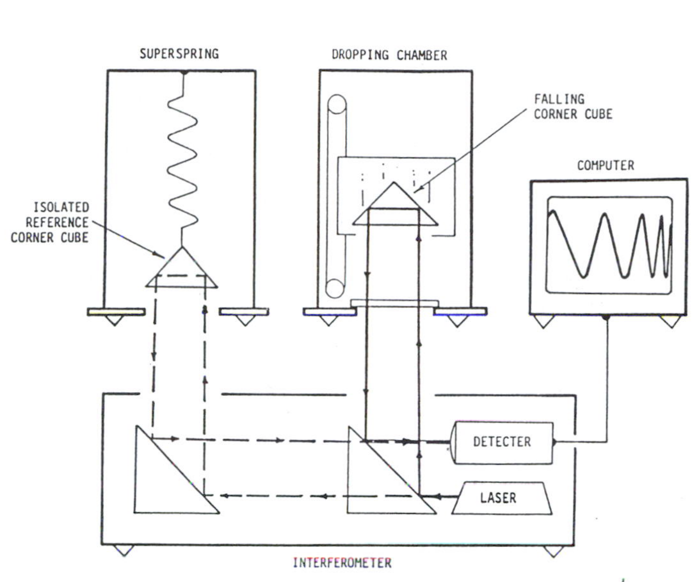

Figure 7–7
Experimental apparatus used to determine the acceleration of gravity on a free-falling corner cube prism. (Courtesy of the National Geodetic Survey of the U.S. National Oceanic and Atmospheric Administration.)

at the other end, then I/mh would equal the length ℓ of that string. Its period would be $\tau = 2\pi\sqrt{\ell/g}$. Of course, such an ideally simple pendulum cannot be constructed.

Let us rearrange Equation 7–8a into the form

$$g = K/\tau^2 \qquad (7\text{–}8b)$$

where $K = 4\pi^2 I/mh$ is called the pendulum constant. Now we should be able to find g by simply timing the swinging pendulum at some location to determine the period τ. Unfortunately, no one has discovered how to find the value of K accurately enough to calculate g within a precision of 0.1 mgal. Two different instruments, however carefully constructed, simply do not yield the same value of g, and we have no way of knowing which value is closer to being correct.

Nevertheless, pendulums have proved to be very useful for measuring the change in gravity Δg between different locations. If we can

TABLE 7–1 Gravity Values from Measurements of a Free-Falling Mass

SITE	LOCATION	NUMBER OF SETS*	FINAL VALUE (mgal)
National Bureau of Standards, Gaithersburg, Md.	Room 01, Bldg. 202 (NBS-3)	5	980,102.394 ± 0.055
National Physical Laboratory, Teddington, England	Room B-17, Bushy House (BH)	19	981,181.930 ± 0.042
Bureau International des Poids et Mesures, Sèvres, France	Salle 1, BIPM Laboratory (Sèvres A)	27	980,925.960 ± 0.041
Air Force Cambridge Research Laboratories, Bedford, Mass.	Pier 1, Seismic Facility	22	980,378.671 ± 0.042
Geophysics Institute, University of Alaska, Fairbanks	Room 1, Patty Bldg.	16	982,234.953 ± 0.042
Universidad Nationale de Colombia, Bogota, Colombia	Quarto 111, Edificio Matematica y Fisica	38	977,390.015 ± 0.087
University of Denver, Denver, Colo.	Room 8, Science Hall	30	979,597.708 ± 0.042
Scott Laboratory of Physics, Wesleyan University, Middletown, Conn.	Room 7B, Scott Lab. (Middletown A)	100	980,305.306 ± 0.041

*Each data set consists of 50 drops.
From J. A. Hammond and J. E. Faller, Results of absolute gravity determinations at a number of different sites, *Journal of Geophysical Research,* v. 76, n. 32, p. 7852, November 10, 1971.

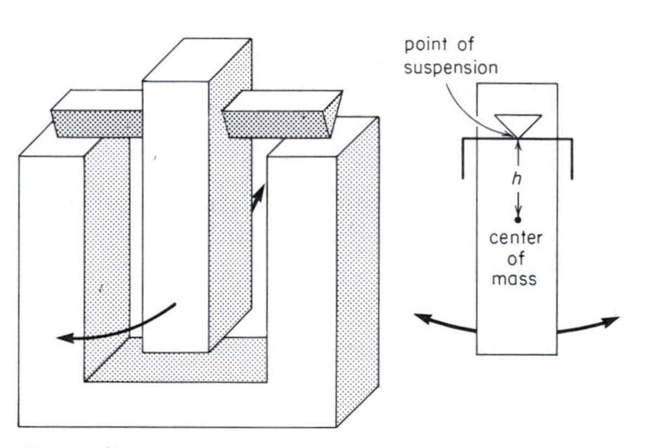

Figure 7–8
Pendulum for measuring gravity. The moving mass is attached to a wedge that rests on a stationary platform.

assume that the value of K for a particular pendulum does not change when it is moved from place to place, we can use the periods measured at these places to find Δg. Consider measurements from two locations. According to Equation 7–8b, we can write:

$$g_1\tau_1^2 = g_2\tau_2^2 \quad \text{or} \quad g_2 = g_1\tau_1^2/\tau_2^2$$

Now suppose that we can determine a relatively imprecise value of K that allows us to calculate gravity within a ± 20-mgal accuracy. For a century, instrument makers have been able to achieve this precision. From this pendulum constant we obtain a somewhat imprecise value of g_1 at the first location, which we use to determine an equally imprecise value of g_2 at the second location. These values could be in error by as much as 20 mgal. But both are incorrect by practically the same amount because the same instrument was operated at the two locations. This means that the change in gravity $\Delta g = g_2 - g_1$ is very accurately determined. Tests show that precision of considerably better than 0.1 milligal is possible for locations separated by large distances.

Portable pendulums were used for geophysical exploration during the 1930s. Standard apparatus designed and constructed by the Gulf Research and Development Company is illustrated in Figure 7–9. The pendulum was a glass bar with wedge-shaped supports on either side resting on glass plates which were attached to a heavy frame. Its period was slightly less than one second. Because tests showed that the swinging of a single pendulum produced a small motion of the frame, the Gulf apparatus consisted of two oppositely moving pendulums. Their supports could be adjusted to produce nearly identical periods, so that their opposing movements greatly reduced any frame vibration. The pendulums were operated in a vacuum chamber to assure accurate data. Thermostatically controlled heating coils maintained a stable internal temperature, which minimized thermal expansion or contraction of the pendulums.

The operation of this equipment ordinarily required a few hours on location. After operators carefully leveled the frame, connected the heating and timing units, and checked the vacuum, the pendulums were allowed to swing continuously for a fixed interval of time lasting at least one-half hour to perhaps more than one hour. A light beam reflected from a mirror on the swinging pendulum for each cycle of motion was photographically recorded. The average period was then calculated from the total number of swing cycles during the carefully measured interval of time. A high-quality tuning fork was used for time control; later a more accurate crystal clock was used for timing.

Perhaps the real value of the Gulf pendulums was not recognized until two or three decades after they were used for oil exploration. Then Professor George P. Woollard (1908–1978) and his associates at the University of Wisconsin used them in a world gravity survey. During several years around 1960, they made pendulum measurements in easily accessible places at many large airports, seaports, and large cities throughout the world. This survey was completed at about the same time that free-falling-mass measurements were yielding useful results at a few well-known observatories. By adding or subtracting from these few falling-mass values of g the gravity differences Δg found from pendulum data, the survey finally made it possible to calculate accurate values of g for a large number of locations around the world. These values comprise a worldwide network of gravity base stations which are described in a publication we refer to later in this section. These base stations are used to adjust the very detailed surveys that can be done with gravimeters or gravity meters, which are much easier to transport and operate than pendulums or falling-mass apparatus.

Spring Gravimeters

Gravimeters are designed to measure the differences in gravity between observation sites. Most of these instruments consist of some arrangement of masses supported by springs. The basic principle is simple. Suppose that a mass m is attached to a spring as in Figure 7–10. The length x that a spring stretches depends on the force pulling on that spring. In this case, the force mg depends on the gravi-

Figure 7–9
Gravity pendulum apparatus developed by the Gulf Research and Development Company. Two pendulums pivoted on quartz glass wedges are mounted to swing simultaneously in opposite directions so that they are able to suppress the movement of the frame induced by the pendulum motion.

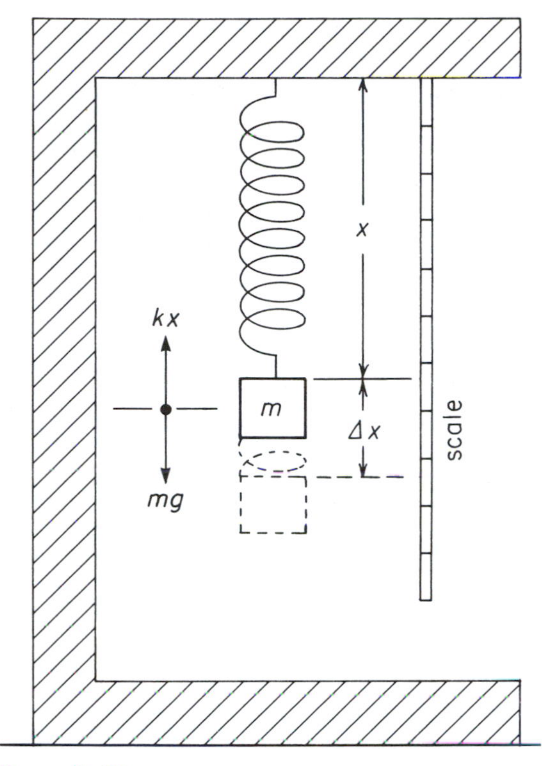

Figure 7–10
Mass supported by a vertical spring. A change in gravity will produce a change in the distension of the spring.

tational attraction of the earth. It is balanced by an upward supporting force kx exerted by the spring, so that

$$kx = mg \qquad (7\text{–}9a)$$

where k is called the spring constant. When this system is moved from place to place, any change in gravity Δg should produce a proportional change Δx in the stretch of the spring:

$$k\,\Delta x = m\,\Delta g \qquad (7\text{–}9b)$$

We can see that this system is similar to a seismometer (Figure 2–24), even though it is being used in quite a different way. Therefore, we know that if the mass is disturbed, it will bob up and down. The natural period τ of this motion is related to the mass and the spring constant:

$$\tau = 2\pi\,\sqrt{m/k} \qquad (7\text{–}10)$$

Now the real challenge is to construct a practical instrument based on this principle. Such an instrument should be portable and easy to operate. It must be sensitive to gravity differences that are at least as small as 0.1 mgal and, for some purposes, 0.01 mgal. Obviously, the mass cannot be too heavy, and the spring cannot be too long if we are to have a compact portable gravimeter. Equations 7–9 and 7–10 provide some guidance for designing the parts of such a gravimeter, but they also point out formidable problems that have to be solved before a useful instrument can be constructed. For example, suppose that we want the gravimeter to have a period of less than 10 seconds so that it does not require too much time to come into balance before a reading can be made. In addition, we want the mass to be less than 10 grams. By rearranging Equation 7–10 and inserting these numbers, we can find the spring constant k that will be required. But what about the length of this spring? We already know that gravity on the earth's surface is close to 980 gals and that we want to detect changes of, say, 0.1 mgal (0.0001 gal). By using a modern micrometer device, we can expect to measure changes in spring length of about 0.001 cm. Rearranging and combining Equations 7–9a and 7–9b shows that $x/g = \Delta x/\Delta g$. We can insert the numbers just given to get $\Delta x/\Delta g = 1$, and taking $g = 980$ gals, we find

$$x = \frac{\Delta x}{\Delta g}\,g = 980 \text{ cm} \cong 32 \text{ feet}$$

This equation tells us that the system illustrated in Figure 7–10 would require a spring at least 32 feet long to have the necessary sensitivity. Clearly, such a system is impractical! Is there any way to achieve the required sensitivity in a more compact and practical instrument?

One way is through a system that consists of a hinged beam supported by a relatively short spring. Such a system (Figure 7–11) has a natural period that can be adjusted by changing the angles of the spring and the beam. To understand the importance of adjusting the natural period, let us rearrange Equation 7–10 to obtain $m/k = \tau^2/4\pi^2$, and then rearrange Equation 7–9b into the form $m/k = \Delta x/\Delta g$. Now, combine these results to get

$$\frac{\Delta x}{\Delta g} = \frac{\tau^2}{4\pi^2} \qquad (7\text{–}11)$$

Thus, by increasing the natural period of a system, we can make the lengthening of the spring produced by a certain change in gravity even greater.

The sensitivity of the system in Figure 7–11 can be significantly increased by using a *zero-length spring* to support the beam. Theoretically, this kind of spring would recoil to a length of zero when not stretched by some force. Because of the finite diameter of the wire in its coils, such a spring cannot actually reduce its length to zero. But by twisting a wire at the same time as it is being coiled, we can produce a spring that behaves this way within the range that it can be stretched. When no force is applied to this spring, its coils press back against one another. By twisting the wire more than is needed to achieve zero length, we can make an *inverted spring*. In theory, such a spring has a negative initial length. It will turn inside out when not stretched by some force. Springs made in this way can be shorter than would otherwise be needed to obtain the required sensitivity.

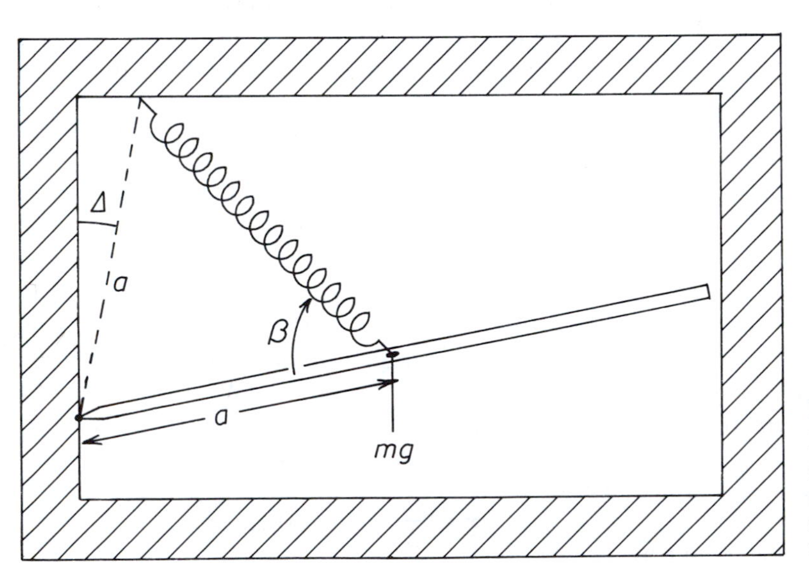

Figure 7–11
Hinged beam supported by an inclined zero-length spring. In this example, the spring is attached at the center of mass of the beam.

Now let us analyze the sensitivity of a hinged beam supported by a zero-length spring. To keep the system as uncomplicated as possible, let us attach the spring to the frame and to the beam at equal distances from the hinge. In addition, the spring is attached at the center of mass of the beam. This arrangement is illustrated in Figure 7–11. Notice that the line between the hinge and the point where the spring attaches to the frame is inclined at an angle Δ from the vertical. The length a of this line is the same as the distance from the hinge to the center of mass of the beam. The spring makes an angle β with the beam, which has a mass m. The natural period of this system is

$$\tau = 2\pi \sqrt{1/[mg \cos(2\beta - \Delta) - (ka^2 \cos 2\beta)]} \tag{7–12}$$

where k is the spring constant. Now suppose that the beam and spring are arranged so that $\Delta = 0$ and $\beta = 45$ degrees. Then we see that $\tau = \infty$, and the system would be so sensitive that it could never come into balance. Obviously, if we want a gravimeter that will come into balance so that a reading can be made, other values for β and Δ must be chosen.

How can we reduce the time required to get a reading and yet retain a sensitivity of better than 0.1 mgal in a gravimeter? Much experimentation has gone into solving this problem. Instrument makers have thoroughly tested many different configurations, and they have developed ingenious ways to make very sensitive springs. Through their efforts we now have gravimeters with natural periods of between 5 and 10 seconds that are sensitive to gravity differences smaller than 0.04 mgal. They weigh less than 10 pounds and can be carried in suitcase-sized containers.

Temperature control is very important in the construction of a reliable gravimeter. Expansion or contraction of its different parts could upset the delicate balance of the beam. Therefore, modern gravimeters are carefully insulated. Some contain battery-operated heating coils. They are controlled by precision thermostats so that internal temperature is maintained within less than 0.002°C. Other instruments are mounted in thermos flasks to minimize temperature effects.

All gravimeters must be calibrated experimentally to find out exactly what change in gravity is indicated by a measured deflection of the beam. This can be done in one of two ways. One method requires repeated measurements on a platform that can be tilted through a range of small but accurately known angles. This changes the direction in which gravity acts on the beam. The effect is the same as changing gravity by an amount $\Delta g = g - g \cos \theta$, where the platform is inclined at an angle θ from the horizontal. From a series of measurements, we can find the *gravimeter constant* that relates beam deflection and change in gravity. A gravimeter can also be calibrated from readings of beam deflection made at different locations where gravity has already been measured with falling mass or pendulum apparatus.

Many kinds of spring gravimeters have been used in all parts of the world to measure gravity differences. The LaCoste–Romberg, Worden, Sodin, and Scintrex gravimeters are instruments currently being manufactured.

LaCoste–Romberg Gravimeter The basic design of the LaCoste–Romberg gravimeter is illustrated in Figure 7–12. It consists of a hinged beam supported by a zero-length spring and is similar to the system shown in Figure 7–11. Notice that the spring is attached to a micrometer screw in the frame. To operate the gravimeter, the observer turns the mi-

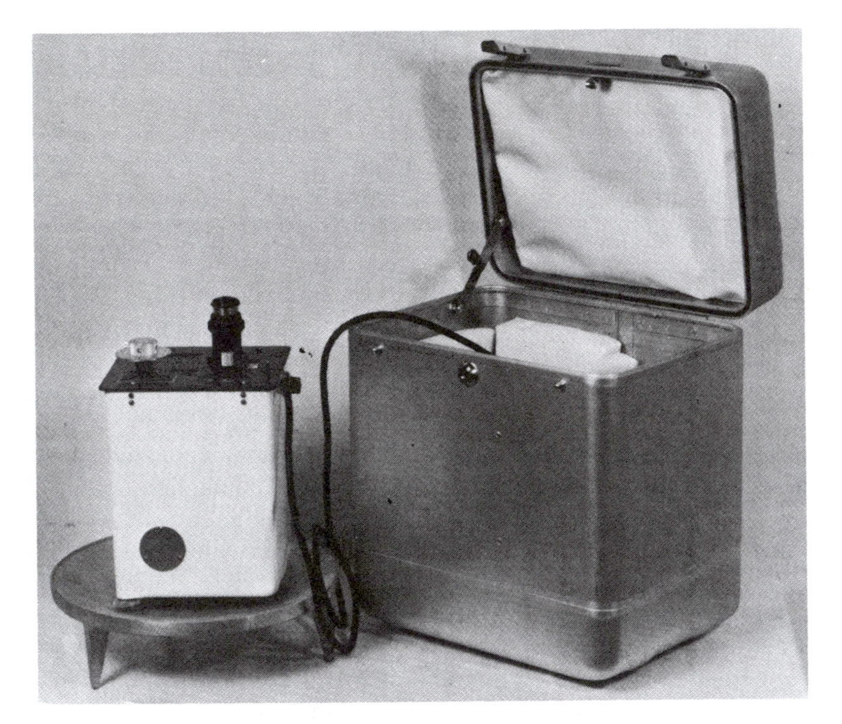

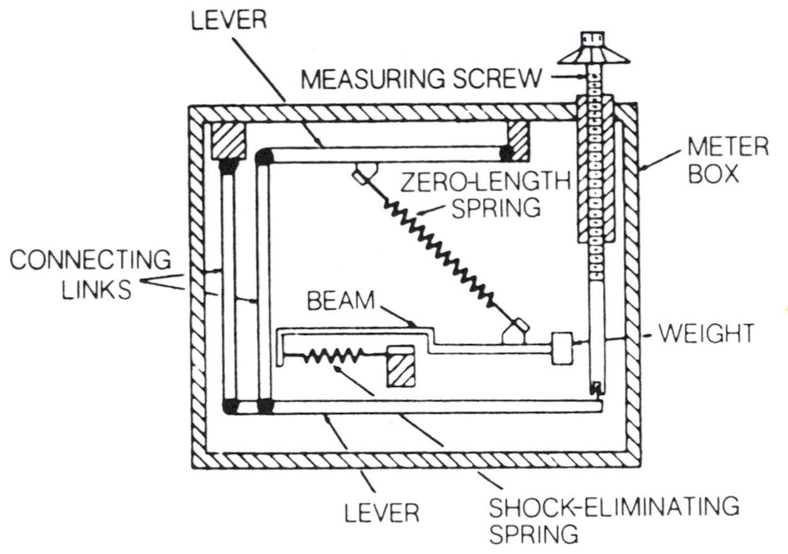

Figure 7–12
Photograph of the LaCoste–Romberg G Series Gravimeter and a cross section showing the basic components. (Courtesy of LaCoste–Romberg, Inc.) The point of attachment of the spring to the frame is adjusted by means of a micrometer screw; this adjustment maintains the beam in the same position for all readings.

crometer screw by means of a dial. This adjusts the upper end of the spring which pulls the beam up or down into a level position. A point of light reflected from a mirror on the beam into an eyepiece is centered when the beam is in the correct position. The reading on the dial indicates the amount of turn on the micrometer screw required to level the beam. All readings are made with the beam in the same position by adjusting the spring to balance the changes in gravity. These gravity changes are then calculated by multiplying differences in dial readings by the gravimeter constant. The instrument can be set up and read in less than two minutes. If it is operated carefully, gravity differences as small as 0.01 mgal can be detected.

LaCoste–Romberg gravimeters are constructed from metals with low coefficients of thermal expansion. The frame with the spring and beam apparatus is placed in an insulated container where temperature is maintained within 0.002°C by electrical heating coils. The first instruments were manufactured before 1940 and weighed nearly 80 pounds (30 kg). Improvements in the quality of available materials and in spring technology have made it possible to reduce the weight to about 6 pounds (2 kg).

Worden Gravimeter The ingenious Worden instruments are constructed entirely from quartz glass rods, fibers, and springs. The mechanism is shown in Figure 7–13. Zero-length springs and inverted springs support the small beam. All readings are made with this beam in the same position. The observer adjusts it by turning a micrometer dial screw to which the dial spring is attached. A pointer viewed through an eyepiece indicates when the beam is properly centered. The dial shows the turn of the micrometer screw required to balance the beam. Differences in dial readings from various locations are converted to gravity differences by multiplication with the gravimeter constant. Gravity changes smaller than 0.04 mgal are easily detected. The system of springs can extend enough to compensate for gravity differences of more than 5200 mgal, depending on the setting of the reset dial spring.

The system of springs and rods is smaller than a fist. It is enclosed in a glass thermos flask to reduce temperature fluctuations. A thermal compensation fiber attached to the inclined rod that supports the main spring is an additional means of controlling temperature. The rod and fiber expand or contract differently in response to a small change in temperature. They combine to move the point of connection of the main spring up or down to counteract the effects of temperature change on the system. One model of the Worden gravimeter is also equipped with a battery-operated temperature control system. The entire gravimeter is packaged in a cylindrical stainless steel container which is about 14 inches (35 cm) high, 7 inches (18 cm) in diameter, and weighs less than 10 pounds (4 kg).

Sodin Gravimeter Quartz glass rods and springs are used throughout the Sodin gravimeter. Figure 7–14 shows how these components are arranged. The small horizontal beam is supported by a zero-length main spring. A pointer attached to its end, which can be seen through an eyepiece tells the position of the beam. Turning a dial on a micrometer screw centers the beam and adjusts the reading spring to counteract the attraction of gravity. The difference in dial readings between two locations is multiplied by the gravimeter constant to find the change in gravity.

Temperature in the Sodin gravimeter is

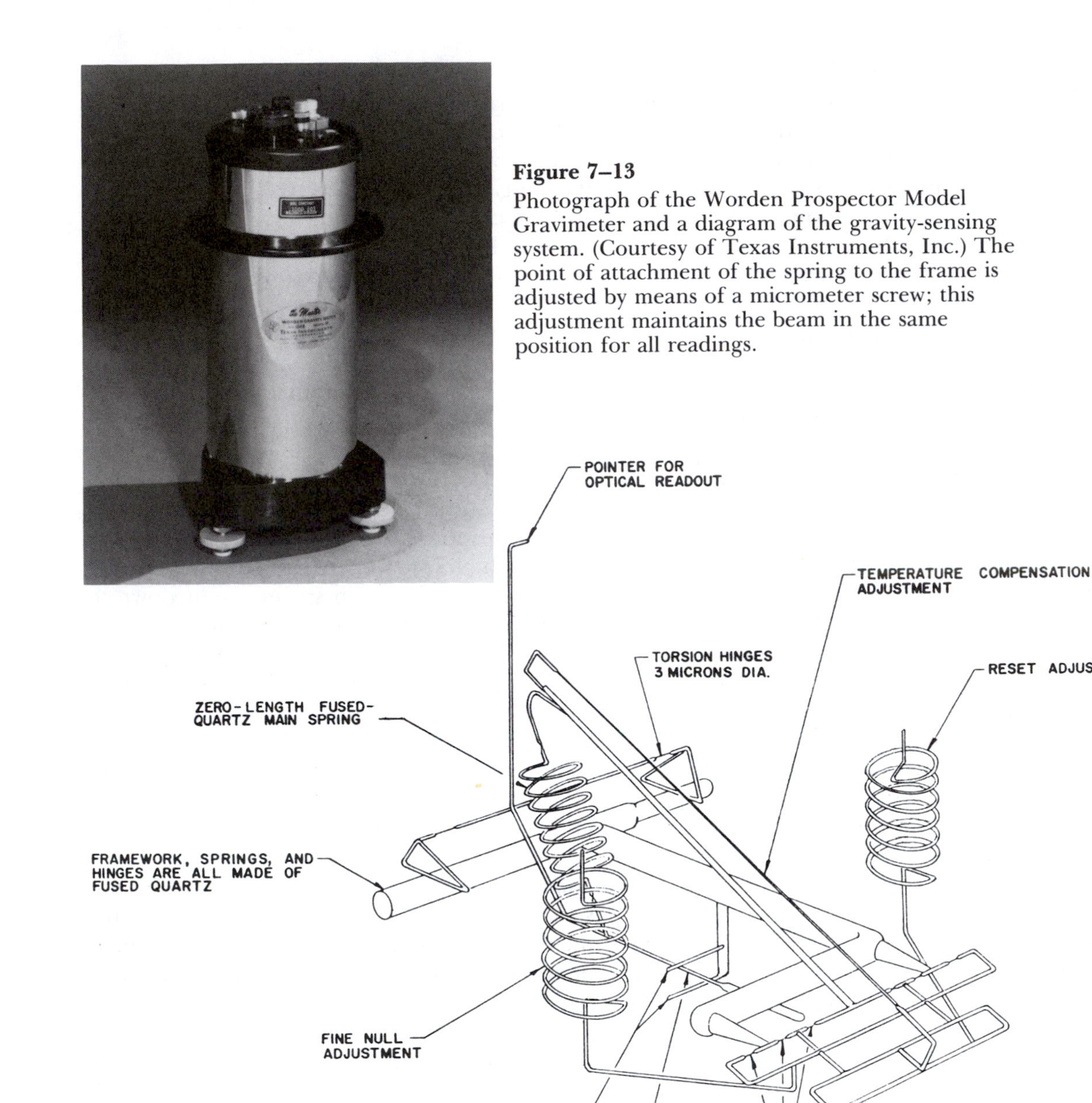

Figure 7–13
Photograph of the Worden Prospector Model Gravimeter and a diagram of the gravity-sensing system. (Courtesy of Texas Instruments, Inc.) The point of attachment of the spring to the frame is adjusted by means of a micrometer screw; this adjustment maintains the beam in the same position for all readings.

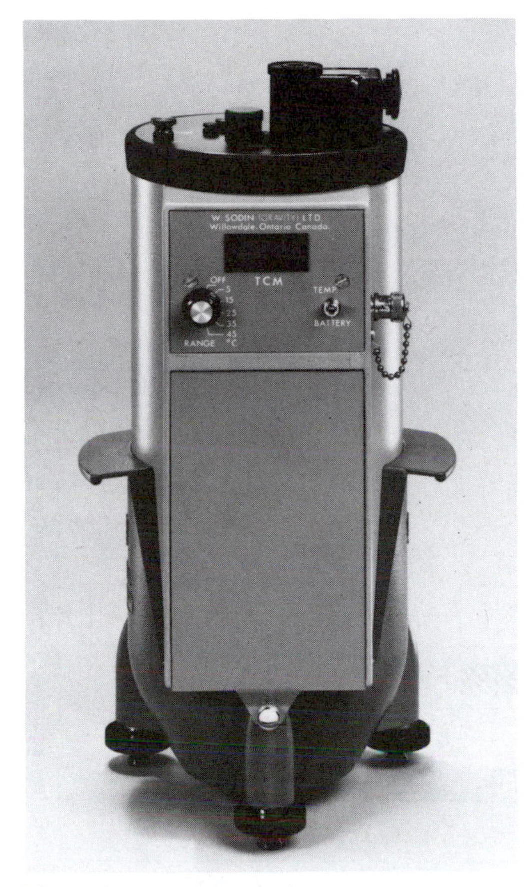

Figure 7–14
Photograph of the Sodin WS-410T Series Gravimeter and a diagram of the gravity-sensing element. (Courtesy of W. Sodin, Limited.) The point of attachment of the spring to the frame is adjusted by means of a micrometer screw; this adjustment maintains the beam in the same position for all readings.

controlled in two ways. A compensator frame supporting the main spring rotates in response to small temperature changes because of the different ways that the rods and the fiber in the frame expand or contract. The entire system is covered by an insulated package with thermostatically controlled heating elements. This package is placed in a thermos flask encased by a stainless steel cover. The gravimeter is 14 inches (35 cm) high, 5½ inches (14 cm) wide, and weighs about 10 pounds (4 kg).

Scintrex Gravimeter The model CG-3 gravimeter, shown in Figure 7–15, consists of a quartz glass mass and spring assembly mounted in a vacuum chamber. The mass also acts as one part of a variable-capacitor trans-

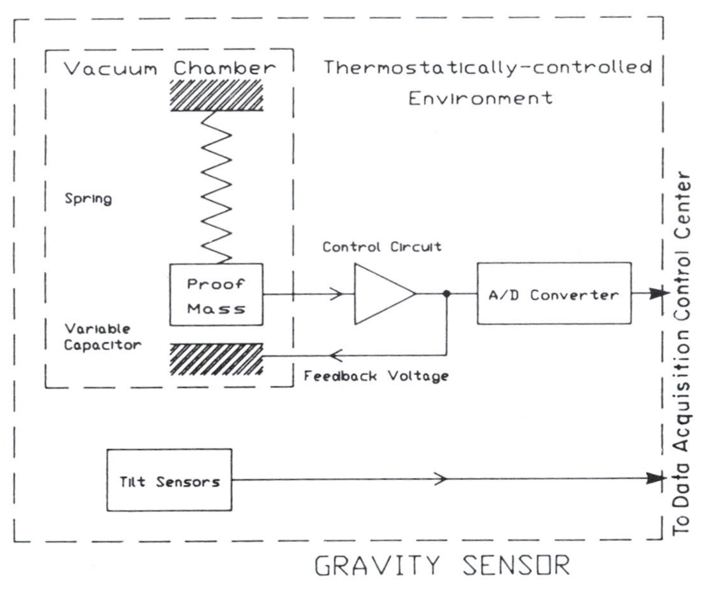

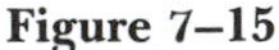

Figure 7–15
Photograph of the Scintrex Model CG-3 Gravimeter and a diagram of the gravity-sensing element. (Courtesy of Scintrex Limited.) The beam is maintained in a null position by a direct-current feedback circuit. A change in gravity produces a change in feedback voltage, which is recorded.

ducer. The attraction ot gravity on the mass is balanced by the spring and by the electrostatic force associated with the capacitor.

A small displacement of the mass produced by a change in gravity is detected by the capacitor transducer, which activates a feedback circuit. By supplying direct-current voltage to the capacitor plates, the feedback circuit alters the electrostatic force, which returns the mass to the null position. The direct-current feedback voltage, which is proportional to the change in gravity, is then recorded.

The entire system is controlled by a microprocessor that uses electronic tilt sensor signals to level the instrument, adjusts the feedback circuit, and converts the feedback voltage into a relative gravity value which is stored in the computer memory. Gravity changes as small as 0.01 mgal can be detected with the Scintrex CG-3 gravimeter.

Development of all the gravimeters we have described has severely tested the ingenuity of instrument makers. The construction of each instrument requires extreme care, and only the finest materials can be used. But all materials undergo a slow process of aging. Gradually, the physical properties change, and the stretch of the springs varies slightly. This causes gravimeters to *drift,* which means that when measurements are repeated at the same location over a period of days, weeks, or months, the readings slowly change. The high-quality gravimeters now in use tend to have uniform drift properties which can be determined by repeating measurements at the same location from time to time. In Chapter 8 we will discuss gravity survey methods in more detail, including the common ways of making drift corrections. For now, it is important to recognize that drift is a feature of all gravimeters.

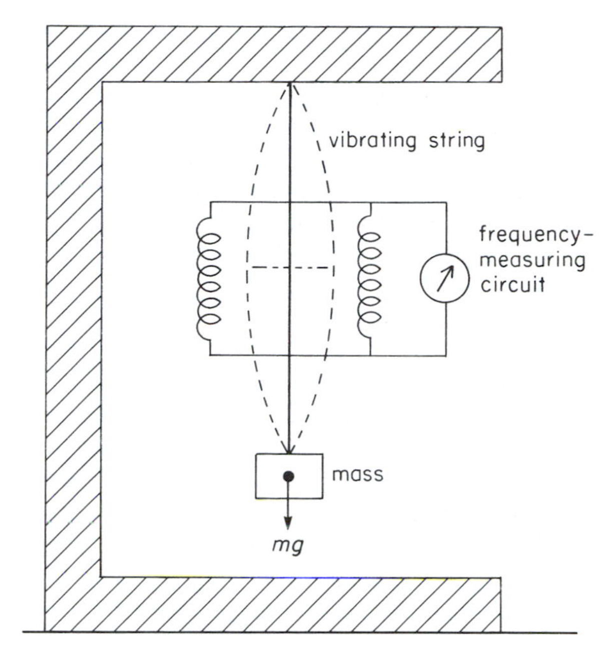

Figure 7–16
Basic components of a vibrating-string gravimeter. A change in gravity will produce a change in the vibration frequency of the fiber supporting the mass.

Vibrating-String Gravimeters

Gravity can be measured with a device consisting of a mass suspended on a string or fiber, as seen in Figure 7–16. If the string is disturbed, it will vibrate with frequency f, which is related to the suspended mass M, the length of the string x, the mass per unit length of the string m, and the attraction of gravity g according to the formula

$$f = \sqrt{Mg/4x^2m} \quad \text{or} \quad g = 4x^2f^2m/M \qquad (7\text{–}13)$$

Moving such a system from place to place allows changes in gravity to be detected from differences in string vibration; these differences can be measured electronically.

Several experimental vibrating-string gravimeters have been constructed. They have not yet reached the standards of precision and stability that are available in spring gravimeters. Nevertheless, this principle might eventually become the basis of a very compact instrument if certain design problems can be solved. Therefore, research on vibrating-string gravimeters continues.

Worldwide Network of Gravity Base Stations

Before 1950 there was no reliable basis for accurately comparing gravity measurements from different parts of the world. Nowhere was gravity known to a precision of better than a few milligals. Measurements of gravity differences between widely separated locations were commonly in error by several milligals. In the 1950s plans for a standardized global gravity survey were made largely through the efforts of Professor George P. Woollard and his colleagues at the University of Wisconsin. The purpose was to obtain accurate measurements at locations throughout the world; these locations could then be used as control points for adjusting detailed gravity surveys in widely separated areas. For the first time, we would be able to examine accurately measured gravity variations on a global scale.

To accomplish the task, geophysicists needed (1) a measurement of gravity at some reference point that was accurate within a few tenths of a milligal; and (2) measurements of the differences in gravity between many widely separated locations, also accurate within a few tenths of a milligal. Geophysicists knew that a suitable falling-mass apparatus was not then available, but it could probably be developed within one or two decades. In the meantime, they could proceed immediately with the survey of gravity differences, which could eventually be adjusted when falling-mass measurements became available.

The global survey of gravity differences included a program of pendulum measurements and a concurrent program of gravimeter measurements. This work was supervised principally by Professor Woollard. Two sets of Gulf pendulums were used, and some measurements were made with pendulums constructed at Cambridge University in England. The first sites were in North America. Gulf pendulums were operated at several airports along a profile extending from Mexico City to Fairbanks, Alaska. Instruments were carried on commercial and military flights between these airports, and each observation site was visited several times. Similar measurements were made along a midcontinent profile reaching from Winnipeg, Canada, to Houston, Texas, using the Cambridge pendulums. During the beginning stage of this program, comparisons of repeated pendulum measurements indicated intolerable errors. Then better ways of controlling temperature and vacuum were adopted, and timing systems were upgraded. Eventually, the equipment and operating procedures were improved so that gravity differences could be measured to a precision of better than 0.3 mgal. Work continued for more than a decade and the survey was extended to more than 100 sites located throughout the world. We have used pendulum measurements from the basic network of sites shown in Figure 7–17 to standardize the global survey of gravity.

There was an important difficulty in using the instruments available in 1950 for a global gravimeter survey. Although these gravimeters were sensitive to gravity changes smaller than 0.05 mgal, they could not measure changes larger than about 100 mgal without making spring adjustments. Thus, if gravity between two places changed by more than 100 mgal, the gravimeter would have to be read at additional intermediate sites within this range. At each intermediate site, readings were made before and after resetting the spring. Therefore, an observer could not simply carry one of these gravimeters on a flight between widely separated observation sites. The additional time required for overland travel to numerous intermediate reset sites, and uncertainties about the effect of these resets on calibration and drift characteristics, made use of these gravimeters impractical.

Some instrument makers were challenged

Figure 7–17
Observation sites for a worldwide network of pendulum gravity measurements used in the preparation of the International Gravity Standardization Net 1971. (From C. Morelli, The International Gravity Standardization Net 1971 [IGSN 71], *Publication Speciale No. 4*, International Association of Geodesy, Bureau Central de l'Association Internationale de Géodésie, 19, rue Auber, 75009 Paris, France.)

by the requirements of the proposed global survey. At first, they modified the springs to accommodate gravity changes of several thousand milligals. But with these springs, the gravimeters were much less sensitive to small changes in gravity. Furthermore, when tested at pendulum control sites, the relation between beam deflection and gravity change was not constant. After much effort, springs with adequate range and sensitivity were produced. Accurate calibration tables were then prepared by reading the gravimeters at pendulum control points covering large variations in gravity.

With these advances, it was possible to complete the global gravimeter survey. It was finished during the same time that the pendulum survey was underway. The basic gravimeter survey included about 1400 observation sites. The principal locations are shown in

Figure 7–18. This worldwide network of gravimeter observations together with the pendulum measurements provides the global standard for adjusting modern gravity surveys.

Shortly after the global gravity survey was completed in 1963, falling-mass experiments provided the reference values given in Table 7–1. By combining the value of g_r from one of these falling-mass sites and the gravity difference Δg between this and one of the global survey sites, we can easily calculate gravity g at the survey site:

$$g = g_r + \Delta g \qquad (7\text{–}14)$$

In this way, we can find g at all the sites in the global survey. Before the values in Table 7–1 were available, another value of g_r was used as an international reference value. This value was measured with pendulum apparatus in

Figure 7–18
Locations of the main gravimeter stations of the International Gravity Standardization Net 1971. (From C. Morelli, The International Gravity Standardization Net 1971 [IGSN 71], *Publication Speciale No. 4,* International Association of Geodesy, Bureau Central de l'Association Internationale de Géodésie, 19, rue Auber, 75009 Paris, France.)

1906 at a site in Potsdam, a city now situated in the German Democratic Republic (East Germany). Later measurements indicated that the Potsdam value was in error by several milligals. Nevertheless, for purposes of international standardization, geophysicists agreed to continue using this reference value until accuracy of about 0.1 mgal could be achieved. As indicated in earlier discussion, this accuracy became an accepted reality in about 1965. After this time, it was possible to adjust gravity differences to values given in Table 7–1. Earlier measurements of Δg were adjusted to the Potsdam value.

The results of the global pendulum and gravimeter survey were presented by Professor Woollard and his associate Professor John C. Rose in a volume called *International Gravity Measurements,* published by the Society of Exploration Geophysicists in 1963. This volume presents values of gravity together with a complete description of the observation sites. But all these values of gravity are based on the Potsdam reference value of g. Our knowledge of modern falling-mass and gravimeter measurements has revealed that the Potsdam value is too high by 14 mgal. Therefore, the values of g given in the Woollard and Rose book can be corrected by subtraction of 14 mgal. In this way, we have obtained accurate values of gravity at a worldwide distribution of observation sites. These sites make up what is called the International Gravity Standardization Net 1971 (IGSN 71).

The results of this global survey of gravity indicate values close to 978.0 gals at the equator. As we would expect, gravity increases with increasing latitude. At the North and South Poles, the values are approximately 983.2 gals. We can attribute the difference of 5.2 gals almost entirely to the flattening and rotation of the earth. Actually, this general conclusion was reached long before we had the results of this modern survey. But the improved accuracy is of critical importance in refining our knowledge of the shape and structure of the earth.

NORMAL GRAVITY

We are now ready to put together a formula for calculating gravity on an idealized model of the earth. This model will be the ellipsoid, which most closely matches the sea-level surface of the earth. We call it either the *reference ellipsoid* or the *normal ellipsoid.* Geodesists, who study the shape of the earth, use this ellipsoid as a basis for describing irregularities in the true sea-level surface. They have discovered that these irregularities extend a few tens of meters above or below the normal ellipsoid in most parts of the world. In Chapter 8, we discuss how geophysicists compare *normal gravity,* which is gravity on this ellipsoid, with measured values to discover density irregularities associated with geologic features.

We can describe the normal ellipsoid quite simply by specifying its equatorial radius R_e, its flattening f, and its speed of rotation ω. These numbers are found from astronomical observations and orbits of artificial satellites. They are combined to obtain the constants C_1 and C_2 in Equation 7–6, which expresses normal gravity. To complete this equation, we must find the value of gravity g_e on the equator of the normal ellipsoid.

Gravity on the equator of the normal ellipsoid should correspond to the average value of gravity at sea level on the earth's equator. Ideally, the measurements used to get g_e should be made along coasts crossing the equator. Only these locations are exactly at sea level. But a quick look at a world map reveals only about a dozen such locations, hardly enough to get a reliable worldwide average value of g_e. Therefore, we must also use gravity measurements from other places. Corrections that account for latitude and elevation must be applied to these additional gravity values so that they will be equivalent to measurements made exactly at sea level where a coast crosses the equator.

For obvious reasons, most geodesists and geophysicists should use the same normal ellipsoid so that they can compare the results of their studies. Therefore, an organization called the International Union of Geodesy and Geophysics (IUGG), after careful study to find reliable measurements, chooses the most appropriate normal ellipsoid. This was first done in 1930 when the following values were adopted:

$$R_e = 6{,}378{,}388 \text{ meters}$$
$$f = 1/297$$
$$\omega = 7.2921151 \times 10^{-5} \text{ radian/second}$$
$$g_e = 978.049 \text{ gals}$$

These values were combined in Equation 7–6 to obtain

$$g_N = 978.049(1 + 0.0052884 \sin^2 \phi - 0.0000059 \sin^2 2\phi) \quad (7\text{–}15)$$

where g_N is normal gravity at latitude ϕ. This particular equation was named the *international gravity formula*. Measurements of gravity used to get g_e were adjusted to the Potsdam reference value. Because the measurements were made before 1928 with pendulums, they cannot be considered very accurate by modern standards. Nevertheless, some geophysicists continued using this formula as late as 1980 in analyses of gravity surveys.

Our capability of launching satellites into orbits around the earth opened the way for making much more precise geodetic measurements. The path of a satellite is very sensitive to the shape of the earth. By carefully observing satellite orbits, we can find more precise values of R_e and f than those adopted in 1930 by the IUGG. While we were developing these methods of satellite geodesy, important advances were also made in methods of measuring gravity. Enough information was available by 1967 to justify revision of the normal ellipsoid. At that time, the IUGG adopted the following values:

$$R_e = 6{,}378{,}160 \text{ meters}$$
$$R_p = 6.356774.5 \text{ meters}$$
$$f = 1/298.247$$
$$\omega = 7.2921151467 \times 10^{-5} \text{ radian/second}$$
$$g_e = 978.031846 \text{ gals}$$

The normal gravity formula obtained by combining these values in Equation 7–6 is

$$g_N = 978.031846(1 + 0.005278895 \sin^2 \phi + 0.000023462 \sin^4 \phi) \quad (7\text{–}16)$$

This is called the *1967 geodetic reference system formula* (GRS67 formula). Gravity measurements used to obtain g_e for this formula were adjusted to the standard values given in Table 7–1.

Further refinements prompted the IUGG to revise the normal ellipsoid again in 1980, but changes in the normal gravity formula are too small to be of importance in exploration geophysics. The GRS67 formula is adequate for analysis of gravity survey results. This formula tells us what we need to know about large-scale variations in gravity caused by flattening and rotation of the earth.

STUDY EXERCISES

1. At 45 degrees latitude, the radius of the GRS67 normal ellipsoid is 6,367,467 m. What is the difference between normal gravity of this latitude and gravity on a perfect stationary sphere having the same mass and radius?

2. Gravity on the normal ellipsoid is the result of the mass gravitational attraction and the centrifugal acceleration. Calculate the mass gravitational attraction of the GRS67 normal ellipsoid at the equator and at the North Pole. Is the difference between these two values due only to the difference between the equatorial and polar radii, or is it also effected by other factors? Discuss!

3. Suppose that gravity is to be measured to a precision of 10 microgals by means of a mass falling a distance of 1 m from a rest position. Calculate how accurately the time of the fall must be known to achieve this precision.

4. If the period of a pendulum is 0.5 s on the equator of the GRS67 normal ellipsoid, what is its period at the North Pole?

5. Suppose you plan to construct a gravimeter according to the diagram shown in Figure 7–11, using a spring with a constant k = 1000 dynes/cm to support a 10-cm-long beam with a mass m = 1 g in a horizontal position. The spring must be attached at some point to the upper part of the frame which is 10 cm above the beam. Determine a set of values for the angles Δ and β for which the natural period of the gravimeter will be about 10 s. Then calculate the length of spring that would have to be retracted by a micrometer screw to hold the beam in a horizontal position if gravity changed by 10 mgal.

6. What are two important differences between a LaCoste–Romberg gravimeter and a Worden gravimeter?

7. What important development made during the 1950s in the construction of gravimeters made it possible to use these instruments efficiently in worldwide surveys?

8. The GRS67 normal gravity formula is an improvement on the 1930 international gravity formula because of improvements in what two kinds of measurements?

SELECTED READING

Heiskanen, W. A., and H. Moritz, *Physical Geodesy*. San Francisco, W. H. Freeman and Co., 1967.

LaCoste, Lucien J. B., A new type long period vertical seismograph, *Physics*, v. 5, pp. 178–180, July 1934.

Morelli, C., The International Gravity Standardization Net 1971 (IGSN 71). *Publication Speciale No. 4,* International Association of Geodesy, Bureau Central de l'Association Internationale de Géodésie, 19, rue Auber, 75009 Paris.

Nettleton, L. L., *Gravity and Magnetics in Oil Prospecting.* New York, McGraw-Hill, 1976.

Torge, Wolfgang, *Geodesy.* Berlin, Walter de Gruyter and Co., 1980.

Woollard, George P., The new gravity system—Changes in international gravity base values and anomaly values, *Geophysics,* v. 44, n. 8, pp. 1352–1366, August 1979.

Woollard, George P., and John C. Rose, *International Gravity Measurements.* Tulsa, Okla., Society of Exploration Geophysicists, 1963.

EIGHT

Gravity Surveying

Geologic features make imprints on the earth's gravity field. We can recognize these imprints by comparing measurements of gravity with corresponding values of normal gravity that account for earth flattening and rotation. Before making these comparisons, we must first introduce certain adjustments for the form and elevation of the landscape where gravity was measured, which obviously is different from the normal ellipsoid surface. Then we can subtract from measured gravity the properly adjusted normal gravity to obtain a value that we call a *gravity anomaly*. The word "anomaly" implies a deviation from the normal. Of particular interest in this chapter are the deviations in gravity related to geologic features that make the earth different from the perfect normal ellipsoid. These gravity anomalies are the imprints that the exploration geophysicist interprets to get a clearer idea about the structure of the earth.

This chapter describes the basic gravity surveying and data processing practices used in exploration geophysics. We begin with a discussion about how to plan and carry out gravity surveys on land. Then we explain the fundamentals of calculating gravity anomalies. The chapter ends with a discussion of gravity at sea and the newly developed airborne operations using helicopters. Gravity continues to be the subject in Chapter 9, where procedures for geologic interpretation of gravity anomalies are presented.

GRAVITY SURVEYING ON LAND

A gravity survey is conducted by making gravimeter readings at many locations in an area of interest. Other measurements and observations must also be made to describe the position of each observation site. Especially important are latitude and elevation, because these values enter directly into the computation of gravity anomalies. Furthermore, we must note the time when each reading is made so that corrections for gravimeter drift can be determined. In the following discussion the important aspects of a gravity survey are first described separately. Then examples of survey practices are presented to illustrate how we can deal with these important aspects.

Gravimeter Observation Sites

Our choice of observation sites depends on the geologic features that are of particular interest and on the accessibility of the area. To detect anomalies caused by relatively small structures such as a buried reef or a salt dome, which are less than a few kilometers wide, we would need gravimeter readings at intervals closer than one or two kilometers. But the anomaly patterns caused by larger geologic features may be evident from measurements spaced several kilometers apart. In some areas, these readings can be made at roadside locations, and the gravimeter and other surveying equipment can be carried by automobile. Elsewhere, we may have to make more costly and time-consuming off-road traverses on foot, in special overland vehicles, or by helicopters.

The Gravity Base Station

The object of the survey is to measure the values of gravity g at the observation sites. We proceed by finding the changes in gravity Δg between these places and a conveniently located reference site, called a *base station*, where gravity has already been determined. The base station should be one of the sites in the 1971 International Gravity Standardization Net (IGSN71), which is described in Chapter 7. If none of these sites is close to the survey area, it may be necessary to determine the reference value of gravity g_r at a more conveniently located base station using gravimeter measurements adjusted to the IGSN71.

Tide and Drift Corrections

The difference in gravimeter dial readings at an observation site and at the base station is used to find the change in gravity Δg. But we cannot simply multiply the dial reading difference by the gravimeter constant to get Δg, for this difference is also affected by the times when the readings were made. First, we must make a *tide correction* that accounts for the time-varying gravitational attraction of the sun and moon, and a *drift correction* that accounts for time-dependent mechanical changes within the gravimeter.

We can find out how gravimeter readings change with time by graphing frequent readings made at the same location with the times of these readings. To determine both tide and drift effects properly, we should make gravimeter readings at one to three-hour intervals. The example in Figure 8–1 shows both cyclic and noncyclic changes. The cyclic variations are caused by lunar and solar gravity forces. These same forces produce the ocean tide. They vary in a cyclic way because the positions of the sun and moon are continually shifting relative to an observation site on the earth.

Observe how groups of points in Figure 8–1 plot alternately above and below an inclined line. This line indicates the noncyclic drift of the gravimeter. As different parts of the instrument age, rods and fibers slowly bend, and the springs gradually stretch. These slow mechanical alterations cause the gravimeter dial reading to change with time.

The dial readings were multiplied by the gravimeter constant to obtain the relative gravity values in milligal units plotted in Figure 8–1. The cyclic displacement of these values above and below the noncyclic drift line shows that the tidal variations of gravity are smaller than a few tenths of a milligal. Notice that high and low points on the tidal cycle occur at intervals of about 6¼ hours and that the amplitudes change from day to day. The tidal cycle also changes from place to place,

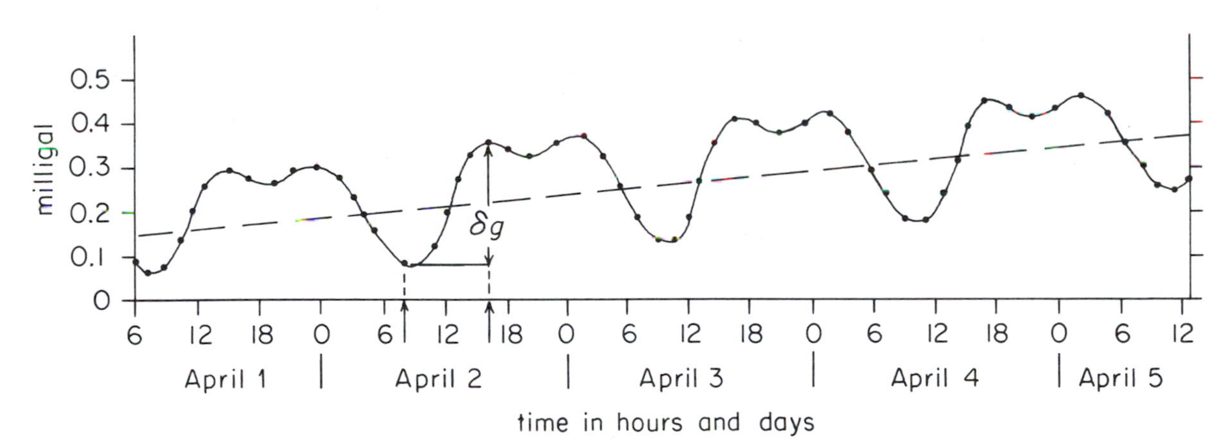

Figure 8–1
Graph showing how gravimeter readings (dots) change with time because of cyclic tidal variations and noncyclic instrument drift. On April 2, the change occurring between the times of 800 and 1600 hr is δg.

but this change is too small to be important if observation sites are less than a few hundred kilometers apart. If it is inconvenient to find the tidal variation of gravity from frequent gravimeter readings, as in Figure 8–1, this fluctuation of gravity δg_T can be calculated using an equation based on the masses of the moon and sun and their positions relative to an observation site,

$$\delta g_T = \frac{3}{2}\sigma GR \left[\frac{M_m}{r_m^3}\left(\cos 2\theta_m + \frac{1}{3}\right) + \frac{M_s}{r_s^3}\left(\cos 2\theta_s + \frac{1}{3}\right)\right] \quad (8\text{–}1)$$

where R is the earth's radius, M_m and M_s are the lunar and solar masses, and their distances from the earth's center are r_m and r_s. The value $\sigma = 1.16$ accounts for the way that the earth itself is stretched elastically by the tidal force. Angles θ_m and θ_s between a line from the earth's center to the observation site and lines from the earth's center to the moon and sun change with time. These angles can be calculated for any particular time from formulas based on astronomical measurements of the relative motions of the earth, moon, and sun. Because these formulas are long and complicated, a computer program is ordinarily used to make the calculations.

Noncyclic gravimeter drift can be determined only from gravimeter readings made in the same location at different times. Before beginning a survey, it is good practice to learn something about the drift characteristics of the gravimeter. The gravimeter is usually read one or more times a day for several days. From such readings we know that some gravimeters drift less than one milligal per week in a regular way. Other instruments may display high and irregular drift rates of close to one milligal per day. Occasionally, an abrupt change, called a *tare*, of several tenths of a milligal or more may occur. Some gravimeters undergo a tare once every few months, whereas others never change in this way.

During a gravity survey, certain observation sites must be reoccupied from time to time to obtain the information for drift and tide corrections. Reoccupations at least every three hours are needed to determine both of these effects properly. The repeated readings are used to prepare a graph similar to the one shown in Figure 8–1. For some surveys, reoccupations more than once or twice a day are impractical. Then it is necessary to combine these few repeated readings with tidal fluctuations computed from Equation 8–1 to prepare a graph of time variations.

After a graph similar to Figure 8–1 is prepared for the duration of a gravity survey, corrections can then be found by interpolation at the times when readings were made at all the observation sites. An example based on Figure 8–1 illustrates how these corrections are applied. Suppose that readings on April 2 were made at the base station at 8:00 A.M. and at observation site X at 4:00 P.M. Refer to Figure 8–1 to see that a change of δg was caused by tide and drift between these reading times. This change must be subtracted from the value $\Delta g'$ found by multiplying the difference in dial readings by the gravimeter constant to get the change in gravity Δg between these two locations:

$$\Delta g = \Delta g' - \delta g \quad (8\text{–}2)$$

We can find the value of gravity g_x at observation site X from this gravity change and the reference value of gravity g_r at the base station by using Equation 7–14:

$$g_x = g_r + \Delta g \quad (7\text{–}14)$$

Positions of Gravimeter Readings

The choice of ways to find the positions of gravimeter readings depends on the survey requirements for site spacing and precision. For some surveys, we can determine these positions directly from a good reference map that is already available for the survey area. Best for this purpose are topographic maps. Insofar as possible, we should read the gravimeter close to features that can be easily identified on the map. Locations of observation sites near road intersections, benchmarks, stream crossings, and prominent landscape features can then be determined from the map coordinates. Compass bearings may be useful for locating other sites that are not close to such obvious features.

Some gravity surveys require more precise positions than can be found on existing reference maps. An engineering transit survey may be needed to locate the gravimeter observation sites. With care, the positions can be found in this way to an accuracy of better than one meter, but a transit survey usually proceeds much more slowly than the time needed to make gravimeter readings. For traverses along curving roads and hilly terrain, the transit and rod may have to be moved several times to measure the distance and bearing between two gravimeter observation sites.

In remote areas for which no reliable reference maps are available, methods of celestial navigation and dead reckoning, familiar to sailors, have been used to locate gravimeter observation sites. Modern electronic satellite navigation systems have proved very useful for this purpose. These systems are discussed in more detail in Chapter 11.

Elevations of Observation Sites

There are three basic ways to find the elevation of a gravimeter observation site: the conventional engineering survey, reference topographic maps, and the aneroid altimeter. The most accurate method is the conventional engineering survey in which differences in elevation are found from transit sightings on a stadia rod through a telescope that is horizontal or inclined at carefully measured angles. Because several sightings from different transit and rod positions may be required to find the elevation difference between two gravimeter observation sites, this is often the most time-consuming and expensive part of a gravity survey. Nonetheless, it is the only practical way to obtain measurements of elevations that are accurate to closer than 30 cm. Later in this chapter, we explain that this accuracy is required to obtain gravity anomaly precision of within 0.1 mgal.

For purposes of some gravity surveys, elevations can be obtained from reference topographic maps already available for the survey area. Elevation values are noted on such maps at benchmarks and many road intersections. These values were determined to an accuracy considerably better than 30 cm by engineering transit surveying for use in preparing the topographic maps issued by many government agencies. Gravimeter readings should be made at these sites. Elsewhere, the elevations of observation sites can be estimated from the map by interpolation between topographic contours. The accuracy of such estimates is ordinarily assumed to be one-half the contour interval. The 7½-topographic quadrangle maps available for many areas of North America have a 20-feet (6.1 m) contour interval. Elevations interpolated between these contours should be accurate within ±10 feet (3.05 m).

It may be necessary to do a gravity survey of an area when reference topographic maps are not available and engineering transit surveying is impractical. An aneroid altimeter can be used to measure elevation. This instrument is sensitive to atmospheric pressure, which tends to decrease with increasing elevation. Atmospheric pressure also changes with time as weather systems move past a survey area, and it is influenced by temperature and relative humidity, both of which can be expected to change with time and location. Therefore, aneroid altimeter readings must be repeated frequently at certain observation sites in order to determine time-dependent corrections. These corrections are applied in much the same way as tide and drift corrections are applied to gravimeter measurements. It is difficult to account for all these influences on atmospheric pressure. For this reason, elevations measured with an aneroid altimeter can be several meters or even tens of meters in error, depending on the terrain of the survey area, weather conditions, and the frequency of reoccupation of certain observation sites.

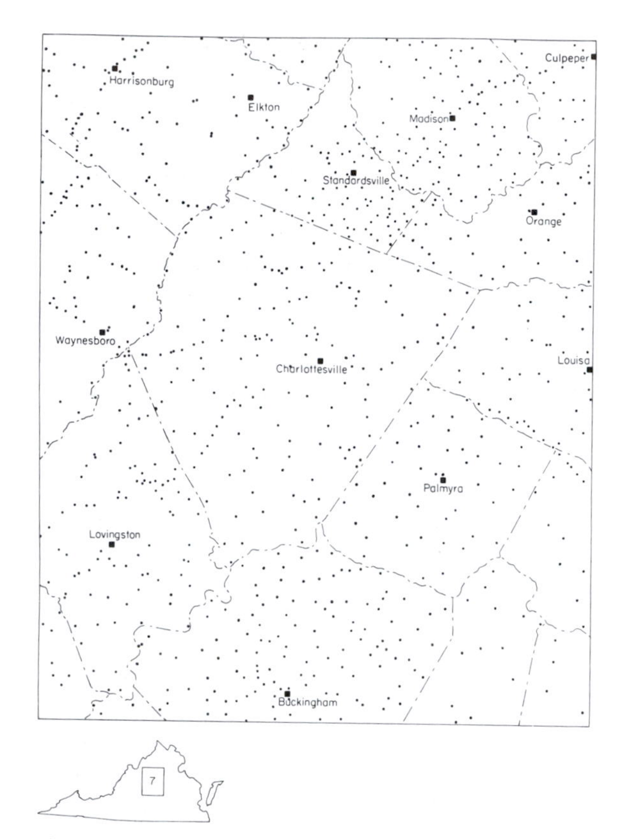

Figure 8–2
Observation sites where gravimeter readings were made during a gravity survey of central Virginia. (From S. S. Johnson, Sheet 7, Report of Investigations No. 27, Virginia Division of Mineral Resources, Charlottesville, Va., 1971.)

Survey Operations

Now let us consider the day-to-day activities of a gravity survey. Common practices can be described by examples. Our first example is a survey of the area in central Virginia shown in Figure 8–2. This survey was part of a larger effort by the Virginia Division of Mineral Resources to make gravity measurements throughout the state. Reference topographic maps of this area were already available for use in planning the survey and for finding locations and elevations of the observation sites. The nearest IGSN71 gravity station was in Washington, D.C. Previous tests with the gravimeter indicated more or less steady drift of less than one milligal per week. A computer program was available for calculating tidal variations of gravity using Equation 8–1.

The survey was planned by marking prospective observation sites on the reference topographic maps at benchmarks and road intersections where elevations were given.

Figure 8–2 shows 896 roadside sites 1 to 10 km apart. Next, a convenient site in Charlottesville, Virginia was selected for the gravity base station, because the IGSN71 station in Washington, D.C. was too far from the area for daily reoccupation. Then, daily routes were proposed for traveling by automobile from Charlottesville to observation sites close enough to one another that gravimeter readings could be made at an average of two to three sites per hour.

Field operations began with measurement of the change in gravity between the IGSN71 gravity station in Washington and the Charlottesville base station. Two consecutive one-day round trips were required. On each day, the gravimeter was read at Charlottesville in the morning, then at the Washington site about midday, and again at Charlottesville in late afternoon. Observation times were noted for purposes of gravimeter drift control.

The base station gravity value was found in the following way. First, the gravimeter dial readings were converted to milligal units by multiplication with the gravimeter constant. Then Equation 8–1 was used to compute tidal gravity at the times of all the gravimeter readings, as well as at hourly intervals for the two days. A curve connects these values to show the tidal variation of gravity. This curve must be adjusted to fit the four instrument readings at Charlottesville so that the change δg with time described in Equation 8–2 can be accounted for. The adjusted curve in Figure 8–3a indicates the combined tide and drift effects. The two readings at the Washington IGSN71 site can also be plotted on this graph. The difference in gravity Δg can then be taken as the average of the offsets Δg_1 and Δg_2 of these readings from the curve. Gravity g at the Charlottesville base station was calculated from Equation 7–14 based on this average value of Δg and the attraction of gravity g_r at the IGSN71 site.

After the base station was established, gravity surveying could commence. Operations each day started with a morning gravimeter reading at the base station. The observer then proceeded along the route planned for that day, reading the instrument as close as possible to the proposed observation sites. Any change in location was marked on the reference topographic map. Data obtained at each site included the gravimeter dial reading, time and date, elevation, latitude and longitude read from the topographic map, and a site identification number. One of the sites was reoccupied every two to three hours to check for tares and drift. After visiting 15 to 20 sites, the observer returned to Charlottesville for a final base station reading.

Gravity values at the observation sites were calculated by the same procedure that was used to find the base station value. Dial readings were converted to milligal units and plotted at the times of observation on the graph presented in Figure 8–3b. A curve of tidal gravity variations adjusted to the morning and afternoon base station readings was drawn on the same graph. Offsets of readings above this tide–drift curve indicate the differences in gravity, Δg, which were used with base station gravity, g_r, in Equation 7-14 to obtain values of g at the observation sites. After this step was completed, field operations for the gravity survey were finished. The results of this survey are presented later in this chapter, following the discussion about how to calculate gravity anomalies.

Now let us consider a second example of gravity surveying practices. This survey was done in northern Illinois to explore for structures where oil and natural gas might be trapped. The countryside in this area is nearly

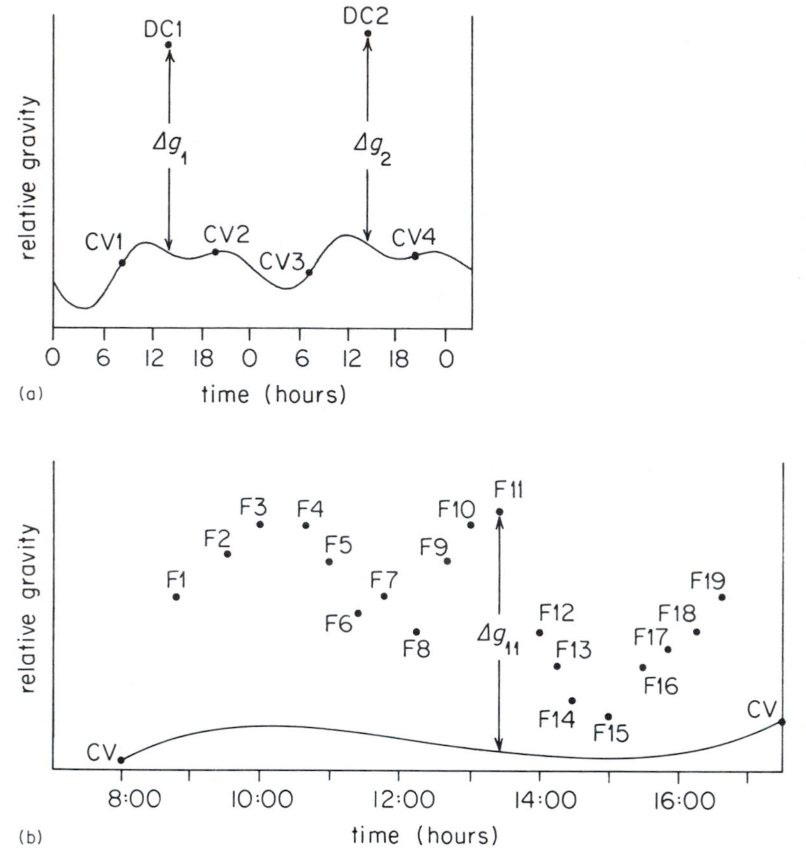

Figure 8–3
(a) Graph of gravimeter measurements used to find the difference in gravity between a base station in Charlottesville, Virginia, and IGSN71 reference station in Washington, D.C. Four readings in Charlottesville, CV1, CV2, CV3, and CV4, are connected by a tide–drift curve, and two readings in Washington, DC1 and DC2, are plotted above this curve to obtain two estimates, Δg_1 and Δg_2, of the gravity difference. (b) Graph of gravimeter measurements used to find the differences in gravity between the Charlottesville base station CV and several survey observation sites, F1, F2, F3, and so on. The two base station readings are connected by a tide–drift curve, and gravity differences, for example, Δg_{11}, are found from the vertical offsets from this curve of the readings at the observation sites.

flat, and bedrock is hidden under a blanket of glacial outwash deposits. Much of the area is crossed by a system of country roads spaced at one-mile intervals in north–south and east–west directions.

Maps showing roads and benchmark locations were used to plan the gravity survey. Experience has shown that readings accurate within 0.05 mgal should be made at intervals of 0.5 to 1.0 km along the roads. The plan called for observations at each benchmark and at intermediate sites where transit surveying would be needed to measure accurately positions and elevations.

During field operations, automobiles were used for transportation. A few convenient sites were chosen as temporary gravity base stations. Repeated gravimeter readings were made at these sites and at an IGSN71 station located some distance away. At the same time, transit surveyors began measuring positions and elevations at all the other observation sites. These sites are shown in Figure 8–4. Each site was marked by a small numbered flag so that it could be found later by the gravimeter operator. Depending on the number of transit and rod relocations between sites, this part of the survey can proceed more

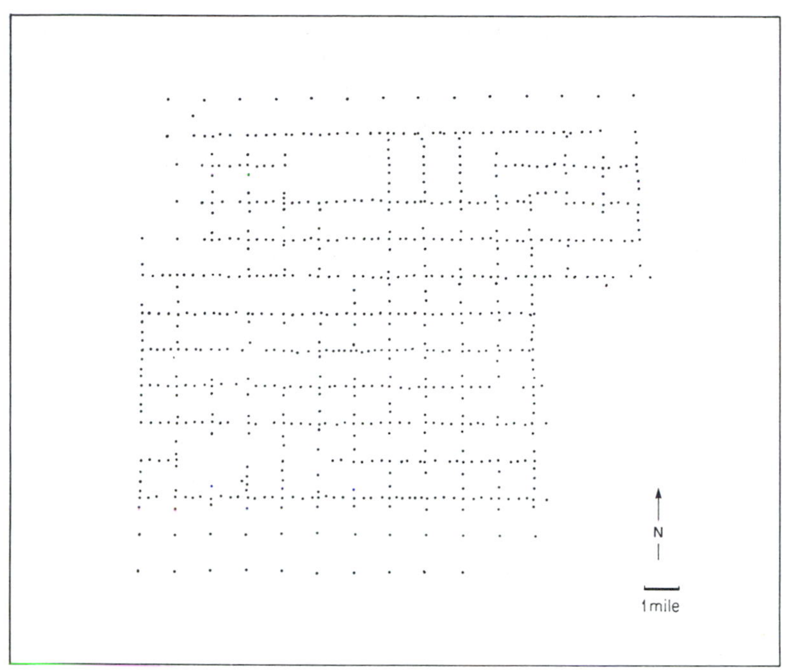

Figure 8–4
Observation sites where gravimeter readings were made during a gravity survey near Sibley, Illinois. (From J. W. Mack, Ph.D. dissertation, University of Wisconsin, 1963.) Elevations of these sites were measured by transit leveling.

slowly than the gravimeter measurements. Sometimes, more than one transit–rod team is employed to keep pace with the gravimeter operator. The elaborate methods of triangulation used in precise engineering surveying are not usually required. A single series of transit–rod sightings adjusted to benchmarks along the route is sufficient.

As a means of ensuring a gravity precision within 0.05 mgal, field operations were planned so that at least every two hours some site was reoccupied. Drift and tide changes during this length of time should be smaller than 0.1 mgal. During a typical day of operation, gravimeter readings could be made at the sites numbered in Figure 8–5 in the following sequence: B1, 2, 3, 4, B1, 5, 6, 7, 4, 8, 9, 10, 7, 11, 12, 13, 10, 14, 15, 16, 17, 13, B1. If we assume that the observer reads a gravi-

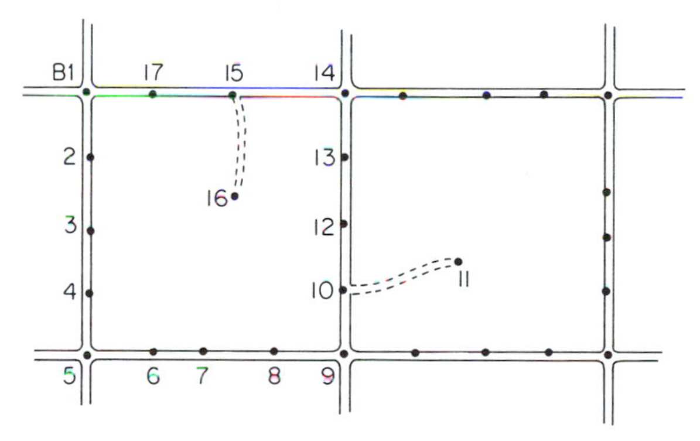

Figure 8–5
Close view of some roadside observation sites (numbered) where gravimeter readings could be made during an average workday.

meter at one site and travels to the next site of this sequence in 10 to 20 minutes, we can conclude that every two hours a reoccupation would be made, in sequence, at sites B1, 4, 7, 10, and 13. In this way, 16 observation sites in addition to the base station B1 could be occupied during one day.

Gravity at the observation sites can then be obtained as follows. First, the dial readings are converted to milligal units by multiplication with the gravimeter constant. These values are plotted at their times of observation in Figure 8–6, and lines are drawn to connect the pairs of readings at the sites that were reoccupied. These connecting lines show accurately enough for purposes of this survey the combined drift and tide variations during these one- to two-hour intervals of time. Gravity differences between one of these reoccupied sites and intermediate sites can now be found from the departure of the intermediate readings above or below the tide–drift line. Each of these departure values must be added to the gravity difference between that reoccupied site and the base station where the reference value of gravity g_r is known. For example, from Figure 8–6 we can see that at site 6 the value of gravity is

$$g_6 = g_r + \Delta g(4, \text{B1}) + \Delta g(6, 4)$$

In this way, gravity values can be determined for all the observation sites without using Equation 8–1 to compute the tidal variations. The results of this exploration gravity survey are presented later in this chapter, following the discussion of gravity anomaly computations.

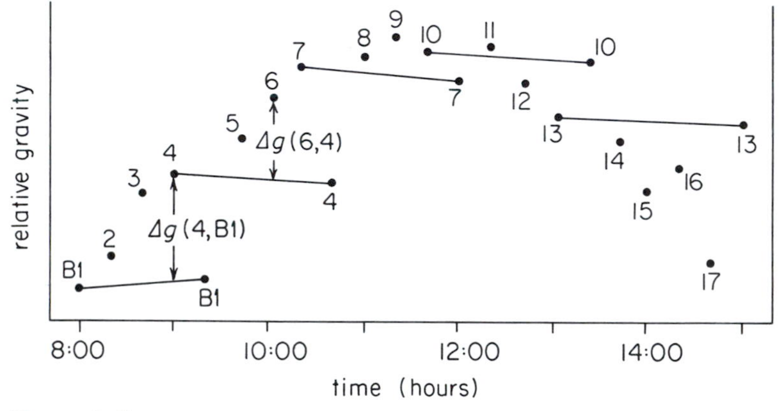

Figure 8–6

Graph of gravimeter readings used to find differences between the gravity values at a base station and at survey observation sites. Certain sites were reoccupied to determine short-term tide–drift curves, which are represented by the lines connecting these pairs of readings. Gravity differences can then be found from vertical offsets above or below these lines of readings at other observation sites. For example, Δg (4, B1) is the difference between the gravity at site 4 and that at site B1, which are shown in Figure 8–5.

GRAVITY REDUCTIONS

The values of g that we have found by gravity surveying are different from place to place. Some reasons for these differences are obvious. We already know that earth flattening and rotation produce latitude-dependent changes in gravity that can be calculated by using Equation 7–16. In addition, we know from the universal law of gravitation (Equation 7–1) that gravity depends on the distance from the earth's center, which changes from place to place according to the elevations of the observation sites. Furthermore, there would appear to be more mass beneath a high observation site than a lower one, causing the gravity values measured at these sites to differ. After we have accounted for these obvious effects, any remaining differences in gravity from place to place indicate density contrasts related to geologic features. How do we account for the effects related to the latitude, elevation, and excess mass at different observation sites? We can begin by imagining a very simplified earth that possesses only features that have an obvious influence on gravity. Then we calculate gravity on this idealized earth for comparison with measured values.

The Latitude Adjustment

Let us start by choosing the normal ellipsoid as a simplified representation of the earth. Then we can use Equation 7–16 to calculate normal gravity g_N at ellipsoid locations corresponding to all the observation sites of a gravity survey. For each observation site, g_N is subtracted from the observed value of gravity g_{obs} to obtain a number we can call the *observed-gravity anomaly* Δg_{obs}:

$$\Delta g_{obs} = g_{obs} - g_N \qquad (8\text{–}3)$$

Now, when we compare the values of Δg_{obs} from different places, we know that dissimilar values cannot result from the latitude-dependent effects of flattening and rotation. These effects are combined in the values of g_N, which have been subtracted from our measurements.

The Elevation Adjustment

If the earth were a perfect nonrotating sphere, gravity on its surface would be given by Equation 7–3:

$$g = GM/R^2$$

Choosing $R = 6.37 \times 10^8$ cm and $M = 5.97 \times 10^{27}$ grams yields the value of $g = 981.78545$ gals. Now let us move 1 meter above the surface, which puts us slightly farther from the center of the sphere. Because the distance has increased to 6.370001×10^8 cm, we calculate a slightly smaller value of $g = 981.78514$ gals. Therefore, we find that by moving 1 meter higher in elevation, gravity decreases by 0.31 mgal. As we continue moving higher and higher, the change in gravity will not remain exactly the same for each 1-meter increase in elevation, but while we are still quite close to the sphere, this rate of change remains very close to 0.31 mgal/m.

The formula for calculating the gravity of a rotating ellipsoid is more complicated than Equation 7–3, but it can be used to make the same kind of calculations to find how gravity changes with elevation above the normal ellipsoid. The results indicate that quite near to the normal ellipsoid surface, the change in gravity with elevation is very close to 0.3086 mgal/m. We will use this value to make elevation adjustments in the following way.

First, let us continue using the normal ellipsoid to predict gravity at an observation site.

Ordinarily, the value of g_{obs} measured at elevation h will be different from the value of g_N computed on the normal ellipsoid surface, which closely matches the earth's sea-level surface. We can improve our prediction by finding the gravity of the normal ellipsoid at the elevation h where we measured g_{obs}. This is done by subtracting from g_N the amount that gravity would change by moving a distance h above the normal ellipsoid. We can find this amount by multiplying h, given in meters, by 0.3086 mgal/m, which is the amount of change for each meter of elevation. Subtraction gives us the value of $g_N - 0.3086h$ for the gravity of the normal ellipsoid at that elevation.

Next, for each observation site, we subtract from measured gravity the normal ellipsoid gravity at the same elevation to obtain a number we call the *free-air gravity anomaly* or simply the *free-air gravity* Δg_{FA}:

$$\Delta g_{FA} = g_{obs} - (g_N - 0.3086h) \quad (8\text{–}4a)$$

where Δg_{FA}, g_{obs}, and g_N are in milligal units, and h is in meters. By using the standard conversion factor of 1 foot = 0.3048 m, we find that gravity above the normal ellipsoid changes at the rate of 0.09406 mgal/foot. We use this rate in the free-air gravity formula,

$$\Delta g_{FA} = g_{obs} - (g_N - 0.09406h) \quad (8\text{–}4b)$$

where elevation h is given in feet. Notice that a value of normal gravity calculated from Equation 7–16 is in gal units and must be converted to milligal units for use in Equations 8–4a and 8–4b.

Now, when we compare values of Δg_{FA} from various observation sites, we know that dissimilar values cannot come from the different elevations of the observation sites. Neither do they result from different latitude-dependent rotation and flattening effects. Each free-air gravity value has already been adjusted for all these effects.

The Excess-Mass Adjustment

Up to this point, we have used the normal ellipsoid as an idealized representation of the earth, but it lacks an important feature that contributes to the gravity measured at an observation site. We must add to the surface of the normal ellipsoid material equivalent to the part of the earth reaching from sea level to the land surface where an observation site is situated, as illustrated in Figure 8–7. The gravitational attraction of this land mass above sea level should then be added to the value of normal ellipsoid gravity at elevation h, which we have already computed.

How do we calculate the gravitational attraction of the land mass above sea level? This task is not difficult when the gravity survey is in an area of subdued topographic relief. The land mass above sea level in Figure 8–7a can be represented by the horizontal plate in Figure 8–7b, which is tangent to the normal ellipsoid at the observation site. The gravitational attraction of a flat plate, which has thickness h and density ρ and extends to infinity, is given by the product

$$\text{gravity of a plate} = 2\pi G\rho h \quad (8\text{–}5a)$$

When thickness is expressed in meters and density in grams per cubic centimeter, the gravity in milligals is

$$\text{gravity of a plate} = 0.04193\rho h \text{ mgal} \quad (8\text{–}5b)$$

If thickness is given in feet, the formula becomes

$$\text{gravity of a plate} = 0.01278\rho h \text{ mgal} \quad (8\text{–}5c)$$

An infinite plate might not be a very real-

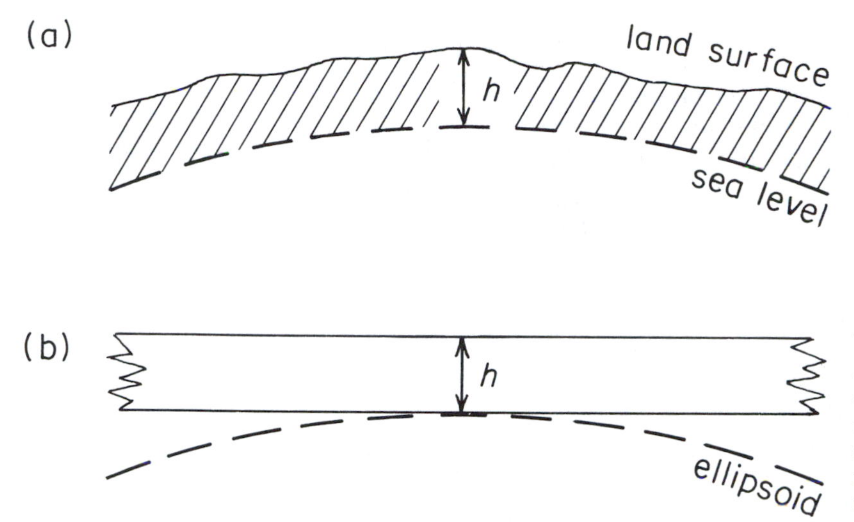

Figure 8–7
Representation of (a) land above sea level by (b) a flat plate placed on the normal ellipsoid for the purpose of Bouguer gravity calculations. The gravitational attraction for such a plate is very close to the attraction of the actual land mass above sea level at a similar observation site in regions of subdued topography.

istic representation of the land above sea level. But its gravitational attraction is very close to the value we would obtain by using Equation 7–3 to calculate, point by point, the combined attraction of all particles of the land above sea level. The reason is that most of the gravitational attraction is due to the small part of the plate situated close to the observation site. Because the more distant parts of the plate contribute very little, we can use the very simple Equation 8–5 to calculate a reasonably accurate value for the gravitational attraction of land mass above sea level.

This infinite plate approximation is all we need to adjust gravity measurements for the effects of land above sea level if the survey is in an area of subdued topographic relief. At each observation site, we make this adjustment as well as the adjustments that account for elevation and latitude-dependent flattening and rotation to obtain a number we call the *Bouguer gravity anomaly* or simply the *Bouguer gravity* Δg_B. For values of g_{obs} and g_N given in milligals, density ρ given in grams per cubic centimeter, and elevation given in meters, Bouguer gravity in milligals is obtained from the formula

$$\Delta g_B = g_{obs} - (g_N - 0.3086h + 0.04193\rho h) \tag{8–6a}$$

Where elevation is in feet, the formula is

$$\Delta g_B = g_{obs} - (g_N - 0.09406h + 0.01278\rho h) \tag{8–6b}$$

Some thought must be given to the density chosen for calculating the gravitational attraction of the land mass above sea level. The same value is generally used for adjusting all the gravity measurements in the survey. The best practice is to select a value close to the average density of this mass. To make an appropriate selection, we should have some idea of the proportions of different kinds of rock, unconsolidated sediment, and soil that make up the mass. It may be possible to estimate these proportions from geologic maps and soil maps if they are available. In addition, we must have some knowledge of the densities that are representative of these different materials. This knowledge is based on measurements of the densities of typical specimens. Ranges of density for common rocks, unconsolidated sediment, and soil are given in Table 8–1.

Another way is to use the gravity measurements for estimating the average density of the land mass above sea level. A method for doing this was proposed by L. L. Nettleton, a well-known geophysicist. A simple example given in Figure 8–8 helps to explain how it is done. Here we see that the density of the land beneath the profile where gravity has been measured is everywhere 2.5 g/cm^3. By using this value in Equation 8–6, we would correctly adjust for the effect of the land mass above sea level. Because there are no density irregularities along this profile, the Bouguer gravity values at all the observation sites should be the same.

TABLE 8–1 Density of Common Rocks

ROCK	DENSITY RANGE (g/cm^3)
Shale	1.95–2.70
Limestone and dolomite	2.50–2.85
Sandstone	2.10–2.60
Soil and alluvium	1.65–2.20
Rock salt	1.85–2.15
Felsic igneous rocks	2.55–2.75
Mafic igneous rocks	2.70–3.00

Suppose, however, that we chose a different density for computing the Bouguer gravity. Then we would be making incorrect adjustments, and the result would be different

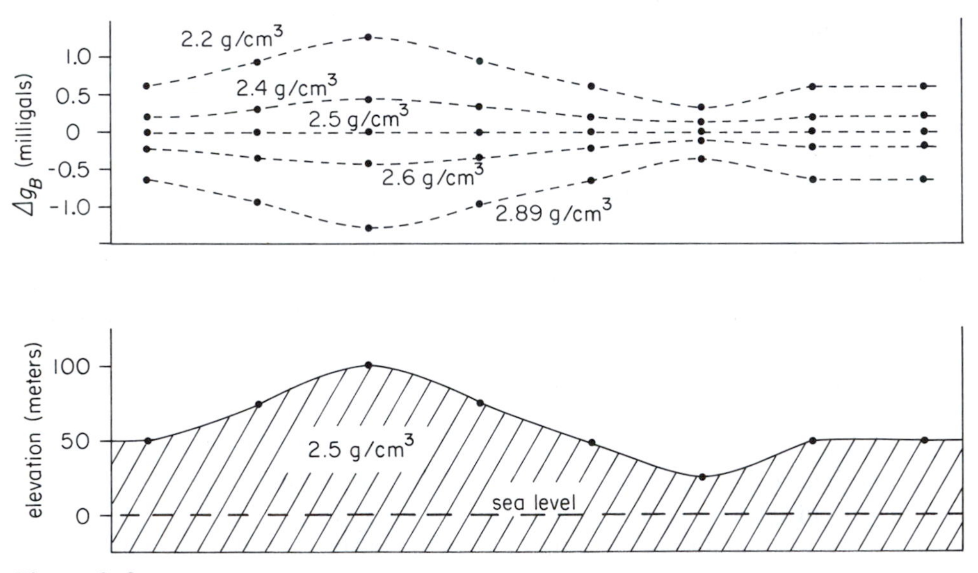

Figure 8–8
Graph illustrating the Nettleton method of density estimation. Dashed curves show Bouguer gravity calculated for several trial values of density.

Bouguer gravity values at different sites. See Figure 8–8 where Bouguer gravity values calculated using several densities are plotted. Observe that by choosing a density that is larger than the correct value, the Bouguer gravity values decrease with increasing elevation of the land surface. The adjustment term, $0.04193\rho h$, in Equation 8–6 becomes disproportionately larger as elevation increases. But if the density used in Equation 8–6 is smaller than the correct value, Bouguer gravity values increase as the land surface rises. This indicates that the adjustment term, $0.04193\rho h$, becomes disproportionately smaller as elevation increases. Only the correct density of 2.5 g/cm^3 produces Bouguer gravity values that do not vary in response to changes in elevation of the land surface along the profile.

Now suppose that you do not know the correct density of the land above sea level in the example given in Figure 8–8. Given the gravity measurements along this profile, we could make a series of trial calculations using different densities. Then we would be able to identify the correct density as the one producing the same Bouguer gravity at all the observation sites. In this way the Nettleton method is used to estimate the average density of the land above sea level in the area of a gravity survey. Because of the density irregularities that exist in nature, we cannot expect to find a trial value that will yield the same Bouguer gravity at all the observation sites of an actual gravity survey. But we will find that one value of density produces less Bouguer gravity variation from place to place than the other trial values. This one value should provide the best adjustment for the gravitational attraction of the land mass above sea level.

For many gravity surveys, the effort has not been made to find the average density of the land above sea level. Instead, a standard density of 2.67 g/cm^3 has been used to compute Bouguer gravity. Insofar as this value is not too different from the correct average value, errors will be quite small. But to compare the patterns of Bouguer gravity variation found from different gravity surveys, we should consider the values of density used in the computations.

Thus far we have discussed the adjustment for land above sea level in a survey area of subdued topographic relief. What must be done in rugged mountainous terrain? It is not sufficient to represent the land above sea level simply by a flat plate. An additional terrain adjustment must be made, and an example illustrates why this adjustment is crucial. A gravimeter situated at the observation site in Figure 8–9a is attracted by the mass of the earth beneath it. But it is also sensitive to a small, upward attraction of the nearby hill that rises above it. To account for this hill, in Figure 8–9b we add a similar mass to the flat plate that has already been placed on the surface of the normal ellipsoid. The gravitational attraction of this hill at the observation site can be represented by tc, which stands for terrain correction. Later, we will describe how the value of tc is calculated. Observe that it is an upward attraction that is opposite to the downward attraction of the flat plate. Therefore, it must be subtracted from the effect of the plate to determine the gravitational attraction of the land above sea level, which can now be expressed as $2\pi G\rho h - \text{tc}$.

Now look at the valley close to the observation site in Figure 8–9a. To account for this topographic feature, we must create a similar valley by removing some mass from the flat plate in Figure 8–9b. Let us again use tc to represent the gravitational attraction of this mass at the observation site. By subtracting it from the attraction of the plate, we obtain the

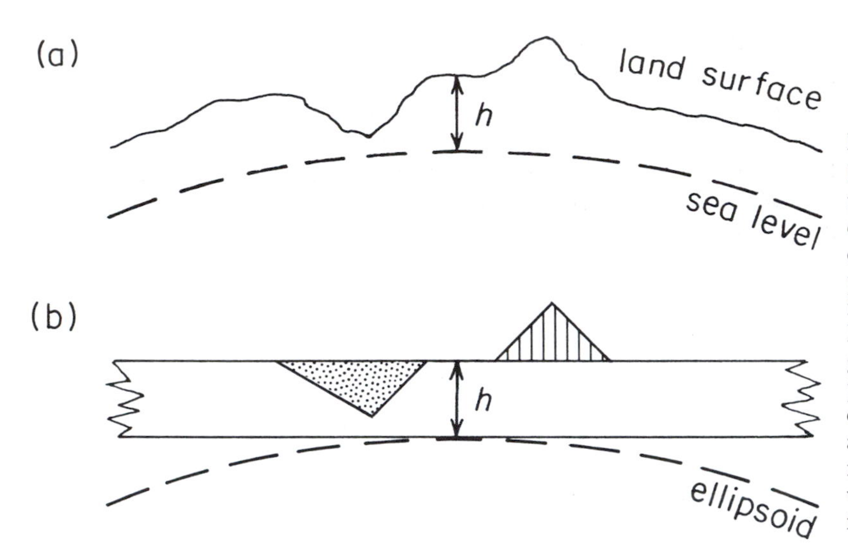

Figure 8–9
Representation of (a) topographic features by means of (b) mass increments added to or removed from a flat plate placed on the normal ellipsoid for the purpose of calculating a Bouguer gravity value. The gravitational attraction of a flat plate modified in this way more closely reproduces the attraction of the actual land mass above sea level compared with the attraction of a single flat plate.

expression $2\pi G\rho h$ − tc for the gravitational attraction of the land above sea level.

To represent a rugged terrain of land above sea level, we can create hills by adding mass on the top of a flat plate, and we can create valleys by removing mass from the plate. The effect of both kinds of topographic features is to diminish the gravitational attraction expressed by Equation 8–5. Now let us use TC to represent the combined gravitational attraction of all nearby hills and valleys that rise above or reach below the observation site. Then the Bouguer gravity is given by the formula

$$\Delta g_B = g_{\text{obs}} - (g_N - 0.3086h + 0.4193\rho h - \text{TC}) \quad (8\text{–}7a)$$

where Δg_b, g_{obs}, g_N, and TC are in milligal units, ρ is in grams per cubic centimeter, and h is in meters. If elevation h is given in feet, the formula is

$$\Delta g_B = g_{\text{obs}} - (g_N - 0.09406h + 0.01278\rho h - \text{TC}) \quad (8\text{–}7b)$$

How can we calculate TC? Perhaps the best-known method is the one developed by the geophysicist Sigmund Hammer. It uses a circular disk like the one in Figure 8–10 and a formula expressing the gravitational attraction, g_{disk}, at the center of its surface,

$$g_{\text{disk}} = 2\pi G\rho\left(\text{H} + r - \sqrt{H^2 + r^2}\right) \quad (8\text{–}8)$$

where r is the radius of the disk, H is its thickness, and ρ is its density. Now use this formula to subtract the attraction of a small disk of radius r_1 from that of a larger disk of radius r_2, where H is the thickness of both. This yields a formula that expresses the gravitational attraction, g_{ring}, at the center of a ring, like the one shown in Figure 8–11, at a point level with its surface:

$$g_{\text{ring}} = 2\pi G\rho\left(r_2 - r_1 + \sqrt{H^2 + r_1^2} - \sqrt{H^2 + r_2^2}\right) \quad (8\text{–}9)$$

Next, the ring in Figure 8–11 is divided into equal segments. The gravitational attraction of each segment at the central point in the di-

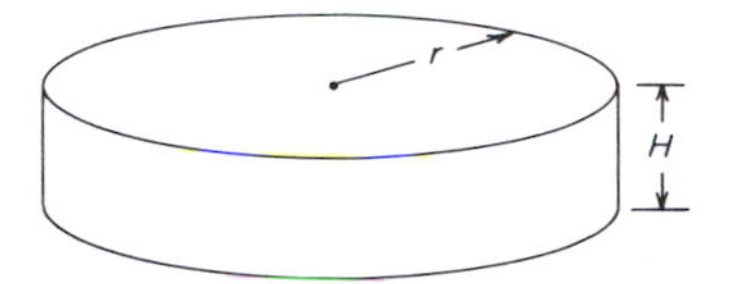

Figure 8–10
The circular disk is the basic figure used for calculating gravity terrain corrections.

rection of the ring axis is found by dividing the gravity of the complete ring by the number N of segments:

$$g_{seg} = g_{ring}/N \qquad (8\text{–}10)$$

For the example in Figure 8–11, we can see that $N = 8$.

Equations 8–9 and 8–10 are used in the following way to estimate the gravitational attraction tc of an increment of terrain that stands higher or lower than a gravimeter observation site. On the topographic map in Figure 8–12, a ring with inner and outer radii of r_1 and r_2 is centered on an observation site where elevation $h = 927$ m. The ring is divided into eight equal segments. Look carefully at the

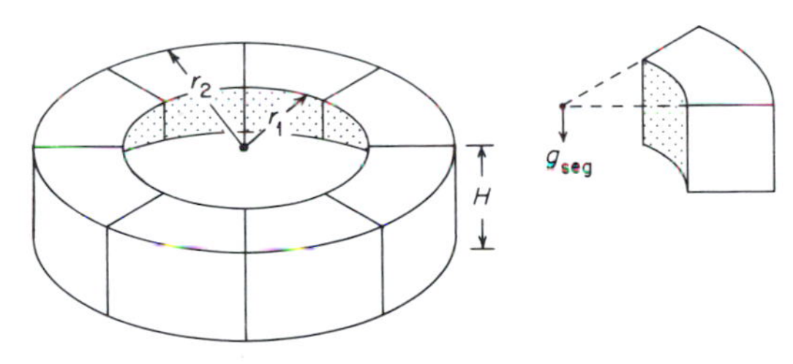

Figure 8–11
A circular cylindrical ring separated into eight equal segments for the purpose of calculating gravity terrain corrections. The component of gravitational attraction of each segment in the direction of the ring axis is equal to the total attraction of the ring divided by eight.

segment enclosed by heavy lines. The average elevation $\bar{h}$ can be estimated from the topographic contours within this segment. In the example, $\bar{h} \cong 950$ m. Now let the difference between this value and the elevation h of the observation site be the thickness H of the ring segment, that is, $H = \bar{h} - h$. Use this value of H in Equation 8–9 to calculate g_{ring}, which is then divided by the number of segments to compute g_{seg} in Equation 8–10. This value is the gravitational attraction tc of the increment of terrain.

To determine the complete terrain correction TC, we must carry out the procedure just described for all increments of terrain surrounding the observation site. This time-consuming task can be simplified by using the set of rings with standard dimensions shown in Figure 8–13, which was described by Sigmund Hammer. He first specified the inner and outer radii and the number of segments for this set of rings. He then calculated tables of values of tc for each segment of each ring using a series of thicknesses H. This information was published in the scientific journal *Geophysics* (volume 4, pages 184–194, 1939). To make use of it, we must draw on a clear plastic sheet the pattern of rings illustrated in Figure 8–13 at the same scale as the topographic map of the gravity survey area using radii specified by Hammer. The plastic sheet is then centered over an observation site marked on the topographic map. Now we can estimate values of average elevation $\bar{h}$ for all ring segments from the contours seen through the plastic sheet. Then values of H must be computed, which are used to find values of tc in Hammer's tables. All values of tc are summed to determine TC, which is needed in Equation 8–7 for calculation of the Bouguer gravity value.

The time required to complete the Hammer terrain correction procedure is approxi-

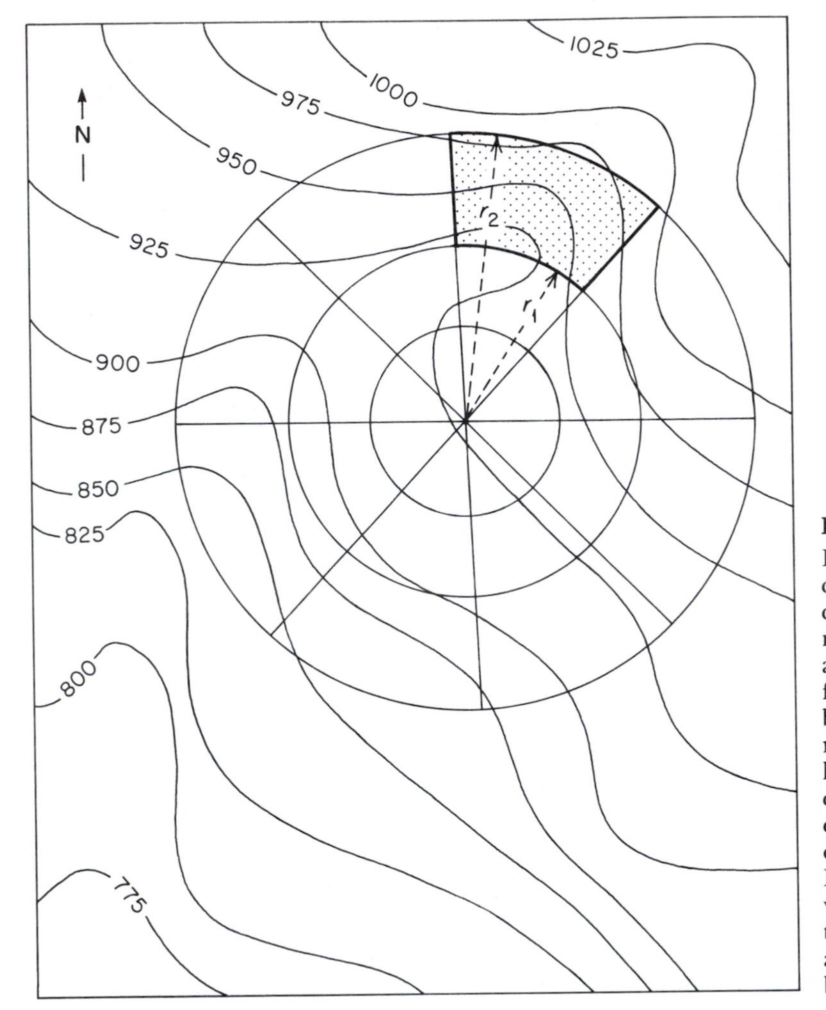

Figure 8–12
Illustration of the procedure for calculating gravity terrain corrections by the Hammer method. The vertical gravitational attraction of the topographic features within a small area bounded by two circular arcs of radii r_1 and r_2 and two radiating lines (shaded area) can be calculated by using the average elevation of this area, $\bar{h}$, and the observation site elevation h in Equations 8–9 and 8–10. The value of $\bar{h}$ must be estimated from the contours crossing this small area. In this example, $\bar{h}$ is seen to be approximately 950 feet.

mately one-half hour for each observation site. For a gravity survey with several hundred sites, this is, indeed, a time-consuming task. The time can be reduced by using one or another of several computer methods that utilize elevations read from a topographic map at regular intervals. Because it is impractical to read these elevations at close enough intervals, these computer determinations are usually supplemented by the Hammer method modified to include only the innermost rings which are very close to each site.

We have already stated that for an area of subdued topography, Equation 8–6 can be used to calculate Bouguer gravity. But in an area of rugged relief, Equation 8–7, which in-

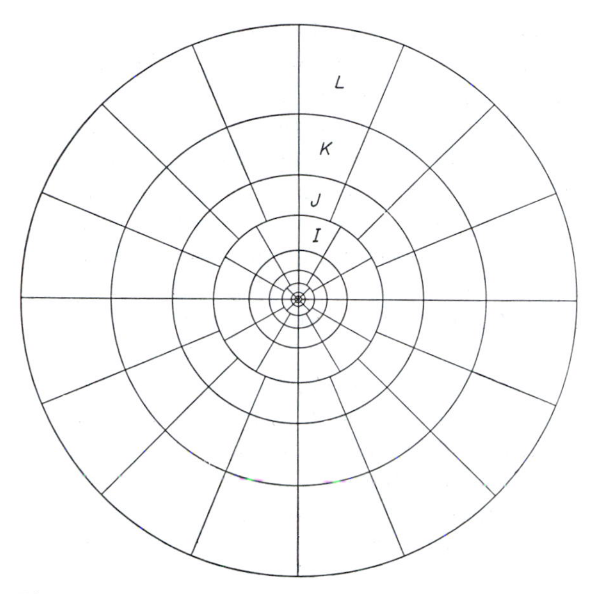

Figure 8–13
The segmented rings used for calculating gravity terrain corrections by the Hammer method. The ring dimensions were specified by Hammer. This pattern is drawn to scale on clear plastic and laid over a topographic map. It is first centered on an observation site. Next, the average elevation in each compartment is estimated from the contours that are visible through the clear plastic. Increments of gravitational attraction are then found from tables of values prepared by Hammer. (From Terrain corrections for gravimeter stations, *Geophysics,* v. 4, pp. 184–194, 1939.)

cludes the adjustment for terrain, must be used. How do we find out whether the land is sufficiently rugged to justify the terrain correction? This judgment is based on the desired accuracy of the Bouguer gravity values. Suppose the desired accuracy is, say, within 0.2 mgal. Then the values of TC should be computed for a few typical observation sites to find out whether the gravitational attraction of the terrain is likely to exceed this limit of accuracy. Trial calculations should also be made to determine how far from an observation site the terrain continues to contribute significantly. For many survey areas, these tests indicate that terrain corrections are small enough to neglect. For other areas, however, they are found to be necessary to achieve the desired Bouguer gravity accuracy.

Bouguer Gravity Accuracy

The measured values used in Equation 8–7 determine the accuracy of a Bouguer gravity value. First is the observed gravity, g_{obs}, which can be measured to within ± 0.05 mgal with most modern gravimeters. If a particularly stable instrument is used and drift corrections are carefully done, an accuracy close to 0.01 mgal is possible.

Next is normal gravity, g_N, which varies with latitude according to Equation 7–16. Values calculated at latitudes of 45 and 46 degrees indicate that g_N changes by 90.26 mgal for this 1-degree change in latitude. By dividing the earth's circumference of approximately 40,000 km by 360 degrees, we find that a 1-degree change in latitude corresponds to a distance north or south of about 111 km. The normal gravity change of 90.26 mgal over this distance indicates the rate of 0.813 mgal/km, or about 0.0008 mgal/m. Thus, for each meter error north or south in the measured position of an observation site, an error of 0.0008 mgal will be introduced into the value of g_N. The rate diminishes at latitudes closer to the equator and the poles. For example, near latitudes of 30 and 60 degrees, it is about 0.0007 mgal/m, and very near the equator and the poles it is negligible.

Two terms in Equation 8–7 depend on elevation. Let us choose a density of 2.67 g/cm^3

for the land above sea level and combine these terms:

$$0.3086h - 0.04193 \times 2.67h = 0.1966h$$

This equation tells us that for each meter of error in the measured elevation, an error of approximately 0.2 mgal is introduced into the Bouguer gravity value.

It is not usually difficult to determine the gravimeter drift characteristics and the position of an observation site accurately enough to obtain g_{obs} and g_N within 0.05 mgal. Much more effort involving transit surveying must go into measuring elevations within ± 25 cm to maintain a Bouguer anomaly accuracy of within 0.05 mgal. Otherwise, if elevation is interpolated, say, within ± 3 meter from a topographic contour map, the Bouguer anomaly accuracy would be closer to within ± 0.6 mgal.

Bouguer Gravity Maps

Finally, we have reached the point where the effects of geology on the earth's gravity field can be detected. These effects are indicated by patterns of Bouguer gravity variation over the survey area. Bouguer gravity values have already been adjusted for the effects of latitude-dependent flattening and rotation, elevation, and land mass above sea level. Therefore, any differences between the Bouguer gravity from place to place must be caused by density irregularities associated with geologic structures.

The patterns of Bouguer gravity variations are usually displayed on contour maps. We can prepare a Bouguer gravity contour map in the same way that we would prepare a topographic map, except that Bouguer gravity values rather than elevations are contoured. An example is shown in Figure 8–14. This is the final result of the gravity survey of central Virginia which we described earlier in this chapter. Bouguer gravity was calculated for all the sites marked in Figure 8–2. The contours in Figure 8–14 illustrate how the Bouguer gravity values vary from place to place over the survey area. What do these patterns of variation indicate? Basically, they tell us about density variations. Bouguer gravity is low over areas where average rock density is relatively low. Higher values indicate areas of relatively high average density. Another Bouguer gravity map is shown in Figure 8–15. It is the final result of the detailed exploration survey in northern Illinois, which was also described earlier in the chapter. The closed contours indicate relatively high Bouguer gravity in the northeastern part of the area. A zone of relatively high-density rock must underlie this area. Much more specific information about this subject is presented in Chapter 9, which deals with the geologic interpretation of Bouguer gravity.

GRAVITY SURVEYING AT SEA

Gravity surveys at sea can be done in two ways: on the ocean floor and aboard ship. The most precise but time-consuming measurements are made on the ocean floor with more or less conventional gravimeters, sealed in special containers and operated by remote electronic controls. Less accurate, but more rapid, shipboard surveying is done with gravimeters in special mountings that suppress accelerations caused by the pitching and rolling of the ship.

Ocean Floor Measurements

Gravity measurements on the ocean floor have been made at depths exceeding 1000 meters

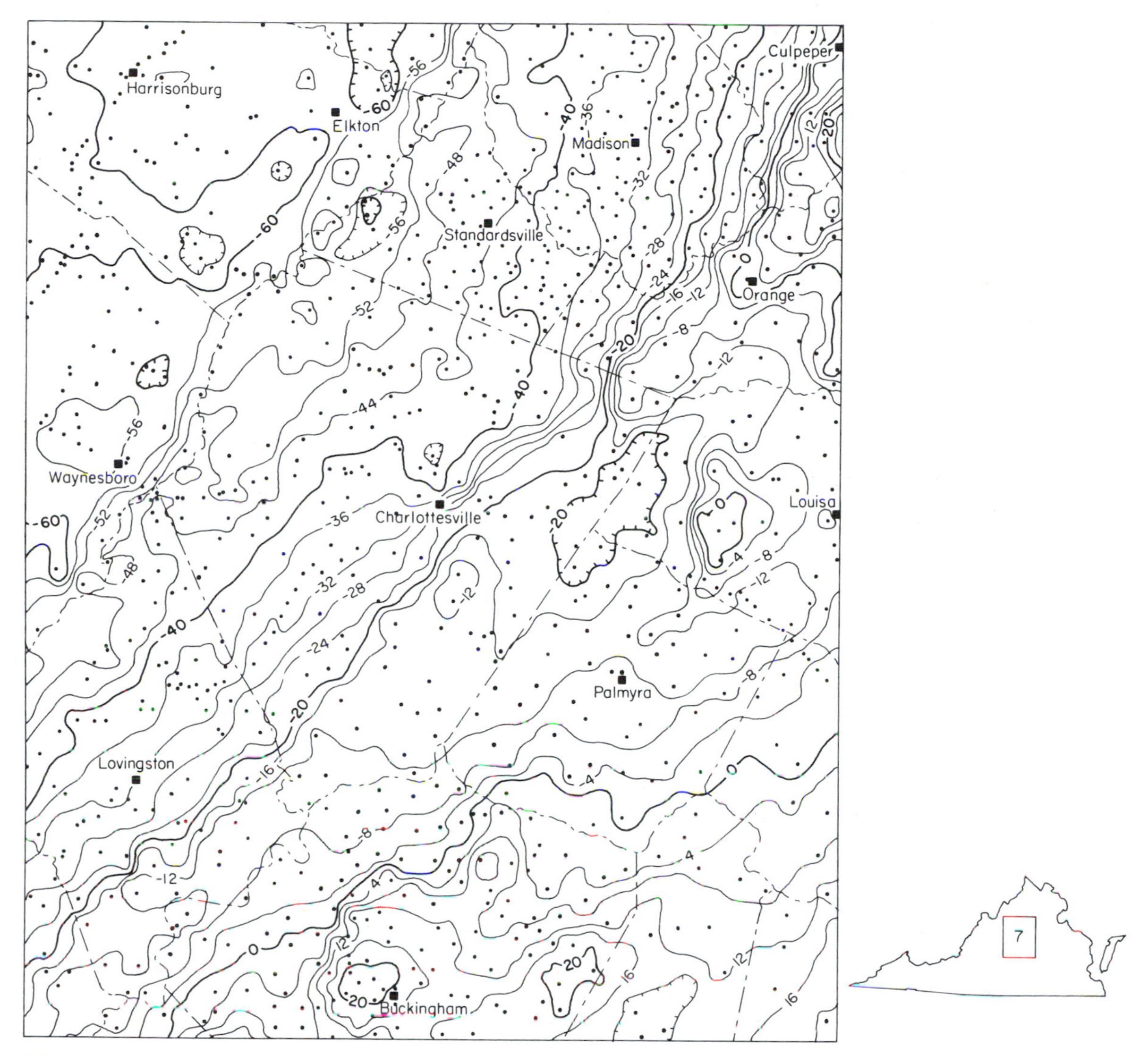

Figure 8–14

Bouguer gravity contour map of central Virginia. The contour interval is 4 mgal. (From S. S. Johnson, Sheet 7, Report of Investigations No. 27, Virginia Division of Mineral Resources, Charlottesville, Va., 1971.) Points indicating observation sites are the same as those in Figure 8–2.

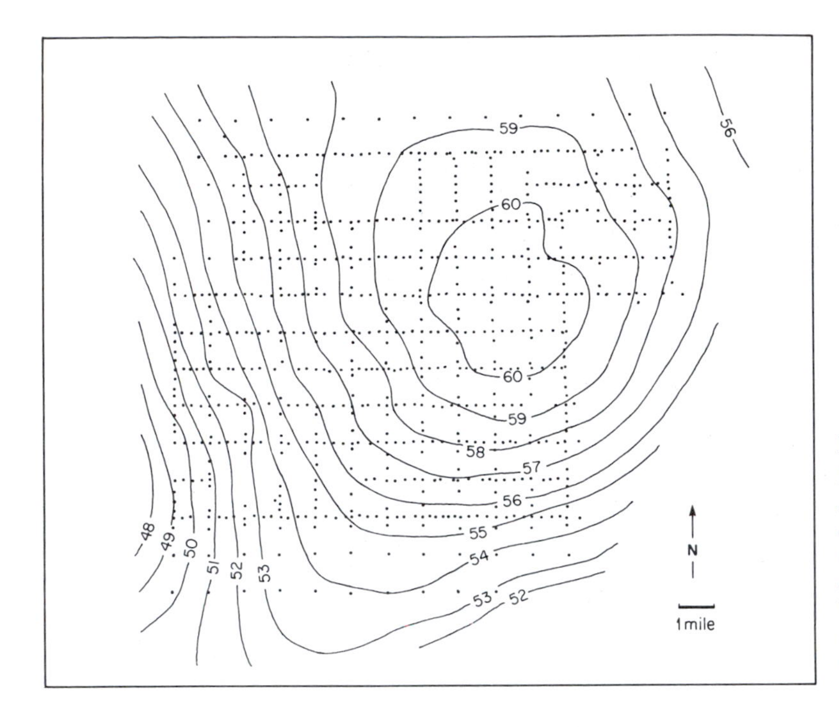

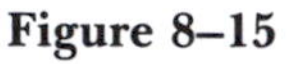

Figure 8–15
Bouguer gravity contour map of an area near Sibley, Illinois. The contour interval is 0.1 mgal. (From J. W. Mack, Ph.D. dissertation, University of Wisconsin, 1963.) Points indicating observation sites are the same as those in Figure 8–4.

(about 3200 feet), but most surveying has been done on the continental shelves where water is less than 200 meters (about 650 feet) deep. The most widely used modern ocean floor gravimeter is the LaCoste–Romberg instrument pictured in Figure 8–16. Signals from a shipboard control unit are transmitted through an electric cable to the gravimeter to operate the leveling system and to move the beam into position by adjusting the spring.

One complication encountered in shallow-water surveying is wave agitation. Motion of the water can rock the instrument slightly. The waves can also disturb the covering of very soft sediment on the bottom, which also moves the gravimeter. The effects of such an unstable environment can be reduced by an automatic leveling system. Motors controlled by a position indicator continuously adjust the frame to maintain beam movements in a vertical plane. Periodic motions of the beam control motors that raise and lower the frame, thereby compensating for the slight vertical movement of the ocean floor caused by waves. In shallow water, readings accurate within approximately 0.2 mgal can be made in 10 to 15 minutes, including the time needed to lower and raise the instrument. Problems of wave agitation diminish in deeper water, but more time is needed to lower and raise the gravimeter. While measurements are underway, the ship propellors are used to hold it in position.

Water depth at an observation site is measured with an echo sounder. At regular intervals of a few seconds, sound pulses emitted from the ship travel to the ocean floor and echo to a transducer on the hull. The travel

Figure 8–16
The LaCoste–Romberg submersible ocean floor gravimeter. (Courtesy of LaCoste–Romberg, Inc.)

time is recorded on a chart. This travel time together with the speed of sound waves in seawater are used in Equation 4–11 to calculate a value h, equal to the water depth d. The position of an observation site is determined with a shipboard navigation system. Modern methods of navigation are discussed in Chapter 11.

How do we calculate the Bouguer gravity at an ocean floor observation site? We subtract from measured gravity g_{obs} a value of normal ellipsoid gravity that has been adjusted for effects of elevation and the mass of the water layer. The first adjustment involves subtracting from normal gravity g_N the attraction of an outer layer of the ellipsoid that has a thickness equal to the ocean depth d. It is common practice to choose the density $\rho_r = 2.67$ g/cm^3 for this outer layer of the normal ellipsoid. Then, according to Equation 8–5b, its gravitational attraction is $0.04193\rho_r d$. Second comes an elevation adjustment that accounts for the increase in gravity caused by moving downward from the normal ellipsoid surface a distance equal to the ocean depth. This adjustment is similar to the one used in Equation 8–4a to obtain a free-air anomaly. The increase in gravity is $0.3086d$. The final adjustment accounts for the upward gravitational attraction of the water layer on the gravimeter, which is opposite to the downward attraction of the underlying solid earth. An equivalent attraction must be subtracted from normal gravity. When a value for seawater density of $\rho_w = 1.03$ g/cm^3 is used, an attraction of $0.4193\rho_w d$ can be found from Equation 8–5b.

The three adjustments just described are combined in the following formula for calculating the Bouguer gravity Δg_B on the ocean floor:

$$\Delta g_B = g_{obs} - (g_N - 0.04193\rho_r d + 0.3086d - 0.04193\rho_w d) \quad (8\text{–}11a)$$

Here Δg_B, g_{obs}, and g_N are in milligals, densities ρ_r and ρ_w are in grams per cubic centimeter, and water depth d is in meters. If $\underline{d}$ is in feet, the formula is

$$\Delta g_B = g_{obs} - (g_N - 0.01278\rho_r d + 0.09406d - 0.01278\rho_w d) \quad (8\text{–}11b)$$

Shipboard Gravity Measurements

Operating a gravimeter on the deck of a moving ship is an attractive alternative to the time-consuming ocean floor measurements. For this method to become a practical reality, it was necessary to solve some formidable problems caused by the motion of the vessel. A gravimeter is indiscriminately sensitive to acceleration. Like a seismometer, it responds to the pitch and roll of the ship as well as to variations in gravity. It is not unusual for the nongravitational accelerations to exceed 100,000 mgal. In such an environment how can we expect to detect gravity anomalies of a few milligals? Before this work can proceed, we must account for the effects of three kinds of motion: (1) pitch and roll that would tilt the gravimeter from a level position; (2) vertical rise and fall of the ship caused by passing waves; and (3) forward movement of the ship over the earth's surface that alters the centrifugal acceleration related to rotation.

A shipboard gravimeter must be operated on a level platform, but a platform on conventional free-swinging gimbals is not sufficient. Additional gyroscopic stabilization is required. As you may know, the principle of the gyroscope is to use the angular momentum of a spinning mass to maintain the spin axis in a fixed position. A properly designed gyroscope can be attached to a platform in a gimbal suspension to keep it level while the ship pitches and rolls. The most widely used shipboard gravity surveying system includes a specially designed LaCoste–Romberg gravimeter mounted on a gyrostabilized platform.

A shipboard gravimeter must be constructed to compensate for the vertical accelerations caused by the rise and fall of waves passing the vessel. It is designed with two basic features that are different from a conventional land gravimeter. First, the natural period of the spring-suspended beam must be considerably longer than the 5- to 10-second periods common to land meters. Because much of the ocean wave motion is at periods of between 5 and 20 seconds, a gravimeter beam with a similar natural period would swing wildly in response to the waves. In contrast, a beam with a much longer natural period would respond too slowly to be affected significantly by a single passing wave. Instead, it would be sensitive to the combination of several waves. In this sequence of waves, the intervals of rising water would produce increments of acceleration that tend to be nullified by the opposing increments of acceleration produced by falling water.

The response of the gravimeter beam to passing waves can also be controlled by a damping system. The mechanical system used in the LaCoste–Romberg instrument consists, basically, of two cylindrical cups. One cup attached to the beam fits as a cover to the other cup, which is attached to the frame. This method is illustrated in Figure 8–17. Any motion of the beam relative to the frame either

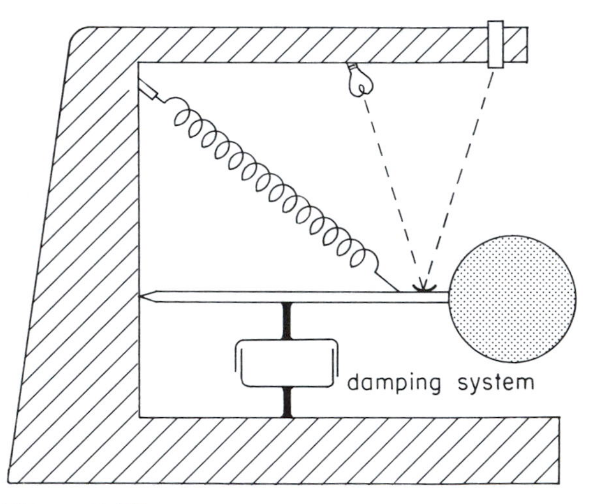

Figure 8–17
Mechanical damping of a gravimeter by means of two closely fitted cups, one acting as a lid covering the other. The restricted flow of air into or out from the chamber retards any effort to draw the cups apart or press them together.

draws the cups apart or pushes them together. This motion will be slowed because of the time required for air to flow into or out from the space enclosed by the closely fitting cups.

A heavily damped gravimeter with a long natural period is too sluggish to be significantly affected by the rapid variations in vertical acceleration caused by ocean waves. How can such a gravimeter be sensitive to gravity anomalies of a few milligals? To answer this questions, we must consider how much time is used to measure gravity variations along a profile. Suppose that we want to detect a variation in gravity that is about 3 km wide. A ship traveling at a speed of 10 knots (18.5 km/hr) would cross this feature in about 10 minutes. The gravimeter would detect it as a variation in acceleration occurring during this 10-minute period of time. This variation would last much longer than the fluctuations of between 5 and 20 seconds that are expected from passing ocean waves. Therefore, the instrument beam would have time to respond. This explanation indicates how a shipboard gravimeter can measure changes in gravity of a few milligals over distances of a few kilometers and yet remain insensitive to rapidly changing accelerations that may exceed 100,000 mgal.

A shipboard gravimeter is equipped with an automatic beam adjustment system. Any movement of the beam is detected by a sensing device. It activates a servomotor that adjusts the spring to compensate for the movement. A continuous record of gravity variations is obtained by measuring the servomotor current required to maintain the beam position.

The gravitational acceleration detected by the gravimeter is profoundly influenced by the speed and direction of the moving vessel. An adjustment called the *Eötvös correction* must be made to a shipboard gravity reading in order to account for this motion. It was first described by the Hungarian geophysicist Baron von Eötvös (1848–1919) about a century ago. This correction can be divided into two parts: (1) the outward centrifugal acceleration related to movement of the ship over the curved surface of the earth; and (2) the change in centrifugal acceleration caused by movement of the ship around the earth's axis at a speed different from the earth's rotation speed.

Consider first the effect of movement along the earth's curved surface. Let the circle in Figure 8–18 be a great circle surrounding the earth. As the ship follows the curved path from A to B in time t, the gravimeter detects an outward centrifugal acceleration $a_m = \Gamma^2 R$, where R is the earth's radius and Γ is the angular speed of the ship. To find Γ, we divide the angle δ between the radii OA and OB by the time t, that is, $\Gamma = \delta/t$. The speed of the ship along its curved path is $V = \mathrm{AB}/t$, and the arc distance $\mathrm{AB} = R\delta$. Therefore, the angular speed $\Gamma = V/R$, which allows us to express the acceleration

$$a_m = V^2/R \qquad (8\text{–}12)$$

This is one way that the forward movement of the vessel influences the value of gravity measured with a shipboard gravimeter.

Next, let us examine how motion of the ship around the earth's axis of rotation affects a gravity measurement. As the earth turns, a point on its surface moves eastward along a circle with a radius d that is the distance from the axis. According to Equation 7–5, the centrifugal acceleration at this point is $a_c = \omega^2 d$, where ω is the earth's angular speed of rotation. But the position of the ship is not fixed. Depending on its speed and direction, its eastward component of motion around the axis can be faster or slower than for a fixed point.

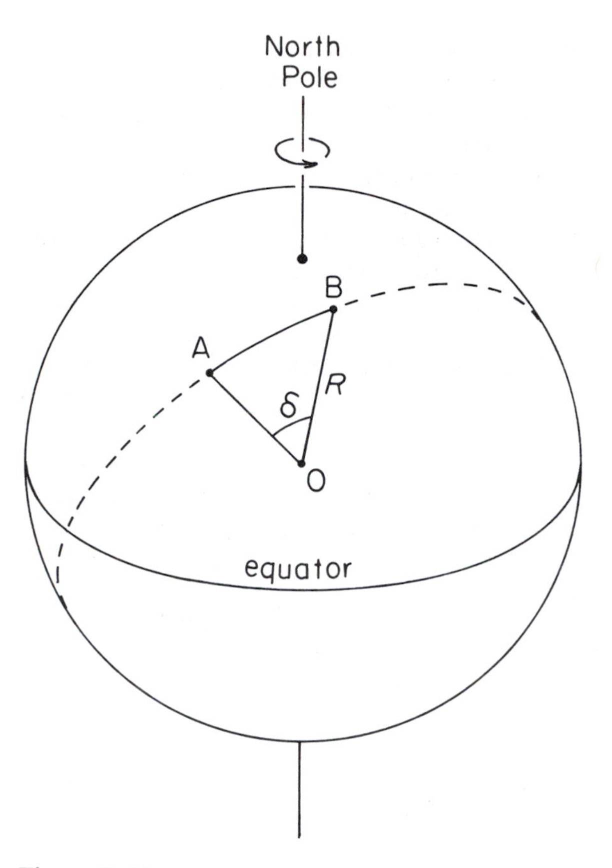

Figure 8–18

A great circle path between points A and B on the ocean surface and the geocentric angle δ between radii from the earth's center to these points.

This affects the centifugal acceleration acting on a shipboard gravimeter. We know that distance d from the axis changes with latitude φ. In addition, the eastward component of motion around the axis depends on speed V and the heading direction that can be specified by the angle β measured from geographical north. These are combined in the formula

$$\Delta a = 2V\omega \cos \varphi \sin \beta \qquad (8\text{–}13)$$

where Δa expresses the difference between a_c at a point on the earth and the centrifugal acceleration acting on a shipboard gravimeter moving over that point.

The two parts, a_m and Δa, make up the Eötvös correction E. Let us combine Equations 8–12 and 8–13 and choose appropriate values so that we can compute E in milligals if the speed of the ship is expressed in knots. For this purpose, $R = 6.371 \times 10^8$ cm, $\omega = 7.2921 \times 10^{-5}$ rad/s, 1 knot = 1.85325 km/hr = 51.479 cm/s, and 1 gal = 1000 mgal were used to obtain the equation

$$E = 7.508V \cos \varphi \sin \beta + 0.00415V^2 \qquad (8\text{–}14)$$

The value of E is positive for all easterly bearings (0 degrees < β < 180 degrees) and is negative for all westerly bearings (180 degrees < β < 360 degrees). This correction must be algebraically subtracted from a shipboard gravity measurement to obtain the value that could be read with a stationary instrument at the same location.

Some numerical examples illustrate the importance of the Eötvös correction. Suppose a ship heading northeast (β = 45 degrees) crosses a point at 45 degrees north latitude at a speed of 10 knots. The correction would be E = 37.97 mgal. If the ship crosses at the same speed on a northwest heading ($\beta = -45$ degrees), the correction would be $E = -37.13$ mgal. This indicates that gravity values measured at the same place along these two headings would differ by 75 mgal. Next consider how small changes in the speed and heading affect the Eötvös correction. According to Equation 8–14, a 1½-degree change from 45 to 46.5 degrees would change E by about 1 mgal. If the ship increases or decreases speed by ¼ knot, there will be a 1-mgal change in E. These numerical examples tell us that high standards of navigation are necessary to achieve a 1-mgal accuracy on a shipboard gravity survey. It is common practice to carry out the survey with some intersecting profiles. Thus, two gravity measurements at each crossing point can be used to estimate the accuracy of the survey results.

Gravity anomalies are calculated from shipboard gravimeter measurements in the following way. Because the vessel is situated at sea level, no elevation adjustment is made. The free-air gravity is identical to the observed gravity anomaly given by Equation 8–3. But to obtain the Bouguer gravity, we must make an adjustment to account for the water beneath the ship. The nature of this adjustment is illustrated in Figure 8–19. The normal ellipsoid is presumed to be solid. The gravitational attraction of its outer shell down to a depth equal to the ocean depth d is different from the attraction of seawater. Therefore, we must subtract from normal gravity g_N the attraction of a solid plate of thickness d and density ρ_r and then add the attraction of a plate of seawater with the same thickness and density ρ_w. We can use Equation 8–5 to find the gravitational attraction of these plates. These adjustments are combined in the following formula to obtain the Bouguer gravity value,

$$\Delta g_B = g_{\text{obs}} - [g_N - 0.04193d(\rho_r - \rho_w)] \tag{8–15a}$$

where Δg_B, g_{obs}, and g_N are in milligals, ρ_r and ρ_w are in grams per cubic centimeter, and d is in meters. If the ocean depth d is given in feet, the formula is

$$\Delta g_B = g_{\text{obs}} - [g_N - 0.01278d(\rho_r - \rho_w)] \tag{8–15b}$$

The value of observed gravity g_{obs} used in this formula is found by applying the Eötvös correction to a shipboard measurement. Then the conventional practice is to choose ρ_r = 2.67 g/cm^3, and seawater density ρ_w = 1.03 g/cm^3.

The Eötvös correction is usually the greatest limitation to Bouguer gravity accuracy. Errors of a few milligals in this correction are common because of the difficulty in controlling small variations in the direction and speed of a ship traveling through a wave-tossed sea. It is impractical to expect the pre-

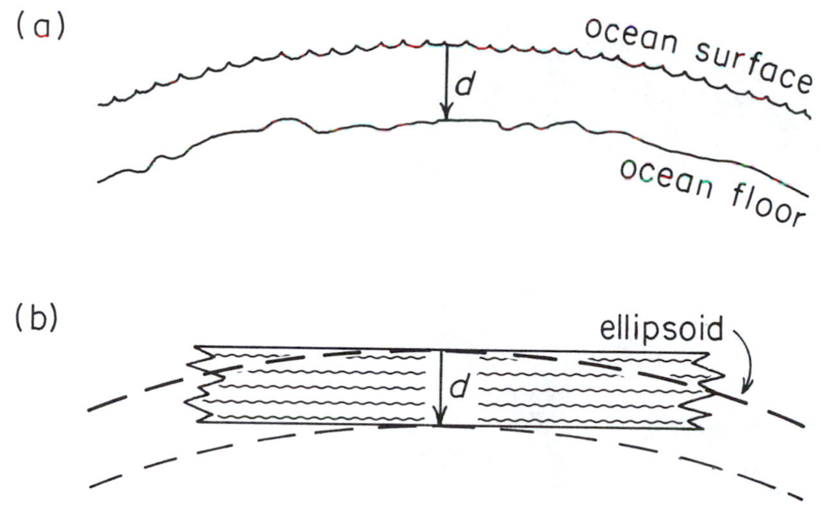

Figure 8–19
Representation of (a) the layer of ocean water by (b) a flat plate placed below the normal ellipsoid for the purpose of calculating a Bouguer gravity value from a shipboard gravity measurement.

cision of Bouguer anomalies obtained from shipboard surveys to be better than one or two milligals.

AIRBORNE GRAVITY SURVEYING

If we could put wings on a gravimeter, no region would be too remote to survey. But some difficult problems of design and operation must be solved to make airborne gravity surveying a reality. These problems are similar to ones encountered in shipboard operations, but they are exaggerated by the greater speed of the aircraft. For example, the Eötvös correction can be more than 1000 mgal in an airplane traveling at a speed of 200 knots.

The first airborne gravimeter measurements were made in 1958. Flights were made along a test path in California where aircraft position could be repeatedly measured from

Figure 8–20
Gyrostabilized gravimeter mounting and the helicopter used for airborne gravity surveying. (Courtesy of Carson Geoscience, a division of Carson Helicopters, Inc.)

ground-based observation stations. The accuracy of the gravity measurements obtained on these carefully controlled flights was within 10 mgal. In subsequent tests made with improved navigation equipment and gyrostabilized gravimeters, accuracy of within a few milligals was obtained during calm weather along well-marked paths.

Experimentation with helicopter transport began in 1971. Flights could be made at much slower speeds, thereby reducing the Eötvös correction and the navigation errors. Still, at speeds of 50 knots, the Eötvös correction can be as much as 375 mgal, depending on the latitude and the heading of the flight. According to Equation 8–14, gravimeter measurements made while crossing an area near the equator at this speed on eastward and westward headings would differ by about 750 mgal without appropriate Eötvös corrections. Because of the difficulty in accounting for moment-by-moment variations in heading and speed, errors of several milligals in the Eötvös correction are likely except under particularly ideal surveying conditions.

Two other effects impose limitations on airborne gravity surveying. The first is the effect of moving the gravimeter farther from the sources of gravity anomaly variation. A few milligals of Bouguer gravity variation on the land surface can diminish to less than a few tenths of a milligal at the level of an airborne survey perhaps 1000 meters higher. The second is the effect of terrain. On the land surface nearby topographic features have a much stronger attraction than similar features farther away. This difference diminishes at the higher level of an airborne survey, so that greater effort must be made to account properly for more distant terrain.

Despite these difficulties, helicopter gravity measurements have proved useful for reconnaissance surveying when accuracy within a few milligals is acceptable. For modern operations, a heavily damped gravimeter is mounted on a gyrostabilized frame and carried by a large cargo helicopter, as indicated in Figure 8–20. Electronic navigation equipment provides a continuous record of position changes, and a radar altimeter continuously measures the height above the land surface.

STUDY EXERCISES

1. Suppose that the following data were obtained on a gravity survey in Virginia.

STATION	TIME (A.M.)	DIAL READING
Base	8:10	2896.31
F1	8:26	2925.93
F2	8:45	2907.89
F3	9:00	2908.92
Base	9:17	2897.03
F4	9:40	2906.63
F5	9:57	2921.65
F6	10:20	2920.49
Base	10:35	2898.26
F7	10:56	2911.94
F8	11:15	2905.05
F9	11:47	2905.60
F10	12:10	2904.26
Base	12:45	2900.26

Measurements were made using a Worden gravimeter with a dial constant of 0.3802 mgal per dial division. Relative gravity values can be obtained by multiplying each dial reading by the instrument dial constant. Observed gravity at the base station is 979701.18 mgal.

a. Use the data to prepare a graph similar to the one in Figure 8–3.

b. What is the value of observed gravity at site F6?

c. What is the difference between the gravity at site F3 and that at site F7?

d. What is the approximate drift rate of the gravimeter?

2. The base station in Exercise 1 is located at latitude 36 degrees 38.67 minutes north, longitude 82 degrees 2.82 minutes west where the elevation is 1798 feet above sea level. Calculate the free-air gravity and the Bouguer gravity values for this station.

3. Antarctica is a continent covered by a vast blanket of ice. The ice has a density of 0.92 g/cm^3. At a location of 80 degrees south, 120 degrees west, the ice surface is 1525 m above sea level, and the ice layer is 2470 m thick. Here the value of observed gravity is 983.061 gals. Density of the rock beneath the ice is assumed to be 2.67 g/cm^3. Calculate free-air gravity and the Bouguer gravity values for this site.

4. The planet Mercury turns very slowly on its axis. For the purpose of calculating gravity anomalies, let us use a stationary sphere of radius 2418 km rather than a rotating normal ellipsoid. Suppose that normal gravity is 376.000 gals on the surface of this sphere. Now suppose that we have a measurement of 376.100 gals at a site on the surface of Mercury that is at a height of 1000 m above the reference sphere. Calculate free-air gravity and the Bouguer gravity values. (Note: We cannot use 0.3086 mgal/m to calculate the free-air correction. We must determine the change of gravity with elevation appropriate for Mercury.)

5. Suppose that at three locations along a road, the following information is obtained.

SITE	ELEVATION (meters)	FREE-AIR GRAVITY (mgal)
1	12	13.45
2	17	13.96
3	27	14.98

a. What would be the best density (not 2.67 g/cm^3) to use for calculating Bouguer gravity along this road?

b. What is the Bouguer gravity at site 2 based on the density value just calculated?

6. Suppose that Bouguer gravity values are calculated based on Equation 8–6 for a site on a mountain top and a site on the floor of a valley. Which of the following statements is correct: (a) values at both sites would be too low; (b) values at both sites would be too high; (c) one value would be too low and the other too high. Explain!

7. Suppose that shipboard gravity measurements were made at a location 40 degrees north, 30 degrees west where the water depth is 2000 m. A value of 980011.81 mgal was obtained with the ship traveling due east at a speed of 12 knots, and a value of 980080.83 mgal was obtained with the ship traveling due south at the same speed. Calculate the true value of observed gravity and the Bouguer gravity at this site.

SELECTED READING

Bott, M. P. H., The use of electronic digital computers for the evaluation of gravimetric terrain corrections, *Geophysical Prospecting,* v. 7, pp. 45–54, 1959.

Dobrin, Milton B., *Introduction to Geophysical Prospecting.* New York, McGraw-Hill, 1976.

Hammer, Sigmund, Terrain corrections for gravimeter stations, *Geophysics,* v. 4, pp. 184–194, 1939.

Kane, M. F., A comprehensive system of terrain corrections using a digital computer, *Geophysics,* v. 27, pp. 455–462, 1962.

LaCoste, Lucien J. B., Measurement of gravity at sea and in the air, *Reviews of Geophysics,* v. 5, pp. 477–526, 1967.

Nettleton, L. L., Determination of density for reduction of gravimeter observations. *Geophysics,* v. 4, pp. 176–183, 1939.

Nettleton, L. L., *Gravity and Magnetics in Oil Prospecting.* New York, McGraw-Hill, 1976.

Telford, W. M., L. P. Geldart, R. E. Sheriff, and D. A. Keys, *Applied Geophysics.* Cambridge, England Cambridge University Press, 1976.

NINE

Bouguer Gravity and Geology

What does Bouguer gravity tell us about geology? The reasoning used to interpret the geologic meaning is the subject of this chapter. We begin by looking at some important features of Bouguer gravity maps and profiles. The typical patterns that will be discussed can be seen on the contour map in Figure 9–1 and the profiles in Figure 9–2. These patterns are the imprints on the earth's gravity field that are produced by contrasts in the density of rock units making up different geologic structures.

These imprints reveal the combined gravitational attraction of several structures. How can we extract from such patterns of overlapping gravitational effects the attraction of certain geologic features of particular interest? We can approach this problem at least partially, by numerical processing methods. They are designed to separate anomaly patterns of different dimensions and bring certain features into clearer focus. Afterward, we can begin a trial-and-error analysis to learn about the shapes and depths of the anomaly-producing structures. This analysis requires calculation of the theoretical gravitational attraction of masses with shapes that represent different geologic features. The particular mass that could produce a gravity field closely resembling a measured anomaly pattern gives us the best indication of the structure that is making this imprint on the earth's gravity field.

REGIONAL BOUGUER GRAVITY FEATURES

Differences over Continents and Oceans

We usually refer to Bouguer gravity patterns that extend many tens or hundreds of kilometers as *regional anomalies.* Superposed on these regional anomalies are *local anomalies* that are less than a few kilometers to a few tens of kilometers wide. Some typically regional features are evident in Figures 9–1 and 9–2. Most dramatic is the difference between Bouguer gravity patterns over the continent and those over the bordering ocean basins. Observe that almost everywhere over the continent Bouguer gravity values are negative numbers. Over the oceans the values are positive. Close to the coast values are near zero. What is evident in Figures 9–1 and 9–2 is true worldwide. We measure negative Bouguer gravity values over most of the area of continents and positive values over the ocean basins of the world.

Bouguer Gravity and Elevation

Examine the Bouguer gravity and the elevations along the west-to-east profile in Figure 9–2. Observe the tendency for an inverse relationship in which the lowest Bouguer gravity, that is, the most strongly negative values, is typical of the regions of highest elevation. In the high reaches of the Rocky Mountains, Bouguer gravity diminishes to between −200 and −300 mgal. In the lowland regions along the coasts and in the Mississippi Valley, higher Bouguer gravity values in the range of −30 to 0 mgal are typical. Over the ocean basins where the earth's solid surface reaches far below sea level, Bouguer gravity rises to more than +200 mgal. This is a worldwide tendency, but it is not an exact inverse relationship. There are many exceptions to the pattern. To understand why Bouguer gravity over large regions tends to vary inversely with elevation, we must look at how the thickness of the earth's crust changes from place to place.

Bouguer Gravity and Crustal Thickness

During the first part of the twentieth century, seismologists discovered the principal interior zones of the earth. Refracted and reflected earthquake waves indicated the four concentric zones shown in Figure 9–3: a solid inner

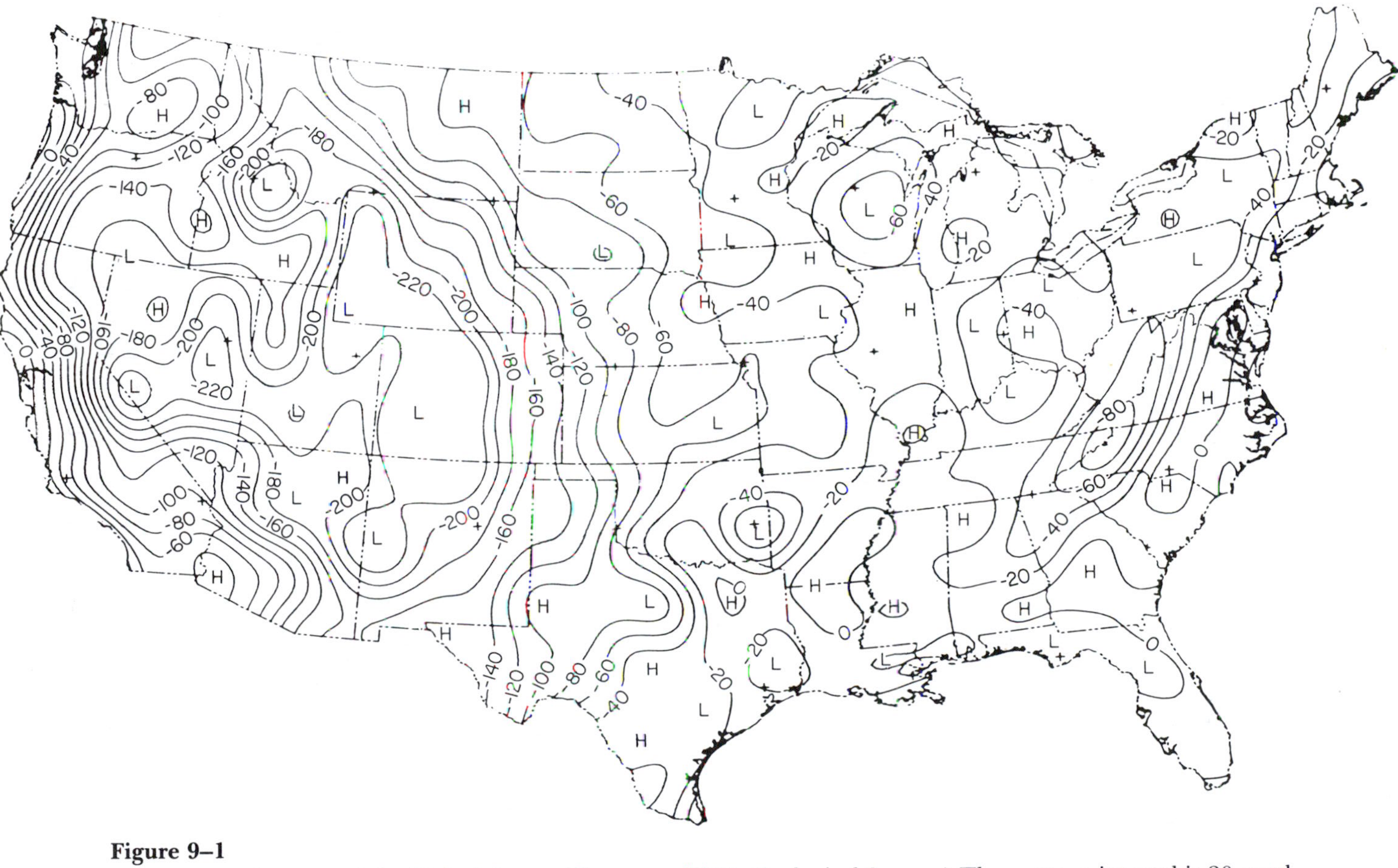

Figure 9–1
Bouguer gravity map of the United States. (Courtesy of U.S. Geological Survey.) The contour interval is 20 mgal.

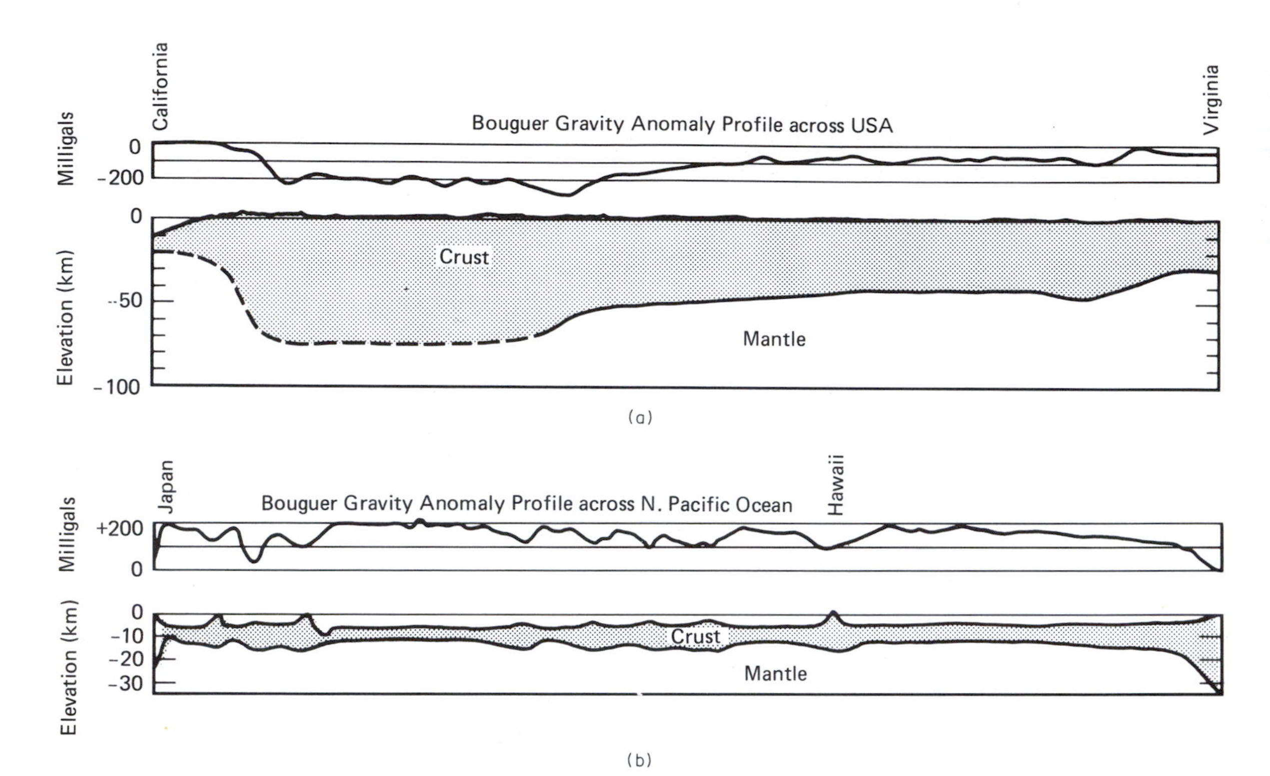

Figure 9–2
Profile of Bouguer gravity variation and crustal thickness across (a) the United States. (Modified from L. C. Pakiser and I. Zietz, p. 513, Fig. 5, *Reviews of Geophysics,* November 1965), and (b) the North Pacific Ocean. (Modified from G. P. Woollard and W. Strange, Monograph 6, p. 72, Fig. 8, American Geophysical Union, 1962).

core, a molten outer core, a solid mantle, and an outer shell we call the crust. Variation in the thickness of the crust is central to understanding regional Bouguer gravity features.

Although our first ideas about the crust came from studies of earthquake waves, most of our knowledge about its thickness and physical properties originated from seismic refraction experiments. At many continental and oceanic locations around the world, multiton explosions have been recorded along geophone lines extending to distances of a few hundred kilometers. Very few reflection measurements were made until the mid-1970s when deep vibroseis surveys became a reality. Since then, a more refined picture of the earth's crust has been emerging. Basically, what we have learned is the following. Under the continents, the crust is mostly between 30 and 60 km thick. Rocks making up the upper part tend to be of felsic composition and have densities between 2.6 and 2.8 g/cm^3. Deeper in the crust, the rocks are more mafic in composition and have densities mostly between 2.8 and 3.0 g/cm^3. The boundary between the crust and the mantle is the Mohorovičić dis-

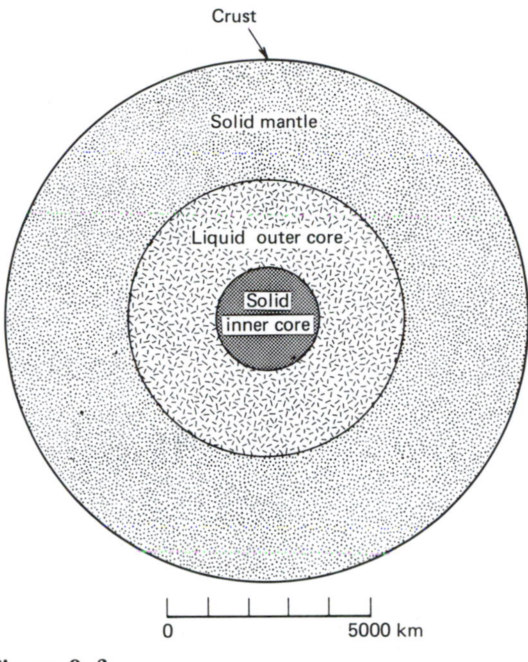

Figure 9–3
Interior zones of the earth.

continuity, often called the Moho. Below is the mantle where ultramafic rocks occur with densities of more than 3.3 g/cm^3. The crust beneath the ocean basins is mostly between 5 and 10 km thick and consists almost entirely of mafic rock.

One important feature of the crust is the correlation between its average thickness and the average elevation of the land surface. This is evident on the profile in Figure 9–2, which shows that a thick crust is typical of elevated regions. The crust is somewhat thinner in lowland areas and becomes much thinner under the oceans where the earth's solid surface extends several kilometers below sea level. Observe that near the coast, the crust is close to 32 km thick. The features we see in Figure 9–2 are typical of the crust beneath continents and oceans in other parts of the world.

Now let us look into how Bouguer gravity is influenced by changes in crustal thickness. It is obvious in Figure 9–2 that low Bouguer gravity (large negative values) is typical of regions where the crust is thick. Higher values that are much closer to 0 mgal exist in lowland areas and close to the coast. Strongly positive values are characteristic of the oceanic regions where the crust is very thin.

The idealized profile in Figure 9–4 helps to explain why Bouguer gravity tends to vary inversely with crustal thickness. Here, the crust is depicted as a layer in which density increases from 2.67 g/cm^3 at the land surface to 3.0 g/cm^3 at the Mohorovičić discontinuity. It rests on a homogeneous mantle where density is 3.3 g/cm^3. First, look at sites A and B which are situated at different elevations. In the process of calculating Bouguer gravity from gravity measurements at these sites, we make adjustments that account for any differences caused by flattening and rotation, elevation, and mass above sea level. But we do not make any adjustment for differences in average density, which depends on the proportions of low-density crustal rock and higher-density mantle rock beneath each site. Vertical lines extend downward from sites A and B to the same depth below sea level into the mantle. A larger proportion of the line extends through crustal rock at A than at B, indicating a lower average density beneath A. This suggests that the gravitational attraction of the combination of crust and mantle should be smaller at A than at B. For this reason the Bouguer gravity at A is lower than the value at B.

Next, look at site C on the oceanic part of the profile in Figure 9–4. Because the crust is so thin, the vertical line extends through a much larger proportion of high-density man-

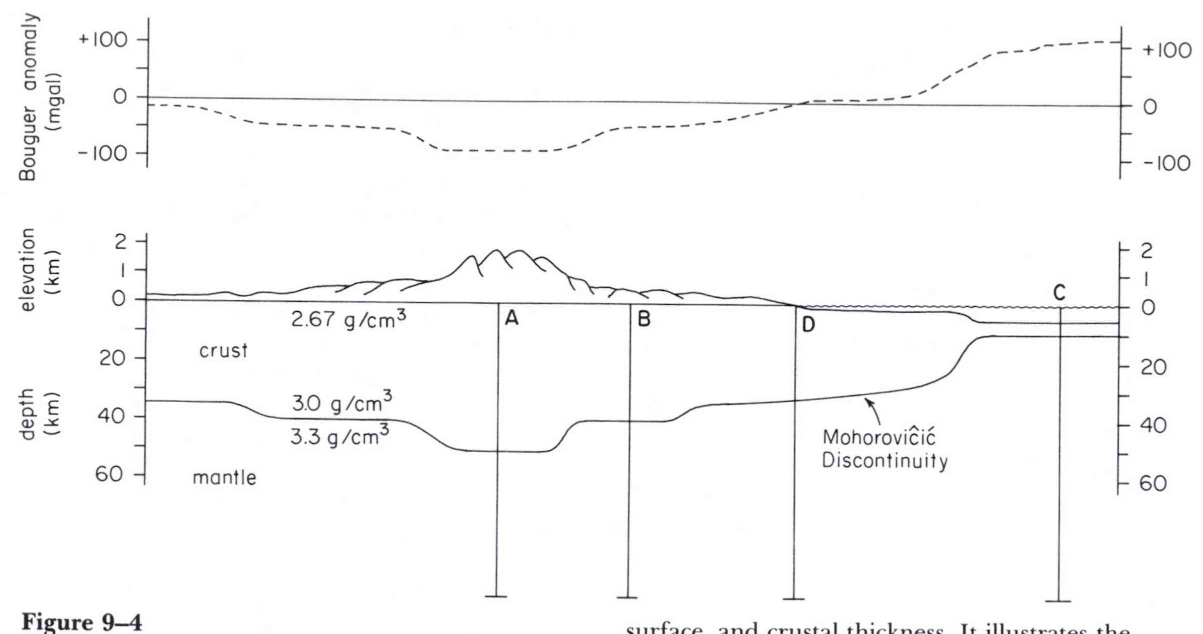

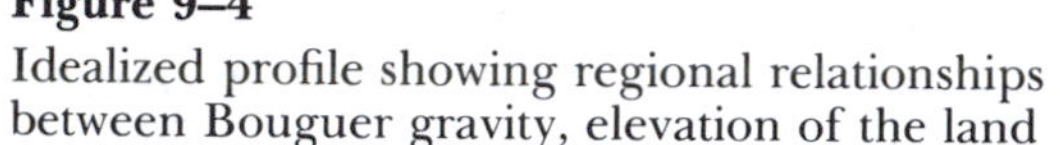

Figure 9–4
Idealized profile showing regional relationships between Bouguer gravity, elevation of the land surface, and crustal thickness. It illustrates the tendency over large regions for Bouguer gravity to vary inversely with elevation and crustal thickness.

tle. The high average density of this crust–mantle combination produces a stronger gravitational attraction and a higher Bouguer gravity at site C compared with that at continental sites A and B.

Finally, there is site D, situated on the coast where the crust is about 32 km thick. Why is the Bouguer gravity equal to zero here? Because the land surface reaches sea level at the coast, the elevation and mass terms disappear from Equation 8-6. Furthermore, for the Bouguer anomaly to be zero, normal gravity g_N must turn out to equal the measured gravity g_{obs}. Look again at the discussion of normal gravity in Chapter 7. Recall that the values of equatorial gravity in Equations 7-15 and 7-16 were determined from coastal gravity measurements. For this reason, we can expect to calculate a value of normal gravity that is close to the global average of coastal gravity measurements. Seismic surveys indicate that in most coastal areas of the world, the crust is close to 32 km thick. This explains why Bouguer gravity values between +20 and −20 mgal, and averaging about zero, are found in coastal areas and in other places where the crust is approximately 32 km thick.

The crust, then, has a profound influence on regional Bouguer gravity variation. Where the thickness of this outer shell departs from a standard coastal thickness of about 32 km, the Bouguer gravity departs from values near zero. These values decrease to lower than −200 mgal in high inland areas where the crust reaches a thickness of as much as 60 km. But over the thin 5-km oceanic crust, values of Δg_B rise to higher than +200 mgal. Large-scale geologic features in the crust also contribute to regional Bouguer gravity variations. Density contrasts related to large faults and regional differences in the thickness of the sediment that has accumulated in broad basins can produce Bouguer gravity patterns that are several tens of kilometers wide.

LOCAL BOUGUER ANOMALY PATTERNS

Anomaly over a Buried Pluton

We can find indications of smaller geologic structures in the local patterns of Bouguer gravity variation. Consider some examples of

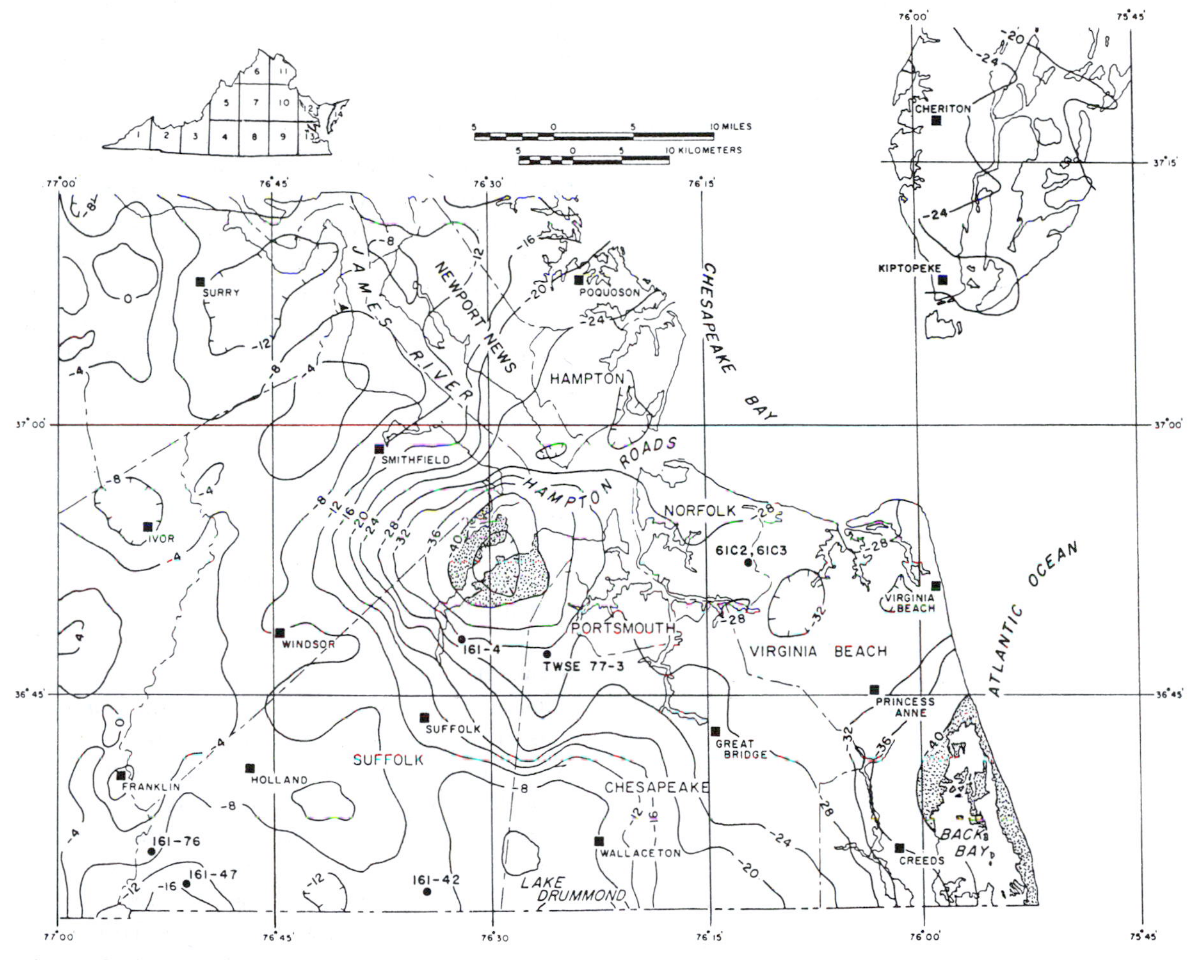

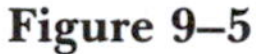

Figure 9–5
Bouguer gravity map of southeastern Virginia. (From S. S. Johnson, Sheets 9 and 13, Report of Investigations No. 39, Virginia Division of Mineral Resources, Charlottesville, Va., 1975.) The contour interval is 4 mgal.

anomalies ranging in width from a few tens of kilometers to a few tens of meters. First, let us look at results of a gravity survey that was done in southeastern Virginia by geophysicists from the Virginia Division of Mineral Resources. The Bouguer gravity contour map is shown in Figure 9–5. Notice the local anomaly pattern about 25 km wide, situated near the center of the map just northwest of Portsmouth. It is clearly indicated by several closed contours that show Bouguer gravity decreasing to lower than −40 mgal. Farther away from this local pattern, the Bouguer gravity values are mostly between 0 and −20 mgal.

What causes the Bouguer gravity to be unusually low in the local area northwest of Portsmouth? This pattern can be explained using the cross section in Figure 9–6. The cross section shows the earth's crust to be about 33 km thick in this region. Here, the upper part of the crust consists mostly of metamorphic rocks with densities between 2.80 and 2.85 g/cm^3. But a granite mass with a density of about 2.67 g/cm^3 is shown just northwest of Portsmouth. The entire area is covered by a blanket of unconsolidated sand and clay deposits about 400 m thick. The local Bouguer gravity values are low because of the density contrast between the granite and the bordering metamorphic rock. If the relatively low-density granite had not intruded the heavier metamorphic rock, the local Bouguer gravity values would not be low. Instead, all the values would be in the range of 0 to −20 mgal typical of the crustal thickness seen in Figure 9–6. Later, we will discuss how to calculate the approximate size and shape of the granite mass required to explain the local anomaly.

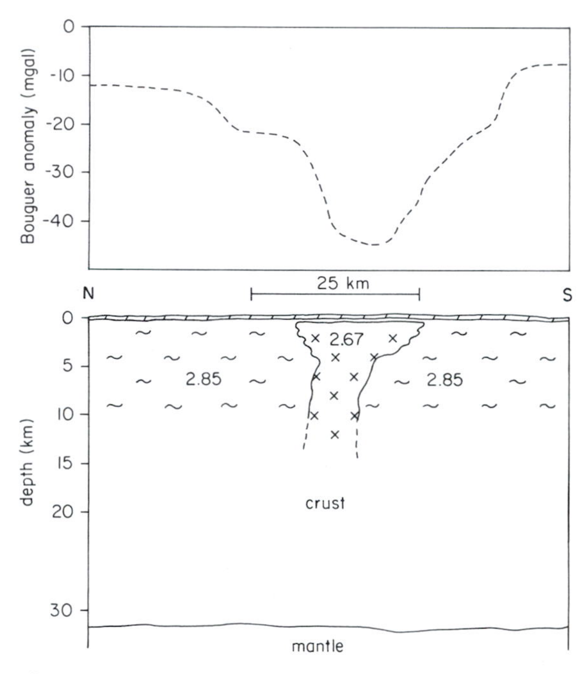

Figure 9–6
Bouguer gravity profile over a buried granitic pluton that intrudes higher-density metamorphic rocks.

Basin and Range Anomalies

Look at the patterns of Bouguer gravity variation in central Nevada that are shown in Figure 9–7. Here, the Bouguer gravity is mostly between −175 and −225 mgal. The relatively thick crust in this part of North America is the reason for such large negative values over the entire area. But superposed on this very low regional field are local variations of as much as 50 mgal. These are caused by density contrasts in the upper part of the crust. Look closely at the Bouguer gravity along profile A–A′ in Figure 9–7, which ranges from lower than −220 mgal in Pine Valley to higher than −180 mgal in the bordering Cortez Moun-

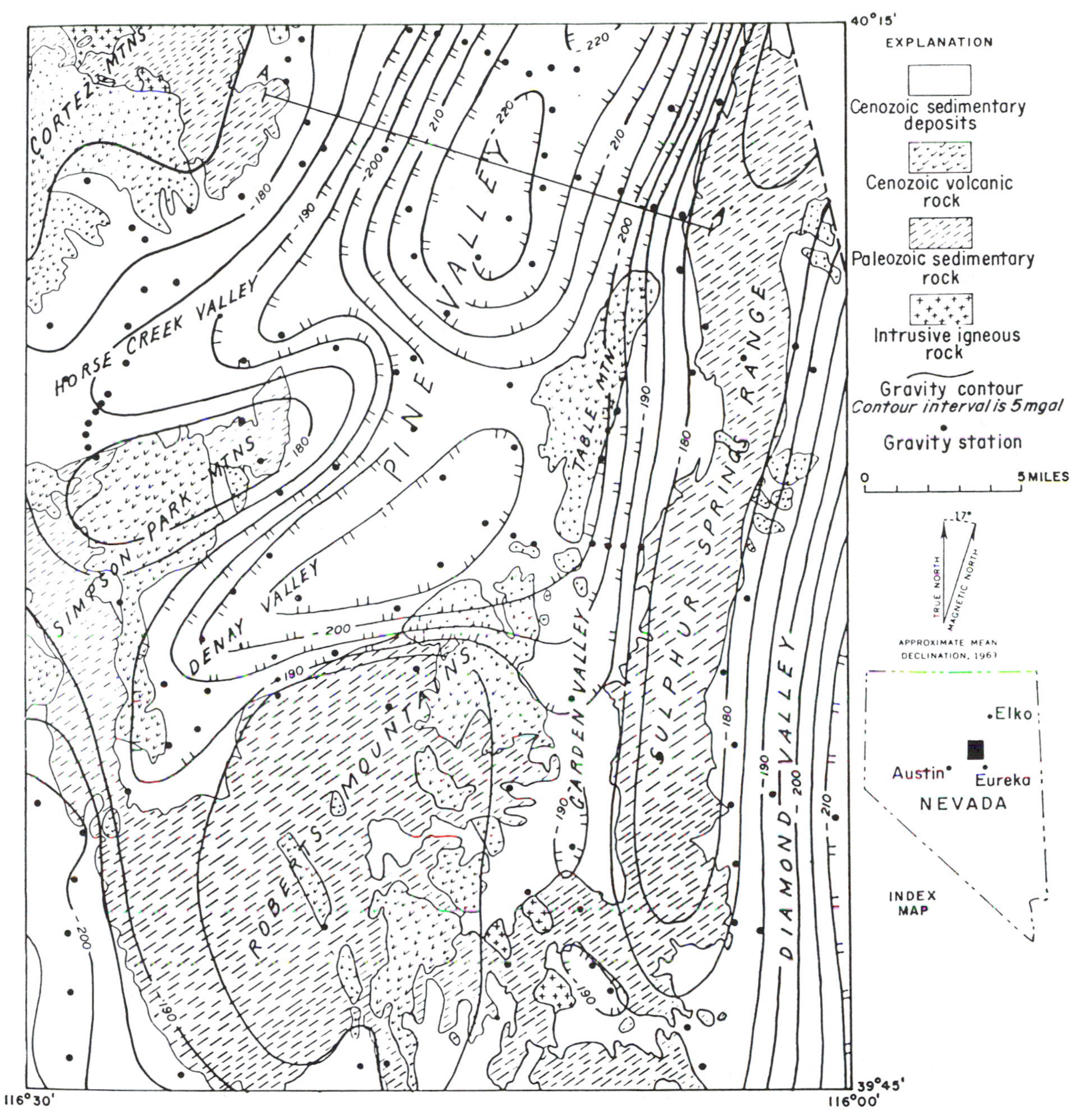

Figure 9–7

Bouguer gravity variation and generalized geology in central Nevada. (Courtesy of D. R. Mabey, U.S. Geological Survey.)

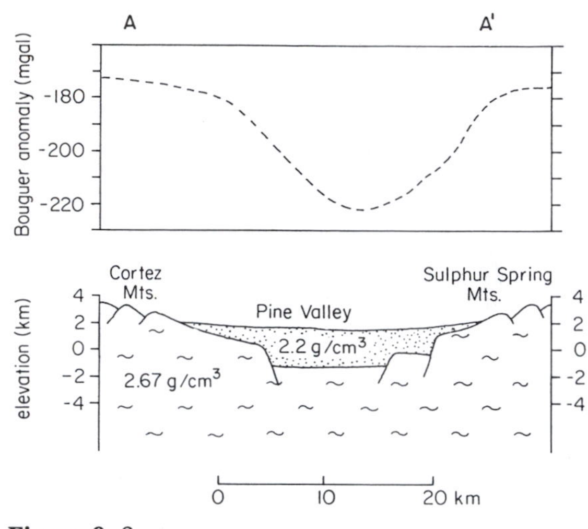

Figure 9–8
Bouguer gravity variation across Pine Valley in central Nevada, along profile A–A′ in Figure 9–7.

tains and Sulphur Spring Range. Because these values have already been adjusted for differences in elevation and mass above sea level, why are values in the valley so much lower than those in the nearby mountains? It must be that the average density beneath the valley is lower than the average density beneath the mountains. The cross section in Figure 9–8 explains this difference in average density. The valley is partly filled by a layer of unconsolidated sedimentary debris eroded from the nearby mountains. The density of this sediment is about 2.2 g/cm^3, which is considerably lower than the deeper bedrock density that is probably close to 2.67 g/cm^3. Therefore, beneath the valley, the combination of sediment and the deeper bedrock has a lower average density than the material making up the mountains, where the sediment is absent. The ways of calculating the thickness of sediment that could produce a Bouguer gravity variation of 40 mgal are discussed later in this chapter.

Anomaly over an Ore Deposit

Consider another example that comes from a precision gravity survey in a chromite mining district in Cuba. Here the gravity measurements were carefully made at sites 20 to 40 meters apart, where positions and elevations were found by transit surveying. The Bouguer gravity map is shown in Figure 9–9. In this small area, Bouguer gravity values differ by less than 0.5 mgal. The prominent anomaly near the center of the area is indicated by contours that enclose an exposure of chromite. The density of the chromite is close to 4 g/cm^3. The chromite deposit is emplaced in rocks with densities generally less than 2.7 g/cm^3. Obviously, the local pattern of high Bouguer gravity values is caused by the heavy chromite deposit.

These three examples illustrate why local patterns of Bouguer gravity variation might be expected when there are density contrasts associated with particular geologic structures. We see that these local variations are superposed on broader regional variations. A principal goal of the geophysicist is to estimate the depths and shapes of different geologic structures from an analysis of local anomalies.

THE GRAVITATIONAL ATTRACTION OF STRUCTURES WITH VARIOUS SHAPES

We use a trial-and-error method to find the approximate shape and depth of the structures that produce Bouguer gravity variations. This method is quite simple. We compare our

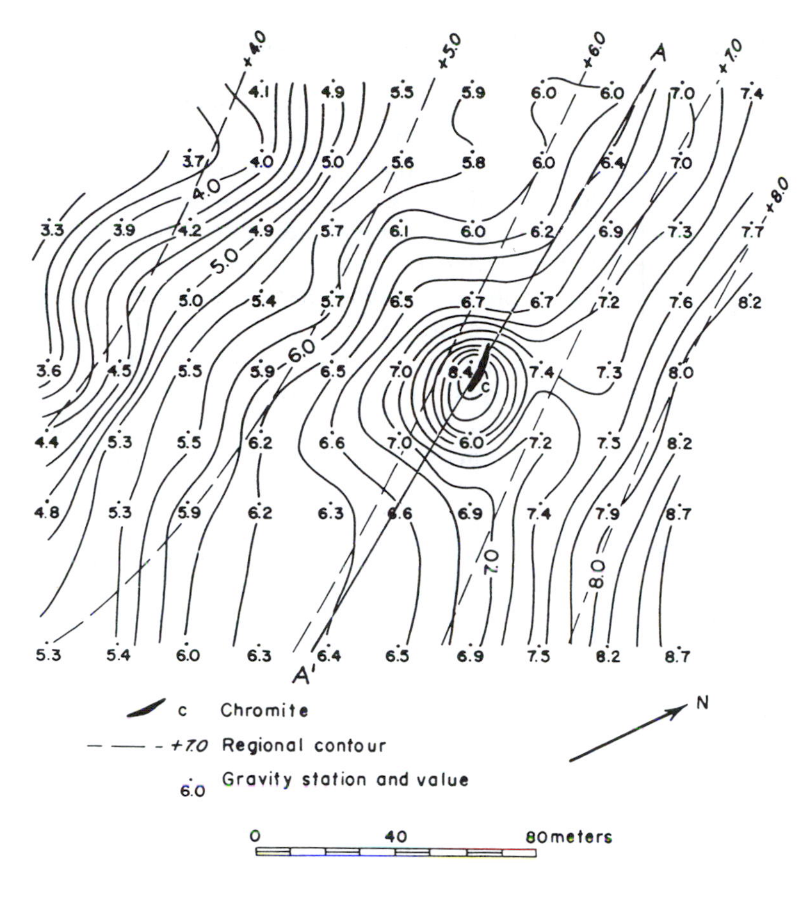

Figure 9–9
Bouguer gravity map over a buried chromite deposit in the Camagüey mining district of Cuba. Values are in gravity units (1 gu = 0.1 mgal). The contour interval is 0.2 gu. (From W. E. Davis, W. H. Jackson, and D. H. Richter, *Geophysics,* v. 22, n. 5, p. 857, Fig. 3, October 1957.)

measured Bouguer gravity variations with patterns of variation which we have calculated for structures having different shapes and situated at different depths. The calculated pattern that compares most closely indicates the shape and depth of the particular structure that best explains the measured variation. Our purpose in this section is to present formulas for computing the gravity anomalies over some structures with quite simple shapes. In a later section, we will discuss how to select and find the dimensions of one or another of these structures to explain the results of a gravity survey.

The Sphere

How do we calculate patterns of gravitational attraction for different structures? Some examples illustrate the procedure. Consider first a spherical structure that has a density different from the rock in which it is embedded. We can see in Figure 9–10 how a sphere could be used to represent, say, a localized accumulation of ore. Although the ore body itself might have quite an irregular shape, the spherical form is a satisfactory approximation of its volume and location. The sphere, then, can be used as a *model* of the ore body for the pur-

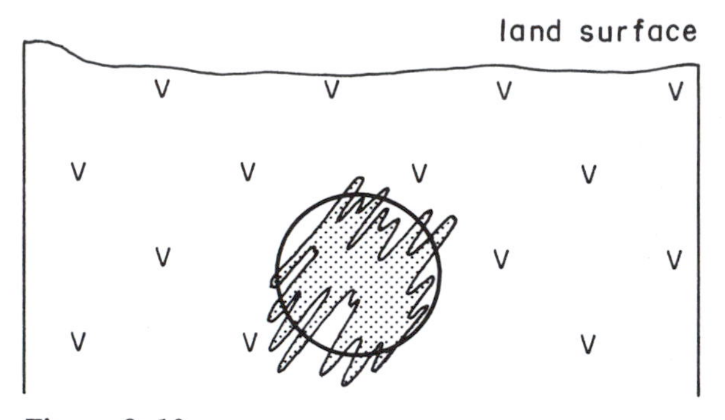

Figure 9–10
A sphere representing an irregular mass for purposes of gravity anomaly analysis.

pose of calculating its gravitational attraction.

To make this calculation, let us assume that a sphere of density ρ_1 and radius d is embedded in rock of density ρ_0, as shown in Figure 9–11. The center of this sphere is at depth z beneath the observation surface. Suppose now that gravimeter readings were made at points along the observation surface. If the sphere did not exist, the density everywhere would be ρ_0, and the gravimeter readings would be the same at all points. The presence of the sphere, however, will cause the gravimeter readings to change along the observation line. Let g_s be the gravity anomaly caused by the sphere, which is the change in gravity that would be detected by the gravimeter. Recall from Equation 7–3 that the total gravitational attraction of a sphere is Gm/r^2 at the distance r away from the center of the mass m. To calculate g_s at some particular observation point, we have to modify this ex-pression.

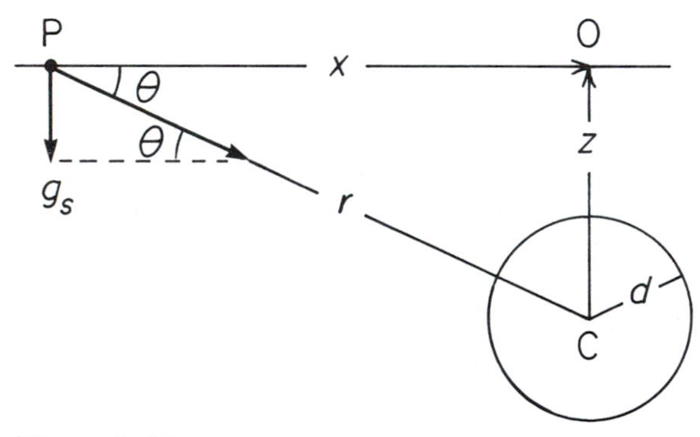

Figure 9–11
Distances and angles used for calculating the vertical gravity anomaly g_s produced by a buried sphere.

First, we do not use the total mass m of the sphere that is found from the product of its volume, $\frac{4}{3}\pi d^3$, and its density ρ_1. Rather, we use the change in mass Δm resulting from the presence of the sphere. This is found from the volume and the density difference $\Delta\rho = \rho_1 - \rho_0$:

$$\Delta m = \frac{4}{3}\pi d^3 \, \Delta\rho \tag{9–1}$$

We must also account for the fact that a gravimeter located at, say, point P in Figure 9–11, is not sensitive to the gravitational attraction of $G\Delta m/r^2$ which is directed toward the center of the sphere. Instead, it responds to the complete gravitational attraction of the earth that acts in the vertical direction. Therefore, the value of g_s must express the effect of the sphere in this direction. This vertical component of attraction is the contribution of the sphere to the gravimeter reading. At location P, it can be calculated using the equation

$$g_s = \frac{G\Delta m}{r^2} \sin\theta \tag{9–2}$$

where the angle θ is indicated in Figure 9–11. We can tell from the triangle OPC in Figure

9–11 that $\sin\theta = z/r$. Substituting this together with the term $Q_s = G\Delta m$, we get

$$g_s = Q_s\frac{z}{r^3} \qquad (9\text{–}3)$$

To calculate g_s in milligals when density is in grams per cubic centimeter, we can combine the terms in Equation 9–1 with appropriate conversion factors to obtain

$$Q_s = 0.02794d^3\Delta\rho \qquad (9\text{–}4a)$$

when sphere radius d, distance r, and depth z are in meters; and

$$Q_s = 0.00852d^3\Delta\rho \qquad (9\text{–}4b)$$

when d, r, and z are in feet.

Now we can calculate the gravity anomaly produced by a buried sphere at points along a profile. First, we must specify the density contrast $\Delta\rho$ and the radius d, so that we can obtain Q_s from Equation 9–4. Then, after we specify the depth z and the horizontal distance x of each point from the center of the sphere, we can obtain $r = \sqrt{x^2 + z^2}$. The values of Q, z, and r are then used in Equation 9–3 to get g_s. The calculations were repeated for each point along the profile in Figure 9–12 to obtain the curve that shows the gravity variation caused by the buried sphere. We see that the gravity anomaly is strongest at the observation point directly over the sphere and diminishes with increasing distance from this point.

The Vertical Cylinder

Another useful model for representing certain geologic structures is the vertical cylinder. The salt dome in Figure 9–13 is approximately cylindrical, so that this model can be used for calculating its gravitational attraction. At a point on the axis of a buried vertical cylinder, the gravity anomaly g_{vc} can be obtained from the formula

$$g_{vc} = 2\pi G\,\Delta\rho\left(z_2 - z_1 + \sqrt{z_1^2 + d^2} - \sqrt{z_2^2 + d^2}\right) \qquad (9\text{–}5)$$

where z_1 and z_2 are depths to its top and its base, and d is its radius. This formula can be

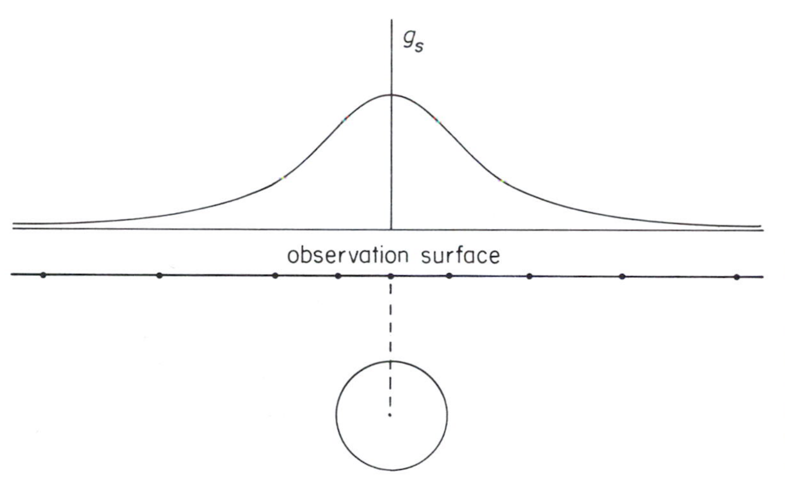

Figure 9–12
Gravity anomaly profile over a buried sphere. In this example, a sphere with a 400-meter radius is centered at a depth of 1000 meters, and the density contrast $\Delta\rho$ is 0.5 g/cm^3. Using these values in Equations 9-3 and 9-4a yields g_s = 0.894 mgal directly over the sphere.

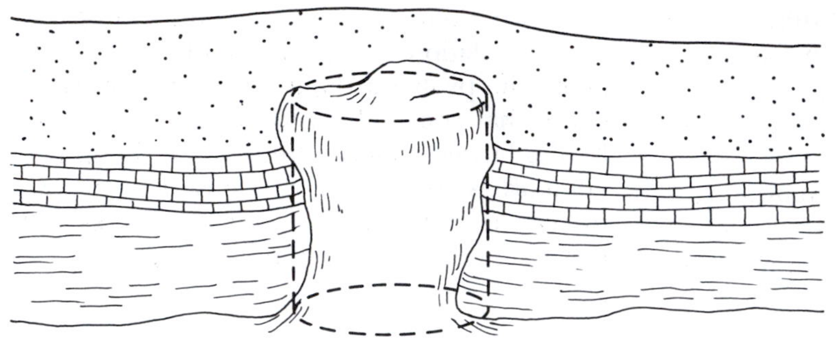

Figure 9–13
A vertical cylinder representing a salt dome for purposes of gravity anomaly analysis.

used to find g_{vc} at the observation point O in Figure 9–14, which is directly above the center of the cylinder. However, it cannot be used to calculate the gravitational attraction anywhere else, say, at point P. There is no simple formula for doing this, but we can use another formula which, while not yielding an exact value, will provide a reasonably good approximation. Imagine that all the mass of the cylinder could be compressed into its axis. By relocating this mass to the line TB in Figure 9–14 and by noting that a line is a continuous succession of points, we can calculate and add together the point-by-point increments of gravitational attraction of the line. Because each point can be viewed as an infinitely small sphere, we can use Equation 9–2 to find its vertical increment of attraction. All these in-

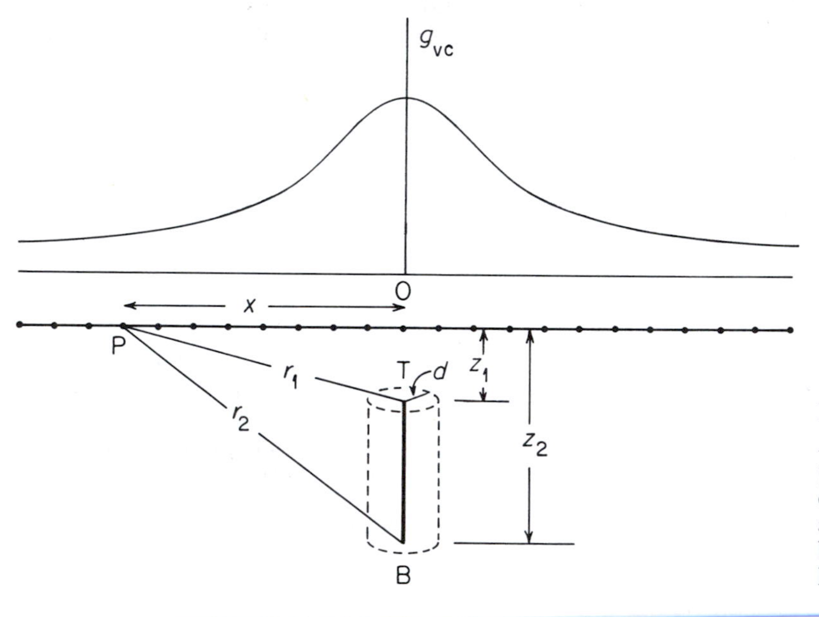

Figure 9–14
Approximate gravity anomaly profile over a vertical cylinder calculated by making the assumption that all its mass has been compressed to the axial line TB.

crements can be summed by mathematical integration to obtain the formula

$$g_{vc} \cong \frac{Q_{vc}}{r_2 - r_1} \tag{9–6}$$

Let d be the radius of the original cylinder, and let $\Delta\rho$ be the difference between its density ρ_1 and the density ρ_0 of the rock in which it is embedded. The distances r_1 and r_2 from the observation point (P) to the top (T) and base (B) of the cylinder axis are found from the horizontal distance x and the depths z_1 and z_2 using $r_1 = \sqrt{x^2 + z_1^2}$ and $r_2 = \sqrt{x^2 + z_2^2}$. If all the terms are in cgs units, then

$$Q_{vc} = \pi G d^2 \, \Delta\rho \tag{9–7a}$$

But if g_{vc} is to be in milligals, and $\Delta\rho$ in grams per cubic centimeter, for d, r_1, and r_2 in meters,

$$Q_{vc} = 0.02096 d^2 \, \Delta\rho \tag{9–7b}$$

For d, r_1, and r_2 in feet,

$$Q_{vc} = 0.00639 d^2 \, \Delta\rho \tag{9–7c}$$

To calculate the gravity anomaly along the profile in Figure 9–14 that crosses over a buried vertical cylinder, we first specify values for d and $\Delta\rho$ so that we can obtain a value for Q_{vc} from Equation 9–7. Then we specify the depths z_1 and z_2 so that a value of g_{vc} can be obtained from Equation 9–6 for each observation point along the profile. The gravity anomaly variation shown in Figure 9–14 was calculated in this way. As mentioned before, the values of g_{vc} are only approximations. To derive the simple Equation 9–6, we had to imagine that all the mass of the original cylinder was compressed into its axis. The accuracy of the approximation is improved by specifying a cylinder with a radius that is considerably smaller than the length of its axis and its depth of burial. Accuracy also improves as the distance to an observation point increases.

Horizontal Cylinder

Another model is the horizontal cylinder, which has been helpful for explaining some long, narrow patterns of variation seen on Bouguer gravity contour maps. The profile in Figure 9–15 extends in a direction perpendicular to a buried horizontal cylinder. To obtain a simple formula for calculating the gravity anomaly over this structure, we will assume that it is infinitely long. We know that infinite structures cannot be buried in the earth. But we can make this assumption because only the close portion of the cylinder contributes significantly to the gravity anomaly along the profile. Its more distant extremities have a negligible effect. If we represent the cylinder by a succession of spheres placed side by side, the increments of gravitational attraction can be expressed by Equation 9–3. They can be summed by the mathematical integration to obtain the formula

$$g_{hc} = \frac{Q_{hc} z}{r^2} \tag{9–8}$$

where the terms indicated in Figure 9–15 include the depth z of the cylinder axis and its distance r from an observation point. Note that $r^2 = x^2 + z^2$. If all terms are in cgs units, then

$$Q_{hc} = 2\pi G d^2 \, \Delta\rho \tag{9–9a}$$

But to get the vertical gravity anomaly g_{hc} in milligals when the density difference $\Delta\rho$ is in grams per cubic centimeter, and the cylinder

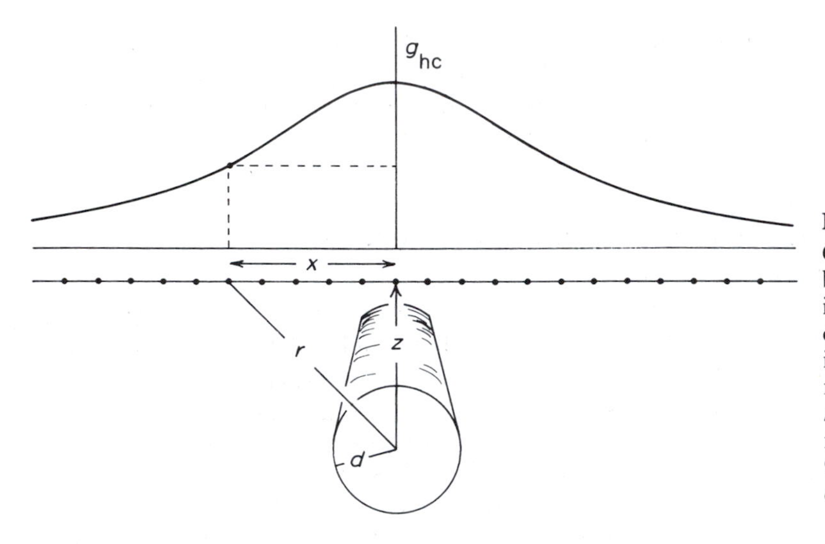

Figure 9–15
Gravity anomaly profile over a buried horizontal cylinder of infinite length. In this example, a cylinder with a 400-meter radius is centered at a depth of 1000 meters, and the density contrast $\Delta\rho$ is 0.375 g/cm^3. Using these numbers in Equations 9–8 and 9–9b yields $g_{hc} = 2.52$ mgal directly above the cylinder.

radius d and z and r are in meters, we also need a conversion factor to obtain

$$Q_{hc} = 0.04193 d^2 \, \Delta\rho \qquad (9\text{–}9b)$$

When d, r, and z are in feet, this term becomes

$$Q_{hc} = 0.01278 d^2 \, \Delta\rho \qquad (9\text{–}9c)$$

The gravity anomaly profile in Figure 9–15 was determined by specifying d and $\Delta\rho$ and calculating Q_{hc}. This value was then combined with a specified depth z to calculate g_{hc} at each observation point.

The Semi-infinite Horizontal Plate

One more simple model to consider is the buried horizontal plate with a vertical edge. It has been used to explain Bouguer gravity variations produced by rock layers that are offset along faults. For the purpose of gravity anomaly calculations, imagine that such a horizontal plate consists of a large number of very small, parallel horizontal cylinders, as shown in Figure 9–16. The increment of vertical gravitational attraction for any particular small cylinder is expressed by Equation 9–8. It can be integrated to obtain the formula expressing the sum of these increments, which is the vertical gravity anomaly g_{hp} caused by the horizontal plate,

$$g_{hp} = Q_{hp} \left[x \ln \frac{r_2}{r_1} + \pi(z_2 - z_1) - z_2\theta_2 + z_1\theta_1 \right] \qquad (9\text{–}10)$$

where the different terms are indicated in Figure 9–16. Notice that distances $r_1 = \sqrt{x^2 + z_1^2}$ and $r_2 = \sqrt{x^2 + z_2^2}$, and that the angles $\theta_1 = \tan^{-1}(z_1/x)$ and $\theta_2 = \tan^{-1}(z_2/x)$. If all terms are in cgs units, then

$$Q_{hp} = 2G \, \Delta\rho \qquad (9\text{–}11a)$$

But if g_{hp} is to be in milligals when $\Delta\rho$ is in grams per cubic centimeter and x, a_1, z_2, r_1, and r_2 are in meters, we use

$$Q_{hp} = 0.01335 \, \Delta\rho \qquad (9\text{–}11b)$$

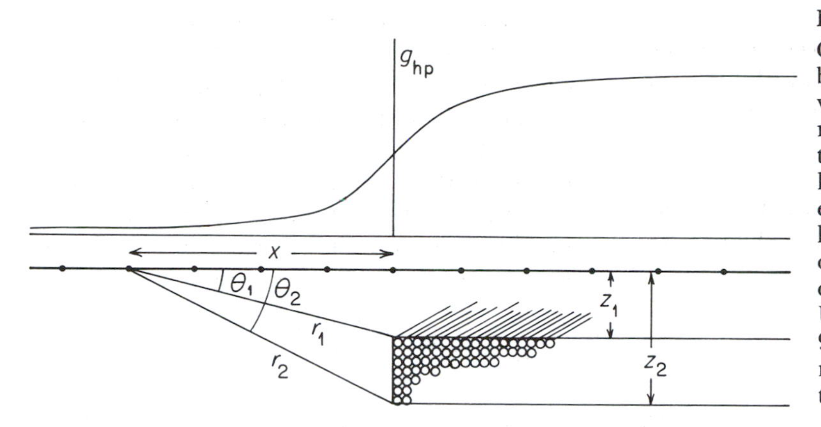

Figure 9–16
Gravity anomaly profile over a buried horizontal plate with a vertical edge. Such a plate can be represented by a large number of thin, parallel rods grouped horizontally side by side. In this example, a 600-meter-thick plate lies 600 meters below the observation surface, and the density contrast $\Delta\rho$ is 0.4 g/cm^3. Using these numbers in Equations 9–10 and 9–11b yields $g_{hp} = 5.03$ mgal directly above the edge of the plate.

For these latter terms in feet, we use

$$Q_{hp} = 0.004068\ \Delta\rho \qquad (9\text{–}11c)$$

The gravity anomaly profile in Figure 9–16 was calculated from Equations 9–10 and 9–11 after specifying values for $\Delta\rho$, z_1, and z_2.

Models of Irregular Shape

So far, we have described four models that have very simple shapes. How can we calculate the gravity anomaly over a structure of more irregular shape? To a limited extent, we can use an appropriate combination of simple models. For example, the three spheres in Figure 9–17 are sufficient to represent the main features of the irregular mass. To find the gravity anomaly, we must calculate the effects of the spheres separately and then sum those for each observation point. This approach becomes impractical when too many simple models are required to represent a structure of complicated shape. There are two kinds of models that are practical for this purpose. For both of these models, the gravity anomaly formulas are complicated and must be solved on a digital computer. Nevertheless, they are widely used for analyzing Bouguer gravity patterns.

Consider first a horizontal plate with a shape that can be described by a polygon with many sides. An example is shown in Figure 9–18. The method for finding the gravitational attraction of this plate was developed by the geophysicists Manik Talwani, Donald Plouff, and Maurice Ewing. Imagine that this plate consists of a very large number of vertical lines. Each line is a vertical cylinder with a

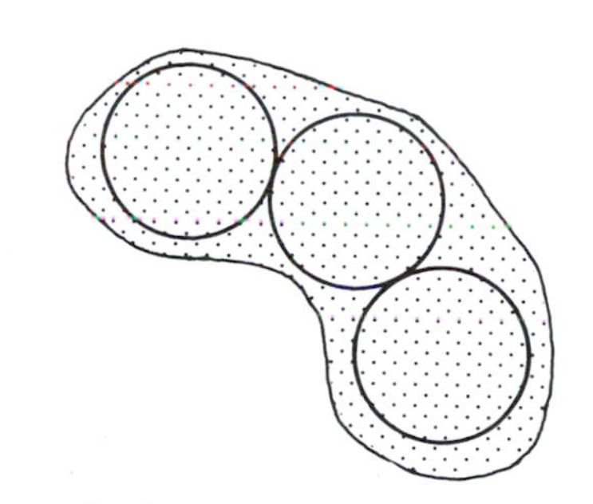

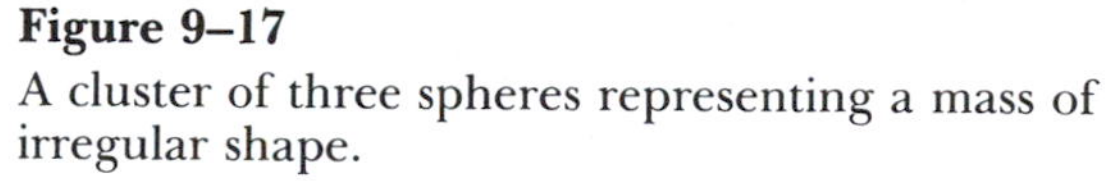

Figure 9–17
A cluster of three spheres representing a mass of irregular shape.

very small radius and a length equal to the thickness of the plate. The increment of vertical gravitational attraction at an observation point produced by each line is expressed by Equation 9–6. These increments can be summed by integration to obtain formulas for calculating the gravity anomaly of the polygonal plate. One formula is used to calculate the factor V_{12} for each side of the polygon:

$$V_{12} = A(z_2 - z_1) + z_2\left(\tan^{-1}\frac{z_2 d_1}{PR_{12}} - \tan^{-1}\frac{z_2 d_2}{PR_{22}}\right)$$
$$-z_1\left(\tan^{-1}\frac{z_1 d_1}{PR_{11}} - \tan^{-1}\frac{z_1 d_2}{PR_{21}}\right) - P\ln\left(\frac{R_{22} + d_2}{R_{12} + d_1}\cdot\frac{R_{11} + d_1}{R_{21} + d_2}\right) \quad (9\text{–}12)$$

where all the terms are as indicated in Figure 9–18. Subscripts 1 and 2 refer to points at the ends of each side of the polygon. If we specify by x_1,y_1 and x_2,y_2 the positions of both ends of the polygon side and by z_1 and z_2 the depths to the top and the base of the plate, other formulas can be obtained for each of the terms in Equation 9–12. For example, the angle $A = \tan^{-1}(x_1/y_1) - \tan^{-1}(x_2/y_2)$, and $R_{11} = \sqrt{x_1^2 + y_1^2 + z_1^2}$. Obviously, Equation 9–12 is complicated, and several other formulas are also needed to calculate each term in it. Nonetheless, all these equations can be included in one computer program that also sums the values of V_{12} for each side and multiplies the result by $G\,\Delta\rho$ to obtain the gravity anomaly g_p of a plate at some observation point:

$$g_p = G\,\Delta\rho(V_{12} + V_{23} + V_{34} + \ldots + V_{n1}) \quad (9\text{–}13)$$

Here the numbers 1, 2, 3, . . ., n refer to polygon corners in Figure 9–18. To use such

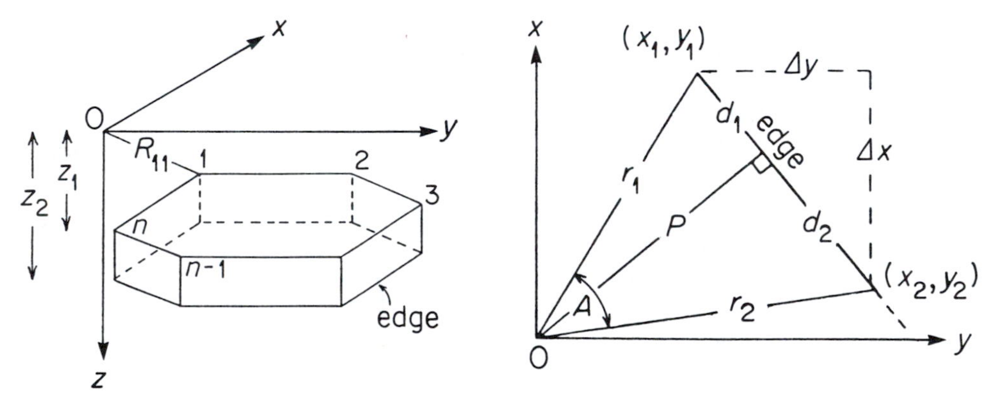

Figure 9–18
Horizontal plate with polygonal shape showing the distances and angles used to calculate its vertical gravitational attraction at the origin of the coordinate system by means of Equations 9–12 and 9–13. (From D. Plouff, Geophysics, v. 41, n. 4, p. 728, Fig. 2, August 1976.)

a computer program, we need specify only the density contrast $\Delta\rho$, the depths z_1 and z_2, and the positions x_1, y_1, x_2, y_2, and so on, of the polygon corners. The advantage of this kind of model is that we can create a plate of any desired shape simply by choosing enough sides to produce this shape. Equation 9–12 must be solved separately for each side; then Equation 9–13 is used to sum all values of V, however many there may be. In a later section, we will discuss how to combine several polygonal plates to reproduce structures of still more complicated form.

The last model to consider is a more complicated form of the horizontal cylinder. It, too, is chosen to be infinitely long, but its cross-sectional shape is described by a polygon. Such a model can be used to represent a geologic structure with a horizontal length that is considerably greater than its width and depth. The example in Figure 9–19 illustrates how a polygon with enough sides reproduces the cross-sectional shape of a folded rock layer. The geophysicists Manik Talwani, J. L. Worzel, and Mark Landisman developed the method for calculating the gravitational attraction of this kind of model. Imagine that such a model consists of a very large number of closely packed horizontal cylinders, all having the same very small radius. The increments of vertical gravitational attraction found from Equation 9–8 can be summed by integration to obtain formulas for calculating the gravity anomaly produced by a structure that is infinitely long in one horizontal direction and has a polygonal cross section. A factor U is calculated for each side of the polygon. For the side between points 1 and 2, the factor is

$$U_{12} = a_1 \sin \varphi_1 \cos \varphi_1 \left[\theta_1 - \theta_2 + \tan \varphi_1 \ln \frac{\cos \theta_1(\tan \theta_1 - \tan \varphi_1)}{\cos \theta_2(\tan \theta_2 - \tan \varphi_1)} \right] \quad (9\text{–}14)$$

where the terms are as indicated in Figure 9–20. Positions of the ends of the side, which are x_1, z_1 and x_2, z_2, can be used to solve for all terms in the equation. For example, $a_1 = x_2 - z_2(x_2 - x_1)/(z_1 - z_2)$, $\theta_1 = \tan^{-1}(z_1/x_1)$, and so on. The vertical gravitational attraction, g_{hp}, of the model is the sum of the factors for all the sides multiplied by $2G\ \Delta\rho$, which is

$$g_{hp} = 2G\ \Delta\rho(U_{12} + U_{23} + U_{34} + \ldots + U_{n1}) \quad (9\text{–}15)$$

where the numbers 1, 2, 3, and so on, refer to polygon corners. A computer program can be prepared to calculate values for all the terms involved, then to solve Equation 9–14 for each

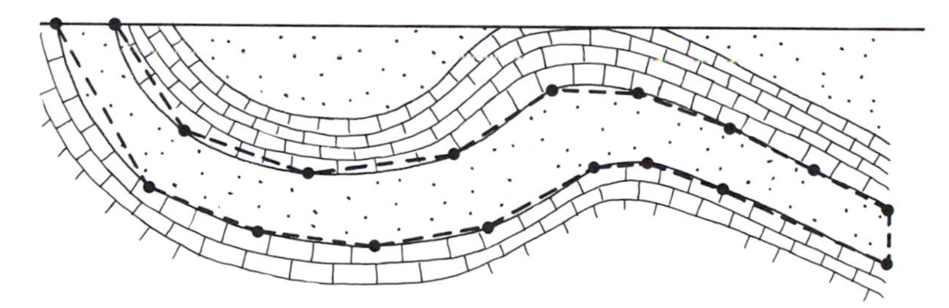

Figure 9–19
Representation of a folded rock layer by a horizontal cylinder that has an irregular polygonal cross section.

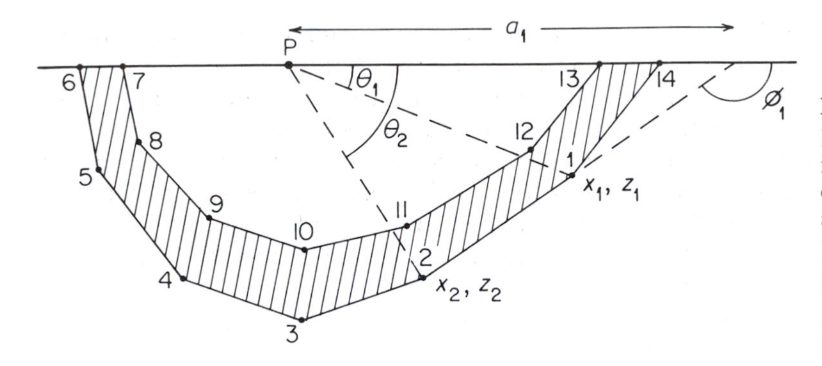

Figure 9–20
Horizontal cylinder with an irregular polygonal cross section described by 14 corner points and showing the distances and angles used for calculating its vertical gravitational attraction at point P by means of Equations 9–14 and 9–15.

side, and finally to sum the results using Equation 9–15. To use such a program, we need specify only the density contrast and the positions of the polygon corners relative to an observation point. A polygon can be designed with as many sides as are needed to reproduce the desired shape.

Any model such as this one, which is assumed to be infinitely long in one horizontal direction, is usually called a *two-dimensional model.* We need specify only two dimensions, the depth z and the horizontal distance x from an observation point, to describe each corner of its polygonal cross section. In a later section, we will discuss how to combine several two-dimensional models to represent more complicated geologic cross sections.

Gravity anomalies produced by these various simple models have two important features: (1) the largest values occur at the closest observation point; and (2) anomaly values diminish with distance, but they reach zero only at infinite distance. These features indicate that local anomaly sources most strongly influence the earth's gravitational field at nearby observation points. Farther away they have a more subdued, but still possibly important, effect.

SEPARATION OF LOCAL AND REGIONAL BOUGUER GRAVITY PATTERNS

The effects of many geologic features are combined in a Bouguer gravity value. Any pattern of variation seen on a Bouguer gravity map is the sum of the attractions of local sources and broader or more distant regional sources. Nowhere can we measure an anomaly from one source that is not distorted by overlapping anomalies from other sources. Certainly, in some places one source is so dominant that the distortion of its anomaly by other anomalies is minor. In other places, however, anomalies indicating structures of particular interest are almost completely hidden. How can we isolate individual anomalies from the patterns of variation seen on Bouguer gravity profiles and maps?

Geophysicists have given much thought to the problem of separating local and regional anomalies. Although there is no way to accomplish perfect separation, geophysicists have developed methods for isolating the principal features of different anomalies. These methods are quite useful for bringing obscure and hidden anomalies into clearer focus. We will discuss three kinds of separation schemes that involve (1) graphical smoothing of profiles and contours, (2) computation of weighted averages, and (3) computation of anomalies caused by sources we already know about.

Graphical Smoothing

Regional and local anomalies can be separated by graphical smoothing in the following way. First, the geophysicist must judge from the appearance of a profile or a contour map how the regional part would look were it not distorted by local irregularities. This is done by sketching lines that bypass the local variations and connect only those more broadly curving parts of a profile or contour line. Observe how the dotted line in Figure 9–21 smoothly connects only the parts of the profile that represent a regional variation according to the judgment of the geophysicist. A regional gravity value Δg_R can then be read from the dotted line where it passes each observation point. These values are subtracted from the Bouguer gravity values at these locations to obtain the local gravity values Δg_L along the profile

$$\Delta g_L = \Delta g_B - \Delta g_R \qquad (9\text{–}16)$$

These values are then plotted, as in Figure 9–21, to illustrate the local gravity anomaly separately.

How can we tell exactly where the dotted line should be drawn in Figure 9–21? There is no way to tell, and no two geophysicists would make the same judgment about this. Insofar as the regional pattern is quite broad and simple and local anomalies are easily recognized, their judgments would probably be similar. But where local anomalies are small and sources of intermediate dimensions add

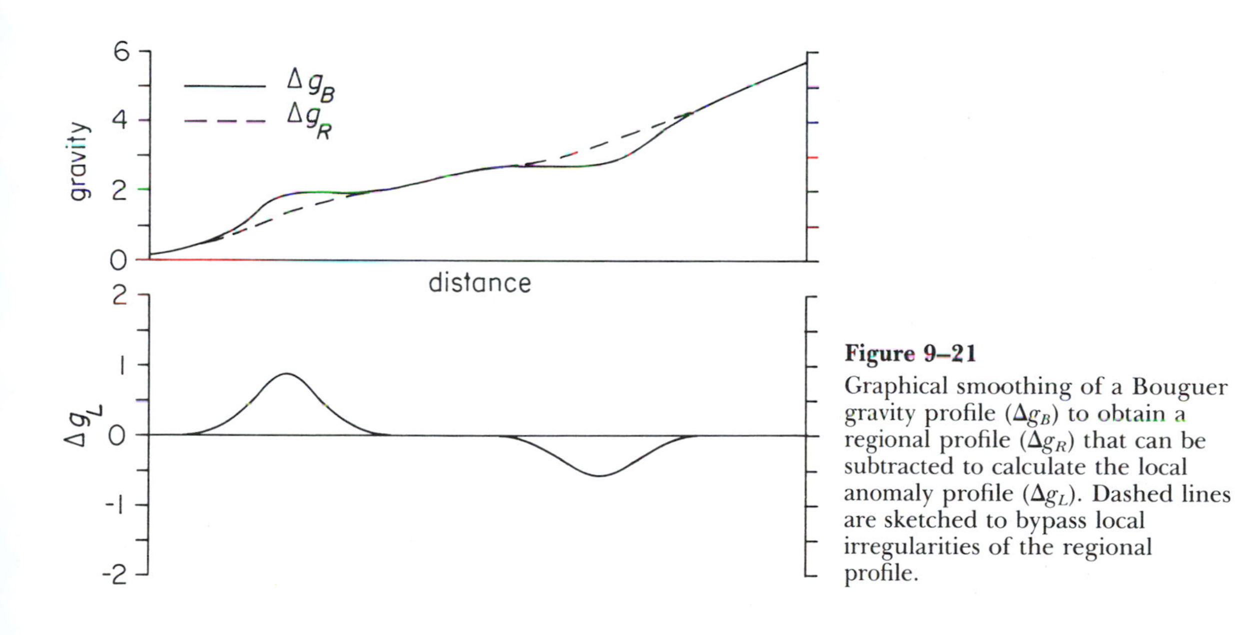

Figure 9–21
Graphical smoothing of a Bouguer gravity profile (Δg_B) to obtain a regional profile (Δg_R) that can be subtracted to calculate the local anomaly profile (Δg_L). Dashed lines are sketched to bypass local irregularities of the regional profile.

to the regional pattern, we can expect larger differences of opinion.

Graphical smoothing of contours on a Bouguer gravity map is illustrated in Figure 9–22. Here we can see that some parts of each contour are straight or gently curving. But these contours also have sharply curving local irregularities. Now look at the dotted contours that have been sketched on Figure 9–22a. They bypass the local irregularities but merge smoothly with the broader, gently curving parts of each Bouguer gravity contour. Only the dotted contours are reproduced in Figure 9–22b to show what we believe is the regional anomaly field. By interpolation between these dotted contours, we can find the regional anomaly value Δg_R for each observation site. Then we can use this value and the corresponding Bouguer gravity Δg_B to calculate the local gravity Δg_L from Equation 9–16 for each site. These values were contoured in Figure 9–22c to produce the map showing local gravity anomalies.

Contour smoothing, like profile smoothing, cannot be expected to separate exactly regional and local anomalies. No one knows precisely where to sketch each regional contour, and, as with profile smoothing, two geophysicists would be unlikely to make the same judgments. Where there are relatively simple patterns, such as those in Figure 9–22, the contour maps made by several individuals would not differ greatly from one another. Only when contour patterns become quite complicated are significant differences of opinion likely.

Gridding or Digitizing Data

Separation of regional and local anomalies by graphical smoothing of contours is time-consuming. This method is impractical for analysis of large areas with complicated patterns of Bouguer gravity variation. More useful are the schemes for estimating the regional anom-

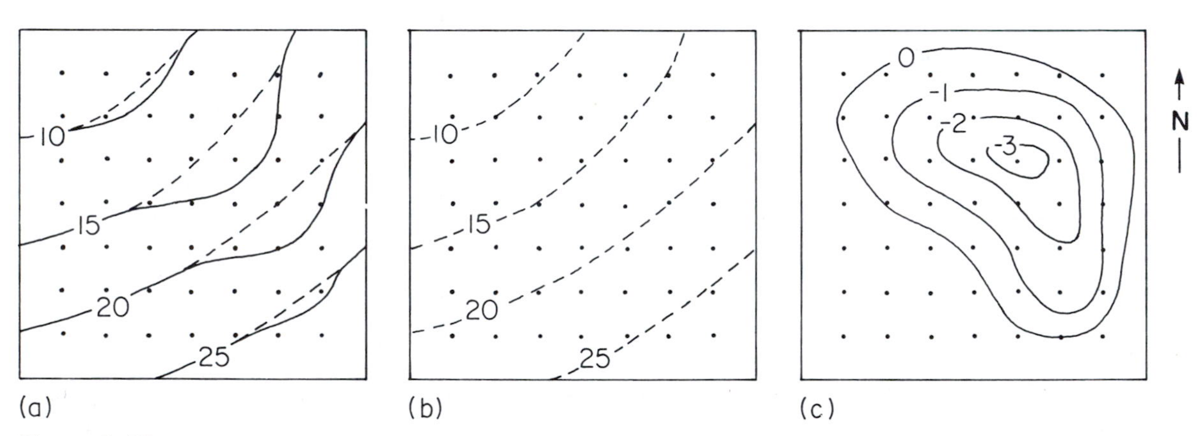

Figure 9–22
Graphical smoothing of Bouguer gravity contours (solid contours in map a) to obtain regional gravity contours (dashed contours in maps a and b). By subtracting regional values from corresponding Bouguer gravity values interpolated at many points on these maps, we can calculate the local anomaly values and plot them on another map c where they are contoured to indicate the patterns of local anomaly variation.

aly value at a location from some kind of weighted average of values from the surrounding area. Because these schemes have been programmed for computer processing, the necessary calculations can be made rapidly even for maps covering large areas.

The various averaging schemes can be applied with much less difficulty if Bouguer gravity values are known at evenly spaced intervals over an area. For most gravity surveys, however, the measurements we make along winding roads and at other convenient places turn out to be at irregular spacings. Therefore, it is common practice to determine Bouguer gravity values at a square grid of points before trying to separate regional and local anomalies. This can be done by drawing or overlaying on a Bouguer gravity contour map an appropriate grid of points or intersecting lines. The value at each point or line intersection is then found by interpolation. The spacing must be chosen so that all the contour patterns could be reproduced from the grid values. Because of the time it takes a technician to interpolate each grid value, computer programs have been prepared for this purpose. When given the Bouguer gravity values and the positions of irregularly spaced observation sites, the computer rapidly searches for sites nearest each grid point and then carries out the interpolation. The process of finding evenly spaced values is called *gridding* or *digitizing* the Bouguer gravity map.

Smoothing by Averaging

We can estimate the regional anomaly at some location from the average of Bouguer gravity over the surrounding area. This is possible because the regional attraction is significant at all the locations, whereas a local attraction is important only at a few closely grouped sites. Even at these few sites, the large number of more distant Bouguer gravity values dominate the average. For an example, see the profiles in Figure 9–23 where two estimates of regional anomaly variation are plotted together with the Bouguer gravity values from which they were obtained. First, an average was calculated for each observation point using five Bouguer gravity values including the one from that point and the ones from the two nearest points on either side. These averages are connected by the dashed line. Such a profile is called a *running average,* which means that we repeat the averaging calculation as we move from point to point.

The other regional anomaly profile, shown by the dotted line in Figure 9–23, is an 11-point running average. Note that it has a lower value at site A than the five-point running average. Because Bouguer gravity values from several more distant points are also used, the effect of the local anomaly is a smaller part of the total. Does this imply that a better estimate of the regional gravity is obtained by averaging a larger number of Bouguer gravity values? To answer this question, we must consider the extent to which the more distant points are affected by other local anomalies. In Figure 9–23, we can improve our estimate of the regional gravity at site A by increasing the number of points, as long as the points we use are not significantly affected by the local anomaly near site B. The use of points near site B would diminish the accuracy of our estimate. The two local anomalies in this example are widely separated so that it is not difficult to decide the optimum number of points to include.

For most gravity surveys, the decision is more difficult. Where several closely spaced, overlapping local anomalies are superposed

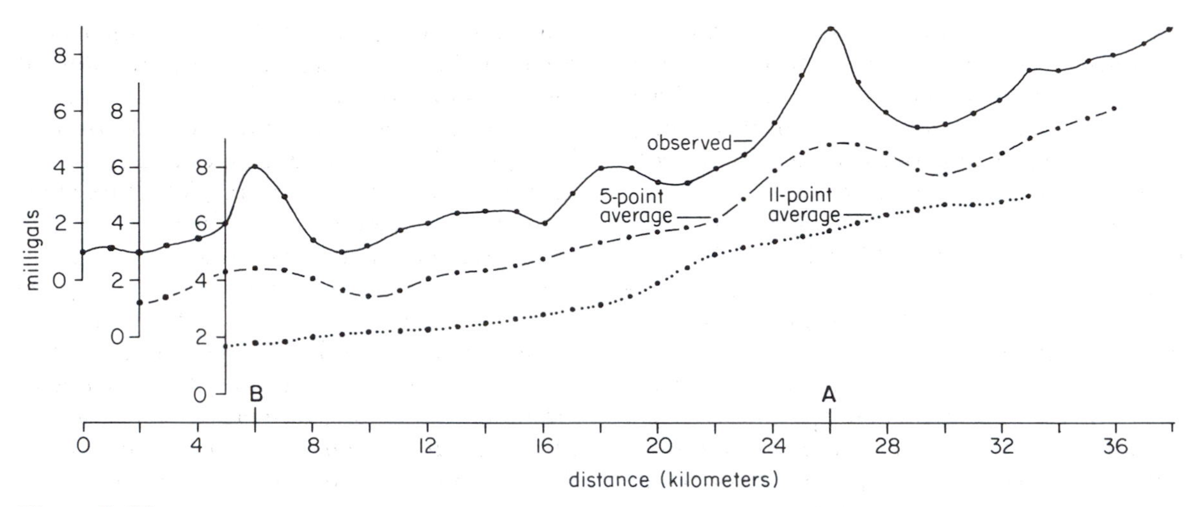

Figure 9–23
Bouguer gravity profile (solid line) and profiles of 5-point (dashed line) and 11-point (dotted line) running averages of the values measured along the Bouguer gravity profile.

on the regional field, there may be no obvious choice. Then it is best to test different running averages, such as the two in Figure 9–23, to see whether one yields a more geologically meaningful anomaly pattern. After an appropriate averaging number has been chosen, the regional anomaly is calculated for each point along the profile and is then subtracted in Equation 9–16 from the Bouguer gravity at that point to obtain the local gravity.

We must consider another practical limitation. To be objective when we calculate a regional anomaly value, we should use the same number of equally spaced points reaching in opposite directions away from the site. Therefore, the total number of points along the profile imposes a limit on the number we can include in any particular average. Look again at Figure 9–23. By choosing a five-point running average, we can calculate regional anomaly values everywhere between the third points from the ends of the profile. If an 11-point average is used, however, regional anomalies can be calculated only along the shorter distance extending between the sixth points from the ends of the profile. Increasing the averaging number, then, decreases the length of the regional anomaly profile that can be obtained. Again, the geophysicists must choose a practical number so that reasonably good estimates of Δg_R can be calculated along a reasonably long profile.

Consider next how running averages are calculated for a map when we have Bouguer gravity values at evenly spaced grid points. We can calculate Δg_R at one point by averaging all Δg_B values within a square area centered on that point, as illustrated in Figure 9–24. Here we have chosen an area that includes Bouguer gravity within two grid lines above and below and two grid lines to the left and right of a point. From the 25 values in this area, we can calculate an average to get our estimate of the regional gravity.

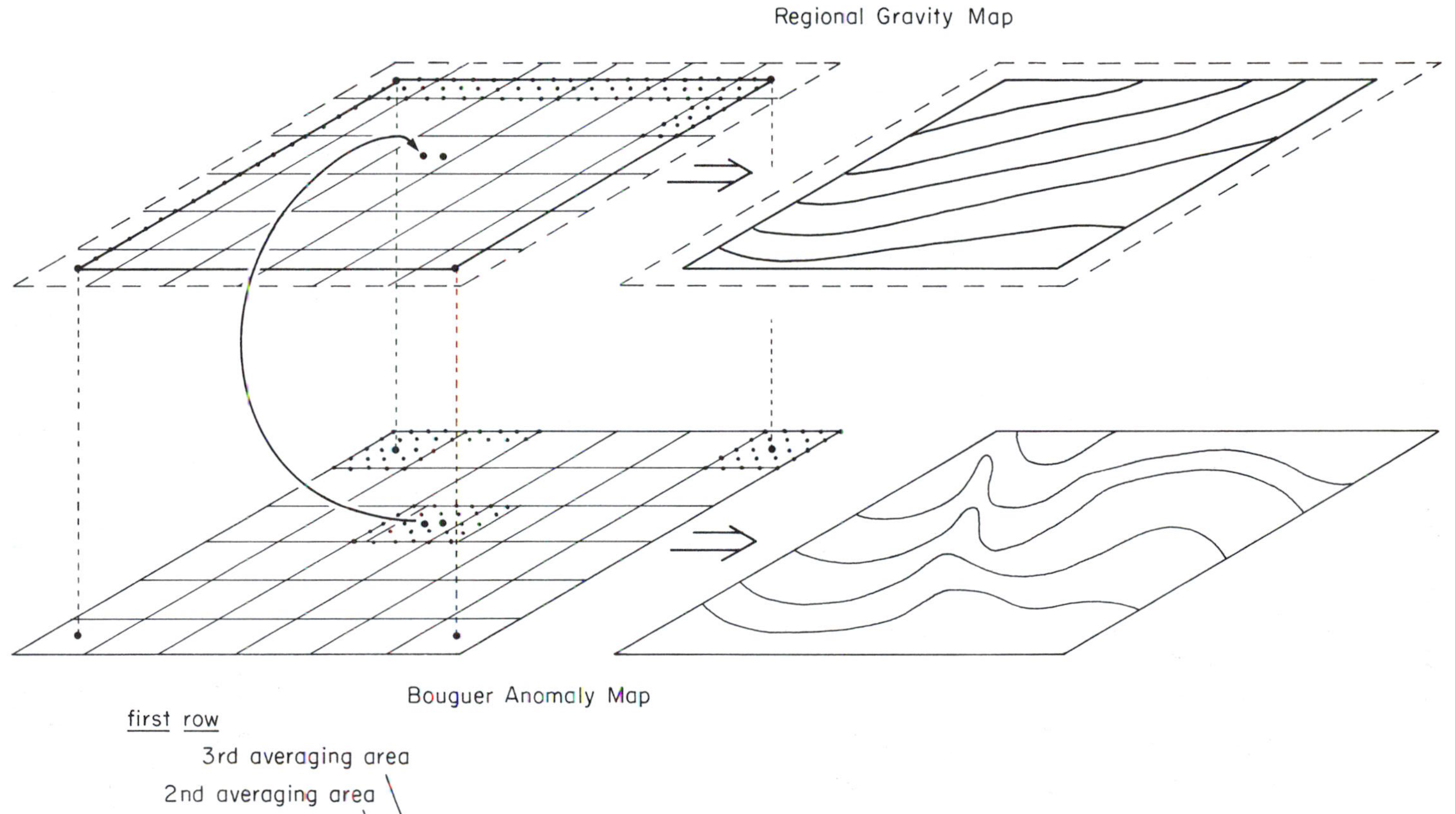

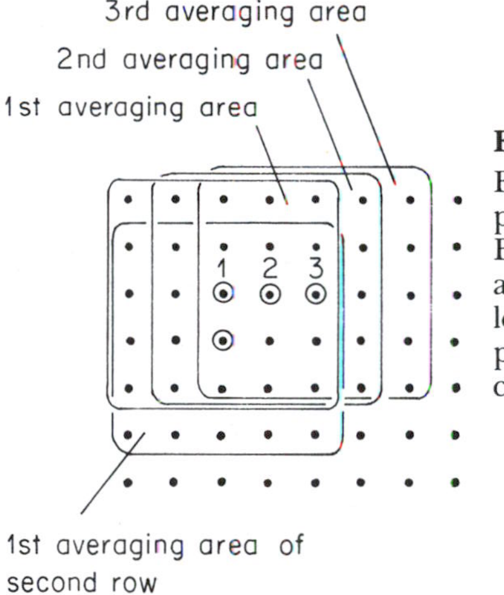

Figure 9–24
Bouguer gravity map and regional gravity map prepared from 25-point running averages of Bouguer gravity values obtained in turn by averaging areas of 5 × 5 grid units. Maps on the left show the grid points used in the averaging process, and maps on the right show the corresponding contours over the same area.

As we advance point by point over the map, the location of the square averaging area changes. For the example in Figure 9–24, we could calculate Δg_R at all points bounded by the third grid lines from the edges of the Bouguer gravity map. Suppose we chose a larger averaging area, say, of four grid lines above and below and to the left and right of a point. Then we could calculate the average of 81 values of Δg_B. But we could obtain these estimates of Δg_R only with the smaller area bounded by the fifth grid lines from the edges of the map. As before, the geophysicist must decide on a practical size for the averaging area. For processing large maps, a computer can be used to search and average the Δg_B values in the area surrounding each point.

Weighted Averaging

For our example in Figure 9–24, we calculated Δg_R at each point simply by adding several Δg_B values and then dividing by the number of values used:

$$\Delta g_R = \frac{1}{25}(\Delta g_{B1} + \Delta g_{B2} + \Delta g_{B3} + \ldots + \Delta g_{B25}) \quad (9\text{–}17)$$

Here we have assumed that all the Bouguer gravity values are equally important. Next, let us consider other methods that involve *weighted averaging*. To obtain a weighted average, we first assign different importance, or weight, to a Bouguer gravity value that depends on its distance from the point where we intend to calculate Δg_R. This is done by multiplying each Δg_B value by an appropriate weighting factor f. A set of weighting factors is displayed in Figure 9–25 for a 25-point averaging area. Note that all points of the same distance from the center are assigned the same factor. At the center point it is f_A, at the nearest four points it is f_B, and so on, to the most distant corners where it is f_F. The weighted average is then found by summing the products, $\Delta g_B \times f$, and dividing by the number of values used:

$$\Delta g_R = \frac{1}{25}[(\Delta g_{B1} \times f_F) + (\Delta g_{B2} \times f_E) + (\Delta g_{B3} \times f_D) + \ldots + (\Delta g_{B25} + f_F)] \quad (9\text{–}18)$$

Now suppose we decide to give greater consideration to Bouguer gravity closer to the center. This could be accomplished by choosing $f_A = 1$, $f_B = 0.8$, $f_C = 0.6$, $f_D = 0.4$, $f_E = 0.3$, $f_F = 0.2$. A contrary decision to assign greater regional importance to the more dis-

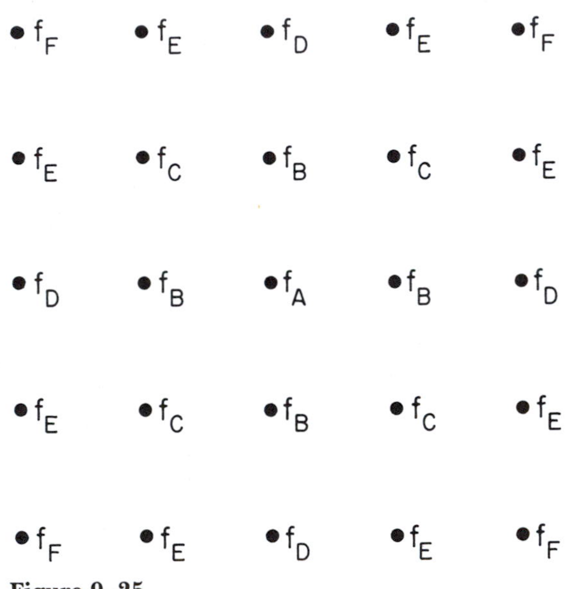

Figure 9–25
An array of weighting factors that can be multiplied with Bouguer gravity values at corresponding locations in 5 × 5 unit averaging areas to obtain weighted averages.

tant values could be implemented by setting $f_A = 0.2, f_B = 0.3, f_C = 0.4, f_D = 0.6, f_E = 0.8, f_F = 1$. But why would we choose these particular values for the weighting factors? Would another choice of factors be more effective? Actually, the geophysicist is free to select any set of values that seems to produce geologically meaningful results. Several methods are available for determining the particular weighting factors needed to carry out certain useful mathematical operations. Without going into much mathematical detail, we can describe two of the most widely used methods: (1) upward continuation and (2) wavelength filtering.

Upward Continuation

Upward continuation is an interesting operation, enabling us to calculate Bouguer gravity on a surface above the earth without making gravity measurements on that surface. What is needed to accomplish this calculation is a map of Bouguer gravity on the underlying land surface. Of what use is this operation? Shifting to a higher surface diminishes the gravitational attraction of the anomaly sources, which are then farther away. But the attraction of local sources tends to be more drastically reduced, because the change in distance is proportionally larger for these sources than for the deeper and broader regional sources. Therefore, on a higher surface, the regional patterns of variation tend to be more clearly displayed, and the local anomalies are less evident.

The basis for upward continuation is a theorem developed by the mathematician G. G. Stokes (1819–1903). This theorem states that if values of the earth's gravitational attraction are known everywhere on its surface, the attraction at any higher point can be calculated from these values. Practical application of Stokes's theorem is expressed in the formula

$$\Delta g_{BP} = \frac{1}{N}\left[\left(\Delta g_{B1} \times \frac{h_1 A}{2\pi R_1^3}\right) + \left(\Delta g_{B2} \times \frac{h_2 A}{2\pi R_2^3}\right) + \left(\Delta g_{B3} \times \frac{h_3 A}{2\pi R_3^3}\right) + \ldots\right] \qquad (9\text{–}19)$$

where, according to Figure 9–26, Δg_{BP} is the Bouguer gravity at the point P above the earth, h and R represent the height and distance of P from each of the Bouguer gravity values Δg_{B1}, Δg_{B2}, . . ., at N locations evenly spaced over the area A of the earth's surface. We can see that this formula is similar to Equation 9–18 if we choose weighting factors of the form

$$f = \frac{hA}{2\pi R^3} \qquad (9\text{–}20)$$

Obviously, the weighting factors decrease as the distance R increases. For this reason, we do not need to use Bouguer gravity from everywhere on the earth's surface to calculate an accurate value of Δg_{BP}. In fact, the weighting factors become so small that values for all points farther from P than about ten times the height of P above the land surface can be neglected. This means that a Bouguer gravity map covering only the nearby area beneath P is all we need for the calculation. To solve Equation 9–19, we let N be the actual number of Bouguer gravity values used, and we let A be the area over which these values are evenly distributed. By repeating the calculation for a grid of points on a surface above the earth, we

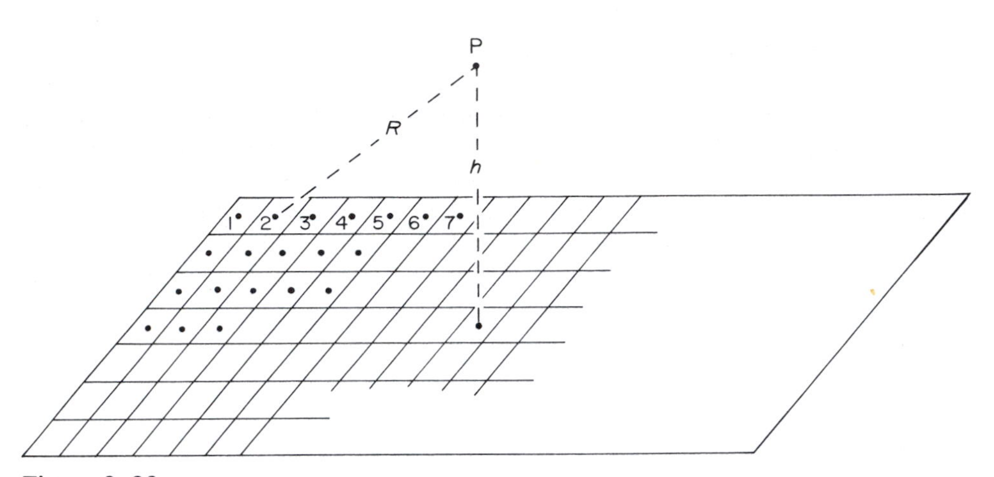

Figure 9–26
Bouguer gravity values evenly spaced over an area A are used with their corresponding height h and distances R to calculate the Bouguer gravity at location P by the method of upward continuation using Equation 9–19.

obtain the upward continuation of our original Bouguer gravity map.

The upward continuation reveals more clearly the regional anomaly pattern. For some purposes, a value of Δg_{BP} can be used as an estimate of Δg_R at the location of the Bouguer gravity value directly beneath it. Then these values are used in Equation 9–16 to obtain an estimate of the local anomaly. But we must bear in mind that the result is only an approximation, because the upward continuation operation does not separate regional and local attractions. Instead, it projects them in a way that makes regional anomalies become the dominant part.

Wavelength Filtering

Wavelength filtering is a mathematical operation that can be used to enhance the appearance of either regional anomaly patterns or local anomaly patterns. It is based on the idea that any pattern of variation along a profile can be reproduced by adding together an appropriate selection of cyclic curves. The example in Figure 9–27 illustrates this idea. Here each cyclic curve has a different wavelength L, amplitude H, and its first peak is offset a different distance d from the left end of the profile.

How closely can we reproduce a gravity anomaly with cyclic curves? A local anomaly is shown in Figure 9–28. Observe how it is duplicated approximately by adding two curves with short but slightly different wavelengths L_1 and L_2. These curves are arranged so that two peaks line up to produce a single large local anomaly. Farther away, the other peaks are not aligned and act to cancel one another. By adding a third curve of considerably longer wavelength L_3, we can improve the synthetic anomaly. This third curve widens it, which makes it more closely resemble the original.

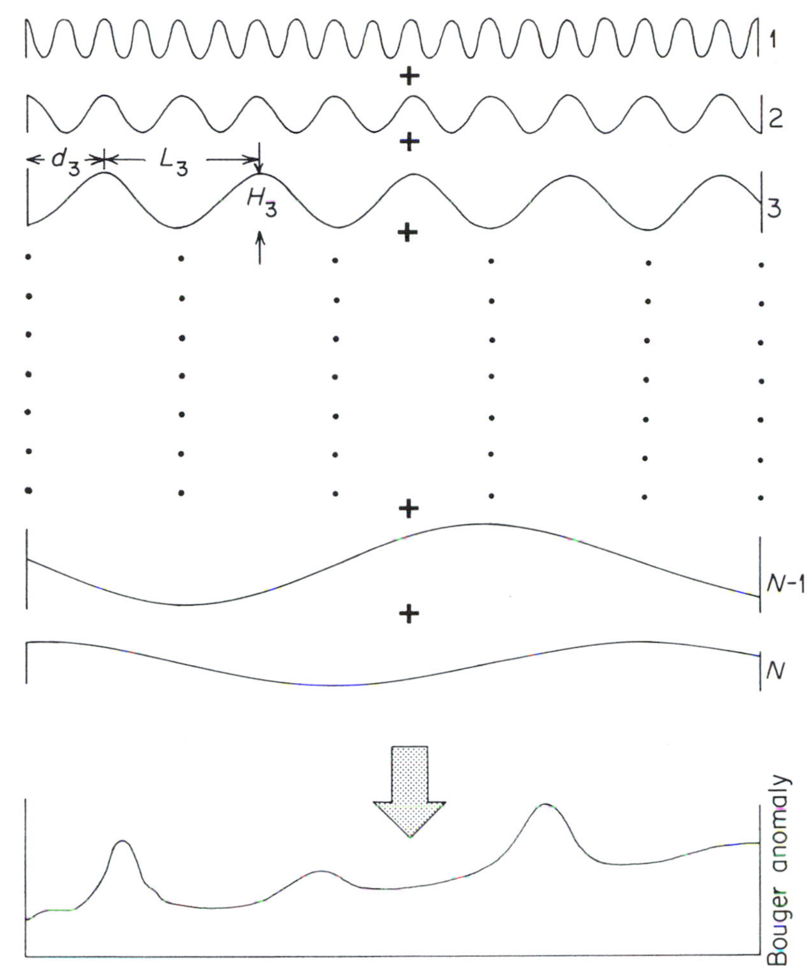

Figure 9–27

A Bouguer gravity profile can be duplicated by summing an appropriate selection of cyclic curves with different wavelengths *L*, amplitudes *H*, and offset distances *d*, which are the distances from the left end of the profile to the nearest wave crests.

Still, the synthetic anomaly is imperfect. Away from the local anomaly peak, the curves do not completely cancel one another. Further improvement would require several more cyclic curves with short, intermediate, and long wavelengths.

The profile in Figure 9–29 displays broad regional variation. This can be reproduced by a cyclic curve with a long wavelength. In fact, the wavelength is considerably longer than the profile, so that only a small part of the curve is used to represent the regional gravity anomaly. Curves with short wavelengths are not ordinarily needed to reproduce regional variations.

These examples point out two important aspects of wavelength filtering. (1) In most instances, no single cyclic curve can be used to represent a local anomaly, for local anomalies are usually distributed irregularly over the

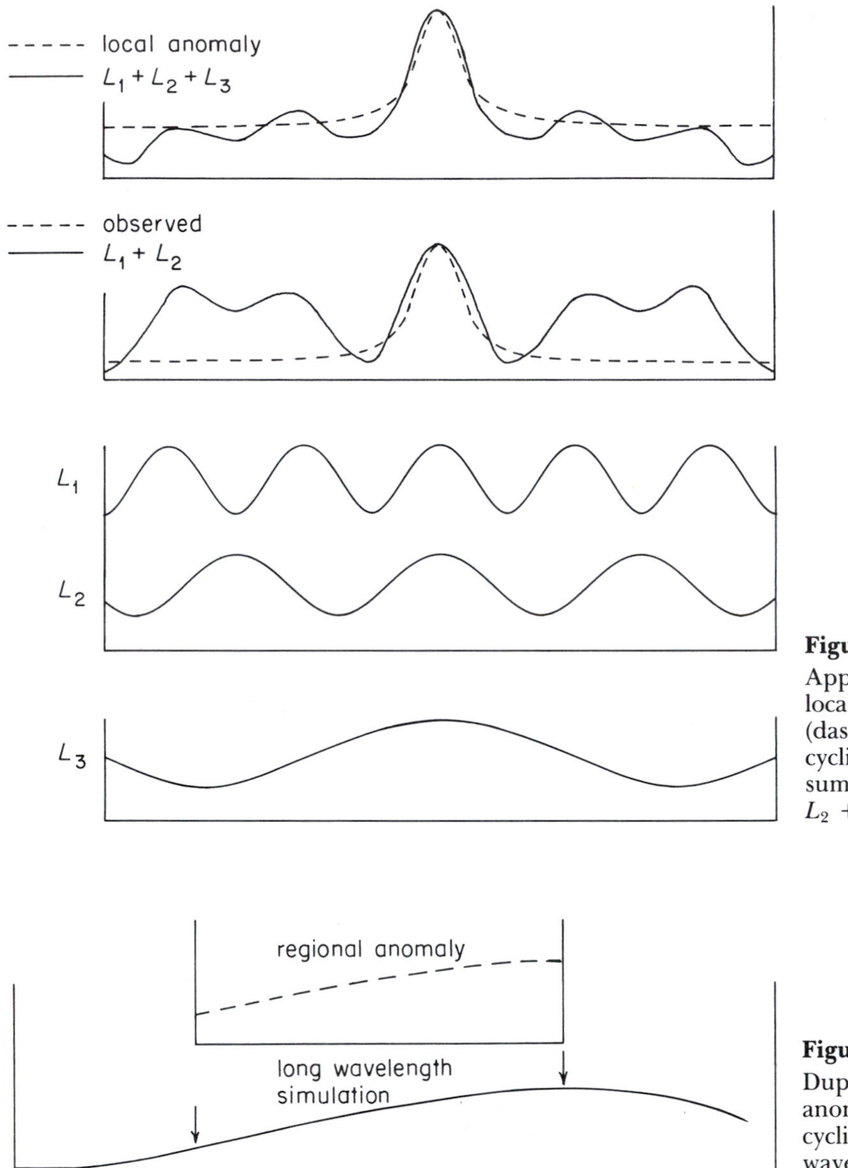

Figure 9–28
Approximate duplications of a local gravity anomaly profile (dashed line) by summing two cyclic curves L_1 and L_2 and by summing three cyclic curves $L_1 + L_2 + L_3$.

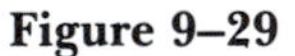

Figure 9–29
Duplication of a regional gravity anomaly profile by a portion of a cyclic curve that has a long wavelength.

area of a survey, and they all have different shapes and amplitudes. To reproduce them, we need several curves that add together in some places and cancel one another elsewhere. (2) Regional anomalies can be reproduced by curves with long wavelengths but local anomalies always require a combination of short and long wavelength curves. Even though the short wavelength curves have larger amplitudes, the long wavelength curves play an important role in the duplication of a local anomaly.

Now comes the important question. How do we determine the appropriate curves for reproducing a gravity profile? For this task we use a mathematical operation called *harmonic analysis*. To begin this operation, we use a formula with a large number of terms to represent the Bouguer gravity value at any distance x from the left end of the profile:

$$g_B(x) = A_1 \cos\left(2\pi\frac{x - d_1}{L_1}\right) + A_2 \cos\left(2\pi\frac{x - d_2}{L_2}\right) + A_3 \cos\left(2\pi\frac{x - d_3}{L_3}\right) + \dots \quad (9\text{–}21)$$

Each term, $A \cos \{2\pi[(x - d)/L]\}$ represents one cyclic curve. Harmonic analysis is an efficient way to test a very large number of possible curves to find the particular curves with amplitudes large enough to be important. The results give us the values of A, L, and d for these particular cyclic curves. We will not attempt to describe exactly how this is done except to say that Δg_B values at evenly spaced points along the profile are used in the calculations.

The search for the particular cyclic curves needed to duplicate a gravity profile is the first step in wavelength filtering. The second step is the attempt to separate regional and local anomalies, which is done by deleting some of the terms in Equation 9–21. When we delete them, we are removing, or filtering out, effects represented by certain wavelengths—hence, the name wavelength filtering. By summing only the terms for short wavelengths and deleting those for long wavelengths, we obtain an approximation of the local anomaly profile. We can see from Figure 9–28 that this will be an imperfect representation of the local anomalies, because they too require that some terms for long wavelengths be closely reproduced. However, the main features of local anomalies are revealed by short wavelength terms. Therefore, this kind of wavelength filtering has helped bring into clearer focus the approximate sizes and shapes of otherwise obscure local anomalies.

In a similar way, we can reproduce only the regional anomalies and filter out local anomalies by deleting terms for short wavelengths from Equation 9–21. Again, the result is imperfect because some of the remaining long wavelength terms may be related to local anomalies. Nevertheless, a clearer view of regional patterns of variations is produced.

So far, we have described wavelength filtering of a Bouguer gravity profile. How can this method be applied to a map that shows variations over an area? Although this operation is more complicated, it is done in a similar way. Rather than cyclic curves, we use surfaces with cyclic corrugations. Two sets of surfaces having corrugations that are at right angles to each other in direction are needed. Each set includes corrugations of many different wavelengths, as the example in Figure 9–30 shows. When combined, surfaces such as these add in some places to reproduce anomalies and cancel

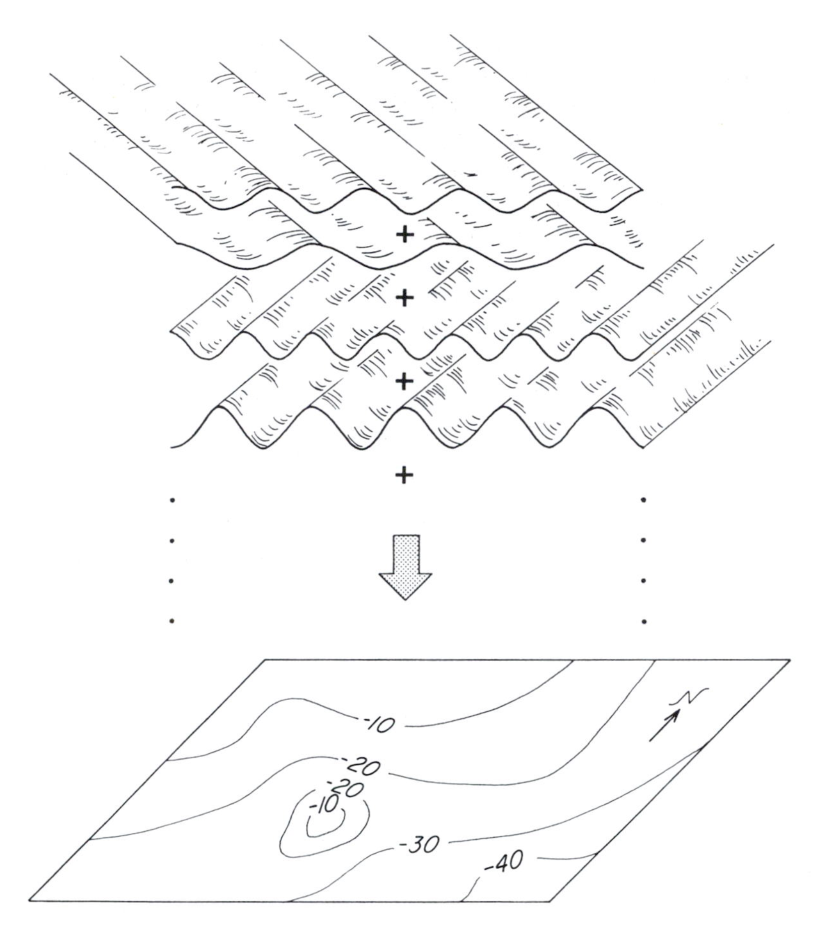

Figure 9–30
A Bouguer gravity map can be duplicated by summing an appropriate selection of cyclic corrugated surfaces. Two sets of surfaces with corrugations whose directions are at right angles to each other must be used.

one another over other areas of a map. The particular corrugation wavelengths needed to reproduce the features of a Bouguer gravity map are determined from harmonic analysis. Filtering is done by deleting certain of these terms while retaining others.

We have described wavelength filtering as a process that involves (1) determination of the wavelengths required to reproduce the features of a profile or map, and (2) deletion of the wavelengths that make up undesired features. These two steps can be combined into a single operation by means of weighting factors designed to enhance certain wavelengths and to suppress others. These weighting factors are applied in the same manner that we described with Figure 9–25 and Equation 9–18, but they do not vary in a simple way with distance from the center of the averaging area. They may first increase and then decrease with distance. Some have positive values, and others are negative. The set of weighting factors used to filter out regional effects is different from the set needed to remove local anomalies. Still other sets can be designed to filter out anomalies of intermediate dimensions. None of these filtering operations perfectly separates desired anomalies from undesired ones, but they have proved to be quite useful for bringing features of interest into clearer focus.

Effects of Known Sources

One final aspect of separating regional and local anomalies should be mentioned. If we have enough independent information about certain features in an area, we may be able to calculate their gravitational attraction. We can then subtract this attraction from the Bouguer gravity to obtain gravity anomaly values caused by other unknown features. For example, in some areas, we already have seismic measurements of the thickness of the crust. By using some combination of two- and three-dimensional models to represent the measured changes in crustal thickness, we could calculate the regional gravity effect from equations introduced in the previous section. In a similar way, local gravity effects of salt structures, igneous intrusions, and other features known from seismic surveys, drilling, and geologic mapping can be calculated. After these known contributions are subtracted, other local anomalies and regional patterns emerge more clearly and with less distortion.

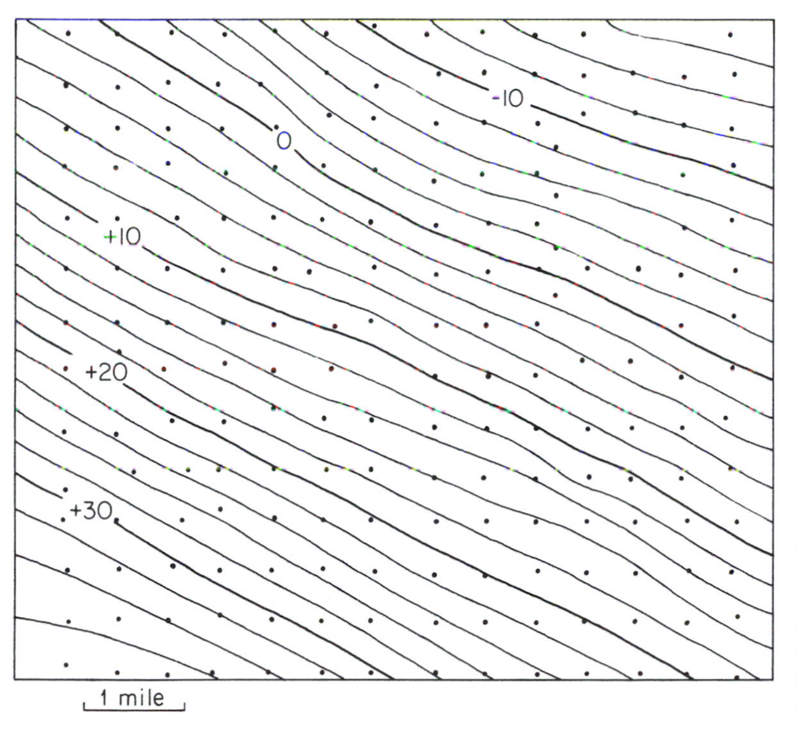

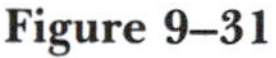

Figure 9–31
Bouguer gravity map of the Los Angeles Basin in California. (Modified from Otto Rosenbach, Geophysical Prospecting, v. 5, n. 2, p. 180, June 1957.) The contour interval is 2 mgal.

SEARCH AND DISCOVERY

We have now introduced the basic tools used by geophysicists to analyze Bouguer gravity maps and profiles. These tools include the formulas for calculating the gravitational attraction of bodies with different shapes and the methods for separating regional and local anomalies. How we use them will depend on the aims of a gravity survey. A weighted averaging operation may be all that is needed to find the local anomalies partly hidden by more dominant regional features. If more specific information about density distribution is of interest, it will be necessary to calculate the attraction of some combination of bodies with different shapes for comparison with the measured Bouguer gravity patterns.

Focussing on Anomalies

Let us look at two examples that illustrate the value of weighted averaging operations for bringing obscure local anomalies into clearer focus. The first is an analysis of a gravity survey of the Los Angeles Basin in California, which it was carried out to identify sites for more intensive oil exploration surveys. The Bouguer gravity map is shown in Figure 9–31. Here, the principal feature is the regional decrease from +37 to −17 mgal in a northeast-

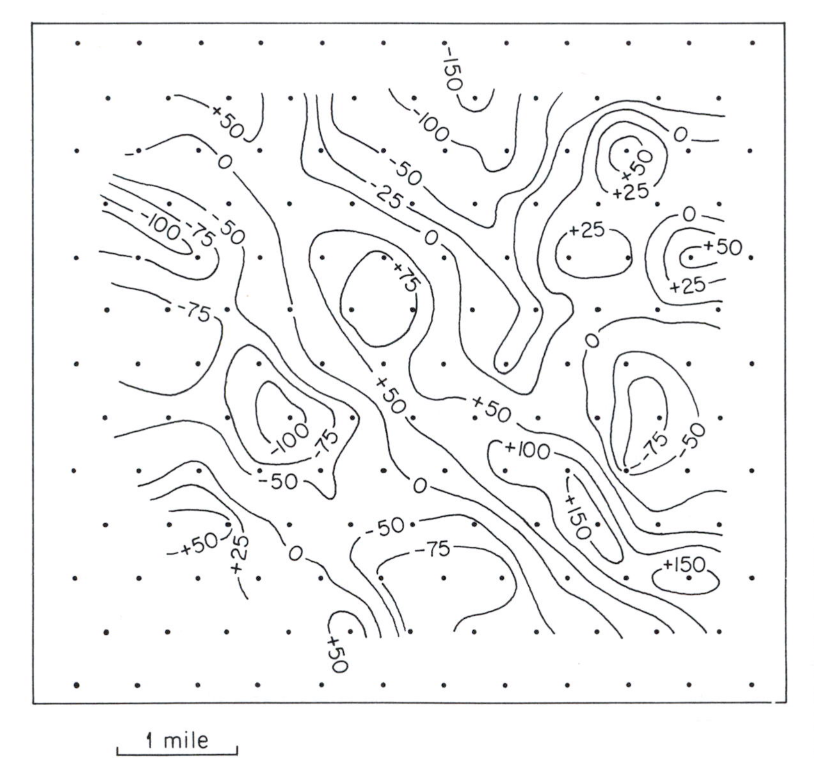

Figure 9–32
Local gravity anomaly map of the Los Angeles Basin in California obtained by processing the data from Figure 9–31 to remove the strong regional effect. (Redrawn from Otto Rosenbach, Geophysical Prospecting, v. 5, n. 2, p. 184, Fig. 6, June 1957.)

erly direction. Except for slight flexures in the contours, there are no obvious local anomalies that indicate sites for further study. However, after we remove the dominant regional variation by a weighted averaging operation, a completely different pattern emerges. Several local anomalies, shown in Figure 9–32, are easily recognized.

Our second example comes from a survey in the Val d'Or mining district of Quebec. Gravity measurements were made to discover the subsurface extent of sulfide ores exposed at the land surface. The Bouguer gravity map in Figure 9–33 reveals several local anomalies that are somewhat distorted by broader regional trends. The regional variation shown in Figure 9–34 was calculated by means of a smoothing operation. After this regional trend was subtracted from the Bouguer gravity, the local anomaly map in Figure 9–35 was prepared. Here the shapes and positions of local anomalies are more clearly evident. The anomaly situated near the eastern margin of the map cannot be identified on the original

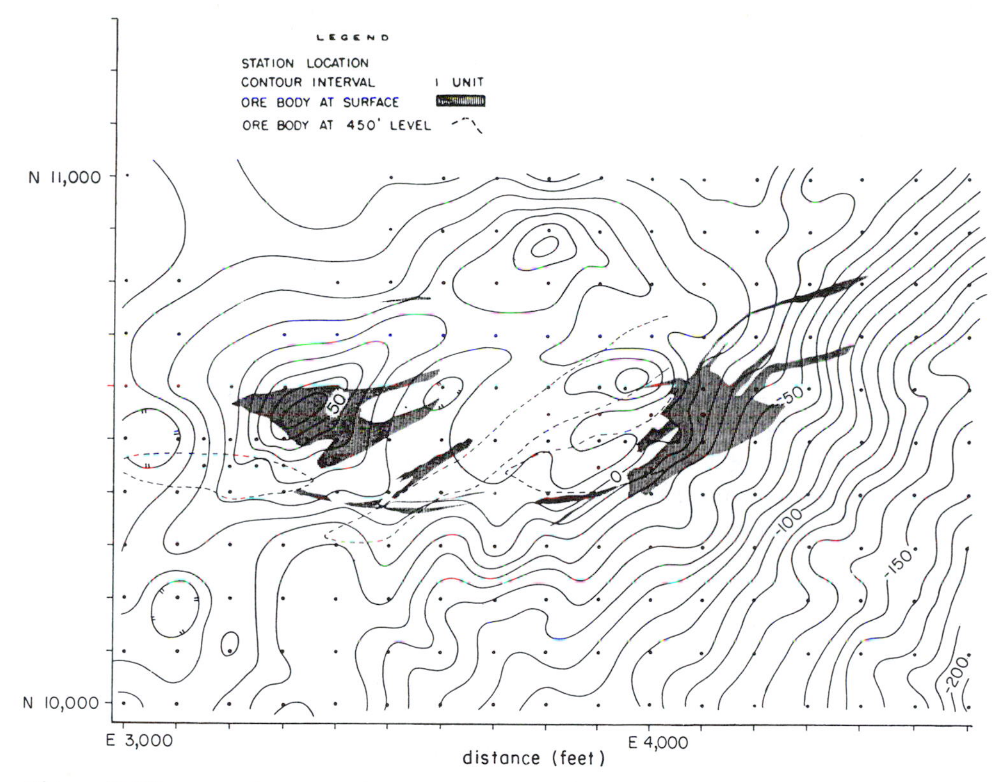

Figure 9–33
Bouguer gravity map of part of the Val d'Or mining district in Quebec, Canada. (From Fraser Grant, Geophysics, v. 22, n. 2, p. 318, Fig. 1, April 1957.) Bouguer gravity is in units of 0.01 mgal, and the contour interval is 0.2 mgal.

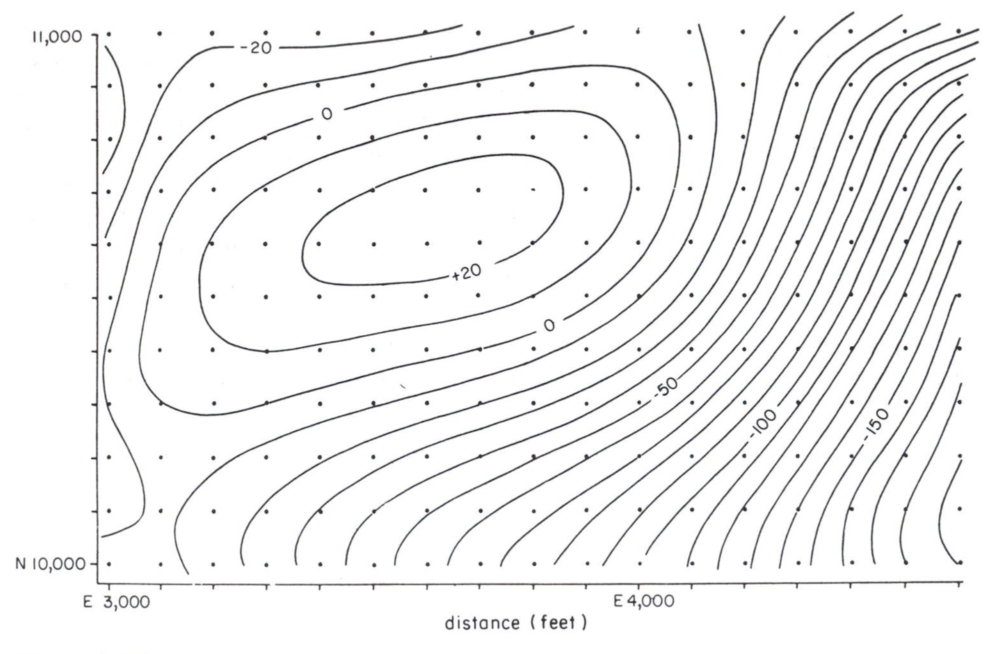

Figure 9–34
Regional gravity map of the Val d'Or mining district in Quebec, Canada. (From Fraser Grant, Geophysics, v. 22, n. 2, p. 335, Fig. 4, April 1957.) Gravity is expressed in units of 0.01 mgal.

Bouguer gravity map. Near the center of the map, the largest anomaly values are slightly to the west of the surface exposure of sulfide ore. The implication is that the ore zone may slant downward in a western direction. Farther to the west is another high local anomaly situated directly over the sulfide ore exposure. The implication here is that the ore may extend more or less vertically downward.

Trial-and-Error Model Analysis

Suppose that we want to learn more about the size and location of particular anomaly sources. We can do so by testing different models by trial and error to find the one that best explains an anomaly. The trial-and-error effort can be time-consuming, but we can follow certain shortcuts if a satisfactory explanation can be found with a relatively simple model. Some examples illustrate these practices.

Look again at Figure 9–9, which shows the gravity anomaly over a chromite deposit in Cuba. After the regional trend was separated out by graphical smoothing, the local anomaly along profile A–A′ was plotted in Figure 9–36. Let us analyze this anomaly by supposing that it is produced by a zone of chromite deposits which has a spherical shape, like the mass in Figure 9–10. We are not really sure

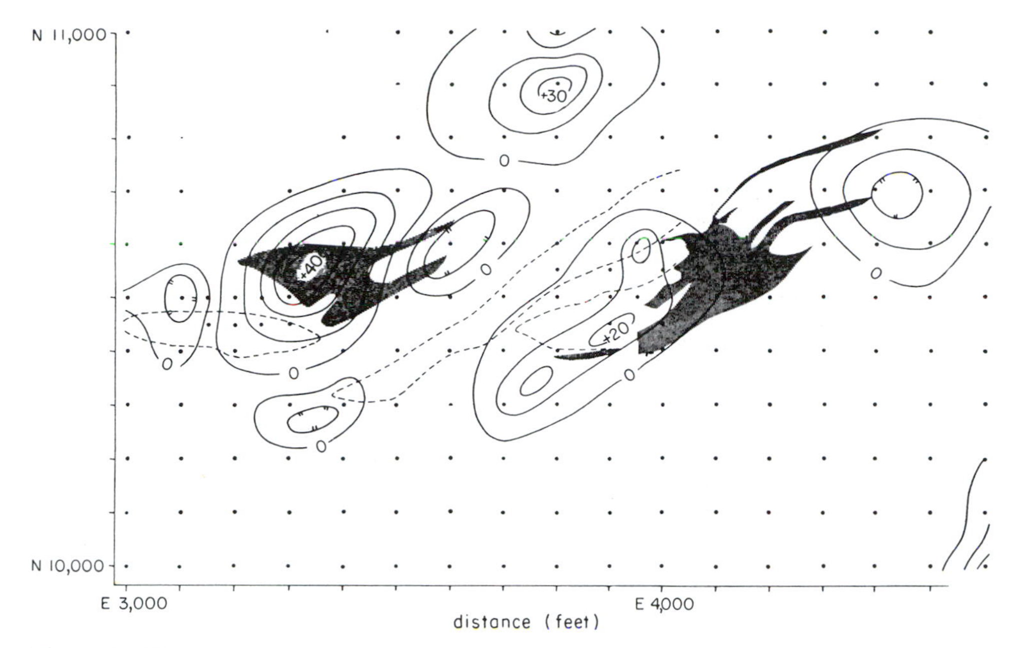

Figure 9–35
Local gravity anomaly map of the Val d'Or mining district in Quebec, Canada, obtained by subtracting the regional field in Figure 9–34 from the Bouguer gravity field in Figure 9–33. (From Fraser Grant, Geophysics, v. 22, n. 2, p. 336, Fig. 5, April 1957.) Gravity is expressed in units of 0.01 mgal.

whether a sphere will prove to be an appropriate model, but it is a simple one to test.

How do we choose the size and depth of a sphere? One way is to make a guess at the radius d and the depth z, and then use these values in Equations 9–3 and 9–4 to calculate a gravity profile for comparison with the measured local anomaly. For example, let us guess that $d = 10$ m and that $z = 25$ m. Recall that chromite density is about 4 g/cm^3 and that the chromite is embedded in rock with an approximate density of 2.7 g/cm^3, which means that the density difference $\Delta\rho = 1.3$ g/cm^3. These values of d and $\Delta\rho$ are substituted into Equation 9–4a to calculate $Q_s = 36.322$. Then Equation 9–3 can be solved for several points to obtain a profile showing the variation of g_s. The lower profile in Figure 9–36 was calculated in this manner. It lies below the measured curve, which tells us that we have chosen a spherical model that is too small and buried too deeply.

Next, let us try another set of values, say, $d = 15$ m and $z = 20$ m. They were used to calculate the upper profile in Figure 9–36. This profile lies above the measured anomaly, indicating that the sphere in our second trial is too large and is not buried deeply enough. By continuing these trial calculations, we could eventually discover the values of $d = 13.4$ m and $z = 21$ m which produce the cal-

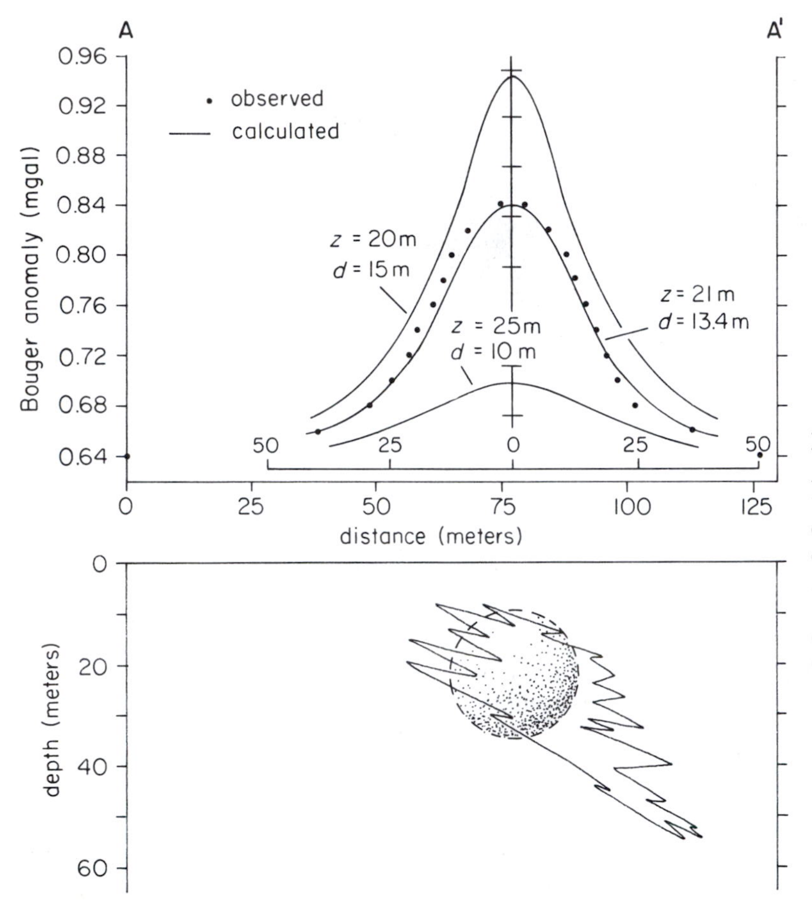

Figure 9–36
Analysis of the Bouguer gravity profile over a buried chromite deposit in the Camagüey mining district of Cuba. Dots indicate Bouguer gravity values along profile A–A′ in Figure 9–9. Solid curves indicate gravity anomalies over three trial models of spherical shape, with depth *(z)* and radius *(d)* given for each curve. The closest fit is given by the sphere of radius $d = 13.4$ m and depth $z = 21$ m, which is compared with an estimate of the true shape of the ore body determined by drilling.

culated curve that closely matches the measured curve in Figure 9–36.

Inversion Methods of Model Analysis

The trial-and-error procedure can be time-consuming and tedious. Is there some way to reduce the number of calculations that must be done repeatedly to duplicate a measured anomaly satisfactorily? Information from Figure 9–37 can be used to simplify the procedure. Observe how gravity diminishes from a maximum value, g_{max}, directly over the sphere to the small value of $0.35g_{max}$ at the horizontal distance x, which equals the depth z. The same anomaly pattern exists for any sphere regardless of its depth or size. Knowing this, we can estimate the depth simply by inspecting the measured profile to find the distance at which the gravity value $g_s = 0.35g_{max}$. For the local anomaly in Figure 9–36, the maximum is 0.2 mgal. The value of $0.2 \times 0.35 = 0.07$ mgal is found at a distance of about 21 meters from

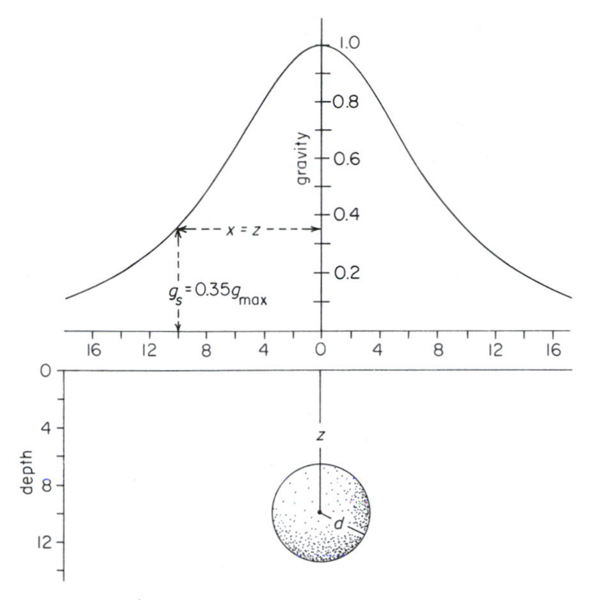

Figure 9–37
Gravity values and distances used in an inversion analysis of the gravity anomaly profile caused by a buried sphere. The depth to the center of the sphere is the same as the distance from the point of maximum gravity, g_{max}, to points where gravity has diminished to 0.35 g_{max}.

the center of the anomaly. Therefore, this must also be the depth of a spherical anomaly source.

Next, we will calculate the size of the sphere. At the observation point directly over a sphere, the distance r from its center is equal to the depth z, so that Equation 9–3 becomes

$$g_{max} = g_s = Q_s/z^2$$

which can be rearranged into the form

$$Q_s = z^2 g_{max} \qquad (9\text{–}22)$$

We can substitute the term on the right-hand side into Equation 9–4a and rearrange the result to obtain the formula

$$d^3 = \frac{z^2 g_{max}}{0.02794\ \Delta\rho} \qquad (9\text{–}23a)$$

where g_{max} is in milligals, $\Delta\rho$ is in grams per cubic centimeter, and z and d are in meters. If z and d are in feet, the formula becomes, from Equation 9–4b,

$$d^3 = \frac{z^2 g_{max}}{0.00852\ \Delta\rho} \qquad (9\text{–}23b)$$

Now we can solve for the radius of the spherical model using values of z and g_{max} obtained from the measured profile and a reasonable estimate of the density difference. Continuing with our analysis of the measured profile in Figure 9–36, we use the values of z = 21 m, g_{max} = 0.2 mgal, and $\Delta\rho$ = 1.3 g/cm^3 to calculate d = 13.4 m.

It is clear that we do not have to make a large number of trial calculations to discover the particular spherical model that best explains a local gravity anomaly. Instead, we can find the depth directly from the measured profile and then calculate the size using Equation 9–23. This modified procedure is called an *inversion* procedure, which means that it is a way to "work backward" from a measured pattern of variation to discover the size and location of its source.

It is highly unlikely that we would find a spherical chromite deposit. Nevertheless, we can use this idealized model to make preliminary estimates of the amount of ore and the depth where we might expect to begin mining it. The density of chromite is about 4 g/cm^3, which means that a spherical zone with a radius of 13.4 meters would contain about 40,000 metric tons of ore. If the center is 21 meters deep, the top of the sphere would be about 8 meters below the land surface. How do these estimates compare with the results of exploratory drilling? A hole drilled at the center of the anomaly reached chromite at 8 meters, the depth predicted from the gravity analysis. About 24,000 tons of ores were discovered, but they were found to be interlayered with peridotite, which has a density close to 3.4 g/cm^3. Therefore, a reasonably accurate estimate of the tonnage of high-density material was obtained from the analysis of a spherical model. The fact that chromite was not the only high-density material present could not have been predicted, because gravity anomalies depend on density variations alone. Drilling revealed that the chromite deposit was a steeply dipping plate, quite different from a sphere. But it is interesting to see how useful this model proved to be for estimating the depth and tonnage of the deposit.

A spherical model is often chosen for preliminary analysis of local anomalies indicated by more or less circular closed contours like the one shown in Figure 9–9. Suppose, however, we had reason to believe that the source of such an anomaly extended vertically downward much deeper than could be represented by a sphere. Perhaps the vertical cylinder in Figure 9–14 would be a more appropriate model. If so, we could attempt to find its dimensions by trial-and-error calculations or by means of an inversion analysis. This operation would involve the same kind of reasoning that we used in the analysis of a sphere. However, the calculations are more complicated because depths to both the top and the base as well as the radius of the cylinder can be adjusted.

Elongate anomaly patterns like those over Pine Valley in Figure 9–8 cannot be explained by a sphere or a vertical cylinder. Instead, a horizontal cylinder might prove to be a more appropriate simple model for preliminary analysis. What we want to find out is the thickness of the low-density alluvium that has been deposited in the valley. We can estimate this thickness by analyzing the local anomaly profile in Figure 9–38, which was prepared by subtracting from Bouguer gravity along profile A-A' in Figure 9–7 a regional trend determined by graphical smoothing.

Can the sediment filling Pine Valley be represented approximately by a horizontal cylinder? Rather than testing this model by trial and error, let us begin immediately with an inversion analysis to find its depth z and its radius d. In Figure 9–15, we can see that at a horizontal distance x which equals z, the gravity g_{hc} is one-half of the maximum attraction g_{max} directly over the cylinder axis. In other words, $g_{hc} = 0.5g_{max}$ where $x = z$. Inspection of Figure 9–38 reveals that $g_{max} = -50$ mgal, and at an approximate distance of 7 km, the local anomaly $g_{hc} = -25$ mgal. Therefore, the depth of the cylinder axis should be $z =$ 7000 m.

Next, we can solve for the cylinder radius. At a point directly over the axis, the distance r equals z, so that Equation 9–8 can be rearranged to obtain the formula

$$Q_{hc} = g_{max}z \qquad (9\text{–}24)$$

Then, by substituting the term $g_{max}z$ into Equation 9–9, we get

$$d^2 = \frac{g_{max}z}{0.04193\ \Delta\rho} \qquad (9\text{–}25a)$$

where g_{max} is in milligals, $\Delta\rho$ is in grams per cubic centimeter, and d and z are in meters. If d and z are in feet, the formula is

$$d^2 = \frac{g_{max}z}{0.01278\ \Delta\rho} \qquad (9\text{–}25b)$$

In Pine Valley, the local anomaly profile consists of negative values, indicating that the relatively low-density alluvium (2.2 g/cm^3) has a weaker gravitational attraction than the sur-

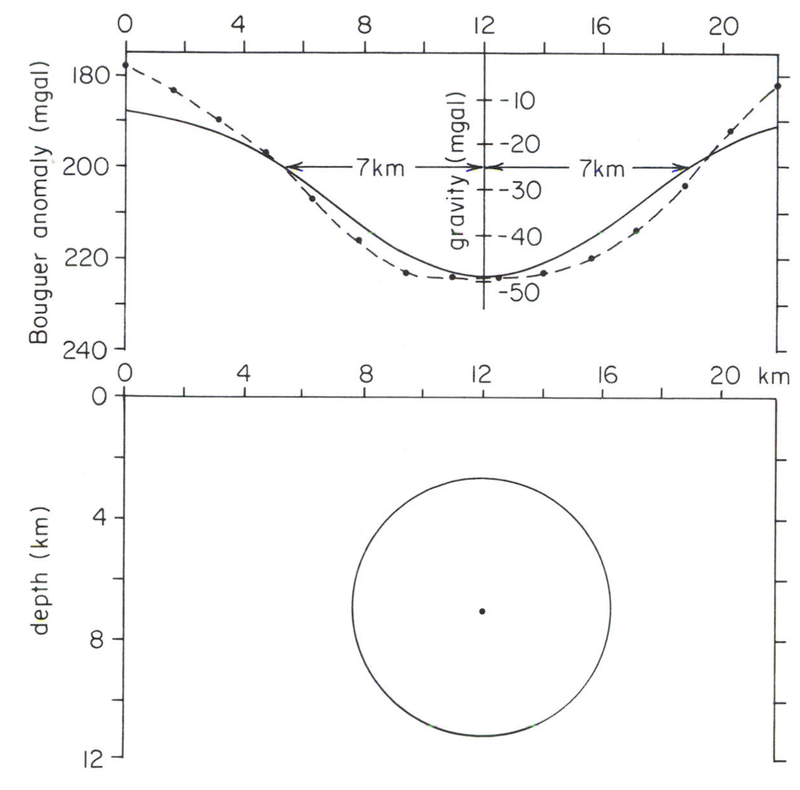

Figure 9–38
Bouguer gravity profile (dots) across Pine Valley in central Nevada is compared with a gravity profile (solid curve) over a single horizontal circular cylinder. A density contrast of $\Delta\rho = -0.47$ g/cm^3 was assumed. The Bouguer gravity is from profile A–A′ in Figure 9–7.

rounding bedrock (2.67 g/cm^3). We therefore choose the negative density difference of $\Delta\rho = -0.47$ g/cm^3 and the values obtained for g and z to calculate $d = 4215$ m. Now we can calculate values of g_{hc} at several points to prepare a profile for comparison with the measured local anomaly profile. The results in Figure 9–38 show how closely the two profiles match. But it is obvious that the horizontal cylinder is a poor representation of the alluvium filling the valley. The cylinder would have to be completely buried at considerable depth, whereas we know that the layer of alluvium extends downward from the land surface. Clearly, the cylinder does not give us an indication of the thickness of this layer.

Perhaps two cylinders side by side close to the land surface would more realistically represent the alluvium in Pine Valley. For this compound model, we must calculate the combined vertical gravitational attraction of two separate bodies. Let us choose the same radius for both cylinders and set the depth $z = d$ so that the cylinders reach the land surface. Because we cannot derive simple formulas for an inversion procedure, we will use the method of trial and error. As before, the density difference appropriate for Pine Valley is $\Delta\rho = -0.47$ g/cm^3. For the first trial, let us guess that $z = d = 3000$ m and then solve Equations 9–8 and 9–9 at several points. The results plotted in Figure 9–39 indicate that the cylin-

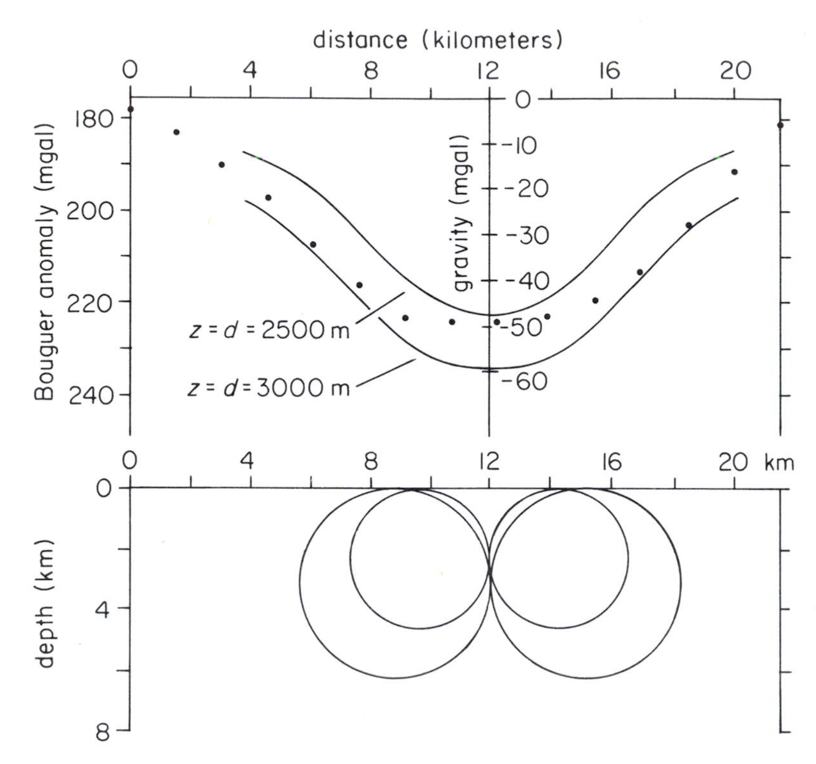

Figure 9–39
Bouguer gravity profile (dots) across Pine Valley in central Nevada is compared with gravity profiles over two compound models, each consisting of two identical parallel, horizontal cylinders. Depth *(z)* and radius *(d)* of the cylinders are indicated for each curve, and a density contrast of $\Delta\rho = -0.47$ g/cm^3 was assumed. The Bouguer gravity is from profile A–A′ in Figure 9–7. Observe that the Bouguer gravity values plot between the gravity profiles calculated for the two compound models.

ders are too large. Therefore, for the second trial, we choose a smaller value of $d = z =$ 2500 m and repeat the calculations. These results, which are also plotted in Figure 9–39, indicate that the cylinders in the second model are too small. Even though we have not succeeded in duplicating the local anomaly, we can conclude from the profiles in Figure 9–39 that the layer of alluvium might be a few thousand meters thick.

We cannot expect to obtain a close match for the local anomaly across Pine Valley by using a compound model consisting of only two parallel cylinders of the same size. The best way of duplicating this profile is to devise a model with a complicated cross section that can be represented by a polygon with many corners. Then, Equations 9–14 and 9–15 can be used to calculate its vertical gravitational attraction at points along a profile. By trial and error, the positions of the polygon corners are adjusted, and the calculations are repeated until a comparable gravity profile is obtained. This was done to match the Pine Valley local anomaly. The result of this analysis is the cross section in Figure 9–40 which shows how alluvium reaching a thickness of more than 3500 meters blankets an irregular bedrock surface.

Analysis of Compound Models

Compound models must be used if we want a detailed explanation of complicated gravity anomaly patterns. There is no limit to the

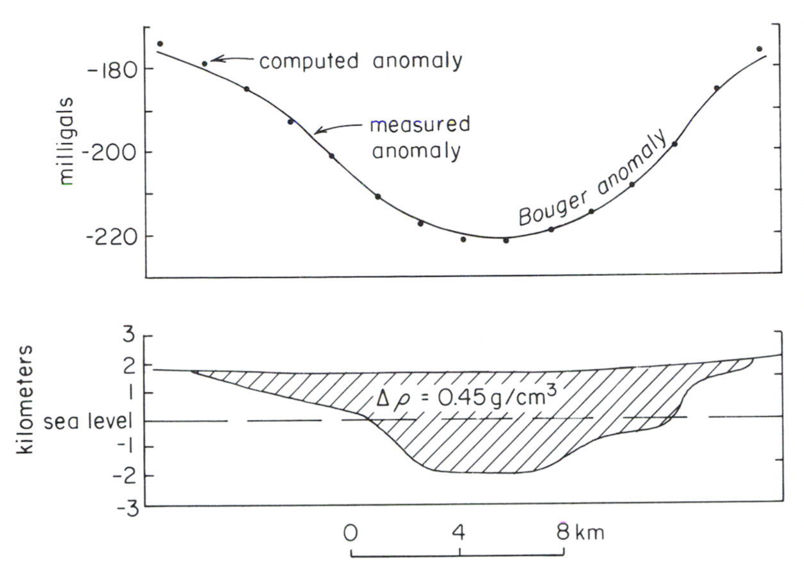

Figure 9–40
The Bouguer gravity profile (solid curve) across Pine Valley in central Nevada is compared with a gravity profile (dots) over a two-dimensional model with an irregular polygonal cross section which represents the relatively low-density layer of alluvium that covers the relatively high-density bedrock floor of the valley. The density difference is $\Delta\rho = 0.45$ g/cm^3 for this model. (From D. R. Mabey, U.S. Geological Survey.) The Bouguer gravity is from profile A–A′ in Figure 9–7. The gravity profile for the two-dimensional model can be calculated by means of Equations 9–14 and 9–15.

number of irregularly shaped bodies that can be included in such a model. However, for most analysis of this kind, the shapes and arrangement of these bodies must be worked out by trial and error. The explanation of the anomaly near Portsmouth, Virginia, seen in Figure 9–6, was worked out by trial-and-error analysis. Recall that this anomaly results from the intrusion of granite into metamorphic rocks of higher density.

How can we describe an irregular mass of granite for the purpose of calculating its gravitational attraction? Specifically, we may combine several horizontal plates, each one having a different irregular polygonal shape. Equations 9–12 and 9–13 are used for calculating the vertical gravitational attraction of each plate, and the results are summed to find the gravity anomaly that would be produced by the compound model.

The Portsmouth gravity anomaly was analyzed in the following way. First, the regional trend determined by weighted averaging was subtracted from the Bouguer anomalies, and the local anomaly contour map presented in Figure 9–41a was prepared. Next, the first trial model was tested. This beginning model consisted of a few polygonal plates chosen by guess. Its gravitational attraction was calculated at several overlying points so that a contour map could be prepared for comparison with the local anomaly map of the Portsmouth area. The calculated map turned out to be a poor representation, but the comparison did provide indications of how the model could be improved. In a succession of trials, more plates were added, and their shapes were adjusted until a satisfactory comparison was obtained. A density difference of $\Delta\rho = -0.15$ g/cm^3 was used in the calculations. The final compound model consisted of the ten plates illustrated in plan view and along cross sections in Figure 9–42. The variations in gravity over this model are shown by the dashed contours in Figure 9–41a, plotted together with the local anomaly contours for easy comparison. You can see that the model explains the main features of the local anomaly. Further

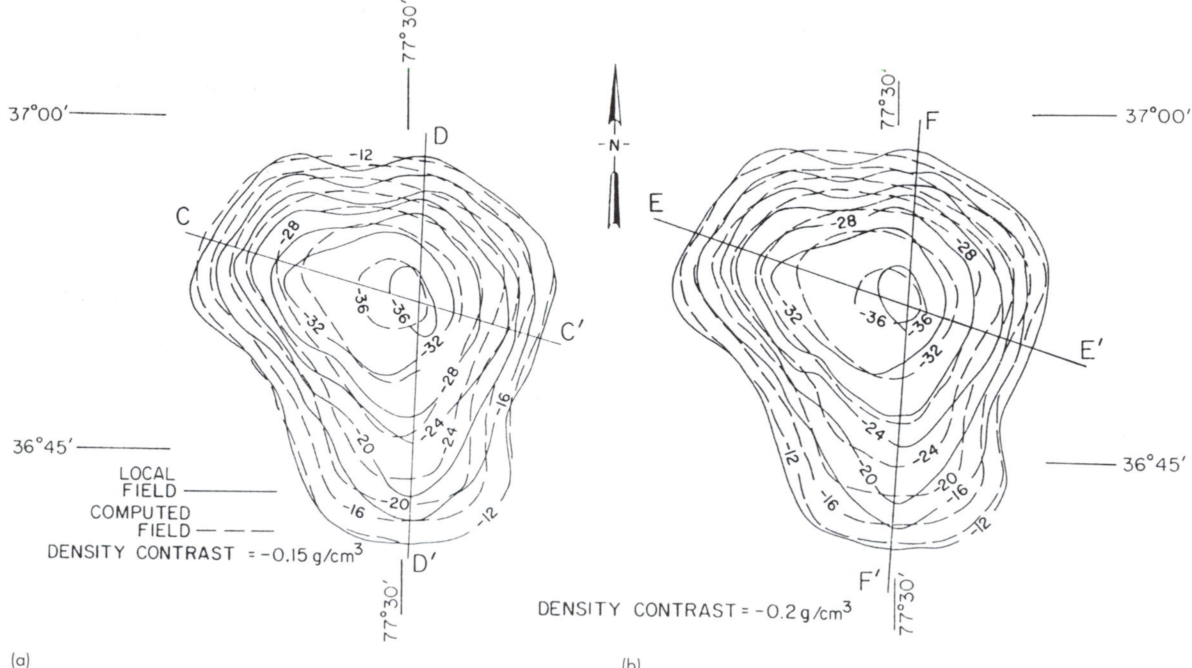

Figure 9–41
Maps of the area near Portsmouth, Virginia, showing local gravity anomaly contours (solid contours) and theoretical gravity anomaly contours (dashed contours) calculated by means of Equations 9–12 and 9–13 (a) for the compound model illustrated in Figure 9–42 and (b) for that illustrated in Figure 9–43. (From J. K. Costain, L. Glover, and A. K. Sinha, Evaluation and Targeting of Geothermal Energy Resources in the Southeastern United States—ERDA Progress Report, p. C-26, 1977.) Gravity is expressed in milligals. Observe that the gravity fields computed for the models reproduce the basic Bouguer gravity pattern, but not all the minor flexures of the contours.

refinements that would be needed to reproduce more closely the minor flexures of the contours would probably not yield much additional information about the granitic intrusion.

In this area, the granite mass and the surrounding metamorphic rocks are all buried beneath a 400-meter-thick blanket of sediments, and so there is no practical way to be certain about the correct density difference. Therefore, a second trial-and-error analysis was done in which a value of $\Delta\rho = -0.20$ g/cm^3 was used. The result is the alternate compound model illustrated in Figure 9–43. The corresponding variations in gravity are shown by the dashed contours in Figure 9–41b, plotted together with local anomaly contours. Both models explain the local anomaly equally well. Rock specimens obtained by drilling indicate that a density difference between -0.15 and -0.20 g/cm^3 is realistic. We may thereby conclude that the shape of the granitic intrusive is probably intermediate between the shapes of the two compound models.

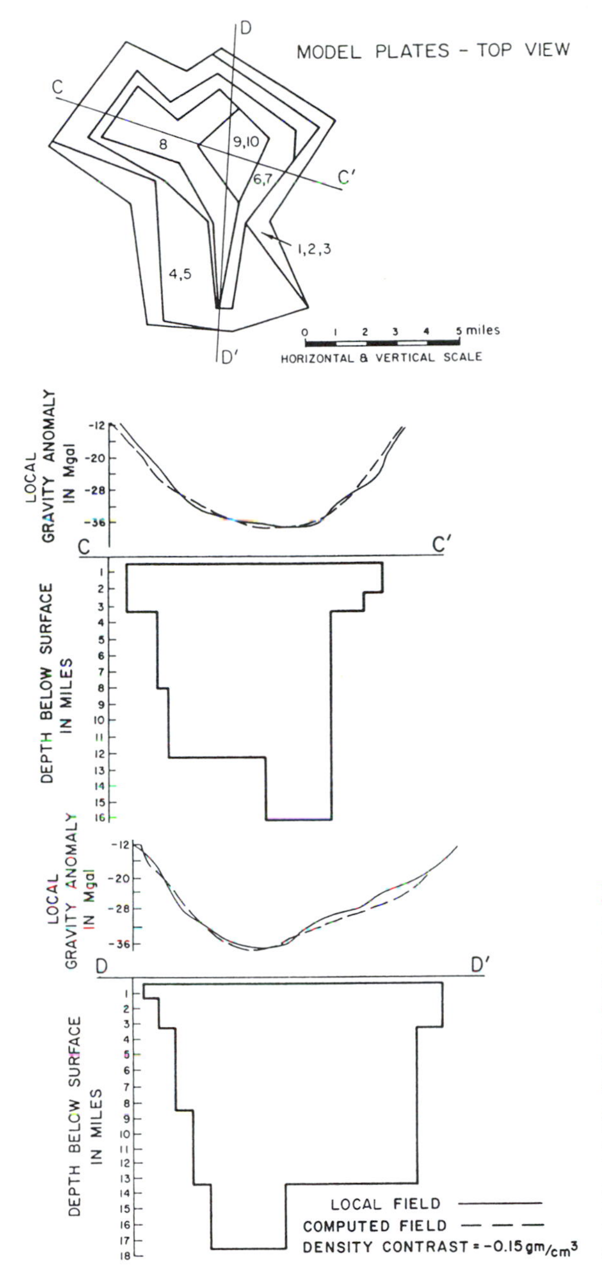

Figure 9–42

Local gravity anomaly profiles C–C′ and D–D′ (solid curves) from Figure 9–41 are compared with gravity profiles (dashed curves) calculated for a compound model consisting of ten horizontal plates with different polygonal shapes. The plates represent the basic shape of a relatively low-density granitic pluton that intrudes higher-density metamorphic rocks. (From J. K. Costain, L. Glover, and A. K. Sinha, Evaluation and Targeting of Geothermal Energy Resources in the Southeastern United States—ERDA Progress Report, p. C-27, 1977.) The density difference is $\Delta\rho = -0.15$ g/cm^3. Gravity is in milligals.

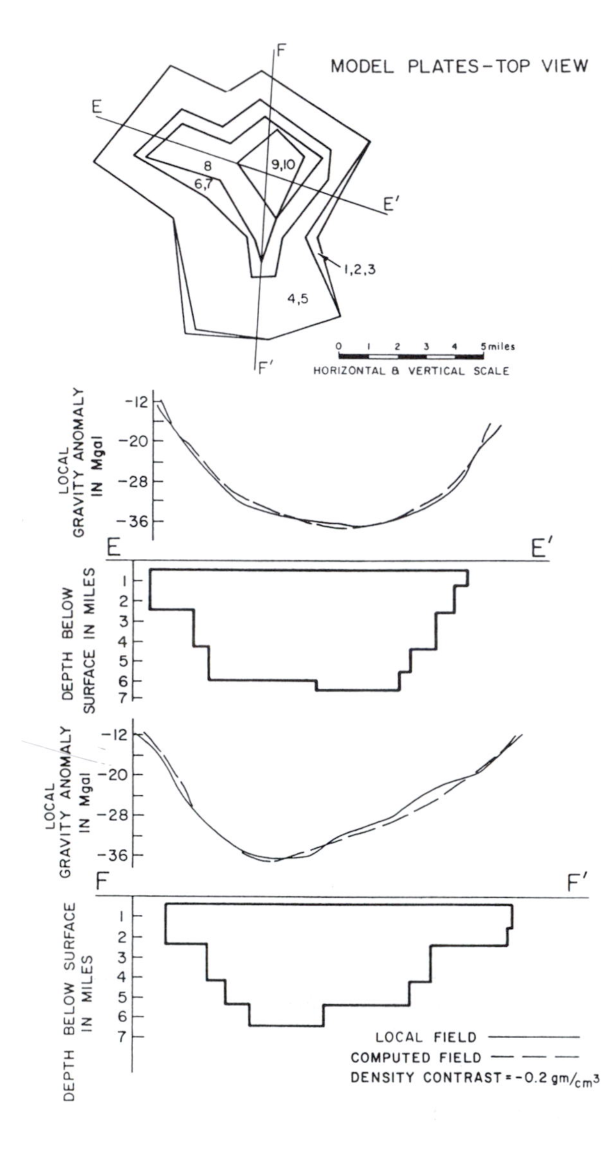

Figure 9–43

Local gravity anomaly profiles C–C′ and D–D′ (solid curves) from Figure 9–41 are compared with gravity profiles (dashed curves) calculated for a compound model consisting of ten horizontal plates with different polygonal shapes. The plates make up another model, different from the one in Figure 9–42, for the granitic pluton that intrudes higher-density metamorphic rocks. (From J. K. Costain, L. Glover, and A. K. Sinha, Evaluation and Targeting of Geothermal Energy Resources in the Southeastern United States—ERDA Progress Report, p. C-29, 1977.) The density difference is $\Delta\rho = -0.20$ g/cm^3. Gravity is in milligals.

Finally, one last example illustrates how gravity measurements can be used in conjunction with seismic surveying and geologic mapping. The area of interest is situated between Richmond and Charlottesville, Virginia. The different rock units in the area were discovered by geologic mapping. These igneous and metamorphic rocks turned out to be so complexly deformed that additional information was needed before a reliable cross section

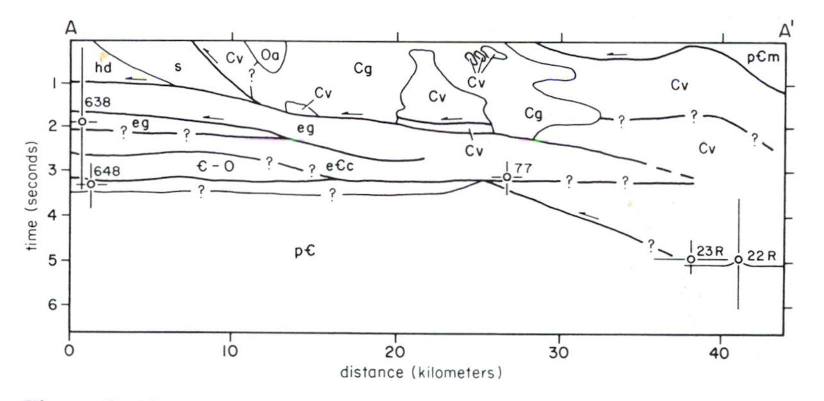

Figure 9–44

Profile in Goochland County, Virginia, showing positions of faults and geologic formation boundaries interpreted from a seismic time section and surface geologic mapping. (Modified from M. R. Keller, E. S. Robinson, and L. Glover, Bulletin of the Geological Society of America, v. 96, n. 12, p. 1581, December 1985, and unpublished data of C. Coruh and J. K. Costain.) Symbols represent different geologic formations.

could be prepared. This information was obtained by seismic surveying. A geologic interpretation of the seismic reflections is presented in Figure 9–44. Here we see a complicated pattern with some reflections coming directly from boundaries between different rock units, and others coming from within individual units. How certain are we that the boundaries have been correctly identified?

Gravity measurements were used to test the interpretation. To do this, the compound, two-dimensional model in Figure 9–45b was prepared. The boundaries indicated in Figure 9–44 are closely reproduced by the polygons in this model. Densities characteristic of the different rock units were estimated from values found for rock specimens collected during geologic mapping. Then Equations 9–14 and 9–15 were used to calculate the vertical gravitational attraction of each polygonal body at several observation points. The results were summed to get the profile of gravity variation, which is plotted as a dashed line in Figure 9–45a, along with the local anomaly profile. The fact that this compound, two-dimensional model explains the local gravity anomalies gives us more confidence that the boundaries given in Figure 9–44 were accurate interpretations of the seismic reflections. No doubt other models with different polygon shapes and densities could produce a similar pattern of gravity variation, but the particular model in Figure 9–45 not only explains this gravity pattern, but also conforms to the independent results of seismic surveying and geologic mapping.

In this chapter we have introduced the fundamentals of geologic interpretation of gravity measurements. Every interpretation must be-

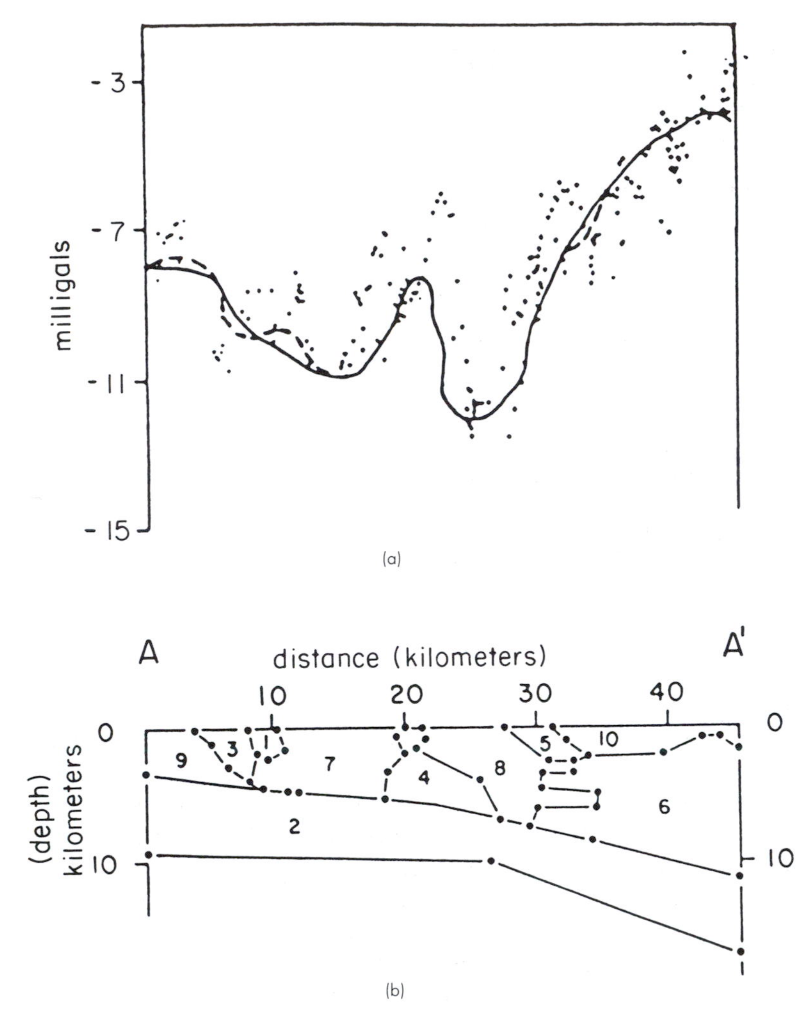

Figure 9–45
(a) Local gravity anomaly variation (dashed line) along the profile in Figure 9–44, and individual local anomaly values (dots) projected to this profile from distances up to one kilometer are compared with the theoretical gravity variation (solid line) calculated by means of Equations 9–14 and 9–15 from (b) the two-dimensional model which consists of units that correspond to the rock units represented in Figure 9–44. (From M. R. Keller, E. S. Robinson, and L. Glover, *Bulletin of the Geological Society of America,* v. 96, n. 12, p. 1584, December 1985.)

gin by recognizing that patterns of Bouguer gravity variation are the superposition of effects from many anomaly sources. To a certain extent, we can separate broader regional anomalies from the local anomalies. Sometimes the process of separation is helpful for locating anomalies that were hidden by the overlapping effects of more dominant features. Additional information about anomaly sources comes from model analysis. Some anomalies can be satisfactorily explained by models of simple geometrical form. For more complicated anomaly patterns, compound models consisting of many irregular bodies of different density can be used. There is always more than one model for reproducing any pattern of gravity variation. For this reason, in interpreting gravity measurements we must make judgments in deciding which explanation is the most geologically meaningful.

STUDY EXERCISES

1. Examine the Bouguer gravity contour map in Figure 9–46. Assume that the regional variation is related to the thickness of the earth's crust and that a local anomaly is caused by an intrusion.

   a. Prepare a regional anomaly map by graphical smoothing and then prepare a local anomaly map.

   b. In what direction is the crust becoming thicker? Explain!

   c. Which of the following explanations of the local anomaly is the most probable? Sediments with density of 2.5 g/cm³ are intruded by (1) rock salt in central Colorado; (2) rock salt in southern Texas; (3) granite with a density of 2.68 g/cm³ in the Rocky Mountains; (4) granite with a density of 2.68 g/cm³ in central Virginia. Discuss your answer!

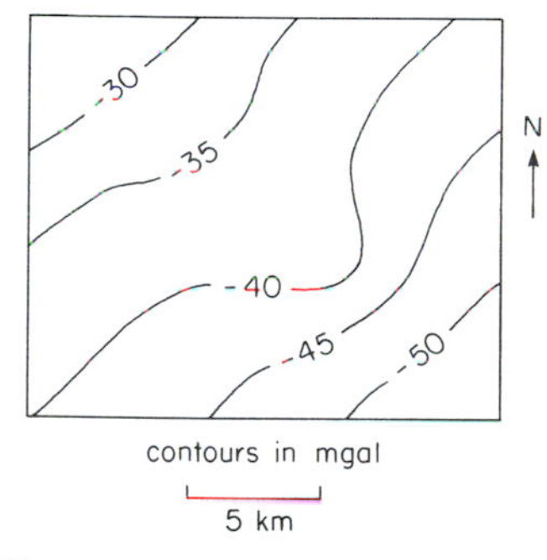

Figure 9–46
Bouguer gravity map contoured at 5-mgal intervals.

2. Analyze the local gravity anomaly pattern evident on the Bouguer gravity map in Figure 9–47 in terms of (a) a sphere embedded in rock with density of 2.5 g/cm³ and (b) a vertical rod embedded in rock of density 2.5 g/cm³. Estimate the depth of the anomaly source, using both models. Discuss which model best reproduces the shape of the anomaly and solve for the density contrast. Was this gravity survey done in Maryland or in Utah? Explain briefly!

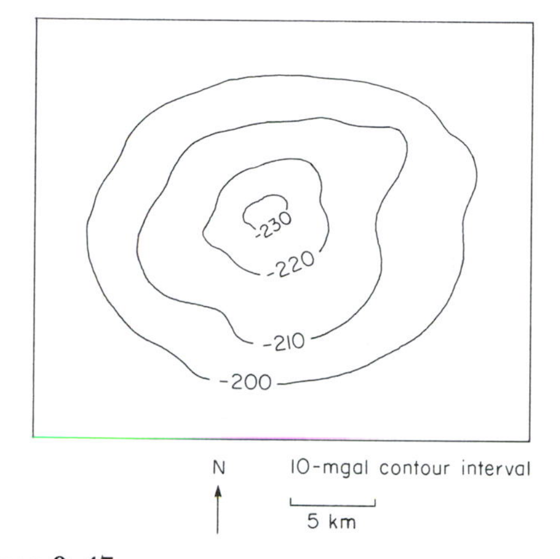

Figure 9–47
Local gravity anomaly map contoured at 10-mgal intervals.

3. The cross section in Figure 9–48 shows a gabbro sill with density of 2.8 g/cm³ that intrudes volcanic rocks with density of 2.55 g/cm³. Calculate a profile showing the variation of gravity along the upper surface of the cross section.

4. The high-density rock layer seen on the cross section in Figure 9–49 is offset along a vertical fault. Sketch the shape of the gravity anomaly that you would expect to observe along a profile on the land surface crossing over this structure.

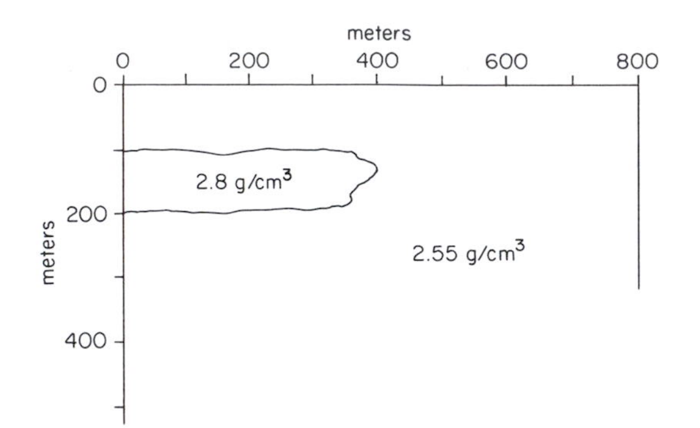

Figure 9–48
Cross section showing gabbro sill with density of 2.8 g/cm^3 intruding rock of density 2.55 g/cm^3.

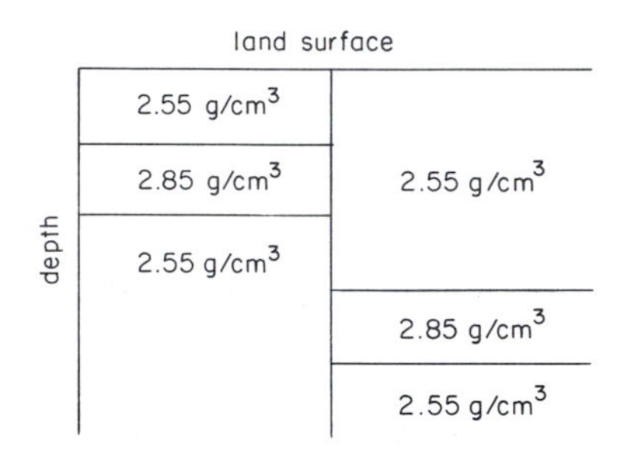

Figure 9–49
Horizontal rock layers offset along a vertical fault.

5. Suppose that a long, straight horizontal passage of a cave passes under a gravity survey area. The cross section of this passage is shown in Figure 9–50. Calculate the gravity variation that you would expect to measure along a profile on the land surface passing over this cave passage. Use a horizontal cylinder to represent the cave.

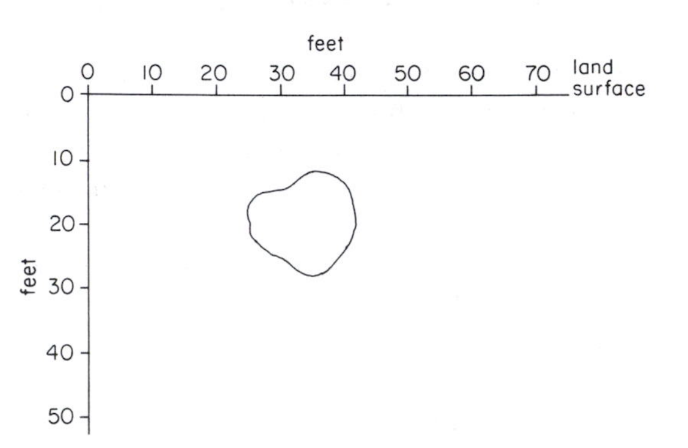

Figure 9–50
Cross section showing a cave passage.

6. Devise an inversion procedure for estimating the position of a model consisting of two horizontal cylinders with the same density and diameter and arranged side by side horizontally as illustrated in Figure 9–51.

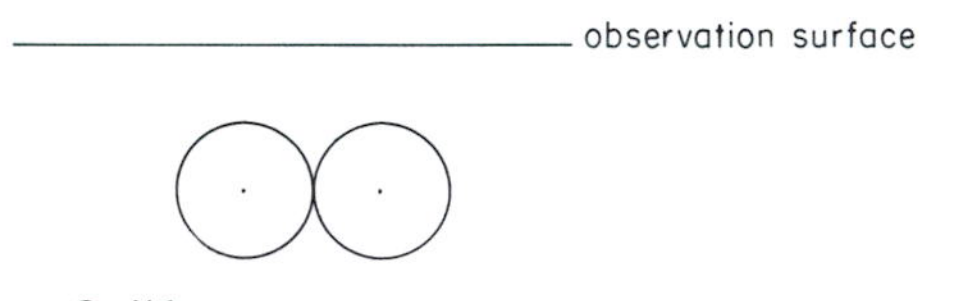

Figure 9–51
Two parallel horizontal cylinders tangent to one another at the same depth.

SELECTED READING

Fuller, Brent D., Two-dimensional frequency analysis and design of grid operators, *Mining Geophysics,* Volume 2. Tulsa, Okla., Society of Exploration Geophysicists, pp. 658–708, 1967.

Grant, F. S., and G. F. West, *Interpretation Theory in Applied Geophysics.* New York, McGraw-Hill, 1965.

Henderson, Roland G., A comprehensive system of automatic computation in magnetic and gravity interpretation, *Geophysics,* v. 25, n. 3, pp. 569–585, June 1960.

Hinze, William J. (editor), *The Utility of Regional Gravity and Magnetic Anomaly Maps.* Tulsa, Okla., Society of Exploration Geophysicists, 1985.

Mack, John Wesley, *A Least Square Method of Gravity Analysis and Its Applications in the Study of Subsurface Geology.* Ph.D. diss., University of Wisconsin, Madison, 1963.

Nettleton, L. L., Gravity and magnetic calculations, *Geophysics,* v. 7, n. 3, pp. 293–310, July 1942.

Nettleton, L. L., Regionals, residuals, and structures, *Geophysics,* v. 19, n. 1, pp. 1–22, January 1954.

Nettleton, L. L., *Gravity and Magnetics in Oil Prospecting.* New York, McGraw-Hill, 1976.

Plouff, Donald, Gravity and magnetic fields of polygonal prisms and applications to magnetic terrain corrections, *Geophysics,* v. 41, n. 4, pp. 727–739, August 1976.

Talwani, M., and M. Ewing, Rapid computation of gravitational attraction of three-dimensional bodies of arbitrary shape, *Geophysics,* v. 25, n. 1, pp. 203–225, February 1960.

Talwani, M., J. L. Worzel, and M. Landisman, Rapid gravity computations for two-dimensional bodies with application to the Mendocino submarine fracture zone, *Journal of Geophysical Research,* v. 64, pp. 49–59, January 1959.

TEN

Earth Magnetism

Why does a compass point toward north? About four centuries ago, the English scientist William Gilbert (1544–1603) explained this phenomenon by proposing that the earth itself is a large magnet. His experiments with a magnetized globe revealed that its effect on nearby test magnets was similar to the earth's effect on a compass. Later, it was learned that most of the earth's magnetism could be explained in terms of a large bar magnet embedded in the core. Even though such a bar magnet does not actually exist, it is a convenient way to represent the magnetism produced by electric currents in the core. But a single bar magnet does not account for the irregularities we call *magnetic anomalies.* They were discovered first from small variations in compass readings, and later with instruments called *magnetometers,* which measure the strength of earth magnetism. Some very broad magnetic anomalies can be explained by additional magnets in the core, but the smaller anomalies can be understood only in terms of magnetism in the crust. These are the targets of exploration geophysicists, who search for them by magnetometer surveying and then interpret them as differences in rock magnetism associated with geologic structures.

It is convenient to separate the earth's magnetism into two parts. By far the strongest part is the *main magnetic field,* which is produced in the core. Much weaker and more irregular is the *anomalous magnetic field,* which originates in the crust. A magnetometer survey measures their combined effect. Therefore, a geophysicist must account for the dominant influence of the main magnetic field before proceeding with a geologic interpretation of the anomalous field. In some ways, this is the counterpart to the practice in gravity exploration of accounting for the very strong effect of earth flattening and rotation before interpreting the gravity anomalies. The role of the main magnetic field in magnetic exploration is similar to the role of the normal gravity field in gravimetric exploration.

This chapter is concerned with (1) basic principles of magnetism, (2) the design of magnetometers, and (3) the nature of the main magnetic field. With this background, we can proceed in Chapter 11 with magnetic surveying practices and the geologic interpretation of magnetic anomalies.

THE NATURE OF MAGNETISM

Magnetic Poles

Our understanding of earth magnetism is based on ideas about how magnets interact with one another and about how magnetism is produced. The eighteenth-century French physicist Charles Coulomb (1736–1806) described the interaction of magnets in terms of forces acting at points called *magnetic poles.* Every magnet possesses a *positive* pole and a *negative* pole, so named because of their opposite effects on the poles of another magnet.

Like poles of two magnets exert a repelling force on one another, whereas unlike poles exert a force of attraction (Figure 10–1). The force f acting on two poles having values of *pole strength* P_1 and P_2, and separated by a distance r, is expressed by Coulomb's law,

$$f = \frac{1}{\mu}\frac{P_1 P_2}{r^2} \qquad (10\text{–}1)$$

where μ represents the *magnetic permeability,* which is a property of the medium where the magnets are located. In a vacuum, $\mu = 1$, which is very close to its value in the earth's atmosphere. We describe the pole strength in terms of the force exerted by a magnetic pole at a specified distance. A pole that exerts a force of one dyne on another pole one centimeter away is said to possess one *unit of pole strength* ($P = 1$ ups). At this same distance, a two-dyne force would be exerted by a pole with two units of pole strength ($P = 2$ ups).

The two poles of a magnet act oppositely

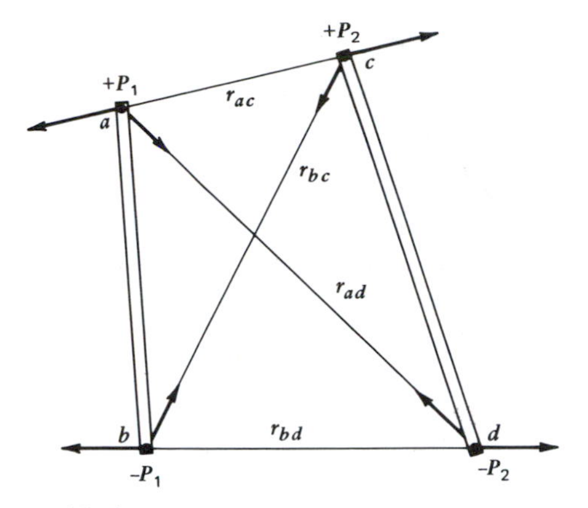

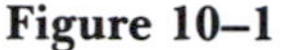

Figure 10–1
Forces acting on the poles of two magnets. These forces are directed along lines connecting pairs of poles that are located near the ends of each magnet and depend on the pole strength and the distance between the poles.

but with equal pole strength. It is not possible to separate or extract either of these poles. To break a magnet is to immediately create two new magnets, each with a positive pole and a negative pole. For this reason, we commonly use the word *dipole* to describe a magnet.

Magnetic Fields

Because a magnet has the capacity to exert force on other magnets or iron objects, it is said to be surrounded by a "field of force," which is called its *magnetic field.* How can we describe a magnetic field? Consider the effect of a large magnet of pole strength P_1 on the positive pole of a small test magnet of pole strength P_2 (Figure 10–2). To keep things simple, we assume that the magnetic permeability $\mu = 1$. According to Coulomb's law, the

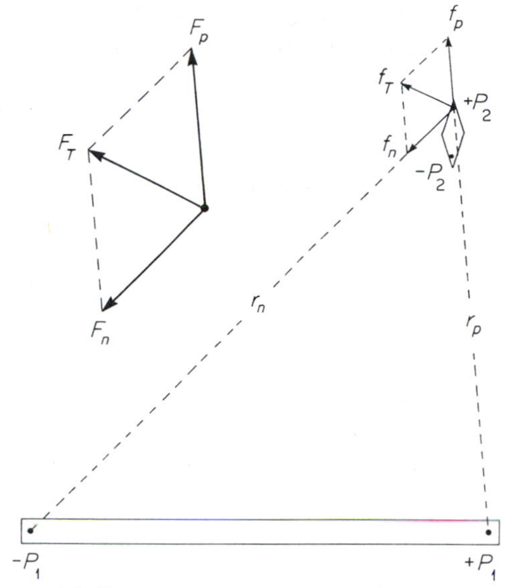

Figure 10–2
Force acting on the positive pole of a test magnet. The force acting on $+P_2$ is the vector sum of the effects of the $-P_1$ and the $+P_1$ poles.

force of attraction from the negative pole of the large magnet is

$$f_n = \left(\frac{P_1}{r_n^2}\right)P_2 \qquad (10\text{–}2a)$$

and the repelling force from its positive pole is

$$f_p = \left(\frac{P_1}{r_p^2}\right)P_2 \qquad (10\text{–}2b)$$

Vectors in Figure 10–2 show the strengths and directions of these forces and how they combine to produce a total force f_T. We can see that these forces depend on the pole strengths of both magnets.

Next, let us introduce a new term called the *magnetic field intensity* to represent separately the effect of the large magnet. The magnetic field intensity from its negative pole is

$$F_n = \frac{P_1}{r_n^2} \qquad (10\text{–}3a)$$

and the magnetic field intensity from its positive pole is

$$F_p = \frac{P_1}{r_p^2} \qquad (10\text{–}3b)$$

Another vector diagram in Figure 10–2 shows how F_n and F_p combine to produce the total magnetic field intensity F_T of the large magnet. We can use this value to express the total force acting on the positive pole of the test magnet:

$$f_T = F_T P_2 \qquad (10\text{–}4)$$

Observe how the effect of the large magnet can be represented by (1) the direction of f_T and (2) the field intensity value of F_T. We use these two features to describe its magnetic field. We can tell from Equations 10–3a and 10–3b that both direction and intensity change as we move the test magnet from place to place. Therefore, the nature of the magnetic field in different places can be illustrated by vectors (Figure 10–3) that point in the directions of f_T at these places. The lengths of the vectors represent the values of F_T.

The units of magnetic field intensity can be found by rearranging Equation 10–4 to get

$$F_T = \frac{f_T}{P_2} \qquad (10\text{–}5)$$

This shows that field intensity must be expressed in units of force divided by pole strength. The basic cgs unit of field intensity is the *oersted,* named for the Danish physicist Hans Christian Oersted (1777–1851), which is defined as

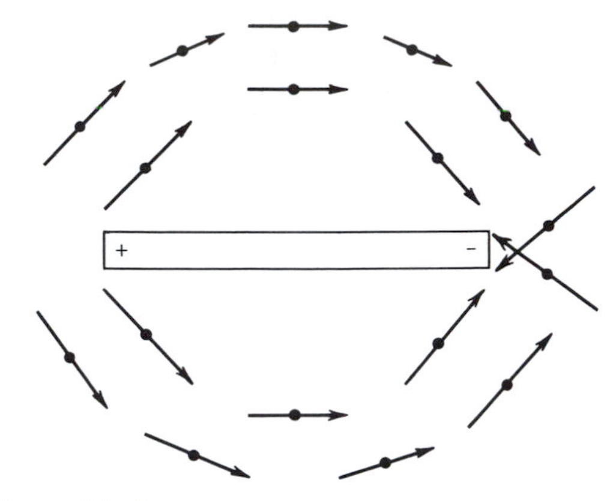

Figure 10–3
Vector representation of the magnetic field surrounding a bar magnet. The field intensity and direction are represented by the length and direction of vectors at different points in the region surrounding the magnet.

$$1 \text{ oersted} = \frac{1 \text{ dyne}}{1 \text{ unit of pole strength}}$$

Because most magnetic anomalies of interest to exploration geophysicists amount to a very small fraction of one oersted, a subunit called a gamma (γ) is more convenient to use. It is defined as

$$1 \text{ gamma} = 10^{-5} \text{ oersted}$$

Another way to display the direction and intensity of a magnetic field is by a pattern of lines that appear to converge toward the poles of the magnet (Figure 10–4). These are called *magnetic lines of force* or *magnetic flux lines.* For a magnet that is a long thin rod, the lines of force converge very close to its ends. But if the magnet is a more equidimensional bar or a sphere, the lines of force point toward interior poles. The orientation of these lines of force indicates the direction of f_T at all locations surrounding the magnet. Intensity is represented

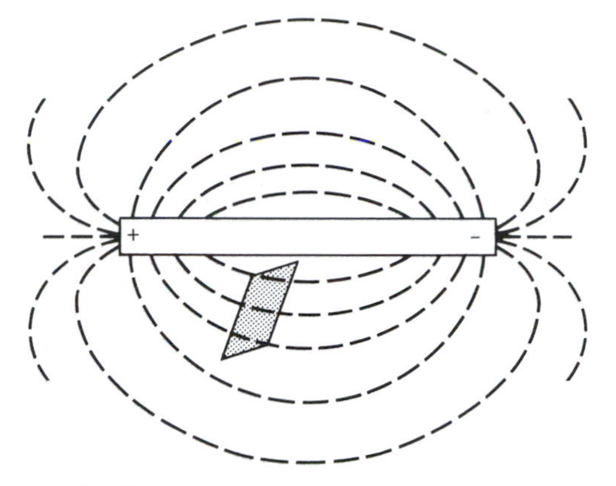

Figure 10–4
Representation of the field of a bar magnet by lines of force. Field direction at any point is indicated by the direction of the line of force at that point, and field intensity is indicated by the spacing of the nearby lines. Field intensity can be illustrated by the number of lines passing perpendicularly through a unit area, for example, the shaded area.

by the spacing of lines. The closely spaced lines near the poles illustrate high field intensity. Farther away, where they are more widely spaced, a weaker field is indicated.

So far, we have described the field of a single dipole magnet. Now let us consider the field produced by a combination of magnets. To find the intensity and direction of this composite field at some location, we can combine the vectors representing the fields of the individual magnets (Figure 10–5). There is no limit to the number of individual dipole fields that can be combined. Depending on how the magnets are arranged, lines of force illustrating the composite field can display complicated patterns.

It is interesting to note some similarities in the ways we describe and analyze magnetism and gravity. The magnetic force at two points said to possess pole strength is expressed by Coulomb's law (Equation 10–1); the equation for this law has the same form as that for the universal law of gravitation (Equation 7–1), which describes the force at two points possessing mass. Our way of expressing the separate effect of one magnet in terms of its magnetic field (Equations 10–2 and 10–3) is equivalent to our description of the separate gravimetric effect of a particle in terms of its field of acceleration (Equations 7–2 and 7–3). The combined effect of many magnets can be found by adding vectors (Figure 10–5), which is the same thing we do to find the gravity field produced by many particles (Figure 7–3). But magnetism tends to be more complicated because the basic element is a dipole which exerts a combination of attracting and repelling

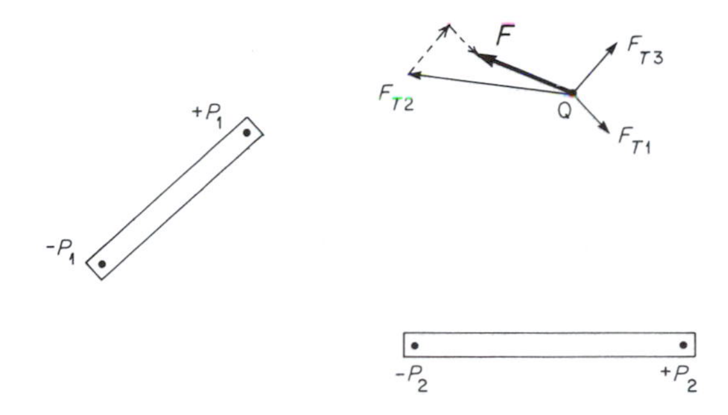

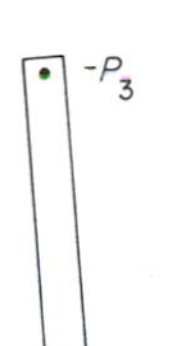

Figure 10–5
Magnetic field **F** at location Q is the vector sum of the individual magnetic fields F_{T1}, F_{T2}, and F_{T3} of three bar magnets.

forces, whereas the basic gravitational element is a particle that exerts only a force of attraction.

Electromagnetism

Early in the nineteenth century, Hans Christian Oersted observed that a wire carrying electric current could change the bearing of a compass needle. Additional experiments revealed that the magnetic field of a dipole of pole strength P and length ℓ could be duplicated by passing electric current through a coil of wire. This could be done by choosing the current i, the number of turns of wire n, and the cross-sectional area A of the coil so that

$$P\ell = niA \qquad (10\text{–}6)$$

It is convenient here to introduce a vector **M**, called the *magnetic moment*, to describe the magnetism of a dipole or an equivalent coil:

$$\mathbf{M} = \mathbf{P}\ell = n\mathbf{i}A \qquad (10\text{–}7)$$

The direction of this vector is in line with the two poles of the magnet or along the axis of the coil. Be sure to understand that the magnetic moment vector is different from a vector that is used to represent the magnetic field intensity and direction at some location. It represents the magnet that produces such a field.

Because the inside of the coil is open, it is possible to see what happens to converging lines of force that cannot be followed into the solid interior of a magnet. The lines of force do not converge exactly to the points we call poles. Instead, they extend through the interior as continuous lines that remain closely spaced and nearly parallel (Figure 10–6). The apparent poles do not actually exist, but for practical purposes, they can be assumed to be situated near the ends of a magnet where the apparently parallel lines of force noticeably begin to spread apart.

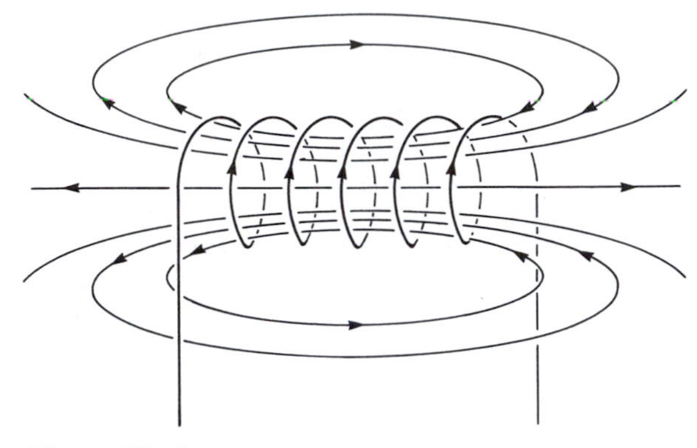

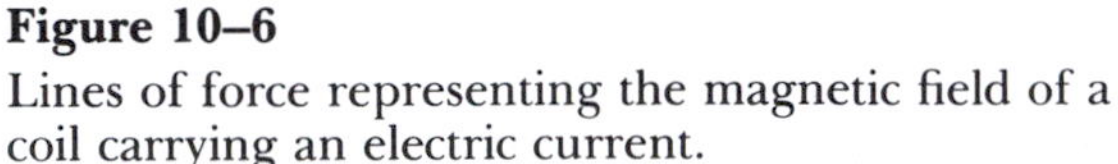

Figure 10–6
Lines of force representing the magnetic field of a coil carrying an electric current.

Let us now reexamine the units of magnetic field intensity in a way that makes use of these concepts of electromagnetism. By rearranging Equation 10–7, we obtain

$$P = niA/\ell$$

which tells us that pole strength can be expressed as the product of current and length. The SI units are ampere-meters. We know from Equation 10–5 that field intensity can be expressed in terms of force divided by pole strength. Therefore, substitution from Equation 10–7 into Equation 10–5 shows that field intensity can be expressed by units of force divided by the product of current and length. The basic SI unit of field intensity is the *tesla* (T), which is defined as

$$1 \text{ tesla} = 1 \frac{\text{newton}}{\text{ampere-meter}}$$

In the cgs system, the oersted, which we have already introduced, can be defined as

$$1 \text{ oersted} = 1 \frac{\text{dyne}}{\text{abampere-centimeter}}$$

where 1 abampere equals 10 amperes. By comparing these two units of field intensity, we find that

$$1 \text{ tesla} = 10^4 \text{ oersteds}$$

For purposes of exploration geophysics, it is convenient to introduce an SI subunit called the *nannotesla* (nT), which equals 10^{-9} tesla. Recalling that 1 gamma (γ) equals 10^{-5} oersted, we see that 1 gamma is numerically equal to 1 nannotesla. These two subunits are currently used interchangeably for describing magnetic field intensity.

Ferromagnetism

Why does a magnet possess a field of force? Basically, this field is created by internal electric currents associated with the atomic structure of the substance. An electric current consists of charged particles in motion, principally electrons. Each negatively charged electron in an atom orbits the nucleus and at the same time appears to spin on an axis like the earth turns on its axis while following its orbit around the sun. Positively charged protons in the nucleus also appear to be spinning. Because of these orbital and spin motions, each electron and proton produces its own dipole magnetic field. However, most atoms do not seem to be magnetized because the electron and proton magnetic moments are randomly oriented and cancel one another. This is not true for iron, the only abundant element in the earth that possesses significant atomic magnetism. Nevertheless, most substances containing iron are nonmagnetic because of the random orientations of the iron atoms in their crystal structures. Pure iron and the minerals *magnetite* (Fe_3O_4), *ilmenite* ($FeTiO_3$), and *pyrrhotite* ($Fe_{1-x}S$) are different. These substances and a few others can acquire and retain magnetism. They are called ferromagnetic substances.

The orientation of iron atoms in a ferromagnetic substance depends on the balance of chemical bonding energy and heat energy. The bonding energy is used to align their magnetic moments, but the heat energy causes the atoms to vibrate out of line. At high temperatures, the heat energy is dominant so that the atomic moments remain randomly oriented and neutralize one another, making the substance nonmagnetic. But at temperatures below a critical point called the *Curie temperature,* the bonding energy prevails. Groups of iron atoms become more or less locked in position with their magnetic moments parallel to one another. Such a group occupies a zone, usually a few microns in diameter, which is called a *magnetic domain.* The entire ferromagnetic substance consists of magnetic domains (Figure 10–7), each of which is a small but strongly polarized magnet.

The first magnets were made from a naturally magnetized substance called *lodestone,* which turned out to contain mostly magnetite. Later, ways were discovered for making magnets from iron. These objects are magnets because a large proportion of their domain moments are aligned. But most iron objects do not appear to be magnetized, for they possess randomly oriented domain moments that neutralize one another.

What influences the orientation of domain moments? Most important is the effect of another magnetic field which acts to strengthen the moments of favorably oriented domains. The fields of neighboring domains, a nearby

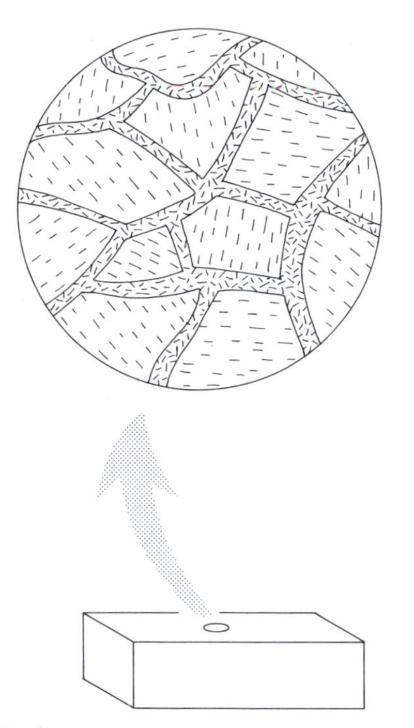

Figure 10–7
Magnetic domains in a ferromagnetic substance. Within each domain, the atomic magnetic moments tend to be aligned. Walls separating the domains are disordered zones of nonaligned atomic moments.

magnet, and the earth's magnetic field all contribute.

Perhaps the strongest alignment of domain moments can be produced by first heating a ferromagnetic substance above its Curie temperature. The heat destroys the existing domains, and then the substance is cooled in the presence of a magnetizing field. As it cools through the Curie temperature, new domains form with a large proportion of the moments in the direction of this field. This preferred alignment produces a *permanent magnetism* in the substance which persists even if the initial magnetizing field is removed.

It is possible to magnetize the substance without raising its temperature. Again, this is done by placing the substance in a magnetizing field. The result is that domains with moments more or less parallel to that field tend to grow at the expense of less favorably polarized domains. To see how this works, we must look closely at the domain structure. Within the domains illustrated in Figure 10–7, the uniformly oriented atomic magnetic moments represent a highly ordered structure. The domains are separated by "walls" of highly disordered atomic moments. One domain grows by capturing ferromagnetic atoms from its neighbors. Energy is required to disrupt the ordered structure in such a neighbor, even if it is unfavorably polarized. Still more energy is needed to reorient these additional atoms into the proper direction in the growing domain. During this process, the domain walls appear to move. The magnetizing field introduces the energy used to move domain walls. The extent to which the substance acquires and retains magnetism depends on the distances that the domain walls are moved.

Suppose we use a coil of wire carrying electric current to produce a magnetizing field F around an unmagnetized bar of iron or some other ferromagnetic substance. The immediate growth of favorably polarized domains creates magnetism in that bar. Let us introduce a property called the *magnetic moment per unit volume, I,* to describe that magnetism. As this new term implies, the value of I is the magnetic moment possessed by each unit of volume in the bar, for example, each cubic centimeter.

The intensity of I depends on the intensity of the magnetizing field F that can be changed by varying the coil current. Values of I corresponding to *small* changes in F are seen in Figure 10–8 to plot along a straight line. This tells us that

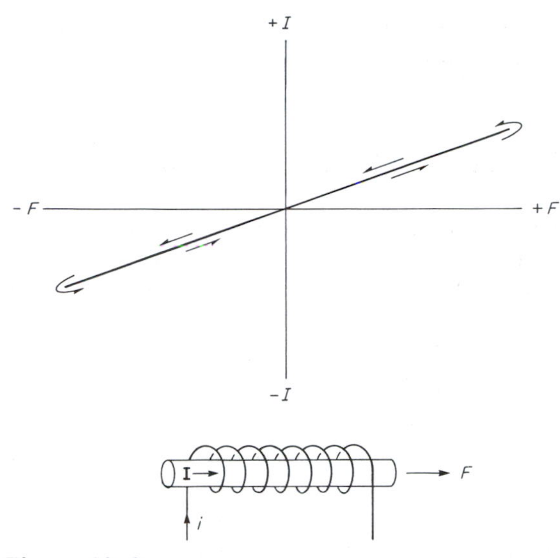

Figure 10–8
Induced magnetic moment per unit volume **I** in a bar that varies in direct proportion to the intensity of the magnetizing field F produced by electric current in the coil. The slope of the straight line is the magnetic susceptibility k of the bar.

$$I = kF \tag{10–8}$$

where the constant of proportionality k is the slope of the line. This constant is called the *magnetic susceptibility* of the substance and indicates its capacity to acquire magnetism. Apparently, the domain walls can adjust easily to small changes in the magnetizing field. Observe, however, that by reducing F to zero, I also becomes zero. This means that no permanent magnetism is retained. This kind of magnetism, which is directly proportional to the magnetizing field, is called *induced magnetism*.

Now let us test how the magnetism of the bar changes in response to *large* variations of the magnetizing field. Suppose that we begin with an unmagnetized bar and a coil carrying no current so that $I = 0$ and $F = 0$. Then, by increasing the current, we create an increasing magnetizing field F. The corresponding change of I in the bar is illustrated in Figure 10–9 by the curve between points 1 and 2. The value of I reaches a maximum at point 2 where the bar is said to become *magnetically saturated*. Further increase of F will not change I because no more growth of favorably oriented domains is possible.

Next, we reduce the magnetizing field which causes I to decrease according to the curve between points 2 and 3 in Figure 10–9. Be sure to observe that at point 3, where $F = 0$, some permanent magnetism I remains in the bar. To demagnetize the bar completely, we must reverse the coil current to produce a magnetizing field in the opposite direction. This step is illustrated by the curve between points 3 and 4 in Figure 10–9.

Continuing along the curve in Figure 10–9 from point 4 to point 5, we see how by increasing this oppositely directed magnetizing

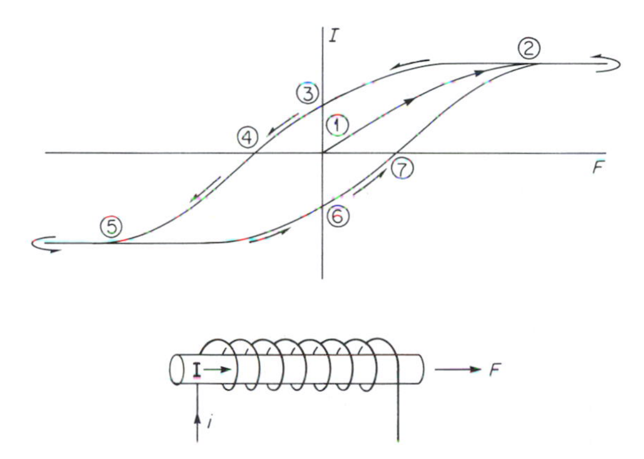

Figure 10–9
Hysteresis curve showing how the magnetic moment per unit volume **I** in a ferromagnetic substance varies with the magnetizing field F produced by electric current in the coil. The sequence of numbers indicates the magnetization cycle.

field, the value of I in that direction increases until magnetic saturation is reached. Then the decrease in F between points 5 and 6 causes I to decrease. Again, permanent magnetism remains at point 6 where $F = 0$. This magnetizing field must then be increased in the opposite direction to point 7 to demagnetize the bar.

These steps illustrate that the magnetism in the bar does not change in a reversible way. We see that I appears to "lag" behind F so that the bar retains some permanent magnetism when the magnetizing field is diminished to zero. This lagging effect is called *hysteresis*. Why does hysteresis occur for large variations of F, whereas for small variations the corresponding changes of I plot along a straight line (Figure 10–8)? Apparently, the balance of bonding energy and thermal energy can produce freely reversible movements of domain walls in response to small changes in F. But larger changes require much greater shifts of the domain walls, and these shifts are impeded by impurities and imperfections in the crystal structure of the substance. These impediments to free movement of domain walls cause the hysteresis effect.

MEASURING THE EARTH'S MAGNETISM

The first scientific surveys of earth magnetism were done with compasses and other delicately balanced test magnets. The purpose was to find the direction of the magnetic field at a large number of locations on the earth's surface. During the past century, much more attention has been focused on measuring field intensity. Magnetometers developed for this purpose are examples of remarkable ingenuity. There are several kinds, including instruments designed for continuous operation in magnetic observatories and portable magnetometers intended for exploration surveying.

The Earth's Magnetic Elements

A vector is used to represent the earth's magnetic field at an observation site. This vector is described by a combination of seven quantities we call the *magnetic elements*. They are illustrated in Figure 10–10. Magnetic *declination, d,* is the angle between geographical north and the direction of magnetic north that is indicated by a compass. *Inclination, i,* is the angle between the field direction and a horizontal plane. The total field intensity F can be separated into a vertical component Z and a horizontal component H in the magnetic north direction. That horizontal component can be separated into an intensity component X in the geographical north direction and an intensity component Y in the geographical east direction. These seven magnetic elements are related in the following ways:

$$F^2 = H^2 + Z^2 = X^2 + Y^2 + Z^2$$

$$F = \frac{H}{\cos i} = \frac{Z}{\sin i} \qquad (10\text{–}9)$$

$$\cos d = X/H$$

$$\sin d = Y/H$$

Any combination of three magnetic elements is sufficient to completely describe the earth's magnetic field vector at an observation site. Several different kinds of instruments have been invented to measure one or another of these basic elements.

Mechanical Magnetometers

Elements of the earth's magnetic field can be measured with test magnets in various me-

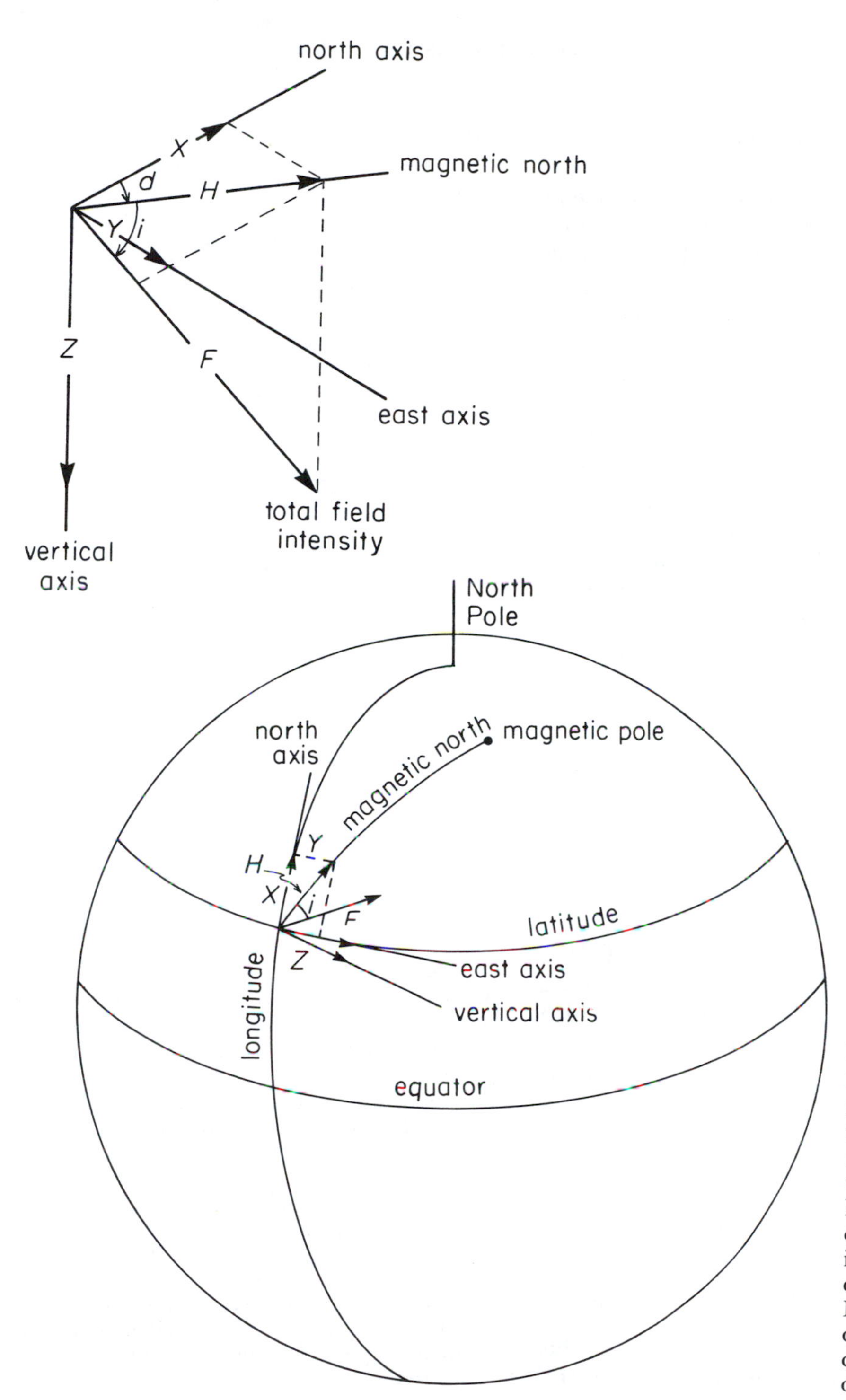

Figure 10–10
Elements of the earth's magnetic field include total field intensity (*F*), north (*X*), east (*Y*), vertical (*Z*), and horizontal (*H*) intensity components, and inclination (*i*) and declination (*d*) angles. Each element can be determined from any combination of three other elements.

chanical mountings. The compass was the first instrument of this kind. Another instrument, the *dip needle,* consists of a magnetized needle mounted on a horizontal axis so that it can rotate freely in a vertical plane. These simple devices are direction indicators. More is needed to measure field intensity. Field intensity can be measured by introducing some system of compensating forces to counteract the effect of the earth's field on the test magnet. Several systems have been used successfully in different magnetometers developed during the past two centuries. By now, most of these instruments have been replaced by electronic magnetometers, which achieve equal or better precision and are easier to operate. Nevertheless, mechanical magnetometers are still used in some permanent observatories and occasionally for exploration surveys.

Perhaps the most conventional procedure in observatories has been to measure the elements H, d, and i (Figure 10–10), from which the other elements can be calculated. A two-step method for finding H was developed by the well-known German scientist and mathematician Carl Friedrich Gauss (1777–1855). First, an experiment is done with a test magnet suspended from a fiber so that it is free to rotate horizontally like a compass (Figure 10–11a). The magnet is disturbed, which causes it to oscillate back and forth before coming to rest in the direction of magnetic north. The period of oscillation τ depends on the earth's horizontal component of field intensity H, the magnetic moment M_T of the test magnet, and the moment of inertia L_T of the test magnet, which can be calculated from its dimensions and point of suspension:

$$\tau = \pi\sqrt{L_T/HM_T} \qquad (10\text{–}10)$$

After τ is measured in the experiment, this equation is used to calculate the product HM_T.

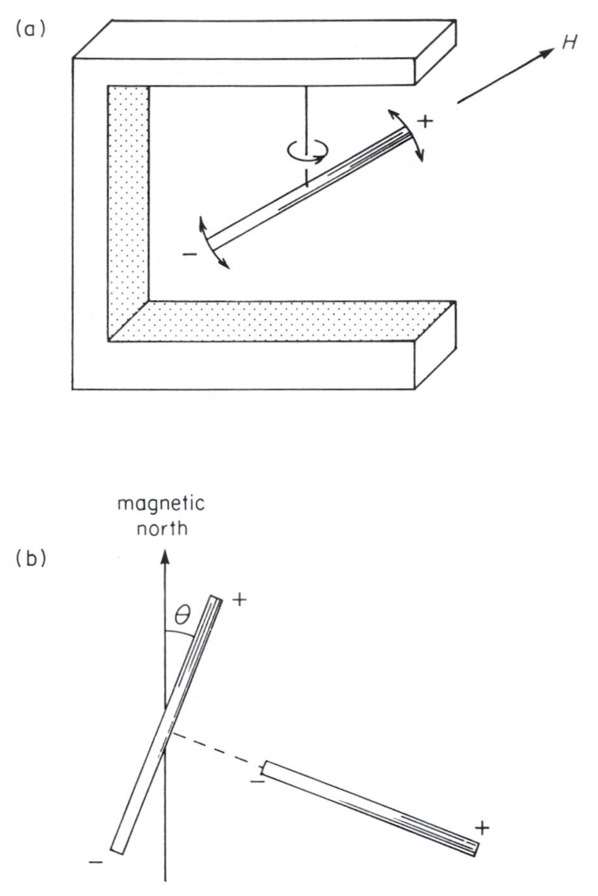

Figure 10–11
Measurement of the horizontal component (H) of the earth's magnetic field intensity by the method of Gauss involves two experiments. (a) In the oscillation experiment, a test magnet is suspended on a fiber so that it is free to swing in the horizontal plane. It is put into motion, and its period of oscillation τ is measured. (b) In the deflection experiment, a second test magnet is mounted on a pivot, like a compass, to rotate in the horizontal plane. The original test magnet is then moved by trial and error into a position in this plane so that it deflects the second magnet from its magnetic north orientation into a direction perpendicular to the alignment of the original magnet. The angle of deflection θ is measured.

A second experiment is conducted to separate the value H from the product HM_T. This experiment involves another test magnet mounted to rotate horizontally. The first test magnet, originally used in the oscillation experiment, is removed from its suspension. It is relocated by trial and error until a position is found where it is perpendicular to the second magnet and points at its center (Figure 10–11b). The purpose is to find the angle θ which is the deflection of the second magnet from magnetic north caused by the balance of forces acting on it. For this configuration of test magnets, we have the relation

$$\frac{H}{M_T} = \frac{C}{\sin\theta} \qquad (10\text{–}11)$$

where the constant C can be calculated from the dimensions and positions of the two magnets. We can combine and rearrange Equations 10–10 and 10–11 to obtain

$$M_T = \frac{\pi^2 K}{H\tau^2} = \frac{H \sin\theta}{C} \qquad (10\text{–}12a)$$

and

$$H^2 = \frac{\pi^2 CK}{\tau^2 \sin\theta} \qquad (10\text{–}12b)$$

To determine the magnetic elements, we measure τ and θ experimentally and then calculate H using Equation 10–12b. The angle of declination d can be found from a compass bearing and the direction of geographical north established from transit sightings on the sun or appropriate stars. Finally, the inclination i can be determined by observing the angle of a dip needle that is oriented so that its vertical plane of rotation contains the magnetic north direction. Inclination can also be measured with an electronic device called an *earth inductor*. Current induced in a spinning coil by the earth's field is reduced to zero by adjusting the spin axis, which is in the plane of the coil, so that it is aligned with the earth's field.

The method of Gauss is impractical for exploration geophysical surveying. The instruments are difficult to transport, and too much time is required to operate them. Other mechanical magnetometers are more efficient for this work. Like gravimeters, these magnetometers measure changes in field intensity between different observation sites rather than absolute values of intensity. Of those still in use, the *torsion magnetometer* is the most convenient to operate. It is sensitive to change in the vertical component, ΔZ, of the earth's field intensity.

In design, the torsion magnetometer is quite simple (Figure 10–12). A small test magnet is attached to a horizontal fiber so that it can be rotated in a vertical plane. One end of the fiber is fixed to the frame of the magnetometer, and the other end is connected to a shaft that can be turned. Observe in Figure 10–12 that if the magnet is inclined, both vertical and horizontal components of the earth's field act to rotate it. But the magnet can also be rotated mechanically by twisting the fiber. To operate the instrument, we turn the shaft until the magnet is horizontal. In this position, only the vertical field intensity component can act to rotate it. This effect is exactly balanced by the twist of the elastic fiber. A pointer fixed to the shaft indicates on a dial the angle of twist in the fiber required to maintain the magnet in the horizontal position. This angle depends on the elastic shear modulus of the fiber. When the instrument is moved from one site to another, the change in the angle of twist is proportional to the change, ΔZ, in vertical field intensity.

The torsion magnetometer is calibrated by

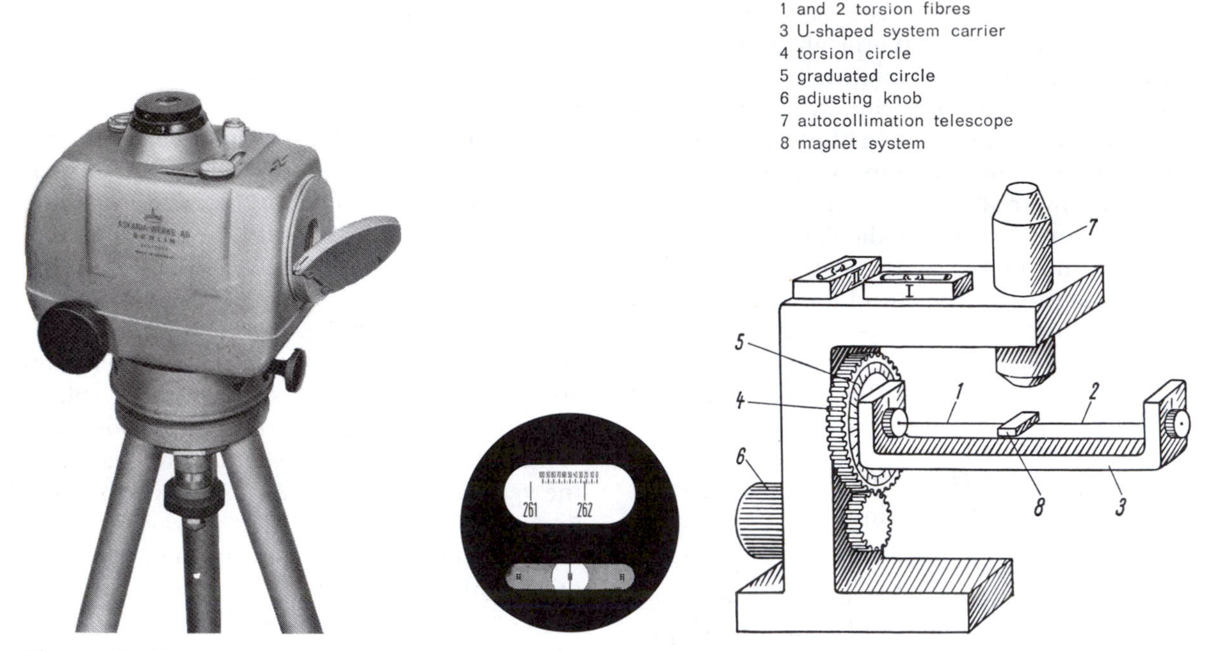

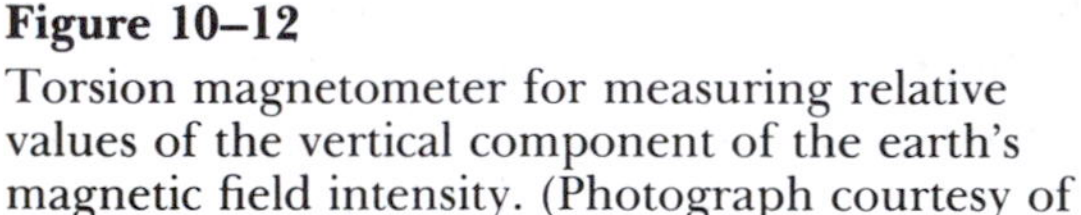
Figure 10–12
Torsion magnetometer for measuring relative values of the vertical component of the earth's magnetic field intensity. (Photograph courtesy of Askania Werke.) The small test magnet is maintained in a horizontal orientation by twisting the fiber to which it is attached.

placing a coil of wire around it. The coil axis is vertical so that a vertical magnetic field can be produced by passing current through the coil. Now the vertical field intensity close to the magnetometer can be changed simply by varying the coil current, and the corresponding change in the twist of the torsion fiber can be observed. Dial readings *(R)* for different intensities can be plotted. The slope *(Q)* of a straight line fitted to the points is the constant of proportionality for converting differences in dial reading (ΔR) into differences in vertical field intensity:

$$\Delta Z = Q\,\Delta R \qquad (10\text{–}13)$$

A tripod-mounted torsion magnetometer is simple to operate. The instrument can be leveled and a reading obtained in one to two minutes. The instrument can detect field intensity changes of less than one gamma.

Saturation Induction Magnetometers

During World War II, a submarine detection device was invented that could be operated in low-flying aircraft. It was sensitive to the small perturbation of the earth's magnetic field intensity produced by the iron in a vessel submerged at shallow depth. This device is called a *flux-gate magnetometer.* It is an electronic instrument consisting of two metal rods that are oppositely magnetized by coils wound around them and that can induce current in another coil when the magnetic balance is altered by the earth's field. The potential of such a device for geophysical exploration was recognized even before its application for submarine detection. Shortly after the war, some design modifications were introduced so that the flux-gate magnetometer could be used for mapping the earth's field intensity. A very important advantage over mechanical magnetometers is that it can be mounted for continuous operation in an aircraft, ship, or any other moving vehicle. Unlike a gravimeter, this kind of magnetometer is not affected by motion or changes in elevation.

The design of the flux-gate magnetometer is based on the principle of electromagnetic induction. Recall how Hans Christian Oersted recognized that a magnetic field is produced by passing electric current through a wire conductor. Experiments also revealed that current could not be induced to flow in a coil of wire simply by placing it in an existing magnetic field. For current to flow, the magnetic field intensity around the coil must be changing. This can be accomplished by placing it in another coil carrying an alternative current that produces an alternating magnetic field β. This time-varying field induces alternating current in the first coil that is proportional to the time rate of change of the cyclic magnetic field, $\Delta\beta/\Delta t$.

The important parts of a flux-gate magnetometer are shown in Figure 10–13. Primary coils connected in series to an alternating-current source are wound around each of two identical parallel rods. These coils are oppositely wound but otherwise similar. The rods are made of an iron–nickel alloy such as Mumetal, which can be magnetized to saturation by a weak magnetizing field that is not much stronger than the earth's field. A secondary coil connected to a source of direct current is wound around this set of rods and their primary coils.

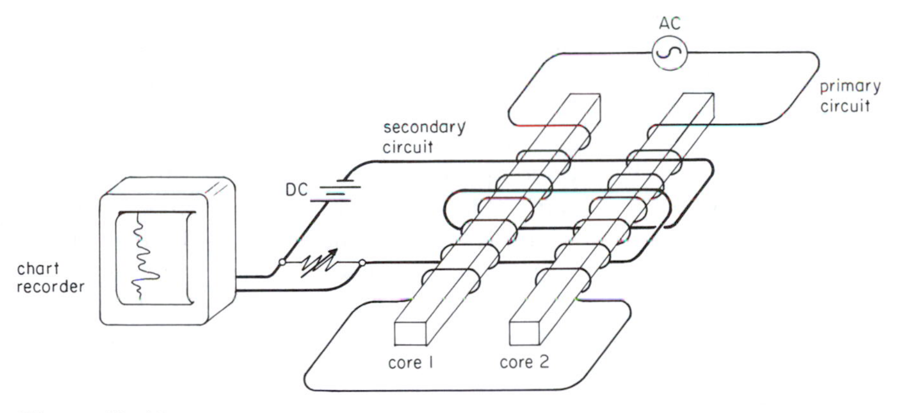

Figure 10–13

Principal components of a flux-gate magnetometer. Oppositely wound primary coils carrying alternating current together with the earth's magnetic field induce magnetism into the two metal cores. Magnetism of the cores induces current in the secondary coil. Direct current produced in the secondary coil by the battery is adjusted to balance the effect of the earth's field.

The flux-gate magnetometer can be operated by adjusting the direct current in the secondary coil to produce a magnetic field that cancels the earth's field inside the coil. The adjustment is controlled by means of the rods with their primary windings that function as a sensing unit. This unit induces pulses of alternating current in the secondary coil that activate a direct-current adjustment circuit. When the direct current is exactly sufficient to cancel the earth's field, the sensing unit ceases to induce alternating-current pulses. A reading of this direct current can be used in Equation 10–7 to find the magnetic moment of the secondary coil that exactly balances the earth's field. The earth's field intensity can be obtained from this value.

How does the sensing unit function? Graphs in Figure 10–14 illustrate what happens when the rods are aligned with the earth's field **F.** Graph a shows the time-varying magnetizing fields, β_1 and β_2, produced by alternating current in the primary coils. Notice that these cyclic curves will be offset vertically for a system situated in the earth's magnetic field, because the steady field **F** reinforces and opposes alternate half cycles of β_1 and β_2. The curves cycle oppositely because of the reverse windings of the coils.

These magnetizing fields induce in each rod time-dependent magnetism that varies according to the hysteresis curve in graph b. For each cycle of β_1, the magnetic moment per unit volume I_1 in rod 1 follows the sequence of points a, b, . . ., o, p. The corresponding cycle of β_2 induces the magnetic moment per unit volume I_2 in rod 2 in the sequence of points i, j, k, . . ., g, h. Notice that the sequence k, l, m, n, o, p indicates magnetic saturation in the direction of **F,** whereas the shorter sequence of d, e, f, g indicates saturation in the opposite direction. Curves of I_1 and I_2 in graph c show that each half cycle in which **F** reinforces the primary magnetizing field reaches saturation sooner and remains saturated longer than the opposite half cycle in which **F** and the primary magnetizing field are reversed.

The combined magnetism of the two rods induces current in the secondary coil. This combination in graph d exhibits pulses of magnetism in the direction of **F,** which are asymmetrically shaped because of the hysteresis effect. These pulses are interspersed with intervals when the rods are magnetically saturated in opposition and balance one another. Alternating-current pulses in graph e are produced only when the magnetic field is changing. Observe how current flows in one direction while the field increases during the first half of a pulse, and then reverses direction as the field decreases.

The alternating-current pulses activate a circuit that introduces direct current in the secondary coil. This direct current is increased until the magnetic field it produces is sufficient to nullify **F.** As a result, the curves of time-varying magnetism in graphs a and c shift into exact opposition so that the alternating-current pulses disappear from the secondary coil. At this point, the adjustment circuit is no longer stimulated to change the direct current. The strength of this direct current, which is proportional to F, is then recorded. If a high alternating-current frequency of about 1000 hertz is used, the balance is reached quickly. We can therefore measure the earth's field intensity almost continuously with a flux-gate magnetometer.

So far, we have described how to measure the intensity F by aligning the sensing unit in the direction of the earth's field. But a very useful feature of the flux-gate magnetometer is its capacity to measure any component of

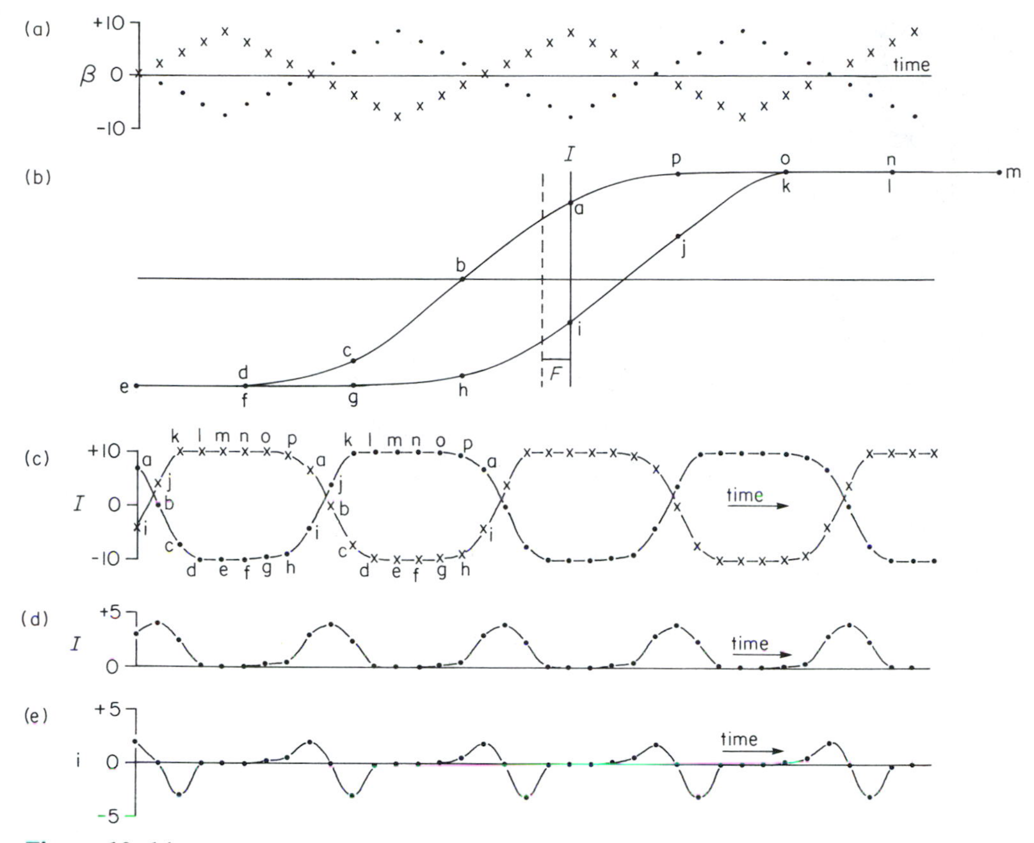

Figure 10–14

Operation of a flux-gate magnetometer. (a) Cyclic magnetic fields produced by alternating current in primary coils 1 (•) and 2 (*x*). (b) Hysteresis curve for magnetization of ferromagnetic cores. (c) Cyclic magnetic moment per unit volume in core 1 (•) and core 2 (*x*). (d) Combined cyclic magnetic moments of cores 1 and 2. (e) Current induced in secondary coil by variation of the combined magnetic fields of cores 1 and 2.

field intensity simply by orienting the sensing unit in the appropriate direction. In a magnetic observatory, we can mount north, east, and vertically directed units to measure the elements X, Y, and Z. For exploration purposes, several portable instruments are available in which the sensing unit is suspended to hang vertically. The one shown in Figure 10–15 can be set up and read in a few moments to obtain a measurement of vertical field intensity that is accurate within one gamma.

For airborne surveying in which the flux-gate magnetometer is in continuous operation, a mechanism is needed to maintain the orientation of the sensing unit. It is most practical to align with the earth's field direction,

Figure 10–15
Hand-carried flux-gate magnetometer for measuring the vertical component of the earth's magnetic field intensity. (Courtesy of Sharpe Instrument Co.) The sensing element is suspended in a vertical orientation.

which means that total field intensity F will be measured. Proper alignment can be maintained automatically by suspending three mutually perpendicular sensing units in a gimbal mounting. When one unit is aligned with **F**, the other two units must lie in the perpendicular plane along which the magnetic field intensity is zero. As they shift out of this plane, the magnetism they begin to detect activates a system of servomotors that realign the units. The earth's field intensity is determined by recording the direct current in the secondary coil of the unit aligned with **F**.

The earth's magnetic field is distorted by metal parts in an aircraft. A magnetometer must be mounted so that these distortions can be accounted for or otherwise reduced to tolerable levels. The simplest practice is to mount the instrument in a streamlined container called a "bird" which is trailed on a long cable behind the aircraft. The magnetometer can also be operated in a long, hollow nonmagnetic cone, called a "stinger," which extends out from the tail or nose of the aircraft. Ordinarily, this does not displace it far enough from the disturbed field surrounding the aircraft, so that properly mounted compensating magnets must be used to nullify the larger distortions. The total field intensity can be measured within an accuracy of one gamma with a modern airborne flux-gate magnetometer.

Proton Precession Magnetometers

A proton possesses distinctive properties that make it useful as a test magnet in an entirely different kind of magnetometer. Every proton is a spinning mass carrying a positive electric charge. This motion of a charged particle creates a magnetic moment in the direction of the spin axis. Because it is spinning, the proton also possesses angular momentum.

The property of angular momentum is what keeps a spinning top from falling over. If it is tilted, the top remains upright, but its axis moves around a cone (Figure 10–16a). This motion is called precession. It results from the balance between gravity, which would cause the top to fall over, and the angular momentum, which would hold the axis in a fixed position.

A proton responds to a different balance, but the result is the same. The earth's magnetic field, rather than gravity, would cause the proton magnetic moment to fall in line. This effect is counteracted by the proton angular momentum, which would maintain the position of the spin axis. As a result, the proton axis precesses around a cone pointing in the direction of the earth's magnetic field (Figure 10–16b). The number of cycles of the axis around the cone in a unit of time is called the precession frequency f. It depends on the magnetic moment M and the angular momentum L of the proton and on the intensity F of the earth's field,

$$f = \frac{M}{2\pi L}F = \frac{GF}{2\pi} \qquad (10\text{–}14a)$$

where the *gyromagnetic ratio* G is a constant for all protons. Nuclear physicists have devised experiments from which they have obtained the very accurate measurement of G =

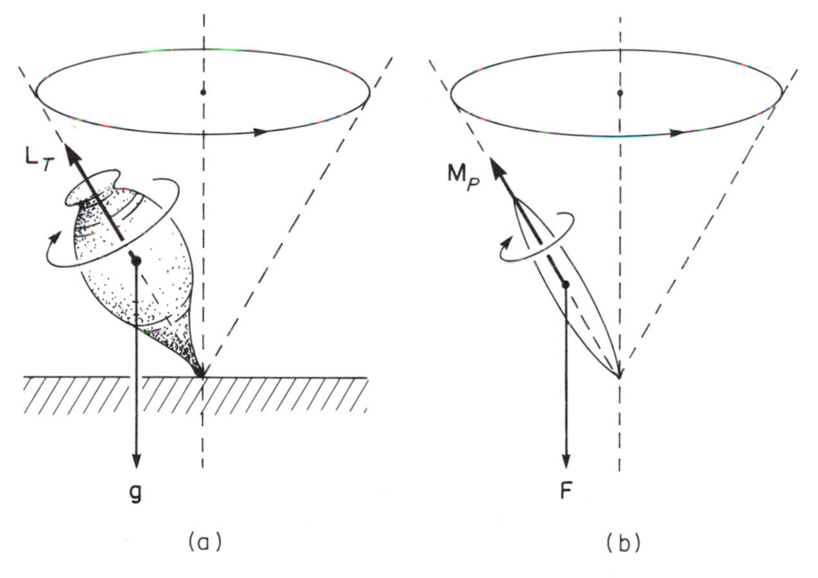

Figure 10–16
(a) Precession of a top in response to the balance of its spin angular momentum vector $\mathbf{L}_T$ and the gravity vector g acting at its center of mass.
(b) Precession of a proton in response to the balance of the magnetic field vector $\mathbf{F}$ and of the proton magnetic moment vector $\mathbf{M}_P$, which results from its electric charge and its spin angular momentum.

0.267513/gamma-s. Knowing this factor, we can rearrange the formula to find the field intensity in terms of the precession frequency:

$$F = 2\pi f/G = 23.4874f \quad (10\text{–}14b)$$

This formula tells us that an instrument designed to measure the precession frequency could be used to measure the earth's field intensity. Such an instrument is called a proton precession magnetometer.

The protons used in a magnetometer exist in water and in all the hydrocarbon fluids. Quite simply, they are the small proportion of dissociated hydrogen ions that have separated from H_2O molecules or other hydrocarbon molecules. Water and heptane ($CH_3[CH_2]_5CH_3$) have proved to be the most suitable proton sources for reasons we mention later. The apparatus illustrated in Figure 10–17 can be used to harness these protons for field intensity measurements. A coil wound around a container of water or heptane can be connected by means of a two-position switch to either a direct-current source

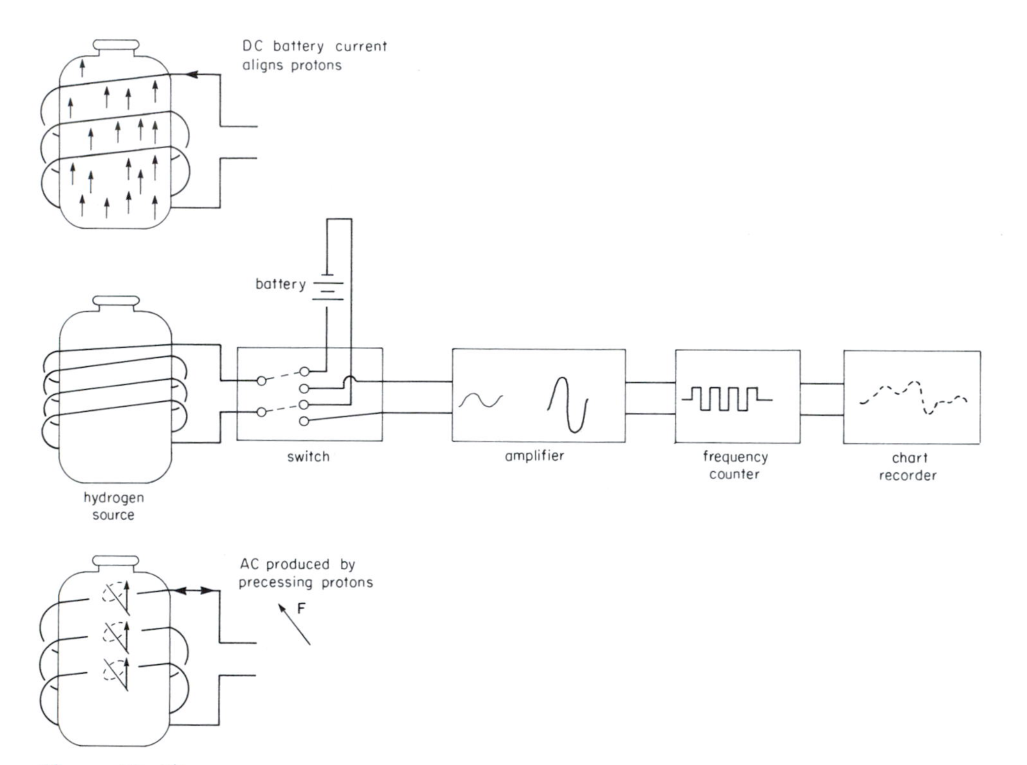

Figure 10–17
Principal components of a proton precession magnetometer. The switch alternately connects the coil to a source of direct current needed to align magnetic moments of protons in the container, and then to a circuit that measures the frequency of alternating current induced into the coil by the magnetic fields of precessing protons.

or a frequency-counting circuit. The coil can be oriented in any direction other than that of the earth's field, but the strength of the signal that is processed in the frequency counter is enhanced by a more nearly perpendicular orientation.

Field intensity is measured in a two-step procedure. First, the coil is switched to the direct-current source, which produces a magnetic field much stronger than the earth's field. It strongly polarizes the protons in the direction of the coil. After 1 to 2 seconds of operation, the direct-current source is disconnected, and the coil is switched to the frequency-counting circuit. This eliminates the strong polarizing field so that the protons immediately begin to precess around the direction of the earth's field. The motion of the proton magnetic moments produces a cyclic magnetic field that induces alternating current in the coil. This alternating current will persist for 2 to 3 seconds before the protons become disoriented by ordinary molecular motions in the water or heptane. This is sufficient time for the frequency-counting apparatus to measure its frequency, which is the same as the proton precession frequency. The value is converted to field intensity units and then transmitted to a recorder or a meter that can be read visually.

Proton precession magnetometers have been designed for operation in aircraft, ships, and other moving vehicles. Hand-carried instruments are also available for ground surveying. Examples are pictured in Figure 10–18. In all models, the sensing unit is connected by a long cable to an electronic package containing the polarizing current source, the frequency-counting apparatus, and a recorder. The sensing unit consists of the coil wound around the container with the proton source. It can be trailed behind an aircraft or mounted in a stinger. No elaborate mechanism is needed to orient the sensing unit. It can be clamped in any position that ensures an angle with the earth's field direction. For hand-carried models, the sensing unit is attached to a short pole which can be held in front or over the head of the observer.

Although dissociated hydrogen exists in all hydrocarbon fluids, the proton precession is too rapidly disrupted in most of them. In water and heptane, this motion persists long enough to obtain accurate frequency measurements. But water is unsuitable for use in subfreezing temperatures encountered in the polar regions and by high-flying aircraft. The best alternative is heptane, which has a very low freezing temperature. However, it is extremely volatile and can evaporate from a seemingly tightly sealed container. In an appropriate container, heptane can be used under all conditions normally encountered in geophysical surveying.

The precession frequencies encountered in geophysical surveying range from 1000 to 3000 hertz. The modern frequency-counting apparatus measures f within a precision of ± 0.02 hertz. According to Equation 10–14b, then, the earth's field intensity can be measured to an accuracy of about one-half gamma. It is important to realize that measurements cannot be made continuously. Because of the time required to polarize the protons and then analyze the precession frequency, a reading is made every 2 to 4 seconds. If the instrument is operated in an aircraft flying 150 mph, readings would be at points spaced at least 450 feet apart. The proton precession magnetometer is sensitive only to total field intensity F and cannot be focused to measure other elements of the field.

Figure 10–18
Hand-carried and airborne proton precession magnetometers. (Courtesy of E. G. and G. Geometrics, Inc.) The sensing element is held on a pole or trailed on a cable to remove it from electromagnetic fields produced by other instrument components or by the aircraft.

Optically Pumped Magnetometers

The most sensitive magnetometers yet invented measure electron precession to find the earth's magnetic field intensity. The spin motion of an electron, like that of a proton, produces angular momentum and a magnetic moment. If the spin axis is not aligned with the earth's field, it will precess at a frequency proportional to the total field intensity. Magnetometers have been designed to measure the electron precession frequency by a procedure quite different from that used in a proton precession magnetometer. The distribution of electrons in vaporized cesium has proved most suitable for these measurements.

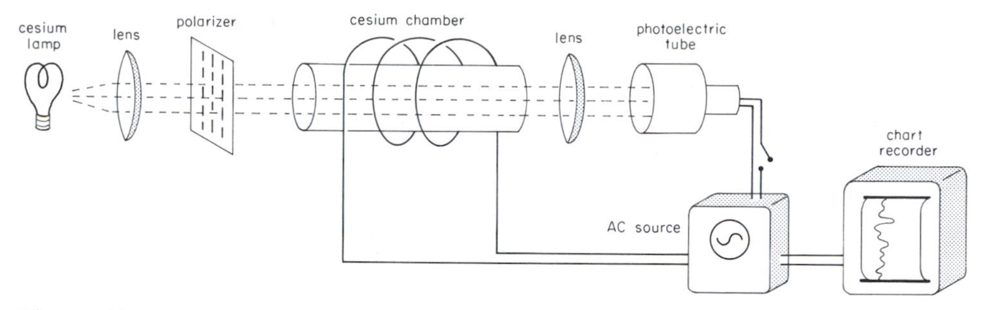

Figure 10–19
Principal components of a cesium vapor magnetometer. The precession of electrons of cesium atoms caused by the earth's magnetic field affects the intensity of a light beam reaching a photoelectric tube.

The instruments in which cesium is used are called *cesium vapor magnetometers.* Vaporized rubidium has also been used in similar instruments called *rubidium vapor magnetometers.* Although both substances are now in use, cesium has proved to be more satisfactory.

The basic magnetometer design is illustrated in Figure 10–19. A coil connected to an alternating-current source is wound around a chamber containing cesium vapor. Polarized light from a lamp with a cesium filament is passed through the chamber to a photoelectric cell that is sensitive to variations in light intensity. The distribution of electrons in the vapor, which affects the light intensity, can be adjusted by tuning the alternating current in the coil to the frequency of electron precession.

The valence electrons of cesium atoms are distributed in different orbits depending on the energy they possess. Consider the four orbits represented in Figure 10–20. Electrons with the least energy follow the lowest orbit A2, whereas those with more energy follow the higher orbits A1, B2, and B_1. By changing the energy of an electron, we can shift it to another orbit. According to the laws of quantum mechanics, only certain discrete shifts are

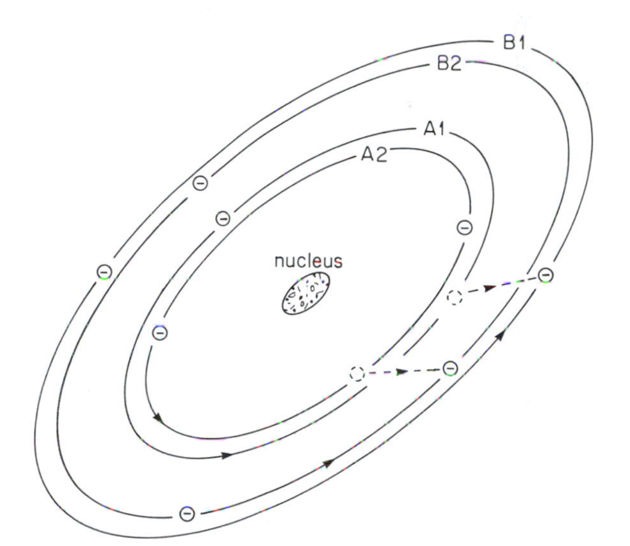

Figure 10–20
Some features of electron motion in four of the orbits of a cesium atom. When electrons in the A1 orbit are energized, they can shift to the B1 orbit. Similarly, energized electrons from the A2 orbit shift into the B2 orbit.

possible. Absorption of photons of light energy can produce shifts from A1 to B1 orbits or from A2 to B2 orbits. Much smaller shifts from A2 to A1 or from B2 to B1 orbits can be accomplished by adding the energy required to change the orientation of electron spin axes.

Light from a cesium lamp possesses exactly the wavelength that can be absorbed by the electrons of cesium atoms. By passing this light through an appropriate filter, we can polarize it in such a way that it can be absorbed only by electrons in A1 orbits. A cesium vapor magnetometer operates by passing the beam of polarized cesium light through the vapor chamber. There some of the photons are absorbed by electrons, which are then shifted from the A1 orbit to the B_1 orbit. This process is called *optical pumping,* because light energy is being "pumped" into the chamber where it is retained by the newly energized electrons. Hence, the cesium vapor magnetometer and the similar rubidium vapor magnetometer are classed as optically pumped magnetometers.

Because photons that are absorbed by electrons do not pass through the cesium vapor chamber, the intensity of the light beam reaching the photoelectric cell is relatively low. When all the electrons in the A1 orbits have been energized and shifted to B1 orbits, no more photons will be absorbed. Thereafter, the light beam reaching the photo cell becomes brighter because all the photons are now passing through the chamber.

When the photoelectric cell detects a light beam of maximum intensity, it activates the alternating-current source so that an alternating magnetic field is produced in the coil surrounding the chamber. The frequency of the field is varied automatically until it is tuned to the precession frequency of the electrons. At this frequency, it provides the energy for shifting some electrons from A2 orbits into the now vacant A_1 orbits. To make this shift, the orientation of the spin axes relative to the earth's field must increase from the small angle typical of low-energy electrons in A2 orbits to the larger angle of electrons in A1 orbits.

Let us examine Figure 10–21, which shows the precession of an electron axis that makes the small angle α with the earth's field direction. Also shown is a rotating vector which illustrates how the direction of the alternating magnetic field changes during each cycle of oscillation. Observe that this vector always points in directions different from the direction of the earth's field. When the frequency of the alternating magnetic field is adjusted to match the electron precession frequency, the vector turns at the same rate as the spin axis is moving around its cone of precession. The result is a steady outward force which acts to rotate the spin axis into the larger angle characteristic of an A1 orbit. For any other frequency, the force exerted by the alternating magnetic field would cycle from an inward to an outward direction, causing the electron to wobble rather than to shift permanently the angle of the spin axis. Therefore, only when the current in the coil is tuned exactly to the precession frequency will the electrons shift from the A2 to the A1 orbits.

Upon reaching an A1 orbit, the newly shifted electron acquires the capacity to absorb light energy. The result is an immediate reduction in the intensity of the beam detected by the photoelectric cell. This change in intensity signals the system to record the current frequency that now matches the electron precession frequency. This value is multiplied by the electron gyromagnetic ratio to obtain the earth's field intensity F.

Optically pumped magnetometers have been designed for airborne surveying and for

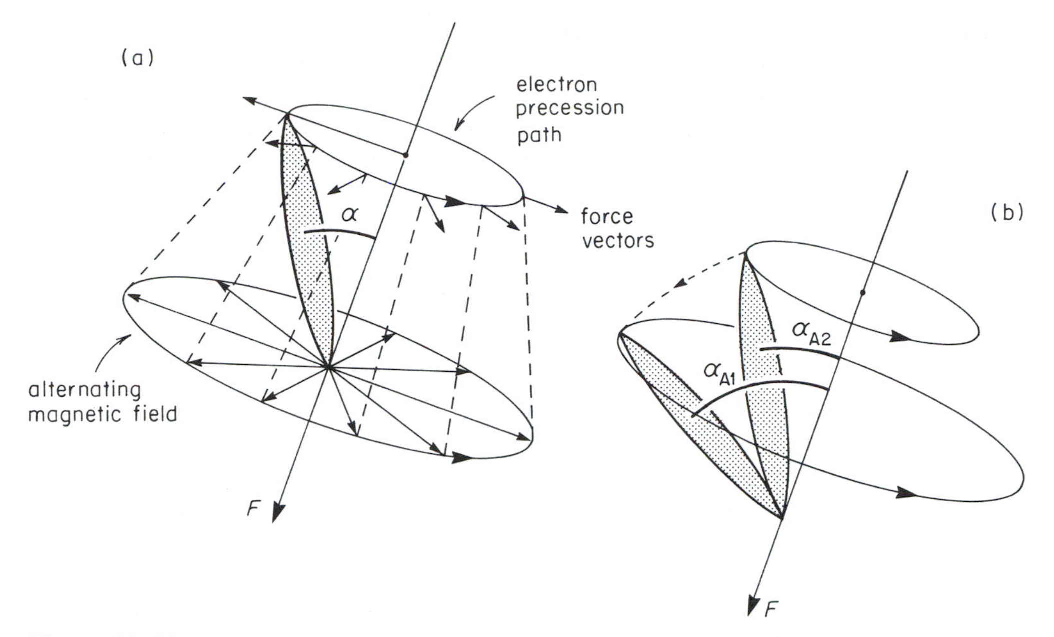

Figure 10–21

(a) Time-varying force exerted by an alternating magnetic field on a precessing electron in a cesium vapor chamber. (b) The force acts to rotate the electron spin axis from low-energy angle α_{A2} characteristic of the A2 orbit into the higher-energy angle α_{A1} characteristic of the A1 orbit. The angle can be changed from α_{A2} to α_{A1}, providing the frequency of the alternating magnetic field is the same as the electron precession frequency, so that a constant outward force acts to rotate the electron spin axis.

operation in permanent observatories. These instruments measure the total field intensity but none of the other elements of the earth's field. Orientation of the sensing unit is not critical, but the alternating magnetic field must not be aligned with the earth's field. With this apparatus, a measurement can be made in less than one second. Because of the very small mass of an electron, its precession frequency is much higher than that of a proton. In the earth's field, the approximate range is 90 to 245 kilohertz. The system can be tuned closely enough to measure field intensity to an accuracy of better than 0.01 gamma.

World Surveys

Our purpose in this chapter is to describe the large regional variations and the time variations of the earth's magnetic field. The exploration geophysicist must account for these features before attempting to analyze local magnetic anomalies caused by geologic structures. Information about the large-scale properties of the earth's field comes principally from geomagnetic observatories. Today, there are more than 140 geomagnetic observatories in the world. They are equipped to obtain continuous records of three elements of the magnetic field. In addition, more than 2000 sites have been selected for measurements of one or more elements that are repeated at intervals of from 5 to 20 years. Insofar as possible, the geomagnetic observatories and the repeat stations are located where there are no large local anomalies. This ensures that the regional properties of the field are measured as accurately as possible.

THE MAIN MAGNETIC FIELD

Before the time of William Gilbert, there was considerable dispute about whether a lodestone compass was attracted by the earth or by poles of force in the heavens. Gilbert presented convincing arguments that the attraction came principally from within the earth. In 1838, Carl Friedrich Gauss published an elegant mathematical proof that approximately 95 percent of the field must originate inside the earth, but that the remaining 5 percent must be produced by external processes. Further research revealed that a large part of the interior field is produced in the core of the earth, but that the remaining part has its source in the crust. We can now separate the earth's magnetic field into the following three parts.

1. The *main magnetic field,* which is produced in the core of the earth and accounts for the very large regional variations in field intensity and direction.
2. The *external magnetic field,* which is produced by electric currents in the earth's ionosphere consisting of particles ionized by solar radiation and put into motion by the solar tidal force.
3. The *anomalous magnetic field,* which is produced by ferromagnetic minerals in the earth's crust.

In this chapter we will concentrate on the main magnetic field, which is, by far, the strongest of the three parts. We will discuss briefly the external field that fluctuates in daily cycles. The anomalous magnetic field, which is of principal interest to exploration geophysicists, is the subject of Chapters 11 and 12.

Geomagnetic Maps

Regional features of the main magnetic field are best illustrated on contour maps. A typical set of maps in Figures 10–22, 10–23, and 10–24 display field intensity, inclination, and declination for the year 1975. Maps such as these are compiled by government agencies of several nations. Because the field changes slowly with time, updated maps are issued every ten years. These geomagnetic maps are made using measurements from the worldwide network of geomagnetic observatories and repeat stations that have been adopted by an international commission to represent the International Geomagnetic Reference Field (IGRF). Values used for contouring are first corrected for effects of the anomalous field and the external field and adjusted to a common time.

The geomagnetic maps show that the field on the earth's surface is strongest in the polar regions where maximum intensity is close to 70,000 gammas. This figure diminishes to an average of about 30,000 gammas near the equator. Inclination is vertical (i = 90 degrees) at the location of 75°N, 101°W in northern Canada, and at the location of 67°S, 143°E in Antarctica. These locations are called the north and south *magnetic dip poles.* Notice that a line connecting the dip poles does not pass exactly through the center of the earth. The *magnetic equator* is the contour of zero inclination.

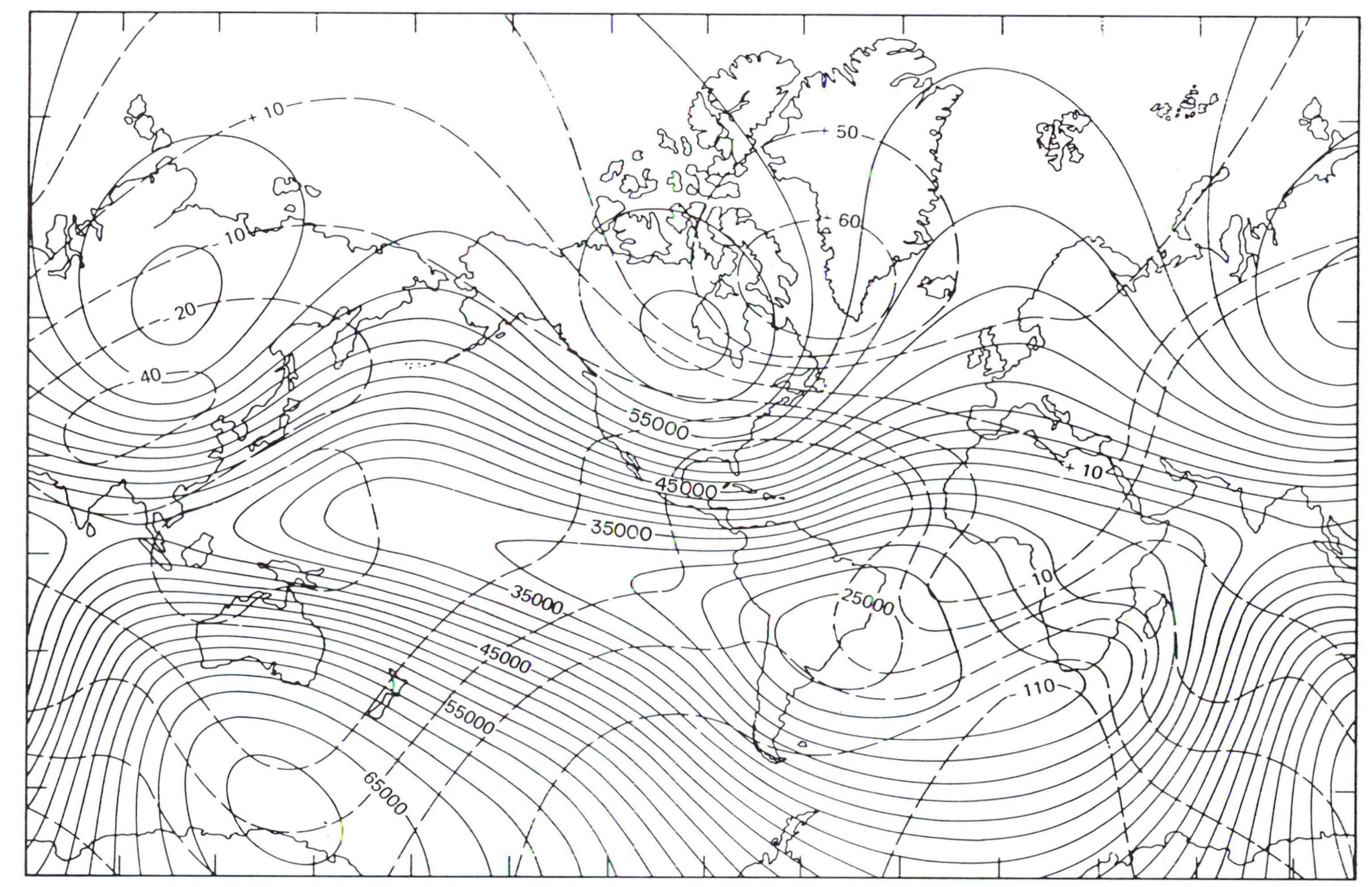

Figure 10–22
Main magnetic field intensity (solid contours) and annual secular change (dashed contours) on the surface of the earth expressed in gammas. (Modified from "Total Intensity of the Earth's Magnetic Field," Epoch 1975.0, chart published by Defense Mapping Agency Hydrographic Center, Washington, D.C.)

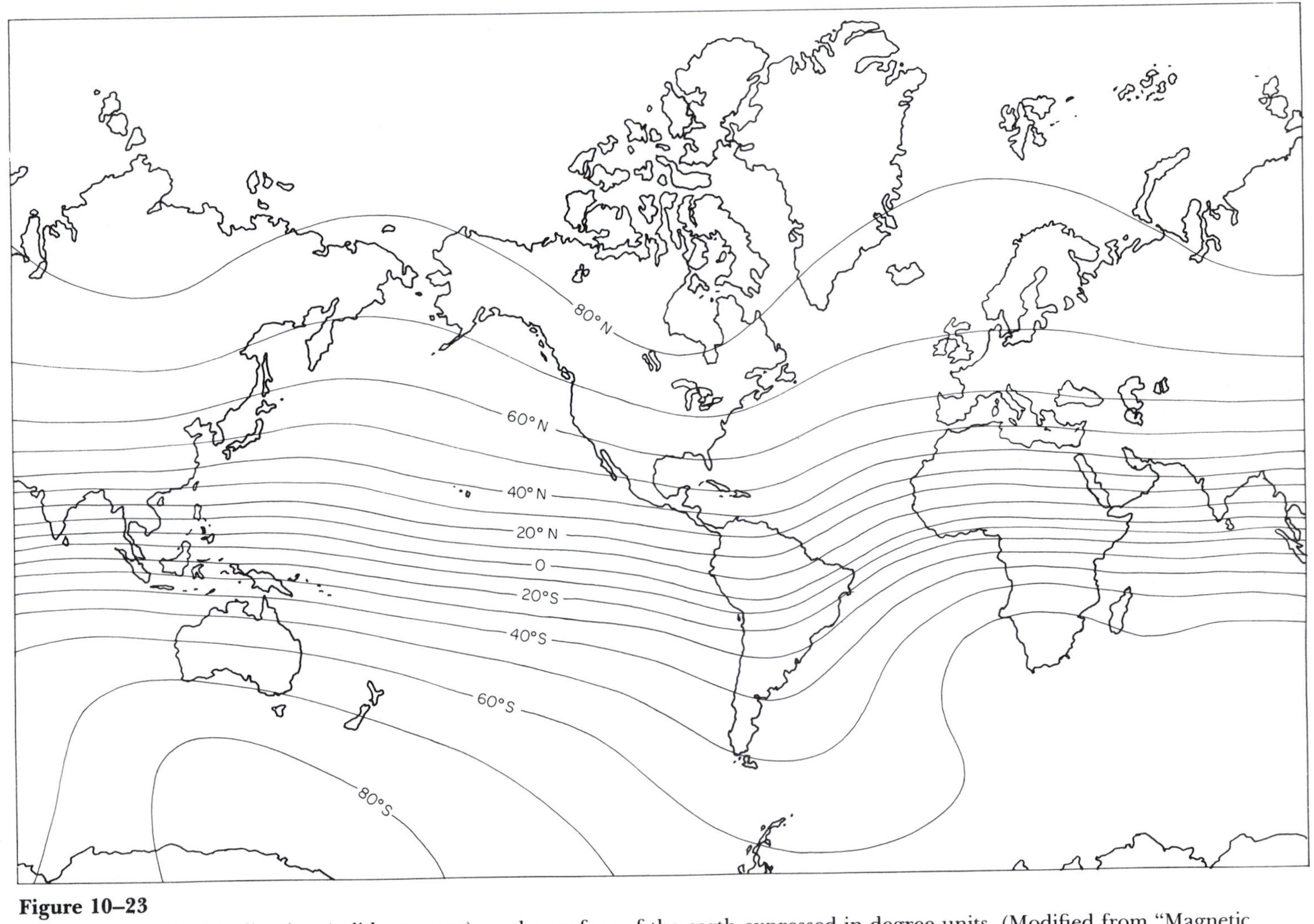

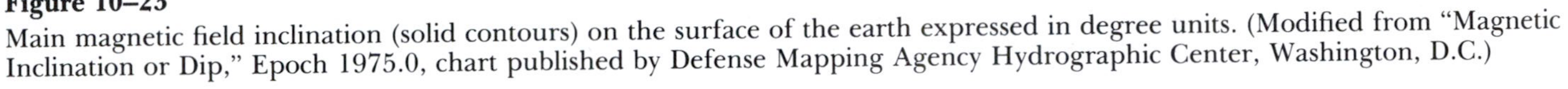

Figure 10–23
Main magnetic field inclination (solid contours) on the surface of the earth expressed in degree units. (Modified from "Magnetic Inclination or Dip," Epoch 1975.0, chart published by Defense Mapping Agency Hydrographic Center, Washington, D.C.)

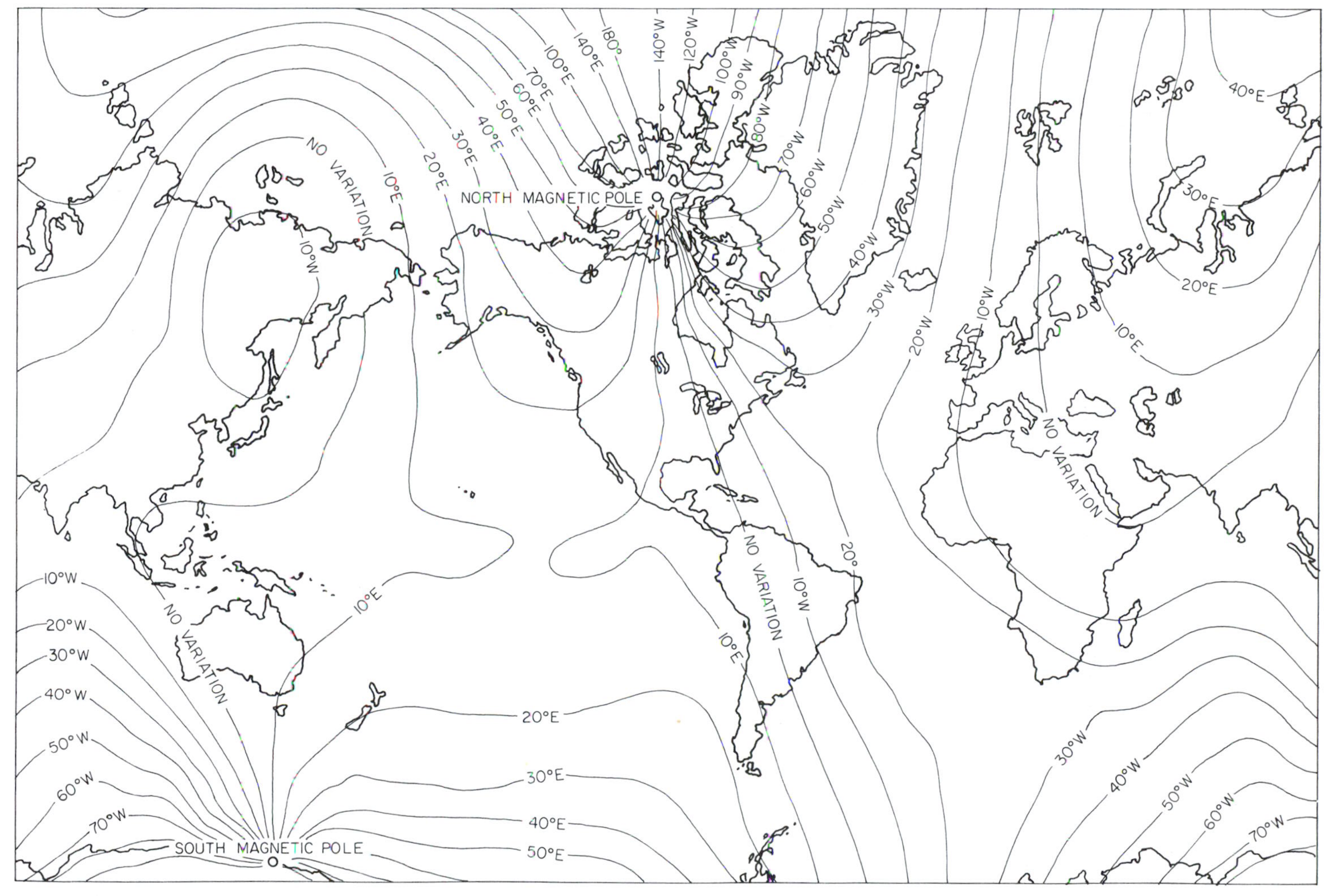

Figure 10–24
Main magnetic field declination (solid contours) on the surface of the earth expressed in degree units. (Modified from "Magnetic Declination," Epoch 1975.0, chart published by Defense Mapping Agency Hydrographic Center, Washington, D.C.)

Dipole and Nondipole Constituents

We could account for a large part of the main magnetic field in terms of a short bar magnet placed at the center of the earth. This fictitious magnet is called the *centered geomagnetic dipole*. Its magnetic moment would have to be 8×10^{25} ups-cm, and its axis would have to be inclined 11½ degrees from the earth's axis of rotation to reproduce the principal features of the main field. A line extended along its axis would intersect the earth's surface at 78°34′N, 69°20′W in the Northern Hemisphere and at 78°34′S, 110°40′E in the Southern Hemisphere. These positions are called the *geomagnetic poles* of the earth.

The centered geomagnetic dipole does not account for regional irregularities in the main field which are indicated by the difference in dip pole and geomagnetic pole positions, variation in field intensity along the magnetic equator, and asymmetrical contour patterns. We could account for these departures from a simple dipole field in terms of about 12 additional bar magnets placed near the boundary of the core with their axes directed toward the center of the earth. These magnets would be much weaker, each having a moment about one-eightieth that of the centered geomagnetic dipole. Their collective field makes up the *nondipole* part of the main field, which is contoured on the earth's surface in Figure 10–25.

Spherical Harmonic Analysis

A distribution of magnets in the core may help to visualize some features of the main field, but it is inconvenient for obtaining a mathematical description. This is done more effectively with a set of three equations which can be used to calculate the vertical and two horizontal elements of field intensity at any location. Geophysicists use these equations to determine the main field in somewhat the same manner that the normal gravity formula (Equation 7–16) is used to determine global variation in gravity. But the main field does not vary in a simple symmetrical way as the normal gravity field does. Rather, it has a far more irregular form. For this reason, the three equations turn out to be very complicated and impractical for everyday use by the exploration geophysicist.

Each equation consists of about 80 separate terms which are called *spherical harmonics*. Our discussion of wavelength filtering in Chapter 9 is useful for explaining, qualitatively, the nature of these spherical harmonic terms. Look again at Figure 9–30, which shows how corrugated surfaces can be superposed to reproduce a pattern of anomalies. Spherical harmonics can be thought of as corrugated surfaces modified to fit on a globe. Each "modified corrugation" illustrated in Figure 10–26 represents a cyclic variation over the surface of the globe. For one set the cyclic corrugations are in the direction of parallels of latitude, and for the other set they are in the direction of meridians of longitude. Superposing an appropriate combination of such terms makes it possible to reproduce all the irregularities of the main field.

Like the terms in Equation 9-21, each spherical harmonic term consists of a constant amplitude factor and a trigonometric factor that depends on position, which is given by latitude and longitude rather than by the distance x. A method of spherical harmonic analysis, which we will not attempt to describe, is used to calculate the amplitude factors from field intensity values measured at the geomagnetic observatories and repeat stations. Once

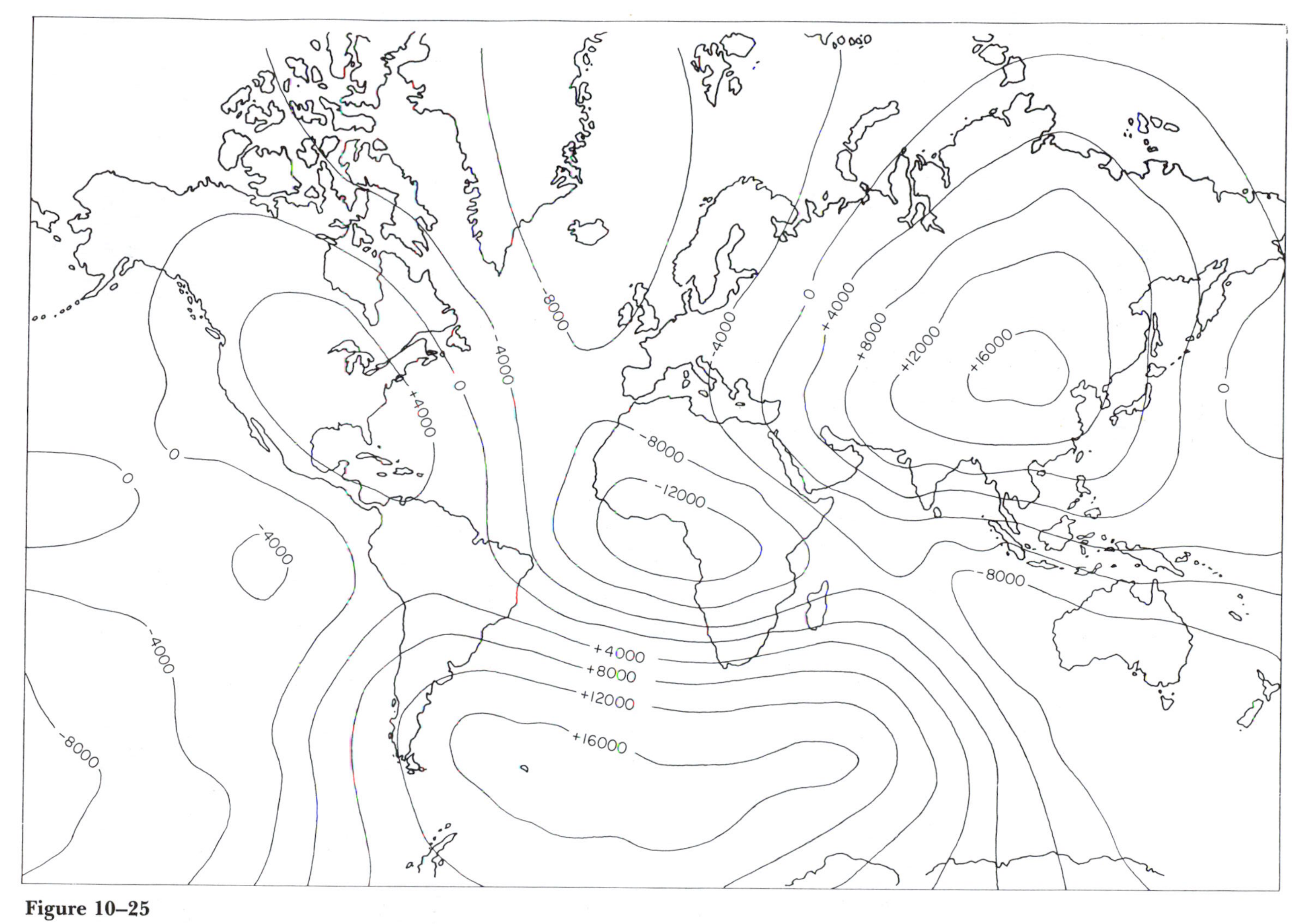

Figure 10–25
Nondipole part of the main magnetic field intensity contoured at 4000-gamma intervals. (Modified from E. C. Bullard, C. Freedman, H. Gellman, and J. Nixon, *Philosophical Transactions of the Royal Society*, Series A, v. 243, pp. 67–92, 1950.)

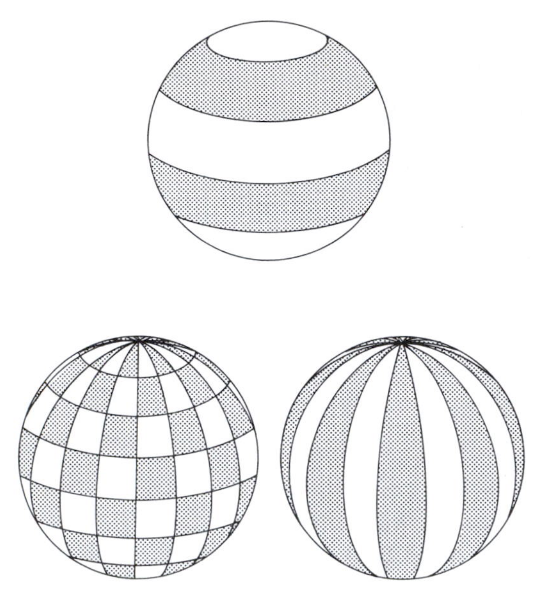

Figure 10–26
Zones on a globe representing functions that vary cyclically with latitude and longitude. Values of the functions are positive in the shaded zones and negative in the unshaded zones. Such functions are called spherical harmonics.

this calculation has been made, it is possible to determine the intensity and direction of the main field at any location. In this way the International Geomagnetic Reference Field (IGRF) is obtained.

Modern computers have greatly facilitated the task of solving the long and complicated spherical harmonic equations. For most geophysicists who are not prepared to undertake the necessary computer operations, tables are available from agencies responsible for issuing IGRF maps (see Table 11–1). In these tables values for elements of the main field are given on a grid of latitude and longitude intersections that extends over the entire surface of the earth.

Secular Variation

The main magnetic field changes with time. This process is called *secular variation.* The first evidence of secular variation was found from magnetic inclination and declination measurements that go back almost four centuries. Some of these measurements are plotted in Figure 10–27. More recently, we have been able to compile a comprehensive picture of secular variation using information from the global network of geomagnetic observatories and repeat stations. Contour maps are prepared to illustrate the annual change over the surface of the earth of all the magnetic elements. For example, the dashed contours in Figure 10–22 indicate an annual rate of change of total intensity. Observe that for most parts of the world, the change in field intensity is between 50 and 150 gammas per year. This information is particularly useful to exploration geophysicists for adjusting magnetic field surveys that were done at different times. These adjustments are discussed in more detail in Chapter 11. Secular variation appears to be related in some manner to the nondipole part of the main field. Observe the general similarity of the dashed contour patterns in Figure 10–22 and of contour patterns in Figure 10–25, which indicates a relationship. There is also evidence that the moment of the centered geomagnetic dipole is decreasing at the approximate rate of one two-thousandth of its value each year. This, too, would contribute to secular variation.

Perhaps the most important aspect of secular variation is called *westward drift.* Most of the large regional features of the main field, indicated by the contour patterns in Figures 10–22 and 10–25, are shifting slowly in a westward direction. The rate of this westward drift is approximately 0.2 degree of longitude per

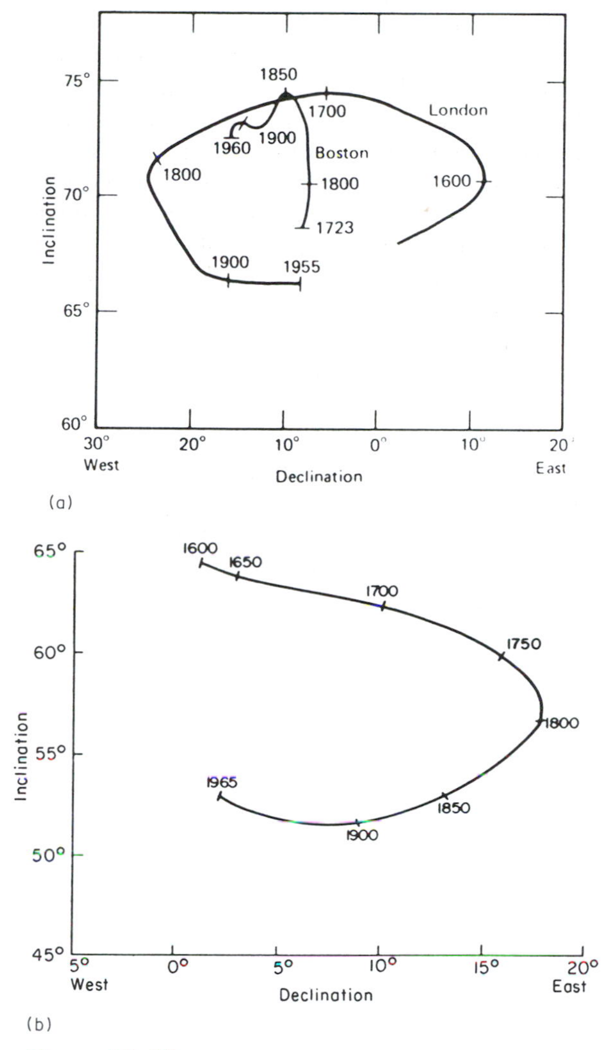

Figure 10–27
Secular change in magnetic inclination and declination at locations (a) near London and Boston and (b) in Sicily. (Modified from Nelson et al., *Magnetism of the Earth*, Publication 40-1, U.S. Department of Commerce, Coast and Geodetic Survey, 1962, and D. W. Strangway, *History of the Earth's Magnetic Field*, New York, McGraw-Hill, 1970.)

year. This explains why field intensity is increasing at some locations while decreasing at other places (Figure 10–22). As a large regional anomaly approaches a location, the field intensity at that location will slowly grow stronger. Later, when the anomaly is drifting away, the field intensity will then decrease. The counterclockwise progression of points plotted in Figure 10–27 is the result of westward drift. We know that the main field must originate in the earth's liquid core. Westward drift of this field tells us that the core is rotating at a slightly slower rate than the solid mantle and crust.

External Field Variations

Geomagnetic observatory records display short-term fluctuations. Prominent on all these records is a daily cycle we called the *diurnal variation*. Samples in Figure 10–28 show that this daily fluctuation amounts to a few tens of gammas. It is strongest in the equatorial region and diminishes at higher latitudes. There is also a seasonal change, the diurnal variation being somewhat larger during the summer than during the winter months.

Careful examination of observatory records reveals that the diurnal variation occurs mostly during the daylight hours. This fact, together with consideration of the latitude and seasonal differences, tells us that the sun must play a leading role in producing this daily cycle. Actually, the sun contributes in two ways. (1) Electromagnetic radiation emanating from the sun, mostly ultraviolet rays and X-rays, ionizes the otherwise electrically neutral particles of the earth's atmosphere in the outer zone we call the ionosphere; and (2) the tidal force associated with the sun produces cyclic wind currents in the ionosphere in the same

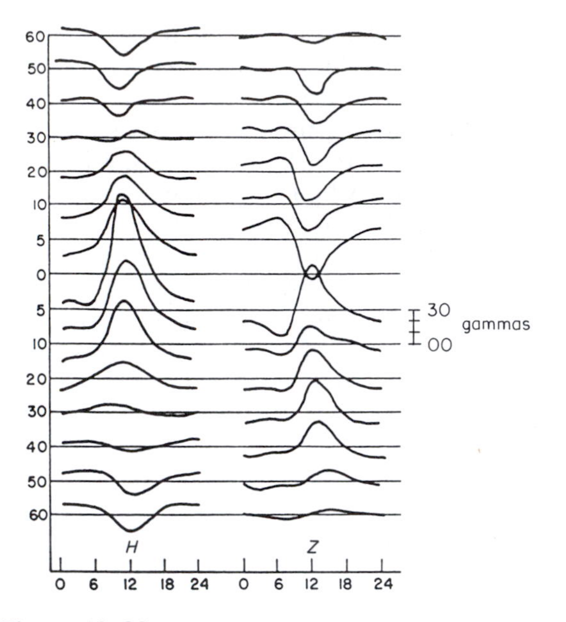

Figure 10–28
Diurnal variation of the H and Z elements of the earth's magnetic field at different latitudes. (From S. Matsushita and W. H. Campbell (editors), *Physics of Geomagnetic Phenomena*, New York, Academic Press, 1967.)

way that it produces tidal currents in the ocean. Therefore, the ionospheric wind, consisting of charged particles in motion, is an electric current. Associated with it is a fluctuating magnetic field. The tidal force from the moon also affects the ionospheric wind, but the moon does not help to ionize the atmosphere. The lunar effect therefore amounts to about 5 percent of the complete solar contribution to the diurnal variation. Ionization occurs during the daylight hours when a portion of the ionosphere is exposed to direct solar radiation. Consequently, there is only one cycle of magnetic field fluctuation, even though the tidal force causes two cycles of wind fluctuation each day.

The cyclic magnetic field produced by the ionospheric wind accounts for only about two-thirds of the diurnal variation measured on the earth's surface. But this field excites electric current in the earth's interior by electromagnetic induction. The interior current, in turn, generates a secondary magnetic field. The combination of the ionospheric field and the secondary field accounts for the diurnal variation.

In addition to the regular diurnal variation, other highly erratic fluctuations occur. These are called *magnetic storms*. Typically, a magnetic storm commences with a sudden dramatic change in field intensity followed by a slower recovery. The record in Figure 10–29 shows this pattern. The maximum change in intensity can range from less than 100 gam-

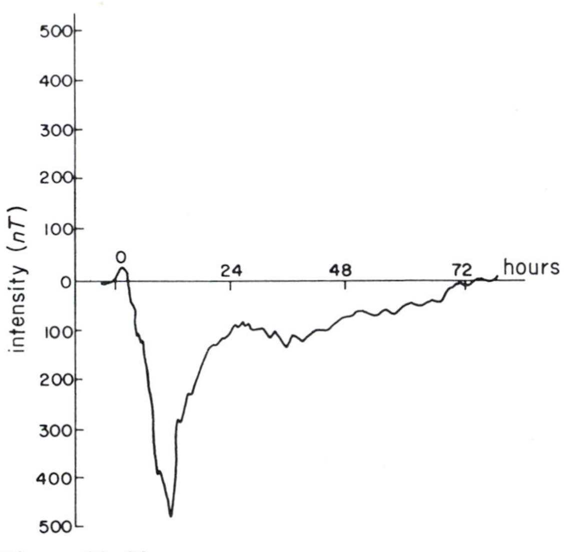

Figure 10–29
Field intensity variation during a typical magnetic storm. (From George D. Garland, *Introduction to Geophysics*, Philadelphia, W. B. Saunders Co., 1979.)

mas to more than 1000 gammas. The duration of a storm is variable, ordinarily lasting anywhere from a few hours to a few days. Magnetic storms are associated with sunspot activity, when streams of charged particles released from the sun bombard the earth.

The exploration geophysicists must adjust magnetic field surveys for the effects of diurnal variation. This adjustment is described in Chapter 11. But magnetic storm fluctuations are usually too strong and erratic to be properly adjusted, and so the most common practice is to suspend survey activities until the storm subsides.

Polarity Reversals of the Main Field

During the 1950s, we were led by ferromagnetic mineral grains to a discovery of fundamental importance. Many times in the history of the earth, the main magnetic field has spontaneously reversed its polarity! Evidence of these reversals is found in the directions of permanent magnetism preserved in grains of magnetite, ilmenite, and pyrrhotite. It is not our purpose in this chapter to present the subject of rock magnetism. Rather, we will describe briefly only the aspects that bear directly on the nature of the main field. Other aspects of rock magnetism will be discussed more fully in Chapter 12.

Most important for learning about main field polarity reversals is the permanent magnetism acquired by ferromagnetic mineral grains when they cool through their Curie temperature. At this instant, magnetic domains are created within these grains. The magnetic moments of these domains tend to align with the direction of the earth's field at their location. Permanent magnetism acquired in this way is called *thermoremnant magnetism.* It is produced in grains of magnetite, ilmenite, and pyrrhotite, which are part of the composition of recently formed igneous rocks in the process of cooling, and of rocks that have been reheated and then cooled following an episode of metamorphism. Thermoremnant magnetism can persist essentially unchanged for indefinite periods of time, preserving a record of the earth's magnetic field direction at the time when it was acquired.

A record of the earth's magnetic field direction can also be produced by sedimentation. As ferromagnetic particles settle to the bottom of the sea or onto the bed of a lake or river, they tend to align with the earth's magnetic field direction and then become buried in the accumulating layer of sediment. This alignment can remain preserved in the sedimentary rock that is eventually produced from such deposits.

We can reconstruct a history of the earth's magnetic field by collecting rock specimens for measurements of the direction of permanent magnetism. Each specimen must be carefully marked to show its original orientation at the location where it was obtained. Corrections are applied to account for any change in orientation that may have occurred after it was permanently magnetized, perhaps by folding and faulting or by compaction of sediment.

Next, the specimen is tested in an instrument called a *spinner magnetometer* to find the direction of permanent magnetism in its ferromagnetic mineral grains. The spinner magnetometer pictured in Figure 10–30 consists of a spindle mechanism, a sensing coil, and a set of coils used to nullify the earth's field in the space surrounding the spindle. This field must be nullified to remove any induced magnetism that would otherwise exist in the specimen.

Figure 10–30
Components of a spinner magnetometer. (Courtesy of Schonstedt Instrument Company.) Compensating coils are adjusted to nullify the earth's field in the region close to the spindle in which a specimen possessing permanent magnetism is mounted. Spinning the magnetized specimen can induce an alternating current in a sensing element.

A small core of the rock specimen is mounted on the spindle, which can be adjusted to spin in any desired direction. Insofar as the core possesses some permanently magnetized grains, its spin motion will induce electric current in the sensing coil. The strength of this current depends on the orientation of the core and the sensing coil relative to one another. When we test it in different orientations, the direction of permanent magnetism can be found from the variation of the induced current.

What do our measurements reveal? Consider the evidence typical of a sequence of basalt layers. Such a sequence is particularly useful because the permanent magnetism is acquired very soon after the rapidly cooling lava congeals to form a basalt layer. Thus, the age of the layer, found by analyzing the decay of radioactive trace elements in the rock, corresponds closely to the time of magnetization. Typically, we find that the sequence consists of some layers that have similar directions of permanent magnetism interspersed with other layers of oppositely directed magnetism.

Initially, many geophysicists viewed these results with suspicion. They believed that errors in the experimental procedure or chemical and physical processes within the layers had produced these opposite directions of permanent magnetism. The idea that they were related to the main field polarity was at first greeted with skepticism. But as evidence continued to accumulate, this idea turned out to be the only generally acceptable explanation. When data were compiled from a worldwide distribution of rocks of the same age, the directions of permanent magnetism were consistent with the lines of force of a dipole field originating in the core of the earth. Rocks of a different age pointed to a dipole field of opposite polarization. Careful sifting of the data revealed the sequence of main field reversals illustrated in Figure 10–31. In the figure, we see that the main field remains in one polarization for an irregular period of time lasting from a few thousand years to several hundred thousand years. Then it undergoes a polarity reversal. The time required for such a reversal appears to be less than a few thousand years.

Reconstructing a history of the positions of the ancient magnetic field is a principal means of confirming the theory of continental drift. We will not attempt to discuss this interesting topic in a book whose specific aim is to introduce the methods of exploration geophysics. For our purposes, it is useful to know that spontaneous reversals of polarity are an important feature of the earth's main magnetic field.

The Source of the Main Magnetic Field

Why does the earth possess a magnetic field of somewhat irregular form and subject to secular variation and polarity reversals? Modern ideas attribute these aspects of the magnetic field largely to the flow of ionized fluid in the core. These ideas are expressed in an adaptation of *magnetohydrodynamic* theory by two well-known scientists, W. M. Elsasser and E. C. Bullard. This theory combines principles of fluid mechanics and the electromechanical dynamo.

Consider first how the simple disk dynamo in Figure 10–32 could, theoretically, generate a self-sustaining magnetic field. Michael Faraday (1791–1867) discovered that a metal disk spinning on an axis aligned with a magnetic field developed a positive electric charge near its rim, whereas its center became negatively charged. Further experiments revealed that electric current flowed in a conducting wire connected between its rim and its center to dissipate the electrical charge difference. Suppose the conductor is wound into a coil surrounding the disk. Current in the coil produces a magnetic field parallel to the original field. Adjusting the spin of the disk to a critical velocity makes this self-generated magnetism exactly sufficient to maintain the current in the coil, so the original field can be eliminated. Now we have a self-sustaining dynamo. Except for mechanical friction and the electrical resistivity of the conductor, this dynamo would operate indefinitely.

The concept of a self-sustaining dynamo helps us to understand how a magnetic field might be maintained indefinitely, but it does not explain spontaneous polarity reversals. An interesting solution to this problem is the combination of disk dynamos in Figure 10–33. The wire connected between the rim and center of each disk is wound into a coil around the other disk. The research of two scientists, T. Rikitake and D. W. Allen, showed that this, too, can be a self-sustaining dynamo system. But the effects of the two spinning disks are very delicately balanced. Even a small perturbation can excite a polarity reversal of the magnetic field generated by this system. Here we have a mechanism for reproducing the fundamental features of the centered geomagnetic dipole that is subject to polarity reversals. Other disk dynamos in different orienta-

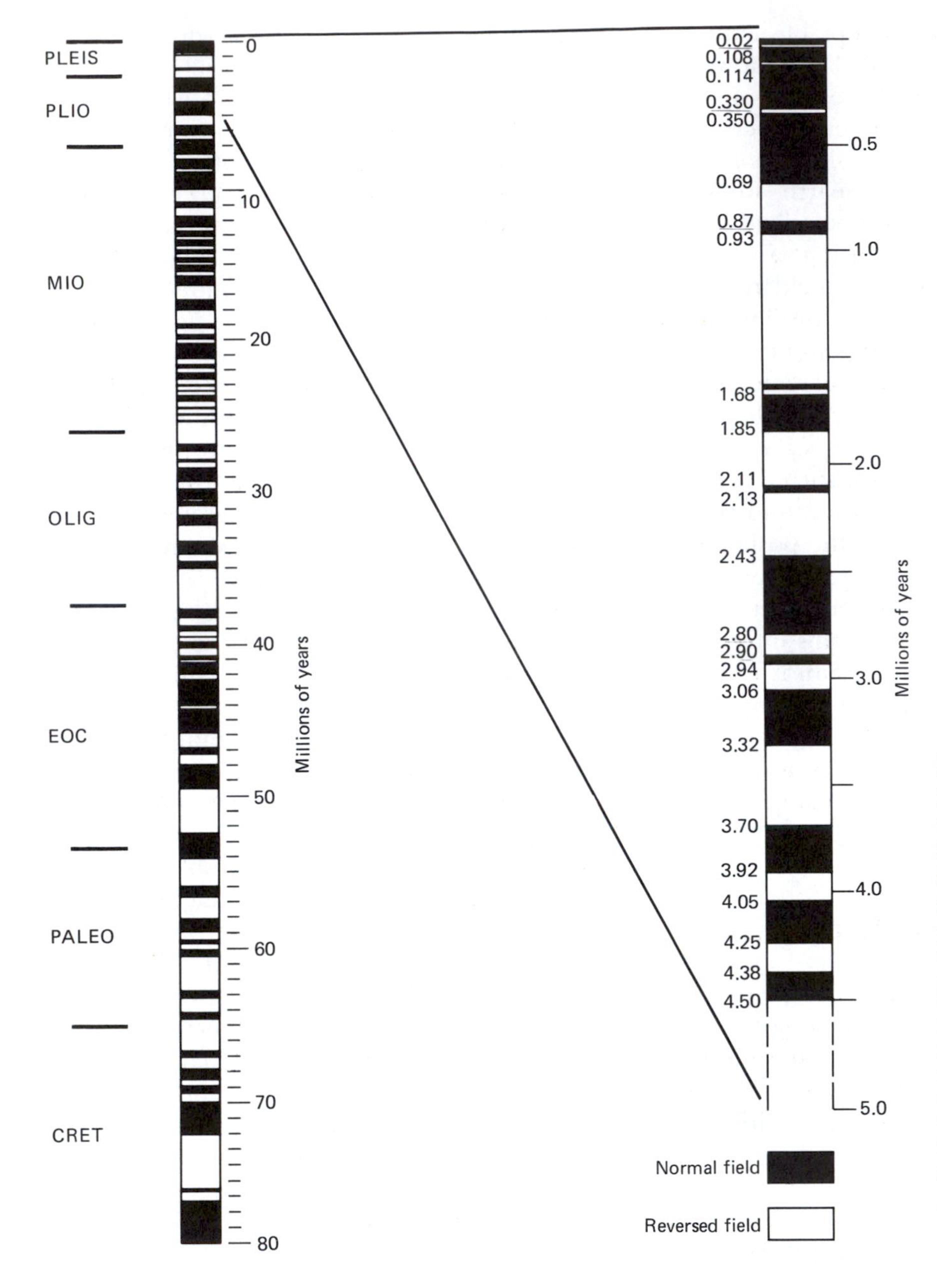

Figure 10–31
Main magnetic field reversals during the past 80 million years. Black indicates polarization in the direction of the present-day field, and white indicates reversed polarization. (From J. R. Heirtzler, G. O. Dickson, E. M. Hebron, W. C. Pitman, III, and X. LePichon, *Journal of Geophysical Research*, p. 2123, March 15, 1968, and Frank D. Stacey, *Physics of the Earth*, p. 176, New York, Wiley, 1969.)

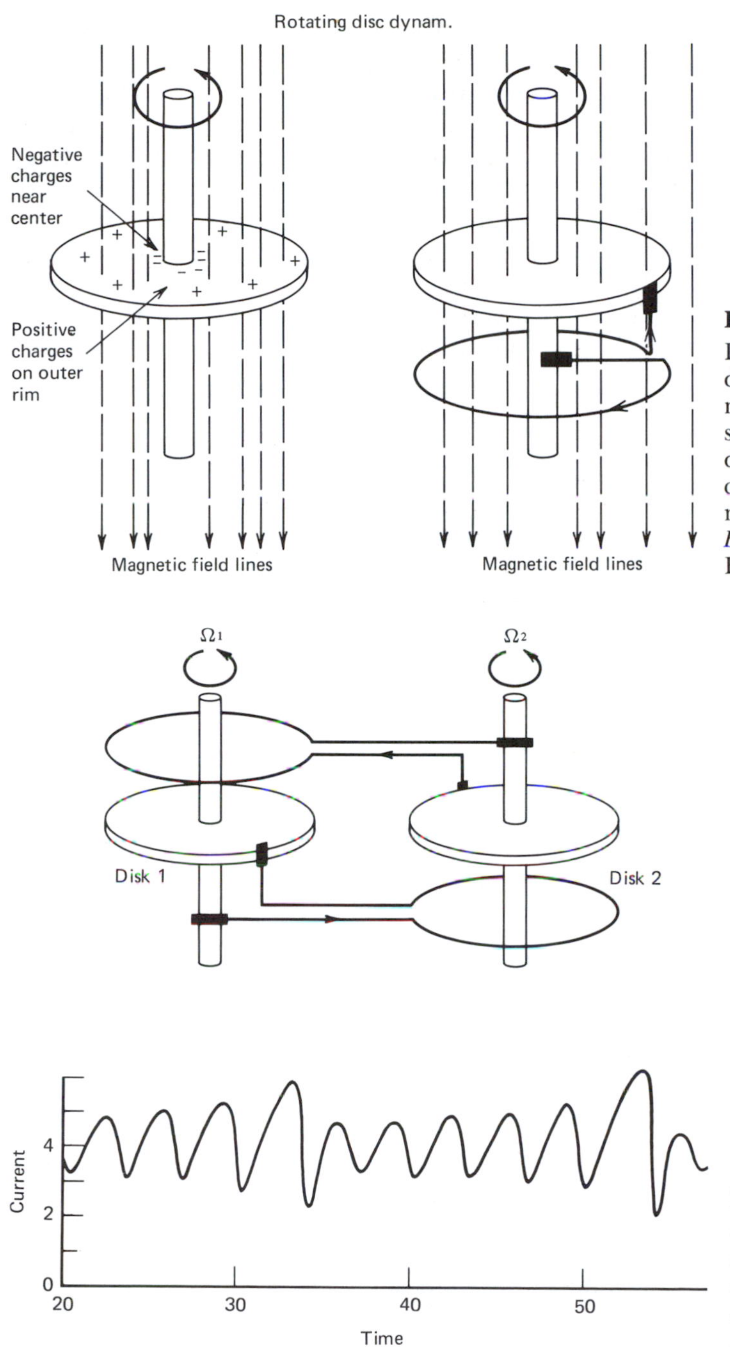

Figure 10–32
Disk dynamo system showing (a) a charge difference caused by rotation in a magnetic field, and (b) generation of a self-sustaining magnetic field by dissipation of the charge as an electric current in the coil surrounding the rotating disk. (From T. Rikitake, *Electromagnetism and the Earth's Interior*, Elsevier North-Holland, 1966.)

Figure 10–33
(a) Coupled disk dynamo system and (b) a graph of spontaneously reversing current flow. (From T. Rikitake, *Electromagnetism and the Earth's Interior*, Elsevier North-Holland, 1966.)

tions can be added to this combination to account for the irregular nondipole part of the earth's field.

A combination of disk dynamos establishes a theoretical basis for reproducing the earth's main magnetic field. It is a system with many parts, quite unlike the earth's liquid core which is simply a rotating spherical shell of ionized fluid. How can the features of a dynamo system be duplicated by motions of this fluid? This is the fundamental magnetohydrodynamic problem, one very difficult to solve.

A system of convection currents is assumed to exist in the earth's liquid core. These convection currents are believed to be produced partly by heat that is released from the decay of radioactive elements and as a consequence of solidification of the inner core. In some ways, they are similar to the convection currents to be seen in a pot of water brought to boiling on a stove. The effect of the earth's rotation is also important, for it imposes axial

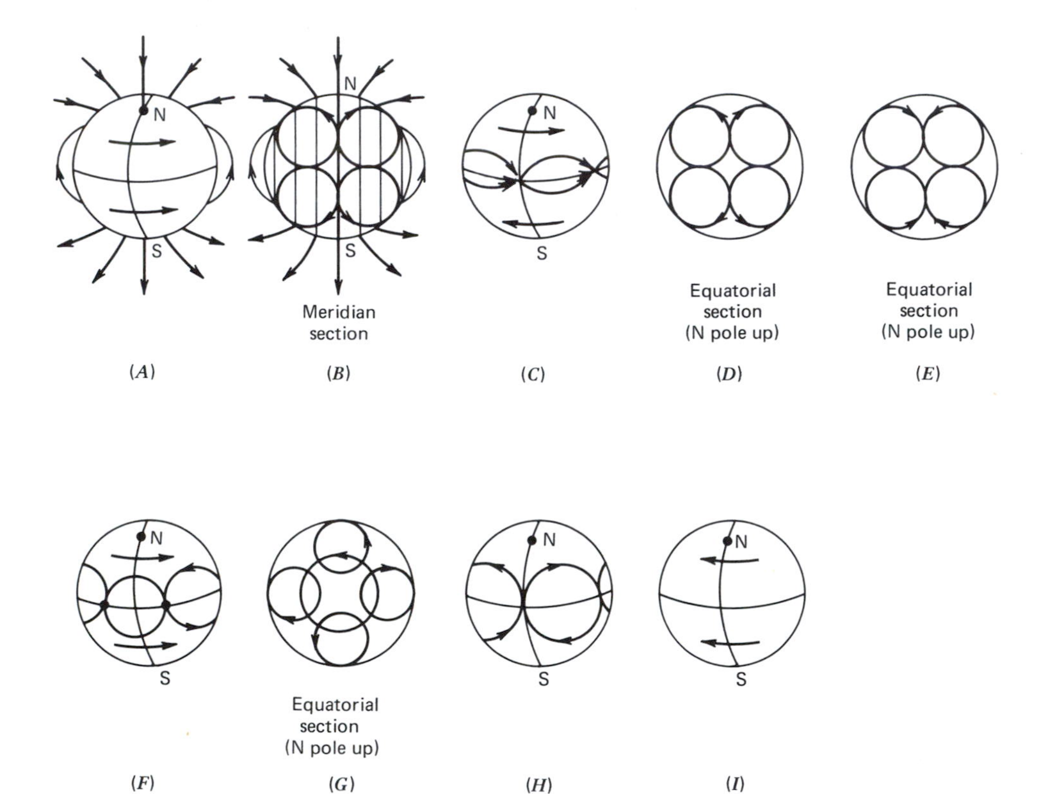

Figure 10–34
Theoretical flow of ionized fluid in the outer core of the earth and the associated magnetic lines of force. (From E. C. Bullard, *Royal Society of London Proceedings*, Series A, v. 197, p. 450, Fig. 6, July 7, 1949.)

symmetry on the system of currents. The mathematical calculations to determine the pattern of currents are formidable. Diagrams in Figure 10–34 illustrate some results. We have learned that such a system can sustain a magnetic field that is more or less aligned with the axis of rotation. Spontaneous polarity reversals are theoretically possible, but we cannot predict when they will occur. Lacking the evidence of direct observation, we can only surmise conditions in the liquid core. Therefore, we cannot expect to explain all the features of the main magnetic field.

STUDY EXERCISES

1. At a point 20 cm from the center of a thin magnetized rod 40 cm long and equidistant from its ends, the magnetic field intensity is 500 nT. What is the pole strength of the rod? Express your answer in cgs units of pole strength and in ampere-meters.

2. Consider a rod 50 cm long with a circular cross-sectional area of 1 cm^3 made of a certain ferromagnetic material. The magnetic moment per unit volume *(I)* produced in this rod by a variable magnetizing field *(F)* is shown by the hysteresis curve in Figure 10-35. Suppose that this rod was first magnetized to saturation and then removed from the magnetizing field.

   a. Calculate the pole strength of the rod that results from the permanent magnetism it retains after being removed from the magnetizing field.

   b. Calculate the magnetic field intensity at a point in line with the rod 50 cm from its nearest end.

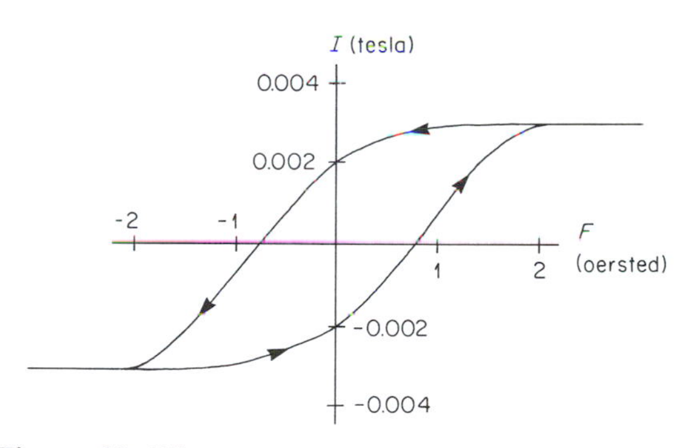

Figure 10–35
Hysteresis curve showing the magnetic moment per unit volume (*I*) produced in a rod by a magnetizing field (*F*).

3. Each of the following magnetometers measures a different specific quantity that is proportional to the earth's magnetic field intensity. What is the specific quantity that is measured by

   a. a flux-gate magnetometer;

   b. a proton precession magnetometer;

   c. a torsion magnetometer;

   d. an optically pumped magnetometer?

4. Use Figures 10–22, 10–23, and 10–24 and the appropriate equations to obtain values for the seven elements of the earth's magnetic field at the location of 40°N latitude, 90°W longitude.

5. Estimate the change in total field intensity at Rio de Janeiro, Brazil (22°50′S latitude, 43°10′W longitude) that westward drift of the earth's magnetic field will bring during the next century. Is this estimated change different from the

total change in field intensity that you can calculate from the annual change indicated on Figure 10–22? If so, what factors other than westward drift might contribute to this difference?

SELECTED READING

Bullard, E. C., C. Freedman, H. Gellman, and J. Nixon, The westward drift of the earth's magnetic field, *Philosophical Transactions of the Royal Society,* Series A, v. 243, pp. 67–92, 1950.

Elsasser, W. M., The earth as a dynamo, *Scientific American,* pp. 44–55, May 1958.

Garland, George D., *Introduction to Geophysics.* Philadelphia, W. B. Saunders Co., 1979.

Nettleton, L. L., *Gravity and Magnetics in Oil Prospecting.* New York, McGraw-Hill, 1976.

Rikitake, T., *Electromagnetism and the Earth's Interior.* Elsevier North-Holland, 1966.

Runcorn, S. K. (editor), *Methods and Techniques in Geophysics.* New York, Interscience, 1960.

Strangway, D. W., *The History of the Earth's Magnetic Field.* New York, McGraw-Hill, 1970.

ELEVEN

Surveying the Anomalous Magnetic Field

For more than three centuries, magnetic field readings have played a role in the exploration for minerals. As the modern oil industry emerged in the second half of the nineteenth century, magnetometer surveying was adapted to the exploration of sedimentary basins. Shortly before 1950, the development of practical methods for airborne magnetometer surveying marked the beginning of a dramatic departure from traditional ground-based geophysical exploration. Large areas could now be surveyed much more rapidly, and exploration could be extended into remote and heretofore inaccessible regions. The introduction of aeromagnetic surveying created a pressing need for efficient data processing that would have been impossible to meet had not computer technology been developing concurrently. Modern aeromagnetic and shipboard magnetic surveying also brought stringent demands for accurate navigation. Our purpose in this chapter is to describe the modern methods of magnetometer surveying and data processing. The product of these methods is a magnetic anomaly map. Interpretation of the magnetic anomalies is the subject of Chapter 12.

A magnetometer survey measures variation of the earth's magnetic field intensity over an area of interest. This area may be as small as a mining claim, covering a few thousand square feet, or as large as an entire nation. Most modern surveying is conducted with magnetometers mounted in aircraft. Some detailed work over small areas, however, is best done on the ground with hand-held or tripod-mounted instruments. Surveys at sea have been carried out using magnetometers trailed from ships as well as airborne instruments.

Position measurements are an important part of a magnetometer survey, but the requirements for accuracy of latitude and elevation are different from those for gravity surveys. Magnetic field intensity values do not require any adjustments comparable to the systematic adjustment of a gravity value to account for the effects of latitude, elevation, and mass above sea level. Nevertheless, the quality of a magnetometer survey depends on how accurately we can plot the intensity values on maps and profiles. To obtain this quality from an airborne magnetometer survey, we must give considerable attention to navigation procedures.

For all magnetometer surveys, we must record the time when each field intensity measurement is made. These times will be used later to make adjustments for diurnal variation of the external field and secular variation of the main field.

LAND SURVEYS

At the present time, land magnetic surveying is used mostly to map anomalies that are too narrow for adequate detection by airborne operations. The field procedure, which consists of reading a magnetometer at many observation sites, is not unlike that of land gravity surveying. In addition to readings along easily accessible roads, the detailed coverage of a typical land magnetometer survey requires numerous off-road traverses where observation sites are surveyed by compass or transit.

While taking readings, the observer should not wear ferromagnetic objects such as belt buckles, metal buttons, and metal eyeglasses; these objects will disturb the magnetometer. Test readings must be made to find an observation site that is an appropriate distance from automobiles, power lines, wire fences, bridges, metal culverts and pipelines, and railways. Depending on the desired accuracy, the magnetometer should be a few tens of feet to more than 100 feet from such objects. The presence of these and other metal objects makes it almost impossible to obtain reliable data in urban areas and along many highways.

The portable magnetometers described in Chapter 10 are designed for land surveys. Simplest to operate is the proton precession magnetometer (Figure 10–18), which is a drift-free instrument that measures total field intensity. Other instruments include hand-held and tripod-mounted flux-gate magnetometers (Figure 10–15) and tripod-mounted torsion magnetometers (Figure 10–12), which measure relative values of vertical field intensity. Some care must be taken to level these vertical magnetometers, but readings are not time-consuming, ordinarily being made in less than one minute. Because flux-gate and torsion magnetometers are sensitive to temperature changes, temperature measurements should be made at the observation sites. Portable magnetometers have been developed for measuring the relative values of horizontal intensity, but these instruments have not been widely used.

Magnetometer surveys must include measurements of the diurnal and secular variation of field intensity. In addition, mechanical magnetometers are subject to drift. For some surveys a recording magnetometer is operated at a base station while the survey is in progress. The more common practice is to reoccupy some of the observation sites at intervals of several hours, much as is done in gravity surveys. For a description of the field procedure, review the discussion of Figure 8–5.

In an exploration survey performed by means of a proton precession magnetometer, it is useful to establish a base station for the purpose of determining time corrections. Even though this magnetometer reads an absolute value of total field intensity at each observation site, it is necessary to designate a *time base*. Then the corrections for diurnal and secular variation can be applied to readings from all observation sites, and they can be adjusted to the same moment in time.

Most portable flux-gate and torsion magnetometers measure only relative vertical intensity. When these instruments are used for an exploration survey, a base station must be designated. The procedure is the same as for gravity surveying. Differences between readings at observation sites and the base station reading are used to adjust all relative values to a standard base station value. But for most land exploration magnetometer surveys, no effort is made to find the absolute vertical intensity at the base station by means of ties with an international network. Unlike gravity surveying, an absolute value of vertical magnetic field intensity is not ordinarily used in adjusting for the effect of the main field. Most vertical intensity surveys from different areas are not adjusted to one another. Each is done primarily to map variations of field intensity over a particular area. Should comparisons between two areas later become important, additional readings at the base stations of these areas can be made to establish the necessary tie.

AEROMAGNETIC SURVEYS

Flight Operations

At the present time, most exploration surveys of magnetic field intensity are done with magnetometers installed in airplanes or helicopters. The operation is called *airborne magnetic surveying* or *aeromagnetic surveying*. The usual procedure is to operate the magnetometer continuously along equally spaced parallel flight lines covering the survey area (Figure 11–1). Tie lines crossing these parallel flight lines are then flown to obtain corrections for time variations of the field. Compared with ground surveying, which can be done by one or two individuals, aeromagnetic surveying is considerably more complicated. A highly trained staff of several people is needed to fly and maintain the aircraft and to operate the magnetometer, the navigation apparatus, and the computers used for data compilation and

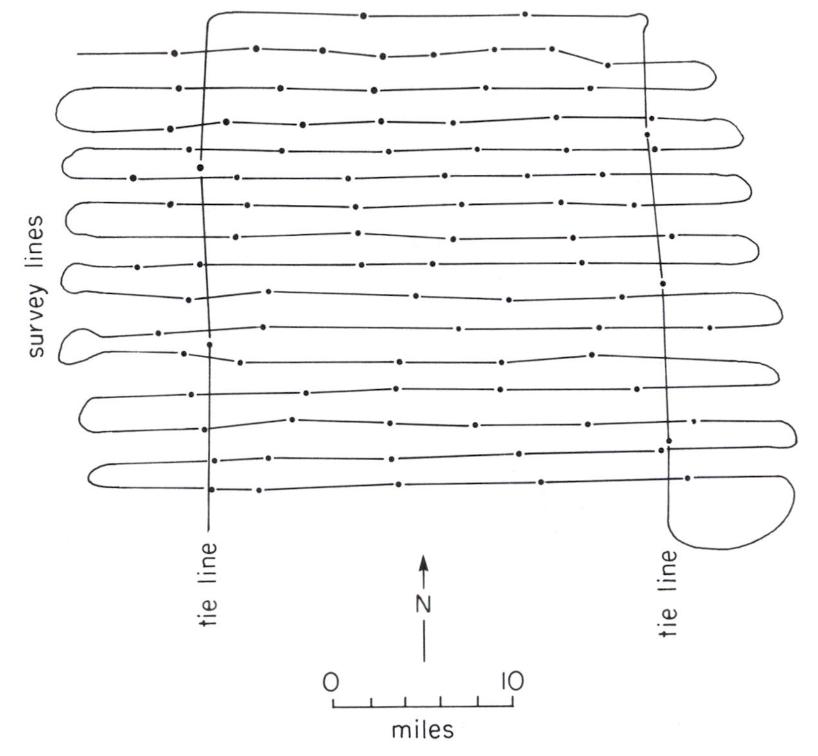

Figure 11–1
Survey flight lines and tie lines of an aeromagnetic survey. Dots indicate position control points.

processing. The initial cost for undertaking this work is much higher than for ground surveying. Nevertheless, aeromagnetic survey costs, usually figured on a line mile or line kilometer basis, are lower than ground survey costs, and much less time is needed to complete the work.

Considerable thought goes into the installation of the magnetometer. It is simplest to place most of the electronics and recording apparatus in the aircraft and to trail the magnetometer sensing unit in a streamline container called a "bird" at the end of a cable 100 to 500 feet long. The only modification of the aircraft necessary is making an opening through which the bird can be lowered and raised on its cable. The magnetometer components (Figure 11–2) are compact and do not occupy much space. Trailing a bird can present problems, however, especially when flying at tree-top level over hilly terrain. Even the best pilots occasionally lose a bird in a tall tree or on a hill top.

Flight operations are less difficult with a magnetometer sensing unit attached to the aircraft, but the sensitivity of the instrument to nearby ferromagnetic parts poses a severe problem. Not only do these parts produce strong interfering magnetic fields, but the fields can change depending on flight direction and speed, and sometimes even on the way that the aircraft is parked when not in use. These effects are partly overcome by placing the sensing unit at a wing tip, in a

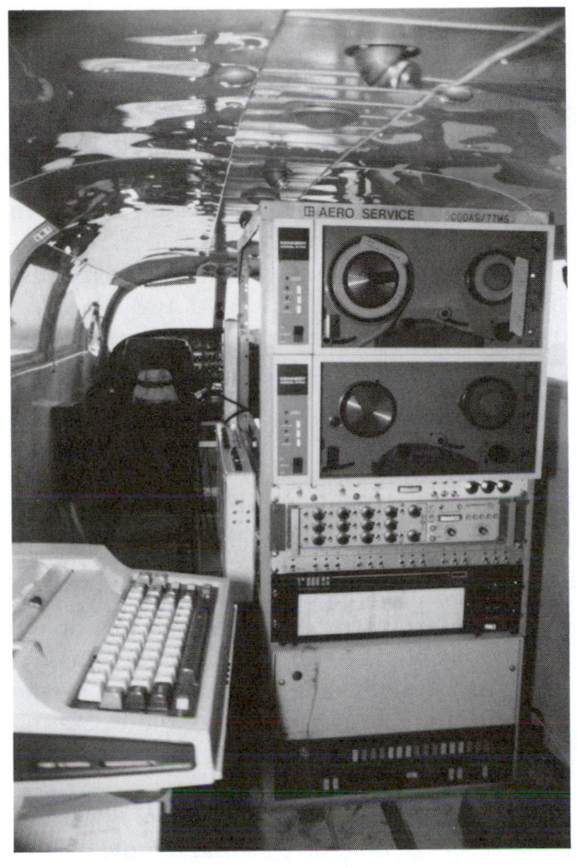

Figure 11–2
Magnetometer controls and electronics and recording apparatus installed in an airplane. (Courtesy of Aero Service, Inc.)

long probe reaching forward from the nose, or in a long "stinger" reaching back from the tail (Figure 11–3). In addition, compensating magnets and coils can be mounted to nullify to some extent the aircraft magnetism close to the sensing head. Costs of instruments, installation, and related structural modifications can almost double the initial cost of the aircraft. Even so, some effects of aircraft magnetism remain. It is necessary to determine *bearing corrections* experimentally by test flights in different directions over the same point. The tests should be repeated frequently to account for changes in aircraft magnetism.

Elevation

Some aeromagnetic surveys are flown at a constant elevation above sea level that is determined with a barometric altimeter. This is the common practice for reconnaissance surveys with flight lines spaced from one mile to several miles or several tens of miles apart. Constant-elevation surveys are also the most practical for rugged mountainous terrain and obviously for surveys over the oceans.

The alternative practice is to maintain constant elevation above the land surface. An accomplished pilot guided by a radio altimeter can maintain elevation over areas of subdued relief and even moderately hilly terrain. This

Figure 11–3
Aircraft with tail "stinger" and wing tip mountings for magnetometer sensing units. (Courtesy of E. G. and G. Geometrics, Inc.) Magnetic fields associated with electric circuits and ferromagnetic parts in the aircraft have less effect on sensing units in such mountings.

kind of survey is called a *draped* survey and is used for more detailed exploration work where lines less than a few miles apart are flown at heights of under 1000 feet. The most detailed aeromagnetic surveying can be done with lines as close as 500 feet apart at heights of about 100 feet. Only the most skillful pilots can fly with this precision.

Navigation

A problem of critical importance in aeromagnetic surveying is keeping track of the aircraft position. It is not simply a question of guiding the aircraft from one location to another some distance away; rather, a record of point-by-point positions along the entire route must be obtained. The positions are recorded by one of several navigation systems.

Aerial photography can be used where it is possible to identify visual reference features on the ground. Before the survey is initiated, proposed flight lines are drawn on aerial photographs and maps that are already available. The pilot then follows these lines as closely as possible, attempting to fly directly over visible landmarks such as roads, buildings, hills, and river bends. At the same time, a camera mounted vertically on the aircraft photographs the ground at constant intervals of time, say, every one to two seconds. Photo numbers are transmitted to the magnetometer recorder for each moment that a picture is taken. Later, these photographs can be compared with the earlier aerial photographs and maps to find the point-by-point aircraft positions. Television cameras can also be used for this purpose. With this equipment, the landscape images are recorded on magnetic tape rather than on film. Aerial photograph navigation is time-consuming and requires several people to process and compare all the pictures. It is not satisfactory for areas with few easily identifiable landmarks and is impractical for surveys over large expanses of Arctic tundra, densely forested lowlands, and oceanic areas.

Electronic navigation is widely used in aeromagnetic surveying. Modified radar systems of the *shoran* type consist of a radar pulse generator in the aircraft and two or more transponders fixed at known locations on the ground. When a transponder receives a radar pulse, it is stimulated to emit a response signal that is detected by an aircraft receiver. Signal travel times between the aircraft and the transponders are accurately measured. Since the wave speed is known, distances from the transponders can be calculated. Circles drawn around the transponder locations at radii corresponding to these distances intersect at the aircraft position. Because shoran systems depend on straight lines of transmission, the curvature of the earth restricts the operating distance to less than about 50 miles.

For a greater distance, a *loran*-type navigation system can be used. It consists of an aircraft receiver that detects radio waves from two or more pairs of transmitters on the ground. Each pair consists of transmitters at two locations some distance apart that broadcast waves of the same frequency. These waves interfere to produce a standing wave pattern on which the lines of maximum constructive interference are indicated by a family of hyperbolas (Figure 11–4). Each pair of hyperbolas defines a corridor. The aircraft receiver detects the strongest signal when crossing one of these hyperbolas. The weakest signal is received along the middle of a corridor where destructive interference is maximum. Distance away from the side of a corridor is interpolated in proportion to signal

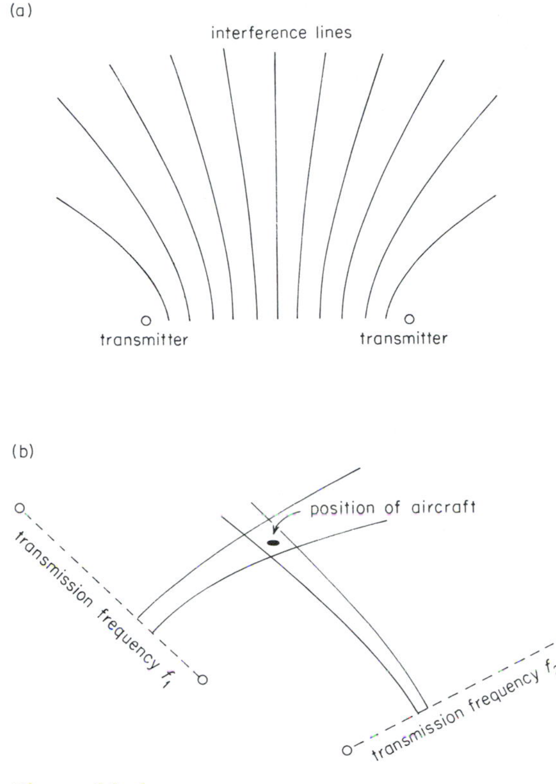

Figure 11–4
(a) This pattern of interference lines represents constructive interference of waves broadcast at the same frequency from two loran transmitters.
(b) Intersecting interference corridors from two pairs of loran transmitters broadcasting at different frequencies are needed to obtain a position.

strength, but the receiver cannot distinguish one corridor from another. The navigator or an appropriate computer makes this distinction by keeping count of each corridor crossing, indicated by a maximum in signal strength, as the aircraft travels away from an established base position. A second pair of ground-based transmitters at two other locations produces another standing wave pattern at different frequency. Aircraft position is found at the intersection of two corridors from these independent standing wave patterns.

A moment-by-moment record of aircraft movement can be obtained with a *Doppler navigation system.* Radar beams from aircraft transmitters are directed to four points on the ground, from which they reflect back to aircraft receivers. Shifts in the frequency of the radar waves, caused by the Doppler effect, are related to transmitter speed. This same kind of frequency shift causes the pitch of a car horn to rise as a car approaches and to lower as the passing car travels away. Frequency shifts of the four radar beams are used to calculate forward and side components of aircraft speed over fixed intervals of time. This information is used to find moment-by-moment vectors of the direction and distance of aircraft movement. These vectors are summed to get the flight path. But the Doppler system is subject to drift. Flight path segments must be adjusted to control positions found by the other navigation methods we have described.

While the pilot is flying an aeromagnetic survey, point-by-point magnetometer readings are recorded, as well as other measurements needed to determine the latitude, longitude, elevation, and time of each reading. The early practice was to have an observer write progress notes about visible landmarks, aircraft speed and direction, and time on the magnetometer strip chart. Upon completion of a flight, the chart was turned over to the processing staff which devoted considerable effort to making graphical adjustments and compiling the results for preparation of profiles and contour maps. This procedure has been largely replaced by integrated data acquisition systems under computer control; these systems encode all measurements simultaneously on digital magnetic tape. Much of the subsequent data processing and even the preparation of profiles and maps can now be done by computers.

Some mention should be made of aeromagnetic survey logistics. Suitable facilities must be available for aircraft maintenance. In remote areas, landing sites must be found, and fuel caches must be established. Survey operations cannot proceed during inclement weather or when ground-based recording magnetometers indicate that magnetic storms are in progress. Calm, clear weather and stable magnetic field conditions are necessary for precision surveying. These are some of the factors that must be considered in plans for a successful operation.

SHIPBOARD MAGNETOMETER SURVEYS

Sea-level magnetic field intensity measurements are usually made in conjunction with other shipboard geophysical operations. For relatively low cost, a magnetometer can be installed in a vessel already equipped for marine seismic and gravity surveying. Except in special circumstances, magnetic field surveying alone cannot justify the considerable expense of operating a ship, which would far exceed the cost of an equivalent aeromagnetic survey.

The most widely used instrument for shipboard surveying is the proton precession magnetometer. Interference from all the ferromagnetic metal parts and electrical apparatus in the ship is reduced by towing the magnetometer in a streamline container 500 to 1000 feet astern. The cable and the container are designed to have enough bouyancy to keep the sensing unit close to the water surface. Thus, it becomes possible to maintain accurate elevation control and to keep the magnetometer from becoming entangled on the bottom in shallow water. The long cable can present problems where the vessel must be maneuvered along closely spaced survey lines or in restricted waters. Operations on sheltered nearshore areas, bays, and lakes can be handled with small power boats equipped with much shorter cables. But aeromagnetic surveys of these areas are usually more cost-efficient.

Tie lines crossing several parallel ship tracks are important for marine seismic and gravity surveys as well as for magnetometer surveys. Measurements along these lines are used to make corrections for diurnal and secular variations of magnetic field intensity.

MAGNETIC GRADIENT SURVEYS

Magnetic field intensity does change with elevation. This change is small and is related mostly to distance from local anomaly sources. The smallness of the change in magnetic field intensity and its dependence on distance from a local anomaly make it unlike the corresponding gravity change, which we already know is quite large and depends mostly on distance from the earth's center of mass. The vertical gradient of magnetic field intensity, $\Delta F/\Delta z$, is the intensity change divided by the small increment of elevation over which it occurs. Until recently, it could not be measured accurately enough for exploration purposes. It can now be measured with high-precision cesium vapor magnetometers.

The first successful surveys of the vertical gradient of magnetic field intensity were done with two magnetometer sensing units in "birds" attached 100 to 200 feet apart on a cable trailed from the aircraft. At relatively low speed with sufficient weight, the cable between the two units can be made to hang almost vertically. Further improvement in instrument sensitivity now allows both units to be mounted about 15 feet apart on a rigid frame attached to the aircraft (Figure 11–5). In this configuration, each sensing unit measures total field intensity independently, as in a conventional aeromagnetic survey, and the gradient is found from the small difference in these measurements. The compensation and bearing corrections for aircraft magnetism that are different for the two sensing units must be done with great care if the results are to be reliable.

Figure 11–5
Dual magnetometer mounting for airborne measurement of the vertical gradient of total field intensity. (Courtesy of Questor Surveys, Ltd.) Small differences in magnetic field intensity can be detected from the readings of these two separated sensing units.

The results of a magnetic gradient survey are independent of diurnal, secular, and main field variations, all of which, for practical purposes, affect both sensing units equally. Broad regional anomalies are almost entirely eliminated by this survey procedure, which turns out to be sensitive only to local magnetic anomalies. These regional anomalies and main field variations are, of course, evident on the separate record of field intensity measurements from either one of the sensing units.

DATA PROCESSING

The processing of magnetic exploration survey data includes all the adjustments and compilation necessary to prepare profiles and anomaly maps suitable for geologic interpretation. Magnetometer measurements are corrected for time-dependent variations, and field intensity values are calculated. Navigation misclosures and elevation errors are adjusted so that positions of field intensity measurements can be determined. The effects of the main magnetic field, and perhaps regional features of the anomalous field as well, are removed. The final products of data processing are profiles and contour maps that display intensity variations of the anomalous magnetic field. For ground magnetic surveys involving relatively few observations, a small calculator may be adequate for making the numerical computations. Larger digital computers have become essential, however, for processing the large volume of data produced by an aeromagnetic survey.

Position

In areas where good base maps are available, some ground survey positions can be accurately determined in the field by specific landmarks such as road intersections and stream crossings. Off-road magnetometer measurements are often made after grid positions have been carefully surveyed by transit or compass and marked by stakes. But for some ground surveys, it is more practical to follow irregular traverses, making all the measurements together. Positions are calculated later from transit or compass sightings.

Navigation measurements from aeromagnetic survey flight lines can require a considerable processing effort. If "in-flight" photography is used, ground points directly beneath the aircraft must be identified on a very large number of pictures. These points must then be plotted on aerial photographs or maps of the survey area and then connected to display flight paths. Electronic navigation requires computer processing of the signals that are transmitted and received in the aircraft. Some on-board processing is necessary to provide preliminary positions needed to keep the aircraft on course. These positions can be refined, and more intermediate positions can be calculated by later reprocessing of the instrument signals. This refinement is important not only for positioning the parallel flight lines, but also for accurate location of the tie line intersections that are so necessary for correcting time-dependent variations of field intensity.

Time-Dependent Corrections

Information for time-dependent corrections is obtained principally by reoccupation of certain observation sites. For ground surveying, the sites most convenient for reoccupation are used, and for airborne surveys they are the intersection points of tie lines that cross a set of parallel flight lines. A magnetometer that records continually at a fixed location can also be used.

Reoccupations should be timed to detect the important features of diurnal variation. Typical records (Figure 11–6) indicate that reoccupation at one- to two-hour intervals is sufficient to record the basic diurnal change but would miss short-term fluctuations of as much as 10 gammas. For ground surveys requiring 1- to 2-gamma accuracy, much more frequent reoccupations, or the record from a base station magnetometer, must be available. Ground magnetometer readings are corrected by the same method we described in Chapter 8 for gravity measurements. Details can be found in the discussions of Figures 8–3 and 8–6.

The idea has been proposed that aeromagnetic field intensity measurements made with drift-free proton or cesium magnetometers can be adjusted simply by subtracting, point by point, the values of diurnal and secular variation obtained from a ground-based recording magnetometer. Tests show that these time-dependent variations ordinarily differ by less than a few gammas over distances of more than 100 miles. But other adjustments must be made to account for the effects of navigation and elevation errors. For many surveys, it is most practical to correct for the combination of these effects and the diurnal and secular variation by adjusting field intensity mis-

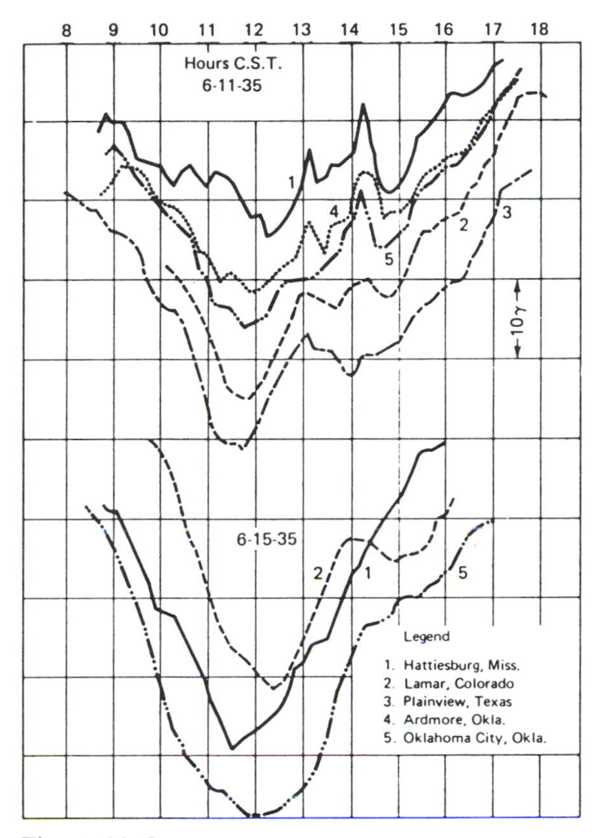

Figure 11–6

Simultaneous records of short-term magnetic field variations at five locations. (Victor Vacquier, Short term magnetic fluctuations of local character, *Terrestrial Magnetism and Atmospheric Electricity*, v. 42, n. 1, pp. 17–28, 1937.) These measured changes in magnetic field intensity indicate for the most part the diurnal variation.

closures around rectangles defined by survey lines and tie lines.

The rectangle ABCD in Figure 11–7 is formed by two aeromagnetic survey lines and two tie lines. Typically, these lines might be spaced at respective intervals of one mile and ten miles. At a flight speed of 150 miles/hour, legs AB and CD would each be flown in four minutes, and legs BC and AD would each be flown in 0.4 minute. Diurnal and secular variation can be assumed to vary linearly with time over these short intervals. Next, look at the change in field intensity along each leg. The values of 132 gammas and −100 gammas are differences between survey line readings at points A and B and points C and D, respectively. Similarly, the values of 15 gammas and −25 gammas are differences between tie line readings at points B and C and points D and A, respectively. In the absence of any errors or time-dependent variations, the sum of the differences along the sides of this rectangle should be zero. Instead, we see that they add up to 22 gammas. This misclosure in field intensity represents the combined effect of diurnal and secular variation, errors in location points A, B, C and D, and differences in aircraft elevation along these flight lines at the places where they cross. Although the pilot tries to hold a constant elevation, unpredictable updrafts and downdrafts, as well as aircraft inertia, cause some discrepancies.

Distributing this 22-gamma misclosure linearly around the rectangle appears to be the next step. For its 22-mile perimeter, this amounts to a correction that increases by one gamma per mile for readings along the path ABCD. If we choose the survey line value at A as a reference value, the correction of 22 gammas for the tie line value at A is calculated for the 22-mile distance around the rectangle. This correction brings the two readings into agreement. But there is a problem if we try the same procedure for the adjacent rectangle DCFE. Here we find a misclosure of 11 gammas that would be distributed at the rate of 0.5 gamma/mile around this rectangle. This

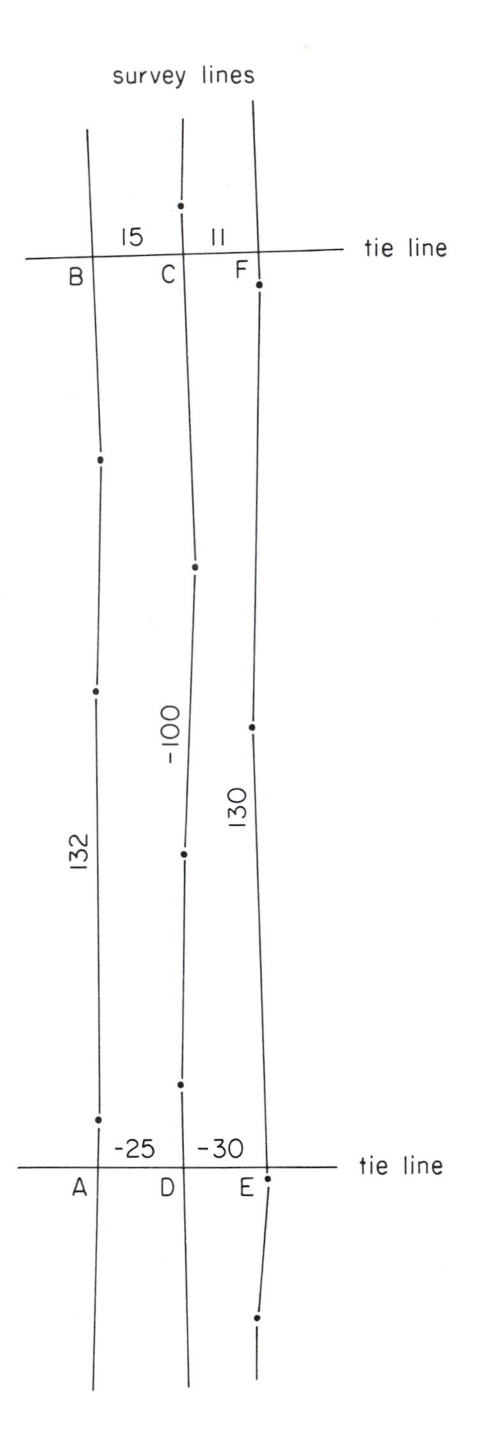

indicates a different correction for the side DC which is common to both rectangles. Should we choose corrections figured on the 0.5-gamma/mile change or the 1-gamma/mile change? Rather than making this choice for these and other rectangles, a numerical scheme can determine intermediate correction values in a way that minimizes the sum of discrepancies around all rectangles used in the analysis. Because the number of values is vast, a digital computer is essential for carrying out these adjustments.

The Main Magnetic Field Adjustment

Magnetic field intensity values that have been corrected for time-dependent variations combine both the main field and the anomalous field. What additional corrections are made before proceeding with a geologic interpretation of these values? No standard correction is ordinarily made to account for the elevation of ground observation sites or aeromagnetic flight lines. Unlike the relatively large standard correction made to gravity values, the corresponding magnetic corrections are small enough to neglect. The small rate of change of the main field intensity with elevation ranges from 0.5 gamma per 100 feet in low latitudes to about 1 gamma per 100 feet near the poles.

It is usually impractical to make topographic corrections by a consistent procedure like the one used for comparable gravity corrections. No correction is needed if the topog-

Figure 11–7
Differences in magnetic field intensity measured between intersection points of aeromagnetic survey lines and tie lines.

raphy is formed from nonmagnetic rocks. But the effects can be significant where magnetized rocks exist. Because of the large differences in rock magnetism from place to place, topographic corrections, rather than a standard data processing procedure, must become part of the geologic interpretation of a survey.

The main magnetic field imposes a broad regional variation over every exploration survey area. A common practice is to remove it by the *scalar subtraction method.* The procedure is basically simple. From each field intensity measurement, the corresponding main field intensity value is subtracted. Main field intensity values can be obtained in different ways. They can be calculated directly from the appropriate spherical harmonic equations, but this method is too complicated for most exploration surveys. The more practical approach is to interpolate values from maps or tables already prepared from solutions to these equations. Tables of values on an appropriate latitude–longitude grid and maps of suitable scale are available from agencies listed in Table 11–1.

TABLE 11–1 Sources of Geomagnetic Information

Branch of Global Seismology and Geomagnetism
U.S. Geological Survey, Mail Stop 967
Box 25046, Federal Center
Denver, Colorado 80225, U.S.A.

Environmental Data Service
National Oceanic and Atmospheric Administration
National Geophysical and Solar-Terrestrial Data Center D62
Boulder, Colorado 80302, U.S.A.

Defense Mapping Agency Hydrographic Center
Washington, D.C. 20390, U.S.A.

Owing to the very large number of total field intensity values (F_T) obtained from an aeromagnetic survey, we must use a digital computer to calculate the corresponding main field values (F_M). A simple procedure can be used, because the main field intensity contours over areas less than 100 miles long and wide are nearly straight and equally spaced. The map in Figure 11–8 is typical. We can use it to find the rates of change in field intensity with distance in the east ($\Delta F/\Delta x$) and north ($\Delta F/\Delta y$) directions. These rates can be expressed in units such as gammas per mile. Now we choose a reference point, perhaps the southwest corner of the map, where the main field intensity is F_0. We can now calculate the main field intensity F_M at another location using its distance east (x) and north (y) of the reference point:

$$F_M(x,y) = F_0 + \frac{\Delta F}{\Delta x} x + \frac{\Delta F}{\Delta y} y \qquad (11\text{–}1)$$

Finally, we subtract this value from the total field intensity F_T at that location to obtain a measure (T) of the anomalous field:

$$T = F_T - F_M \qquad (11\text{–}2)$$

A digital computer is easily programmed to calculate F_M and T for a large number of measurements. The same procedure can be used for vertical field intensity surveys on the ground by substituting the vertical intensity (V_M) of the main field in place of F_M.

Next, we must consider very carefully just what we have obtained by removing the main field in the manner just described. Here it is important to point out a limitation of most magnetic surveys. Airborne magnetometers measure *only* the intensity and not the direction of the magnetic field. Similarly, on the ground only vertical or total intensity is mea-

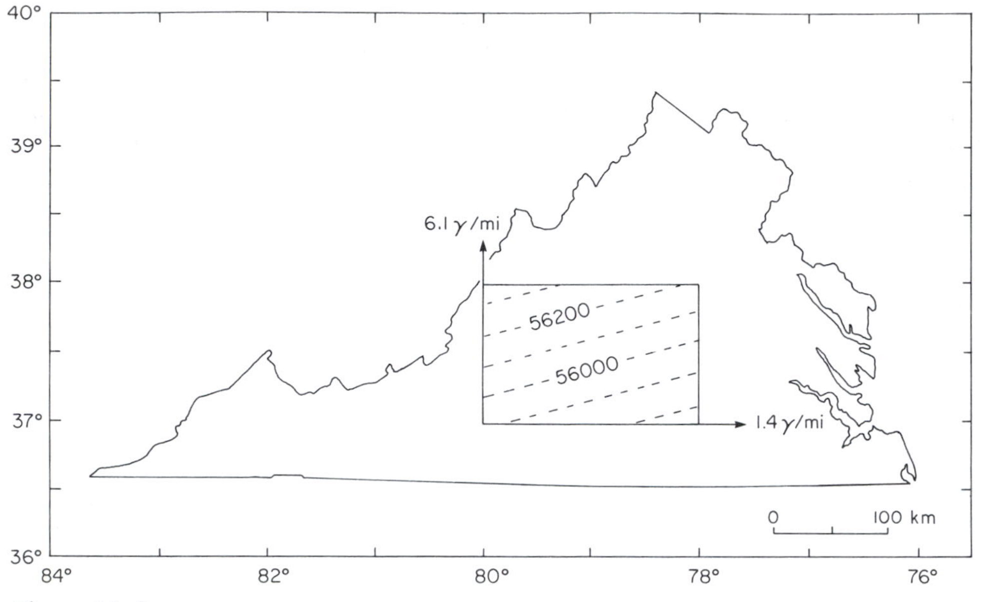

Figure 11–8
Contours of the main magnetic field intensity (F_M) and rates of change of field intensity in north and east directions for an area in central Virginia. These contours were drawn using International Geomagnetic Reference Field values at latitude and longitude intervals of one degree. Sources of these values are listed in Table 11–1.

sured. We know that three elements are needed to describe a field completely (Figure 10–10). Exploration surveys provide values of one element.

Now let us examine how the separate contributions of the main field and the anomalous field are combined in a total field intensity value. We will use the simple example presented in Figure 11–9 in which the source of the anomalous field is a single dipole buried in the crust. Vectors representing its field, together with main field vectors, are shown at five observation points along a profile. We assume that this profile is short, say less than 10 miles long. Over this distance the main field changes so little that we consider its intensity and direction to be constant. At each point the vector representing the total field ($\mathbf{F}_T$) is the sum of the anomalous field vector ($\mathbf{F}_A$) and the main field vector ($\mathbf{F}_M$).

A total-intensity magnetometer measures the length of the $\mathbf{F}_T$ vector, which changes along the profile. This is plotted in Figure 11–9. But the inclination of this vector, which also changes, is not measured. What do we get by a scalar subtraction of main field intensity from total field intensity as in Equation 11–2? Figure 11–10 shows what the resulting value of T represents. Here the vector $\mathbf{F}_M$ is rotated to the vector $\mathbf{F}_T$ along the arc ab. The difference in their lengths is T, which is obviously different from the anomalous field intensity

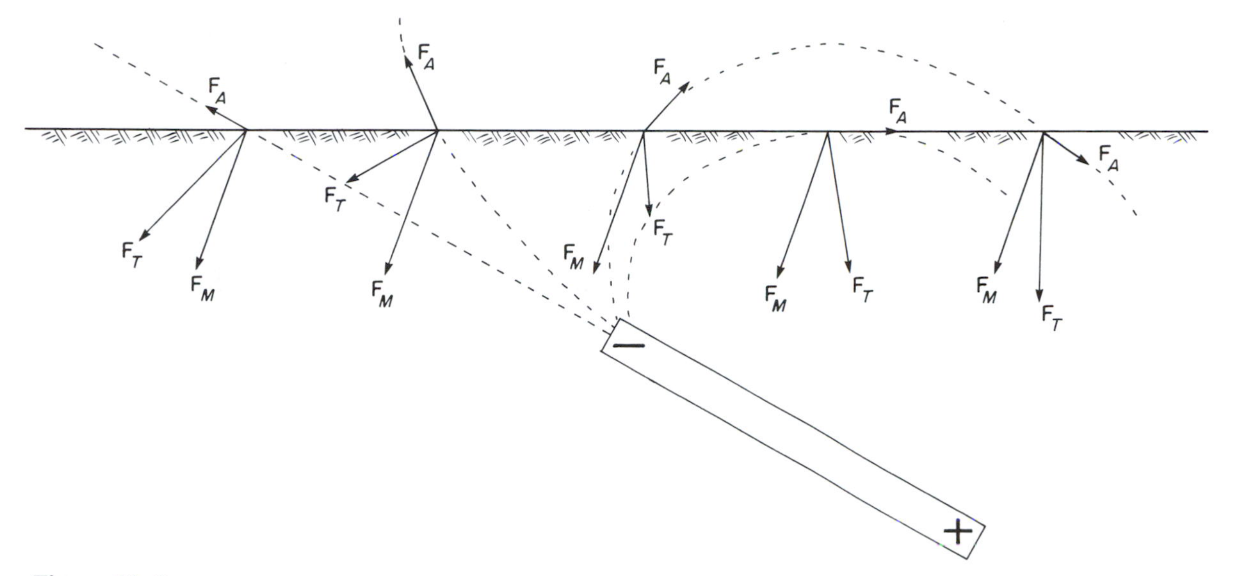

Figure 11–9
Vectors at five locations along a profile crossing a buried dipole that indicate the dipole field intensity ($\mathbf{F}_A$), the main field intensity ($\mathbf{F}_M$), and the total field intensity ($\mathbf{F}_T$) which is the vector sum of $\mathbf{F}_A$ and $\mathbf{F}_M$. Vectors at different points are not drawn to exact scale relative to one another. A total-intensity magnetometer measures the magnitude of $\mathbf{F}_T$.

F_A. This illustrates that the conventional practice of removing the main field by scalar subtraction does not yield values of anomalous field intensity. It is impossible to determine the intensity F_A without knowing the direction of the vector $\mathbf{F}_T$. Nevertheless, we can calculate T over models of magnetic anomaly sources for comparison with measured values. This subject is discussed in the section on interpretation in Chapter 12.

Another common practice is to remove the main field effect by conventional methods of graphical smoothing, averaging, and wavelength filtering. These methods, which are described in Chapter 9, are used on magnetic survey data in the same way that they are used on gravity survey data. Because they cannot separate the specific contribution of the main field from other broad regional anomalies related to the anomalous field, they are less satisfactory than the scalar subtraction method. But they can be quite effective for data processing when the survey objective is to target local anomalies.

Contour Maps and Profiles

The final step in data processing is the preparation of profiles and contour maps that display magnetic anomalies of geologic interest. In recent years, the traditional methods of plotting field intensity values and manual contouring have been yielding to computer-drawn maps and profiles. Although manual methods are satisfactory for some ground-

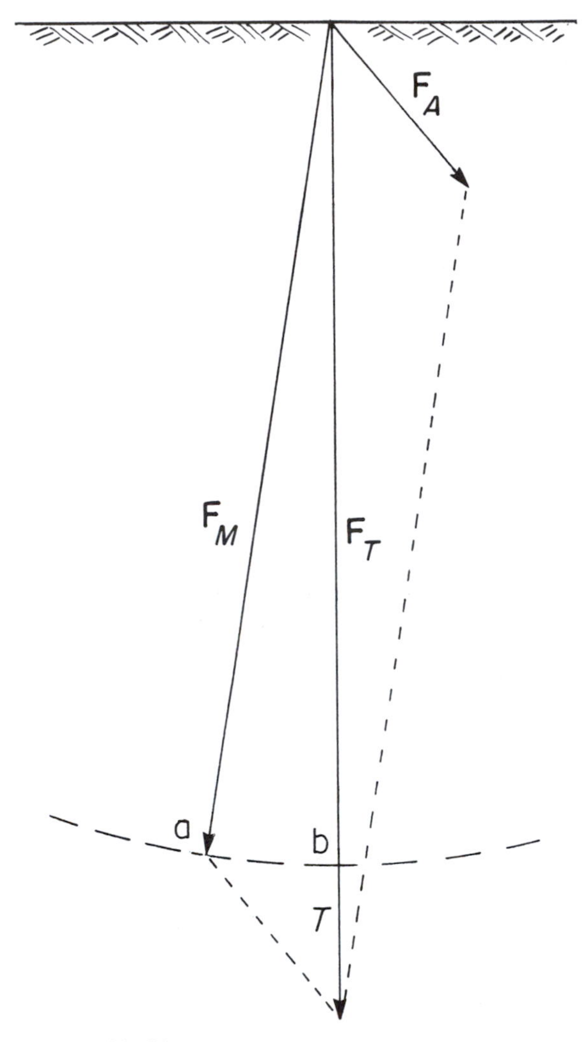

Figure 11–10
Vector diagram of the anomalous field intensity ($\mathbf{F}_A$), the main field intensity ($\mathbf{F}_M$), and the total field intensity ($\mathbf{F}_T$). The total-intensity anomaly T is the difference in the lengths of $\mathbf{F}_T$ and $\mathbf{F}_M$.

based surveys involving relatively few measurements, compilation of most aeromagnetic maps and profiles is at least partially, if not completely, automated.

Aeromagnetic flight lines are always somewhat irregular. Even the best pilot cannot anticipate all the wind variations and other factors that cause the aircraft to deviate from the plan of straight, evenly spaced lines. Therefore, an airborne magnetometer produces field intensity measurements at points irregularly spaced over the survey area. Values on a uniform grid must be interpolated from these measurements before computer contouring can be done.

Several computer gridding schemes can be used with aeromagnetic data. Some involve weighted averaging. The computer stores the intensity values and position coordinates of all flight line observation points. Grid coordinates are also specified. Positions of the observation points are scanned to locate those situated within a predetermined distance from each grid point. Intensity values at these nearby points are multiplied by weighting coefficients that vary according to distance. Field intensity at that grid point is found from the average of these products.

Methods of polynomial interpolation can also be used to find intensity values at grid points. For example, suppose we say that in a small area surrounding a grid point, the variation of field intensity with position can be represented by a quadratic polynomial,

$$F(x,y) = a_1 + a_2x + a_3y + a_4x^2 + a_5xy + a_6y^2 \quad (11\text{–}3)$$

where x and y are position coordinates and $a_1 \ldots a_6$ are constants that can be determined from the six nearest observation points. Six separate equations are then obtained by inserting, in turn, the coordinates x and y and the field intensity F from each nearby point into Equation 11–3. Simultaneous solution of these six equations yields values for a_1, a_2, $\ldots$, a_6. Now they can be used together with the grid point coordinates in Equation 11–3 to calculate F at this location. This procedure is repeated for each grid point.

We can now proceed with automatic contouring of the computer gridded data. A simple scheme for doing so is illustrated in Figure 11–11. In this figure the sides and diagonals of the grid cells form a pattern of triangles. This example shows how the computer searches for points on, say, the 100-gamma contour. It begins on the top grid row by comparing adjacent intensity values, advancing point by point until a place is found where one value is higher and the other lower than the 100-gamma contour level. It calculates by linear interpolation the intermediate position where field intensity is 100 gammas, and then it scans the other two sides of the triangle to locate the other position that must exist where the intensity is 100 gammas. A digital plotter is commanded to draw a line connecting these positions. The second position lies along a side shared with another adjacent triangle, whose other sides are searched to find the third position where intensity is 100 gammas. The digital plotter then extends the line from the second to this third position. The procedure continues in this way through a succession of adjacent triangles. Adjoining line segments drawn by the plotter make up a 100-gamma contour line. After scanning the entire grid, the computer returns to the top row to begin a new search for another contour level.

The sample in Figure 11–11 has diagonal sides that extend from the upper left to the lower right corners of each grid cell. Suppose that the analysis is repeated using diagonals extending from the lower left to the upper right corners. In all likelihood, a slightly different contour pattern would be obtained. Which pattern is correct? Probably neither exactly reproduces the true pattern of field intensity variation. There will always be some ambiguity because only the values from a set of grid points are used, rather than values from all points in the area. Such ambiguity is not restricted to automatic contouring prac-

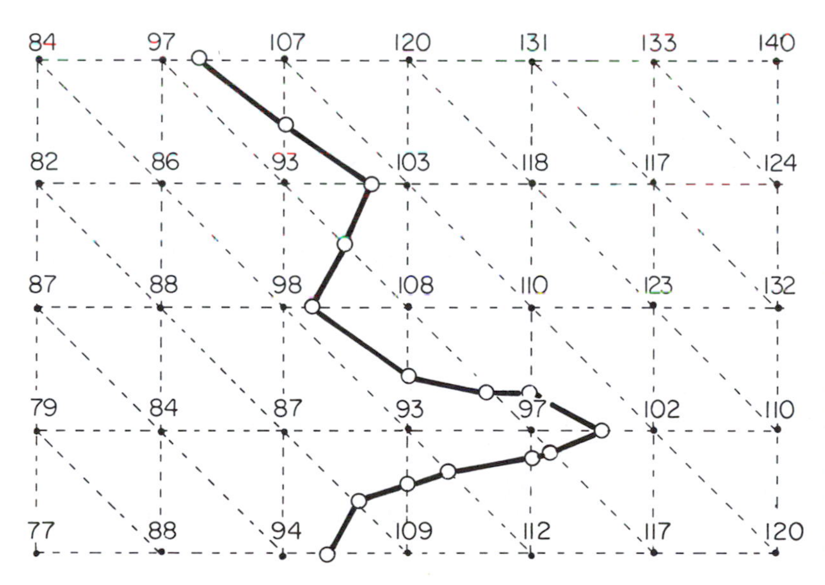

Figure 11–11
Procedure for machine contouring of values given on a grid involves a search for points on a contour line by linear interpolation along the sides of triangles formed by grid lines and diagonals. These points are connected by a computer-controlled plotter to form a contour line.

tices. Differences can also be expected when the same data are contoured manually by different individuals.

A factor that must be considered when contouring aeromagnetic data is the difference in spacing of measurements along and transverse to the flight lines. Suppose that north–south lines are flown at one-mile intervals. Along each flight line, a flux-gate or cesium vapor magnetometer produces an almost continuous sequence of readings. At normal speeds, even a proton precession magnetometer with an operation cycle of one to two seconds produces readings at points only a few hundred feet apart. Therefore, very narrow anomalies can be detected on the north–south flight line profiles.

What about an east–west profile? In this direction, the only available readings are spaced at the one-mile intervals where the profile intersects the flight lines. This spacing is too large to detect anomalies as narrow as some that appear along flight lines. If a contour map is prepared from readings that are spaced differently in the north–south and east–west directions, the anomaly patterns tend to be distorted.

To ensure against such distortion, we must smooth each flight line profile by a method of averaging or wavelength filtering. This operation must be carried out before gridding and contouring in order to eliminate all anomalies that are too narrow to be properly represented by readings at intervals equal to the flight line spacing. Following this procedure, the data can be contoured to obtain undistorted anomaly patterns. The contour map will display no evidence of the narrow anomalies that originally appeared along the flight lines. The best way to present this information is by a separate display of the original flight line profiles.

MAGNETIC ANOMALIES

What typical features can be looked for on a magnetic anomaly map? The contour patterns can be quite complicated because of overlapping anomalies from several sources. But one feature common to most contour patterns is an association of nearby magnetic anomaly highs and lows. This feature is evident on the magnetic anomaly map in Figure 11–12, which was prepared from vertical torsion magnetometer readings at 100-foot intervals on a grid covering the survey area. This survey in southwestern Utah was part of a search for other mineralized zones in the vicinity of a small mine where sulfide ores had been discovered. Notice the numerous highs and nearby lows that make up the magnetic anomaly pattern. This complicated pattern suggests several anomaly sources, each of which possesses a dipolar field. In Chapter 12 we describe how the anomaly produced by each source is high in an area where the field of the positive pole is strongest and low in a nearby area dominated by the field of the negative pole. The dipolar nature of magnetic anomaly sources leads to magnetic anomaly patterns that are more complicated than those found on gravity anomaly maps.

On a larger scale, the magnetic anomaly map of west Texas (Figure 11–13) also displays a pattern of associated highs and lows. This map was prepared from aeromagnetic measurements 400 feet above the land surface along east–west flight lines at 3-mile intervals. Different parts of the survey were done with a proton precession magnetometer and with a flux-gate magnetometer. The magnetic anomalies are much broader than those seen in Figure 11–12, but they make up a similarly complicated pattern. Notice that the anomalies on both maps have amplitudes ranging from a

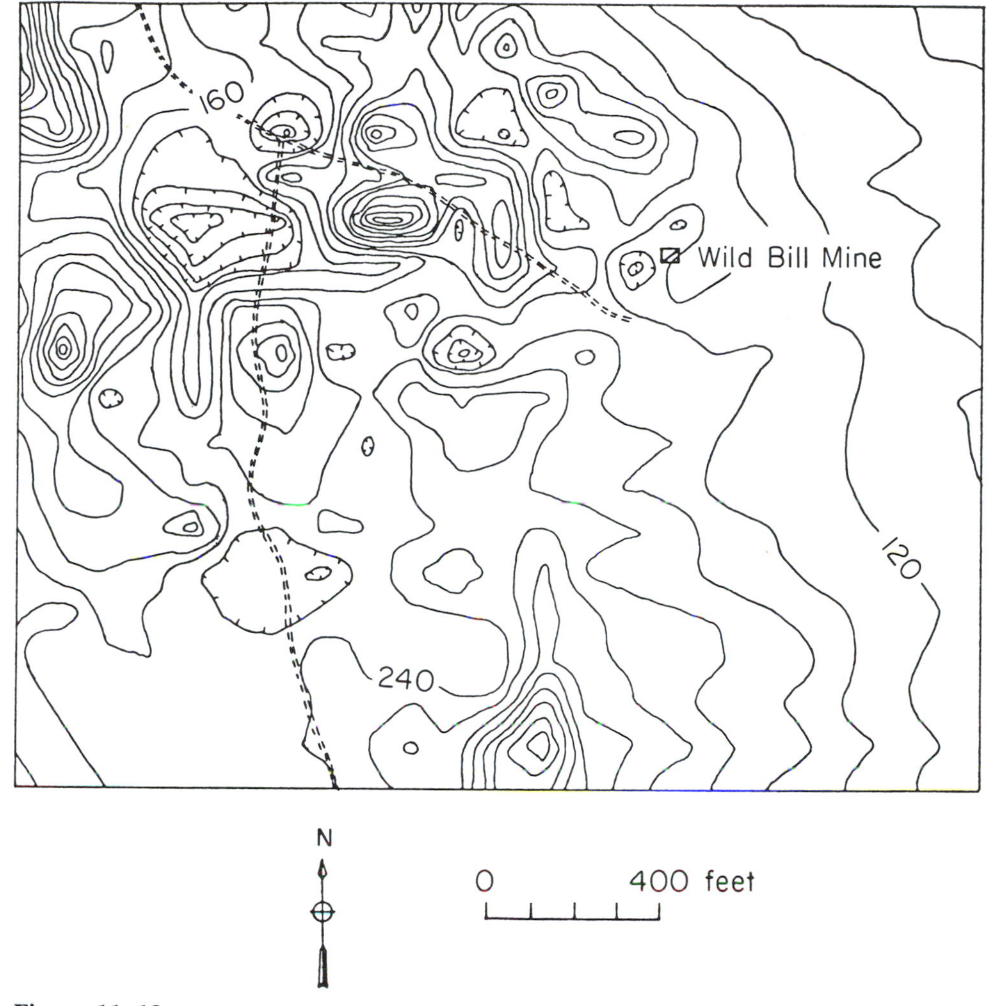

Figure 11–12

Vertical-intensity magnetic anomaly map of an area near the Wild Bill Mine in the San Francisco Mountains of southwestern Utah prepared from torsion magnetometer readings at 100-foot intervals on a grid covering the area. The contour interval is 40 gammas. (From J. W. Schmoker, Interpretation of aeromagnetic, magnetic, and gravity data from the San Francisco Mountains vicinity, southwestern Utah, Ph.D. dissertation, Virginia Polytechnic Institute and State University, p. 98, June 1969.)

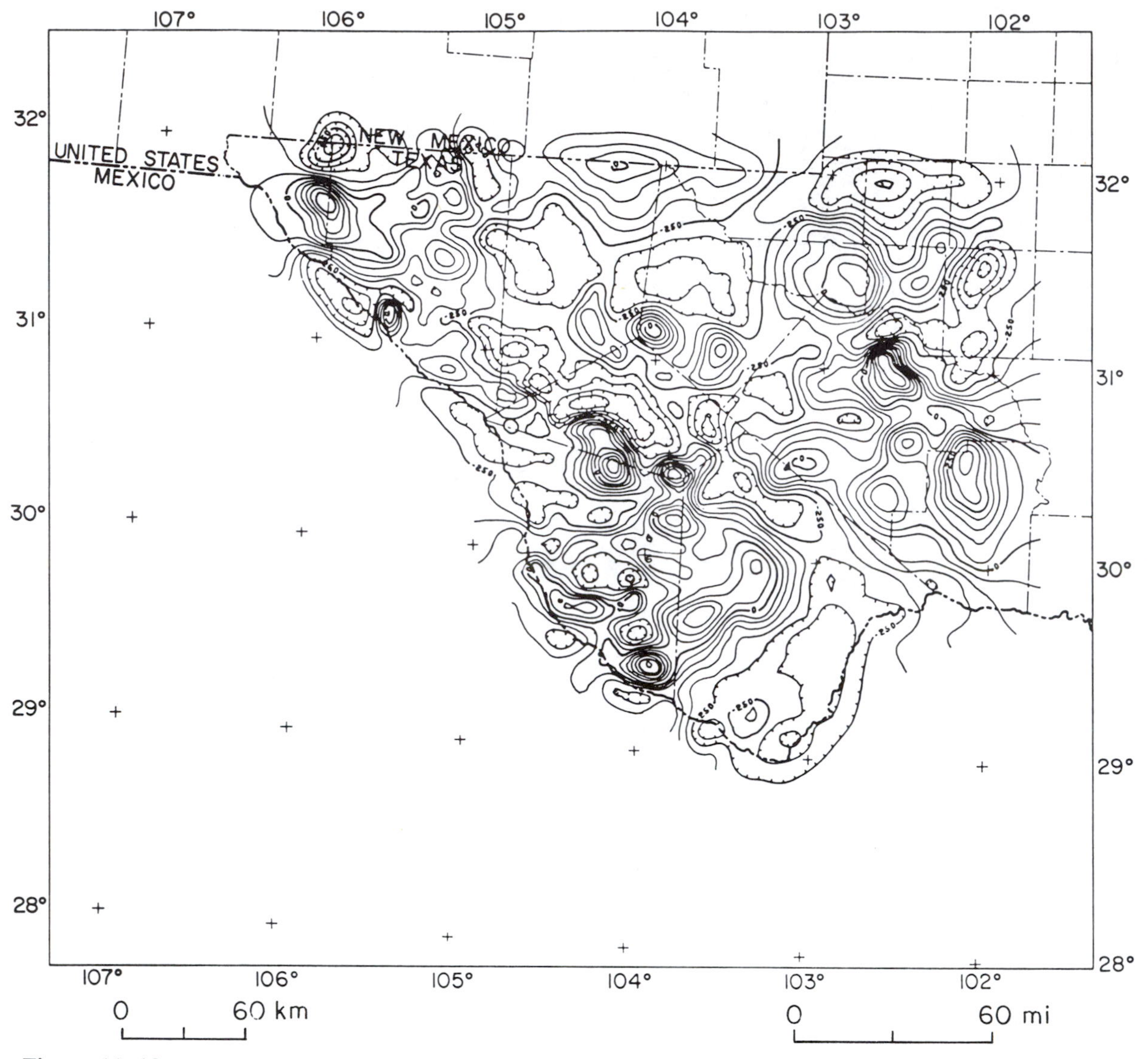

Figure 11–13
Total-intensity aeromagnetic anomaly map of western Texas with a contour interval of 50 gammas. (From G. R. Keller, R. A. Smith, W. J. Hinze, and C. L. V. Aiken, Regional gravity and magnetic study of west Texas. In *The Utility of Regional Gravity and Magnetic Anomaly Maps*, p. 202 (ed. W. J. Hinze). Tulsa, Okla., Society of Exploration Geophysicists, 1985.)

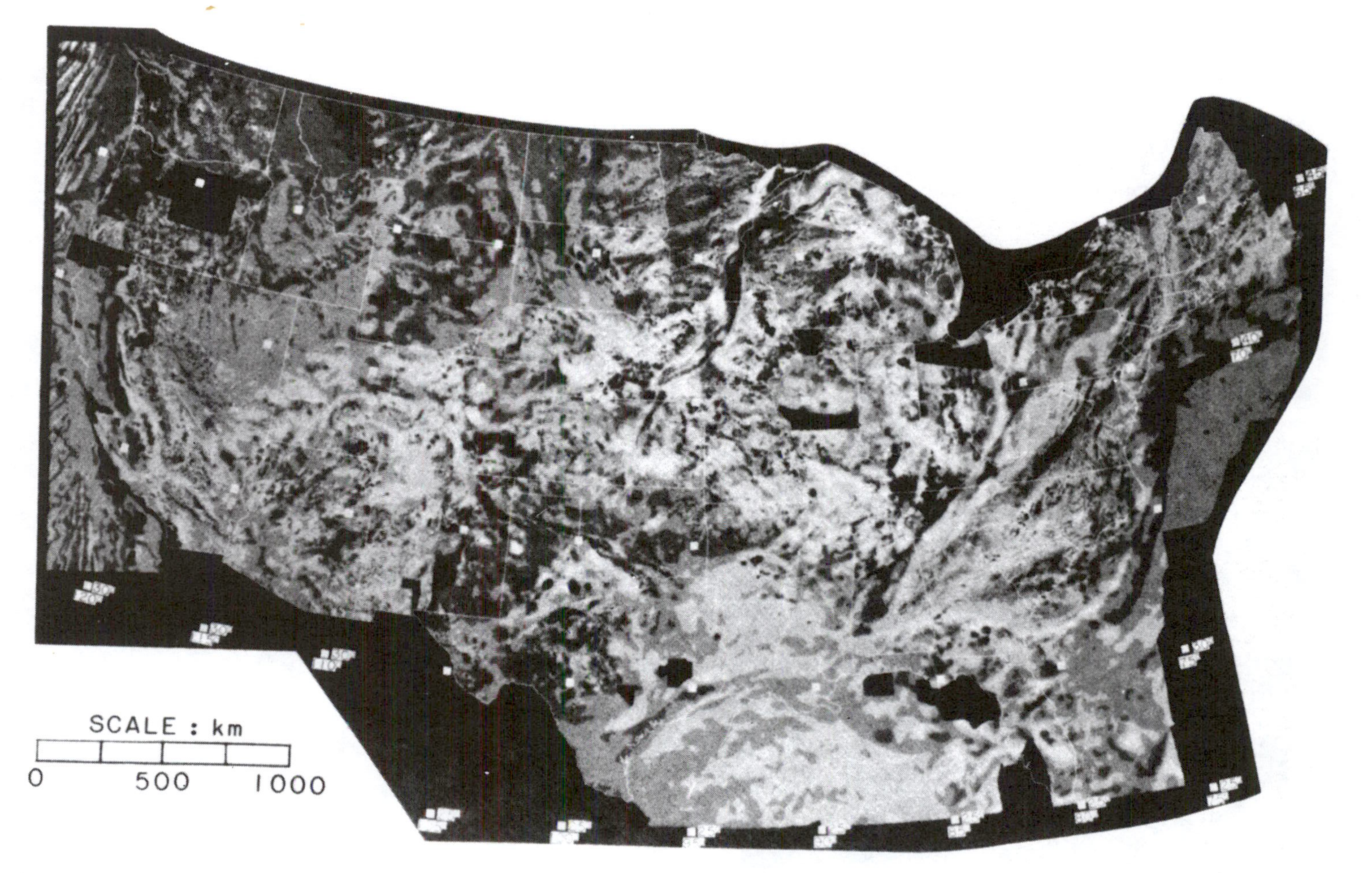

Figure 11–14
Total-intensity magnetic anomaly map of the conterminous United States. Shades of gray, ranging from white (lowest) to black (highest), indicate ranges of field intensity, increasing at 200-gamma intervals. (From W. J. Hinze and I. Zietz, The composite magnetic anomaly map of the conterminous United States. In *The Utility of Regional Gravity and Magnetic Anomaly Maps*, p. 3 (ed. W. J. Hinze). Tulsa, Okla., Society of Exploration Geophysicists, 1985.)

Figure 11–15

Magnetic anomaly maps of the Lockhart River District in the Northwest Territories of Canada. Anomaly patterns are displayed (a) by shading with different tones of gray between contour lines and (b) by shaded relief (From S. D. Dods, D. J. Teskey, and P. J. Hood, The new series of 1:1000000 scale magnetic anomaly maps of the Geological Survey of Canada. In *The Utility of Regional Gravity and Magnetic Anomaly Maps*, p. 79 and p. 81 (ed. W. J. Hinze). Tulsa, Okla., Society of Exploration Geophysicists, 1985.)

few tens to a few hundreds of gammas.

We must use large-scale maps to display some features of the anomalous magnetic field. Contours alone do not always provide the most satisfactory graphical presentation of these features. On some maps, colors or different shades of gray are used to produce a clearer display of anomaly patterns. This second method was used to prepare the magnetic anomaly map of the United States in Figure 11–14. Each shade of gray represents a different range of total field intensity. Several regional features stand out on this map. Especially noteworthy are the parallel linear zones extending in a northeasterly direction along the Appalachian Mountains from Alabama to Maine. When this map is compared with the Bouguer anomaly map of the United States (Figure 9-1), some important differences are evident. There is no obvious relation between magnetic anomalies and elevation or crustal thickness such as exists for Bouguer anomalies. Although regional magnetic anomalies may extend linearly for great distances, they tend to be narrow features consisting of adjacent highs and lows.

Another interesting way to display magnetic anomalies is by means of a shaded relief map. To understand how it is prepared, imagine that magnetic anomalies make up a landscape of hills and valleys. Now suppose that light is beamed on this magnetic landscape from a certain angle. Some parts of the area will be strongly illuminated, and others will be covered by shadows. It is the shadow pattern that is displayed on a shaded relief map. A computer-controlled plotting system is used to prepare such a map. It is a particularly effective map for drawing attention to narrow linear features that are less obvious on other kinds of anomaly maps. Examine the two maps in Figure 11–15. On map a, different tones of gray represent different ranges of field intensity. Map b displays the anomalies in shaded relief, which brings the linear features into much sharper focus. The shaded relief map does not, however, provide the numerical representation of field intensity that is revealed on map a by the tones of gray. Contours, colors or tones of gray, and shaded relief each offer some advantages in helping us to recognize magnetic anomaly patterns.

STUDY EXERCISES

1. A magnetometer sensing element mounted to an aircraft in a "stinger" is sensitive to aircraft magnetism as well as to earth magnetism. Describe two methods that are used to compensate or correct for the effects of aircraft magnetism.

2. Suppose that you want to obtain a contour map of magnetic field intensity for a region in central Greenland where there are no easily recognizable landmarks over the vast expanse of the ice sheet covering it. Describe a procedure for accomplishing this objective which includes a discussion of field operations, the equipment needed for obtaining the most precise measurements, the data processing steps, the sources of error, and preparation of the final contour map.

3. The map in Figure 11–16 shows aeromagnetic survey lines and tie lines. The field intensity values at intersection points, measured along survey lines and tie lines, are the following.

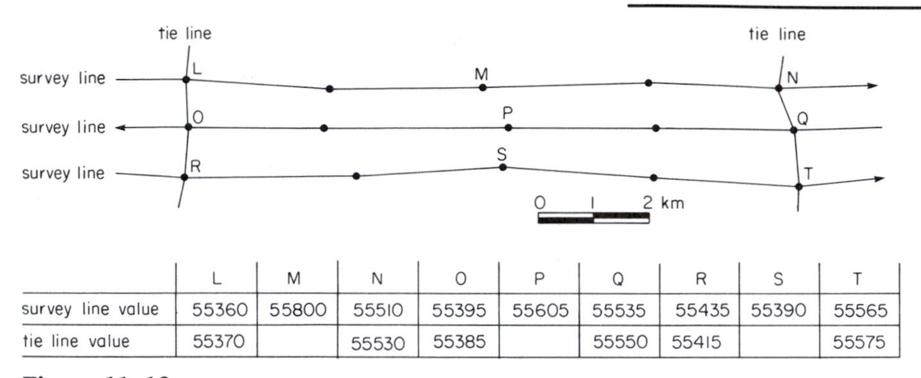

	L	M	N	O	P	Q	R	S	T
survey line value	55360	55800	55510	55395	55605	55535	55435	55390	55565
tie line value	55370		55530	55385		55550	55415		55575

Figure 11–16
Survey lines, tie lines, and measurement points on an aeromagnetic survey flight map.

a. Determine the misclosures for rectangles LNQO and OQTR and the average of these two misclosure values.

b. Using the average misclosure value, determine corrections to be applied to field intensity values measured along survey lines at points M, P, and S. Discuss your procedure.

c. Calculate field intensity at M, P, and S using these corrections.

4. The map in Figure 11–17 shows contours of total magnetic field intensity (F_T). At the southwest corner of this map, the main field intensity (F_M) is 54,000 gammas, and the rates of change of main field intensity are 0.9 gamma/km toward the north and 0.8 gamma/km toward the east. Calculate the scalar difference (T) between the total field intensity and the main field intensity at location A.

5. Using the field intensity values given at grid intersections in Figure 11–11, show contours at 10-gamma intervals that would be obtained by machine contouring based on diagonals crossing from the upper left to the lower right corners of each grid cell. Then show the alternate contours that would be obtained by extending the diagonals from the upper right to the lower left corners of each grid cell. Are there any important differences in the two contour patterns?

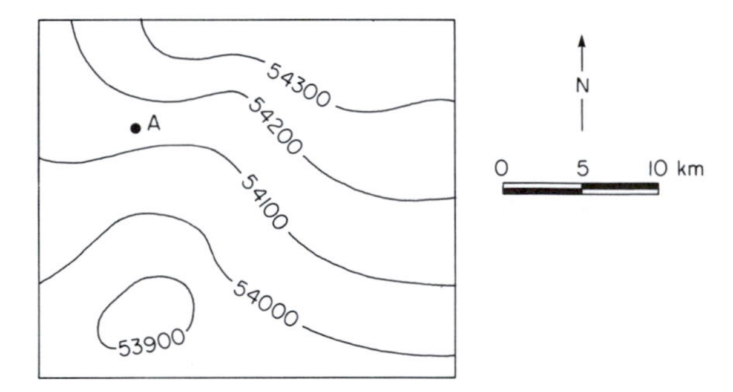

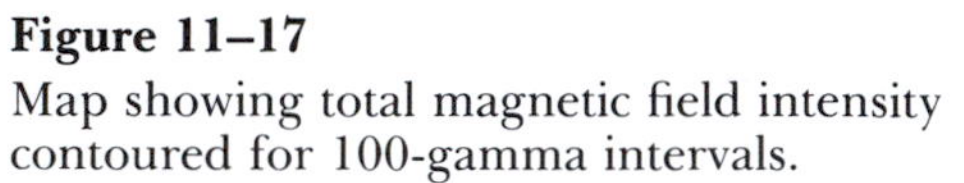
Figure 11–17
Map showing total magnetic field intensity contoured for 100-gamma intervals.

SELECTED READING

Bolandi, G., F. Rocca, and S. Zanoletti, Methods for contouring irregularly spaced data, *Geophysical Prospecting,* v. 25, n. 1, pp. 96–119, March 1977.

Domzalski, W., Some problems of the aeromagnetic surveys, *Geophysical Prospecting,* v. 5, n. 4, pp. 469–479, December 1957.

Hinze, William J. (editor), *The Utility of Regional Gravity and Magnetic Anomaly Maps.* Tulsa, Okla., Society of Exploration Geophysicists, 1985.

Nettleton, L. L., *Gravity and Magnetics in Oil Prospecting.* New York, McGraw Hill, 1976.

Proubasta, Dolores, Airborne surveying: A view from the cockpit, *The Leading Edge,* v. 4, n. 4, pp. 20–25, April 1985.

TWELVE

Magnetic Anomalies and Their Geologic Sources

Ferromagnetic minerals in the earth's crust are the source of the anomalous magnetic field. The contrasting proportions of these minerals in different crustal rocks produce the magnetic anomalies that are the targets of magnetometer surveying. This chapter introduces the procedure for obtaining useful geologic information from magnetic field measurements. We begin with the nature of the rock magnetism that is the basic source of magnetic anomalies. Then we will present methods for interpreting these anomalies in terms of possible ore-bearing zones, or the thickness of sedimentary basins containing oil and gas traps, or other geologic structures. These methods include separation of regional and local anomalies and calculation of the magnetic fields over different models for comparison with measured anomalies.

The exploration geophysicist's methods for processing and interpreting magnetic anomalies are very similar to the methods used for analyzing gravity anomalies. Comparison of Newton's universal law of gravitation (Equation 7–1) and Coulomb's law (Equation 10–1) shows that both the gravitational force and the magnetic force vary inversely with the squared distance from the source. Because of this basic similarity, we are justified in using many of the same procedures for processing and interpreting gravity and magnetic field measurements. Magnetic anomaly patterns usually appear to be more complicated because the sources possess both positive and negative poles, whereas gravity anomaly sources exert only a single positive attraction. Nevertheless, the effects of the two magnetic poles can be calculated separately by procedures similar to those used for gravity analysis. Therefore, we will find that the methods for separating regional and local anomalies and the idea of comparing measured anomalies with anomalies calculated for models, which are presented in Chapter 9, are easily adapted for magnetic field analysis.

ROCK MAGNETISM

Magnetic anomalies are produced by contrasts in rock magnetism. Before attempting to interpret these anomalies, we must first look into the nature of rock magnetism. Let us begin by acknowledging that all minerals are affected in some way by a magnetizing field. The capacity of a mineral to acquire magnetism by induction in a magnetizing field is described by its susceptibility (Equation 10–8), a term we introduced in Chapter 10. Later, we will describe how susceptibility is measured. But first we introduce some representative values that will help you recognize the importance of different minerals in producing magnetic anomalies. We will also discuss the capacity of a few minerals to retain permanent magnetism and how it can be measured.

Paramagnetism and Diamagnetism

The atoms of most chemical elements do not group in ways that form magnetic domains. Nevertheless, they are affected by a polarizing field that causes subtle shifts in the speed and direction of orbiting electrons. The result is an atomic magnetic moment. For atoms with odd numbers of electrons, the atomic moments tend to align with the polarizing field. Minerals in which the magnetic moments of these atoms predominate are said to be *paramagnetic.* They have positive values of susceptibility. Paramagnetic minerals include important rock-forming silicates such as olivine, pyroxene, amphibole, and biotite. Typical values of susceptibility for these minerals are listed in Table 12–1.

In some other minerals, atoms with even numbers of electrons have the strongest effect on the magnetism acquired in a polarizing field. The atomic moments tend to oppose the polarizing field, as is indicated by negative values of susceptibility. Such minerals are called *diamagnetic.* Important species include quartz, calcite, halite, galena, and sphalerite. Their susceptibilities are also listed in Table 12–1.

The values of susceptibility for paramag-

TABLE 12–1 Magnetic Susceptibility of Some Paramagnetic and Diamagnetic Minerals

MINERAL	MAGNETIC SUSCEPTIBILITY (emu/g)
Paramagnetic	
Fayalite (Fe_2SiO_4)	100×10^{-6}
Pyroxene ($FeSiO_3$)	73×10^{-6}
Amphiboles	$13–75 \times 10^{-6}$
Biotite	$53–78 \times 10^{-6}$
Garnets	$31–159 \times 10^{-6}$
Diamagnetic	
Quartz	-0.50×10^{-6}
Calcite	-0.38×10^{-6}
Halite	-0.52×10^{-6}
Galena	-0.34×10^{-6}
Sphalerite	-0.26×10^{-6}

Data from David W. Strangway, *History of the Earth's Magnetic Field,* New York, McGraw-Hill, p. 4, 1970.

netic and diamagnetic minerals are so small that their effects cannot be distinguished by conventional magnetic surveying practices. On rare occasions, very weak negative anomalies have been attributed to large diamagnetic masses of rock salt or marble. For practically all exploration surveys, the effects of paramagnetism and diamagnetism can be ignored. The important magnetization contrasts are produced by a few ferromagnetic minerals, which are the subject of the following discussion.

Ferromagnetic Minerals

In a few crystalline substances, the atoms are so arranged that electron orbital and spin motions combine to produce strong magnetism within small regions of the crystal structure. Some features of these small regions, called magnetic domains, are described in Chapter 10. Let us look more carefully at the nature of these domains. The alignment of atomic moments within each domain has one of three basic patterns identified as (1) true ferromagnetism, (2) antiferromagnetism, and (3) ferrimagnetism.

In a domain of a *true ferromagnetic* substance, all atomic elements tend to align in the same direction (Figure 12–1a). The elements iron, nickel, and cobalt in pure form are true ferromagnetic substances. In nature they rarely occur in pure form. Rather, they form compounds with other elements. Iron plays

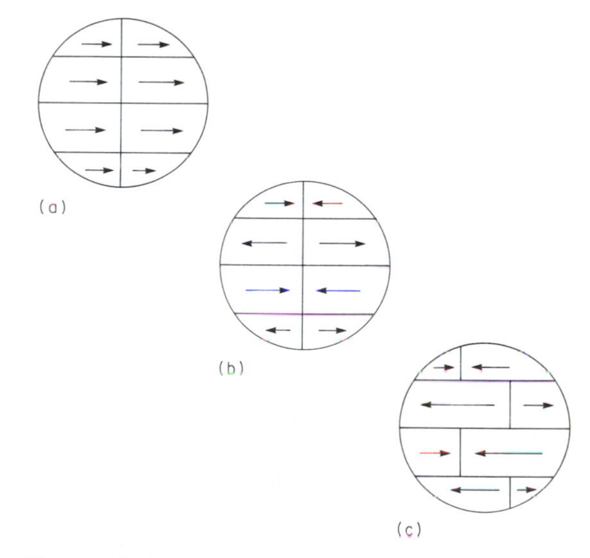

Figure 12–1
Schematic representation of the alignment of magnetic moments of domains formed in a magnetizing field. (a) In an ideal true ferromagnetic substance, all domain moments are in the same direction. (b) In an ideal antiferromagnetic substance adjacent domains tend to have equal but opposing domain moments. (c) In an ideal ferrimagnetic substance, adjacent domains have unequal opposing domain moments, so that the moments in one direction are stronger than those in the opposite direction.

the most important role, nickel is much less abundant in the earth's crust, and cobalt is too rare to be of importance in explaining magnetic anomalies.

A domain in an *antiferromagnetic* substance possesses atomic moments that tend to align equally in one or the other of two opposite directions (Figure 12–1b). Therefore, the domain has no net magnetic moment. For this reason, antiferromagnetic minerals do not contribute to magnetic anomalies. Nevertheless, they are useful in studies of paleomagnetism, because the atomic moments indicate the orientation of lines of force of the earth's field in former times. *Hematite* is the best-known antiferromagnetic mineral. It is the iron oxide Fe_2O_3 in a rhombohedral crystal structure. Although this important iron ore does not produce magnetic anomalies, it is commonly associated with minerals that are anomaly sources. Hydrous iron oxides in the limonite group are antiferromagnetic.

Of principal interest for exploration purposes are the *ferrimagnetic* minerals. Within a ferrimagnetic domain, the atomic moments tend to line up in two opposite directions, one of which is preferred (Figure 12–1c). Because more atomic moments align in the preferred direction, the domain possesses a net magnetic moment. The most abundant ferrimagnetic mineral is *magnetite,* which is the iron oxide Fe_3O_4. Much less abundant but also important is *ilmenite,* which is an iron–titanium oxide $FeTiO_3$. The mineral *pyrrhotite* is a ferrimagnetic iron sulfide. Its composition is between that of *troilite* (FeS) and *pyrite* (FeS_2) and is written as $Fe_{1-x}S$. For values of $0.2 < x < 0.94$, the mineral possesses ferrimagnetism. Minor amounts of nickel commonly occur in pyrrhotite, giving it value as an ore. Finally, there is the mineral *maghemite,* which has the same chemical composition as hematite, but in an inverse spinel crystal structure. It, too, is ferrimagnetic and may contribute to magnetic anomalies.

Magnetization of Ferrimagnetic Minerals

Even though individual domains possess strong magnetism, the various domain moments tend to be randomly oriented in a ferrimagnetic mineral grain. Therefore, the entire grain would display no magnetism. But when the grain is placed in a polarizing field, the domain walls adjust so that the grain acquires a magnetic moment. In the discussion of this subject in Chapter 10, a hysteresis curve (Figure 10–9) describes the relation between the strength of the magnetic moment and the intensity and direction of the polarizing field. In Figure 10-9 the polarizing field reaches intensities high enough to produce magnetic saturation in the specimen.

Suppose that weaker fields are used. Hysteresis curves in Figure 12–2 illustrate magnetization in strong, intermediate, and weak fields. The hysteresis effect, indicated by different paths for increasing and decreasing magnetization, is obvious where the strong polarizing field is used. Although the paths are closer where the intermediate field is used, hysteresis is still evident. In the weak polarizing field, however, the paths are nearly coincident. This last example best represents magnetization of minerals by the earth's main field. The two paths are so close to a straight line that the slope of such a line indicates the mineral susceptibility, *k,* to an accuracy sufficient for exploration purposes. Susceptibility values for the important ferrimagnetic minerals are given in Table 12–2. These values were obtained in weak testing fields similar to the

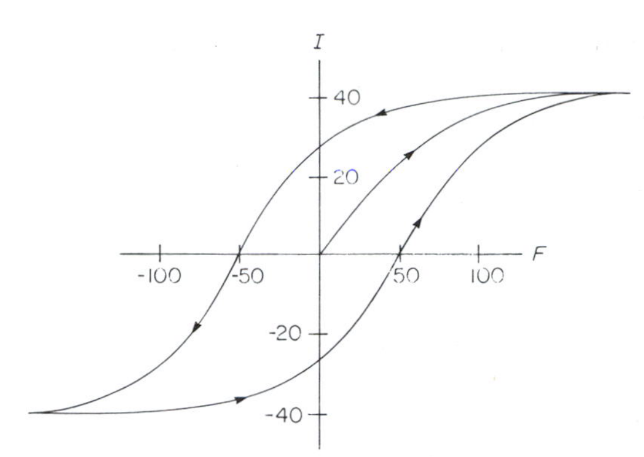

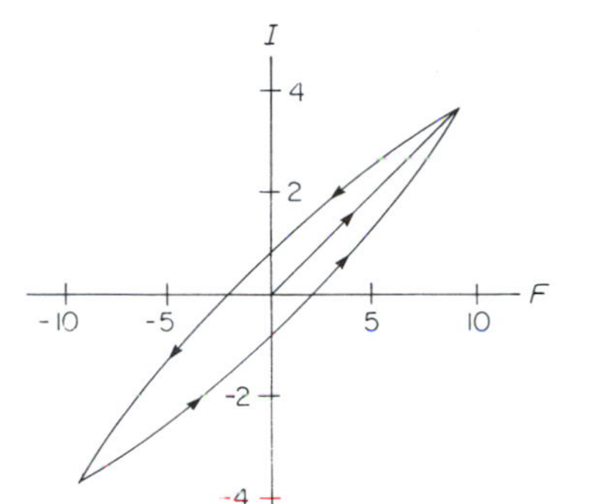

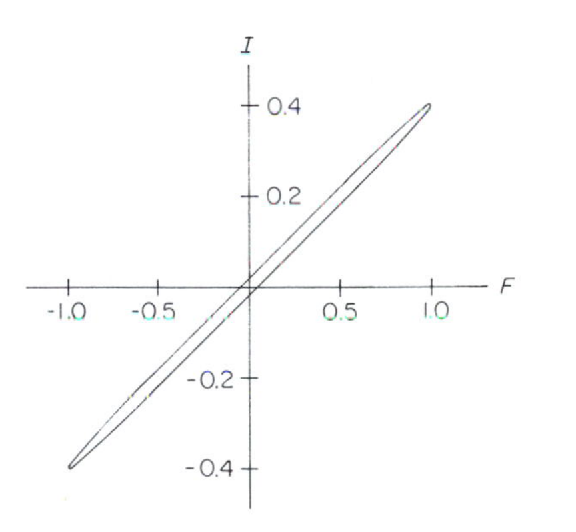

Figure 12–2

Hysteresis curves illustrating magnetization (I) of a ferromagnetic substance by magnetizing fields that undergo strong, intermediate, and small fluctuations. (From Methods in Paleomagnetism, Developments in Solid Earth Geophysics 3, New York, Elsevier, pp. 159–162, 1967.) In a weak magnetizing field such as the earth's main field, the hysteresis effect is small, and the relation between magnetization (I) and the magnetizing field (F) is very nearly linear.

earth's field. Observe how large these values are compared with those in Table 12–1. This difference explains why only these few ferrimagnetic minerals are used to explain the anomalies detected from exploration magnetic surveys.

The magnetization of a ferrimagnetic mineral grain is influenced by the grain shape, an effect not considered in Equation 10–8. A modification must be introduced to ensure that magnetic lines of force are smooth and continuous across the grain boundaries. The induced magnetic moment per unit volume (I) is expressed in terms of the susceptibility k and the magnetizing field F in the following way:

$$I = \frac{k\mathrm{F}}{1 + \eta k} \tag{12-1}$$

where η is the *demagnetization factor*. It is called this because the effect of the term ηk is to reduce the value of I. For a long, thin needle aligned in the direction of magnetization, the value of η is close to zero, but for a sphere, $\eta = \frac{4}{3}\pi$. Values for a thin disk range from $\eta = 0$ if magnetization is directed along the plane of the disk to $\eta = 4\pi$ where magnetization is perpendicular to the disk. These values tell us that a particle of magnetite with a sus-

TABLE 12–2 Susceptibility and Curie Temperature of Some Important Ferrimagnetic Minerals

MINERAL	SUSCEPTIBILITY (emu)	CURIE TEMPERATURE (°C)
Magnetite	0.3–0.8	580
Ilmenite	0.135	50–300
Pyrrhotite	0.125	320
Maghemite	variable	545–675

ceptibility of $k = 0.3$ in a magnetizing field of 50,000 gammas could acquire magnetization of between $I = 4500$ and $I = 1194$, depending on its shape and orientation.

So far, we have been discussing the magnetization acquired by a mineral grain because it is situated in a polarizing field. Recall from Chapter 10 that this is called induced magnetism. If the grain is moved to another location where the polarizing field F is different, its induced magnetism I changes according to Equation 12–1. With the change in position, the mineral grain may retain unchanged another form of magnetism. This form is its permanent magnetism, also called *remnant magnetism,* which we also introduced in Chapter 10. Remnant magnetism exists where there is a preferred alignment of domain moments that is independent of any existing polarizing field.

The strongest remnant magnetism is produced by creating the original domains in a polarizing field. One way is by cooling the mineral from a temperature too high for domains to exist down through the Curie temperature where the domain structure is created, as discussed in Chapter 10. Curie temperatures of ferrimagnetic minerals are listed in Table 12–2. Recall that magnetism acquired in this way is called *thermoremnant magnetism* (TRM).

Another way of producing remnant magnetism, which can be very important in lower-temperature environments, is to create the mineral grain itself in the presence of a polarizing field. This occurs as a result of chemical reactions associated with metamorphism. An example is the creation of magnetite from the decomposition of the iron-rich but nonmagnetic silicate mineral *epidote*:

$$\underset{\text{epidote}}{Ca_2FeAl_2Si_3O_{12}(OH)} \rightarrow \underset{\text{magnetite}}{2Fe_3O_4} + \underset{\text{anorthite}}{12CaAl_2Si_2O_8} + \underset{\text{quartz}}{12SiO_2} + \underset{\text{water}}{6H_2O} + \underset{\text{oxygen}}{O_2} \qquad (12\text{–}2)$$

As magnetic grains form at relatively low temperature by such reactions, the domains are created with moments tending to align with the polarizing field. Magnetism produced in this way is called *chemical remnant magnetism* (CRM).

The total magnetic moment of a ferrimagnetic mineral grain is the vector combination of its induced and remnant magnetic moments. The remnant component depends on the extent to which domain moments were aligned at the earlier time when the domains were originally created. The induced component depends on the ease with which domain walls can shift to favor domains that have moments aligned with the present polarizing field. These components can be measured independently in a mineral specimen. A spinner magnetometer (Figure 10–30) for finding remnant magnetism is equipped with compensating coils to nullify any polarizing field that would produce induced effects. Magnetic susceptibility can be measured in an electromagnetic inductance bridge, which is described in the next section. Alternating current that is insensitive to remnant magnetism is used in the tests, so that the value of k can be used in Equation 12–1 to obtain an independent value of induced magnetism. Relative strengths of the induced and remnant components vary widely for different specimens. Some possess only an induced moment, whereas others are dominated by the remnant component.

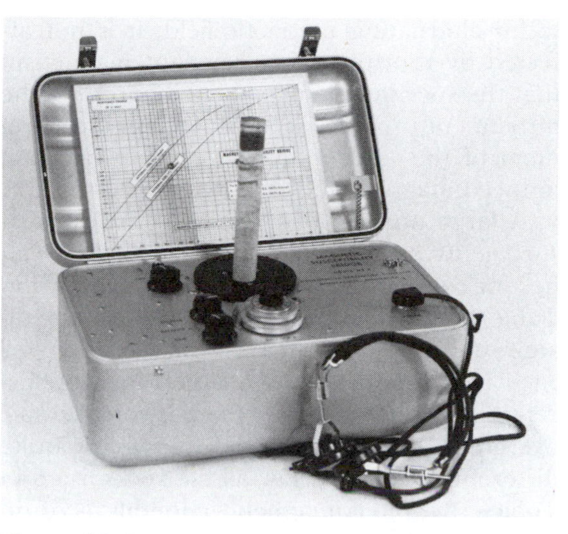

Figure 12–3
Electromagnetic inductance bridge for measuring the magnetic susceptibility of rock specimens. (Courtesy of Geophysical Specialties Company.)

Magnetism in Rocks

For purposes of exploration geophysics, rock magnetism is attributed entirely to the presence of ferrimagnetic minerals. Depending on the volume proportion of these minerals, the way they are disseminated, and the magnetism they possess, the rock itself is said to be magnetized. A rock specimen can be tested to determine, independently, its susceptibility to magnetization by induction in the present main field and its remnant magnetism acquired at a former time.

The most common practice for finding rock susceptibility is to test specimens in an electromagnetic inductance bridge (Figure 12–3). This bridge consists of a set of coils carrying opposing alternating currents that can be adjusted by variable resistors. In this way the alternating magnetic fields of the coils are balanced to nullify each other in the region surrounding a nearby sensing coil. A specimen chamber is located nearer one of the balancing coils, where its alternating magnetic field is not completely nullified. When a rock specimen is placed in the chamber, it acquires induced magnetism in proportion to its susceptibility k. This induced field upsets the balance at the sensing coil, producing an alternating current. The balance is then restored by adjusting one of the variable resistors. The difference in resistance required to balance the system with and without a specimen in the chamber is proportional to the susceptibility of the specimen. The constant of proportionality for the instrument has been previously determined by testing standard specimens of known susceptibility. How is the susceptibility of these standard calibrating substances found? It can be determined through magnet deflection experiments performed in a known magnetic field, experiments not unlike the one illustrated in Figure 10–11.

The electromagnetic inductance bridge is designed to test rock cores or samples of crushed rock, both of which are easily placed in the cylindrical specimen chamber. The system is portable and simple to operate in the field. Because the sensing coil responds only to an alternating magnetic field, it is not affected by the unvarying remnant magnetism that the specimen may possess. Therefore, the instrument provides an independent measurement of the susceptibility needed to calculate induced magnetism.

A large number of rock susceptibility measurements have been published. Typical values for common rock types are summarized in Table 12–3. Two important facts are immediately obvious from this information: (1) for each rock type, there is a large range of susceptibility values; and (2) there is considerable overlap in the susceptibility ranges of quite different rock types. For all the rocks listed in Table 12–3, ferrimagnetic minerals account for only a small volume proportion. Even in the iron-rich mafic rocks, these minerals rarely exceed 5 percent of the volume. This small proportion is highly variable from one specimen to another. For example, some basalts possess 5 percent magnetite, but others possess almost none. Rock susceptibility also depends on the demagnetizing factors of the ferrimagnetic mineral grains that vary with their shape and orientation. Clearly, rock susceptibility has a much larger range than rock density, which is its counterpart in gravity exploration.

Some efforts have been made to estimate rock susceptibility from volume proportions of the ferrimagnetic minerals. Most of this work focuses on magnetite, which is by far the most abundant of these minerals. Results of two studies of typical rocks are presented in Figure 12–4. Curves drawn through the data on these graphs can be used to estimate susceptibility from analyses of the mineral composition of rocks. Because the data do not group closely to these lines, such estimates should be used with caution.

Remnant magnetic moments of ferrimag-

TABLE 12–3 Magnetic Susceptibility of Common Rocks

	SUSCEPTIBILITY × 10^6 emu			SUSCEPTIBILITY × 10^6 emu	
TYPE	RANGE	AVERAGE	TYPE	RANGE	AVERAGE
Sedimentary			Igneous		
Dolomite	0–75	10	Granite	0–4,000	200
Limestones	2–280	25	Rhyolite	20–3,000	
Sandstones	0–1,660	30	Dolerite	100–3,000	1400
Shales	5–1,480	50	Augite–syenite	2,700–3,600	
			Olivine–diabase		2,000
Metamorphic			Diabase	80–13,000	4,500
Amphibolite		60	Porphyry	20–16,700	5,000
Schist	25–240	120	Gabbro	80–7,200	6,000
Phyllite		130	Basalts	20–14,500	6,000
Gneiss	10–2,000		Diorite	50–10,000	7,000
Quartzite		350	Pyroxenite		10,500
Serpentine	250–1,400		Peridotite	7,600–15,600	13,000
Slate	0–3,000	500	Andesite		13,500

Data from W. M. Telford, L. P. Geldart, R. E. Sheriff, and D. A. Keys, *Applied Geophysics*, Cambridge, England, Cambridge University Press, p. 121, 1976.

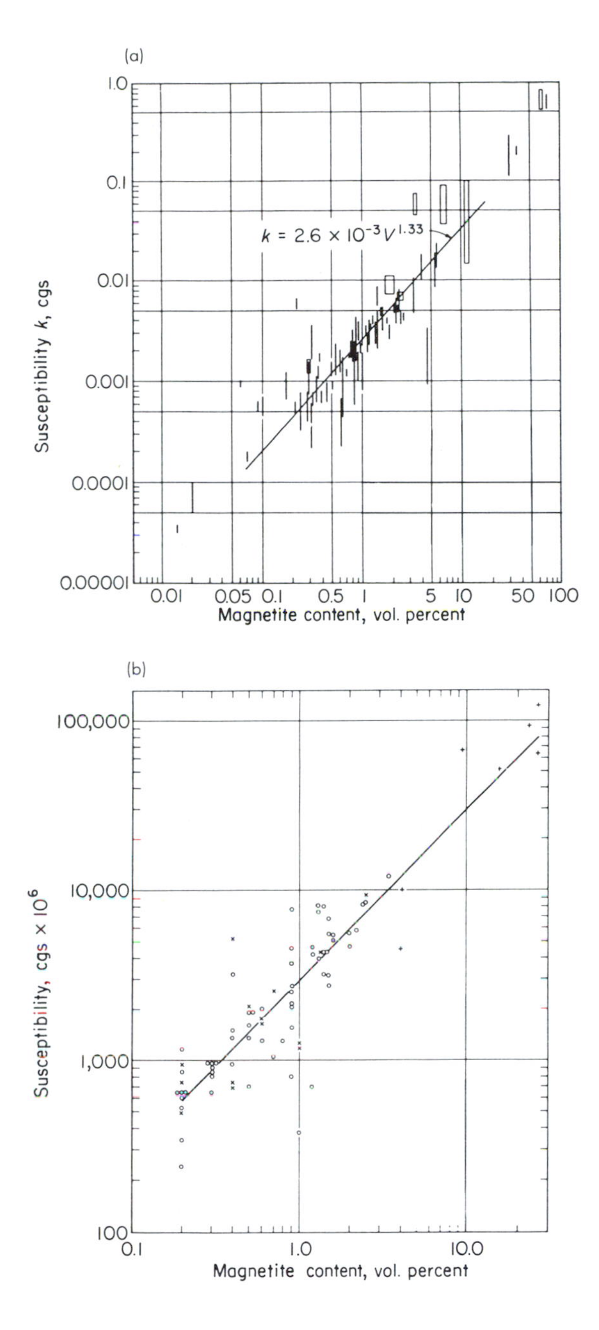

Figure 12–4

Comparisons of magnetic susceptibility and magnetite content of representative rock specimens. (Graph a from J. R. Balsley and A. F. Buddington, Iron-titanium oxide minerals, rocks, and aeromagnetic anomalies of the Adirondack Area, New York, *Economic Geology*, v. 53, pp. 777–805, 1958. Graph b from H. M. Mooney and R. Bleifuss, Magnetic susceptibility measurements in Minnesota, Part II, Analysis of field results, *Geophysics*, v. 18, n. 2, pp. 383–393, April 1953.)

netic grains also contribute to rock magnetism. Most important in igneous and metamorphic rocks is the TRM or CRM acquired when the domains were first created. Sedimentary rock may also possess remnant magnetism because of mineral grains that were formerly magnetized in these ways and now make up part of the mixture of sediment particles. The formerly magnetized grains tend to align with the earth's field at the time of sedimentation. The result is called *detrital remnant magnetism* (DRM). Occasionally, a rock exposure is struck by lightning, which generates a strong temporary magnetic field capable of altering the domain structure of nearby mineral grains. The effect seldom extends beyond a few tens of feet from the point of impact. Because there is no appreciable temperature change during the process, magnetism acquired in this way is called *isothermal remnant magnetism* (IRM). Because it is so localized, it is rarely of importance in the exploration analysis of magnetic anomalies. Even so, the idea helps us to understand the occasional erratic values in a series of rock magnetism measurements. Finally, we should mention the possibility of very gradual shifts in domain structure occurring over the long expanse of time that a rock resides in the earth's field. Alignment of domain moments resulting from the slow change produces *viscous remnant magnetism* (VRM).

We have already discussed laboratory measurements of remnant magnetism with a spinner magnetometer (Figure 10–30). Because this apparatus is impractical for field measurements, another portable instrument has been developed for this purpose. It is a small flux-gate probe (Figure 12–5) which can be used for quick field tests to determine the existence and direction of remnant magnetism in hand-held rock specimens. The flux-gate sensing element is mounted to a stake driven in the ground. Then it is rotated on the mounting into the position where a zero meter reading indicates that it is perpendicular to the earth's field. In this position, it is insensitive to the induced magnetism of a specimen held near its end. Remnant magnetism, however, is detected. By holding the specimen in different orientations, we observe that meter deflection is maximum when the remnant moment is aligned with the flux-gate probe.

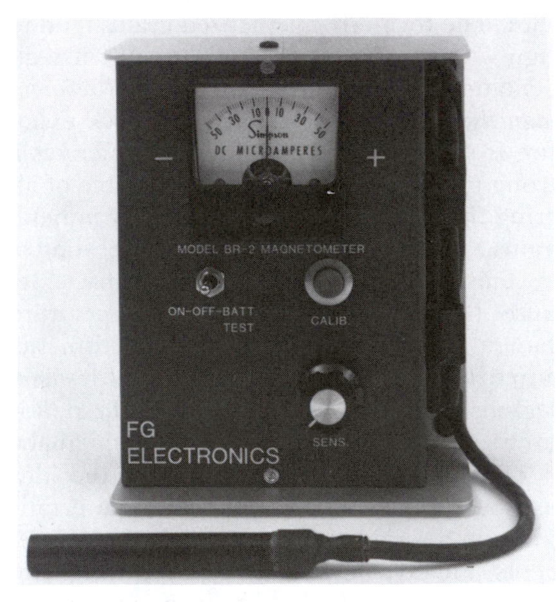

Figure 12–5
Flux-gate probe for detecting permanent magnetism in rock specimens. (Courtesy of FG Electronics, Inc., Menlo Park, Calif.)

Most rocks are identified according to the proportions of a certain few abundant minerals. For example, granite consists mostly of alkali feldspar, quartz, and mica, and the principal minerals in basalt are plagioclase and amphibole. As we mentioned earlier, ferrimagnetic minerals make up only small proportions of these rocks. In a large rock mass in which the proportions of the principal minerals vary only slightly, there can be large local differences in the content of ferrimagnetic minerals. In an otherwise uniform rock magnetite concentrations of as much as 5 percent may exist in zones a few tens of feet in diameter but can be almost absent in other nearby zones. Such a situation can cause considerable local irregularity in rock magnetism.

Evidence of this local irregularity is seen along the profile in Figure 12–6, which was prepared from ground magnetometer measurements over a quartz monzonite mass in Utah. Readings were made at 100-foot intervals along the profile. Along the center part of the profile where the quartz monzonite is exposed at the land surface, local variations in vertical field intensity exceed 600 gammas. But nearer the ends where the quartz monzonite is covered by nonmagnetic sediments, the local variations almost disappear. The strongly magnetized zones are farther from the magnetometer so that their effects merge to produce a more smoothly varying field intensity profile.

Many of the facts about rock magnetism serve to emphasize how variable it can be from place to place. Even rocks most likely to

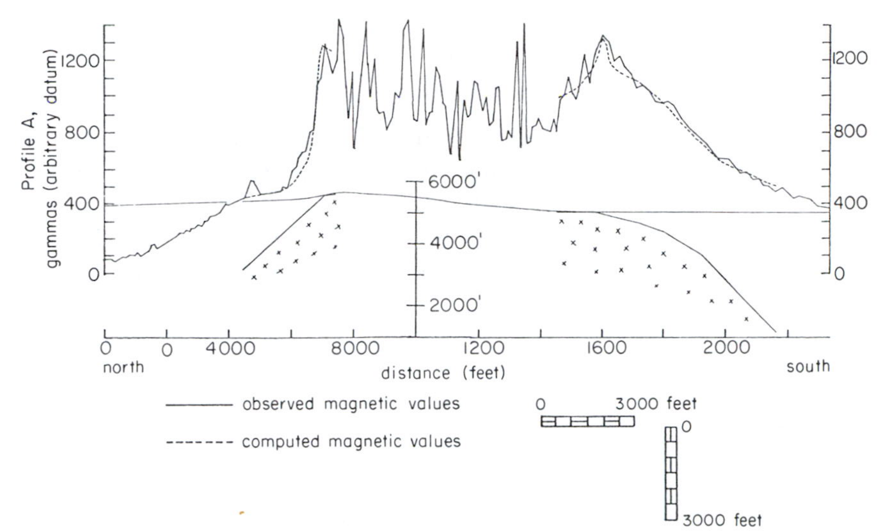

Figure 12–6
Vertical-intensity profile prepared from torsion magnetometer measurements at 100-foot intervals along a profile crossing a magnetized granitic mass in the San Francisco Mountains of southwestern Utah. (From J. W. Schmoker, Ph.D. dissertation, Virginia Polytechnic Institute and State University, p. 90, June 1969.)

produce magnetic anomalies can have large differences in magnetism from place to place. For magnetic surveying purposes, collecting specimens for rock magnetism measurements can be quite useful. But this information is often difficult to use directly in the numerical analysis of anomalies. Rather, it qualitatively guides the estimates we must make about the magnetization of hidden anomaly sources. It is important to be aware of the general nature of rock magnetism so that good judgment can be exercised in interpreting anomalies.

ANOMALIES CAUSED BY MAGNETIZED MODELS

To interpret magnetic anomalies, we can use the same trial-and-error method described in Chapter 9 for interpreting gravity anomalies. Theoretical anomalies calculated for models with different shapes are compared with a measured magnetic anomaly. The one matching most closely indicates the model that best represents the anomaly source. The calculations tend to be more complicated than similar gravity calculations because the effects of both positive and negative poles must be considered, as well as the direction of the magnetizing field.

The simplest magnetic model is a bar magnet. This dipole can be used to represent, approximately, a somewhat irregularly shaped body (Figure 12–7) in the same way that we used a sphere to represent a more irregular gravity anomaly source (Figure 9–10). Other models useful for duplicating more closely the shapes of magnetic anomaly sources include inclined plates, and two- and three-dimensional polygons.

The Dipole

Suppose that the mass in Figure 12–7 is magnetized by induction in the earth's main field. We can represent its magnetism approximately by a dipole aligned with the main field.

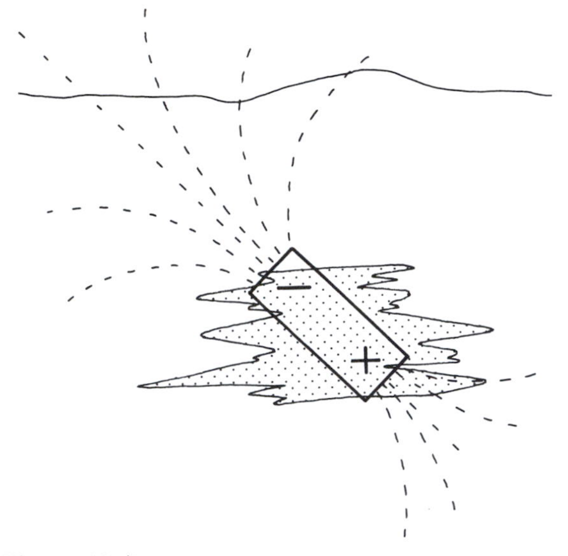

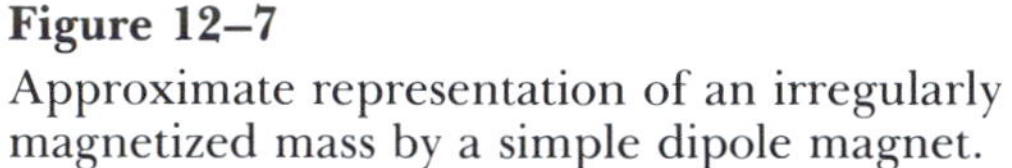

Figure 12–7
Approximate representation of an irregularly magnetized mass by a simple dipole magnet.

How do we calculate the magnetic anomaly produced by such a dipole? This can be done by means of equations first presented in Chapter 10. Let us begin by calculating the field intensity at points along a profile passing over the dipole in the direction of magnetic north (Figure 12–8). According to Equation 10–3, contributions to the field intensity at some point Q from the positive (F_p) and negative (F_n) poles are

$$F_p = P/r_p^2 \quad \text{and} \quad F_n = -P/r_n^2 \qquad (10\text{–}3)$$

where P is the pole strength. Positions measured from the left end of the profile are x_p, z_p for the positive pole and x_n, z_n for the negative pole. Therefore, their distances from a point Q, located at position x_q along the profile, are

$$r_p = \sqrt{(x_q - x_p)^2 + z_p^2} \qquad (12\text{–}3\text{a})$$

and

$$r_n = \sqrt{(x_q - x_n)^2 + z_n^2} \qquad (12\text{–}3\text{b})$$

Next, we must calculate the vertical *(v)* and horizontal *(h)* components of field intensity. Noting that the lines of r_p and r_n make angles α_p and α_n with the profile, we obtain the following components for the positive pole,

$$v_p = F_p \sin \alpha_p = \frac{Pz_p}{r_p^3}$$

and (12–4a)

$$h_p = F_p \cos \alpha_p = \frac{P(x_q - x_p)}{r_p^3}$$

and similarly for the negative pole,

$$v_n = -F_n \sin \alpha_n = \frac{-Pz_n}{r_n^3}$$

and (12–4b)

$$h_n = -F_n \cos \alpha_n = \frac{-P(x_q - x_n)}{r_n^3}$$

These components are then summed and combined to obtain the field intensity F_A caused by the dipole:

$$v_A = v_p + v_n, \quad h_A = h_p + h_n,$$
$$\text{and} \quad F_A = \sqrt{v_A^2 + h_A^2} \qquad (12\text{–}5)$$

By repeating calculations of v_A and h_A along a line of closely spaced points, we obtain vertical and horizontal field intensity profiles over the dipole (Figure 12–8). These profiles can be compared with anomaly profiles measured with magnetometers that read vertical or horizontal field intensity. But a corresponding profile of F_A cannot be compared with a profile of total-field magnetometer readings. The reason why is explained in Figure 11–10, where we see that the total intensity (F_T) is obtained from the vector addition of the anomalous field ($\mathbf{F}_A$) and the main field ($\mathbf{F}_M$).

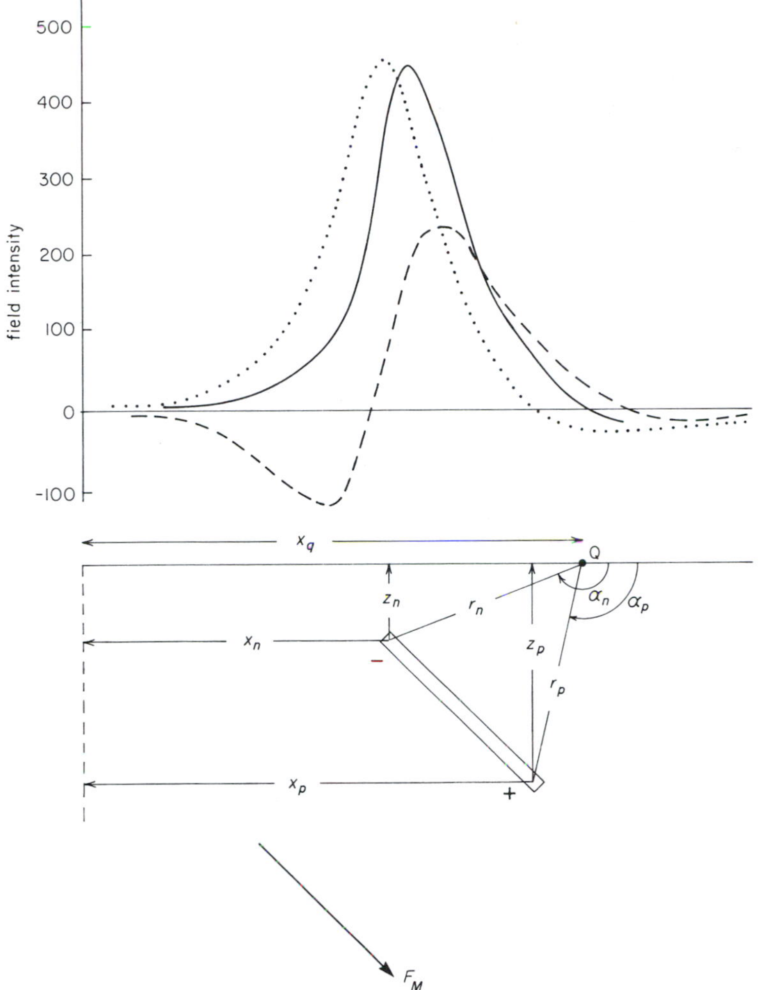

Figure 12–8
Profiles of the vertical-intensity (dotted line), horizontal-intensity (dashed line), and total-intensity (solid line) anomalies over a dipole aligned in the direction of the main magnetic field.

How do we obtain a profile for comparison with total-field magnetometer readings? We sum and combine the vertical (V_M) and horizontal (H_M) elements of the main field intensity with those produced by the dipole model to obtain elements of the total field:

$$Z = v_A + V_M, \quad H = h_A + H_M, \quad \text{and } F_T = \sqrt{Z^2 + H^2} \qquad (12\text{–}6)$$

An anomaly profile obtained by repeated calculations of F_T along a line of closely spaced points can be compared directly with a total-intensity magnetometer profile that has not been corrected for the effect of the main field. But suppose that the magnetic field measurements have been adjusted for the contribution of the main field by the method of scalar subtraction described earlier in Chapter 11. Then

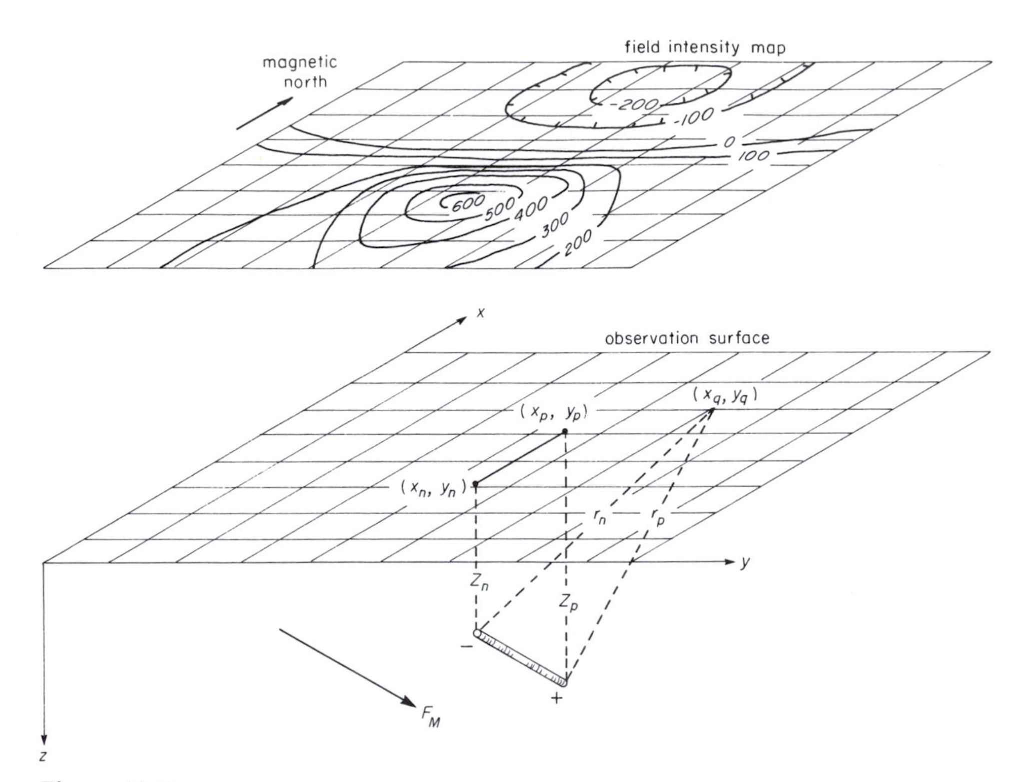

Figure 12–9
Total-intensity anomaly map over a dipole aligned in the direction of the main magnetic field (qualitative scale). Notice that the anomaly pattern consists of a high part and a low part.

we must make a similar adjustment to the values of F_T that were calculated on the basis of the dipole method. For each point along the line, we obtain the value T by scalar subtraction of the main field intensity F_M from our calculated total intensity F_T:

$$T = F_T - F_M \tag{12–7}$$

The profile of T in Figure 12–8 can be compared with a profile of total-intensity magnetometer readings from which the effect of the main field has been removed.

The same basic procedure can be followed to calculate a map of the magnetic anomaly over a dipole model. Field intensity is calculated at points on a surface above the dipole (Figure 12–9). We can use rectangular coordinates to describe the positions of the poles and of points on the surface. Let point Q be specified by x_q, y_q and the positive and negative poles by x_p, y_p, z_p and x_n, y_n, z_n. We see that distances of the poles from Q are given by

$$r_p = \sqrt{(x_q - x_p)^2 + (y_q - y_p)^2 + z_p^2}$$

and

$$r_n = \sqrt{(x_q - x_n)^2 + (y_q - y_n)^2 + z_n^2} \tag{12–8}$$

Field intensity contributions at Q can be calculated separately for each pole using Equations 10–3. These values must be separated into x-, y-, and z-directed components. Geometrical features used to obtain these components for the positive pole are illustrated in Figure 12–10. In terms of these angles and distances, the vertical (v_p) and two horizontal (h_{xp} and h_{yp}) components are

$$v_p = F_p \sin \alpha_p = \frac{P z_p}{r_p^3} \qquad (12\text{–}9\text{a})$$

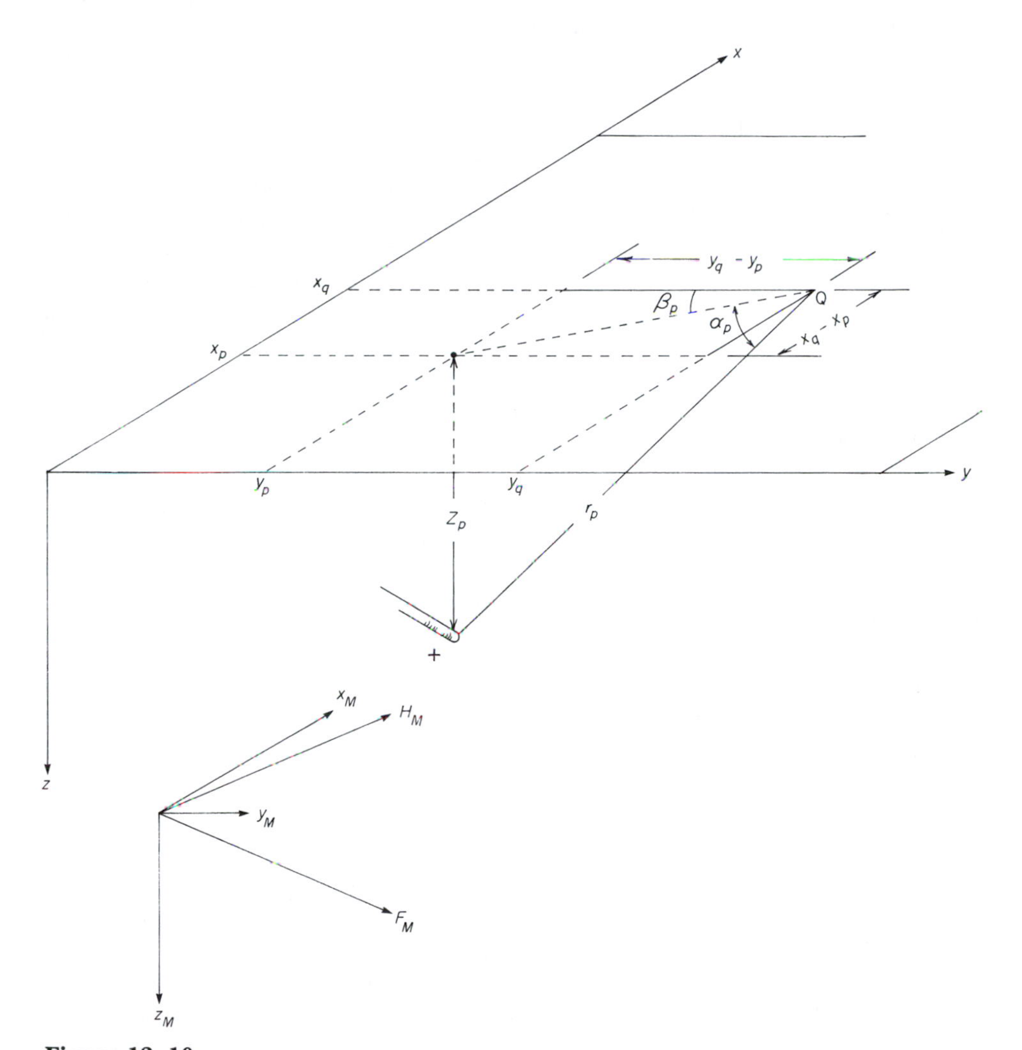

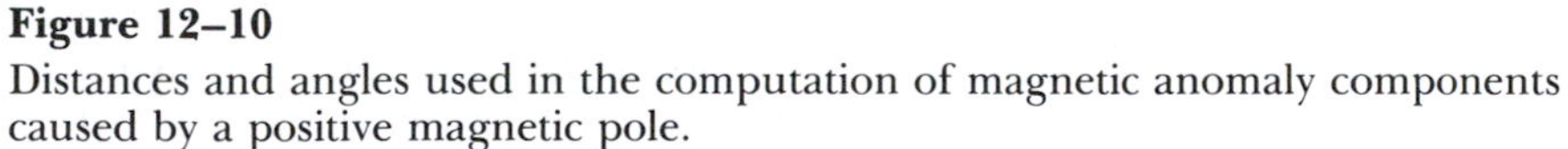
Figure 12–10
Distances and angles used in the computation of magnetic anomaly components caused by a positive magnetic pole.

$$h_{xp} = F_p \cos \alpha_p \sin \beta_p = \frac{P(x_q - x_p)}{r_p^3} \qquad (12\text{–}9b)$$

$$h_{yp} = F_p \cos \alpha_p \cos \beta_p = \frac{P(y_q - y_p)}{r_p^3} \qquad (12\text{–}9c)$$

By similar analysis, we obtain the components for the negative pole:

$$v_n = \frac{-Pz_n}{r_n^3} \qquad (12\text{–}10a)$$

$$h_{xn} = \frac{-P(x_q - x_n)}{r_n^3} \qquad (12\text{–}10b)$$

$$h_{yn} = \frac{-P(y_q - y_n)}{r_n^3} \qquad (12\text{–}10c)$$

The dipole field at Q is then found by summing and combining these components in the following way,

$$v_A = v_p + v_n \qquad (12\text{–}11a)$$

$$h_{xA} = h_{xp} + h_{xn} \qquad (12\text{–}11b)$$

$$h_{yA} = h_{yp} + h_{yn} \qquad (12\text{–}11c)$$

and, finally,

$$F_A = \sqrt{v_A^2 + h_{xA}^2 + h_{yA}^2} \qquad (12\text{–}12)$$

To calculate the total field intensity, we combine these components with the intensity components of the main field. Before this step can be completed, we must specify the orientation of the axes of our coordinate system. One conventional way of specifying it is to direct the x axis toward geographical (true) north and the y axis toward geographical east. For this orientation, we use the components of the main field, H_{xM}, H_{yM}, and V_M, and the dipole components h_{xA}, h_{yA}, and v_A to obtain the elements X, Y, Z, and F_T of the total field, which were first described in Equations 10–9:

$$Z = v_A + V_M \qquad (12\text{–}13a)$$

$$X = h_{xA} + H_{xM} \qquad (12\text{–}13b)$$

$$Y = h_{yA} + H_{yM} \qquad (12\text{–}13c)$$

and

$$F_T = \sqrt{X^2 + Y^2 + Z^2} \qquad (12\text{–}14)$$

As before, the values of F_T can be compared directly with total-intensity magnetometer readings, and values of T from Equation 12–7 can be compared with readings that have already been adjusted by subtraction of the main field.

So far, we have described a way to calculate the magnetic anomaly over a dipole model. How do we relate the pole strength of this model to the magnetic susceptibility of the rock mass that it represents? We can obtain an approximate relationship by the following simple procedure. Assume that the model is a circular cylinder with diameter D and length ℓ equal to the distance between the positive and negative poles. According to Equation 10–7, its magnetic moment $\mathbf{M} = P\ell$, and its magnetic moment per unit volume $\mathbf{I} = \mathbf{M}/\mathcal{V}$, where $\mathcal{V} = \pi D\ell$ is the volume of the cylinder. But Equation 10–8 tells us that $I = kF_M$ if the cylinder has susceptibility k and is magnetized by the main field F_M. We can equate these relationships to obtain

$$M = P\ell = I\mathcal{V} = (kF_M)(\pi D\ell) \qquad (12\text{–}15)$$

and

$$P = \pi DkF_M \qquad (12\text{–}16)$$

After specifying the susceptibility and dimensions of a cylindrical dipole model (Figure 12–7), we can use this formula to estimate its pole strength. This value will be only an approximation because we have not accounted for the difference between the physical length of a cylindrical bar magnet and the distance between its poles. In addition, we have not considered demagnetization, which according

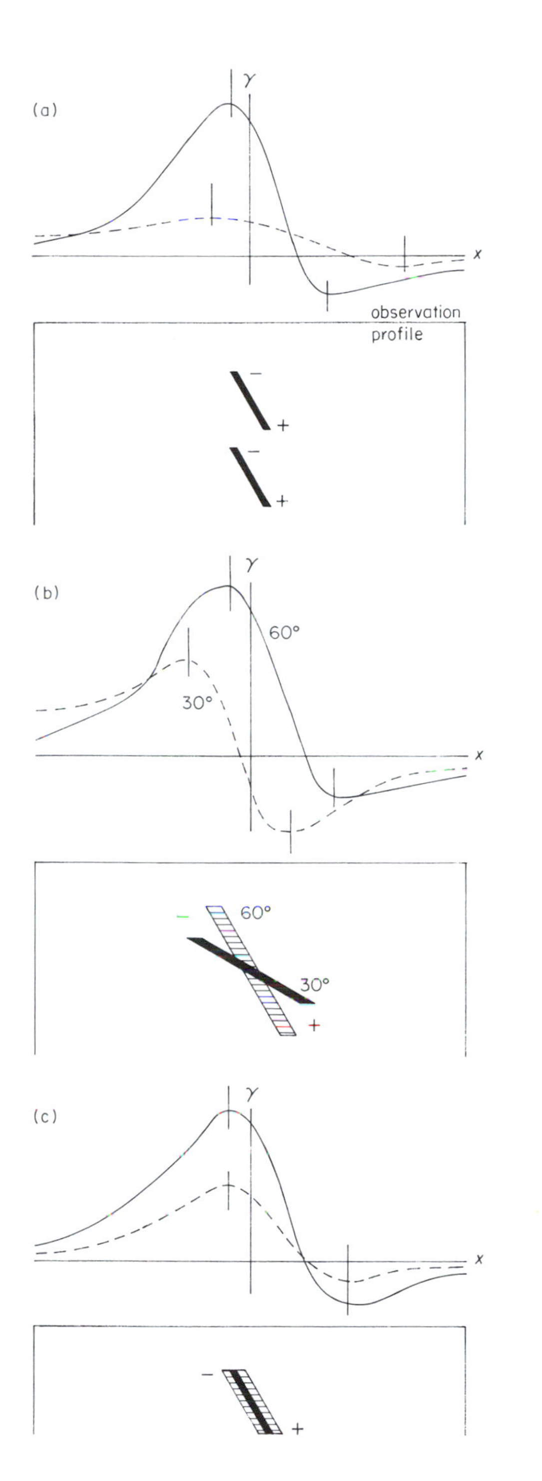

Figure 12–11
Comparison of total-intensity anomalies over (a) identical dipoles at different depths, (b) identical dipoles inclined at different angles, and (c) dipoles of different pole strength in the same position (qualitative scale).

to Equation 12–1 should be small for susceptibilities of most rocks listed in Table 12–3.

Now let us examine some important features of dipole magnetic anomalies. Observe in Figures 12–8 and 12–9 that the anomaly consists of adjacent positive and negative parts. The spacing and relative amplitudes of these high and low parts depend on the depth and inclination of the dipole. This is illustrated in Figure 12–11 which shows anomalies for dipoles in different positions. First, look at the effect of depth by comparing the anomaly profiles over two dipoles of identical inclination that are situated at different depths. Notice the following features, which are related to an increase in depth: (1) the anomaly amplitude diminishes; (2) the distance separating the high and low parts of the anomaly increases; and (3) the ratio of the high and low amplitudes does not change.

Next, look at the effect of dipole inclination by comparing the anomaly profiles over two dipoles which are inclined at different angles but which are situated at the same depth. The following important features can be observed for an increase in the inclination angle: (1) the ratio of the high and low amplitudes increases; and (2) the total difference in field intensity between the high and low parts remains approximately the same.

Finally, let us consider the effect of pole strength by comparing the anomaly profiles over dipoles of different pole strength that are inclined at the same angle and are located at the same depth. Notice that an increase in

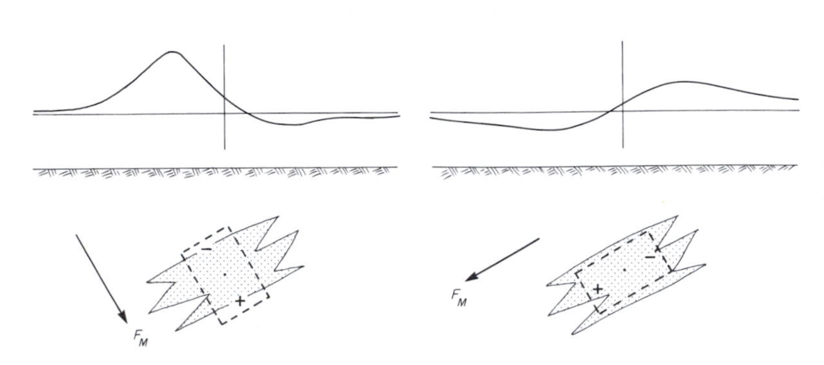

Figure 12–12
Total-intensity anomalies over identical irregular masses situated in different places so that their magnetic fields must be represented by differently oriented dipoles (qualitative scale).

pole strength (1) increases the total amplitude of the anomaly; (2) does not change the ratio of the high and low parts; and (3) does not change the distance separating the high and low parts.

Our analysis of the dipole model shows that relatively simple equations can be used to calculate the magnetic anomaly. However, the procedure requires a series of calculations to (1) determine pole strength after specifying the susceptibility and dimensions of the dipole; (2) determine horizontal and vertical components of field intensity; (3) combine these with corresponding components of the main field and calculate a total field profile; and (4) separate the main field from the total field by scalar subtraction to obtain the total field anomaly T. Although each calculation is quite simple, several more steps are needed to determine the magnetic anomaly than are required to obtain a gravity anomaly over a simple model.

We began this analysis by suggesting that a dipole model could be used to represent, approximately, the magnetism of a more irregular anomaly source (Figure 12–7). Now let us suppose that rock masses of identical shape and susceptibility are located in different places where the intensity and direction of the main field are not the same. Figure 12–12 shows how the dipole models of induced magnetism and the corresponding magnetic anomalies will change from one location to another. Unlike a gravity anomaly that is the same for all locations, the shape and amplitude of the magnetic anomaly over identical rock masses change with location.

Multiple Dipole Models

It is fortunate that the development of aeromagnetic surveying technology has been paralleled by the development of electronic computer technology. Otherwise, there would be no practical way to solve the complicated equations that we routinely use for calculating field intensity over models of magnetic anomaly sources. The theoretical basis for these equations is quite simple. The magnetism of any body, however irregular, can be represented by a sufficiently large number of dipoles. The magnetic anomaly over the body is found by summing the effects of these many dipoles. For relatively simple geometrical forms, the summing is done in part by mathematical integration. The result is a complicated equation, but one that can easily be programmed for computer solution.

The magnetic moment per unit volume **I** of a body specifies the magnetism of each volume unit, for example, each cubic centimeter, inside that body. Each one of these magnetized units in Figure 12–13a is represented by a small bar magnet. By summing the effects of all the bar magnets, we can calculate the anomaly over the body. But there is another way to look at this magnetism. In Figure 12–13b, a set of parallel lines in the direction of **I** is superposed on the body. The small bar magnets inside the body lie along these lines with the positive pole of one magnet close enough to the negative pole of the next magnet in line that they cancel each other. Along each line, only the single negative pole on the upper surface of the body and the single positive pole on the lower surface remain. Therefore, each line represents a single slender bar magnet extending between the two points where it intersects the surface of the body. Now we have a multiple dipole model. By solving Equations 12–9 for each pole on the surface of the body and summing the results, we can obtain the vertical and horizontal components of its magnetic anomaly.

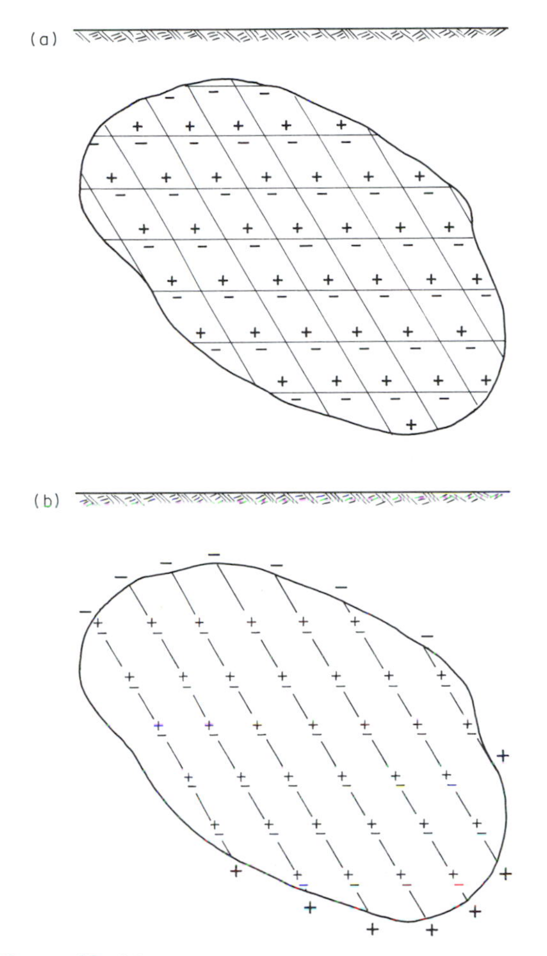

Figure 12–13
Magnetism of a rock mass represented (a) by magnetized volume elements, and (b) by magnetic poles distributed over the surface. Inclined lines indicate the direction of magnetization.

The distribution of poles on the surface of a body depends on the direction of magnetization. Suppose that the body is magnetized by induction in the earth's main field. The set of lines superposed on the body would be in the direction of the main field, and the spacing of the lines would be proportional to the main field intensity. In Figure 12–14 the direction and spacing of lines are illustrated for the magnetization of identical bodies at two different locations.

Calculation of magnetic anomalies over multiple dipole models by repeated solution of Equations 12–9 turns out to be impractical,

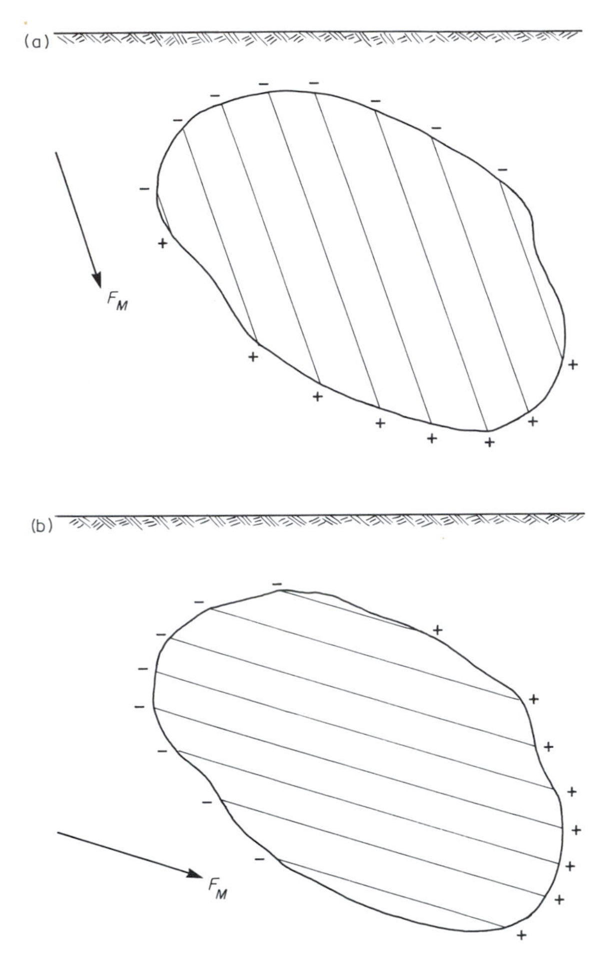

Figure 12–14
Distributions of poles on the surface of a mass that result from different directions of magnetization.

except for very simple models, because of the effort that would be required to specify the large number of pole positions. The more practical procedure is to use horizontal plates in different combinations to represent anomaly sources of more complicated shapes. First, the formulas expressing the magnetic anomaly over a horizontal plate will be presented.

The Three-Dimensional Horizontal Plate

A model that is particularly useful for analysis of magnetic anomalies is a horizontal plate of polygonal shape and vertical sides (Figure 12–15). Observe that this model has the same form as the horizontal plate used in the analysis of gravity anomalies (Figure 9–18). The x and y coordinates of the polygon corners and the depths z_1 and z_2 to the top and bottom surfaces of the plate are specified. These values together with the components of magnetic moment per unit volume I_x, I_y, and I_z are used to determine the field intensity at the origin caused by the magnetic poles on the surfaces of the plate. By mathematical integration, the following equations were obtained by two geophysicists, Manik Talwani and Donald Plouff:

$$\begin{aligned} h_{xA} &= I_xV_1 + I_yV_2 + I_2V_3 \\ h_{hA} &= I_xV_2 + I_yV_4 + I_2V_5 \qquad (12\text{–}17) \\ v_A &= I_xV_3 + I_yV_5 + I_2V_6 \end{aligned}$$

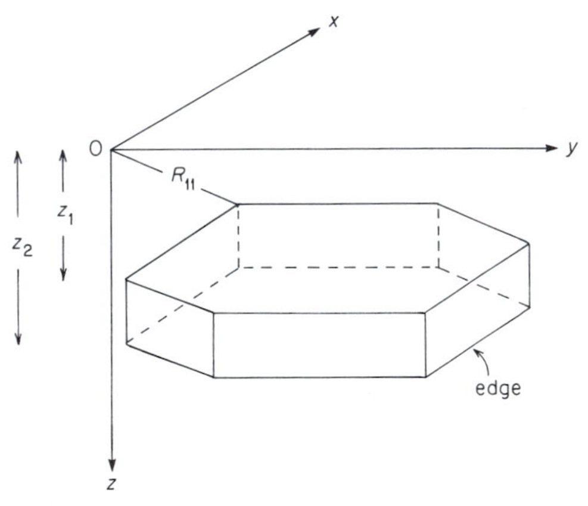

Figure 12–15
Horizontal plate of polygonal shape used to represent a three-dimensional magnetic anomaly source. (From D. Plouff, *Geophysics*, v. 41, n. 4, pp. 727–739, August 1976.)

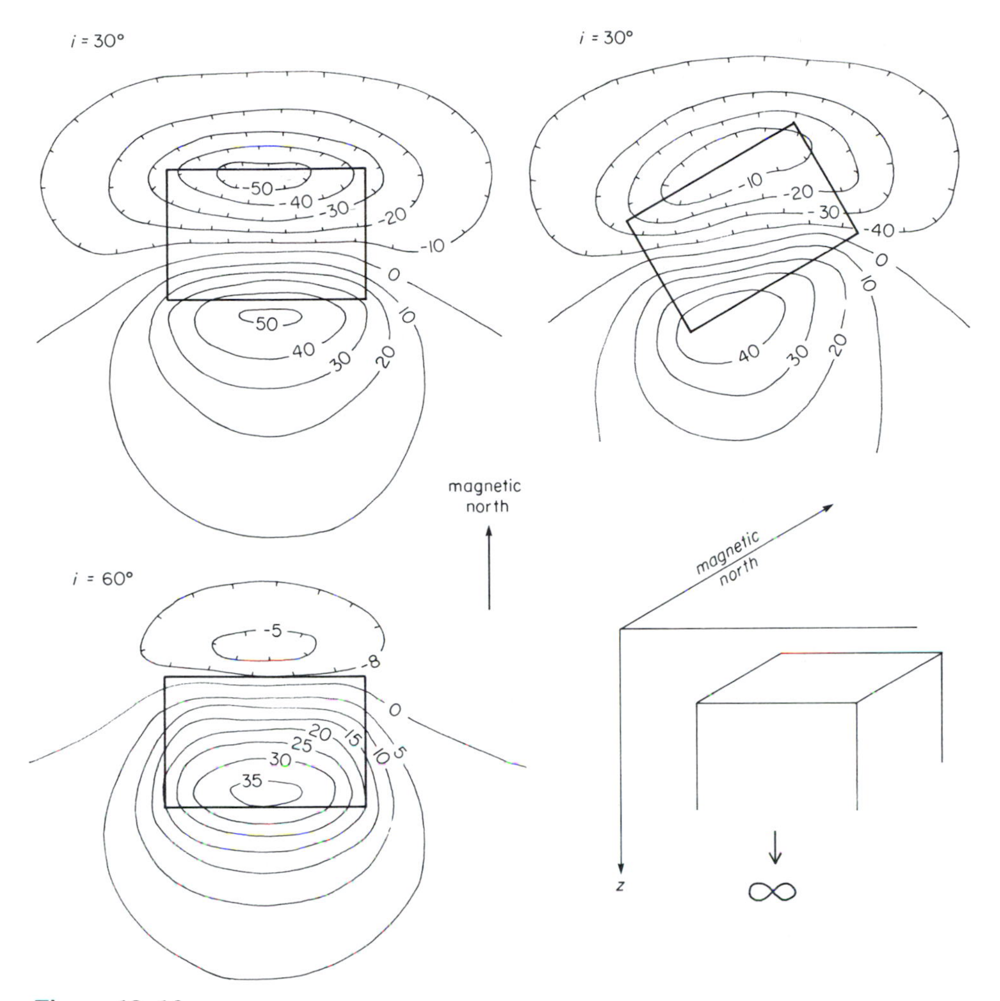

Figure 12–16

Contour maps of total-intensity anomalies over identical rectangular prisms with different orientations and directions of magnetization. The tops of the prisms are at the same depth, and the prisms extend infinitely downward. (Modified from V. Vacquier, N. C. Steenland, R. G. Henderson, and I. Zietz, *Geological Society of America, Memoir 47*, pp. 47, 104, 119, 1951 and reprinted 1963.)

Here $V_1, \ldots, V_6$ are complicated terms[1] that depend on z_1, z_2, and the polygon corner coordinates. After these components are calculated with a digital computer, they are combined with the main field elements in Equations 12–11, 12–12, and 12–14 to obtain total-intensity anomaly values.

Examples of magnetic anomalies calculated by this method for rectangular prism models are shown in Figure 12–16. These examples illustrate how the anomaly pattern differs according to the orientation and direction of magnetization of otherwise identical models. Although different from one another, all the patterns have adjoining high and low parts.

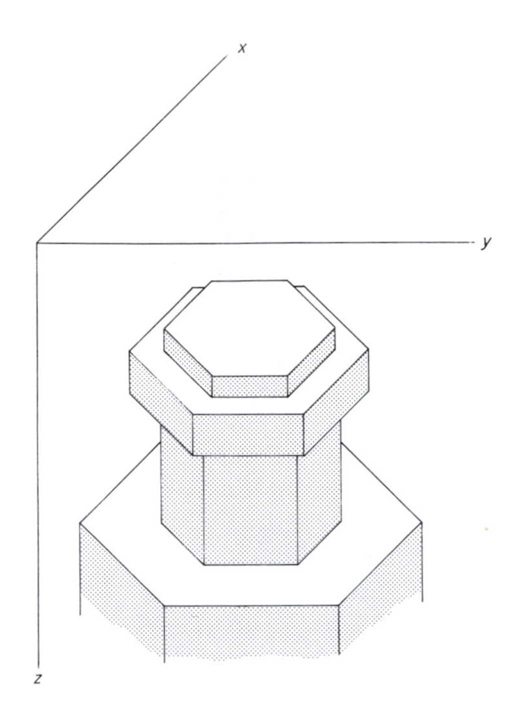

Figure 12–17
Irregular magnetic anomaly source model consisting of several horizontal polygonal plates.

Compound Three-Dimensional Models

Combinations of horizontal polygonal plates can be used to represent magnetized bodies with complicated shapes. The idea is the same as we described in Chapter 9 for three-dimensional gravity anomaly sources. The combination of plates in Figure 12–17 illustrates this procedure. The magnetic anomaly for each plate is calculated separately using Equations 12–17. Results are then summed to obtain the anomaly over the compound model.

Three-dimensional models are described by specifying the positions of polygon corners and the depths to their upper and lower surfaces. A rectangular coordinate system is usually chosen for purposes of describing these positions and depths. One common convention we introduced earlier in this chapter is to direct the x axis toward geographical (true) north and the y axis toward geographical east. For this orientation, the field intensity components obtained from Equations 12–17 can be combined in Equations 12–13 and 12–14 with the corresponding elements of the main field to obtain the total intensity F_T.

[1]Interested readers can find expressions for these terms in the following article by Donald Plouff: Gravity and magnetic fields of polygonal prisms and their application to magnetic terrain corrections, *Geophysics,* v. 41, n. 4, pp. 727–741, August 1976.

The Semi-infinite Horizontal Plate

Linear magnetic anomalies, like linear gravity anomalies, can be analyzed by comparing them with theoretical anomalies over two-dimensional models. For purposes of magnetic anomaly calculations, a semi-infinite horizontal plate with a sloping edge is the most useful geometrical form, because it can be used in various combinations to make up two-dimensional models with more complicated shapes.

Magnetization of a semi-infinite horizontal plate can be represented by a distribution of positive and negative poles on its surfaces. These poles are situated at points where a set of evenly spaced inclined lines, representing the magnetizing field, intersect the plate (Figure 12–18). Observe how these poles lie along the dotted lines on the surfaces of the plate that are parallel to its edge. What magnetic field is produced by a single horizontal line of poles that is infinitely long (Figure 12–19)? Obviously, the field intensity at some observation point is mostly due to the nearby poles,

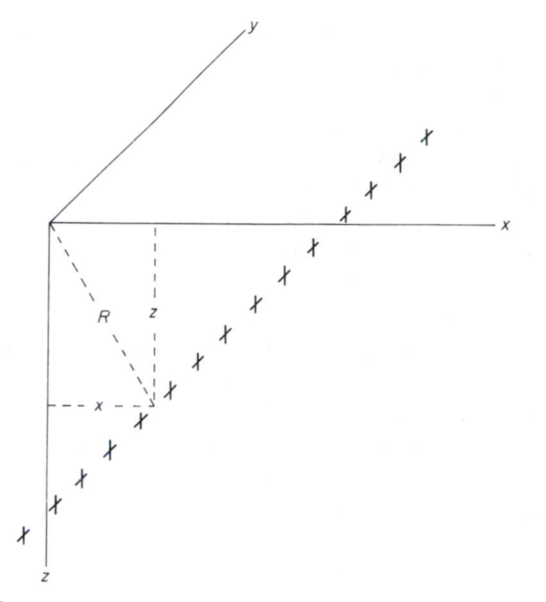

Figure 12–19
Distances used for calculating components of magnetic field intensity resulting from an infinitely long horizontal line of magnetic poles that is parallel to the y axis.

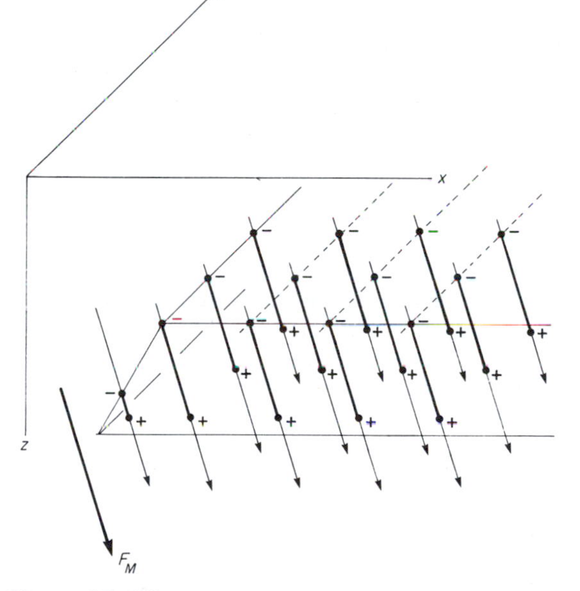

Figure 12–18
Distribution of magnetic poles on the surface of a horizontal semi-infinite plate magnetized by induction in the earth's main field.

whereas those along the more distant parts of the line have a negligibly small effect. Summing the effects of individual poles by means of integration yields the field intensity F_L at distance R from the line,

$$F_L = 2P'/R \qquad (12\text{–}18)$$

where P' is the pole strength per unit length along the line. The value of F_L is positive or negative depending on whether the line consists of positive or negative poles. In terms of a coordinate system with the y axis parallel to the line of poles (Fig. 12–19), the components of field intensity in the x and z directions are

$$h = 2P'x/R^2 \quad \text{and} \quad v = 2P'z/R^2 \qquad (12\text{–}19)$$

where $R = \sqrt{x^2 + z^2}$.

Next, look at the cross section of a semi-infinite horizontal plate (Figure 12–20) which extends infinitely in both the positive y and negative y directions, as well as in the positive x direction. The edge of the plate is parallel to the y axis of the coordinate system. Each pole shown on the cross section represents a line of poles reaching infinitely into and out from the plane of the cross section. The magnetic anomaly is the sum of the fields produced by these line sources. This sum is expressed by formulas for the v and h components of field intensity, which were first presented by Manik Talwani and James Heirtzler. After pole strength is expressed in terms of the magnetic moment per unit volume **I**, their results are

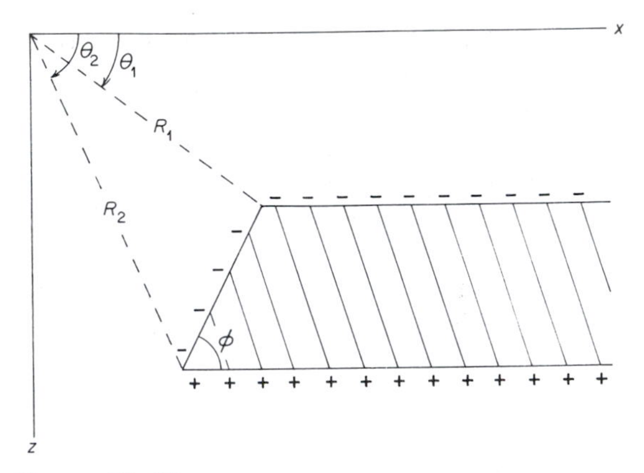

Figure 12–20
Distances and angles used to calculate components of magnetic field intensity resulting from a semi-infinite horizontal plate with a sloping edge.

$$v = 2 \sin \varphi \, I_x \left[(\theta_2 - \theta_1) \cos \varphi + \sin \varphi \ln \left(\frac{R_2}{R_1} \right) \right] - 2 \sin \varphi \, I_z \left[(\theta_2 - \theta_1) \sin \varphi - \cos \varphi \ln \left(\frac{R_2}{R_1} \right) \right] \tag{12–20a}$$

$$h = 2 \sin \varphi \, I_x \left[(\theta_2 - \theta_1) \sin \varphi - \cos \varphi \ln \left(\frac{R_2}{R_1} \right) \right] + 2 \sin \varphi \, I_z \left[(\theta_2 - \theta_1) \cos \varphi + \sin \varphi \ln \left(\frac{R_2}{R_1} \right) \right] \tag{12–20b}$$

where I_x and I_z are components of **I**, and R_1, R_2, θ_1, and θ_2 are the lengths and angles of lines from an observation point to the upper and lower corners of the plate.

We must remember that the coordinate system in this analysis has the y axis parallel and the x axis perpendicular to the edge of the plate. This is different from the convention of north- and east-directed x and y axes described for three-dimensional model analysis. Therefore, we cannot use Equations 12–13 to obtain total field components. Rather, we must find H_{xM} and H_{yM}, the components of the horizontal element H_M of the main field in the x and y directions that correspond to the plate,

$$H_{xM} = H_M \sin \alpha \quad \text{and} \quad H_{yM} = H_M \cos \alpha \tag{12–21}$$

where α is the angle between magnetic north and the direction of the edge of the plate. Now the total field intensity F_T can be found,

$$F_T = \sqrt{(h + H_{xM})^2 + H_{yM}^2 + (v + V_M)^2} \tag{12–22}$$

where h and v are determined from Equations 12–20. This value can then be used in Equation 12–7 to obtain the total-intensity anomaly T.

Examples of total-intensity anomaly profiles are shown in Figure 12–21 for a semi-infinite horizontal plate with a vertical edge. These profiles differ from one another because of

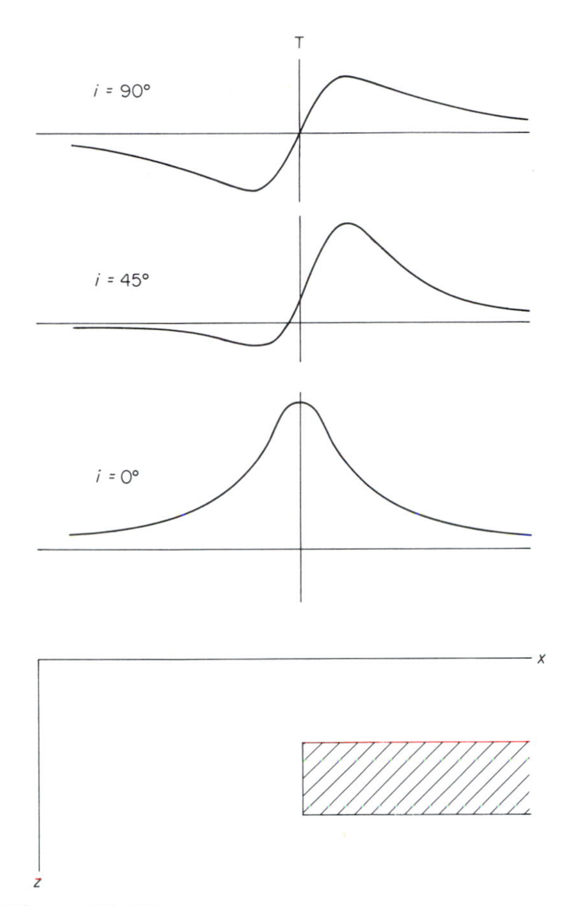

Figure 12–21
Total-intensity anomalies over identical horizontal, semi-infinite plates magnetized in different directions. The x axis extends in the magnetic north direction.

differences in the direction of magnetization of otherwise identical models. All the profiles display adjoining high and low parts. An important difference between the magnetic anomaly and the gravity anomaly (Figure 9–16) over a horizontal plate should be pointed out. The magnetic anomaly exists only near the edge and diminishes over the plate with increasing distance from the edge. In contrast, the gravity anomaly over the plate increases with distance from the edge.

Two-Dimensional Inclined Plates

Many linear magnetic anomalies can be attributed to inclined plates of magnetized rock. For purposes of computing a theoretical magnetic anomaly, this kind of structure can be reproduced by subtracting from one thick horizontal plate another plate with a parallel sloping edge that is offset slightly (Figure 12–22). The magnetic anomaly over the inclined plate is found by means of Equations 12–20. At points along a profile, the values of v and h for one plate are subtracted from those for the other plate that extends farther to the left.

This method was used to calculate the magnetic anomaly profiles over plates that are (1) inclined at different angles; (2) oriented in different directions relative to magnetic north; and (3) magnetized in different directions. The results show that these three factors all influence the distance between the high and low parts of the anomaly and the ratio of the high and low amplitudes. In other examples, we have already pointed out that differences in orientation and direction of magnetization produce different anomalies over structures having the same shape and magnetic susceptibility. The examples in Figure 12–22 also indicate that plates of different inclination, magnetization, and orientation might possibly produce very similar anomalies.

Irregular Two-Dimensional Models

Linear geologic structures with complicated cross-sectional shapes can be represented by

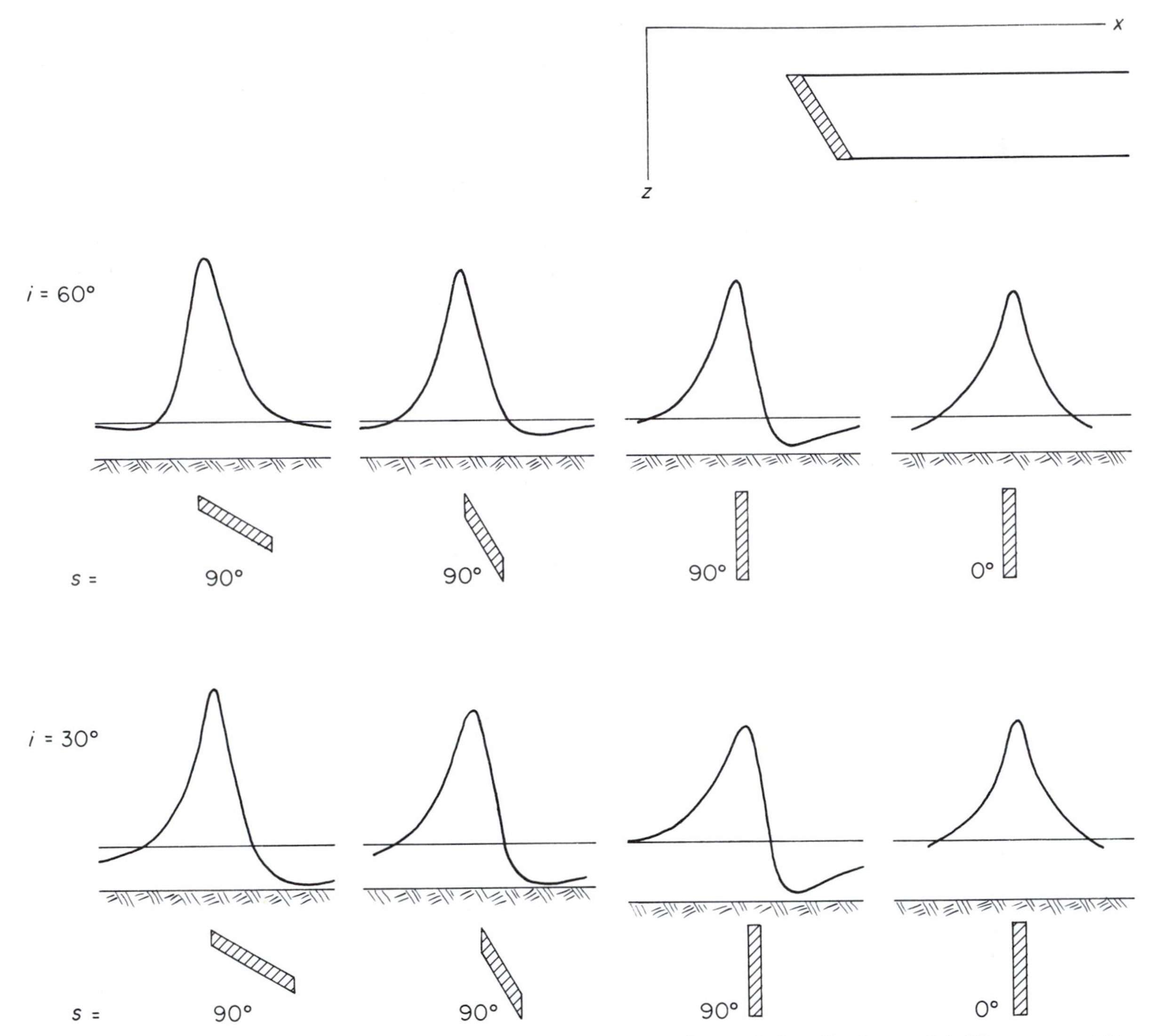

Figure 12–22
Total-intensity anomalies over plates of infinite horizontal extent that are inclined at different angles, magnetized in different directions (*i*), and oriented in different directions (*s*) relative to magnetic north (qualitative scale). These anomalies are calculated by subtracting from the anomaly over a horizontal plate with a sloping edge the anomaly over a similar but slightly offset plate.

two-dimensional models with polygonal cross sections. The idea is the same as the one we described in Chapter 9 for gravity anomaly analysis. To obtain a model of a magnetic anomaly source, we can combine several plates with differently sloping edges to reproduce a complicated polygonal cross section. Try to figure out how several horizontal plates could be added and subtracted to reproduce the polygon in Figure 12–23, which represents a syncline. Magnetic field components can be calculated separately for each plate by means of Equations 12–20 and then added and subtracted in the same way to obtain the magnetic anomaly over the model.

Summary

The procedure for calculating magnetic anomalies over models is more complicated than the corresponding procedure for calculating gravity anomalies over models. After an appropriate coordinate system is chosen, the vertical and horizontal components of field intensity are calculated separately for the model and for the main field. These are then combined to calculate a total field. Finally, the total-intensity anomaly is found by scalar subtraction of the main field from the total field intensity. The result is an anomaly with high and low parts. The distance between the high and low parts and the ratio of their amplitudes differ according to the location, orientation, depth, and direction of magnetization of the model. Practical methods for rapid and accurate calculations of model anomalies depend heavily on electronic computers.

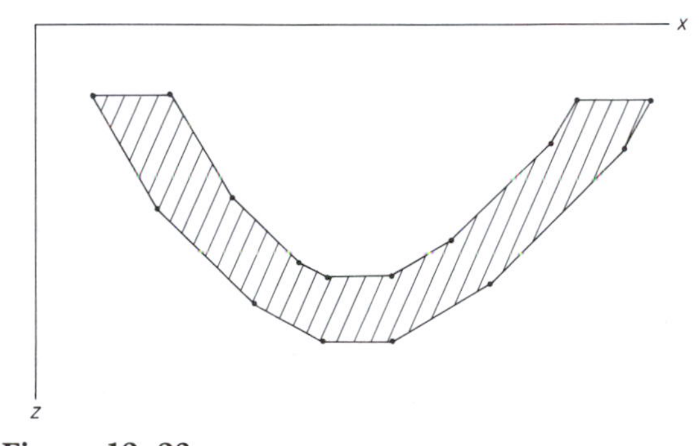

Figure 12–23
Two-dimensional polygon model representing a syncline.

MAGNETIC ANOMALY INTERPRETATION

Survey Objectives

Magnetometer surveys are undertaken for various purposes. Most exploration surveys have at least one of three general objectives: (1) detection of structural trends; (2) mapping the depth to "magnetic basement"; and (3) detection and analysis of specific anomaly sources.

Structural trends are indicated by patterns of linear anomalies and alignments of anomalies. Often, the interpretation involves little more than inspecting a magnetic anomaly map. Some anomaly patterns correlate directly with structural features on a geologic map. What may be more interesting are the linear patterns that cannot be attributed to obvious surface features. They are the basis for inferring the existence of hidden faults or zones of deformation. In mining districts, inspection of magnetic anomaly maps may be all that is necessary to recognize patterns similar to those over known ore deposits, so that exploration drilling can commence without further analysis.

One of the principal uses of magnetometer surveying in petroleum exploration has been to measure the depth to the "magnetic basement." What is meant by this term? It refers to the zone beneath the earth's surface where sources of magnetic anomalies are situated. The magnetic basement is mostly an assemblage of igneous and metamorphic rock possessing large magnetization contrasts. In areas of petroleum exploration, it typically underlies a sequence of nonmagnetic sedimentary rocks. The real aim is to estimate the thickness of this sequence, which may contain petroleum reservoirs. This estimate can be made by calculating the depths to underlying magnetic anomaly sources. These sources can be situated throughout a wide range of depths, but the shallowest ones are presumed to be at or near the upper surface of the magnetic basement. A common practice is to use inversion schemes that yield depth values from measurements of certain features of the magnetic anomalies. Recall that the idea of inversion schemes was introduced in Chapter 9 for the analysis of gravity anomalies. It is entirely possible that a zone of nonmagnetic igneous and metamorphic rock may lie between the base of the sedimentary layer and the depth of the shallowest sources of magnetic anomalies. However, experience shows that this is usually not the case.

The purpose of some magnetometer surveys is to determine specific details about the shape and location of anomaly sources. A trial-and-error method is the best way to find the shapes of two- or three-dimensional models that most accurately reproduce measured anomalies. The procedure can be time-consuming and is not suitable for rapid analysis of a large number of anomalies.

Derivative Methods for Separating Regional and Local Anomalies

The methods described in Chapter 9 for separating regional and local gravity anomalies are also used in the analysis of magnetic anomaly maps. Field-derivative calculations are another viable approach. This procedure is introduced here because it is used in one of the most widely applied magnetic anomaly inversion schemes. However, it is applicable for both magnetic and gravity analysis.

We can use the method of upward continuation to explain the nature of field derivatives. Look again at the discussion of Equations 9–19 and 9–20 to recall how upward continuation is done. Now suppose that we have total-intensity magnetic anomaly values, T_0, at N equally spaced grid points on an observation surface. Let us use upward continuation to find the anomaly T_z at the height z above the center of this area. The equation is the same as Equation 9–19 except that magnetic anomaly values rather than gravity anomaly values are used:

$$T_z = \frac{1}{N}\left[\left(\mathrm{T}_{01} \times \frac{z_1 A}{2\pi R_1^3}\right) + \left(T_{02} \times \frac{z_2 A}{2\pi R_2^3}\right) + \left(T_{03} \times \frac{z_3 A}{2\pi R_3^3} + \ldots\right]\right. \quad (12\text{–}23)$$

Here the terms A and R are as described in Figure 9–26, but z is used in place of h to represent height. The next step is to repeat this calculation at a slightly higher level, $z + \Delta z$, to obtain the anomaly value $T_{z+\Delta z}$. Then we find the difference between these two values, which we divide by the small change in height,

$$\frac{T_{z+\Delta z} - T_z}{\Delta z} = \frac{\Delta T}{\Delta z} = T' \quad (12\text{–}24)$$

where T' is the *first vertical derivative* or the *vertical gradient* of the total field anomaly at this point. It tells us the rate of change of the anomaly with elevation. If we calculated values of T' above each one of the grid points on the observation surface, we could contour the values to obtain a vertical gradient map. The effect of this procedure is to suppress regional anomalies and enhance local anomalies.

Further calculations can be done to bring local anomalies into still sharper focus. First, we obtain values of T' at two slightly different levels. Then we divide the difference between these values by the small difference in height,

$$\frac{T'_{z+\Delta z} - T'_{z}}{\Delta z} = \frac{\Delta T'}{\Delta z} = T'' \qquad (12\text{–}25)$$

where T'' is the *second vertical derivative* or the *curvature* of the total field anomaly at this point. Curvature is a mathematical expression of how sharply curved or distorted the anomaly surface is at the location. A curvature anomaly map is prepared by contouring values of T'' that were calculated above each grid point on the observation surface.

We pointed out in Chapter 9 that upward continuation could be viewed as a weighted averaging scheme in which the weighting coefficients were chosen to achieve this result. It is also possible to determine other sets of weighting coefficients that can be used in equations similar to Equation 9–18 to obtain gradient and curvature values directly from a grid of total-intensity anomaly values. This procedure was followed using total anomaly values from the map in Figure 12–24a to calculate curvature values contoured in Figure 12–24b. The result is a more sharply defined contour pattern which turns out to parallel the edges of the underlying anomaly source more closely than the contour pattern of the total-intensity anomaly.

Inversion Methods for Estimating Depths of Anomaly Sources

Efficient methods for calculating the depths of anomaly sources are essential for purposes of mapping the depth of the magnetic basement. But magnetic anomalies depend on the source's orientation and direction of magnetization as well as on its shape and depth. Is there a simple method for rapid calculation of depth that accounts for all these factors? Depth can be calculated rapidly at the expense of accuracy by using models very simple in shape to represent the anomaly sources. Most inversion schemes make use of an important feature of magnetic anomalies. All sources for which the dimensions and depth are in the same proportions produce anomalies of identical shape. An example illustrates this feature. Consider two vertical cylinders that have the same magnetic moment per unit volume. One has a length of 5 miles, a diameter of 1 mile, and a depth to its top of 2 miles. The other has a length of 5 meters, a diameter of 1 meter, and a depth to its top of 2 meters. Anomaly values at one-mile intervals along a profile over the large cylinder are identical to the values at one-meter intervals along a profile over the small one. Of the numerous inversion methods that have been devised, we will describe three that illustrate the kinds of simplifications that are necessary for efficient estimation of source depths.

One of the most widely used magnetic anomaly inversions is the *Peters half slope method,* which was proposed by the exploration geophysicist Leo Peters in 1949. The method

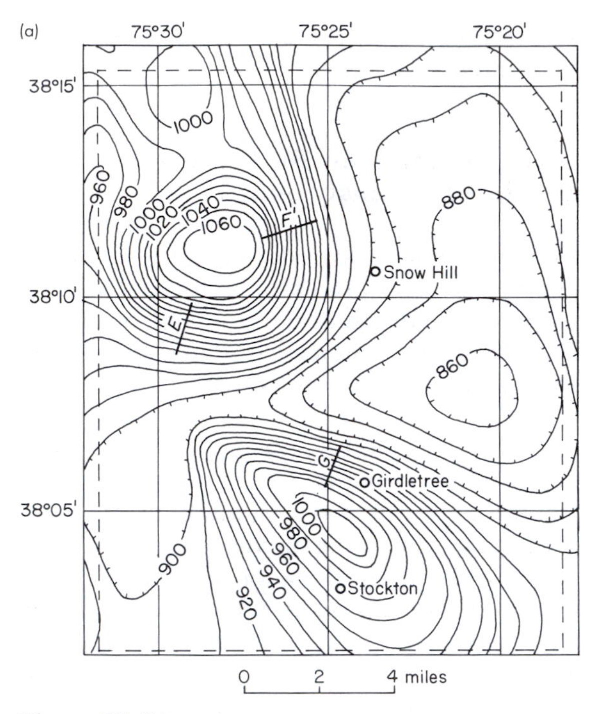

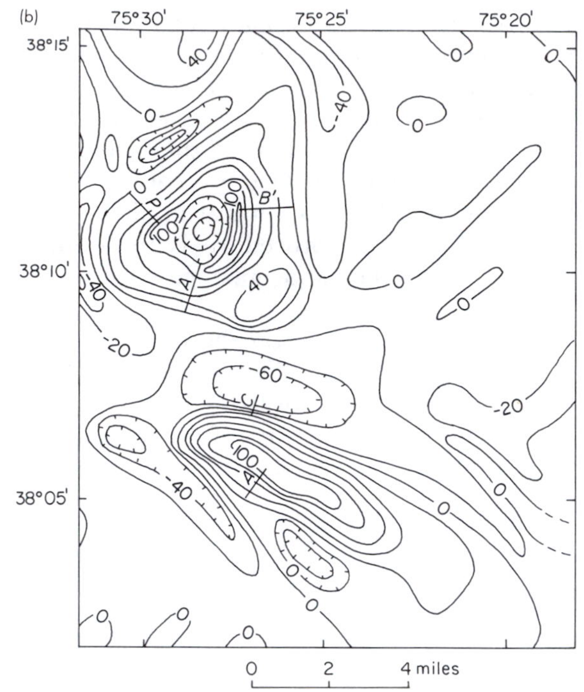

Figure 12–24

(a) A total-intensity anomaly map and (b) a magnetic curvature map of part of Worcester County, Maryland. (From V. Vacquier, N. C. Steenland, R. G. Henderson, and I. Zietz, *Geological Society of America, Memoir 47*, pp. 21–22, 1951 and reprinted 1963.)

is based on analysis of the anomaly over a vertical plate that is vertically magnetized (Figure 12–25). The plate is assumed to extend infinitely downward from the depth z. The magnetic anomaly along a profile crossing perpendicular to such a plate is

$$v = 2I\left(\tan^{-1}\frac{z}{x_1} - \tan^{-1}\frac{z}{x_2}\right) \qquad (12\text{–}26)$$

where x_1 and x_2 are the horizontal distances, and z is the vertical distance of the plate corners from a point on the profile. The formula can be used to calculate anomaly profiles over plates of different thickness and depth. In a study of profiles over many different plates, Peters found that the depth z could be estimated from the shape of an anomaly profile in the following way. First, a line is drawn tangent to the steepest part of the anomaly; this line is identified as line 1 in Figure 12–25. The slope of this line is measured, and then two lines, labeled 2 and 3, are constructed so that they are tangent to the anomaly and have slopes that are one-half the slope of line 1. The horizontal distance d is then measured between the points of tangency of lines 2 and 3. There is an approximate relationship between d and the depth z:

$$z \cong 1.6d \qquad (12\text{–}27)$$

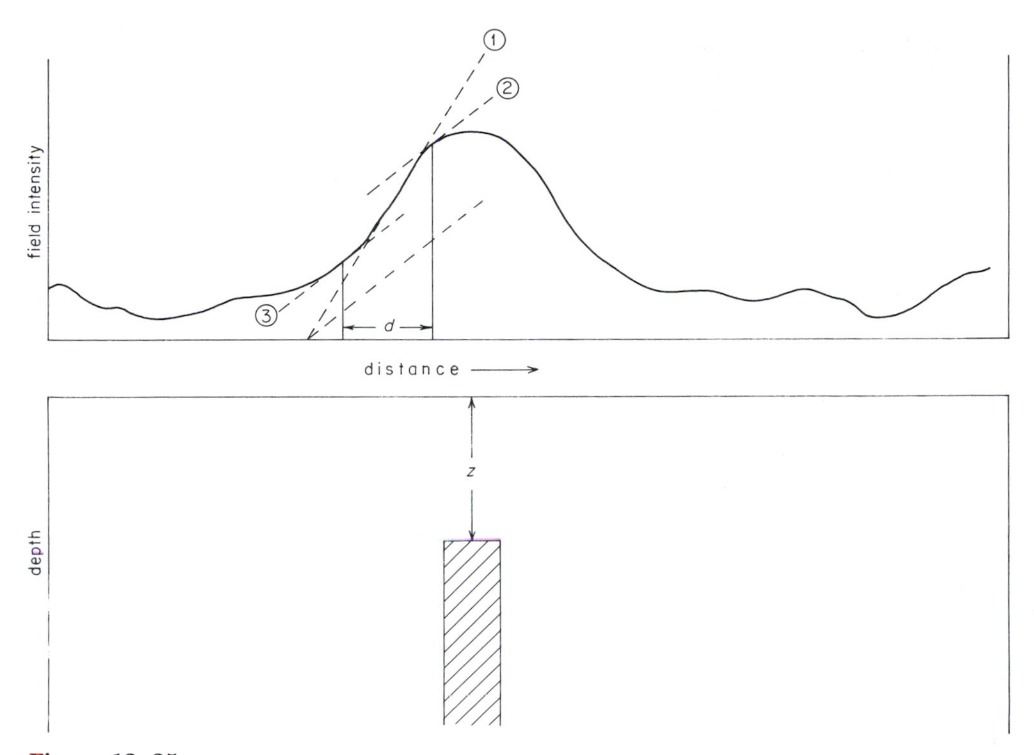

Figure 12–25
Analysis of a magnetic anomaly by the Peters half slope method to estimate the depth z to the top of a semi-infinite vertical plate model. Line 1 is drawn tangent to the anomaly at the point of maximum slope. Lines 2 and 3 are drawn tangent to the anomaly at points where the slope is one-half the maximum slope. The distance d between these points is related to the depth of the anomaly source.

In nature we could expect to find such a vertically magnetized plate only near the north and south magnetic poles of the earth. In these locations, the vertical- and total-intensity anomalies are identical over vertical plates oriented in any direction. At other locations, anomalies with the same symmetrical shape would be produced only by vertical plates oriented in the direction of magnetic north.

Ideally, the Peters half slope method should be used only to analyze narrow symmetrical anomalies that trend close to magnetic north and that are not appreciably distorted by other nearby anomalies. It is not uncommon to calculate depths that are accurate within 10 percent from analyses restricted to these anomalies. With some sacrifice of accuracy, the method has been used widely in the United States and Canada to analyze both vertical- and total-intensity anomalies trending in other directions. Because the main field inclination is larger than 60 degrees over most of this area, the anomalies over vertical plates, though somewhat asymmetrical, are not severely distorted. Careful application of the Peters half slope method ordinarily yields depths that are accurate within 30 percent. Results become much less reliable where there is distortion from nearby overlapping anomalies and where the main field inclination is less than about 60 degrees. When used to analyze appropriate anomalies, the Peters half slope method can provide useful information about depth to magnetic basement.

Another well-known inversion method for analyzing total-intensity anomalies was devised by the geophysicists V. Vacquier, N. D. Steenland, R. G. Henderson, and I. Zietz in 1951. They prepared a total of 83 reference maps that show total-intensity anomalies and curvature patterns over a set of standard models. All these models are rectangular prisms with horizontal tops and vertical sides that extend infinitely downward. These standard models have different horizontal dimensions, orientations, and directions of magnetization. An example is the reference map in Figure 12–26 which shows total-intensity anomaly contours on the east side and curvature contours on the west side. The prismoidal model outlined on the map is eight grid intervals long and wide, and the inclination of its magnetization is 75 degrees. Depth to its top was set equal to one grid interval for purposes of calculating the total-intensity anomaly and curvature values. Because of the proportionality relation between the depth and dimensions of an anomaly source, the spacing of grid lines can be assigned any value without altering the contour patterns. That is, a model 8 miles long and wide, at a depth of 1 mile, would produce the same pattern as a model 8 feet long and wide, at a depth of 1 foot, and so on for other units.

Analysis begins by comparing observed anomalies with reference maps for models

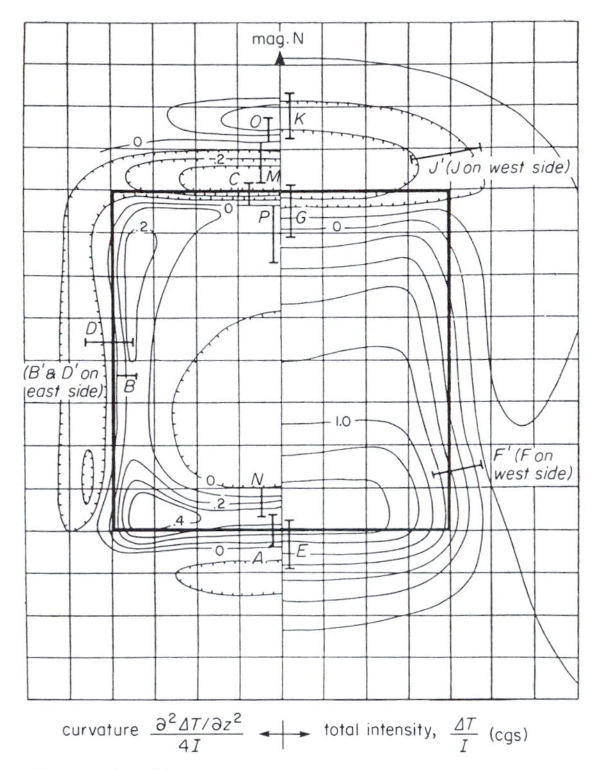

Figure 12–26
Reference map over a rectangular prism showing total-intensity contours on the right and curvature contours on the left. The depth to the prism top is one grid unit, and the prism extends infinitely downward. Index lines are drawn where contours are most closely spaced. (From V. Vacquier, N. C. Steenland, R. G. Henderson, and I. Zietz, *Geological Society of America, Memoir* 47, p. 13, 1951 and reprinted 1963.)

that have a magnetization direction closest to the main field inclination in the survey area. After the map that best reproduces the observed anomaly is selected, index lines such as those marked by letters in Figure 12–26 are drawn at corresponding positions on both maps. These lines are located where contours are the most closely spaced. The lengths of index lines drawn on the observed anomaly are determined from the map scale. Lengths of corresponding lines on the reference map are measured in grid units. These values are used to calculate the grid interval, which is also the depth to the source. Because most anomaly sources do not have perfectly horizontal tops, different index lines usually yield different values for the grid spacing. These values can be averaged to obtain a depth estimate.

In Figure 12–24 this procedure was followed to obtain the depth to the source of the anomaly situated just west of Snow Hill. Index lines labeled E and F' on the total-intensity anomaly map were compared with lines in Figure 12–26 to obtain the same depth estimate of 5600 feet for both locations. Index lines A, B', and P on the curvature map yield values of 6000 feet, 8200 feet, and 6000 feet, respectively. If the unusually high value from line B' is judged to be unreliable, the remaining four values yield an average depth of 5800 feet.

The last inversion method we will introduce is the most versatile. It is perhaps the best known of several computer-based approaches to analyzing total-intensity magnetic anomalies. The idea originated in 1953 with a Swedish geophysicist, S. Werner. It was further developed and adapted for computer processing by R. R. Hartman, D. J. Teskey, and J. L. Friedberg. The basic model is a thin semi-infinite sheet. By attributing all field intensity variations along a profile to some combination of similarly oriented thin sheets, we can view the result as the superposition of many anomalies, all having the same characteristic shape. This notion is not unlike the idea in Chapter 6 that a seismogram is the superposition of many pulses, all of which have the shape of an initial source pulse. Recall that the technique for decomposing a seismogram to extract a pulse shape and a series of points indicating where the pulses occur is called deconvolution. A magnetic anomaly profile, too, can be "deconvolved" to extract a characteristic anomaly shape and a set of points indicating where anomalies with this shape originated. The procedure has been named *Werner deconvolution.* It is applied to magnetic anomaly profiles to obtain the positions of points that presumably indicate the tops of semi-infinite sheet sources.

Equations for calculating the magnetic anomalies over thin semi-infinite sheets inclined at different angles and magnetized in different locations are quite complicated. Nonetheless, Werner recognized that these equations could be put in the general form

$$F(x) = \frac{A(x - x_0) + Bz}{(x - x_0)^2 + z^2} \qquad (12\text{–}28)$$

where F is the anomaly value at position x that is produced by a thin sheet with its top at distance x_0 along the observation profile and depth z below it. The terms A and B represent complicated functions related to the orientation, geometry, and magnetization of the sheet. Rather than using the complicated mathematical expressions for A and B, Werner proposed a simple method for computing their numerical values along short segments of the profile. By substituting the terms $a_0 = Ax_0 + Bz$; $a_1 = A$; $b_0 = -x_0^2 - z^2$; and $b_1 = 2x_0$, we can rearrange Equation 12–28 into the polynomial

$$a_0 + a_1x + b_0F + b_1xF = x^2F \qquad (12\text{–}29)$$

Now, by substituting measured values of F from four points along the profile, we obtain four separate equations. Simultaneous solution of these equations yields values for a_0, a_1, b_0, and b_1. The terms just given are combined so that these values can be used to calculate the position of the top of the thin sheet:

$$x_0 = \tfrac{1}{2}b_1 \quad \text{and} \quad z = \tfrac{1}{2}\sqrt{-4b_0 - b_1^2} \qquad (12\text{–}30)$$

Accurate values of x_0 and z can be obtained only if the four measured values of F are not affected by other nearby anomaly sources. Because overlapping effects of nearby sources are almost always present, Equation 12–29 should be modified to take account of these effects. For practical purposes they can be represented, approximately, by adding more terms to the polynomial

$$a_0 + a_1x + b_0F + b_1xF = x^2F + c_0 + c_1x + c_2x^2 \qquad (12\text{–}31)$$

where the polynomial coefficients c_0, c_1 and c_2 must also be determined. The values of F from seven points along the profile are substituted to obtain seven equations for simultaneous solution. After the seven coefficients are determined, the values of x_0 and z can be calculated. Accuracy of the result depends on how well the additional polynomial terms represent the effects of nearby sources.

These additional terms do not properly account for anomalies associated with contacts between large rock masses of contrasting magnetization. Therefore, the procedure requires further modification, which is based on an interesting feature. Magnetic field intensity along a profile over a thin sheet varies in the same pattern as the field intensity gradient over a plane contact in the same position between two contrasting semi-infinite masses. Therefore, if gradient values are available along the observation profile, they can be used in place of intensity values to obtain depth estimates based on a plane contact model rather than a thin-sheet model. The gradient of total field intensity in the direction of magnetization can be calculated from its vertical component, $\Delta T/\Delta z$, and its horizontal component, $\Delta T/\Delta x$. We have already described how to measure the vertical component of the gradient with two magnetometers mounted one above the other (Figure 11–5) or to calculate it by the method of weighted averaging. The horizontal component is easily calculated from the difference between adjacent values of T along the profile divided by the distance between the two measurements.

For different parts of a profile, it is usually difficult to know which model, the thin sheet or the plane contact, is most appropriate. Therefore, the conventional practice is first to use total-intensity anomaly values in Equation 12–31 to calculate a set of thin-sheet positions, and then to repeat the procedure with gradient values to obtain an alternate set of plane contact positions. The analysis advances along the profile using sets of several alternate values or sets consisting of every third value, and so on. This should be done to ensure that narrow and broad anomalies alike are properly sampled. As the computer completes each calculation, it instructs a mechanical plotter to mark the position on a graph. Different symbols are used for the thin-sheet and the plane contact models. The example presented in Figure 12–27 illustrates the result of this kind of analysis.

How do we interpret the seemingly irregular pattern of marks produced by Werner deconvolution? At this point, human judgments must be made. Experienced interpreters have gained some knowledge about these patterns from analyses of theoretical profiles over different combinations of anomaly source models. Much of the scatter is due to anomaly sources that are not perpendicular to the profile and that are not thin sheets or plane contacts. However, careful examination does re-

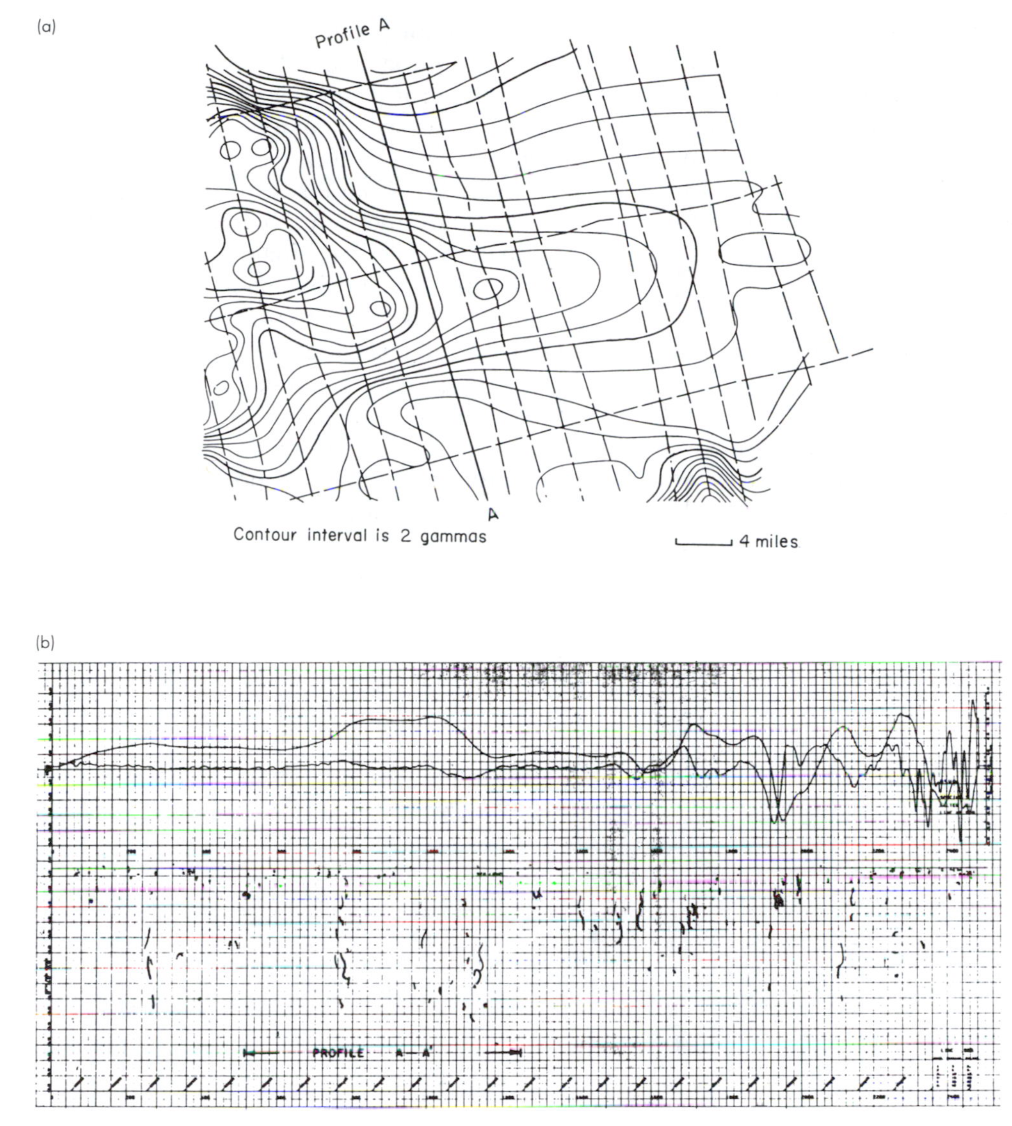

Figure 12–27
(a) Total-intensity anomaly map and (b) profiles of the total-intensity anomaly and the horizontal gradient of total field intensity. Beneath the profiles are marks indicating depth solutions for thin-sheet and plane contact models based on the Werner deconvolution analysis. (From R. R. Hartman, D. J. Teskey, and J. L. Friedberg, *Geophysics*, v. 37, n. 5, pp. 891–918, October 1971.)

veal some close groupings. The interpreter selects only these clusters and marks them with appropriate symbols on another graph (Figure 12–28). Different combinations of these positions are examined, and a judgment is made about which combination indicates the most geologically reasonable picture of the magnetic basement.

Interpreting Shapes of Specific Anomaly Sources

Careful model analysis of individual magnetic anomalies or groups of anomalies can be very productive for interpreting the shapes of hidden geologic features. For elongate anomalies, we use two-dimensional models with polygonal cross sections. Otherwise, models consisting of three-dimensional horizontal polygonal plates with vertical sides are appropriate.

Model analysis of magnetic anomalies is carried out in exactly the same manner that we described in Chapter 9 for gravity anomalies. Although more mathematical steps are necessary to calculate magnetic fields over models, they can be incorporated into computer programs that are no more difficult or time-consuming to operate than their counterparts in gravity analysis. For two-dimensional models (Figure 12–23), the interpretion specifies the corner positions and magnetization components for each polygon. Similarly, for three-dimensional models, the components of magnetization and polygon corner positions are stated along with depths to the

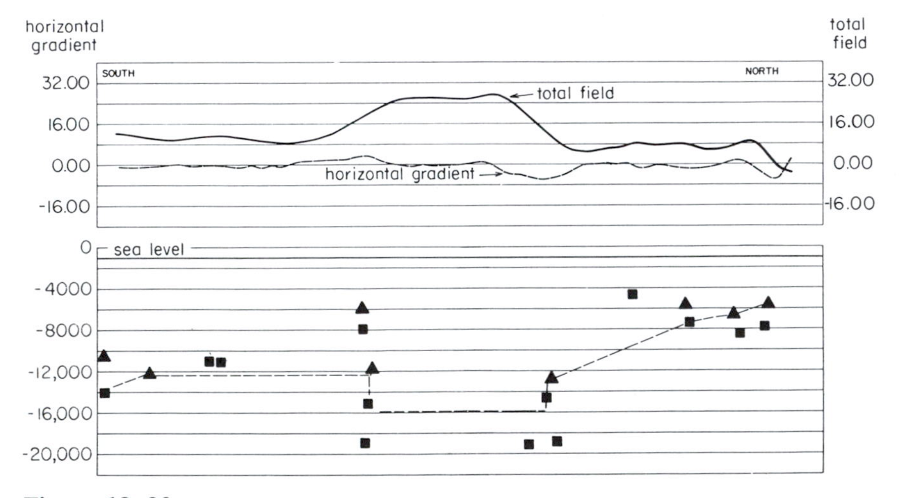

Figure 12–28
The geologic interpretation of clusters of depths from Figure 12–27, indicated by (■) for thin-sheet solutions and by (▲) for plane contact solutions, consists of selecting meaningful patterns that indicate the top of the magnetic basement. The dashed line connecting certain of these clusters represents the interpreter's judgment of this surface. (From R. R. Hartman, D. J. Teskey, and J. L. Friedberg, *Geophysics*, v. 37, n. 5, pp. 891–918, October 1971.)

upper and lower surfaces of the plates. In both cases, models evolve by trial-and-error adjustment of these parameters until observed anomalies are satisfactorily reproduced.

This procedure was followed in the two-dimensional model analysis of anomalies over folded metamorphic rocks in western North Carolina. Total-intensity anomaly contours are superposed on a geologic map of the area in Figure 12–29. Here we find interbedded metasandstone and metabasalt layers folded into a syncline. Although the structure could be recognized from surface geologic mapping, the subsurface shape of the syncline was not evident. The large linear magnetic anomaly appeared to be associated with magnetite

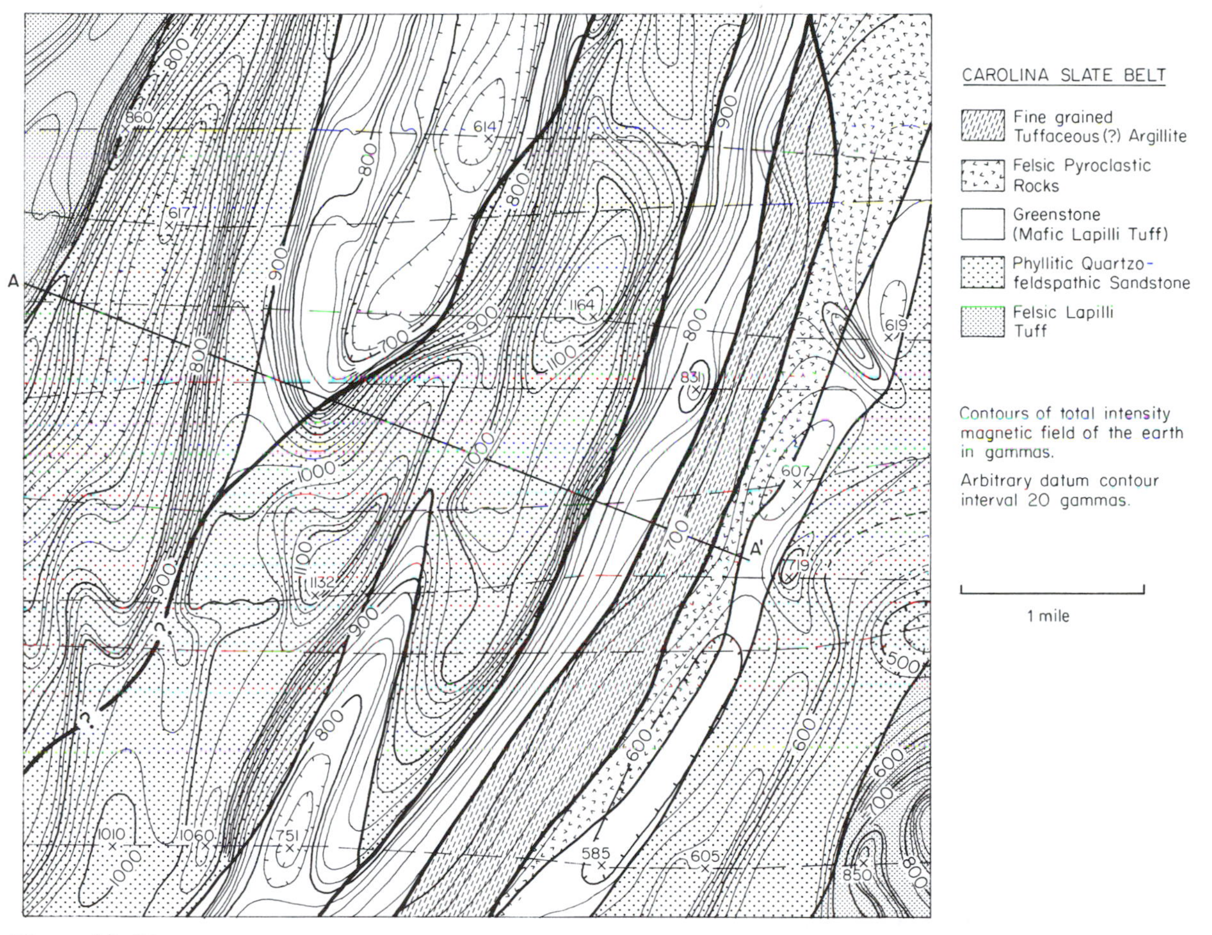

Figure 12–29
Total-intensity anomaly and geologic map of an area near Roxboro, North Carolina. (Courtesy of Lynn Glover, Virginia Polytechnic Institute and State University, Blacksburg, Virginia.)

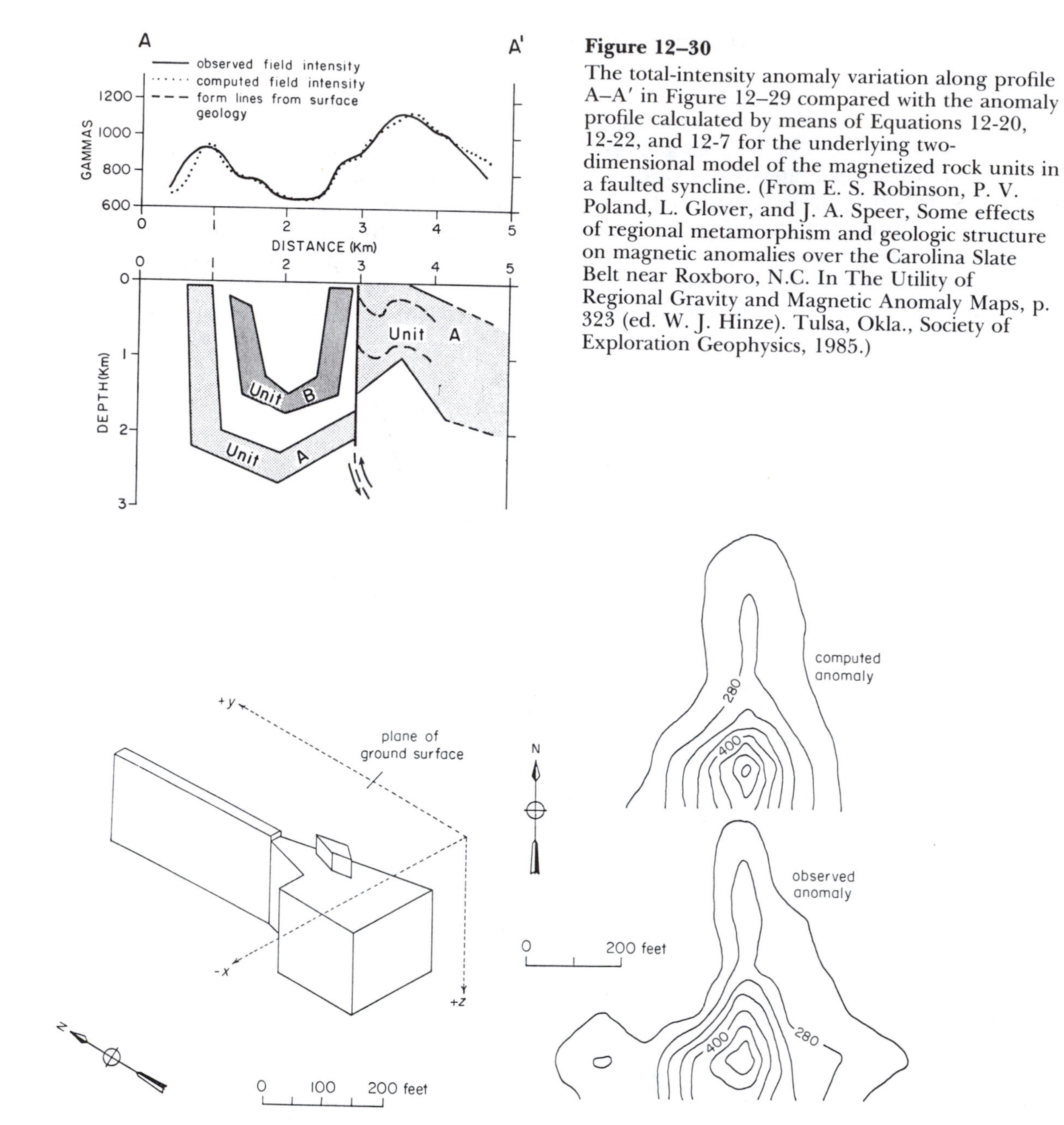

Figure 12–30
The total-intensity anomaly variation along profile A–A′ in Figure 12–29 compared with the anomaly profile calculated by means of Equations 12-20, 12-22, and 12-7 for the underlying two-dimensional model of the magnetized rock units in a faulted syncline. (From E. S. Robinson, P. V. Poland, L. Glover, and J. A. Speer, Some effects of regional metamorphism and geologic structure on magnetic anomalies over the Carolina Slate Belt near Roxboro, N.C. In The Utility of Regional Gravity and Magnetic Anomaly Maps, p. 323 (ed. W. J. Hinze). Tulsa, Okla., Society of Exploration Geophysics, 1985.)

Figure 12–31
Three-dimensional model interpretation of a vertical-intensity anomaly. The contour pattern computed for the model by means of Equations 12–17, 12–14, and 12–7 can be compared with the contour pattern observed along the southern border near the center of the map in Figure 11-13. (From J. W. Schmoker, Ph.D. dissertation, Virginia Polytechnic Institute and State University, June 1969.)

known to be present in the metasandstone. A two-dimensional model (Figure 12–30) was developed to explain the anomaly along a profile crossing the syncline. Comparison of the observed and theoretical anomalies indicates that magnetized rock units in the form of a syncline are the probable source. This interpretation points out the subsurface shape and depth of the syncline.

Look again at the magnetic anomaly map in Figure 11–12. Along the southern border near the center of this map is a north-trending anomaly with an amplitude of approximately 250 gammas. A three-dimensional model con-

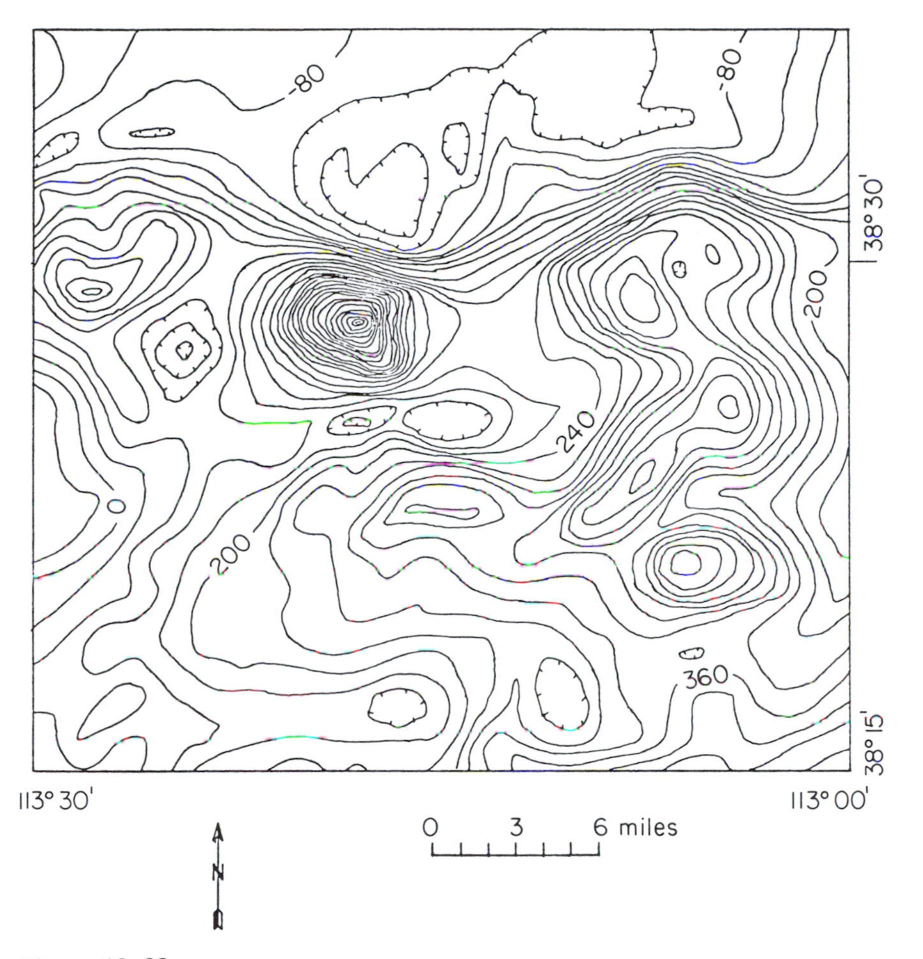

Figure 12–32
Total-intensity anomaly map of the San Francisco Mountains in southwestern Utah. (From J. W. Schmoker, Analysis of gravity and aeromagnetic data, San Francisco Mountains and vicinity, southwestern Utah, *Utah Geological and Mineralogical Survey*, Bulletin 98, p. 3, November 1972.)

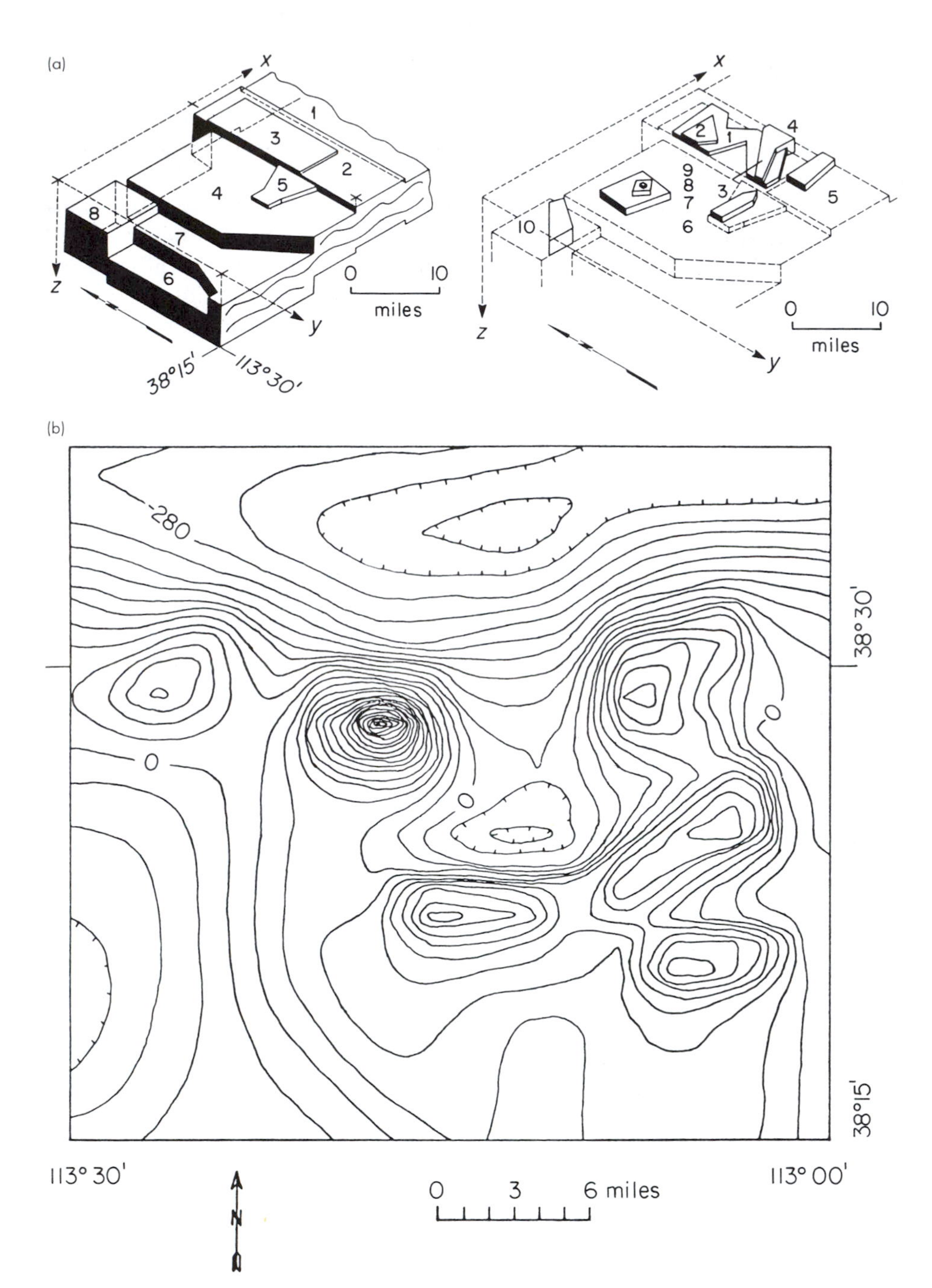

Figure 12–33

(a) Three-dimensional model of magnetized rock and (b) corresponding total-intensity contour map to be compared with the measured total-intensity anomaly map in Figure 12–32. (From J. W. Schmoker, Analysis of gravity and aeromagnetic data, San Francisco Mountains and vicinity, southwestern Utah, *Utah Geological and Mineralogical Survey*, Bulletin 98, pp. 8, 10, November 1972.)

sisting of three plates was developed to explain this anomaly. Figure 12–31 shows this model and the observed and theoretical anomaly maps. Although minor contour flexures are not exactly reproduced, the model satisfactorily explains the principal features of the anomaly.

On a much larger scale is the three-dimensional model analysis of the San Francisco Mountains in southwestern Utah. The ground magnetic survey in Figure 11–12 covers a small part of this area. Results of an aeromagnetic survey covering the entire area are presented in Figure 12–32. Observe how a set of nearly parallel contours extends across the area close to 38°30′N latitude. To the south of this line, magnetic anomalies are mostly positive and occur over granitic rocks. On the north is a broad negative anomaly zone underlain by sedimentary and volcanic rocks. A compound three-dimensional model (Figure 12–33a) to reproduce this pattern of anomalies was developed after many trials. The same principal features are clearly evident on the theoretical anomaly map in Figure 12–33b. Observe that the model of magnetized rock has its northern border near 38°30′N latitude. This model is a somewhat irregular horizontal plate. As we might expect from the profiles in Figure 12–21, such a plate should produce a negative anomaly zone to the north of its edge and a positive anomaly zone to the south. On the basis of the steep northern edge of the model, a fault was interpreted to cross the area near this latitude, which separates magnetized granitic rocks on the south from nonmagnetized sedimentary and volcanic rocks on the north.

Throughout this chapter, we have compared the methods of processing and interpreting magnetometer measurements with the methods of gravity exploration. Many similarities are evident, including the fact that two- and three-dimensional model analyses provide the clearest and most detailed look at the nature of anomaly sources.

STUDY EXERCISES

1. Consider two masses of rock, both having the same magnetic susceptibility and neither of which possesses remnant magnetism. Mass A is more or less equidimensional, and mass B is a thin, horizontal sheet of rock. Suppose that both masses are magnetized by induction in the earth's main magnetic field. From consideration of the demagnetization effect, which mass would possess the strongest magnetic moment per unit volume

   a. If both masses were situated in an Arctic region where the main field inclination is nearly vertical?

   b. If both masses were situated in an equatorial region where the main field inclination is nearly horizontal?

2. The diagram in Figure 12–34 shows a rod that is 1 km long and 10 m by 10 m in cross section and is inclined at an angle of 65 degrees from the *x* axis along a line passing through the origin. This rod has a magnetic susceptibility of 0.01 and is embedded in nonmagnetic rock. The depth to the top of the rod is 50 m. It is magnetized by induction where the earth's main field intensity is 55,000 gammas and the inclination is 65 degrees. The *x* axis points toward

magnetic north. Calculate the anomalous field intensity (F_A) and the total field intensity (F_T) at the origin.

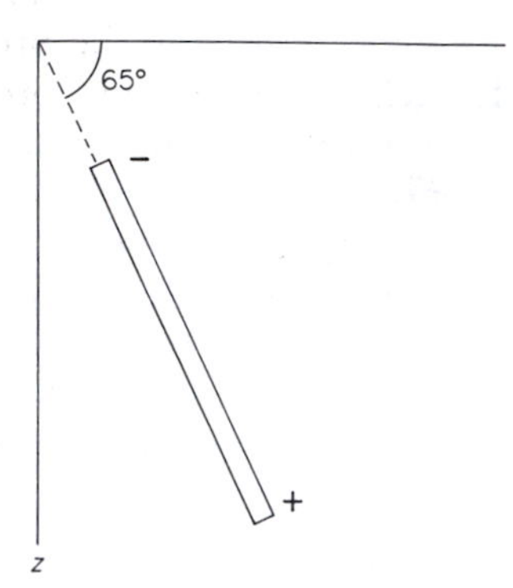

Figure 12–34
Cross section of an inclined magnetized rod.

3. Beginning with Equation 12–20a, obtain an equation expressing the vertical field intensity component (v) over a horizontal plate with a vertical edge. Guided by this equation, prepare profiles similar to those in Figure 12–21, but showing the variation of v caused by a horizontal plate with

   a. A vertical magnetic moment per unit volume.

   b. A horizontal magnetic moment per unit volume.

   Assume that the x-axis is in the magnetic north direction.

4. Total magnetic field intensity values at 0.5-km intervals along a profile 100 m above the land surface are given in column A in the following table. Corresponding values measured at 110 m above the land surface are given in column B. Calculate the vertical gradient for each pair of corresponding field intensity values. Then plot profiles showing the variation of field intensity at the 100-m level and the variation in vertical gradient. Describe different features of these profiles.

DISTANCE (km)	FIELD INTENSITY (gammas)	
	A	B
0	55761.3	55749.9
0.5	55757.3	55746.7
1.0	55746.3	55737.4
1.5	55766.0	55754.7
2.0	55770.8	55759.3
2.5	55773.7	55761.8
3.0	55756.5	55761.8
3.5	55747.9	55737.9
4.0	55760.1	55748.5
4.5	55764.8	55753.0
5.0	55763.2	55751.5
5.5	55761.5	55749.6
6.0	55760.5	55748.8

5. Two magnetic anomalies are evident on the profile in Figure 12–35. Both are caused by sources having the same size, shape, and direction of magnetization. Which source body

Figure 12–35
Magnetic field intensity along a profile.

   a. Is deepest?

   b. Has the strongest magnetization?

   Explain your answers.

6. Use the Peters half slope inversion to estimate the depth to the source of the anomaly shown along the profile in Figure 12–36.

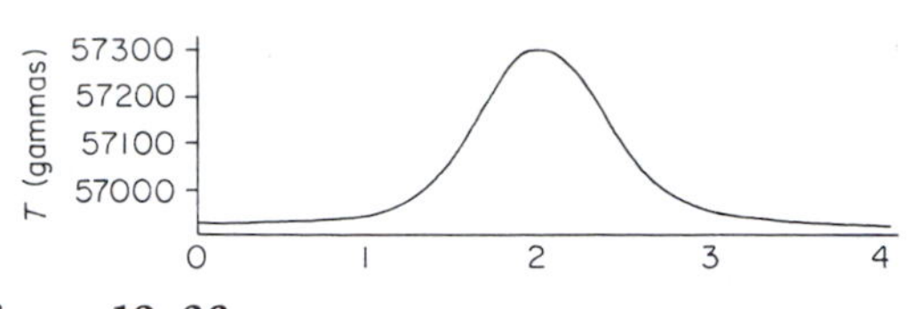

Figure 12–36
Profile of magnetic field intensity.

7. The total-intensity anomaly *(T)* contour map in Figure 12–37 covers an area near the borders of Texas, New Mexico, and Mexico. This map is an enlargement of the northwest corner of the map in Figure 11–13. The contour interval is 50 gammas. Use the method of trial and error to develop a simple dipole model to reproduce the main features of the anomaly pattern, which consists of a positive part and an associated negative part. Obtain the information about the main field by interpolation from Figures 10–22, 10–23, and 10–24. Prepare a drawing that shows the location and orientation of the model. Then draw a contour map of the magnetic field intensity over the model that can be compared with the observed anomaly map.

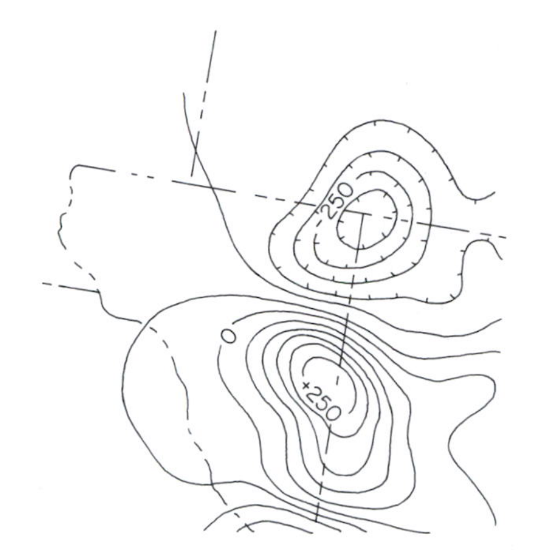

Figure 12–37
Total magnetic field intensity anomaly field contoured at 50-gamma intervals. (From G. R. Keller, R. A. Smith, W. J. Hinze, and C. V. L. Aiken, Regional gravity and magnetic study of west Texas. In *The Utility of Regional Gravity and Magnetic Anomaly Maps*, p. 202 (ed. W. J. Hinze). Tulsa, Okla., Society of Exploration Geophysicists, 1985.)

SELECTED READING

Grant, F. S., and G. F. West, *Interpretation Theory in Applied Geophysics.* New York, McGraw-Hill, 1965.

Hartman, Ronald R., Dennis J. Teskey, and Jeffrey L. Friedberg, A system of rapid digital aeromagnetic interpretation, *Geophysics,* v. 36, n. 5, pp. 891–918, October 1971.

Henderson, Roland G., A comprehensive system of automatic computation in magnetic and gravity interpretation, *Geophysics,* v. 25, n. 3, pp. 569–585, June 1960.

Hinze, William J. (editor), *The Utility of Regional Gravity and Magnetic Anomaly Maps.* Tulsa, Okla., Society of Exploration Geophysicists, 1985.

Nagata, Takesi, *Rock Magnetism.* Tokyo, Maruzen Co., Ltd., 1961.

Nettleton, L. L., Regionals, residuals, and structures, *Geophysics,* v. 19, n. 1, pp. 1–22, January 1954.

Plouff, Donald, Gravity and magnetic fields of polygonal prisms and applications to magnetic terrain corrections, *Geophysics,* v. 41, n. 4, pp. 727–739, August 1976.

Talwani, Manik, Computation with the help of a digital computer of magnetic anomalies caused by bodies of arbitrary shape, *Geophysics,* v. 30, n. 5, pp. 797–817, October 1965.

Talwani, Manik, and James J. Heirtzler, Computation of magnetic anomalies caused by two-dimensional structures of arbitrary shapes, *Computers in the Mineral Industries,* Part 1. Stanford University Publications, Geological Sciences, v. 9, n. 1, pp. 464–480, 1964.

Vacquier, Victor, Nelson C. Steenland, Roland G. Henderson, and Isadore Zietz, Interpretation of aeromagnetic maps, *Geological Society of America, Memoir 47,* 1951 and reprinted 1963.

THIRTEEN

Geoelectrical Surveying

The capacity of the earth to produce and respond to electric fields supports a variety of geophysical exploration procedures. The idea of exploring for ores by means of electrical measurements was introduced in the early 1800s, but its successful application did not follow until nearly a century later. The earliest methods sought to detect ore concentrations that possessed natural electrical polarization. Soon afterward, other procedures that introduced electric current into the earth by means of electrodes placed at the surface were developed. Still other approaches used coils to produce currents in the ground by electromagnetic induction. These developments opened the way for modern airborne as well as ground-based electrical surveying.

The various electrical methods of exploration geophysics all test the flow of electric current in the ground. Electric current, which is the movement of charged particles, can take place in three different ways. In the first mode of conduction, termed *ohmic conduction,* the electrons flow through the crystalline structure of some materials. Ohmic conduction occurs most readily in metals.

Electric current is also conducted by ions dissolved in groundwater. These ions can move through the interconnected pores of a permeable mass of soil, unconsolidated sediment, or rock. This is called *electrolytic conduction.*

In the third mode of conduction, called *dielectric conduction,* an alternating electrical field causes ions in a crystalline structure to shift cyclically. Even though there is no actual flow of charged particles, the cyclic change in the positions of ions is a movement that can be viewed as an alternating current. Dielectric conduction can occur in electric insulators, which are materials that otherwise do not carry electric current.

All substances act to retard the flow of electric current, so that energy must be expended to move charged particles. The extent to which a substance restrains this movement is described by its *electrical resistivity.* One of the principal goals of electrical surveying is to measure this physical property as a basis for distinguishing layering and structures in the earth.

Electrical resistivity in the earth depends on a combination of ohmic and dielectric effects related to lithology and electrolytic effects related to the groundwater. Because of the ways these effects can vary relative to one another, lithologically similar formations can have a considerable variation in resistivity. This variation tends to be much more extensive than the corresponding variation of seismic wave velocity, density, and rock magnetism. Therefore, compared with these other physical properties, it is generally more difficult to distinguish lithology by means of resistivity.

Electrical resistivity surveying procedures that introduce direct current into the ground through surface electrodes provide the best means for focusing on specific zones in the earth. The results of these procedures tend to be the least difficult to understand and interpret. Therefore, we will begin with discussion of these direct current methods for measuring resistivity.

Another electrical surveying method, called induced polarization, is an outgrowth of resistivity surveying practices. Introduction of current into the ground produces an electric field that persists for a short time after the current source is turned off. The duration of this temporary field depends on the capacity of the ground to dissipate concentrations of charge developed by current input. We will discuss how induced polarization surveying measures the persistence of the temporary electric field.

Other electrical surveying methods test the shape of naturally occurring electric fields. The so-called *self-potential* methods aim to map permanent fields that exist near concentrations of electric charge. Physical and electrochemical processes associated with some structures and ore deposits produce these charge concentrations so that the structure acts as a natural battery. Natural electric fields are also related to *telluric currents* produced in the earth by ionospheric and atmospheric electric discharges.

Finally, we will discuss electromagnetic surveying techniques that use alternating current input coils to produce temporary electric fields by induction. The shape of the temporary field is altered by variations in electrical conductivity in the ground. These methods have been adapted for airborne as well as land-based surveying.

Electrical surveying has figured prominently in mining exploration efforts. In recent years, it has played an increasingly important role in groundwater and engineering surveys. It has received some use in preliminary reconnaissance surveys of the oil prospects in sedimentary basins, but difficulties in probing the depths ordinarily tested for oil exploration limit this application of electrical surveying.

ELECTRICAL RESISTIVITY SURVEYING

Measurements of electrical resistivity are made with circuits in which the earth is one component, namely, a resistor. It is connected into the circuit by means of electrodes placed in the ground. To explain how it works, we will begin by examining the basic circuit. Then we will examine ways of focusing on specific zones in the earth to obtain resistivity measurements.

Ohm's Law and Resistivity

Positive and negative electric charges exert forces of attraction on one another. To create an electric battery, we must hold positive and negative charges in separate positions that are the terminals of the battery. Ways of creating batteries are presented in most introductions to physics and electrical engineering. The energy required to keep the positive and negative charges separated is the potential energy of the battery. Because it maintains oppositely charged terminals, we say that there is a *difference in potential* between these terminals.

To dissipate the energy of a battery, we can connect an electric conductor between the terminals. The energy is then used to compel charges to flow through the conductor. Positive charges move from the positive terminal toward the negative terminal, and negative charges move in the opposite direction. Although an electric current can consist of both positive and negative charges, it is conventional to say that the direction of current flow is the direction of positive charge movement. Therefore, we say that the current will flow through the conductor from the positive to the negative terminal.

By connecting a conductor between the battery terminals, we have assembled a very simple electric circuit through which current will flow because of the difference in potential. Now let us add another component, an electric resistor, to the circuit, as in Figure 13–1. Compared with an ideal conductor, which in no way impedes the current, a resistor consists of some material that restrains the movement of charges. Energy is required to push these charges through the resistor. This dissipation of energy reduces the potential energy of the battery. Therefore, we say that there is a change in potential resulting from the flow of current through the resistor. In 1827 Georg Simon Ohm reported the following relationship between *resistance* (r) of the resistor, current (i), and the corresponding change in potential v:

$$v = ir \tag{13–1}$$

This empirical relationship is now known as

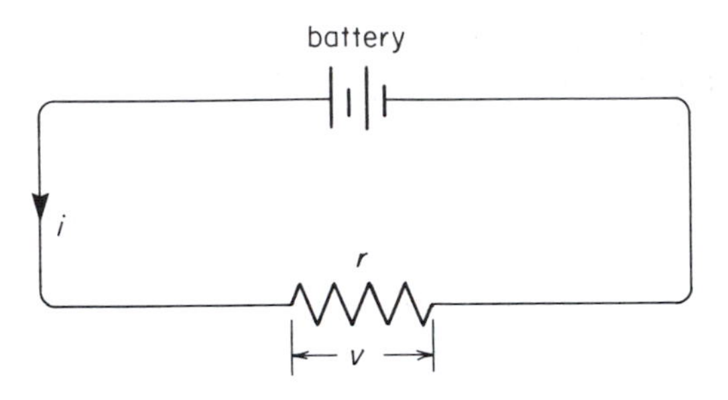

Figure 13–1
Electric circuit consisting of a battery and a resistor. Because the resistor impedes the flow of current, there is a change in potential (v) across the resistor that is proportional to the current (i) and the resistance (r).

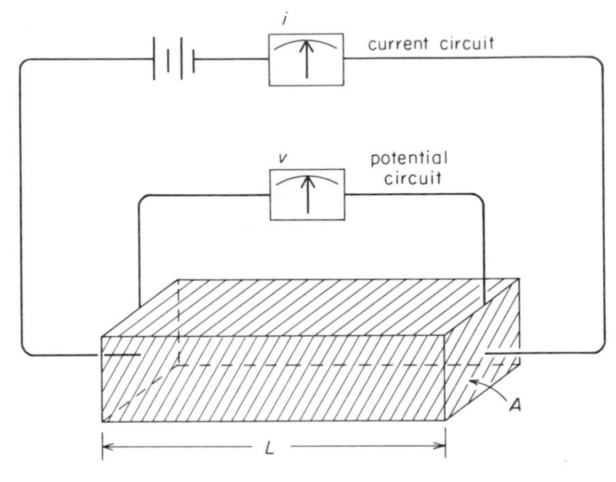

Figure 13–2
Electric resistor consisting of a rectangular bar. The current (i) flowing through the circuit is directly proportional to the resistance of the bar, which is inversely proportional to the cross-sectional area (A) and directly proportional to the length (L) of the bar.

Ohm's law and is limited to current flow below a saturation level. These three quantities can be expressed numerically: the basic unit of potential difference is the *volt,* the basic unit of current is the ampere, and the basic unit of resistance is the ohm, so that

$$1 \text{ volt} = 1 \text{ ampere} \times 1 \text{ ohm}$$

Let us look more closely at the resistor. Suppose that it is made from a rectangular bar of length L and square cross-sectional area A, as in Figure 13–2. We can represent the electric current flowing through this resistor by a uniform distribution of charges moving along parallel paths from one end to the other. The resistance r of this resistor is described in terms of the length L of the path followed by a charge, the cross-sectional area A over which the charges are uniformly distributed, and the *resistivity* R, which is a physical property of the substance used to make the resistor,

$$r = \frac{RL}{A} \qquad (13\text{–}2)$$

By rearranging this expression, we can express resistivity as

$$R = \frac{rA}{L} \qquad (13\text{–}3)$$

which tells us that resistivity can be expressed in units of resistance x length. Common units used in exploration geophysics include the ohm-meter, the ohm-centimeter, and the ohm-foot.

We see from Equation 13–2 that resistance can be increased by lengthening the resistor so that charges must travel longer paths through it. The cross-sectional area can also be decreased, which impedes the movement of charges by crowding them into a smaller volume. The concentration of charge passing through a cross-sectional area of the resistor is expressed by the *current density,* which is

$$\mu = \frac{i}{A} \qquad (13\text{–}4)$$

Current Flow in Three Dimensions

Suppose that we make up an electric circuit in which the earth is the resistor. We can connect electrodes at two different locations to the battery terminals, as in Figure 13–3. These electrodes are simply metal stakes driven into the ground with wires leading to the battery terminals. The electrode connected to the positive terminal is called the *source,* and the electrode connected to the negative terminal is the *sink.*

Because of the difference in potential between these electrodes, current is compelled to flow along paths leading from the source to the sink. To find the directions of these paths, we begin by considering separately the effects of the source and the sink. Then we can combine these effects to determine the pattern of three-dimensional current flow in the earth. We used this same approach in Chapter 10 to find the form of a magnetic field by combining the separate effects of the positive pole and the negative pole of a magnet. To keep things as simple as possible, we will start by assuming that resistivity is constant throughout our model of the earth.

The source electrode is positively charged. Therefore, it repels positive electric charges, pushing them outward into the ground. The implication is that electric current flows outward from the source into the ground. Because the resistivity is uniform, current moves away from the source, radiating outward and uniformly in all directions as shown by the paths in Figure 13–3.

Now let us consider the resistance encountered by current that has flowed a distance *d* from the source. Because it spreads outward in all directions, it has moved through a hemispherical zone. Let us think of this zone as a resistor. The current flows out of this resistor

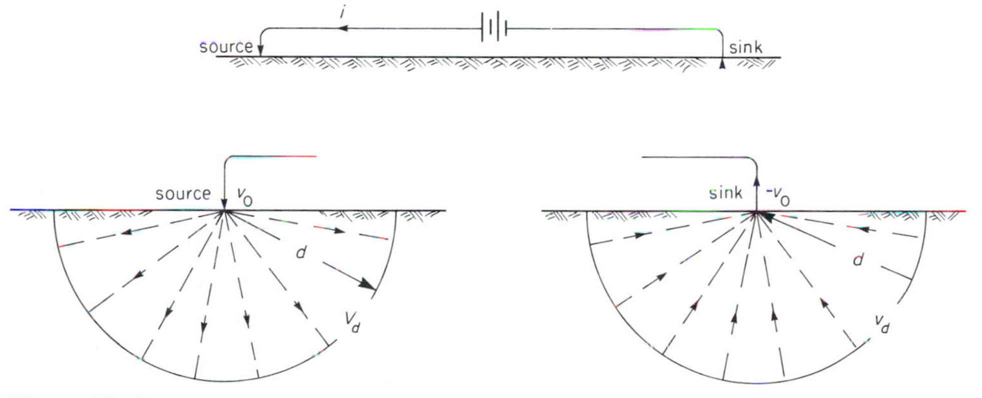

Figure 13–3
Current lines radiating out from a source electrode and converging on a sink electrode. When the source and sink are placed far enough apart, the current lines close to the source radiate outward equally in all directions. Close to the sink the current lines converge equally from all directions if the resistivity of the medium is everywhere the same. Equipotential surfaces, perpendicular to the current lines, are surfaces on which the potential is everywhere the same. Close to the source and sink, these are hemispheric surfaces.

when it moves across the area of $2\pi d^2$, which is the surface of the hemisphere. According to Equation 13–2, the resistance r can be expressed by the product of resistivity R and the distance d that the current has traveled divided by the area $2\pi d^2$ across which it must flow:

$$r = \frac{Rd}{2\pi d^2} = \frac{R}{2\pi}\left(\frac{1}{d}\right) \qquad (13\text{–}5)$$

The change in potential resulting from the flow of current through this hemispherical resistor can be found from Ohm's law:

$$v = ir = \frac{iR}{2\pi}\left(\frac{1}{d}\right) = v_0 - v_d \qquad (13\text{–}6)$$

This equation expresses the difference between the electric potential v_0 at the source and the electric potential v_d at any point in the ground a distance d from the source. The surface of the hemisphere of radius d contains all points at this distance, which means that the electric potential related to current flowing from the source is the same everywhere on that surface. Such a surface is called an *equipotential surface.* If you were to connect a voltmeter, which measures change in potential between the source and any point on an equipotential surface, you would obtain the same reading.

Look next at the effect of the sink electrode. If the potential at the source is v_0, the potential at the sink will be $-v_0$, for it is connected to the negative terminal of the battery. Because the negatively charged sink attracts positive charges, we say that current flows toward the sink. Figure 13–3 shows how these charges follow paths that converge on the sink from all directions.

The resistance encountered by current converging on the sink can be analyzed in terms of the zone that it flows through. Current flowing from all points at a distance d from the sink must move through a hemispherical zone that is identical to a zone with the same radius that surrounds the source. The resistance of this zone, then, is expressed by Equation 13–4. We can use this expression and Ohm's law to find the difference between the electric potential $-v_0$ of the sink and the potential v_d at all points a distance d away from it:

$$-v = ir = \frac{iR}{2\pi}\left(\frac{1}{d}\right) = v_d - v_0 \qquad (13\text{–}7)$$

This equation tells us that hemispherical equipotential surfaces are concentric about the sink.

So far, we have analyzed the effects of the source and sink separately, ignoring interference of the other. But to find the electric potential v at a point in the ground, we must combine the contributions of the source and sink. Using Equations 13–6 and 13–7, we obtain

$$v = \frac{iR}{2\pi}\left(\frac{1}{d_1} - \frac{1}{d_2}\right) \qquad (13\text{–}8)$$

where d_1 and d_2 are the distances to the source and the sink.

Suppose that we use Equation 13–8 to calculate the potential point by point throughout the ground. Then, by connecting points of equal potential, we would obtain the pattern of equipotential surfaces shown in Figure 13–4. Now look again at Figure 13–3, which shows that when source and sink effects are considered separately, electric charges follow paths that are perpendicular to equipotential surfaces. This relationship also holds when these effects are combined. Current always flows in directions perpendicular to equipo-

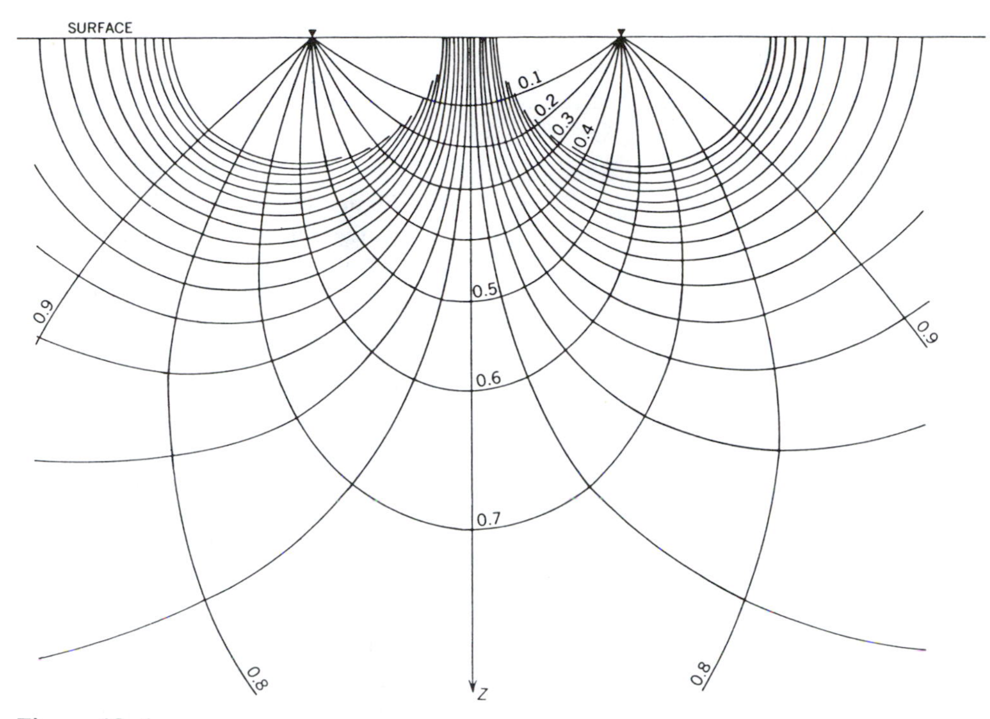

Figure 13–4
Current lines and equipotential surfaces produced by a source and sink in a medium of uniform resistivity. The current lines extend from the source to the sink, and the equipotential surfaces are perpendicular to them. The current lines and the equipotential surfaces are symmetrical about the vertical plane midway along and perpendicular to the line between the source and sink. (From R. G. Van Nostrand and K. L. Cook, Interpretation of Resistivity Data, U.S. Geological Survey Professional Paper 499, p. 31, 1966.)

tential surfaces. Therefore, having determined the positions of equipotential surfaces by means of Equation 13–8, we can construct a set of lines perpendicular to them. This set of lines, also shown in Figure 13–4, represents the flow of current between the source and sink along a plane surface.

The pattern of equipotential lines and current lines in Figure 13–4 applies for any plane surface containing the source and sink, regardless of its inclination. To gain a more complete three-dimensional perspective, we can draw this pattern on a set of planes, each inclined at a different angle, as shown in Figure 13–5. We can thereby visualize more clearly how current spreads out through the ground as it flows between the source and the sink electrodes.

Current Density

Electric charges move between a source electrode and a sink electrode along paths that ex-

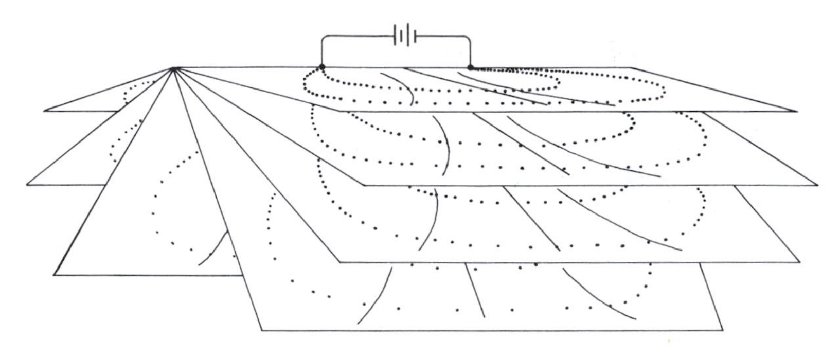

Figure 13–5
Current lines and equipotential surfaces along different planes containing a source and a sink in a medium of uniform resistivity.

tend, in theory at least, through the entire earth, as in Figure 13–6. How is it possible to focus on particular zones in the earth for purposes of measuring electrical resistivity? To find an answer to this question, we begin by looking at the variation of current density in the ground.

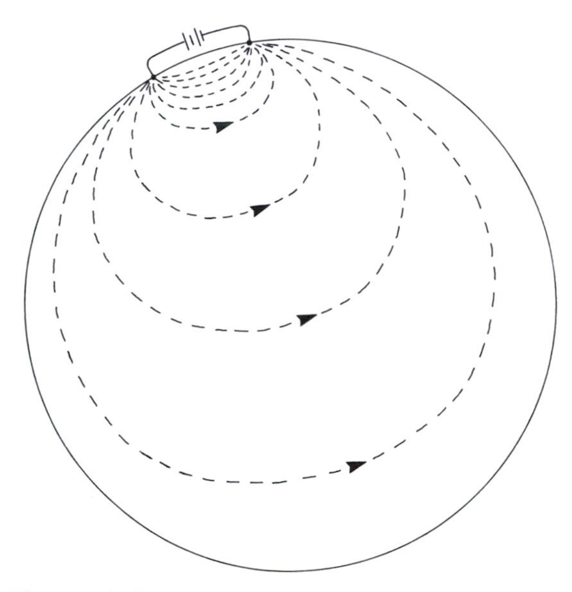

Figure 13–6
Current lines in a sphere of uniform resistivity. The spacing of current lines increases with distance from the source and sink, indicating a proportional decrease in current density.

Earlier in Equation 13–4, we introduced the concept of current density μ. This equation can be rearranged to give

$$i = \mu A$$

By substituting this result and the right-hand side of Equation 13–2 into Equation 13–1, we obtain Ohm's law in terms of current density:

$$v = \frac{RL}{A}\mu A = \mu RL$$

This equation can be rearranged to obtain the expression

$$\mu = v/RL \qquad (13\text{–}9)$$

which we can use to find the variation of current density in the ground.

First, we can determine how current density varies along a current flow line. Figure 13–7 reproduces part of the pattern of current and equipotential lines from Figure 13–4. Look closely at the current line that crosses equipotential lines at points, a, b, c, d, e, f, g, and so forth. The equipotential lines are drawn for equal changes in potential, that is, the change in potential v is the same between all pairs of adjacent lines. This means the

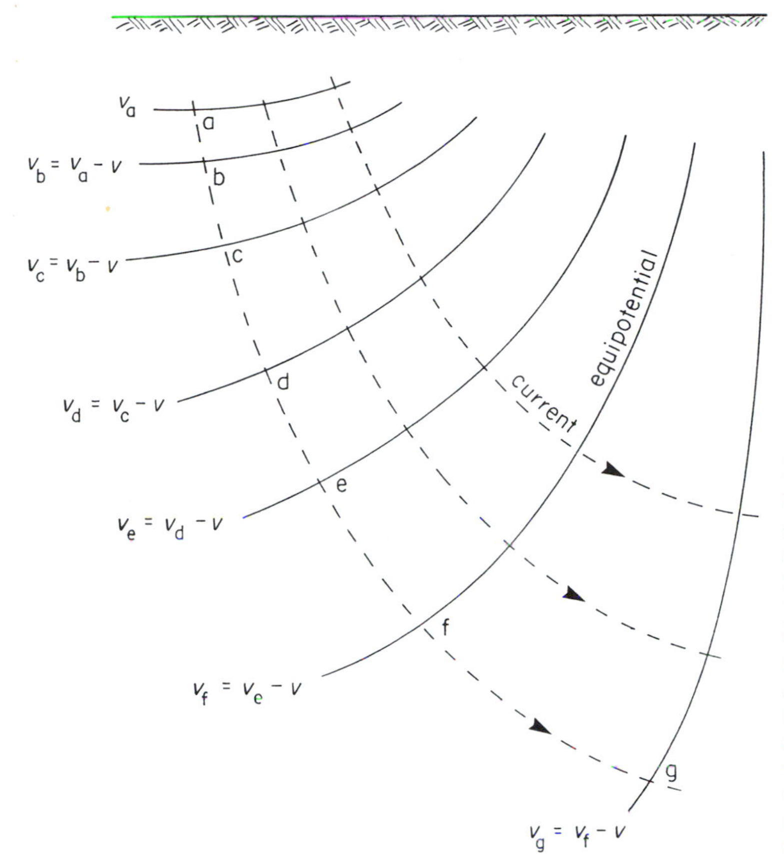

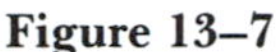

Figure 13–7

Current lines and equipotential surfaces that illustrate how the spacing of intersections varies inversely with current density. The points a, b, c, d, e, f, and g on one of the current lines (dashed) mark intervals of equal potential change along that line. Equipotential surfaces containing these points are shown with solid lines. Observe that the intervals of equal change in potential along the current lines become greater with distance from the source and sink.

value of v will be the same along each of the segments ab, bc, cd, de, ef, fg, and so on, of the current path. Because the spacing of equipotential lines increases with depth, the lengths of successive path segments also increase with depth, so that

$$\mathrm{ab} < \mathrm{bc} < \mathrm{cd} < \mathrm{de}$$

An electric current consists of moving charges. The current density expresses the spacing of these charges. A high current density indicates that the charges are closely spaced, and a lower value tells us that they are farther apart. To determine current density along a segment of the path marked in Figure 13–7 by means of Equation 13–9, we choose the length of that segment as the distance L. Therefore, we have

$$\mu_{\mathrm{ab}} = \frac{v}{R}\left(\frac{1}{\mathrm{ab}}\right), \quad \mu_{\mathrm{bc}} = \frac{v}{R}\left(\frac{1}{\mathrm{bc}}\right),$$

$$\mu_{\mathrm{cd}} = \frac{v}{R}\left(\frac{1}{\mathrm{cd}}\right), \ldots$$

It is clear from this analysis that current density decreases with distance from the source and sink, because the lengths of path segments increase while there are no corresponding changes in potential.

We see that the intersections of a current line with equipotential lines can be used to find the variation of current density. It becomes apparent, then, that the spacing of these intersections should be proportional to the spacing of the electric charges that make up the current along this path. Therefore, in Figure 13–4 the spacing of intersections of the sets of current lines and equipotential lines should vary in proportion to the charge spacing. Points corresponding to these intersections are plotted in Figure 13–8 to indicate the variation of current density in the ground. Near the source and sink, the closely spaced points reveal high current density. Farther away where points are more widely spaced, the current density is lower.

We should draw attention to the similarity in the form of current lines in Figure 13–4 and magnetic lines of force in Figure 10–4. In the same way that the spacing of magnetic lines of force indicates field intensity, the spacing of current lines indicates current density.

To determine how much current flows through different zones of the earth, we can mark the points in Figure 13–8 on each of the inclined planes in Figure 13–5. Next, let us also divide the ground into a series of horizontal layers of equal thickness, as in Figure 13–9. For convenience, we choose the layer thickness to be onethird of the distance a between the source and sink. Now we can see that the numbers of points on all the inclined planes lying within each of the layers indicate the proportions of current flowing through these layers. The results are shown by the bar graph in Figure 13–10, where i_1 is the total current passing above a horizontal plane at depth z_1. In addition, the total numbers of points lying above and below each of the layer boundaries indicate the proportions of current flowing above and below the depth of that boundary. The curve in Figure 13–10 designates these proportions.

We can tell from Figure 13–10 that most of the current is concentrated near the land surface. Fifty percent of the current flows in the

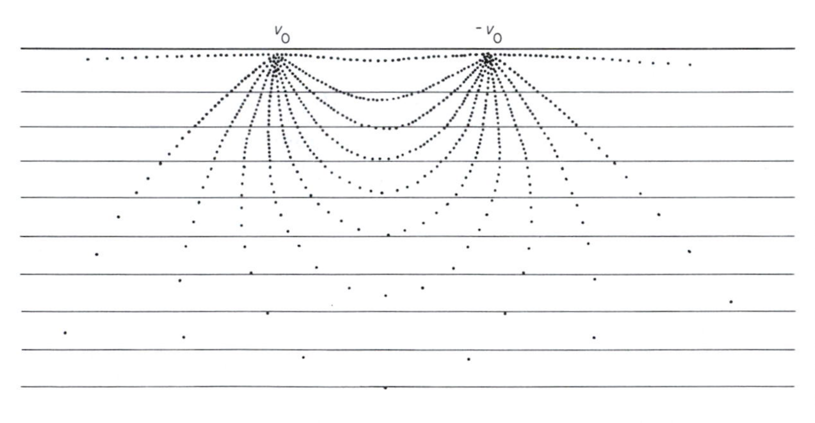

Figure 13–8

Intersection points from Figure 13–4 illustrating variation in current density. Each point corresponds to an intersection of a current line and an equipotential surface.

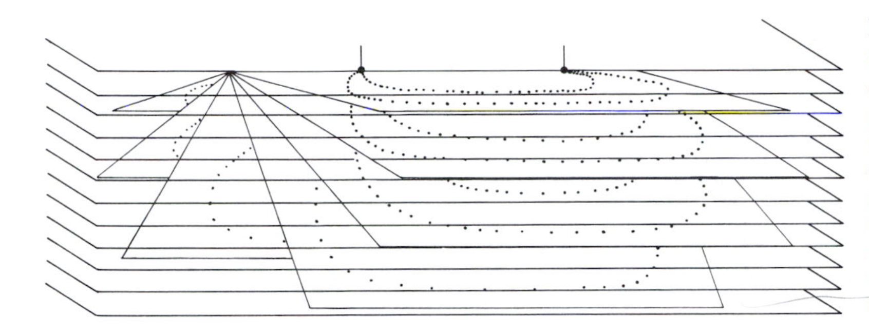

Figure 13–9
Points illustrating current density on several inclined planes. Horizontal planes indicate boundaries of layers of equal thickness. Proportions of points within each layer indicate the variation of current density from layer to layer. Low current density is represented by fewer points.

zone reaching from the land surface to a depth of one-half the distance between source and sink. Ninety percent of the current does not reach deeper than three times this electrode spacing. This analysis of current density tells us that the zone close to the land surface has the largest influence on the flow of current, even though a small proportion of this current, in theory, spreads through the entire earth.

Current Flow across a Boundary

Up to this point, we have assumed uniform resistivity in the ground. Now let us examine what happens to the flow of current across a boundary where resistivity changes from R_1 to R_2. In Figure 13–11 we show two current lines and two equipotential lines that cross a boundary along the x axis. The angles θ_1 and θ_2 indicate the directions of a current line

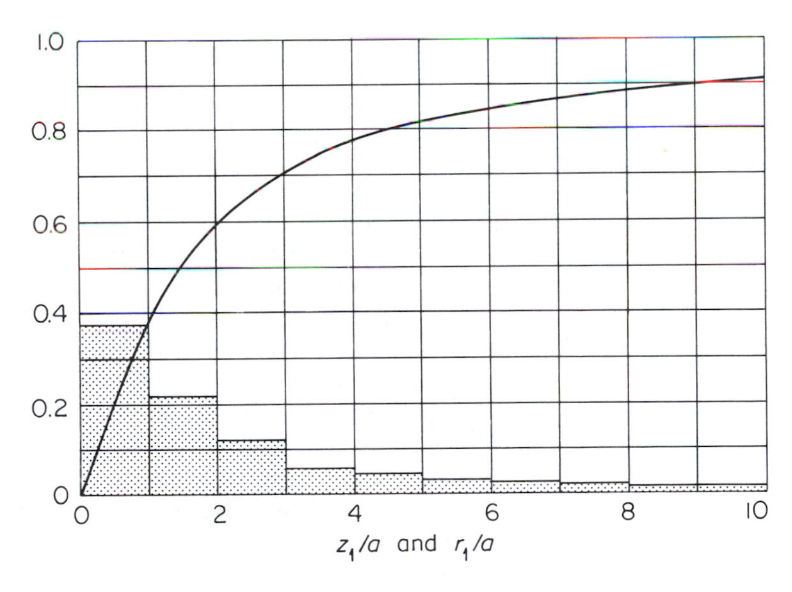

Figure 13–10
The proportion of current i_1 passing above a horizontal plane at depth z_1 and the current applied. The units of z are equal to one-third of the distance ($3a$) between the source and the sink. (From R. G. Van Nostrand and K. L. Cook, U.S. Geological Survey Professional Paper 499, p. 34, 1966.)

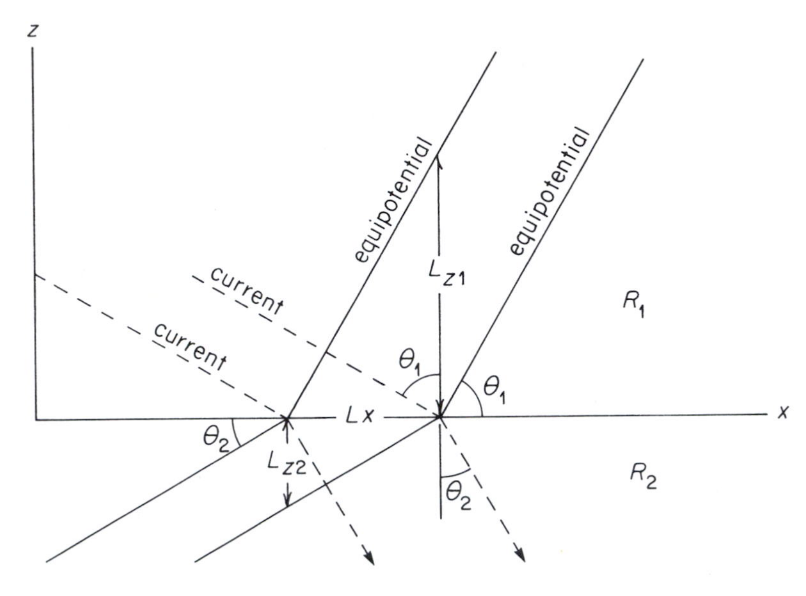

Figure 13–11
Bending of current lines and equipotential surfaces at a boundary between zones of contrasting resistivities R_1 and R_2.

measured from a reference line drawn perpendicular to the boundary.

We know from Equation 13–8 that current density can be determined from the distance between equipotential lines. Observe in Figure 13–11 that the distance L_x measured along the boundary and the distances L_{z1} and L_{z2} measured in the direction of the z axis can be used to express the angles

$$\tan \theta_1 = \frac{L_{z1}}{L_x}, \quad \tan \theta_2 = \frac{L_{z2}}{L_x}$$

These expressions can be rearranged and combined to get

$$\frac{L_{z1}}{L_{z2}} = \frac{\tan \theta_1}{\tan \theta_2}$$

Then, by rearranging Equation 13–9, we see that distance

$$L = \frac{v}{\mu}\left(\frac{1}{R}\right)$$

is inversely proportional to resistivity. This tells us that

$$\frac{L_{z1}}{L_{z2}} = \frac{R_2}{R_1}$$

so that

$$\frac{\tan \theta_1}{\tan \theta_2} = \frac{R_2}{R_1} \qquad (13\text{–}10)$$

This expression is of fundamental importance for determining the pattern of current lines and equipotential lines in a structure consisting of layers or masses of contrasting resistivity. Like the refraction of the paths and wave fronts of seismic waves, the current lines and equipotential lines bend at boundaries where resistivity changes. However, be sure to note the important difference between Equation 13–10, which uses the angle tangents, and Snell's law, which uses the angle sines.

We can use a method of sources and images to calculate electric potential in a structure consisting of different substances. This method is similar to the one we used in Chapter 4 to analyze seismic reflection from an inclined boundary. To explain this method, we will examine the simple structure in Figure 13–12, which consists of zones with resistivities of R_1 and R_2 separated by the boundary of an inclined plane.

We begin by analyzing the separate effect of the source electrode. At a point such as P_A, which is situated on the same side of the boundary as the source, we consider the boundary to be a partial mirror. From the source a line can be constructed perpendicular to the boundary along which the image can be marked at equal distance on the opposite side. Electric current flows from the source to P_A along the direct path d_1 and along the reflected path that has the same length as the distance d_2 from the image. Now we can use the form of Equation 13–6 to express the change in potential v_A between the source and P_A that depends on both the source and its image,

$$v_A = \frac{iR_1}{2\pi}\left(\frac{1}{d_1} + \frac{k}{d_2}\right) \qquad (13\text{–}11)$$

where k is the reflection coefficient. It accounts for the fact that only part of the current reflects from the boundary, and that the other part is transmitted across the boundary.

Next, we can analyze the change in potential v_B between the source and a point such as P_B, which is on the other side of the boundary. Current reaching P_B along the path d_3 has been transmitted through the boundary.

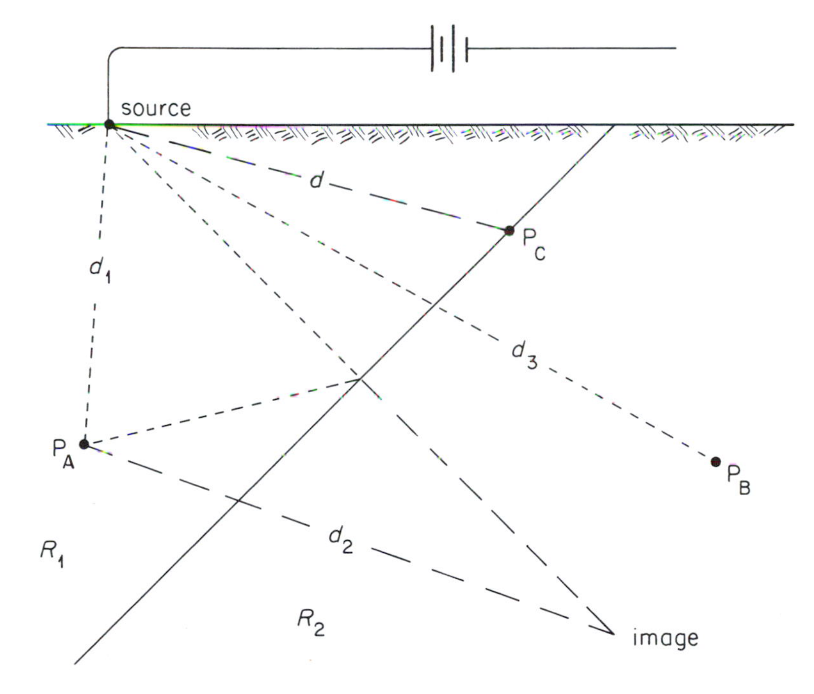

Figure 13–12
Geometry of current reflection and transmission at a boundary between zones of contrasting resistivities. This example illustrates only the effect of the source that can be analyzed separately from the sink and later combined with a similar analysis of the effect of the sink. Current paths to points P_A, P_B, and P_C are shown by dotted lines. Equivalent paths from the image are shown by dashed lines. The boundary shown by a solid line separates the zone of resistivity R_1 from the zone of resistivity R_2.

There is no reflected path to this point because it is situated behind the "mirror." Therefore, the change in potential will be

$$v_B = \frac{iR_2}{2\pi}\left(\frac{1-k}{d_3}\right) \qquad (13\text{–}12)$$

where the term $1 - k$ is the transmission coefficient indicating the proportion of current transmitted through the boundary.

Now let us shift P_A and P_B to the same location P_C on the boundary. Then we have the condition that

$$v_A = v_B = v_C$$

and

$$d_1 = d_2 = d_3 = d$$

By setting Equations 13–11 and 13–12 equal to each other for this case, we get

$$\frac{iR_1}{2\pi d}(1 + k) = \frac{iR_2}{2\pi d}(1 - k)$$

which can be rearranged to obtain

$$k = \frac{R_2 - R_1}{R_2 + R_1} \qquad (13\text{–}13)$$

By means of Equations 13–11 and 13–13, the change in potential can be calculated at any point on the same side of the boundary as the source. Equations 13–12 and 13–13 are used to find the change in potential at any point on the other side of the boundary. Potential differences related to the sink electrode are determined in the same way, except that the negative value $-i$ is used to express current.

We now have a way to determine electric potential and current density in structures with boundaries separating zones of contrasting resistivity. First, values of potential can be calculated for points distributed throughout the structure. Second, points of equal potential can be connected to obtain the equipotential surfaces. Next, current flow lines can be constructed perpendicular to these equipotentials. Results for horizontal and vertical boundaries are illustrated in Figure 13–13. They show that the highest current density, indicated by closely spaced flow lines, is in zones of lowest resistivity. This finding is expected because current flows most readily in the least resistive material.

The simple structures in Figure 13–13 show the basic features of electric current flow across a horizontal layer boundary and across a vertical fault. The method we have described can be used to analyze structures that have additional boundaries. However, because different reflection and transmission coefficients must be introduced for each boundary, this graphical method is impractical for structures with more than a few different materials separated by simple boundaries. Computers must be programmed to carry out the mathematical operations required to analyze more complicated structures.

Measuring Resistivity

The conventional practice in electrical resistivity surveying is to use source and sink electrodes connected to a battery, or some other source of electric power, to compel current to flow in the ground. An ammeter is included in this circuit to measure the current. Two other electrodes connected to a voltmeter are placed in other positions to measure differences in potential. Usually, but not always, the four electrodes are placed in a line.

In Figure 13–14, the source and sink electrodes are A and B, and the so-called potential electrodes are M and N. The M electrode is at

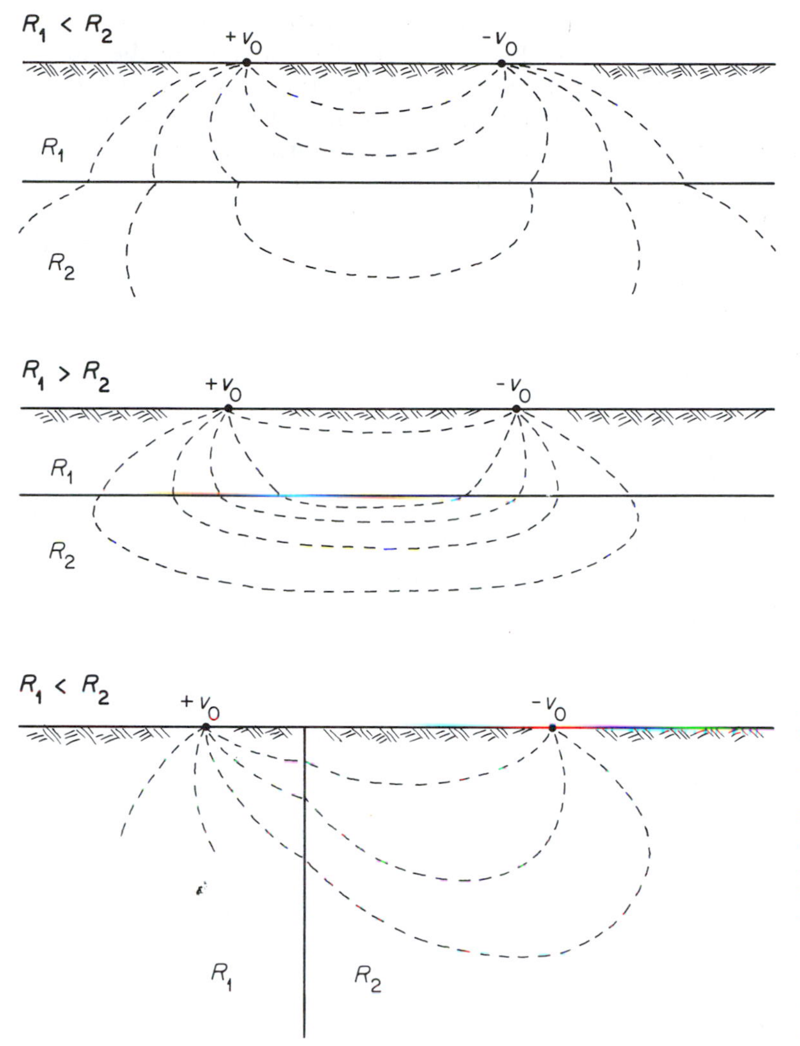

Figure 13–13
Current lines in a medium with two zones of contrasting resistivity. These examples indicate that close to the boundary between zones of different resistivities, current density is higher in the zone of lowest resistivity. In the structures with horizontal boundaries, the current flow patterns are symmetrical. The structure with a vertical boundary has an asymmetrical pattern.

distances d_1 and d_2 from the source and sink, and the N electrode is at distances of d_3 and d_4. According to Equation 13–8, if the resistivity R is uniform, the electric potential v_M at the M electrode will be

$$v_M = \frac{iR}{2\pi}\left(\frac{1}{d_1} - \frac{1}{d_2}\right)$$

and the potential v_N at the N electrode will be

$$v_N = \frac{iR}{2\pi}\left(\frac{1}{d_3} - \frac{1}{d_4}\right)$$

Therefore, the difference in potential v_{MN} measured by the voltmeter will be

$$v_{MN} = v_M - v_N = \frac{iR}{2\pi}\left(\frac{1}{d_1} - \frac{1}{d_2} - \frac{1}{d_3} + \frac{1}{d_4}\right) \quad (13\text{–}14)$$

Observe in Figure 13–14 that the M and N electrodes are situated on two different equipotential surfaces. It is the zone between these equipotentials that is being tested by the electrode array. Equation 13–14 can be rearranged to express the resistivity in this zone:

$$R = 2\pi\frac{v_{MN}}{i}\left(\frac{1}{d_1} - \frac{1}{d_2} - \frac{1}{d_3} + \frac{1}{d_4}\right)^{-1} \quad (13\text{–}15)$$

Suppose that resistivity is not uniform between the two equipotential surfaces. If there are boundaries between materials of contrasting resistivities, the equipotential surfaces would have more irregular form than those shown in Figure 13–14. The value obtained from Equation 13–15 would then be a weighted average of resistivities in the different materials within the zone between these equipotentials. Such a weighted average is called the *apparent resistivity* R_a. Ordinarily, we do not know what different materials are present at the time when the electrical survey is underway. Therefore, it is more appropriate to express a measurement as an apparent resistivity,

$$R_a = \frac{v_{MN}}{i}G \quad (13\text{–}16)$$

where

$$G = \frac{2\pi}{1/d_1 - 1/d_2 - 1/d_3 + 1/d_4} \quad (13\text{–}17)$$

is the *geometrical factor* that depends on the electrode arrangement.

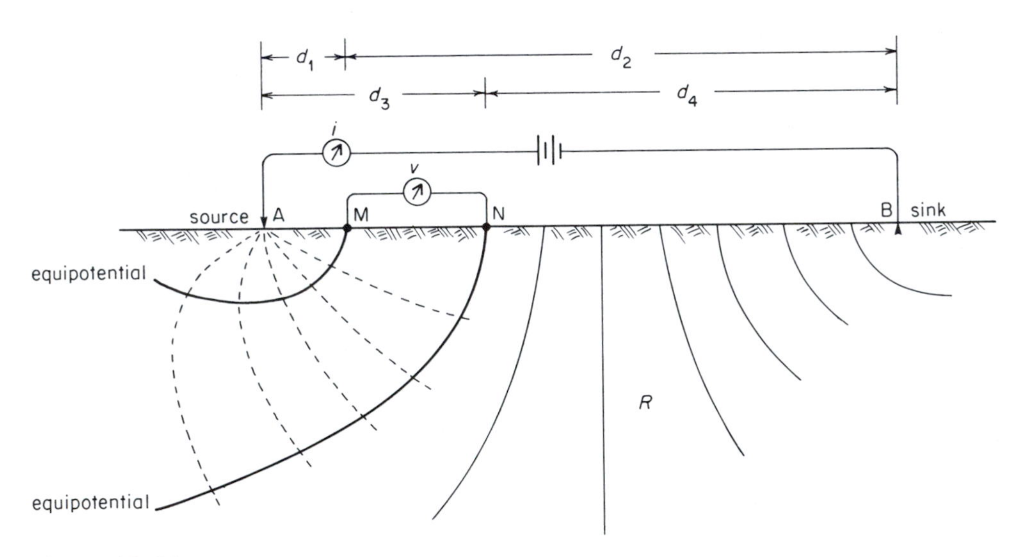

Figure 13–14
Current electrodes A and B and potential electrodes M and N are used to measure potential difference v, which depends on resistivity in the zone bounded by the two equipotential surfaces that reach the land surface at M and N.

Most electrical resistivity surveying is done with one or another of the electrode configurations illustrated in Figure 13–15. In the *Wenner configuration,* electrodes are spaced at equal intervals *(a)* with the potential electrodes M and N in a straight line between the current electrodes A and B. From Equation 13–17, we see that the geometrical factor for this arrangement will be

$$G_w = \frac{2\pi}{1/a - 1/2a - 1/2a + 1/a} = 2\pi a \qquad (13\text{–}18)$$

In the *Schlumberger configuration,* the current electrodes A and B are at equal distances L in opposite directions from the center of the array. The potential electrodes M and N are between A and B at equal distances b from the center of the array. Therefore, from Equations 13–17, the geometrical factor becomes

$$G_s = \frac{2\pi}{\dfrac{1}{L-b} - \dfrac{1}{L+b} - \dfrac{1}{L+b} + \dfrac{1}{L-b}} = \frac{\pi(L^2 - b^2)}{2b} \qquad (13\text{–}19)$$

A third electrode arrangement is called a *dipole configuration.* It is more accurate to call this arrangement a class of configurations in which the potential electrodes are not situated between the current electrodes. Figure 13–15c shows the general pattern, and the usual variations of this pattern are described as follows.

1. Equatorial dipole: $\alpha = \beta = 90$ degrees.
2. Radial dipole: $0 < \alpha < 90$ degrees; $\beta = 180$ degrees.
3. Polar dipole: $\alpha = 0$; $\beta = 180$ degrees.
4. Parallel dipole: $\alpha = \beta$.
5. Dipole–dipole: $a = b$; $c = na$; $\alpha = 0$, $\beta = 180$ degrees.

For the polar dipole configuration, the geometrical factor is

$$G_p = \frac{2\pi}{\dfrac{1}{c - \frac{a}{2} - \frac{b}{2}} - \dfrac{1}{c + \frac{a}{2} - \frac{b}{2}} - \dfrac{1}{c - \frac{a}{2} + \frac{b}{2}} + \dfrac{1}{c + \frac{a}{2} + \frac{b}{2}}} \qquad (13\text{–}20)$$

and for the dipole–dipole configuration, it is

$$G_{dd} = \pi n(n^2 - 1)a \qquad (13\text{–}21)$$

To measure apparent resistivity, then, we begin by choosing an appropriate configuration and placing the four electrodes in their proper positions. The battery or other power source is then switched on, and values of i and v_{MN} are read from the ammeter in the current circuit and the voltmeter in the potential circuit. These values are used with the appropriate geometrical factor in Equation 13–16 to calculate apparent resistivity. Later, we will discuss the advantages of the different electrode configurations used for this purpose.

Equipment for Electrical Resistivity Surveying

To measure electrical resistivity, we must have a suitable source of electric power, meters for measuring voltage and current, and electrodes that can be embedded in the ground. Most electrical resistivity surveying on the land surface is done to probe depths of a few hundred

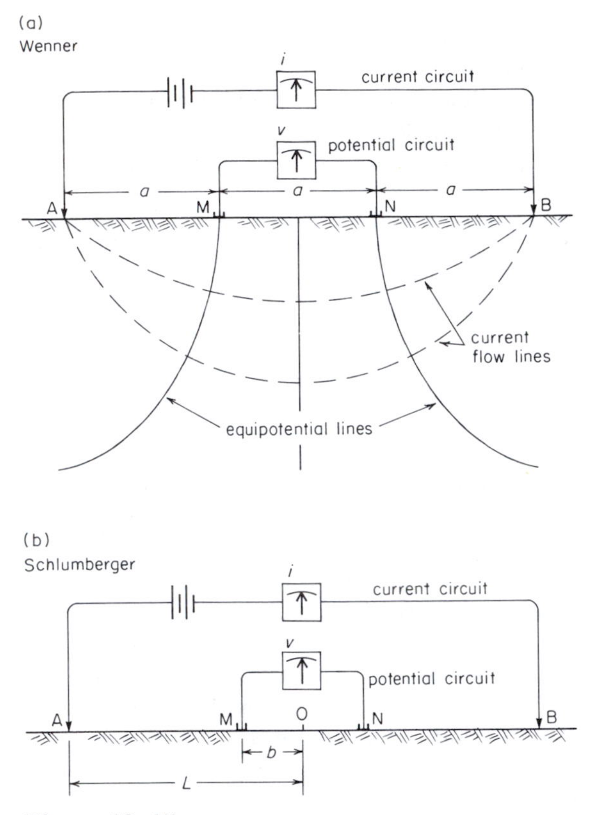

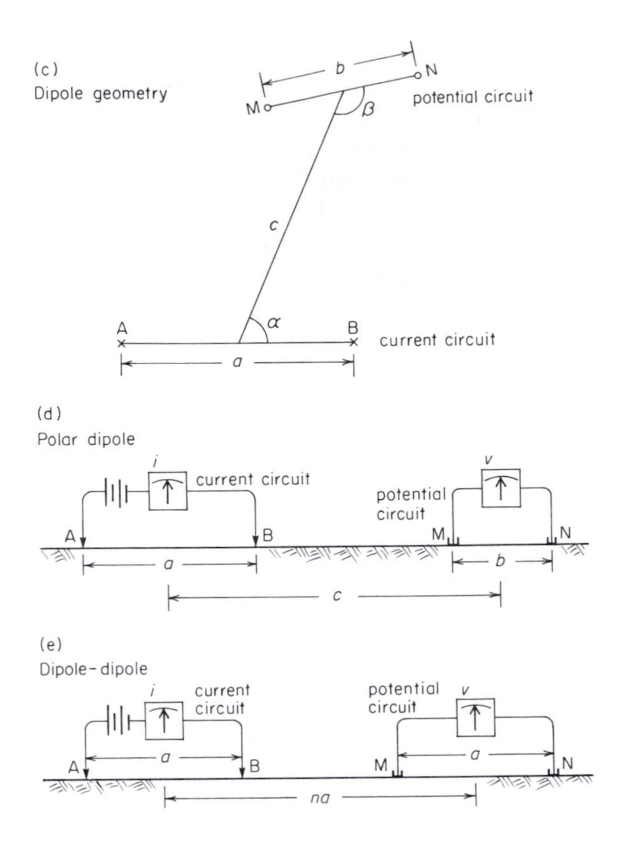

Figure 13–15

Electrode configurations for resistivity surveying. (a) The Wenner electrode system is symmetric with four equally spaced electrodes. In this system, the inner electrodes are used to measure potential difference. (b) The Schlumberger electrode system is also a symmetric four-electrode system. The distance ($2b$) between the inner potential electrodes is kept relatively small to measure the electric field close to the midpoint. The current is introduced by the outer electrodes. (c) The dipole electrode system has the most freedom for arranging electrodes. Lengths a and b of the current dipole (A–B) and the potential dipole (M–N) are kept relatively small compared with the separation (c) of dipoles. Depending on the selection of values for α and β, different dipole systems are obtained. (d) The special form defined by $\alpha = 0$ degrees and $\beta = 180$ degrees is called the polar dipole system. (e) Most commonly used is the so-called dipole–dipole electrode configuration in which $\alpha = 0$ degrees, $\beta = 180$ degrees, and $b = a$.

feet or less. These shallow measurements provide information about groundwater reservoirs, construction sites, and near-surface ores. For this kind of work, portable systems with all the necessary components can be purchased from commercial suppliers. Typical of this portable equipment is the system pictured in Figure 13–16.

Ideal for resistivity surveying is a direct-current source. However, such a source must be

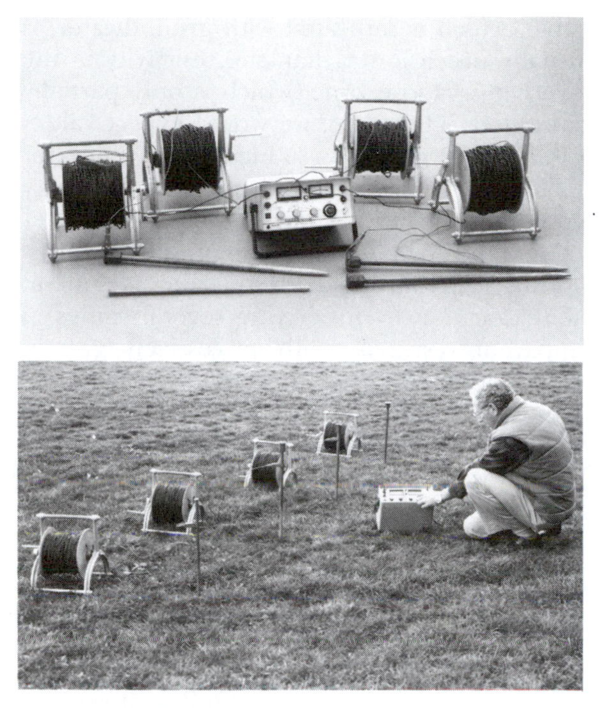

Figure 13–16
A portable system for resistivity surveying includes a control unit with circuits for measuring current and voltage, electrode stakes, electric cables, and cable reels. Survey operations can be carried out by two persons.

equipped with a commutator or a switching circuit that reverses the current direction a few times or more each second. If this is not done, charges from the ground tend to collect near the current electrodes, producing ambiguous readings. Batteries generally do not have sufficient power to sustain a direct-current source. A generator with the capacity of a few watts or a few tens of watts is required for shallow surveying.

Most modern portable systems have replaced direct-current sources with low-frequency alternating-current sources that can be satisfactorily powered by batteries. Good results can be obtained with frequencies lower than 100 hertz. At higher frequencies, the results cannot be interpreted by means of the equations we have developed from Ohm's law, which suppose direct current. But low-frequency alternating current achieves the same result as commutated direct current. It can be produced by a more compact and portable unit.

Meters needed for resistivity surveying include an ammeter sensitive to currents ranging from a few milliamperes to a few hundred milliamperes, and a voltmeter sensitive to potential differences ranging from a few tens of millivolts to a few volts. Some power sources are regulated to produce a fixed current, which eliminates the need for an ammeter.

Metal stakes made of aluminum, copper, or steel usually make suitable electrodes. They are driven several inches to more than a foot into the ground. Under very dry conditions, the ground close to an electrode can be moistened to improve contact. Where charge buildup in the ground has been a problem, another kind of electrode has been used. It is called a *porous pot electrode.* A metal core is placed in a porous ceramic cup that is filled with an electrolyte containing ions of the same metal. For example, a copper core is placed in a copper sulfate solution. Seepage of the electrolyte through the porous pot acts to dissipate charge concentrations in the ground near the electrode. The use of commutated direct-current or alternating-current sources usually eliminates the need for such an electrode.

Insulated low-resistance electric wire carried on reels is used to connect the electrodes in the current and potential circuits. For shallow resistivity surveying, enough wire must be available for electrodes to be spaced at least a few hundred feet apart.

For deep electrical sounding to probe many hundreds of thousands of feet below the surface, special equipment must be assembled. Generators capable of producing hundreds or thousands of watts are needed. Care must be taken to use low-resistance conductors, and electrodes must be carefully embedded to ensure proper contact in the ground.

Surveying Procedures

Most electrical resistivity surveying involves a *profiling* procedure or a *sounding* procedure. Profiling is usually done with a fixed electrode spacing. A series of apparent resistivity measurements is made by moving this electrode array from place to place along a profile. If values are obtained along several parallel profiles, a contour map of apparent resistivities can be prepared. Individual profiles and the contour map shown in Figure 13–17 were prepared from such a survey.

Apparent resistivity is usually profiled with a Wenner or a Schlumberger electrode configuration. The electrode spacing is chosen to ensure that the zone of interest is adequately tested. The curve in Figure 13–10 provides information for making this choice. For example, suppose the aim of the survey is to detect variations in the depth of the water table which is believed to range between 10 and 100 feet below the land surface. According to the curve in Figure 13–10, 37 percent of the current flows above the depth equal to the electrode interval in a Wenner configuration. Therefore, by choosing an interval of about 100 feet, we see that for a zone of interest reaching to a depth of 100 feet, the current density should be quite sufficient to detect changes in the depth to the water table.

The resistivity in the zone below the water table, which is saturated with groundwater, is usually much lower than the resistivity in the overlying vadose zone, which is only partially saturated. Therefore, we would expect values of apparent resistivity to increase as the water table depth increases. An apparent resistivity contour map would indicate changes of water table depth by high and low contour patterns.

An electrical resistivity sounding is done at some location by measuring several values of apparent resistivity with successively greater electrode spacing. As the electrode array expands, the same proportion of current is distributed through an increasingly deeper zone. Therefore, resistivities in deeper and deeper layers have a proportionally larger effect on the apparent resistivity. Alternating high- and low-resistivity layers may be evident on a graph of apparent resistivity and electrode spacing, as in Figure 13–18.

Although the Wenner configuration is commonly used for resistivity sounding, it is probably the least convenient for field operations. All four electrodes must be moved for each additional reading. When a Schlumberger configuration is used, the potential electrodes can remain in the same position. Only the current electrodes need to be moved for additional readings. When a dipole configuration is used, the current electrodes are fixed while the potential electrodes are moved.

For very deep sounding, both the Wenner and Schlumberger configurations turn out to be impractical. Too much wire is required to connect the electrodes. For this purpose, the dipole configurations must be used. The distance c in Figure 13–15c can be as long as necessary, whereas the lengths a and b remain relatively short. When electrical resistivity sounding is done in reconnaissance oil exploration, the measurements are made with expanding dipole configurations.

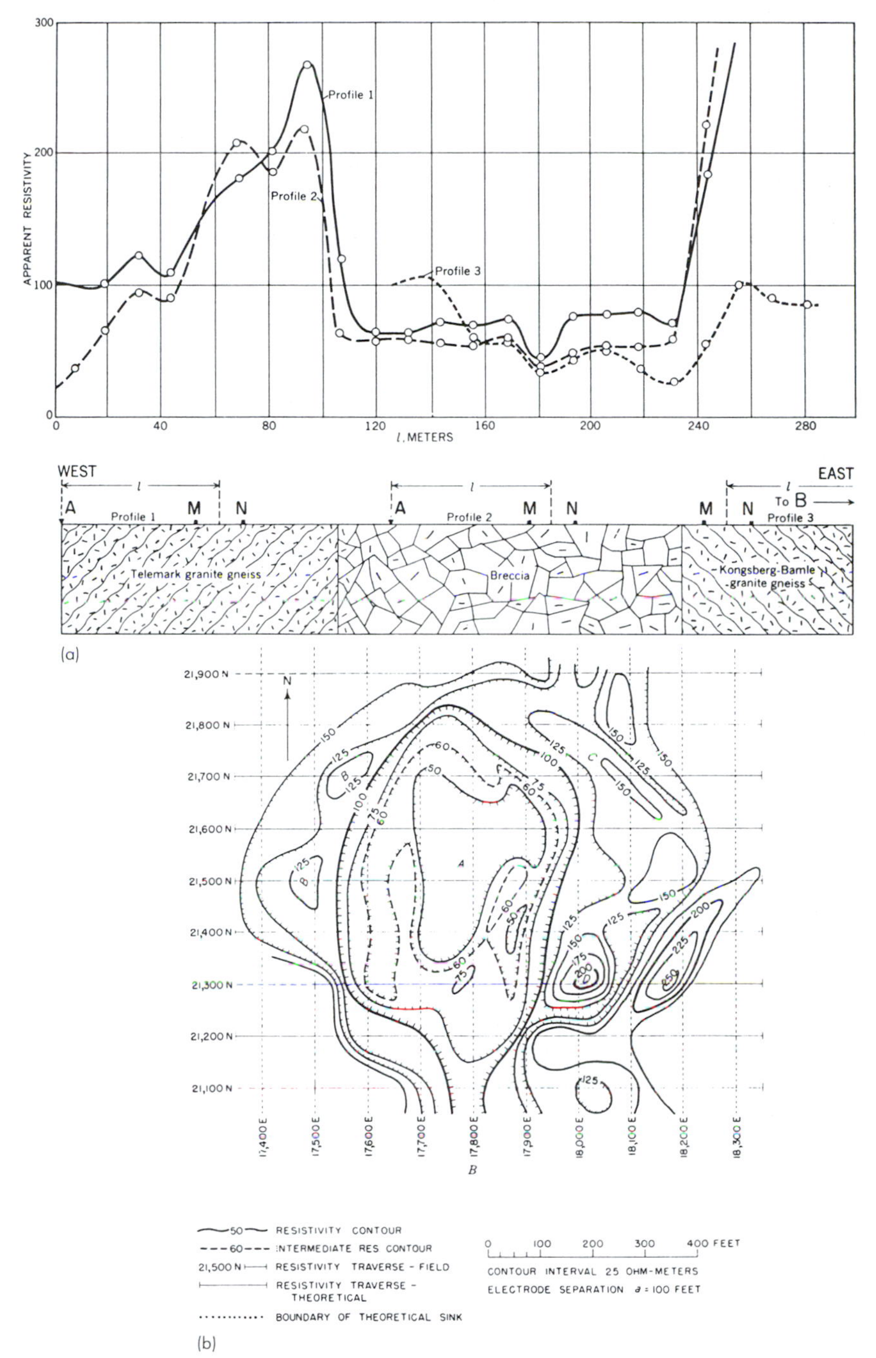

Figure 13–17

(a) Apparent resistivity profiles from an area near Kongsberg, Norway, and (b) an apparent resistivity contour map of an area in the Tri-State Mining District in Cherokee County, Kansas. (From R. G. Van Nostrand and K. L. Cook, U.S. Geological Survey Professional Paper 499, pp. 129 and 240, 1966.)

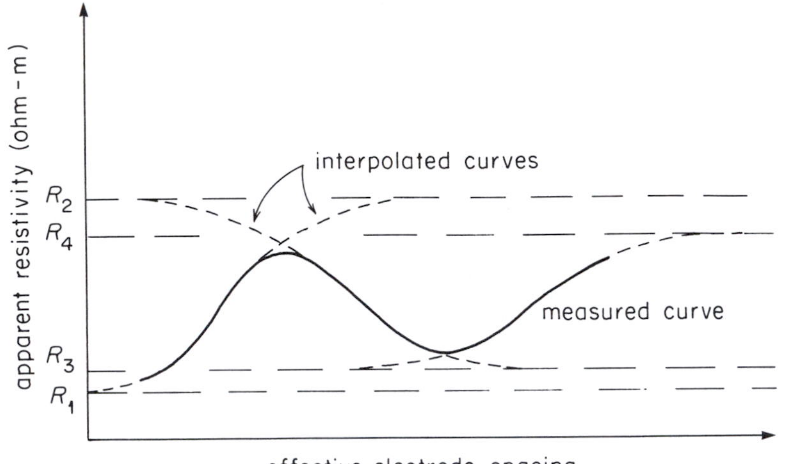

Figure 13–18
Variation of apparent resistivity determined from a Wenner configuration, with increasing electrode spacing indicating four layers with different resistivities. At close spacings, resistivity increases as spacing increases, suggesting a top layer of low resistivity and a second layer of higher resistivity. At intermediate spacings, apparent resistivity decreases with increased spacing, suggesting a third layer of lower resistivity. At wider spacings, apparent resistivity again increases, suggesting a still-deeper zone of higher resistivity. Resistivities are indicated on the vertical axis.

ANALYSIS OF RESISTIVITY MEASUREMENTS

Electrical resistivity surveys provide a series of apparent resistivity values. These values are obtained at different locations along a profile or at one location by means of different electrode spacings. Variation in apparent resistivity indicates the existence of zones of contrasting resistivity.

There are several ways of analyzing this variation to locate the boundaries of these zones and to estimate the values of true resistivity in them. Some of these methods involve direct processing of the measured values. We will begin with a simple inversion method that can be used to estimate depths of horizontal layer boundaries from resistivity sounding data. Other methods involve trial-and-error comparison of measured values of apparent resistivity with values calculated for idealized models of simple geological structures. Filtering methods are also used for interpreting data obtained from resistivity sounding.

The Cumulative Resistivity Inversion Method

The object of electrical sounding is to determine the variation of resistivity with depth. To analyze the measurements, we must access the relative influence of the material at different depths on an apparent resistivity reading. The focus of our analysis will be a four-electrode system consisting of a source and sink with two potential electrodes in line between them.

Let us consider the change in potential v between the source electrode A and a potential electrode at distance d. This change in potential is related to the flow of current through the entire zone within the equipotential surface shown in Figure 13–19. Be sure to understand that the combined effect of source and sink is being considered here, but that we are concentrating our attention on the region surrounding the source.

Suppose that we divide the zone within the equipotential surface in Figure 13–19 into a succession of thin, horizontal laminations.

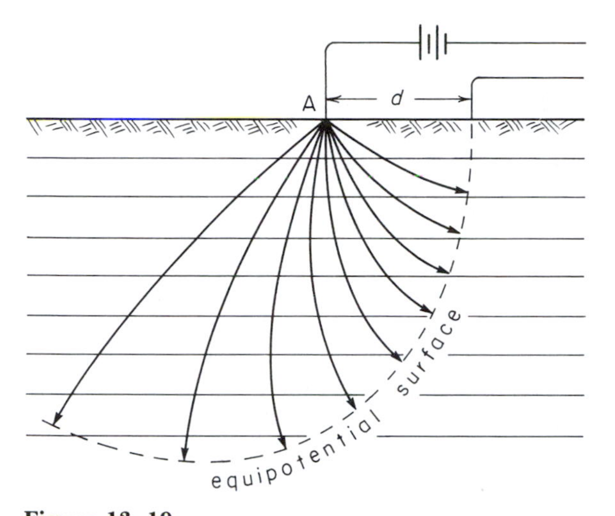

Figure 13–19
The change in potential between the source A and an electrode at a distance *d* away from it depends on the resistivity and the flow of current, shown by solid current lines, in a zone that has a boundary marked by the equipotential surface, shown by the dashed line.

Then we can determine the proportion of v that occurs within each lamination. We use Equation 13–9 to calculate the potential change for the segment of each current line that crosses a lamination and then sum these values. Results obtained in this manner can then be plotted at the corresponding depths of the laminations to obtain the graph shown in Figure 13–20. In this figure the depth is expressed in units equal to the separation d of the two electrodes. The graph indicates that the largest proportion of the change in potential occurs at a depth equal to approximately three-fourths of the electrode spacing.

The area under the curve in Figure 13–20 represents the total change in potential v. Depth zones of thickness $d/3$ are drawn on the graph, and the percentages of the total area occupied by each of these zones are indicated. These values tell us the relative influence of the material at different depths on the total change in potential. For example, the top

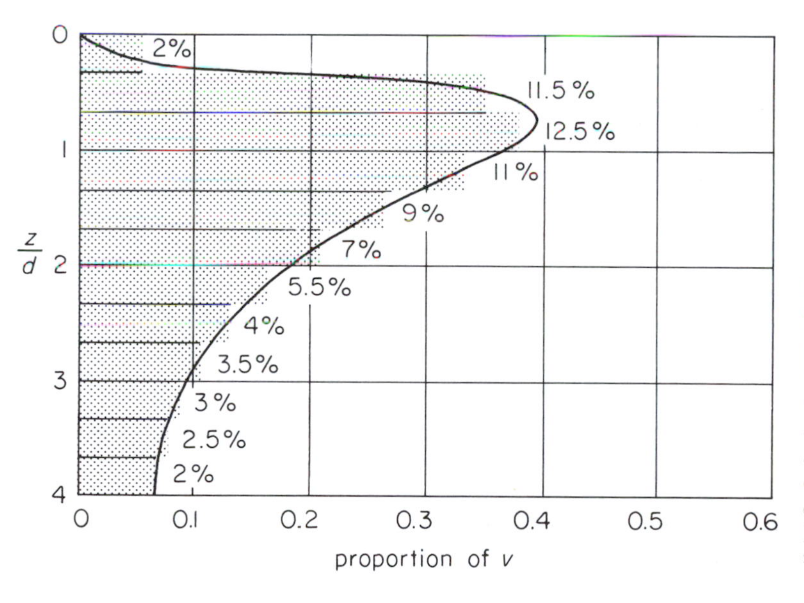

Figure 13–20
Sensitivity curve indicating the proportion of change in potential between the source and an electrode at a distance d occurring at different depths. The depths are expressed in units of d. For example, 12.5 percent of the total change in potential occurs in the horizontal zone reaching from $z = \frac{2}{3}d$ down to $z = d$.

zone that extends from the land surface to a depth of $d/3$ accounts for only 2 percent of v, whereas the zone of greatest influence, which is between $\frac{2}{3}d$ and d, accounts for about 12.5 percent of v.

How is information from Figure 13–20 used to analyze measurements of apparent resistivity? Suppose, for example, that the measurements are made with an expanding Wenner configuration. Let us find the depth zone that most strongly influences each value of apparent resistivity. The measured change in potential v_{MN} depends on the electric potentials v_M at the M electrode and v_N at the N electrode. Therefore, we can replot the curve from Figure 13–20, first setting the distance $d = a$ to account for v_M at the M electrode, and then setting $d = 2a$ to account for v_N at the N electrode. The results are curves M and N in Figure 13–21. These curves are now combined to obtain a curve that indicates the sensitivity of the Wenner configuration to different depth zones. This curve indicates that it is most sensitive to the zone at a depth approximately equal to the electrode spacing.

Now we are in a better position to explain variations in apparent resistivity measured at a sounding site. Consider the structure in Figure 13–22 which consists of three layers. If readings are made with the eight electrode spacings $a_1, a_2, \ldots, a_8$, the depths of maximum sensitivity $z_1, z_2, \ldots, z_8$, which are approximately the same as the spacings, are distributed within the three layers.

Now it should become evident that the apparent resistivities $R_{a1}, R_{a2}, \ldots, R_{a8}$ corresponding to these electrode spacings will tend to vary according to resistivities within the layers where the depths $z_1, z_2, \ldots, z_8$ are situated. There will be relatively small differences between values of R_a corresponding to depths z within the same layer. In contrast, there will be relatively large differences between values of R_a with corresponding depths in different layers.

These variations can be emphasized by a

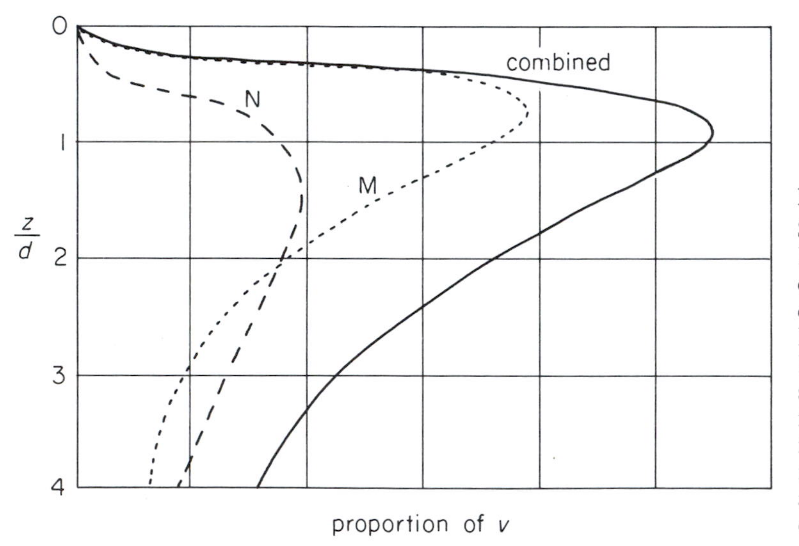

Figure 13–21
Sensitivity curves for the M and N electrodes and their combination in a Wenner configuration. Curves M and N have the form of the curve in Figure 13–20. The curve showing their combined effect indicates that the zone of largest influence on a resistivity measurement is at the depth approximately equal to the electrode spacing.

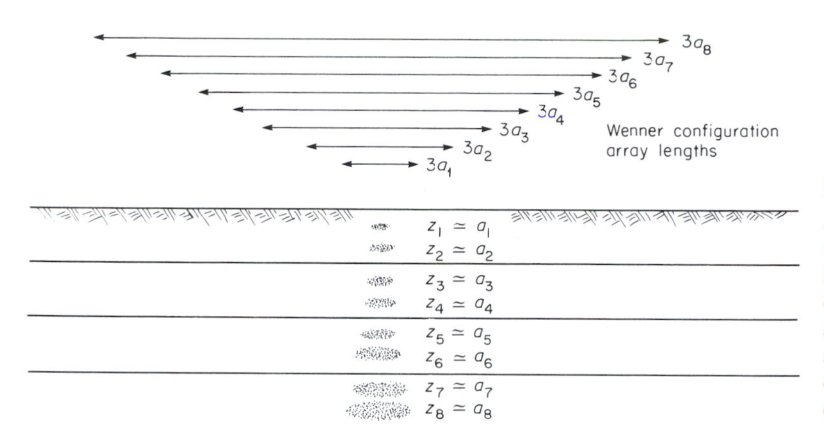

Figure 13–22
Approximate depths of maximum sensitivity in a three-layer structure corresponding to different Wenner configuration array lengths. Each depth is approximately equal to the electrode interval in the corresponding Wenner array.

graph of cumulative resistivity, prepared in the following way. For each electrode interval, a sum of resistivity values is calculated as shown in the following table.

INTERVAL	CUMULATIVE RESISTIVITY
a_1	$\Sigma R_1 = R_{a1}$
a_2	$\Sigma R_2 = R_{a1} + R_{a2}$
a_3	$\Sigma R_3 = R_{a1} + R_{a2} + R_{a3}$
a_4	$\Sigma R_4 = R_{a1} + R_{a2} + R_{a3} + R_{a4}$
.	.
.	.
.	.

Then the cumulative resistivity values are plotted at their corresponding electrode intervals, as shown in Figure 13–23. On such a graph the cumulative resistivities with corresponding depths of maximum sensitivity in the same layer tend to plot in nearly straight alignments. The intersections of lines drawn through these groupings indicate depths of the layer boundaries, and the slopes of these lines provide relative estimates of the layer resistivities. This is the basis of inversion analysis of cumulative resistivities.

We should stress the fact that points on a graph of cumulative resistivities do not have exactly straight alignments. Unlike the arrivals of seismic waves refracted or reflected from exact boundaries, we cannot obtain such specifically focused electric signals. We can estimate only the zones having relatively greater influence on our readings. Electrical surveying does not yield results that are as accurate or reliable as those that can be obtained by seismic surveying.

A Practical Example

The equipment pictured in Figure 13–16 was used to make electrical resistivity soundings in southwest Virginia. Current and potential readings made at one location are listed with the corresponding electrode intervals in Figure 13–24. Because the measurements were made using a Wenner configuration, values of apparent resistivity, also listed, were calculated by means of Equations 13–16 and 13–18. Cumulative resistivity values were then obtained and plotted on the graph, where straight lines were fitted to three alignments of points.

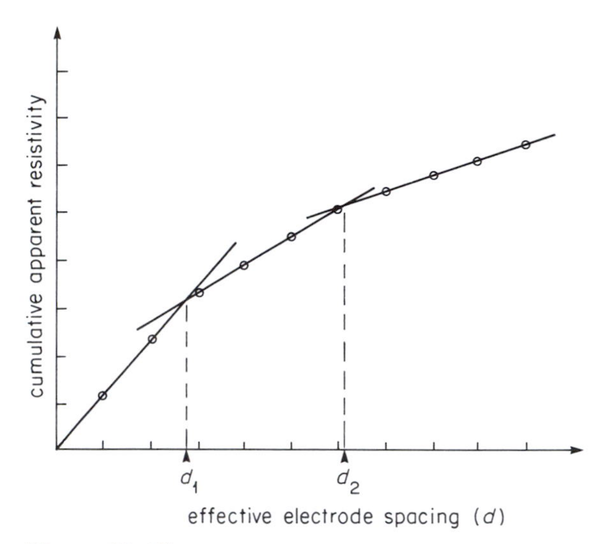

Figure 13–23
Graph of cumulative resistivities. Cumulative resistivity values are plotted at corresponding electrode spacings. Lines drawn through alignments of points intersect at distances that are close to the depths of boundaries between layers of contrasting resistivities.

Line intersections on the graph indicate layer boundaries at depths of about 18 and 37 feet. Slopes of the lines indicate lower resistivity in the middle layer compared with those above and below. From these results, the water table was interpreted to be at a depth of 18 feet, and the boundary between unconsolidated sediment and bedrock was interpreted to be at a depth of 37 feet.

The Barnes Parallel Resistor Method

Another simple inversion method can be used to analyze soundings of apparent resistivity made with the Wenner configuration, if the electrode spacing is expanded in equal intervals. This method, called the *Barnes layer method* or the *Barnes parallel resistor method.* It is based on a simplifying assumption that each apparent resistivity value is the average resistivity $\overline{R}$ in a layer reaching from the land surface to a depth equal to the electrode spacing. This assumption implies that each time the electrode spacing is increased, the depth of the layer being tested increases by an equal increment.

Adding a succession of equal layer increments is equivalent to adding a succession of equal resistors in a parallel electric circuit. This analogy is illustrated in Figure 13–25a. According to electric circuit theory, the relation between the combined resistance $\overline{r}$ and the individual resistances is

$$\frac{1}{\overline{r}} = \frac{1}{r_1} + \frac{1}{r_2} + \frac{1}{r_3} + \ldots$$

or in terms of resistivity,

$$\frac{1}{\overline{R}} = \frac{1}{R_1} + \frac{1}{R_2} + \frac{1}{R_3} + \ldots$$

This equation indicates that we can calculate true resistivity in any layer from two successive values of apparent resistivity:

$$\frac{1}{R_i} = \left(\frac{1}{R_{a,i}} - \frac{1}{R_{a,i-1}}\right) \qquad (13\text{–}22)$$

A graph of values determined in this way indicates the variation of resistivity with depth, as in Figure 13–25b. Because of the initial simplifying assumption, the results are considered only estimates of the true resistivity values.

The Method of Characteristic Curves

A widely used method for analyzing apparent resistivity measurements is to make trial-and-error comparisons of them with curves that il-

a (feet)	i (ma)	v (mV)	R_a (ohm-feet)	ΣR_a
5	17	5.4	9.9	9.9
10	20	2.5	7.8	17.7
15	20	1.7	8.0	25.7
20	23	1.3	7.1	32.8
25	40	0.8	3.1	35.9
30	52	0.8	2.9	38.8
35	70	1.3	4.1	42.9
40	108	2.1	4.9	47.8
45	120	3.0	7.1	54.9
50	125	3.2	8.0	62.9

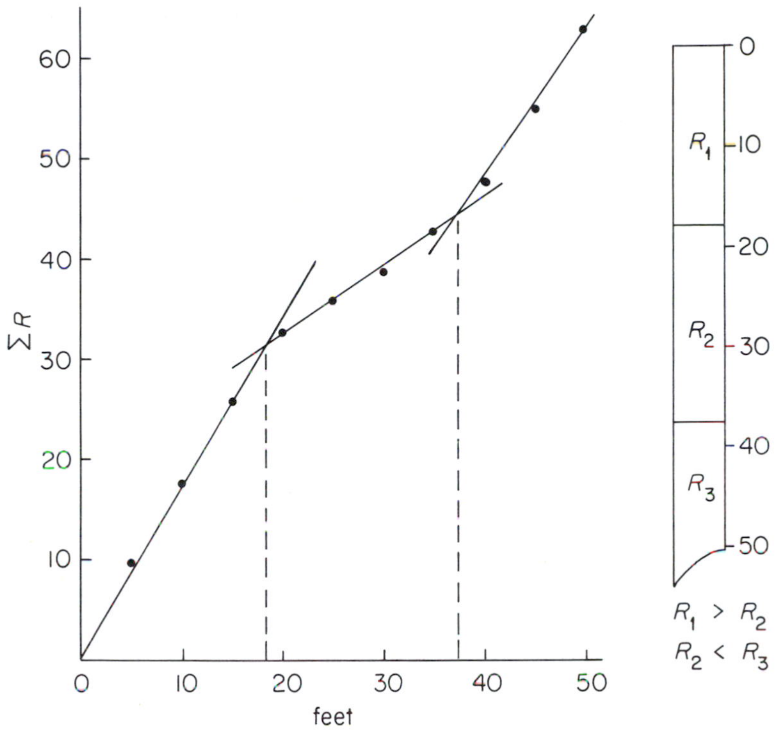

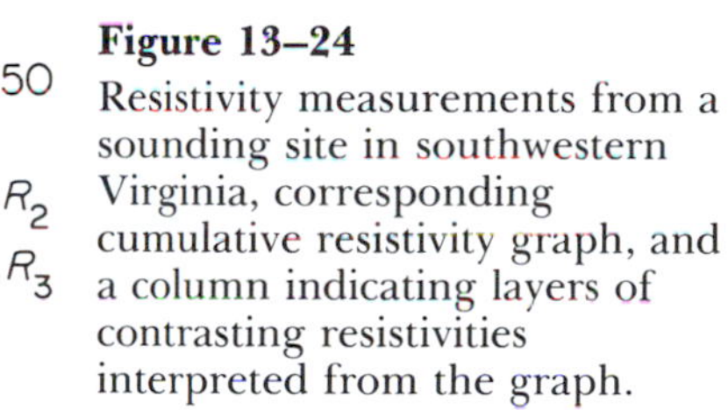

Figure 13–24
Resistivity measurements from a sounding site in southwestern Virginia, corresponding cumulative resistivity graph, and a column indicating layers of contrasting resistivities interpreted from the graph.

lustrate the resistivity variations of simple structures. To illustrate this method, we will look at a set of curves that are characteristic of a structure consisting of two layers separated by a horizontal boundary.

Let us begin with the structure shown in Figure 13–26 in which resistivity is R_1 in the upper layer and R_2 in the lower layer, and the boundary is at depth z_1. First, the electric potential is calculated at points along the surface

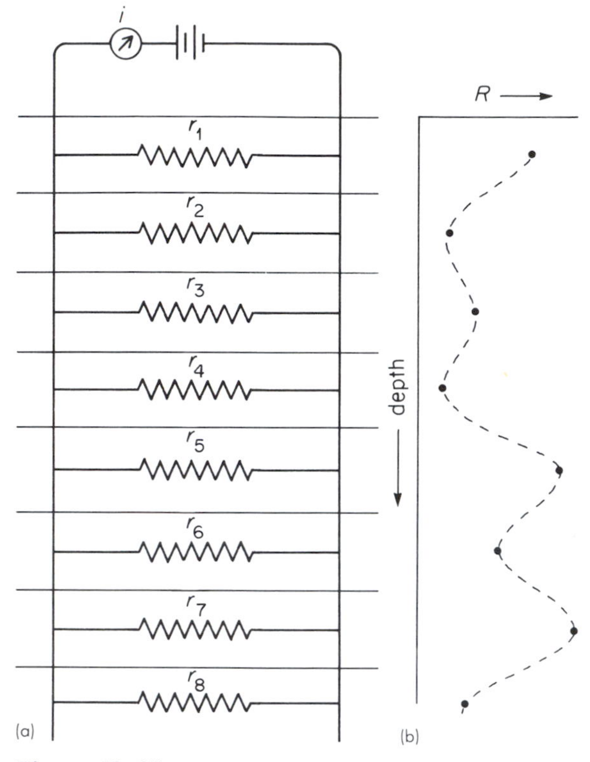

Figure 13–25
Sequence of layers of equal thickness and an equivalent representation consisting of resistors wired in a parallel circuit. The graph shows the resistivity variation with depth determined by the Barnes parallel resistor method for analyzing electrical resistivity data obtained by sounding.

of the structure by means of Equations 13–11, 13–12, and 13–13. This information is then used for determining differences in potential corresponding to different electrode spacings, so that values of apparent resistivity can be calculated from Equation 13–16. A graph of the variation of resistivity with electrode spacing can then be prepared. This procedure is then repeated using different values of R_1 and R_2. In this manner, the set of curves in Figure 13–26, which is characteristic of a two-layer structure, was prepared.

A set of characteristic curves is commonly plotted using the ratio a/z_1 to express electrode spacing in units of layer thickness. Apparent resistivity is usually expressed as the ratio R_a/R_1. When these ratios are used, the curves become general so that they apply to any two-layer structure, regardless of the depth of the boundary or the range of resistivities.

A numerical example illustrates how we can use these characteristic curves to analyze resistivities obtained by sounding. Suppose that we have measured the apparent resistivities and electrode spacings given in Table 13–1. In addition, suppose that by placing the electrodes quite close together, we have measured the value of R_1 = 11 ohm-m. This value is used to calculate the ratios of R_a/R_1 which are listed in Table 13–1.

Our aim is to determine the depth z_1 of the boundary and the value of R_2; we can obtain them by trial-and-error. First, we choose a trial depth, which is perhaps based on a preliminary analysis by the cumulative resistivity method. In our example we begin by choosing z_1 = 18 m. Then we obtain the electrode interval–depth ratios of $a/18$, which are listed in Table 13–1. These values and the corresponding resistivity ratios are then plotted as the symbol x on the graph of characteristic curves in Figure 13–26. Observe that a line connecting these points is not parallel to the adjacent characteristic curves. This means that another depth value must be tried.

Next, we choose a value of z_1 = 24 m and calculate another set of electrode interval–depth ratios, which are also listed in Table 13–1. These values and the corresponding resistivity ratios are plotted as dots on the graph of characteristic curves. This time we see that the

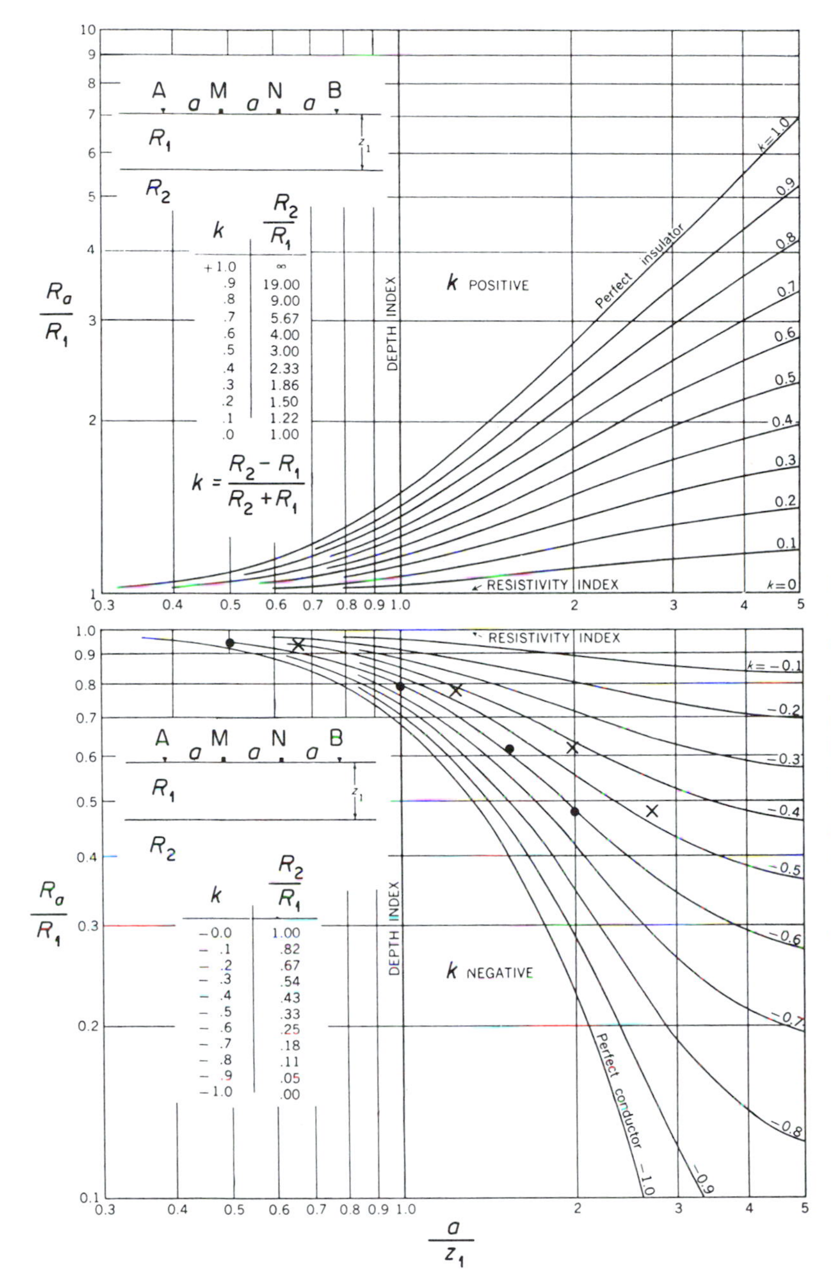

Figure 13–26
Apparent resistivity curves based on the Wenner configuration for a two-layer structure. The top set of curves represents a structure for which resistivity R_1 in the top layer is lower than resistivity R_2 in the lower layer. The bottom set of curves represents a structure for which resistivity R_1 in the top layer is higher than resistivity R_2 in the bottom layer. Each curve corresponds to a different reflection coefficient k. Electrode spacing is expressed in units of the depth of the boundary. The points shown with the symbol × and dots on the bottom curves correspond to analysis of data in Table 13–1. (From R. G. Van Nostrand and K. L. Cook, U.S. Geological Survey Professional Paper 499, p. 91, 1966. Adapted from I. Roman, U.S. Geological Survey Bulletin 927-A, 1941.)

TABLE 13–1 Electrical Resistivity Survey Data

ELECTRODE SPACING a (feet)	APPARENT RESISTIVITY R_a (ohm-feet)	R_a/R_1 (R_1 = 11 ohm-feet)	$a/18$	$a/24$
12	10.3	0.94	0.5	0.67
24	8.7	0.79	1.0	1.33
36	6.8	0.62	1.5	2.0
48	5.2	0.47	2.0	2.67

points plot very close to the particular curve for which the reflection coefficient $k = -0.6$. This tells us that the trial value of $z_1 = 24$ m must be the correct depth. Furthermore, we can use the reflection coefficient and our value of R_1 in Equation 13–13 to calculate the value $R_2 = 2.75$ ohm-m.

This graphical trial-and-error process is a basic procedure for analyzing apparent resistivities obtained by sounding. A computer can be programmed to perform this operation. The computer can also be used to calculate and plot characteristic curves for structures with more layers, so that more complicated patterns of variation in apparent resistivities can be analyzed.

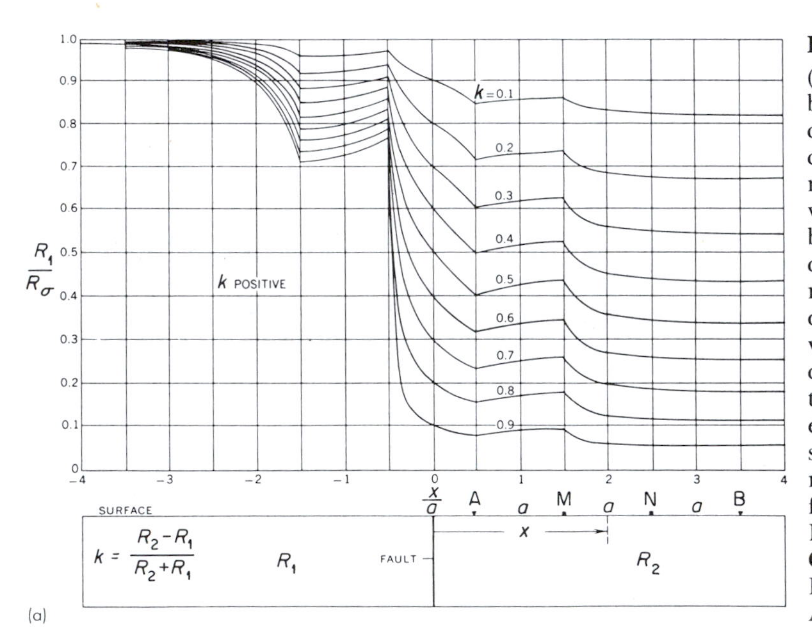

Figure 13–27
(a) Apparent resistivity curves based on the Wenner configuration for structures consisting of two contrasting resistivities separated by a vertical boundary (fault) and (b) by two vertical boundaries. Each curve corresponds to a different reflection coefficient k. The distance from the fault (single vertical boundary) and from the center of the dike (zone between two vertical boundaries) is expressed in units of electrode spacing. The zone of highest resistivity is on the left of the fault. (Modified from R. G. Van Nostrand and K. L. Cook, U.S. Geological Survey Professional Paper 499, p. 125, 1966. Adapted from Tagg, 1930.)

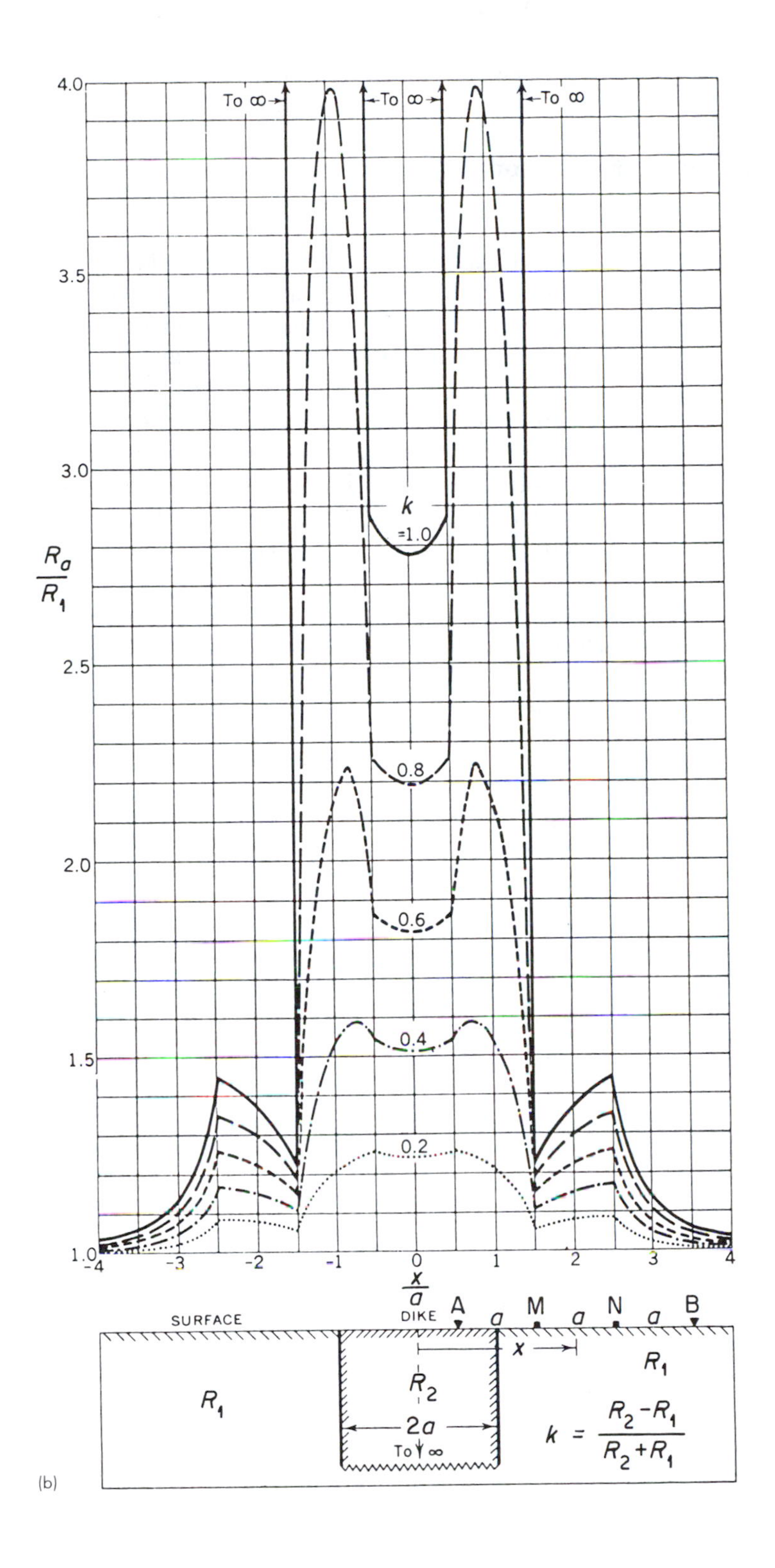

To ∞
To ∞
To ∞
k
=1.0
0.8
0.6
0.4
0.2
4.0
3.5
3.0
2.5
2.0
1.5
1.0
-4
-3
-2
-1
0
1
2
3
4
$\frac{R_a}{R_1}$
$\frac{x}{a}$
SURFACE
DIKE
A
a
M
a
N
a
B
x
R_1
R_2
R_1
2a
To ∞
$k = \frac{R_2 - R_1}{R_2 + R_1}$

(b)

Resistivity Profiles over Faults and Dikes

Up to this point, we have described ways of analyzing resistivities obtained by sounding. Let us now look at readings made with a fixed electrode spacing in different positions along a profile. These readings can also be analyzed by trial-and-error comparisons with characteristic curves.

Resistivity variations along profiles over idealized structures with faults and dikes can be calculated by the same procedure we described for a two-layer structure. Changes in electric potential can be found from Equations 13–11, 13–12, and 13–13 for electrodes in different positions relative to the fault or dike. These potential differences are used in Equation 13–16 to calculate the apparent resistivity at these electrode positions, and the results are plotted on a profile to display the pattern of variation over the structure. Sets of characteristic curves are prepared by repeating this procedure with different values of resistivity.

Characteristic curves over a vertical fault and a vertical dike are given in Figure 13–27. These curves indicate the patterns of variation along a profile of measured values that crosses such structures. As before, a computer can be programmed to calculate and plot profiles over still more complicated structures with inclined faults, dikes, and boundaries between

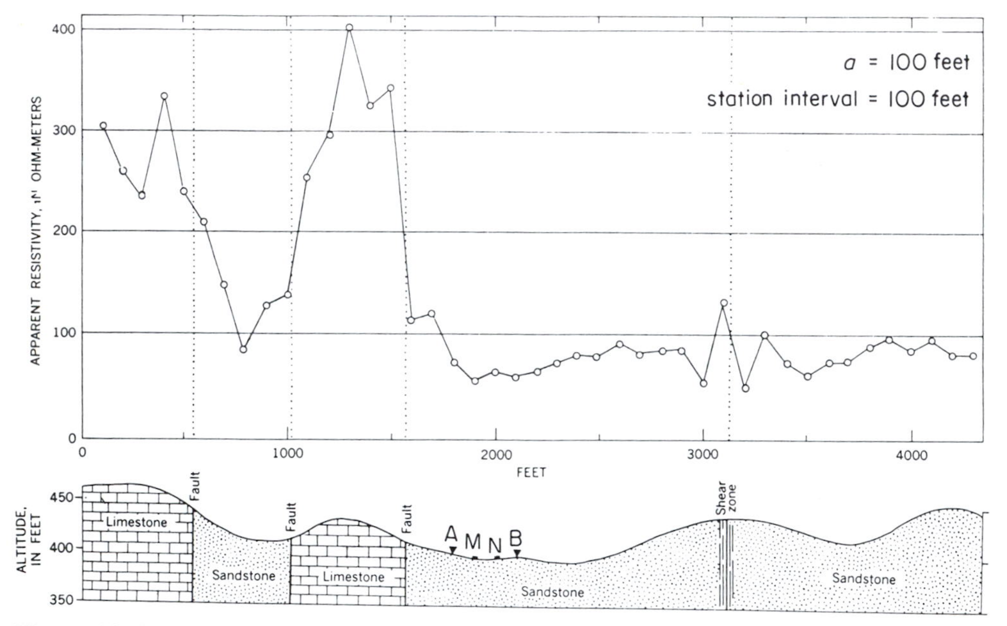

Figure 13–28

Profile of apparent resistivities measured in faulted terrane in Illinois. Both the electrode spacing and the reading interval are 100 feet. (From R. G. Van Nostrand and K. L. Cook, U.S. Geological Survey Professional Paper 499, p. 164, 1966. Adapted from Hubbert, 1952.)

TABLE 13–2 Resistivities of Earth Materials

EARTH MATERIAL	RESISTIVITY, AVERAGE OR RANGE (ohm-m)
Metals	*Average R*
Copper	1.7×10^{-8}
Gold	2.4×10^{-8}
Silver	1.6×10^{8}
Graphite	10^{-3}
Iron	10^{-7}
Lead	2.2×10^{-7}
Nickel	7.8×10^{-8}
Tin	1.1×10^{-7}
Zinc	5.8×10^{-8}
Sulfide Ore Minerals	*Average R*
Chalcocite	10^{-4}
Chalcopyrite	4×10^{-3}
Pyrite	3×10^{-1}
Pyrrhotite	10^{4}
Molybdenite	10
Galena	2×10^{-3}
Sphalerite	10^{2}
Oxide Ore Minerals	*Range of R*
Bauxite	10^{2}–10^{4}
Chromite	1–10^{6}
Cuprite	10^{-3}–300
Hematite	10^{-3}–10^{7}
Magnetite	10^{-5}–10^{4}
Ilmenite	10^{-3}–10^{2}
Rutile	10–10^{3}
Silicate Minerals	*Range of R*
Quartz	10^{10}–10^{15}
Muscovite	10^{2}–10^{14}
Biotite	10^{2}–10^{6}
Hornblende	10^{2}–10^{6}
Feldspar	10^{2}–10^{4}
Olivine	10^{3}–10^{4}
Other Minerals	*Range of R*
Calcite	10^{12}–10^{13}
Anhydrite	10^{9}–10^{10}
Halite	10–10^{13}
Coal	10–10^{11}
Crystalline Rocks	*Range of R*
Granite	10^{2}–10^{6}
Diorite	10^{4}–10^{5}
Gabbro	10^{3}–10^{6}
Andesite	10^{2}–10^{4}
Basalt	10–10^{7}
Peridotite	10^{2}–10^{3}
Schist	10–10^{4}
Gneiss	10^{4}–10^{6}
Slate	10^{2}–10^{7}
Marble	10^{2}–10^{8}
Quartzite	10–10^{8}
Sedimentary Rocks	*Range of R*
Shale	10–10^{3}
Sandstone	1–10^{8}
Limestone	50–10^{7}
Dolomite	10^{2}–10^{4}
Unconsolidated Sediment	*Range of R*
Sand	1–10^{3}
Clay	1–10^{2}
Marl	1–10^{2}
Groundwater	*Range of R*
Portable well water	0.1–10^{3}
Brackish water	0.3–1
Seawater	0.2
Supersaline brine	0.05–0.2

Modified from W. M. Telford, L. P. Geldart, R. E. Sheriff, and D. A. Keys, *Applied Geophysics,* Cambridge, England, Cambridge University Press, pp. 451–455, 1976.

different lithologies. The results help us to understand resistivity variations such as those shown in Figure 13–28, these particular variations are caused by steeply inclined boundaries.

Resistivity and Lithology

The electrical properties of minerals and rocks have been a subject of study for almost two centuries. Laboratory measurements of resistivity and the dielectric constant have been made for most known varieties of minerals, rocks, soil and unconsolidated sediment, and for groundwater. In addition, many field measurements have been made with these substances in their natural settings. Typical values of resistivity are presented in Table 13–2.

Electrical resistivity is the most variable physical property of concern in exploration geophysics. Values range over 15 orders of magnitude, which is much larger, even, than the range of magnetic susceptibility. There is considerable variation between lithologically and mineralogically similar specimens.

For purposes of interpreting electrical resistivity surveys, it is usually suitable to classify substances with resistivities lower than one ohm-meter as electric conductors, and those with resistivities above 10^7 ohm-meters as electric insulators that do not conduct current. The intermediate range describes resistive conductors.

Natural metals and metallic ore minerals tend to be good conductors of electric current. Otherwise, most minerals are insulators or highly resistive conductors. Rocks consisting of metallic ore minerals have broader but generally lower ranges of resistivity, largely because water is present in the pore space of the rock.

In the ground, ohmic conduction accounts for a very small part of the electric current that flows mostly by electolytic conduction. Therefore, the values measured in a survey of electrical resistivity depend much more on rock porosity and the nature of the groundwater than on lithology. For this reason, there are no generally applicable guidelines for interpreting the lithology of the zones of contrasting resistivity delineated by a survey. Interpretation may be possible on a local basis, however, if there is independent information about the nature of the rocks in the area of the survey.

INDUCED POLARIZATION SURVEYING

When electric current is compelled to flow in the ground by means of source and sink electrodes, concentrations of charge may build up at various places. After the imposed current is switched off, these charges shift back to their original distribution in the ground. During the time that the concentrations of charge are being dissipated, electric potential persists. This phenomenon is called *induced potential* and is abbreviated as IP.

For many years, geophysicists had recognized the effect of induced potentials while carrying out electrical resistivity surveys. The voltmeter between the potential electrodes continued to indicate a weak signal after the current from the source and sink had been switched off. The decay of this weak induced potential ranged from a few moments to several minutes. Beginning in the 1940s, efforts were made to measure the decay time for exploration purposes. By the mid-1950s, induced potential, or IP, surveying had become one of the principal tools in the search for base metal ores. More recently, there have been indications that the method may have application in groundwater surveys.

Source of Induced Potential

In the ground a balance is normally established between electric charges on the surfaces of mineral grains and ions in the water that saturates the pores. Positive ions from the water tend to cluster around negatively charged clay particles, and negative ions are attracted by positive charges on the surface of some sand grains. Altogether, the mobile ions maintain positions that act to produce electrical neutrality, as in Figure 13–29a.

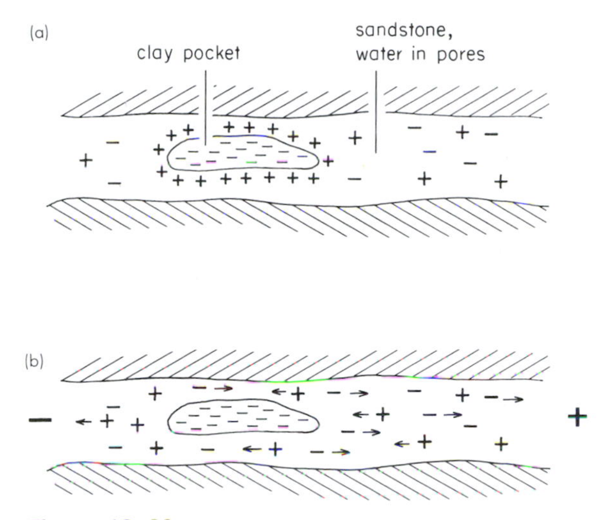

Figure 13–29
Distribution of ions and electrically charged grains of sediment in a natural setting (a) where positive groundwater ions tend to concentrate on surfaces of negatively charged clay particles and negative groundwater ions collect on sand grains carrying positive charges. This natural balance is disrupted when source and sink electrodes are placed in the ground causing (b) positive groundwater ions to migrate in one direction and negative ions to migrate in the opposite direction. Removal of these electrodes allows these ions to return to their original positions. During this interval of time, the returning ions produce voltage of decaying amplitude.

The natural balance of charges, which is like that of a rundown battery, is upset by imposing current in the ground, which acts to recharge the battery. This means that positive and negative charges are drawn apart from one another, as in Figure 13–29b.

The recharging effect is strongest in the zones of highest current density. We know from Figure 13–13 that these are the zones of lowest resistivity. It is in the low-resistivity zones that mobile charges become the most widely separated when they are present. In a sense, these zones are the natural batteries that are the most easily recharged.

Metallic ore minerals tend to have much lower resistivity than other minerals, as is evident from the data presented in Table 13–2. Zones with high concentrations of metallic ore minerals, then, turn out to be low-resistivity zones. We can expect that separations of charges will be abnormally large, and therefore that large induced potentials might develop in these zones if current is imposed in the ground. This idea is the basis for using induced potential (IP) surveys to explore for base metal ores.

Measuring Induced Potentials

The conventional electrode configurations used for electrical resistivity surveying are also used to measure induced potentials. There are sometimes problems with Wenner or Schlumberger configurations because electromagnetic signals are induced by the flow of current in the wires connecting the electrodes. Therefore, the polar dipole–dipole configuration is preferred for IP surveying.

In addition to the equipment normally used for resistivity surveying, an oscilloscope or recording system is required to preserve the IP signal. A typical IP surveying system is illustrated in Figure 13–30.

An IP record obtained with this equipment begins while current is still flowing in the source–sink circuit; it continues after the current is switched off until no additional voltage can be detected. The length of an IP record can range between a few seconds and several minutes. The form of the record is illustrated in Figure 13–31. It shows an induced potential that persists from time t_1 to time t_2 before it becomes too weak to be measured.

Values measured in different ways can be used to represent the strength of the induced potential effect. One or the other of two values can be determined directly from a single IP record. First, we will define the value that we will call the induced potential ratio Q_{IP}. It is the ratio of the voltage v_t measured at a fixed time t after current in the source–sink circuit has been switched off, and the voltage v_c measured while the current is on:

$$Q_{IP} = v_t/v_c \qquad (13\text{–}22)$$

The voltages for this measurement are indicated in Figure 13–32a.

Another measure of induced potential is called the chargeability M. To obtain this value, we must measure the area α under the voltage curve between two times, t_a and t_b, as in Figure 13–32b. The chargeability can then be expressed as

$$M = \alpha/v_c \qquad (13\text{–}23)$$

Because the area under the curve is in units of time multiplied by volts, the unit of M is time, which is usually given in milliseconds. Since the values of Q_{IP} and M are both determined from a single record of voltage variation with time, they are called *time domain* IP measurements.

Other values for specifying IP strength are determined from measurements made using different alternating-current frequencies. These are called *frequency domain* IP measurements. They are determined from apparent resistivity values that tend to vary with the frequency of an alternating current. A value called the *frequency effect*, FE, is given by

$$\text{FE} = \frac{R_{dc} - R_{ac}}{R_{ac}} \qquad (13\text{–}24)$$

where R_{dc} is the resistivity found by means of a direct current or a very low alternating current with a frequency of less than one hertz, and R_{ac} is the resistivity found with a higher-frequency (up to ten hertz) alternating current. Another number expressing IP strength that can be calculated from these resistivity values is called the *metal factor*,

$$\text{MF} = 2\pi \left(\frac{R_{dc} - R_{ac}}{R_{dc}R_{ac}} \right) \times 10^5 \qquad (13\text{–}25)$$

Resistivities used to calculate FE and MF are determined from Equation 13–16 and the appropriate geometrical factor. The polar dipole–dipole configuration is the most commonly used electrode system, and its geometrical factor is found from Equation 13–21.

Results of IP Surveying

The practice in IP surveying is to refer to the circuit with the source and sink electrodes as the *transmitter* circuit and to refer to the circuit

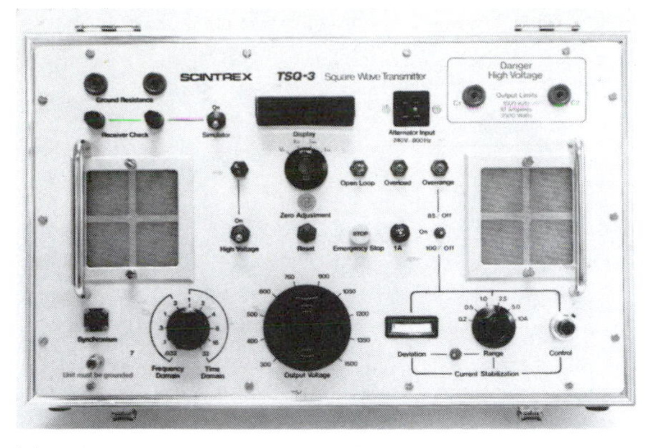

(a)

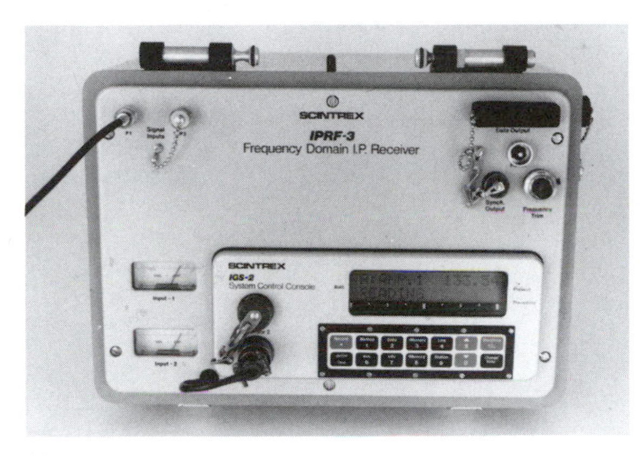

(b)

(c)

Figure 13–30
A portable system for IP surveying. (a) The transmitter unit which controls the voltage and the frequency and duration of voltage to be applied. (b) The receiver unit which measures the decaying voltage after the transmitter is switched off. (c) A field operation showing electrodes in the ground and connecting cables. (Courtesy of Scientrex, Ltd.)

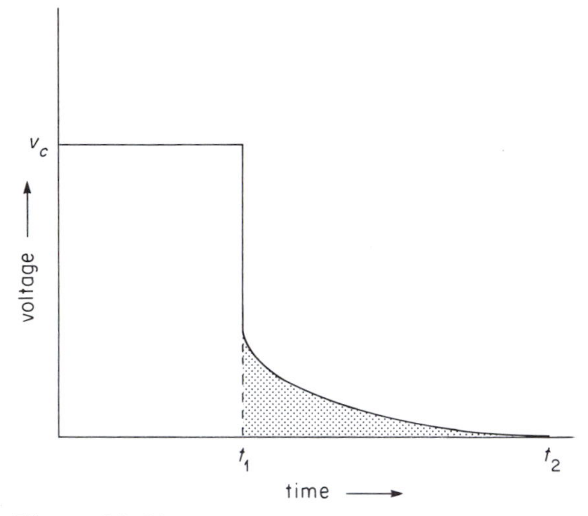

Figure 13–31
An IP record showing variation of voltage with time. For the interval of time before t_1, constant voltage is maintained between source and sink electrodes. The constant voltage separates positive and negative groundwater ions as in Figure 13–29b, producing an induced polarization. At time t_1, voltage is switched off. As the ions return to their original positions, they maintain a decaying voltage that lasts until time t_2.

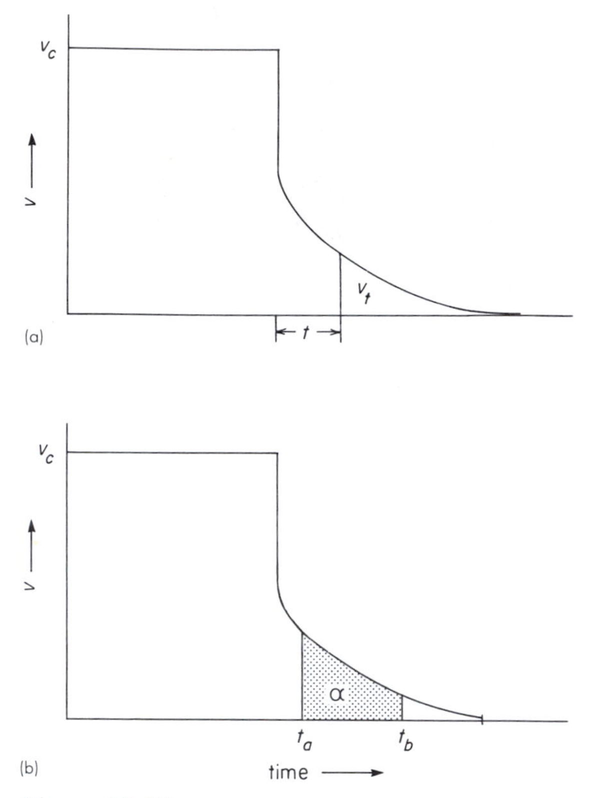

Figure 13–32
IP records showing information used to obtain (a) an induced potential ratio and (b) chargeability. Two kinds of measurements can be made. One is the amplitude v_t of the voltage remaining at a fixed time t after switching off the inducing voltage. The other is the integrated voltage α which is the area under the decay curve between two fixed times t_a and t_b after switching off the inducing voltage.

with the potential electrodes as the *receiver* circuit. In the usual polar dipole–dipole configuration, the transmitter electrodes and the receiver electrodes have the same spacing, and the distance between the mid points of the two arrays is a multiple of this spacing.

IP surveying is usually done along profiles. The most common procedure is to begin with the transmitter electrodes at one end. Readings are then made with the receiver electrodes in a succession of positions at intervals equal to the electrode spacing. The transmitter is then advanced to the next position, and the procedure is repeated. This method, illustrated in Figure 13–33, is similar in some ways to the common midpoint profiling used in seismic surveying.

Values measured along a profile can be plotted in an interesting way; they can create an apparent cross section on which images of the IP sources are indicated by contours. In

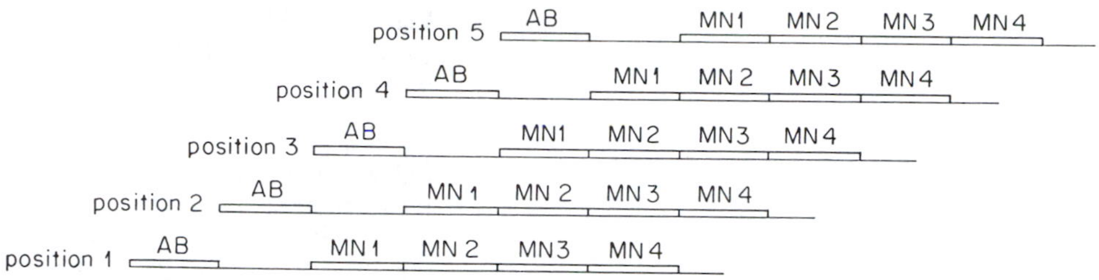

Figure 13–33

Procedure for IP profiling. With the transmitter AB in position 1, readings are made with the receiver MN in locations MN1, MN2, MN3, and MN4. The transmitter is then advanced to position 2, and four more measurements are made in the same way. This procedure is repeated as the transmitter is moved to the other positions along the profile.

Figure 13–34 the transmitter and receiver positions are indicated by the first and second numbers, respectively. The top diagram shows one transmitter position with a line from the midpoint extending downward toward the receivers at a 45-degree angle. It intersects the lines drawn at 45-degree angles from the receiver positions defined by the midpoint of each receiver electrode. Values measured with the receiver in different positions are marked at the corresponding intersections of these inclined lines. Observe that the values plotted in this way lie beneath the mid points of transmitter–receiver lines along which they were measured. In addition, the depths of these points increase as the transmitter–receiver distance increases. Although there is no direct relation between these points and the positions of IP sources, we do expect signals from deeper sources to have a proportionally larger effect on readings from longer transmitter–receiver distances.

The middle diagram in Figure 13–34 shows how the graph is extended by plotting the results from a succession of transmitter positions. For each transmitter position, the procedure illustrated by the top diagram is repeated. This produces an array of IP values which can then be contoured, as shown in the bottom diagram.

Contours on the bottom diagram in Figure 13–34 create patterns that tend to image the relative locations of IP sources. These contour images must not be interpreted as showing actual positions of sources, but they do give an approximation of where along the profile a source may be situated and whether it is relatively shallow or deep.

The usual practice is to obtain electrical resistivity measurements at the same time that IP measurements are made. The resistivity data can be displayed on the same kind of graph that is used for the IP results, as in Figure 13–35. Here it is possible to compare the contour images to see whether there are common sources of IP and resistivity variation.

Most IP surveys are done for reconnaissance purposes to identify zones that can be tested further by drilling or other geophysical surveying. Ordinarily, little effort is made to explain the IP variations in terms of models of specific geological structures.

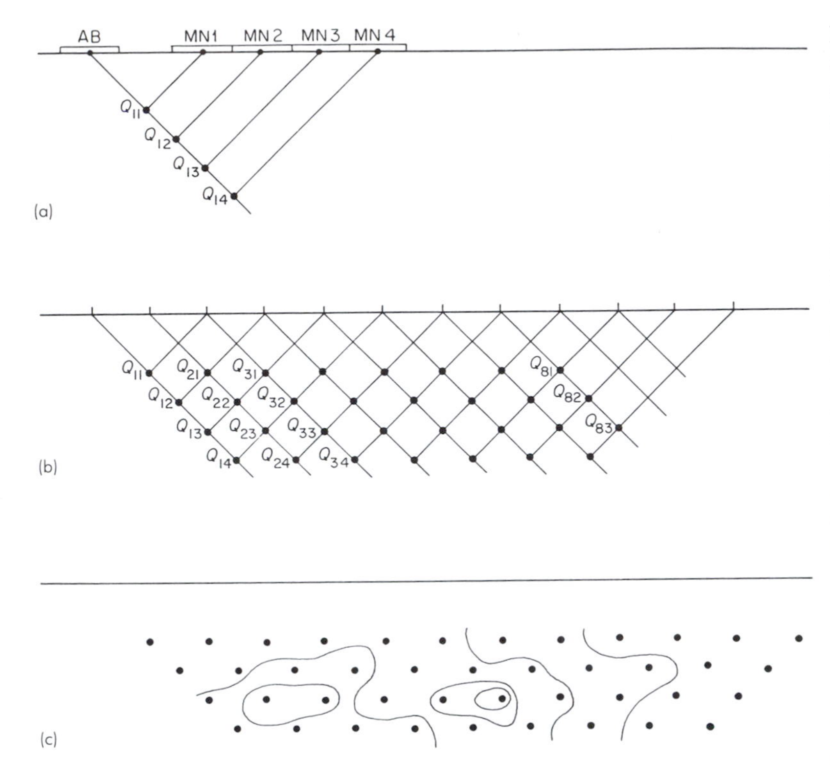

Figure 13–34
Graphic display of IP measurements on a pseudo cross section which reveals relative positions of zones that produce induced potentials. (a) Each measurement Q is plotted at the intersection of lines extending downward at 45-degree angles from the transmitter and receiver midpoints. For a succession of receiver positions and a fixed transmitter position, values of Q_{11}, Q_{12}, Q_{13}, Q_{14}, and so on, all lie along the line extending from the transmitter. (b) By moving the transmitter and repeating the procedure, we can plot an array of values of Q at equal intervals along a set of parallel lines extending downward at 45-degree angles from the transmitter positions. (c) The array of Q values can be contoured, and the contour patterns can be interpreted to identify relative positions of IP source zones.

SURVEYING NATURAL POTENTIALS

One of the first geophysical methods used to explore for metallic ores was to measure natural potentials. As early as 1830, natural potentials were associated with copper ores in the Cornwall district of England. The various explanations of natural potentials that have been proposed infer complicated geochemical reactions. We will describe briefly one process that can produce electric fields around shallow concentrations of metallic ore. Other processes are introduced in Chapter 14 in the discussion of borehole measurements of electrical potentials.

Electric fields are produced on a global scale by interaction of the earth's magnetic field and electric discharges in the atmosphere and the ionosphere. Currents in the earth called telluric currents are associated with these fields. We will present a brief discussion of how they can be used for reconnaissance geophysical surveying.

Exploring Shallow Natural Potentials

Mineralized zones close to the earth's surface may give rise to natural potentials. One explanation has been proposed for metallic ore concentrations that are partly above and partly below the water table. Chemical reactions including oxidation proceed at different rates in the upper and lower parts of such an ore body, producing an upward ohmic con-

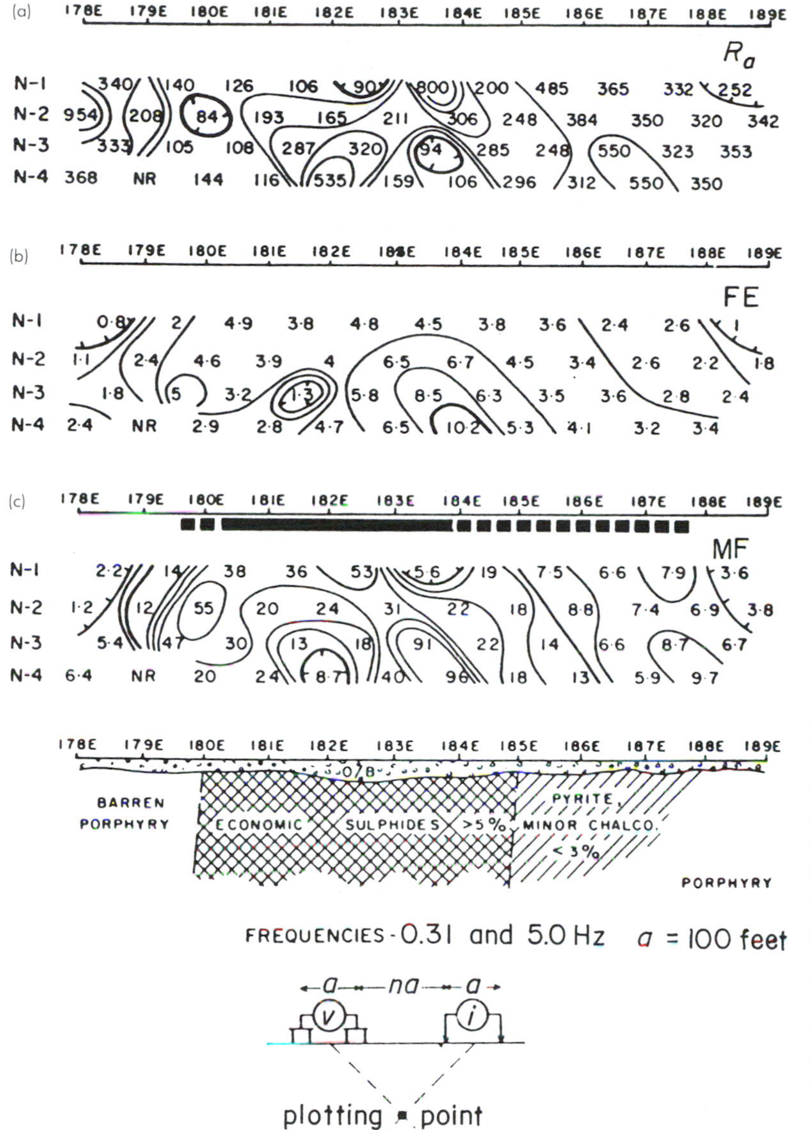

Figure 13–35
Measurements of (a) apparent resistivity, (b) frequency effect, and (c) metal factor contoured on a pseudo cross section along a profile crossing a zone of sulfide mineralization. The locations of massive sulfide ore bodies were then established by test drilling at sites where the frequency effect and metal factor were detected. Notice the corresponding relatively low resistivity values. (Modified from D. K. Fountain, *Geophysics*, v. 37, n. 1, p. 157, February 1972.)

duction of electrons. The concentrations of charges cause electrolytic conduction of groundwater ions in the surrounding zone, as in Figure 13–36. The ore body, then, acts as a natural battery.

Methods for detecting shallow natural potentials can be quite simple. Electrodes connected to a millivolt meter are moved from place to place, and potential differences are measured. Such a survey is usually done in a

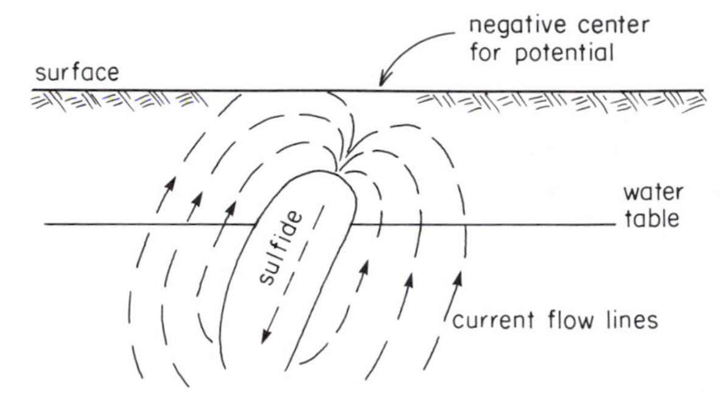

Figure 13–36
A natural potential developed by charge concentrations and current flow in a conducting body situated partially above the water table. Because its lower part extends below the water table, the body acts as a natural battery with ions moving from one side to the other. Potential differences measured on the surface indicate a minimum at the location where the axis of the more or less symmetrical current field interacts with the land surface.

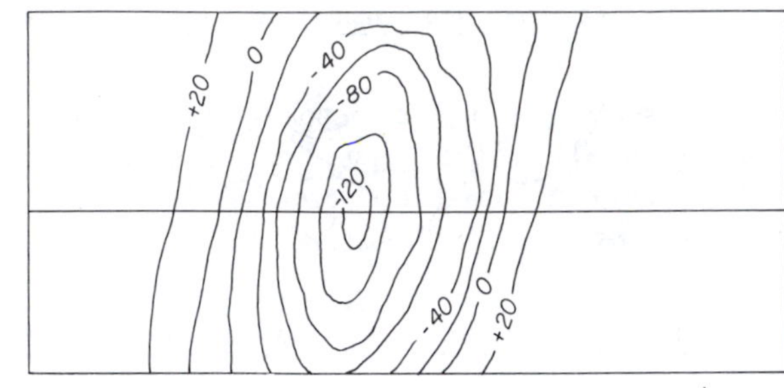

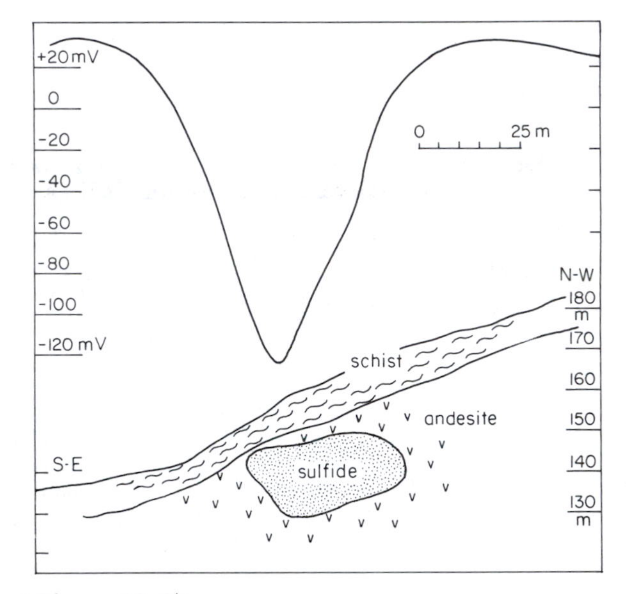

Figure 13–37
A contour map and a profile showing variation of the natural potential produced by a sulfide ore body situated within an andesite mass covered by schist at Istanbul, Turkey. A low value of about -120 mV was detected directly over the sulfide, and a high value of about $+30$ mV was measured on each side along the profile. The overall anomaly of 150 mV is considered a strong one. The contoured map locates the area with a negative potential anomaly. (Modified from S. H. Yüngül, *Geophysics*, v. 19, n. 3, p. 458, June 1954.)

systematic way by moving the electrodes along profiles and maintaining a constant spacing. For this purpose, it is necessary to use porous pot electrodes.

Results of a natural potential survey can be plotted to show potential variations along a profile. Where measurements have been made along several parallel profiles, the results can be contoured on a map to display these variations. Examples in Figure 13–37 show that natural potential differences of more than 100 millivolts are sometimes observed.

Most surveys of shallow natural potentials were done simply to locate targets for further exploration by other means. If a particularly uniform potential field is mapped, however, it may be possible to estimate the depth to the source. One approach is to represent the source as a dipole battery that has a positive charge Q at one end and a negative charge

$-Q$ at the opposite end. The potential v_p at a point on the land surface is, then,

$$v_p = \frac{Q}{r_1} - \frac{Q}{r_2} \qquad (13\text{–}26)$$

where r_1 and r_2 are the distances to the charge concentrations. By a trial-and-error procedure, not unlike that used to explain a magnetic anomaly by means of a dipole magnet, the positions of the charge concentrations can be located.

Telluric Currents

It is well known that the motion of particles ionized by solar radiation produces electric currents in the ionosphere. In Chapter 10 we described the diurnal fluctuations of magnetic field intensity produced from these ionospheric currents by electromagnetic induction. By this same process, an alternating electric current with a variable frequency of one hertz or smaller is induced to flow horizontally in the upper part of the earth's crust. This is the telluric current.

In a horizontal rock layer of uniform thickness, the telluric current density μ_T can be considered to be constant, because the source of the current is so far from the area where measurements are being made. According to Equation 13–9, the difference in potential v between two electrodes that is associated with this current will be

$$v = \mu_T RL \qquad (13\text{–}27)$$

where L is the distance between the electrodes. If the layer thickness changes, the current density will also change, as will the potential difference between the electrodes, as indicated in Figure 13–38.

The depth h that the telluric current penetrates into the earth depends on its frequency f and the resistivity. This depth of penetration is expressed approximately by

$$h = \tfrac{1}{2}\sqrt{R/f} \qquad (13\text{–}28)$$

where R is in ohm-meters, f is in hertz, and h is in kilometers.

Telluric Current Surveying

Reconnaissance surveying of variations in the thickness of a sedimentary basin can be done by measuring potential differences associated with telluric currents. This method has been used in the initial stages of oil exploration in areas where there is little information about the thickness of the sequence of sedimentary rock.

The basic idea of telluric current surveying is to measure differences in potential at a base station, where thickness z_0 and resistivity R of the sedimentary basin are known, and at other satellite stations. Then, by assuming the same value of R for the entire area of the survey, we can compute the current density for each station by means of Equation 13–27. As can be seen from Figure 13–38, the change in thickness Δz is approximately proportional to the change in current density $\Delta\mu$ between the base station and a satellite station. Therefore, the thickness z_s at a satellite station can be calculated:

$$z_s = z_0 + \Delta z = z_0 + \frac{z_0}{\mu_0}\Delta\mu \qquad (13\text{–}29)$$

The direction of the telluric current varies from place to place, and its frequency is irregular. To account for differences in direction, we must measure components of the potential with the electrodes oriented in two perpendic-

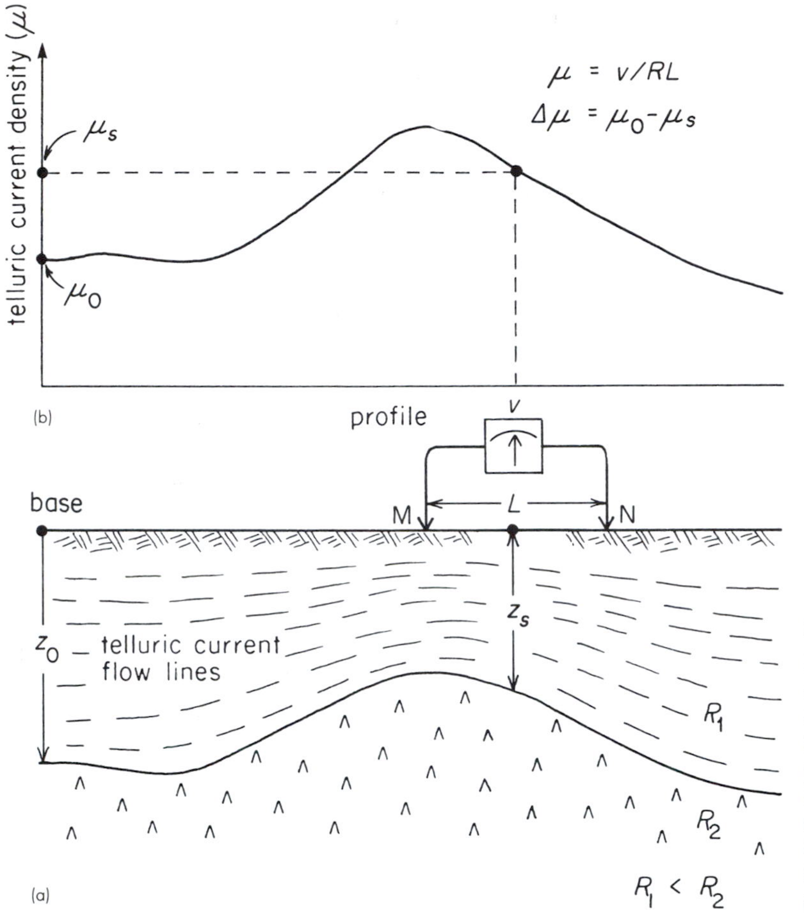

Figure 13–38
(a) Telluric current flow lines in a layer of low resistivity resting on a zone of high resistivity and (b) the potential difference produced by these currents. Lines show how the current density increases because the layer has become thinner where higher potential differences are observed.

ular directions, usually north and east. Because of the frequency variation, the potential fluctuates. Therefore, the output of voltmeters connected to the two perpendicular sets of electrodes must be recorded simultaneously for a certain interval of time, as in Figure 13–39. Voltage components v_x and v_y measured at the same time are then combined to obtain the difference in potential v, which is then used to determine current density:

$$v = \sqrt{v_x^2 + v_y^2} \qquad (13\text{–}30)$$

The accuracy of depths determined from Equation 13–29 is limited by possible variations in resistivity from place to place. The presence of structures and lithologic variations in different locations can introduce serious error. It is important that current penetration be sufficient to reach entirely through the rock sequence being surveyed. Whether pen-

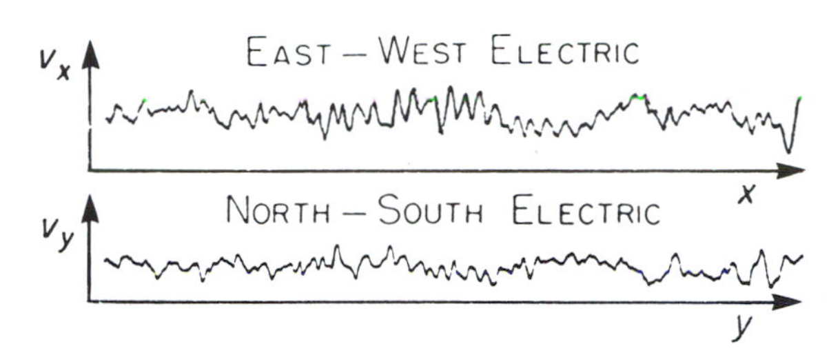

Figure 13–39
Examples of charts obtained during a telluric current survey. They show the variation with time of voltage measured between electrode pairs aligned in north and east directions. (From K. R. Vozoff, R. M. Ellis, and M. D. Burke, *Bulletin of the American Association of Petroleum Geologists*, v. 48, pp. 1890–1901, 1964. Reprinted with permission of the American Association of Petroleum Geologists.)

etration is sufficient can be determined by means of Equation 13–28 and frequencies found from the voltage records such as those in Figure 13–39.

Magnetotelluric Surveying

According to the principle of electromagnetic induction, an alternating electric current has an associated alternating magnetic field. Measurement of the intensity components of this field in the same directions as the components of voltage provides another means of determining the thickness and apparent resistivity of a layered sequence of rocks. Magnetotelluric surveying combines these measurements.

In a zone of thickness z, the apparent resistivity R_a can be estimated from the expression

$$R_a \cong \frac{0.2}{f}\left(\frac{E_x}{H_y}\right)^2 \qquad (13\text{–}31)$$

where f is the telluric current frequency and H_y is the associated magnetic field intensity in a direction perpendicular to the electric field intensity component E_x. The electric field intensity is defined as the change in potential v over a unit increment of distance x:

$$E = v/x \qquad (13\text{–}32)$$

The thickness of the zone is related to the frequency and the apparent resistivity by the

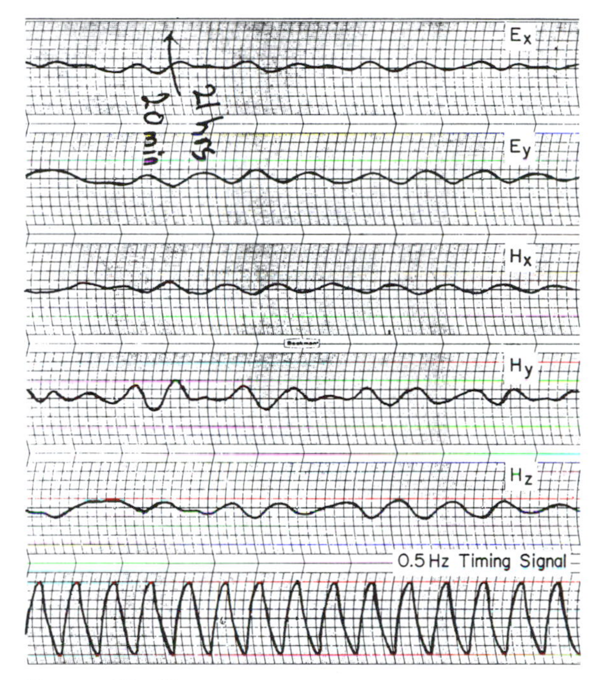

Figure 13–40
Simultaneous records of north and east components of the variation with time in the electric field and the north, east, and vertical components of the variation with time in the alternating magnetic field. Information in these records is used to compute the apparent resistivity by means of Equation 13-31. (K. R. Vozoff, *Geophysics*, v. 37, n. 1, p. 101, February 1972.)

approximation

$$z \cong \frac{1}{2\pi} \sqrt{5R_a/f} \qquad (13\text{–}32)$$

where z is in kilometers when R_a and f are given in ohm-meters and hertz, respectively.

A set of records obtained from measurements of a magnetotelluric sounding is shown in Figure 13–40. Most difficult to measure are the weak magnetic field components that have intensities of a few tens of milligammas. Because conventional magnetometers are insensitive to such weak fields, special instruments must be constructed for this purpose. One limitation of magnetotelluric surveying is the effort required to obtain reliable magnetic intensity records.

The kind of information that can be obtained from a magnetotelluric survey is illustrated in Figure 13–41. Although magnetotelluric surveying is not suitable for detecting discrete thin layers, it definitely indicates a layered structure existing along this profile. Reasonably reliable depths are obtained for the base of the structure. These results show that magnetotelluric surveying can be quite satisfactory for preliminary reconnaissance exploration.

ELECTROMAGNETIC SURVEYING

The electrical surveying techniques most widely used to explore for base metal ores are based on the principle of electromagnetic induction. Rather than electrodes, these techniques make use of coils to compel currents to flow in the ground and to detect their effects. Electromagnetic (EM) surveying is usually done with two coils, one called the *transmitter* and the other the *receiver*. Land-based surveying usually requires two individuals to operate the two coils.

It is possible to mount a coil system on an airplane or helicopter for airborne surveying, a principal advantages of EM methods over electrical methods, which require electrodes. Airborne EM surveying and aeromagnetic surveying are the principal tools for reconnaissance exploration for metallic ores.

The principal value of EM surveying is in targeting zones that are good electric conductors. The EM methods are limited by the depth of the induced current penetration, which is usually not more than a few hundred meters. Because of this depth limitation, EM methods are not used for oil exploration. And because of their insensitivity to resistive and nonconducting structures, they are seldom used for engineering surveys.

The Principle of EM Surveying

To explain how an EM system works, let us suppose that a transmitter coil and a receiver coil are placed at two locations above a body that is a good conductor, as in Figure 13–42. An alternating current flows in the transmitter with a frequency in the range of a few hundred hertz to a few kilohertz. This current induces an alternating magnetic field, called

Figure 13–41
(a) A geologic cross section compiled from drilling data and (b) a resistivity cross section determined from magnetotelluric data across the Anadarko Basin in Oklahoma. The magnetotelluric survey results correlate with a boundary in the depth range of contact between Precambrian crystalline rocks and younger sedimentary rocks. (From K. R. Vozoff, *Geophysics*, v. 37, n. 1, pp. 135, 136, February 1972.)

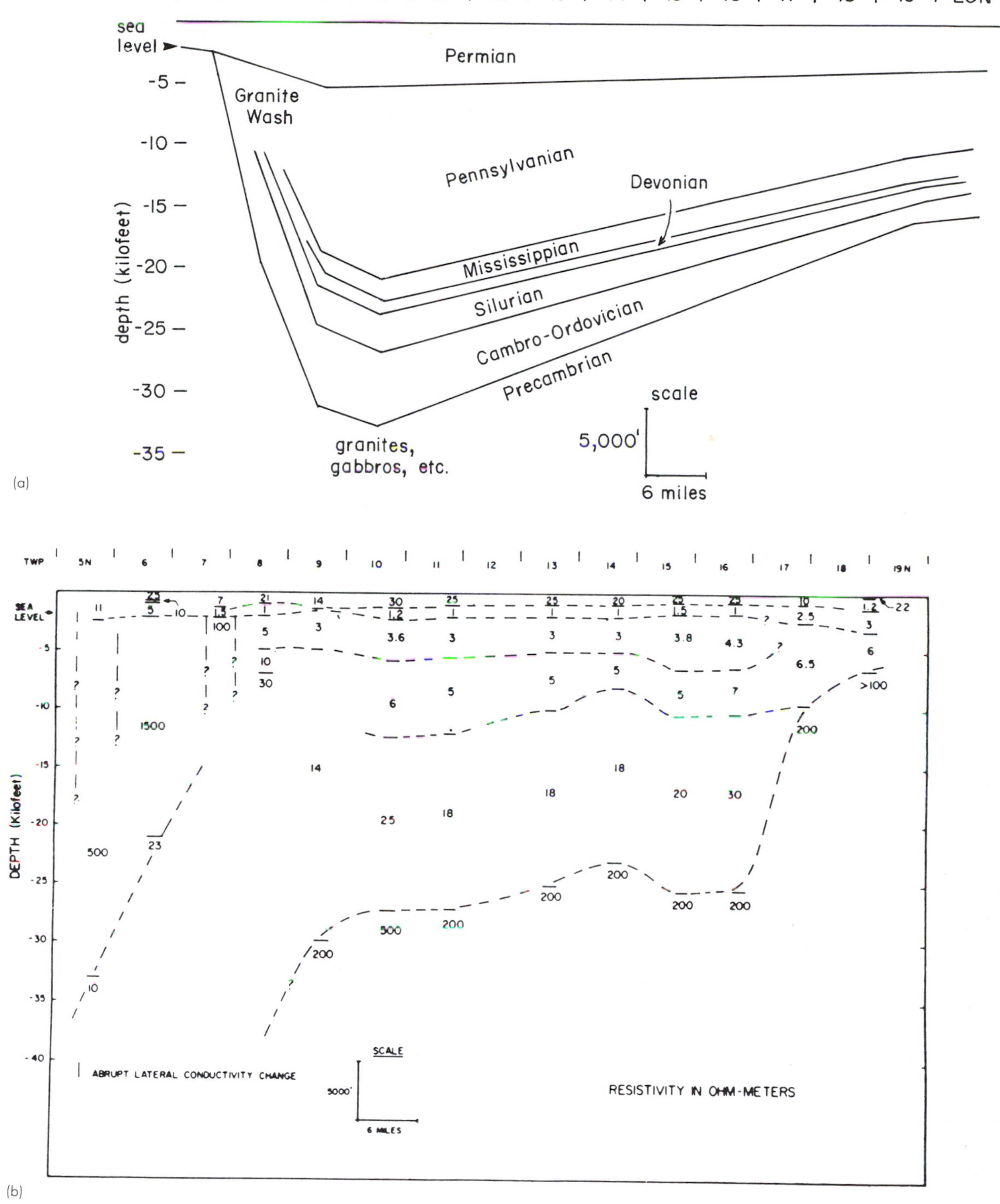

7N
8
9
10
11
12
13
14
15
16
17
18
19
20N
sea level
Permian
Granite Wash
Pennsylvanian
Devonian
Mississippian
Silurian
Cambro-Ordovician
Precambrian
granites, gabbros, etc.
depth (kilofeet)
scale
5,000'
6 miles
(a)
TWP
DEPTH (Kilofeet)
SEA LEVEL
ABRUPT LATERAL CONDUCTIVITY CHANGE
SCALE
5000'
6 MILES
RESISTIVITY IN OHM-METERS
(b)

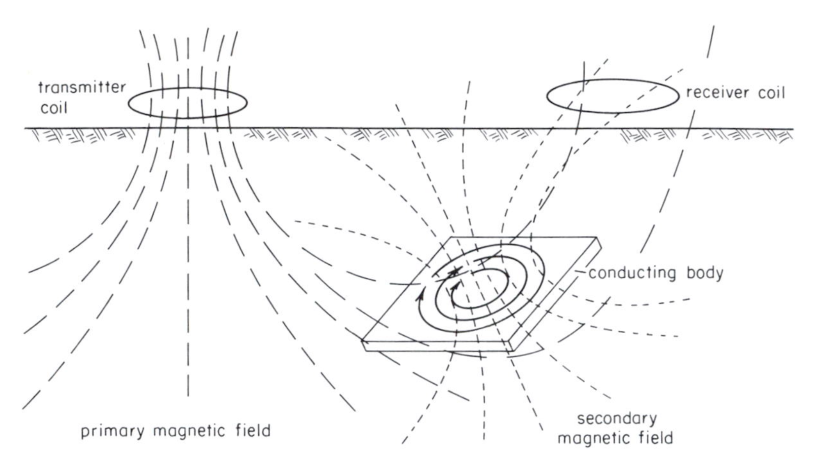

Figure 13–42
Transmitter and receiver coils used for EM surveying, and the corresponding primary and secondary magnetic field lines. The secondary field is induced by current flowing in the conducting body.

the *primary field,* which extends through the surrounding region. The region contains the conducting body and the receiver. The primary field induces an alternating current to flow through a loop within the conducting body. This ground current induces another alternating magnetic field, called the *secondary field,* which also extends through the region that includes the receiver. The complete magnetic field around the receiver, then, is made up the primary and secondary fields.

The combined magnetic field induces an alternating current in the receiver coil. This current, which is measured, can be used to determine the combined magnetic field intensity H_s at the location of the receiver. The usual practice is to compare H_s with a value H_p, which is the intensity of the primary field alone at that location. Because the coil positions and the current in the transmitter coil are known, the value of H_p can be calculated.

EM surveying has many different transmitter–receiver configurations, for measuring various aspects of H_s. For some procedures, the measurements are made with identical coils spaced at a constant interval. Others use coils of different shape and orientation spaced at varying intervals. Coils can vary in size from less than one meter to a few kilometers. Very large coils are formed by placing a loop of insulated wire on the ground in a circular, triangular, or rectangular shape.

We will not attempt to describe and compare all the configurations that are effective for EM surveying, but we will look closely at two of the most widely used procedures. One makes dip angle measurements, which can be done on the ground with lightweight portable equipment. The other, which uses horizontal coils at a fixed interval, is suitable for both ground and airborne surveying.

Parallel-Line Dip Angle EM Surveying

The direction of the combined alternating magnetic field is measured by parallel-line dip angle EM surveying. The transmitter coil (T) is placed in a vertical position in line with the receiver (R_1), which is some distance away as in Figure 13–43. The receiver coil is then rotated on an axis pointing at the source until

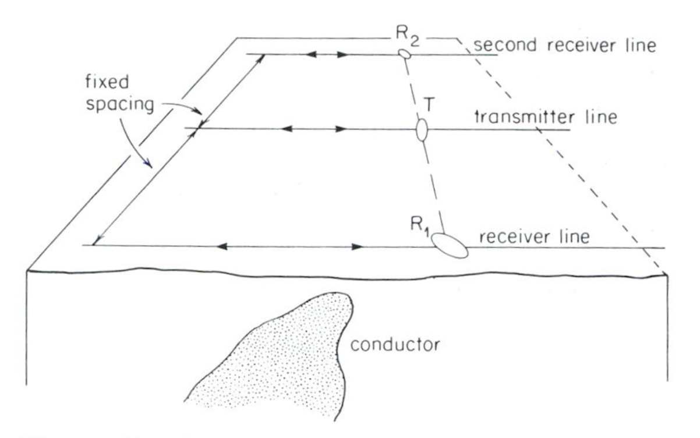

Figure 13–43
Positions of transmitter and receiver coils for parallel-line dip angle EM surveying. After measurements are made from these positions, the coils are moved equal distances along their respective profiles, and the procedure is repeated.

the position is found where no current is induced into it. In this position it is exactly parallel to the direction of the combined magnetic field; that is, it does not cut any of the lines of force. The angle of the receiver coil is then measured.

Next, the transmitter and receiver coils are moved equal distances along parallel profile lines, and another reading is made. The survey proceeds in this way so that readings are obtained at equal intervals all along the profile while the same transmitter–receiver distance is maintained. Sometimes two receiver coils (R_1, R_2) are operated simultaneously along parallel profiles on opposite sides of the transmitter so that the survey can proceed faster, as shown in Figure 13–43.

Because the transmitter coil is in a vertical position, the primary field will be horizontal at the receiver position. Thus, if no conducting bodies are present, no current will be induced in the receiver coil when it is horizontal. Nor will a current be induced if the receiver is situated directly above a conducting body when the secondary field is also horizontal. In other locations, the secondary field will be inclined. Angles of coil inclination along a profile crossing over a conducting body change from positive to negative from one side to the other.

Dip angle EM measurements are usually analyzed by comparing them with characteristic curves for idealized structures. The mathematical operations for expressing these effects are quite complicated and will not be presented here. Characteristic curves in Figure 13–44 show how the dip angle varies over a conductor that is a vertical plane sheet. These curves show that the depth to the top of the conductor can be determined from the range of dip angles, a smaller range corresponding to greater depth. If the depth is very small, the angles measured along a profile can range between ± 30 degrees. However, if the ratio z/L of the depth z and the transmitter–receiver distance L increases to about one-third, the range of angles is less than ± 10 degrees. These curves also show how the results are skewed if the profile is not perpendicular to the strike of the conducting plane.

Many dip angle EM surveys are done simply to locate targets for test drilling or other kinds of geophysical measurements. An indication of the position and strike of a conducting body and an estimate of its depth make up the basic information provided by such a survey.

Horizontal-Loop EM Surveying

For horizontal-loop EM surveying transmitter and receiver coils of approximately the same size are oriented horizontally in the same plane. Readings are made along a profile with the transmitter and the receiver in line and a

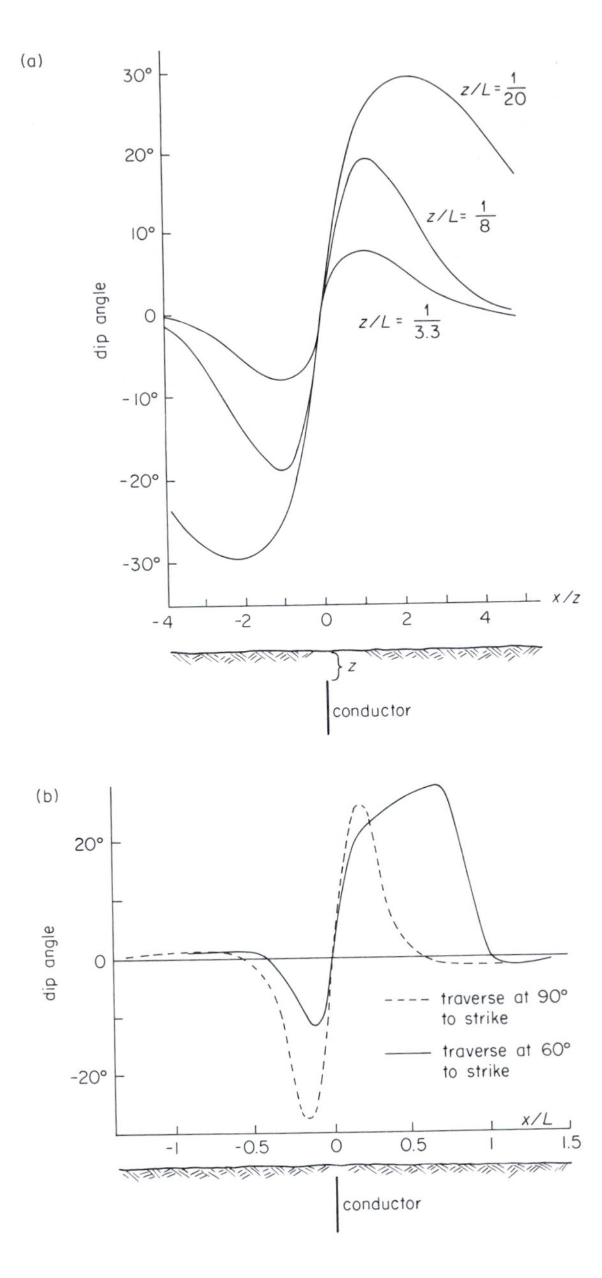

Figure 13–44
(a) Curves showing variation in the dip angle over a vertical sheet conductor at depths of $z/L = 1/20$, 1/8, 1/3.3, where L is the transmitter–receiver spacing. The horizontal distance is in units of depth z to the top of the conductor. (b) Curves illustrating variation of the dip angle along profiles that cross a vertical sheet conductor at different angles. (From W. M. Telford, L. P. Geldart, R. E. Sheriff, and D. A. Keys, *Applied Geophysics*, pp. 560–561, Cambridge, England, Cambridge University Press, 1976.)

fixed distance apart, as in Figure 13–45. For ground surveying, a cable between the two coils is useful for maintaining the same separation of the coils and for conducting a reference signal to synchronize the coils.

As with all EM systems, an alternating current is induced in the receiver coil by the combination of the primary and secondary alternating magnetic fields. This current is used to

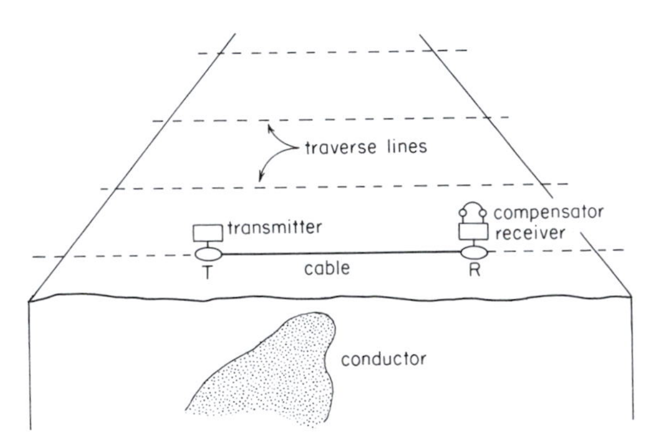

Figure 13–45
Horizontal-loop EM surveying procedure. After a reading is made, the transmitter and receiver are moved farther along the traverse line, another reading is made, and so on. All readings are made with the same transmitter–receiver spacing.

determine the combined magnetic field intensity at a particular location. This measure of field intensity is processed in an interesting way to obtain the reading.

To explain what is done, let us compare the fluctuating field intensity at the receiver with the field intensity of the primary field alone. Figure 13–46 shows that the combined field has a different amplitude, and that its peaks and troughs are shifted with respect to those of the primary field. This shift is called the *phase shift.*

It is possible to reproduce the form of the combined field mathematically by means of two fictitious sine wave curves that have the same frequency but are in phase and 90 degrees out of phase, respectively, with the primary field. This means that the peaks and troughs of one of these fictitious waves, called the *inphase signal,* are in line with those of the primary field. The peaks and troughs of the other fictitious wave, called the *quadrature signal,* are in line with points midway between peaks and troughs of the primary field. The

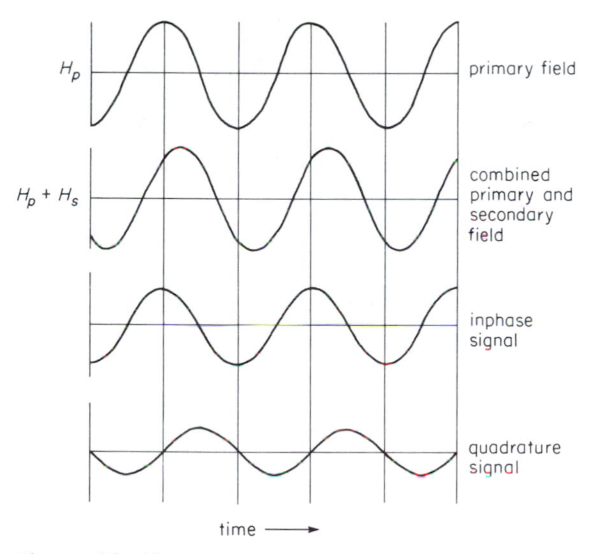

Figure 13–46
Primary and combined primary and secondary alternating magnetic fields at the location of an EM receiving coil, and the inphase and quadrature signals that add together to reproduce the combined field. Amplitudes of the inphase and quadrature signals are measured at each recording site along the traverse lines in the area of the survey.

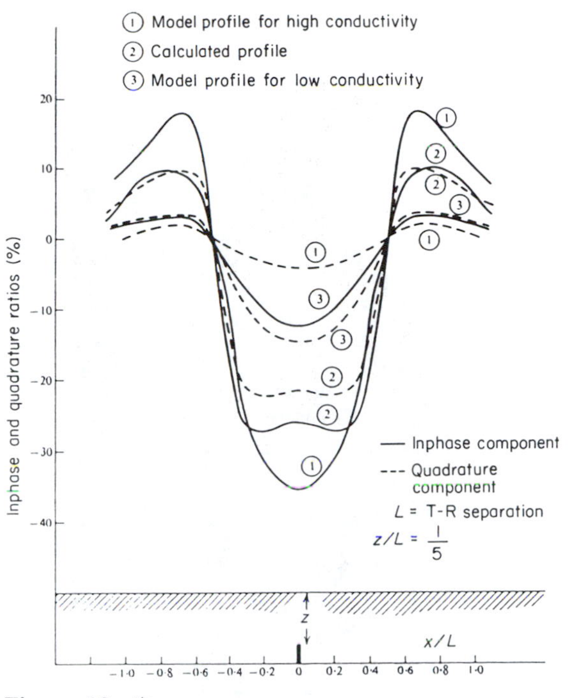

Figure 13–47
Curves showing the variation of inphase and quadrature signals along a profile crossing a semi-infinite vertical sheet conductor at a fixed depth and possessing different values of conductivity. Curves marked 1 and 3 were measured over models, and the curves marked 2 were calculated from theoretical expressions. (From W. M. Telford, L. P. Geldart, R. E. Sheriff, and D. A. Keys, *Applied Geophysics*, p. 566, Cambridge, England, Cambridge University Press, 1976.)

(a)

(b)

Figure 13–48
Coils and electronic controls for ground-based EM surveying: horizontal-loop EM (a) transmitter and (b) receiver. (Courtesy of Gisco.)

inphase and quadrature signals are shown in Figure 13–46.

By means of an electronic analyzing circuit and a primary reference signal conducted through the cable from the transmitter to the receiver, we can determine the amplitudes of the inphase and quadrature signals. These two values make up the reading at a particular location. A succession of readings made in this way are obtained by advancing the transmitter and receiver along a profile, always maintaining the same separation.

Figure 13–49
Transmitter–receiver configurations for airborne EM surveying. The transmitting and receiver coils may be mounted on the wingtips, nose, and tail. (Courtesy of Scientrex, Ltd.)

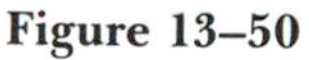

Figure 13–50
Transmitter and receiver instrumentation mounted in a long bird for low-elevation helicopter EM surveying. (Courtesy of Dighem Surveys and Processing, Inc.)

Variations of inphase and quadrature signals are usually analyzed by comparing them with characteristic curves for idealized models of conducting bodies. Figure 13–47 illustrates the patterns of variation over a plane vertical sheet of different conductivities. For simple structures such as vertical or inclined plane sheets, the inphase and quadrature terms can be expressed mathematically. For structures of more irregular shape, the mathematical operations become impractically complicated. Responses obtained from complicated structures can be tested with laboratory models properly scaled to reproduce natural conditions.

Equipment for ground surveying by the horizontal-loop method is lightweight and easily portable. Transmitter and receiver coils about three feet in diameter and the electronic analyzing units can be carried in shoulder harnesses by two individuals. The equipment pictured in Figure 13–48 is typical of that used for ground surveys.

Airborne EM Surveying

A very attractive aspect of EM methods is the opportunity for airborne operations. Different schemes for mounting transmitter and receiver coils on airplanes and helicopters are shown in Figure 13–49. The transmitter coil may be fixed to the aircraft and a receiver mounted in a "bird," which is trailed some distance behind, or both transmitter and receiver may be mounted on the aircraft or in a bird.

Because of the limited depth penetration possible with EM surveying, it is essential to operate the transmitter and receiver as close to the ground as possible. It is also necessary to maintain the same spacing between transmitter and receiver during survey operations. These requirements are much more stringent than those imposed on aeromagnetic surveying.

The earliest airborne EM surveys were done with the transmitter mounted on the aircraft and the receiver trailed in a bird. Because the bird swayed, the receiver distance and angle could not be rigidly maintained. Therefore, it was possible to measure accurately only the quadrature part of the signal. Attempts to improve operations by mounting both elements on the aircraft posed the problems of avoiding electronic noise from the aircraft and maintaining a great enough separation.

Improvements in instrumentation have made it possible to place both transmitter and receiver in a very long bird. In helicopter surveys, this equipment can be trailed close to the ground, as seen in Figure 13–50.

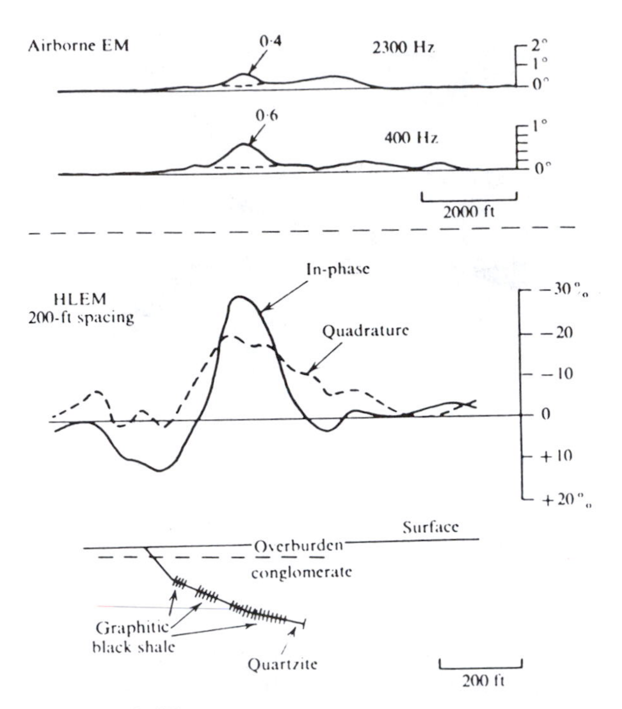

Figure 13–51
Examples of (a) airborne and (b) horizontal-loop EM ground profiles of the inphase and quadrature signals over (c) graphitic ore deposits. (From W. M. Telford, L. P. Geldart, R. E. Sheriff, and D. A. Keys, *Applied Geophysics*, p. 606, Cambridge, England, Cambridge University Press, 1976)

Results of EM surveys are displayed on maps and profiles such as those in Figure 13–51. Both airborne operations and ground surveying have proved to be effective for base metal ore exploration and for delineation of fresh and saline groundwater zones.

STUDY EXERCISES

1. Assume a homogeneous medium of resistivity 120 ohm-m. Using the Wenner electrode system with a 60-m spacing, assume a current of 0.628 ampere. What is the measured potential difference? What will be the potential difference if we place the sink (negative-current electrode) at infinity?

2. Suppose that the potential difference is measured with an electrode system for which one of the current electrodes and one of the potential electrodes are at infinity. Using Figure 13–12 and a current of 0.5 ampere, compute the potential difference between the electrodes at P_A and infinity for $d_1 = 50$ m, $d_2 = 100$ m, $R_1 = 30$ ohm-m, $R_2 = 350$ ohm-m.

3. Suppose that an electrical resistivity survey was done using an expanding Wenner electrode configuration. The current of 0.25 ampere was kept the same for all the readings. Potential differences measured with different electrode spacings are given in the following table. Interpretations of these measurements indicates that a layer of resistivity R_1 lies above another layer of resistivity R_2. Determine the depth of the boundary between these two layers. Estimate the resistivities of the layers.

ELECTRODE SPACING (meters)	POTENTIAL DIFFERENCE (volt)
1	0.800
2	0.420
4	0.280
6	0.155
8	0.125
10	0.120
15	0.105
20	0.100
25	0.098
30	0.086
40	0.076
50	0.064

4. Plot resistivity data as a function of electrode spacing, and determine the particular electrode spacing corresponding to the inflection point for the data given in Exercise 3. Compare the electrode spacing at the inflection point with the depth of the boundary from Exercise 3.

5. If the Schlumberger electrode system with $L/b = 5$, as defined in Figure 13–15, is used to conduct the resistivity survey explained in Exercise 3, what will the potential readings be? Use resistivity values found in Exercise 3 to compute potential differences for each of the electrode spacings L = 1, 2, 4, 6, 8, 10, 15, 20, 25, 30, 40, 50 m while a constant current of 0.250 ampere is applied.

6. For purposes of an IP survey, resistivity values are determined from both direct and alternating current using the same electrode arrangement. If the resistivities for direct and alternating current are $R_{dc} = 50$ ohm-m and $R_{ac} = 40$ ohm-m, respectively, what will the frequency effect and the metal factor values be?

7. Assume that a telluric current survey is to be carried out to outline large-scale features of a sedimentary basin 5 km deep. A resistivity of 50 ohm-m is supposed to represent the sedimentary section. What is the maximum frequency of the telluric current that will penetrate below the basin?

8. Suppose that a magnetotelluric survey indicates an apparent resistivity of 5 ohm-m at a frequency of 1 Hz. What is the thickness of the layer?

SELECTED READING

Dobrin, M. B., *Introduction to Geophysical Prospecting,* 3rd edition: New York, McGraw-Hill, 1976.

Griffiths, D. H., and R. F. King, *Applied Geophysics for Geologists and Engineers.* Oxford, England, Pergamon Press, 1981.

Kearey, P., and M. Brooks, *An Introduction to Geophysical Exploration.* Oxford, England, Blackwell Scientific Publication, 1984.

Keller, G. V. and F. C. Frischnecht, *Electrical Methods in Geophysical Prospecting.* Oxford, England, Pergamon Press, 1966.

Koefoed, O., *Geosounding Principles 1—Resistivity Sounding Measurements.* Amsterdam, Elsevier, 1968.

Parasnis, D. S., *Mining Geophysics.* Amsterdam, Elsevier, 1973.

Parasnis, D. S., *Principles of Applied Geophysics,* 4th edition. London, Chapman and Hall, 1986.

Sharma, P., *Geophysical Methods in Geology.* Amsterdam, Elsevier, 1986.

Summer, J. S., *Principles of Induced Polarization for Geophysical Exploration.* Amsterdam, Elsevier, 1976.

Telford, W. M., L. P. Geldart, R. E. Sheriff, and D. A. Keys, *Applied Geophysics.* Cambridge, England, Cambridge University Press, 1976.

Wait, J. R., *Geo-Electromagnetism.* New York, Academic Press, 1982.

Yungul, S. H., Spontaneous polarization survey of a copper deposit at Sariyer, Turkey, *Geophysics,* v. 19, n. 3, pp. 455–458, June 1954.

FOURTEEN

Geophysical Well Logging

We can gain direct access to rocks beneath the earth's surface by drilling. The best samples are obtained by core drilling, but this method is very costly and is seldom practical for depths of more than a few thousand feet. In addition, the conventional methods of deep drilling produce rock cuttings that are flushed from the borehole by drilling fluids. Because these rock cuttings become mixed and contaminated by the drilling fluid, they can be difficult to interpret. Furthermore, these broken fragments tell us little about the porosity and permeability of the unbroken matrix of rock, or the natural fluids that exist in the pore space. Therefore, instruments have been developed for measuring certain rock properties directly, while they are lowered in a borehole. The operation of such instruments is called *well logging,* and the strip chart on which the measurements are plotted is called a *well log.*

Geophysical well logging was introduced in 1928 by two brothers, Conrad and Marcel Schlumberger. They devised a probe that could be lowered in a well, continuously measuring electrical resistivity as it moved past different rock formations. A strip chart, the electrical well log, was produced while the operation was underway. Variations in resistivity indicated the contacts between different lithologic units more clearly than the rock cuttings obtained while the well was being drilled.

In the decades that followed, the inventory of well logging instruments expanded. The first improvements increased the variety of electrical measurements. Resistivity was measured using different electrode configurations, and natural voltages were recorded. Several values could be recorded simultaneously by means of a probe, called a *sonde,* on which several electrodes were mounted. These values could be analyzed in different combinations to estimate formation porosity and to make inferences about the natural fluids filling the pore space.

Later improvements included the addition of devices to log radioactivity. These tools detect levels of natural gamma radiation, and they measure radiation produced by bombarding a formation with atomic particles. At about the same time, other logging tools were developed for measuring seismic wave speeds close to the wall of the well. This is called sonic logging. Different combinations of radioactivity and sonic logs have proved very useful for determining lithology, density, and porosity. On a more limited scale, borehole gravimeters and magnetometers have been used in well logging operations.

In this chapter, we begin with a discussion of well drilling to point out certain factors that must be considered in interpreting well logs. Then we will describe the conventional well logs and introduce methods for interpreting them. The general objective of well logging is called *formation evaluation.* Well logs are used to discover as much as possible about the properties of rock formations penetrated by a drill.

WELL DRILLING

Most modern well drilling is done with rotary drills and with percussion drills. The rotary drill can be used for both shallow and deep wells. The record depth reached with a rotary drill is more than 40,000 feet. A percussion drill is faster, but it cannot be operated if the well becomes flooded by groundwater while drilling is in progress. This problem restricts its use for the most part to depths of less than a few thousand feet.

Rotary Drilling

The basic parts of a rotary drill are illustrated in Figure 14–1. Drilling is done with a bit attached at the bottom of a column of round hollow pipe. The top of this column connects to a square hollow bar, the *kelly,* which fits through a square hole in the rotary table. An engine turns the rotary table, causing the entire column to rotate at speeds of up to 300 rpm. The teeth on the rotating bit grind up the rock, working the hole deeper into the earth.

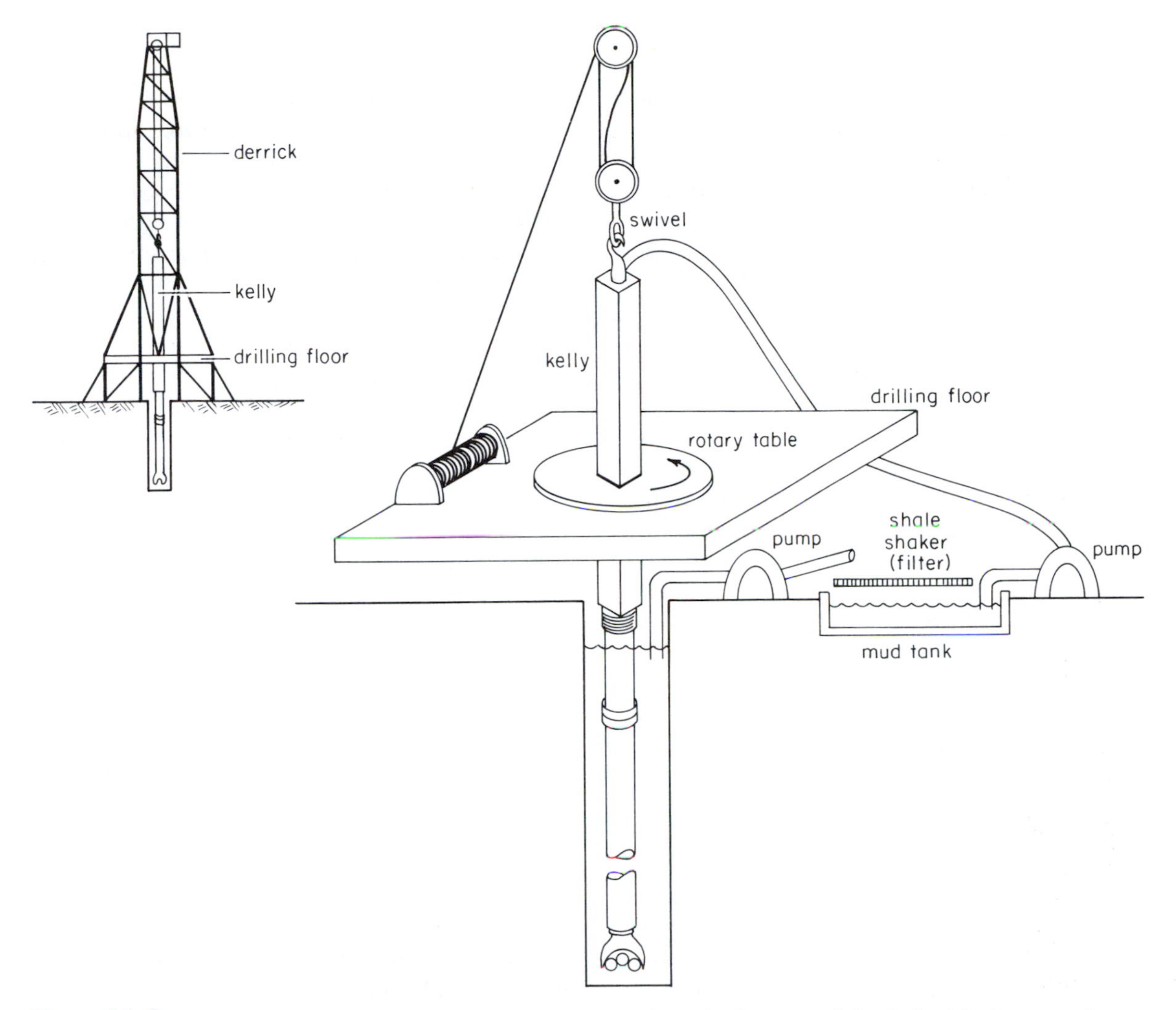

Figure 14–1
Principal parts of a rotary drill. The power system turns the rotary table, which then rotates the entire column of drilling pipe. The bit grinds the rock at the bottom of the hole. Mud pumped down through the column of pipe then rises in the hole, flushing out the rock cuttings.

At the same time drilling fluid, stored in the mud tank, is pumped down the column of pipe. It flows out through holes in the bit and rises in the well, carrying the rock cuttings up to the land surface. On the surface the cuttings are filtered out, and the fluid returns to the mud tank.

As the well deepens, additional sections of round pipe, threaded at the ends, are inserted beneath the kelly to lengthen the column.

Drilling continues until the bit becomes dull. To replace it requires that the entire column be raised by means of the cable and pulleys hooked to the swivel at the top of the kelly. The column is lifted, unscrewed section by section, and stacked in the derrick tower. A new bit is attached and the column is reassembled, after which drilling resumes. The task of disassembling the pipe, changing the bit, and reconnecting the pipe is called "making a trip." In a deep well, this operation can require at least several hours and perhaps more than a day.

Drilling fluid serves several purposes. We have already mentioned that it carries rock cuttings out of the well. For shallow depths water itself is suitable, but in deeper wells a more viscous fluid is desirable to hold the cuttings in suspension. The fluid also lubricates the rotating column of pipe, and it cools the bit. Another important function of the fluid is to seal the rock surface so that pore fluids cannot seep into the well. In this capacity, the drilling fluid acts to prevent gushers and blowouts that would otherwise occur when the bit penetrates high-pressure oil and gas-bearing zones. To seal the rock surface effectively, the drilling fluid must exert pressure against it that exceeds the pressure of the groundwater, oil, or gas which fills the pores in the rock.

Pressure on the wall of the well depends on the weight of the drilling fluid, which increases with depth. To increase the weight at some particular depth, we must increase the density of the drilling fluid. Particles of different substances are added to turn the drilling fluid into a thicker and more viscous mud. Considerable thought goes into the choice of additives to achieve the desirable weight, viscosity, and lubrication.

By sealing the rock surface, the drilling fluid acts to hide the nature of the pore fluids that would otherwise seep or gush into the well. Oil- and gas-bearing zones can be so completely sealed that we have no direct indication they have been drilled into. One of the principal objectives of well logging is to help detect the presence of these hydrocarbons after the well has been drilled. Not only are the natural pore fluids prevented from seeping into the well, but because of its higher pressure the drilling fluid itself seeps into the rock, displacing the natural pore fluids.

The zone surrounding the well in which drilling fluid has infiltrated is called the *zone of invasion;* its structure is illustrated in Figure 14–2. At the side of the well where drilling fluid first begins to seep into the rock, most of the additive particles are filtered out. Only the remaining filtrate with its dissolved constituents passes into the rock. Close to the well the filtrate completely replaces the natural pore fluids, creating the *flushed zone.* Beyond is the *annulus of invasion* where the proportion of filtrate diminishes gradationally with distance from the well. As seepage progresses, the additive particles build up a coating on the side of the well called the *mud cake.* It thickens until it forms an impermeable barrier which prevents further invasion.

In relatively impermeable rock such as shale, the entire zone of invasion reaches only a few inches into the rock. Porous and permeable sandstones and carbonates may be invaded to distances of several feet or more. The mud cake may reach a thickness of a few tenths of an inch, and the flushed zone can extend several inches to more than a foot into the rock. To "see through" the effects of invasion into the undisturbed formation requires combinations of several well logging devices.

Coring is done using a rotary drill equipped with a special hollow and tubular bit

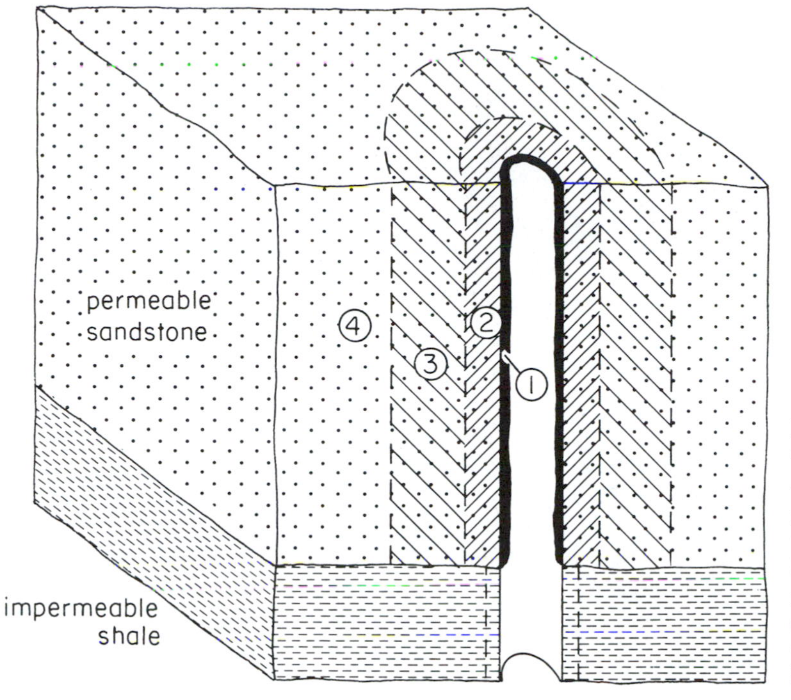

Figure 14–2
The zone of invasion includes (1) the mud cake, (2) the flushed zone, and (3) the annulus of partial invasion. Beyond the zone of invasion is (4) the undisturbed formation. The zone of invasion extends much farther from the well in permeable sandstone than in relatively impermeable shale.

(Figure 14–3). The end of the bit is studded with fragments of industrial diamond or some other very hard substance. As the bit turns, it cuts a tubular hole around a solid core of rock which moves into a container, the *core barrel,* inside the column of drill pipe. When the core barrel is full, it is retrieved by "making a trip." This process is carried out much more frequently in core drilling than in noncore drilling, because of the limitation imposed by the length of the barrel, which is ordinarily less than 20 feet. In addition, drilling must proceed more slowly and carefully to minimize core breakage. For these reasons, core drilling is very expensive compared with other noncoring operations. The core drill produces the best rock samples, but the natural formation fluids are flushed out by the drilling field.

Percussion Drilling

Modern percussion drilling is done with a hammer activated by an air compressor. The principle is not unlike that of a jack hammer commonly used to break up pavement. The drill rig includes a hammer bit, illustrated in Figure 14–4, mounted at the bottom of a column of pipe. Air from a compressor flows through a high-pressure hose into the top of the column and then down the column to drive the hammer.

A simplified sketch (Figure 14–4) illustrates the principal features of a bit assembly; it consists of a hammer attached to a piston that moves up and down in a cylinder. Compressed air, pumped into the top chamber, forces the piston downward. This action com-

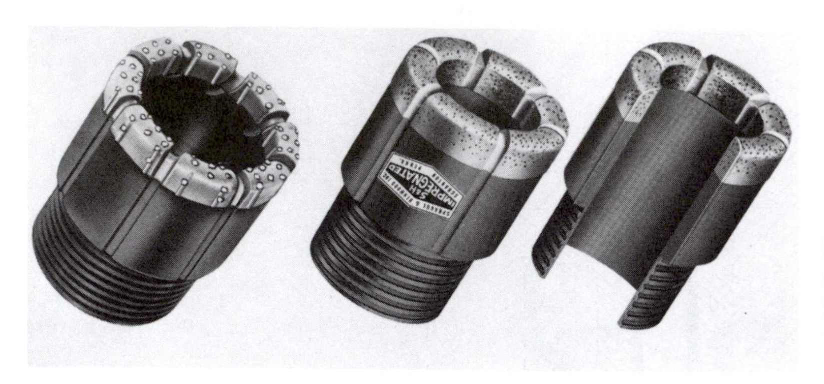

Figure 14–3
Coring bits studded with industrial diamond fragments. (Courtesy of Sprague and Henwood, Inc.)

presses the air that is sealed in the bottom chamber. When the piston is pushed down far enough, the compressed air in the top chamber escapes through holes in the side of the cylinder. Upon release of pressure in the top chamber, the compressed air in the bottom chamber expands, driving the piston upward. This closes the holes, allowing pressure to build up once again in the top chamber. The cycle is repeated several times per second. The rapid pounding of the hammer, which is studded with knobs of tungsten carbide or other hard materials, pulverizes the rock into a fine dust. Air escaping from the upper chamber blows the dust out of the well.

As drilling progresses, lengths of pipe, threaded at the ends, are added to the column. Eventually, the studding on the hammer wears away, and it is necessary to "make a trip" to replace the bit. Percussion drilling usually proceeds faster than rotary drilling. But a depth is reached beyond which the escaping air can no longer remove the rock dust. The depth that can be drilled is also limited by the effect of groundwater seepage, which obstructs the upward flow of air and rock dust.

Casing

Several difficulties in drilling and maintaining wells are overcome by means of *casing.* A well is cased by lining it with a column of pipe, which is lowered into the hole after the drilling pipe has been removed. Cement is then pumped into the well to fill the space outside the casing pipe and to fix it solidly in place. This seals the well, preventing natural formation fluids from seeping into it and keeping loose rock from collapsing and blocking the hole. After the casing has been set, it can be perforated at particular depths to allow water from an aquifer, or oil and gas from reservoir rock, to flow into the well. The casing must sometimes be set before a well is completely drilled, for example, when groundwater seepage floods the well so that a percussion drill cannot be operated. In rotary drilling problems with collapsing rock and excessive loss of drilling fluid into fracture zones can be solved by casing. After the casing is set, drilling can resume with a smaller bit that will pass through the casing pipe.

Casing creates some serious problems for well logging. Electric logs cannot usually be run in wells with steel casing. Sonic logs can

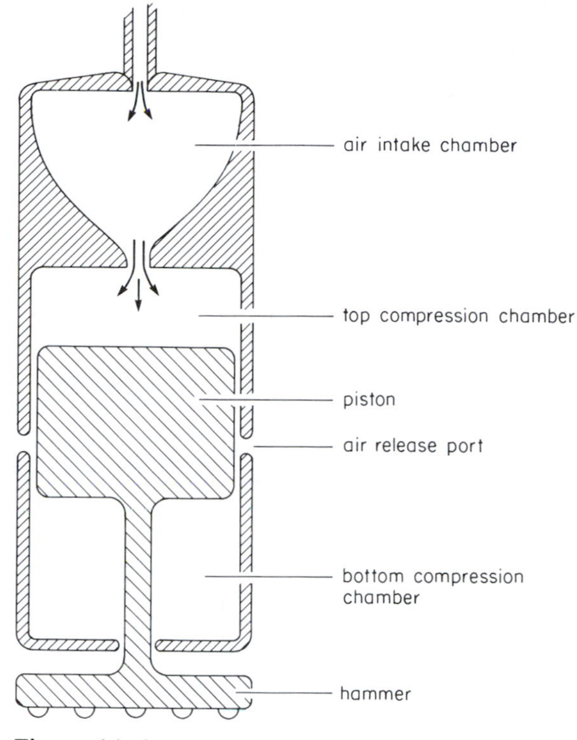

Figure 14–4
Schematic drawing of a bit assembly used for percussion drilling. The hammer is driven by compressed air, which is pumped down through the column of drilling pipe into the air intake chamber. Rock cuttings are blown out of the hole by compressed air escaping from the air release ports.

also be severely distorted. But, certain radioactivity logs can be run effectively where casing prevents other kinds of logging.

FORMATION EVALUATION

The properties of a formation that can be estimated from well log measurements include lithology and bed thickness, porosity, permeability, and the proportions of water and hydrocarbons occupying the pore space. What kinds of information are needed to determine these properties? We will answer this question first, before describing the logging instruments and techniques used to obtain this information.

Lithology and Bed Thickness

Electrical resistivity, natural voltage differences, natural radioactive emissions, and seismic wave speed can all help us to distinguish different lithologies. None of these properties uniquely characterizes particular kinds of rock, however. We already know that there is some overlap in the ranges of values measured in typical sandstones, shales, limestones, and other lithologies. Nevertheless, distinctive differences usually exist for particular sequences of interbedded lithologies and for particular areas. For example, in a section of alternating sandstone and shale beds, values of natural voltage, radioactivity, and *P*-wave travel time over a fixed interval tend to be high in shale and low in sandstone. These patterns (Figure 14–5) exist almost everywhere, even though the actual values measured in one area can be quite different from those in another area.

Cuttings produced by drilling tell us, initially, what kinds of rock are penetrated by the well. Because of the mixing that occurs while these cuttings are carried in suspension in the drilling fluid, they do not provide precise information about bed thicknesses and the locations of bed boundaries. But they do suggest what patterns we should look for, such as those in Figure 14–5, so that we can use logs for accurate measurement of these boundaries and thicknesses.

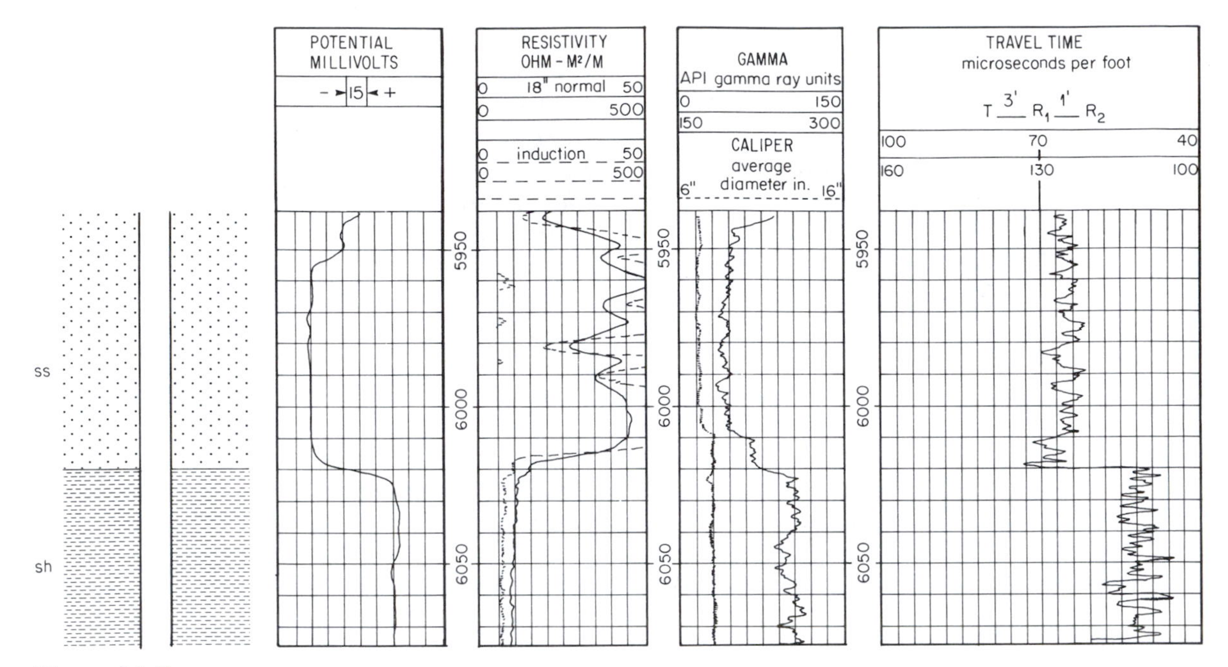

Figure 14–5
Representative well logs illustrating the variation across the boundary between sandstone and shale layers of spontaneous potential (natural voltage), electrical resistivity, natural gamma radiation, and sonic transit time (*P*-wave travel time) between two receivers one foot apart.

The usual practice is to interpret lithology in a well by comparing logs from that well with reference logs from another well in the area in which lithology is already known from careful study of cuttings and, perhaps, rock cores. Patterns of variation corresponding to these lithologic units are recognized by visual correlation of the logs (Figure 14–6). It is useful to know that peaks and troughs on well logs are not usually recorded exactly at bed boundaries. Rather, these boundaries are located by more subtle inflections between the high and low points on a log. The reasons for these subtle indications are discussed in a later section.

For correlation purposes, it is usually not necessary to understand exactly what produces the patterns of variation that characterize different rock layers. More important is the ability to recognize the particular point within a pattern that marks a layer boundary. Because different types of logging devices record different patterns of variation, it can be quite difficult to correlate, say, a resistivity log from one well with a radioactivity log from another well. Accurate identification of the same layer boundary in different wells is best done by comparing logs recorded with the same kind of instrument. No particular kind of log turns out to be generally superior. In some

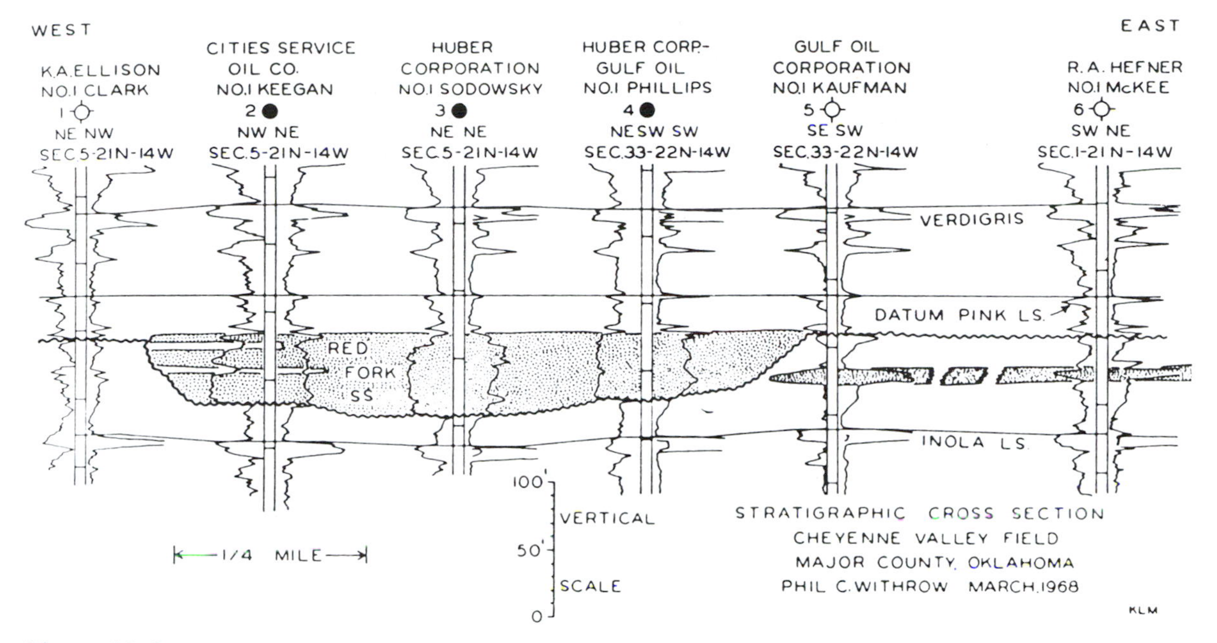

Figure 14–6
Correlation of formation boundaries between different wells based on similar patterns of variation identified on logs from these wells. (From Phillip C. Withrow, Depositional environments of Pennsylvanian Red Fork Sandstone in Northeastern Anadarko Basin, Oklahoma, *Bulletin of the American Association of Petroleum Geologists,* v. 52, n. 9, pp. 1638–1654, September 1968.)

areas, resistivity logs are easiest to compare, but in other places the correlation patterns are clearer on radioactivity logs.

Porosity

In a sample of a rock formation, the proportion of the volume that consists of openings between grains, along fractures, and in cavities is the *porosity*. The porosity of reservoir rocks containing petroleum, natural gas, or fresh water is especially important because it indicates the volume of fluid that can be stored in the rock. Methods for estimating porosity from well log measurements are based mostly on experiments that have tested electrical resistivity, *P*-wave travel time, and nuclear emissions in rock specimens for which the porosity is known from independent measurements.

Insofar as the pore fluid resistivity differs from the solid matrix resistivity, we can expect a relationship between formation resistivity and porosity. In 1942 G. E. Archie proposed such a relationship. It was determined from laboratory measurements of the resistivities of sandstone cores that had been saturated with water containing different concentrations of dissolved NaCl. The porosities of these cores were measured by a method independent of resistivity. Archie introduced a property called the *formation resistivity factor* F, which depends on the formation resistivity R_0 and the pore fluid resistivity R_w:

$$F = R_0/R_w \quad (14\text{–}1)$$

He then plotted values of F and porosity φ that had been found for the different cores on a logarithmic graph (Figure 14–7). Because a straight line could be fitted to this logarithmic plot, he recognized that φ and F could be related by an equation of the form

$$\varphi = aF^{-m} \quad (14\text{–}2)$$

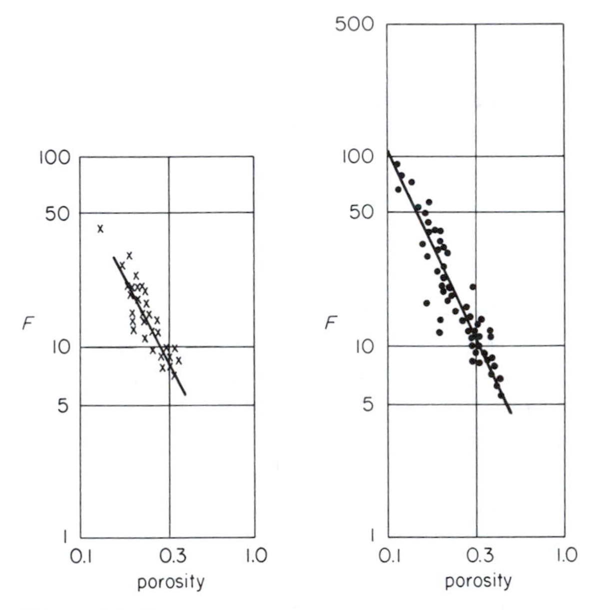

Figure 14–7
Graphs of the variation of formation factor (F) with porosity (φ), based on values measured in sandstone cores from two regions. (From G. E. Archie, *Transactions of the AIME,* v. 146, pp. 54–62, 1942.)

where m is a constant that depends on the cementation of the reservoir rock and a is a constant that must be found by testing the rocks in an area of interest. This equation, which is known as the Archie formula, has become the most widely used equation in formation evaluation. Tests for many different areas indicate that values for a are mostly between 0.62 and 1.0, and values for m are mostly between 2.0 and 3.0.

Another property of a formation that is related to porosity is the compressional wave speed, which well loggers usually call the sonic wave speed. The sonic wave speed (V_w) through the pore fluid is usually quite different from speed (V_m) through the solid matrix. Because the sonic wave speed (V_0) that is characteristic of the formation depends on the proportions of these two parts, it can be related to porosity:

$$\frac{1}{V_0} = \frac{\varphi}{V_w} + \frac{1 - \varphi}{V_m} \quad (14\text{–}3)$$

The formation wave speed (V_0) can be found from the travel time (T_0) between two points that are a known distance apart (Equation 2–8). Because these points remain a fixed distance apart on a sonic logging instrument, the porosity can be related directly to the travel time,

$$T_0 = \varphi T_w + (1 - \varphi)T_m \quad (14\text{–}4)$$

where T_w and T_m are travel times along the same distance through a fluid in which the wave speed is V_w and through a solid in which the wave speed is V_m. This formula can be rearranged to find the porosity:

$$\varphi = \frac{T_0 - T_m}{T_w - T_m} \quad (14\text{–}5)$$

The last porosity relationship that we will consider concerns the formation density ρ_0.

This physical property depends on the proportion of pore fluid, which has the density of ρ_w, and the proportion of solid matrix, which has the density of ρ_m. Therefore, it can be related to the porosity:

$$\rho_0 = \varphi\rho_w + (1 - \varphi)\rho_m \quad (14\text{–}6)$$

This equation can be rearranged to obtain

$$\varphi = \frac{\rho_m - \rho_0}{\rho_m - \rho_w} \quad (14\text{–}7)$$

We know that density of formation water is close to 1 g/cm^3, and that the matrix density is typically between 2.6 and 2.7 g/cm^3. Radioactivity can be used to estimate the formation density in two ways. The first employs a device for bombarding the formation with electromagnetic photons, which become scattered by collisions with electrons in the formation. As a result of the scattering process, a certain proportion of these photons are returned to a detector. This proportion depends on the concentration of electrons, hence the density (ρ_0) of the formation. The other method bombards the formation with neutrons which become absorbed into atomic nuclei in the formation; this causes the nuclei to emit gamma rays. The level of gamma radiation reaching a detector depends on the formation density. Values of ρ_0 measured in these ways can be used in Equation 14–7 to calculate porosity.

Water and Hydrocarbon Saturation

The pore space in rocks is almost everywhere filled with water. Except at shallow depths, this water ordinarily contains some dissolved NaCl and other ions. Because of these dissolved constituents, the water becomes a good electric conductor with low resistivity. In contrast, the common rock-forming minerals, making up the solid matrix of a formation, are poor conductors with high resistivity. Consequently, the flow of electric current in a formation is almost entirely through the pore water.

Very sparsely distributed in the earth's sedimentary rocks are zones containing petroleum and natural gas. These zones are the elusive targets of the exploration geophysicist. Petroleum and natural gas, like the rock-forming minerals, are poor conductors with high resistivity. Therefore, when these hydrocarbons percolate into a reservoir rock displacing some of the pore water, the electrical resistivity of the formation increases.

A typical reservoir rock contains a mixture of water and hydrocarbon fluids. Because of their lower densities, the hydrocarbons tend to migrate upward, becoming more concentrated in the upper part of the reservoir. In no place is the water completely replaced, but it does diminish gradationally to an irreducible minimum. Therefore, a reservoir containing hydrocarbon fluids tends to have a gradational upward increase in electrical resistivity.

For purposes of estimating potential production from a well, it is useful to determine the proportions of water and hydrocarbons in different parts of the reservoir by means of resistivity measurements. In 1942 G. E. Archie proposed a formula for finding the water proportion, otherwise called the *water saturation (S)*. It is based on laboratory measurements of the resistivities of sandstone cores containing different proportions of petroleum and of water having a fixed salinity. Logarithmic plots of S and the fraction R_0/R_t could be fitted by straight lines, indicating that

$$S = (R_0/R_t)^{1/n} \quad (14\text{–}8)$$

where R_t is the resistivity of rock containing a mixture of hydrocarbons and water, and R_0 is the resistivity of the same rock matrix completely saturated with water. Because the formation factor F is defined in terms of rock that is completely water-saturated, the formula can also be written as

$$S = (FR_w/R_t)^{1/n} \qquad (14\text{–}9)$$

where R_w is the resistivity of the water that partly fills the pore space. In a reservoir where the water saturation (S) decreases gradationally upward, the value of R_0 must be measured from well log readings in the lower part that is completely saturated with water.

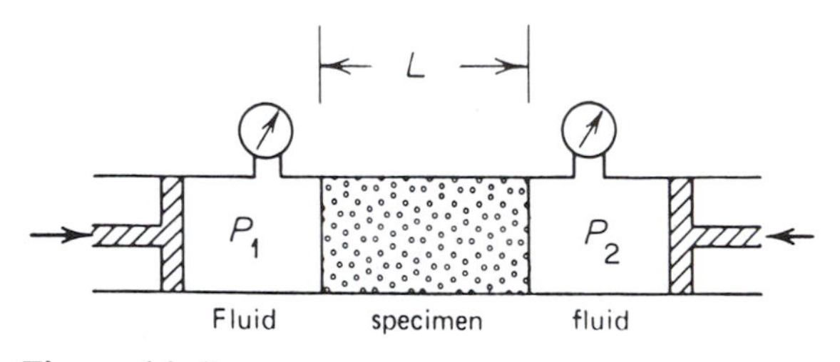

Figure 14–8
Schematic diagram of the apparatus for measuring permeability in rock cores. Pushing in the plunger on the left and drawing out the plunger on the right makes the pressure P_1 in the left chamber higher than the pressure P_2 in the right chamber. This pressure difference causes fluid to flow through the specimen from the left chamber to the right chamber.

Permeability

The capacity of a formation to transmit fluid is called its *permeability*. It depends on the extent to which the pores are interconnected and on the diameter of the openings. It is the most difficult property to estimate from well log measurements.

The permeability of a core sample of rock can be measured with the apparatus shown schematically in Figure 14–8. Fluid is forced to flow through the specimen by applying different pressures at either end. The permeability (k) can be expressed in terms of the discharge (q, volume per unit of time) through the core, the length (L) of the core, the pressure difference $(P_1 - P_2)$, and the viscosity (μ) of the fluid:

$$k = \frac{q\mu L}{P_1 - P_2} \qquad (14\text{–}10)$$

The permeability has units of area, the most common being the *darcy*, which has as its value 1 darcy = 0.987×10^{-8} cm^2; and the *millidarcy*, with the value 1 darcy = 1000 millidarcies.

Using permeability measurements in sandstone cores, G. E. Archie showed that values of k and the corresponding formation resistivity factors (F) plotted close to a straight line on a logarithmic graph (Figure 14–9). However, other factors such as the capillary force, which depends on the diameters of the channels of interconnected pores, also influence k. These other factors cause the inclination of such a straight line to change too much from place to place to obtain a reliable relation depending only on k and F.

Additional experimentation has revealed that capillary force effects can be represented in terms of the capacity of the rock to retain water. We know that migrating hydrocarbon fluids are able to displace all but an irreducible minimum amount of water. This remaining fraction of water is called the *irreducible water saturation* (S_{irr}). The value of S_{irr} can be estimated from resistivity measurements through a reservoir formation in which water saturation is decreased to the irreducible minimum by hydrocarbon displacement.

A relation between permeability and the formation resistivity factor is evident in Figure

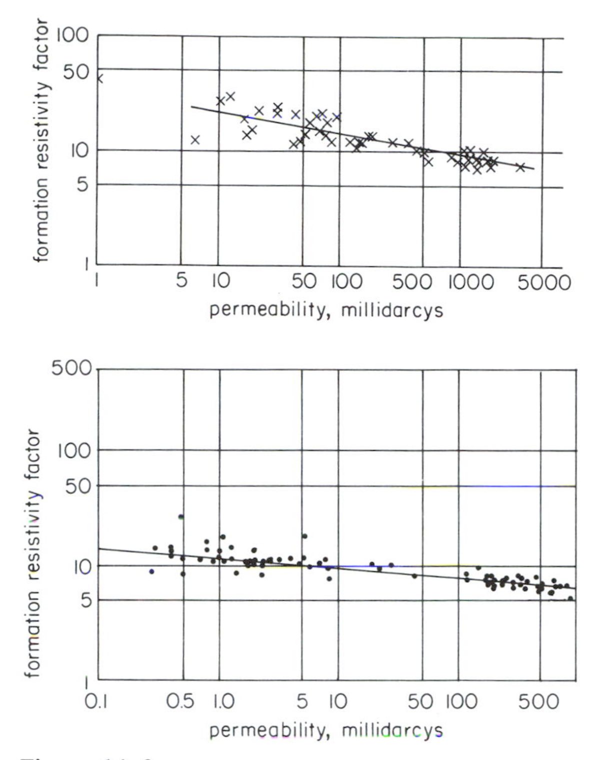

Figure 14–9
Graphs of the variation of formation factor (F) with permeability (k), based on values measured in sandstone cores from two regions. (From G. E. Archie, *Transactions of the AIME*, v. 146, pp. 54–62, 1942.)

14–9. We know from Equation 14–2 that the formation resistivity factor also depends on porosity. These two facts suggest that permeability is influenced by porosity. Measuring permeability in rock cores having different porosity and water saturation tests the combined effects of porosity and capillary force. Graphs of the results indicate the following approximate relation,

$$k \cong (C\varphi^3/S_{irr})^2 \qquad (14\text{–}11)$$

where the constant C depends on the lithology and average grain size of the rock. The problem with using this formula to estimate permeability is that values of φ, S_{irr}, and C must first be estimated from well log measurements. Any error in φ will be multiplied sixfold, and errors in S_{irr} and C will be multiplied twofold in the estimate of k. These facts point out why permeability is a difficult property to measure.

We have now introduced some simple formulas that are used for quantitative formation evaluation. To use these formulas, we must be able to measure electrical properties, radioactivity, and sonic wave speeds in a formation. We turn now to the well logging practices that provide this information.

ELECTRIC LOGGING

Electric logging is the procedure for measuring the resistivity and the natural electric potential in the formations penetrated by a well. The measurements are made with a device, the sonde, which moves in the well suspended on a wire line. Electrodes on the sonde are connected in different circuits that are designed to test the resistivity in different zones of the formation and to detect the natural potential, more commonly called the *spontaneous potential* (SP). It is necessary to test the resistivity in different zones to correct for the effects of invasion by a drilling fluid that has a resistivity different from that of the natural pore fluid. We now describe each of the circuits used in conventional electric logging operations.

Special Forms of Ohm's Law

The basis for resistivity logging is Ohm's law, which was introduced in Chapter 13. Look again at Equation 13–8, which is the form of

Ohm's law used in the analysis of resistivity measurements made with the electrodes placed on the land surface. Recall that in a homogeneous material, current spreads out from the source electrode through concentric hemispherical shells (Figure 13-3). The factor 2π in Equation 13–8 originates from the expression of the area of a hemispherical surface of radius r, which is $2\pi r^2$.

The situation will be different if the electrodes are placed underground so that current can spread out through concentric spherical shells (Figure 14–10) rather than hemispherical shells. Because the area of a sphere of radius r is $4\pi r^2$, the factor 2π in Equation 13–8 must be changed to 4π in the equation we use in the analysis of resistivity measurements made in wells. Otherwise, the same reasoning is used to obtain the following expression of Ohm's law,

$$v_{\mathrm{MN}} = \frac{IR}{4\pi}\left(\frac{1}{\mathrm{AM}} - \frac{1}{\mathrm{AN}} - \frac{1}{\mathrm{BM}} + \frac{1}{\mathrm{BN}}\right) \tag{14–12}$$

where v_{MN} is the change in potential between two electrodes M and N, I is the current flowing between a source electrode A and a sink electrode B, R is the resistivity of the material in which the electrodes are embedded, and AM, AN, BM, and BN are distances between the electrodes (Figure 14–11).

Let us consider the zone that is tested with the electrode arrangement in Figure 14–11. Observe that electrodes M and N are situated on two equipotential surfaces. The potential difference v_{MN} is related to the flow of current I and the resistivity R everywhere in the shaded zone between these two surfaces. If the electrodes M and N were placed at any other points on these surfaces, the same value of v_{MN} would be measured. The entire shaded zone between these equipotential surfaces is tested.

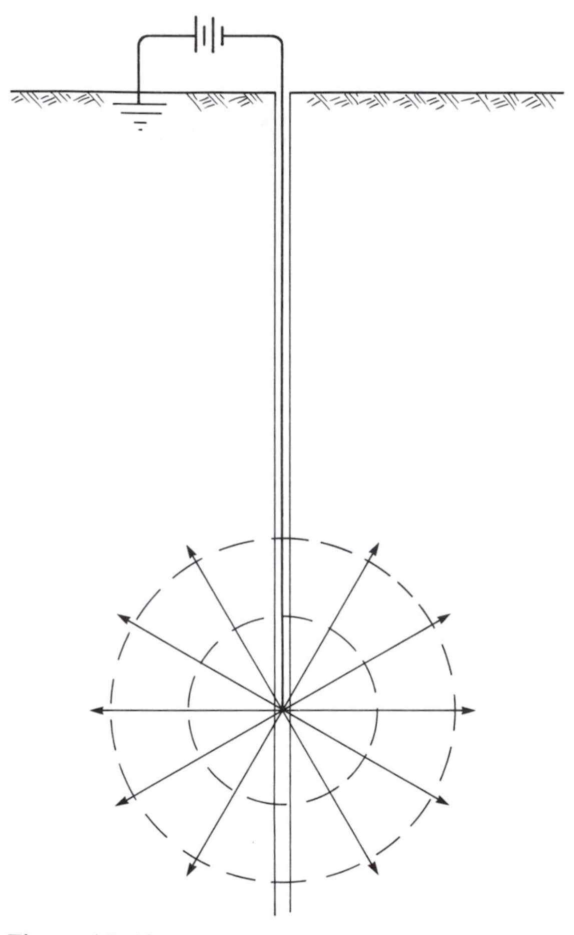

Figure 14–10
Spherical radiation of electric current outward into a homogeneous formation from a source electrode situated underground, far from the sink electrode situated on the land surface.

We can express Ohm's law in other forms that apply to particular electrode arrangements. It is common practice in electric logging to place the sink electrode at the land surface far from the other electrodes. Because this practice makes the distances BM and BN very large, the terms 1/BM and 1/BN become

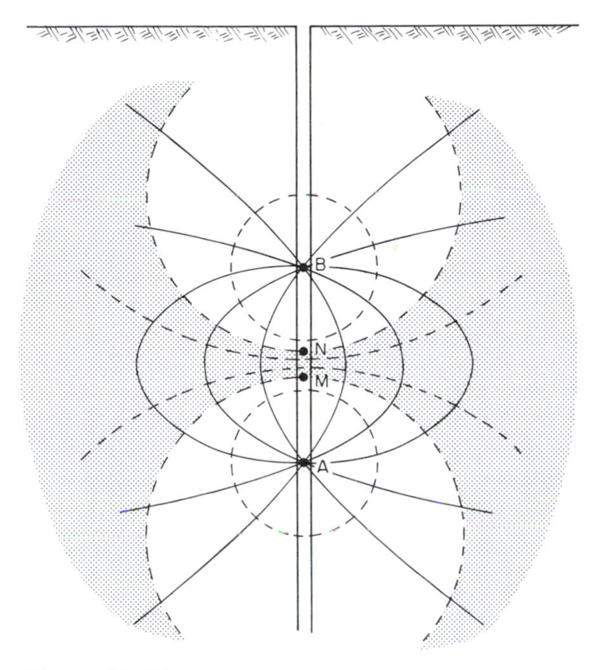

Figure 14–11
Current rays (solid lines) and equipotential surfaces (dashed lines) illustrating the flow of current and the electric field between a source electrode (A) and a sink electrode (B) situated underground in a homogeneous formation. The difference in potential measured between two other electrodes M and N is affected by all the material in the shaded zone between the equipotential surfaces on which M and N are located.

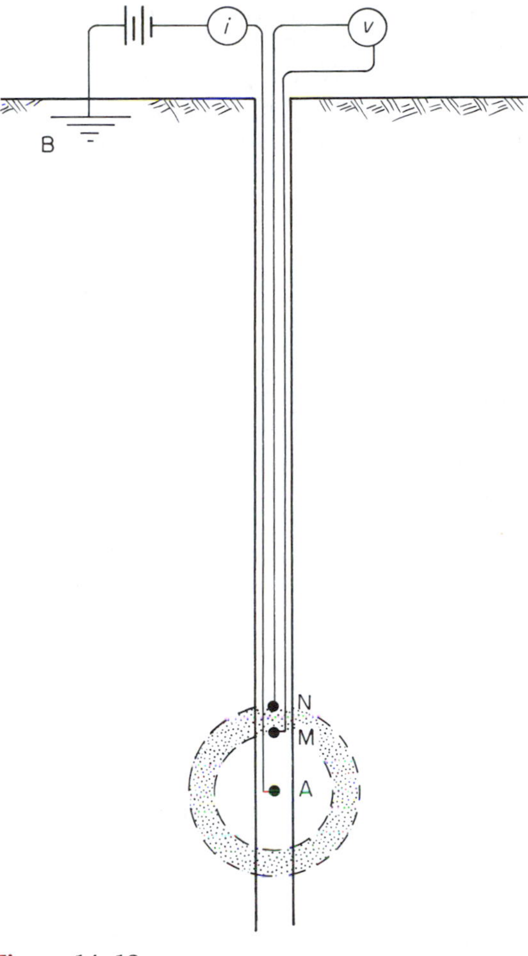

Figure 14–12
Electrode arrangement in which the source electrode (A) is located far from the sink electrode (B), and the difference in potential (v) is measured between the M and N electrodes that are located close to A and far from B. The value of v is affected by all the material in the spherical shell having an inner radius AM and an outer radius AN.

very small and can be neglected. Equation 14–12 can then be modified to the form

$$v_{\mathrm{MN}} = \frac{IR}{4\pi}\left(\frac{1}{\mathrm{AM}} - \frac{1}{\mathrm{AN}}\right) \qquad (14\text{–}13)$$

In this arrangement, the M and N electrodes are situated on nearly spherical equipotential surfaces, with the A electrode at the center. Therefore, the zone being tested is a spherical shell with inner radius AM and outer radius AN (Figure 14–12).

Let us consider another electrode arrangement in which A and M are close together, and both B and N are placed far away. Because AN is very large, the term 1/AN becomes small enough to neglect. Equation 14–13 can then be modified to obtain

$$v_{MN} = IR/4\pi \text{ AM} \qquad (14\text{–}14)$$

The zone tested by this electrode arrangement is a thick shell with inner radius AM and a very large outer radius (Figure 14–13). However, most of the change in potential occurs in the innermost part of this shell. We can understand why by considering the change in potential that would be measured if the N electrode were placed at the distance AN = 2AM. By substituting 2AM for AN in Equation 14–13, we obtain $v_{MN} = 0.5 \times IR/4\pi\text{AM}$, which is one-half the value of v_M given by Equation 14–14. This tells us that one-half of the change in potential found from Equation 14–14 occurs within a distance of 2AM from the source electrode. If the N electrode is placed at the distance AN = 10AM, the same analysis indicates that 90 percent of the potential change found from Equation 14–14 occurs within a distance of 10AM from the source electrode. It is clear, then, that even when the N electrode is placed far from A and M, measurements are the most strongly influenced by rock in a relatively thin spherical shell with the inner radius AM. The gradational shading in Figure 14–13 illustrates how the effect of rock within this zone diminishes with distance from the inner equipotential surface.

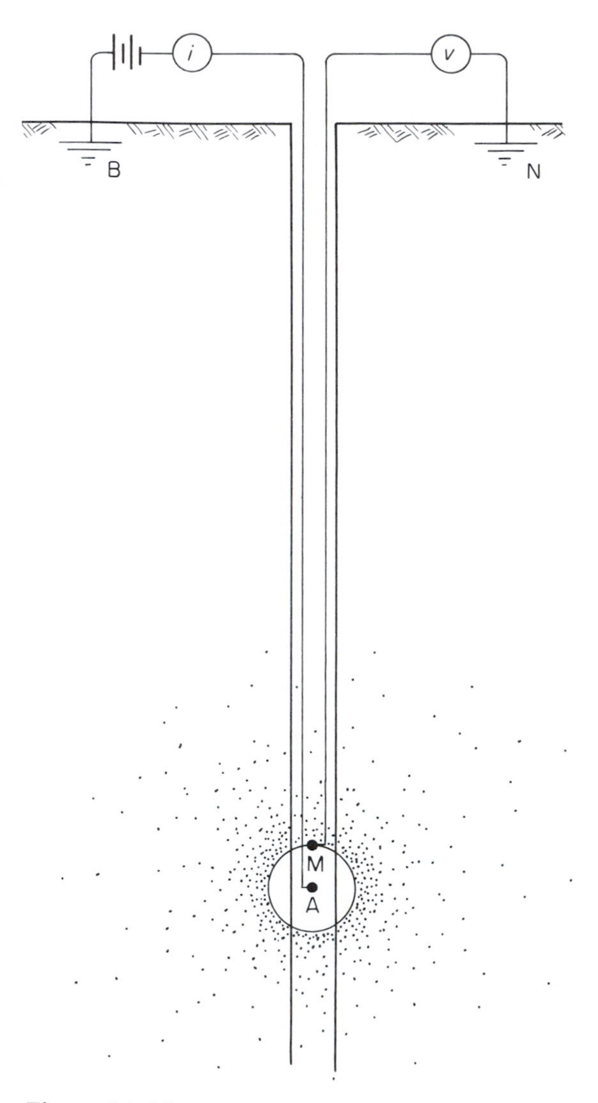

Figure 14–13
Electrode arrangement in which the source electrode (A) is located far from the sink electrode (B), and the difference in potential (v) is measured between the M electrode, placed near A, and the N electrode, which is placed far from M. The value of v is affected by all the material in a spherical shell of inner radius AM and outer radius AN, but material close to M has a much larger influence than material farther from M, as illustrated by the gradational shading.

Normal Logs

One of the most widely used electrode arrangements for logging resistivity is called the *normal* electrode configuration (Figure 14–14). Only the source electrode A and the potential electrode M are mounted on the sonde. The sink electrode B and the other potential electrode N are situated far from the sonde. A generator connected between A and B supplies a constant current I, and the change in potential v_M is measured continuously with a voltmeter connected between M and N. The voltmeter output is plotted on a strip chart, the *normal log,* as the sonde moves in the well. For this electrode configuration, the resistivity R is found by rearranging Equation 14–14 into the form

$$R = \frac{4\pi AM}{I} v_M \qquad (14\text{–}15)$$

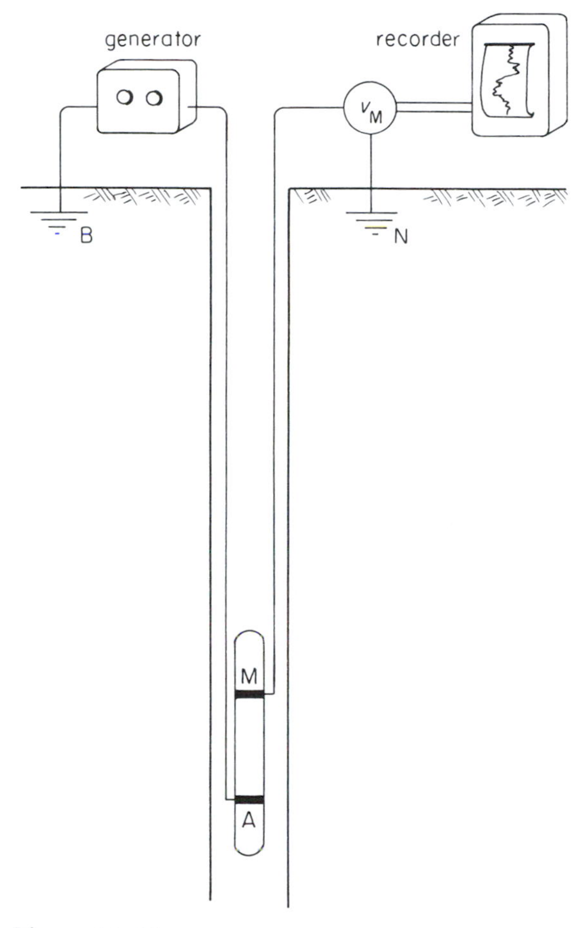

Figure 14–14
Schematic diagram of a normal logging circuit. The source electrode A is far away from the sink electrode B, and the electrode M is far away from the electrode N. The difference in potential v_M is measured between M and N.

Because the A amd M electrodes are a fixed distance apart on the sonde and the current is held constant, we see that R varies directly with v_M. Therefore, the strip chart can be scaled in resistivity units (usually ohm-meters or ohm-feet) rather than in volts. Then it is possible to read resistivity directly from the log.

A normal logging sonde moving through a very thick homogeneous formation would test a continuously changing zone with the form illustrated in Figure 14–13. But actual electric logging is not done in a homogeneous environment. The zone of testing includes beds of contrasting resistivity, and the sonde must be operated in a well filled with drilling mud, which has yet another resistivity. Current emanating from the A electrode first passes through the drilling mud and then into the formation. Rays representing this flow of current refract into different directions at each boundary where the resistivity changes (Figure 14–15). Observe how the corresponding equipotential surfaces, which must be perpendicular to these rays, are distorted into nonspherical shapes. Because these equipotential surfaces indicate the shape of the zone of testing, it is clear that this zone will not be spherically symmetrical, and that its shape will change as the sonde moves through a well. In

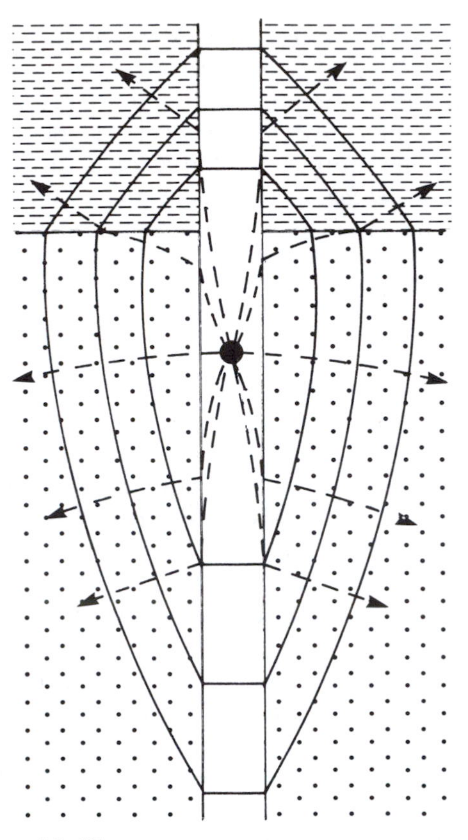

Figure 14–15
Refraction of current rays at the side of the well and at a boundary between two formations of different resistivities. (From Interpretation Handbook for Resistivity Logs, *Schlumberger Document No. 4,* Schlumberger Well Surveying Corporation, 1951.)

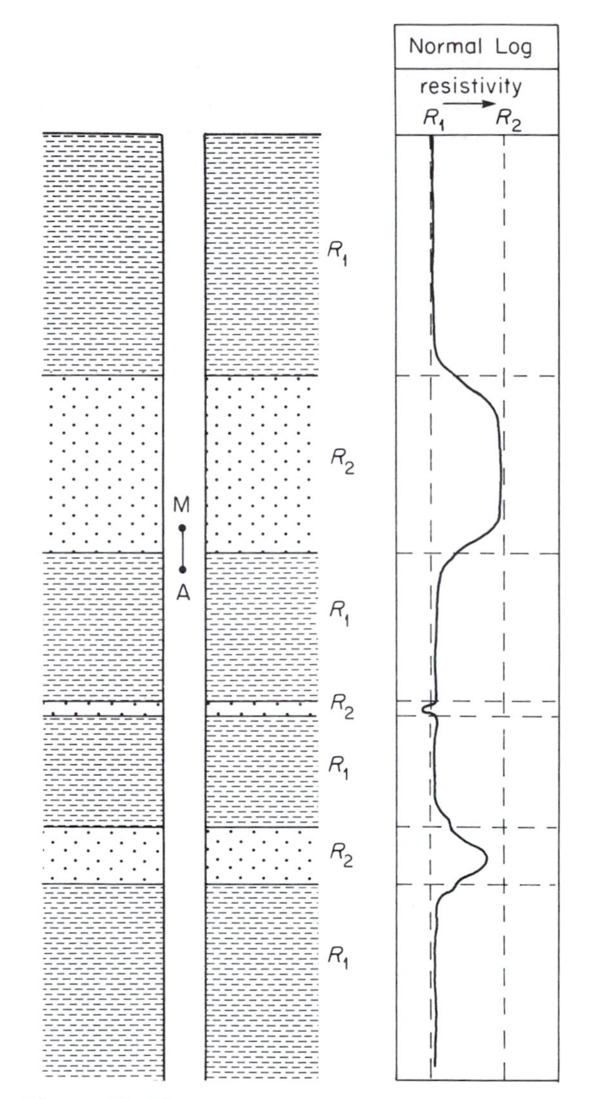

Figure 14–16
Idealized normal log variation at boundaries between beds of different thicknesses and contrasting resistivities. Contrasts between high and low resistivity are properly indicated for beds considerably thicker than the distance between A and M. For the thin bed, the response is inverted, which gives an incorrect indication of the resistivity contrast.

this heterogeneous environment, Equation 14–15 expresses only an apparent resistivity that depends on the proportions of the zone of testing which have contrasting resistivities.

Suppose that we obtain a normal log in a well that penetrates beds of contrasting resistivities. An idealized example (Figure 14–16) illustrates how the apparent resistivity changes as the sonde moves through the well. Observe that there are no abrupt changes in R as the sonde passes bed boundaries. Rather, we see gradational changes. The reason why is that the A and M electrodes mark the inner radius of a zone of testing that reaches beyond the dimensions of the sonde. Therefore, the zone of testing moves through a bed boundary before the sonde actually passes this boundary. Therefore, a nearby bed influences the apparent resistivity measurements, even though the sonde is still moving through another bed.

Observe in Figure 14–16 that the apparent resistivity measured with the sonde near the center of a thick bed comes closest to the true formation resistivity R_t. Here the bed occupies most of the zone of testing, and distortion caused by the beds above and below is at a minimum. For beds that are thicker than 10AM, the measured resistivity should differ from R_t by less than 10 percent near the center of the bed. For thinner beds, the value of R_t is not accurately indicated on a normal log.

There is a particular problem when a bed thinner than AM is encountered. Look at the apparent resistivity variation near the thin layer in Figure 14–16. Small increases appear at its upper and lower boundaries, but low apparent resistivity is indicated near the center where R_t is high. This distortion results from refraction of current rays, which occurs when the A electrode is below the layer and the M electrode remains above it. The proportion of current flowing in the layer is reduced so that an incorrect measure of resistivity is recorded.

Another factor affecting the measurement of apparent resistivity is invasion. Insofar as the resistivity of the drilling mud differs from that of the natural pore fluid it has displaced, the resistivity in the invaded zone will be different from R_t. Therefore, where invasion is deep, even the apparent resistivity measured near the middle of a thick bed can be quite different from the true formation resistivity.

One aim of electric logging is to measure R_t for purposes of calculating the porosity, saturation, and permeability of a formation. Do the problems posed by thin beds and invasion prevent us from obtaining accurate measurements? To overcome these difficulties, we begin by finding bed thickness and estimating the depth of invasion. We do this by comparing electric logs recorded with different electrode spacings. The common practice throughout most of the petroleum industry is to record a short normal log with an AM spacing of 16 inches, and a long normal log with an AM spacing of 64 inches, as well as other logs, which we will discuss later.

What can we learn from long and short normal logs? Consider first the effect of invasion. Suppose that resistivity in the drilling fluid is higher than the resistivity of the natural pore fluid. The effect of invasion will be to increase the resistivity R_i in the invaded zone so that $R_i > R_t$. Next, consider the zones tested by long and short normal logging. For the long normal log, this zone is larger and reaches farther into the undisturbed part of the formation. Invasion affects a smaller proportion of this zone compared with the proportion of the zone tested by the short normal log. Therefore, the long normal log will indicate a lower value of apparent resistivity. This difference can be seen in Figure 14–17 where both long and short normal logs indicate a high-resistivity bed between the depths of 1750 and 1820 feet. The apparent resistivity values indicated by these logs can be compared with values read from standard correction charts that have been calculated for models consisting of beds with different thicknesses, resistivity contrasts, and depths of invasion.[1] The true resistivity R_t and the depth of invasion are found from the particular chart that best reproduces the apparent resistivities on the 16- and 64-inch normal logs.

Even though the short normal log is more adversely affected by invasion, it is useful for detecting thin beds, which are incorrectly represented on a long normal log. Compare the logs in Figure 14–17 near 6677 feet. Here only the short normal log properly indicates the presence of a high-resistivity bed which must be between 16 and 64 inches thick.

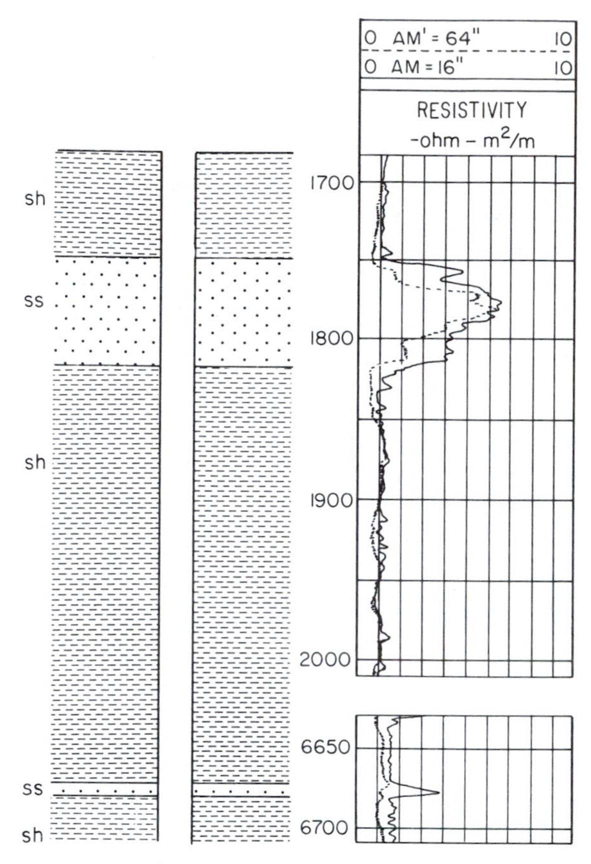

Figure 14–17
The effect of invasion on apparent resistivity measured with 16-inch and 64-inch normal logging devices. The 16-inch normal log is more strongly influenced by the high-resistivity invaded zone in the thick sandstone layer. The 64-inch normal log does not properly detect the thin sandstone bed near 6680 feet.

The Lateral Log

Another way to measure resistivity is with the *lateral* electrode configuration (Figure 14–18). The M and N electrodes are placed near each other on the sonde. Farther away in the well is the A electrode, and very far away at the land surface is the B electrode. In the petroleum industry the M and N electrodes are usually spaced 32 inches apart and the A electrode 18 feet, 8 inches from the point O situated midway between M and N. Because it is impractical to operate a sonde of this length, the A electrode is mounted on the wire line above a shorter sonde. The lateral log is almost always operated in combination with normal logs or other kinds of electric logs.

As with normal logging, the procedure in lateral logging is to supply constant current I from the generator connected between A and B. The potential difference v_{MN} is measured continuously with a voltmeter connected between M and N, as the sonde moves through the well. Because the spacing of A, M, and N remains fixed, this potential difference varies in direct proportion with resistivity according to Equation 14–13. The lateral log is obtained by recording the voltmeter output on a strip chart scaled in resistivity units.

In a very thick, homogeneous formation,

[1]Interested readers can find correction charts for normal logs and other kinds of well logs in George B. Asquith and Charles B. Gibson, *Basic Well Log Analysis for Geologists,* Tulsa, Okla., American Association of Petroleum Geologists, 1982.

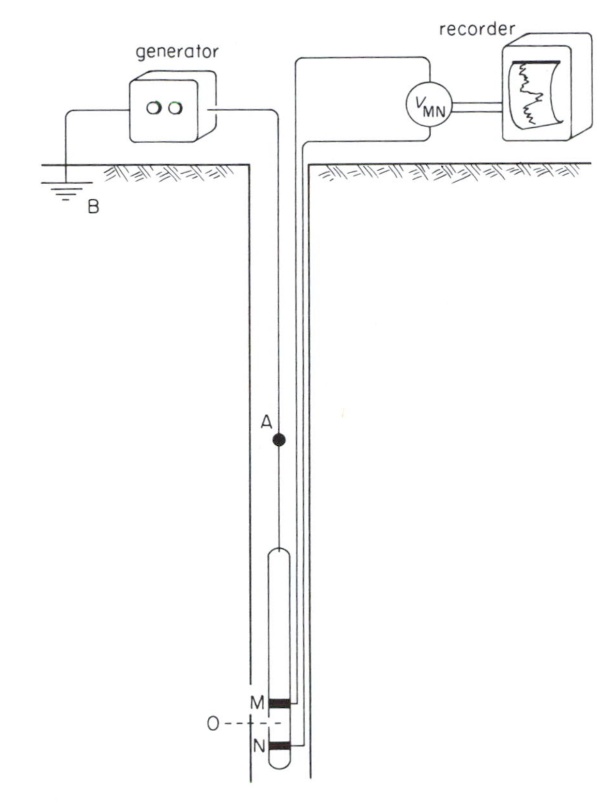

Figure 14–18
Schematic diagram of a lateral logging circuit. The M and N electrodes are quite close to the source electrode A but remain far away from the sink electrode B.

the lateral log would test a spherical shell with inner radius AM and outer radius AN (Figure 14–12). However, in an environment consisting of beds of contrasting resistivities, the zone of testing is distorted into a nonspherical shell because of the refraction of current rays (Figure 14–15). Regardless of this distortion, the zone tested by lateral logging reaches much farther from the well than the zones tested by conventional normal logging. For this reason, a much larger proportion of the lateral zone occupies formation that is not contaminated by invasion. Therefore, the apparent resistivity measured in a thick bed by lateral logging tends to be closer to R_t than the values obtained from normal logs.

Lateral logs have peculiar variations in apparent resistivity near bed boundaries. These distortions are related to the refraction of current rays that occurs where the M and N electrodes are moving through one bed while the A electrode is still moving through another overlying bed. This refraction produces the typical features of a lateral log that are illustrated in Figure 14–19.

First, let us examine the variation in apparent resistivity related to the thick bed having a high resistivity. Observe the small decrease that appears close to its upper boundary. This occurs because the M electrode has moved below the boundary while the N electrode remains above it, where the current rays are oriented differently. Apparent resistivity increases gradually after N moves below this boundary, and then increases more sharply when the A electrode passes below it. This change ceases in the middle part of the bed where the apparent resistivity levels off at a value close to R_t. A sharp rise is recorded near the base of the bed, reaching a maximum very close to its lower boundary as the M electrode moves below. After N passes this boundary, the apparent resistivity decreases to an intermediate level and then diminishes rapidly after A, too, moves below the bed.

A different pattern of variation is related to the thin, high-resistivity layer in Figure 14–19. The small decrease that is recorded as M and then N pass the upper boundary is followed by a sharp increase, reaching a peak value as these electrodes pass the lower boundary. The log decreases sharply after M and N have moved below the bed. Then, at the distance

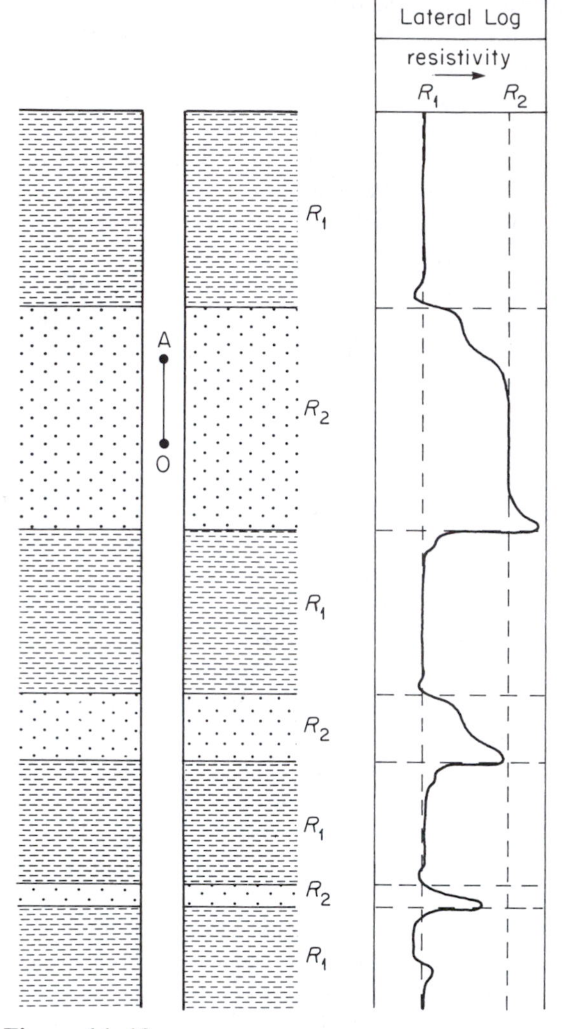

Figure 14–19
Idealized lateral log variation at boundaries between beds of different thicknesses and contrasting resistivities. The spurious effects are measured close to the bed boundaries and at a distance below a thin bed that is equal to the AO electrode spacing.

AO beneath the bed, a small resistivity increase is indicated. This effect is caused by a change in current refraction as the A electrode moves through the thin bed. It is a completely spurious variation that should not be mistaken for the effect of another thin bed.

The advantages of a lateral log compared with normal logs are (1) a more accurate measure of R_t in the middle part of thick beds, and (2) a much sharper indication of the lower bed boundary. A similarly sharp indication of the upper bed boundary can be detected by lateral logging with the A electrode placed beneath M and N. Deep invasion can introduce some distortion in the resistivity read from a lateral log, albeit less severe than that on normal logs. Corrections for the effects of invasion can be estimated from standard charts similar to those used in normal log analysis. The spurious resistivity variations that are recorded below thin beds are a serious disadvantage of lateral logging. When closely spaced thin beds are encountered, it can be very difficult to interpret lateral logs.

The Laterolog

The electric logs that we have been discussing are often referred to as unfocused logs, because no attempt is made to control the direction of current flow. It spreads outward in all directions from the A electrode and is refracted at boundaries where resistivity contrasts exist (Figure 14–15). There is another procedure, called *laterologging,* in which circuits are designed to focus the current in particular ways.

The most widely used laterologging devices direct the current in a sheet that spreads horizontally outward from the well. One way of accomplishing this is illustrated in Figure

14–20. The sonde consists of a short electrode centered between two long electrodes, called the *guard* electrodes, which are connected to each other. Current, which can be controlled separately for the guards and the central electrode, is adjusted automatically to maintain all of these electrodes at the same fixed potential. It then becomes impossible for current to flow from any one of these electrodes toward another. Therefore, current from the central electrode must move outward in a horizontal sheet. Because the potential is fixed, the amount of current from the central electrode varies in proportion to the resistivity of the formation into which it flows. This current is plotted on a strip chart scaled in resistivity units to obtain the *laterolog*.

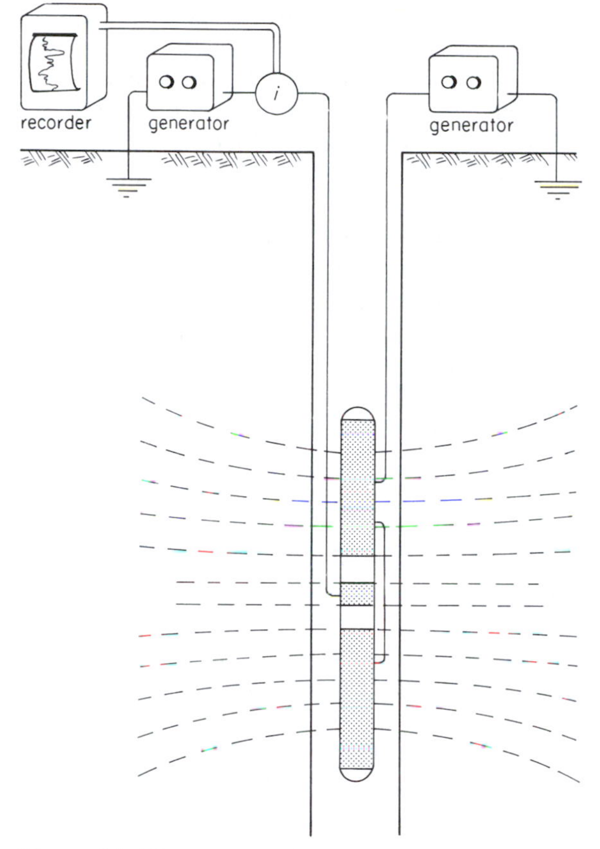

Figure 14–20
Schematic diagram of a laterologging circuit. The effect of the guard electrodes above and below, which are held at the same potential, is to make nearly horizontal current rays (dashed lines) flow outward from the center source electrode.

Laterologging sondes ordinarily have guard electrodes that are more than five feet long. The length of the central electrode ranges from as short as three inches to more than one foot. The zone tested by the laterolog is a circular disk (Figure 14–21) with a thickness approximately equal to the central electrode length. Within this zone, the rock closest to the well has the greatest effect on a resistivity measurement. Closely spaced rays indicate the greatest current intensity. The effect of rock farther from the well diminishes as the rays spread apart.

Compared with unfocused electric logs, the laterolog is much more sensitive to thin beds. It can properly detect a bed as thin as the central electrode length. If the effect of invasion is small, the apparent resistivity read from the laterolog is close to R_t. However, deep invasion can introduce significant distortion. Because the invaded zone is close to the well, it has a proportionally larger effect on the laterolog than on unfocused electric logs. This is the principal disadvantage of the laterolog. Corrections can be determined from standard charts if the depth of invasion can be estimated independently from normal and lateral logs.

Induction Logging

Up to this point, we have been concerned with logging devices that introduce current into the

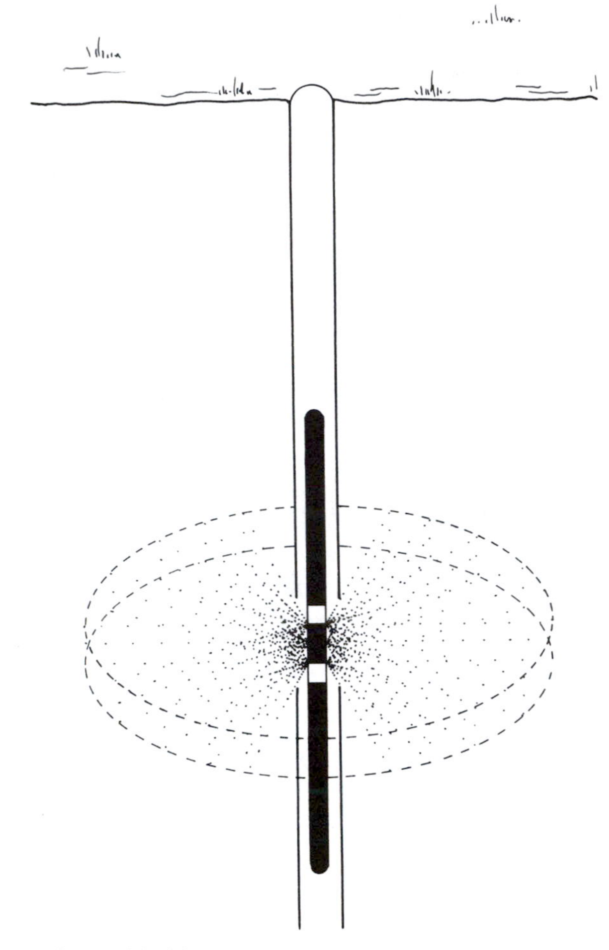

Figure 14–21
The zone tested by a laterolog is a circular disk with a thickness equal to the length of the central electrode. The material closest to the well has the strongest influence on the resistivity measurement. Influence diminishes gradationally with distance from the well, as illustrated by the gradiational shading.

formation by means of electrodes mounted on a sonde. To reach the formation, current must first flow from the electrode through the drilling fluid to the side of the well. In the event that the well is filled with nonconducting fluid, the sonde will be insulated and have no means of introducing current into the formation. This difficulty can be encountered when crude oil rather than water is used for making the drilling mud.

A different procedure, *induction* logging, has been developed for operation in wells filled with nonconducting fluids. It makes use of coils rather than electrodes, which are mounted on the sonde. Electromagnetic induction is the means for introducing current into the formation and for measuring its effects. Unfocused induction logging can be done with a two-coil sonde (Figure 14–22). A generator produces alternating current in the transmitting coil. This alternating current creates an alternating magnetic field in the region surrounding the coil, which reaches into the formation. Electromagnetic induction associated with this field compels current to flow through the formation in circular paths centered on the well. This alternating ground current creates a secondary magnetic field that compels current to flow in the receiving coil, again by electromagnetic induction. The intensity of the ground current varies with resistivity, and so do the secondary magnetic field intensity and the induced current in the receiving coil. The induced current in the receiving coil is recorded on a strip chart to obtain an unfocused induction log.

Suppose that the sonde is placed in a very thick homogeneous formation. Then it is possible to calculate the intensity of the primary magnetic field and the corresponding intensity of the ground current at any point in the formation. Such calculations help us deter-

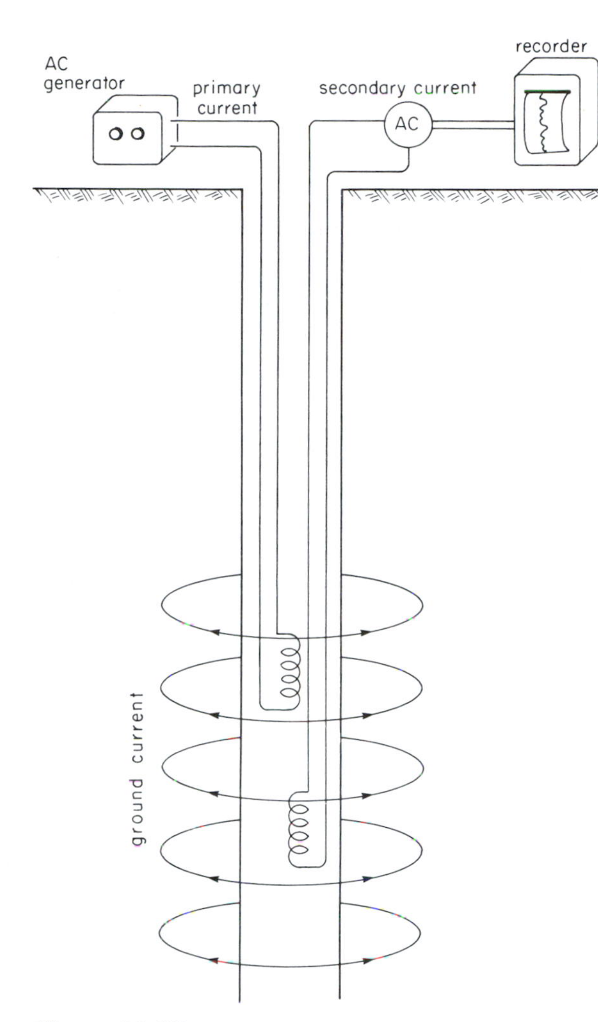

Figure 14–22
Schematic diagram of an unfocused two-coil induction logging circuit, showing the ground currents induced by the alternating magnetic field of the primary coil. The alternating magnetic field associated with these ground currents induces alternating current in the secondary coil.

mine the zone that is tested by this induction logging procedure. Graphs in Figure 14–23 illustrate how the effect of different parts of the formation vary with distance from the sonde. They show that the ground current is concentrated mostly in the zone within a distance of about three times the coil spacing *(L)*.

An unfocused induction log records a gradational change in apparent resistivity as the sonde passes the boundary between two beds of contrasting resistivity. This is shown in Figure 14–24. In a thick bed in which the sonde is more than three times the coil spacing from either the upper or lower boundary, the measured apparent resistivity should be close to R_t unless the results are distorted by invasion. Corrections for invasion can be determined from standard charts, but additional information is required; this information can usually be obtained from other kinds of logs.

Observe in Figure 14–24 that inflection points on the induction log lie close to the bed boundaries. These inflections can be difficult to identify. A sharper indication of bed boundaries can be obtained by means of focused induction logging. This method requires additional control on the flow of ground currents, which is accomplished by adding more coils to the transmitting and receiving circuits. The simplest focusing circuits are shown in Figure 14–25. Two additional coils are used. One is in series with the original transmitting coil but is situated close to the original receiving coil; the other is in series with the original receiving coil but is placed near the original transmitting coil. The log obtained with this arrangement is compared with the unfocused log in Figure 14–24. The focusing coils produce a sharper indication of bed boundaries, but they introduce spurious oscillations in apparent resistivity. Still more focusing coils can be arranged in more com-

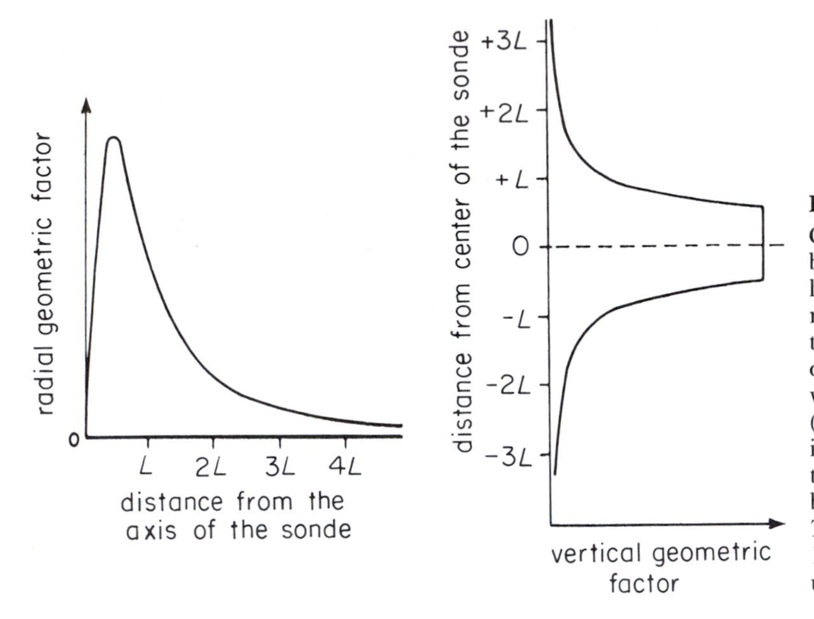

Figure 14–23
Graphs illustrating the zone tested by an unfocused two-coil induction logging circuit. Geometric factors read from these graphs indicate the relative influence of material at different distances away from the well and above or below the sonde. (From H. G. Doll, Introduction to induction logging and application to logging of wells drilled with oil base mud, *Journal of Petroleum Technology,* AIME, p. 16, June 1949.) The distance is expressed in units of coil spacing.

plicated circuits to suppress these oscillations while retaining the steeper change in apparent resistivity at the bed boundary.

The induction log was introduced for the purpose of testing wells filled with nonconducting drilling mud. This logging procedure can also be used in wells containing mud that does conduct current.

Spontaneous Potential Logging

Natural potential differences have been found to exist near bed boundaries. The circuit for measuring these so-called *spontaneous potentials* (SP) is quite simple. A voltmeter is connected between an electrode mounted on a sonde and another electrode fixed at the land surface (Figure 14–26). The SP log is obtained by recording the voltmeter output on a strip chart.

Why do spontaneous potentials exist? Although their origin is not clearly understood, we have some ideas about the cause. Where pore water solutions that have different ion concentrations are in contact, ions tend to move across the boundary to equalize the concentration. Insofar as some ions can move more easily than others, charge imbalances can develop, creating potential differences.

Consider the situation in which a permeable sandstone is in contact with a relatively impermeable shale. Suppose that both beds contain the same saline pore water. Invasion is much deeper in the sandstone than in the shale, however, and the NaCl concentration in the mud filtrate is different from the concentration in the natural pore water. Therefore, near the well, along the boundary between these beds, the mud filtrate in the sandstone is in contact with the natural pore water in the shale.

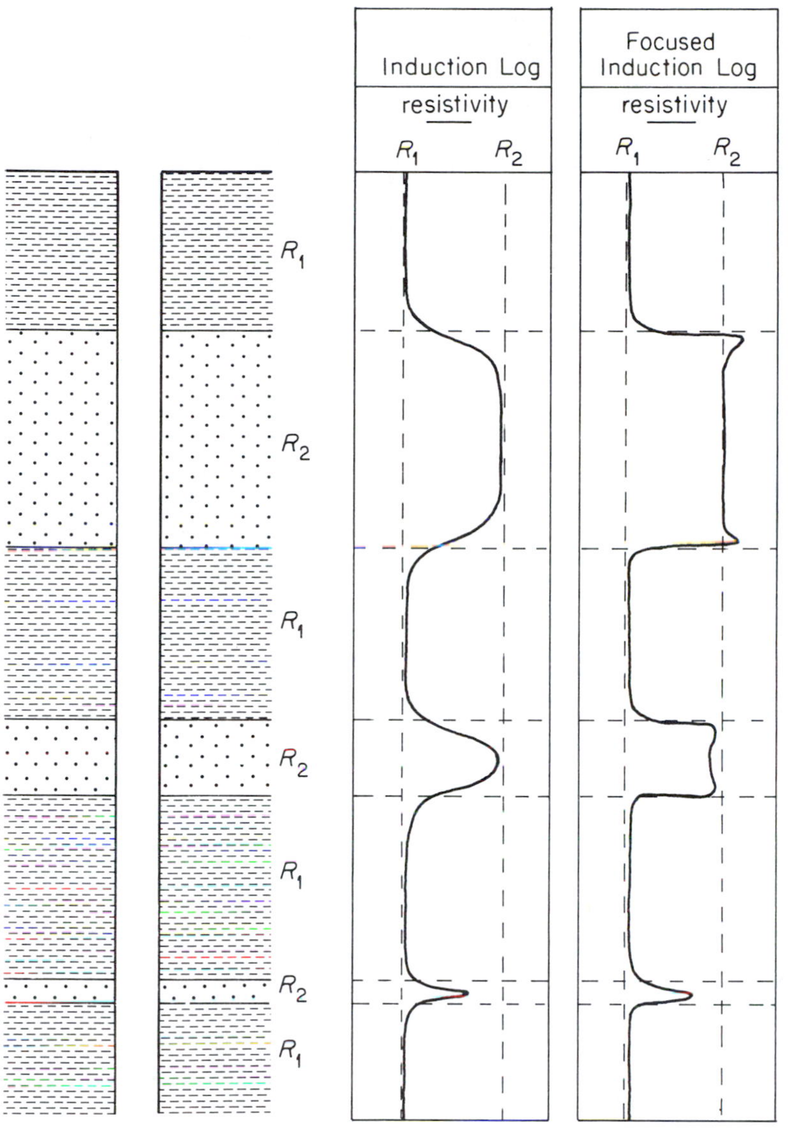

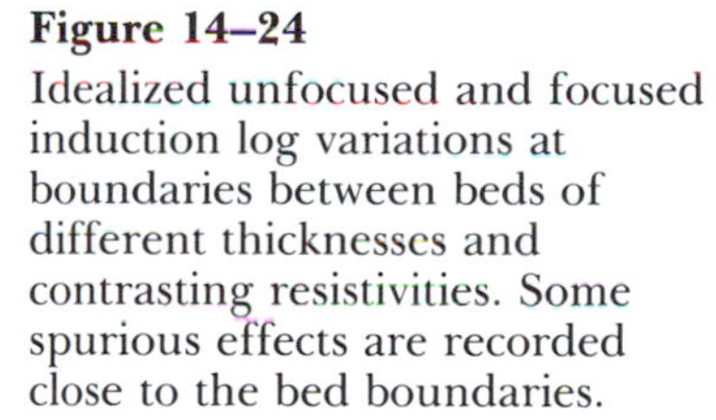

Figure 14–24
Idealized unfocused and focused induction log variations at boundaries between beds of different thicknesses and contrasting resistivities. Some spurious effects are recorded close to the bed boundaries.

The movement of ions to equalize the concentrations in these two solutions is impeded by surface charges on the clay particles in the shale. These charges tend to immobilize the negative chloride ions but have little effect on the positive sodium ions. Therefore, a charge imbalance develops across the boundary, and a spontaneous potential is created.

In the permeable sandstone, both sodium and chloride ions are mobile, but the sodium

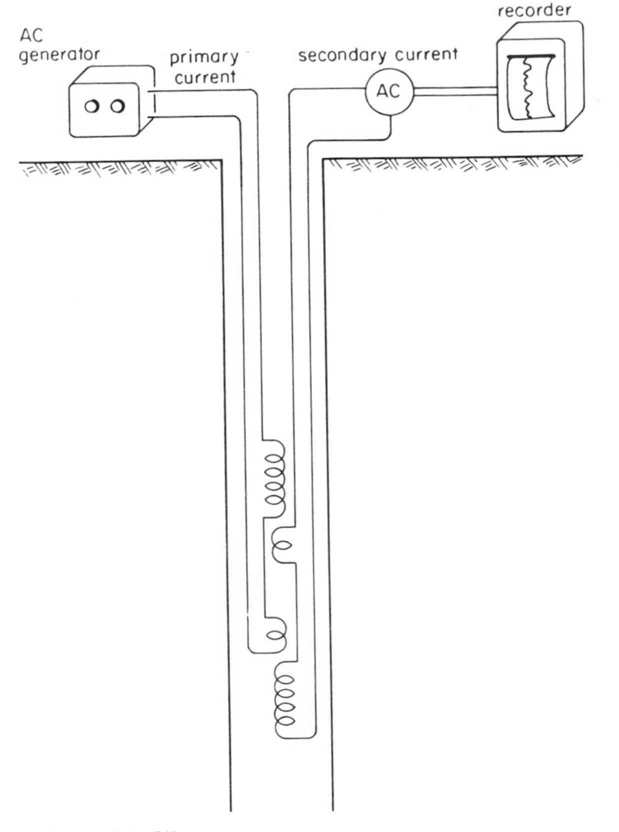

Figure 14–25
Schematic diagram of a focused four-coil induction logging circuit. Both the primary circuit and the secondary circuit contain two coils. The small secondary coil is placed close to the large primary coil, and the small primary coil is placed close to the large secondary coil.

ions tend to move more freely. Therefore, another source of spontaneous potential is the transitional boundary between the mud filtrate near the well and the natural pore water beyond the invaded zone. We do not fully understand the nature of these and other processes that contribute to spontaneous potential variations. But we can measure their combined effect; it produces variations of a few tens to a few hundreds of millivolts.

Some important features of SP logs can be seen in Figure 14–27. In a sequence of sandstone and shale beds, a negative change in potential is recorded as the sonde electrode passes from shale into sandstone. Compared with conventional resistivity logs, the SP logs

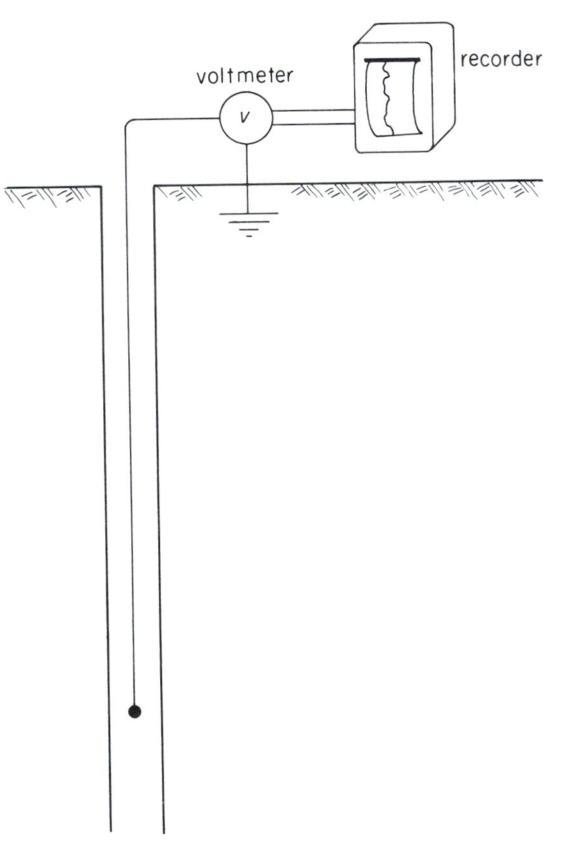

Figure 14–26
Schematic diagram of a circuit for measuring spontaneous potential. The potential difference is measured between an electrode mounted on the sonde and another electrode placed far away.

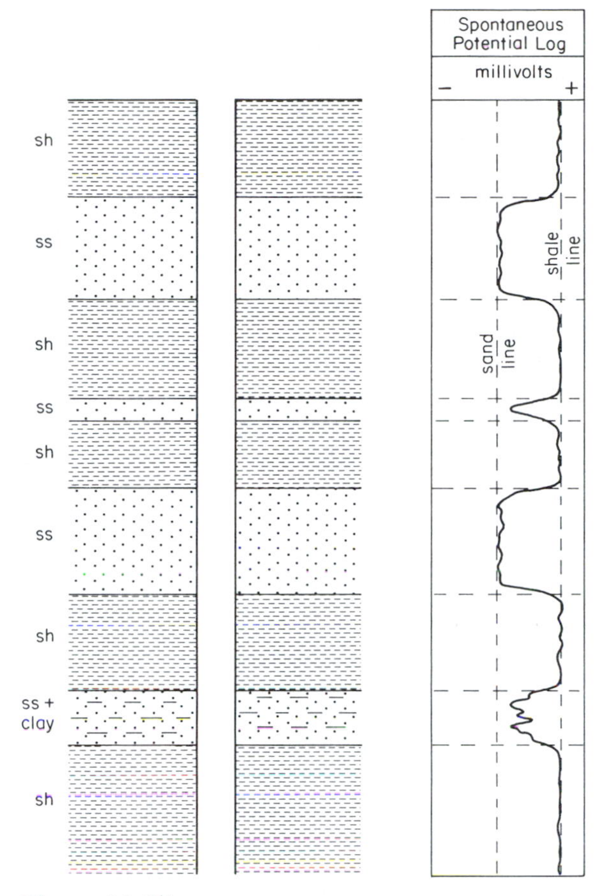

Figure 14–27
Idealized spontaneous potential variations in alternating beds of sandstone and shale. The inflection points on the log occur close to the bed boundaries.

change more steeply, with the inflection point on the curve occurring at the boundary. This feature of the SP log provides an accurate means of locating bed boundaries.

A feature observed in many regions is that all the shale beds are at approximately the same spontaneous potential. Then it is possible to draw a straight line, called the *shale line* (Figure 14–27), along the right-hand side of the SP log, showing the alignment of the parts of the log that were recorded in shale. If the well also encounters relatively thick beds of sandstone uncontaminated by clay particles, it is not unusual to measure the same spontaneous potential in these beds. Then a line, the *sand line* (Figure 14–27), can be drawn along the left-hand side to show the alignment of these parts of the SP log. Having marked these two SP extremes, we can estimate the clay contamination in other beds in which intermediate spontaneous potentials are measured. If an uncontaminated sand bed is too thin, the sharp peak recorded on the log may not reach the sand line. Nevertheless, the bed boundaries are correctly indicated by the inflection points above and below such a peak.

Micrologs

Special sondes have been designed for measuring resistivity with very closely spaced electrodes. These electrodes are placed on a pad which can be pressed firmly against the side of a well (Figure 14–28). These electrodes, which are usually one or two inches apart, are connected into different circuits to obtain *micronormal* logs, *microlateral* logs, and *microlaterologs*.

These micrologs can detect very thin beds. However, their principal purpose is to measure the resistivity of the mud cake lining a well and the resistivity in the invaded zone. These measurements are of critical importance for determining correction factors that are needed to calculate true resistivity R_t from standard normal, lateral, and induction logs.

Because the microlog electrodes must be pressed against the side of the well, the sonde has to move very slowly. It is generally im-

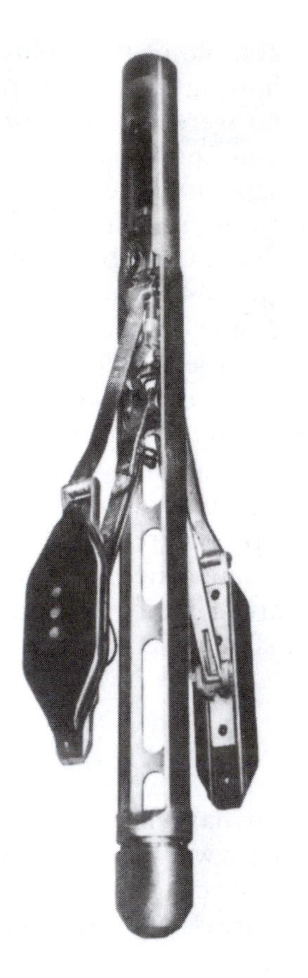

Figure 14–28
A micrologging sonde with electrodes mounted on a pad which can be pressed against the side of the well. (Courtesy of Schlumberger, Limited.)

practical to run micrologs through the entire depth of a well. Only zones of particular interest are logged.

Electric Log Combinations

The apparent resistivity measured by an electric log depends on the true resistivity R_t in a bed, the resistivity in the invaded zone, and the resistivities in other nearby beds. Different electrode configurations test zones having different proportions of these resistivities. Only by combining apparent resistivity values measured with different electrode configurations can we calculate R_t.

The electric logging apparatus typically consists of a sonde with several electrodes that are automatically switched into different circuits, so that several logs can be plotted simultaneously on the same strip chart while the sonde moves through the well. For most operations, the recording equipment and electronic controls are mounted in a truck which also carries the winch and wire line for lowering and raising the sonde (Figure 14–29). For offshore operations, the same equipment is installed in a small vessel or is airlifted to an offshore drilling platform. Different sondes can be attached to the wire line, each one providing a different combination of logs.

The combination most widely used in the petroleum industry includes the SP log, 16-inch and 64-inch normal logs, and the 18-foot, 8-inch lateral log (Figure 14–30). Another very common combination includes the SP log, the laterolog, and two induction logs. In both of these combinations, the SP log can be used to locate bed boundaries, and the other logs provide data for estimating R_t. If additional resistivity measurements are needed, different sondes equipped for micrologging can be attached to the wire line.

The trucks used for electric logging are usually equipped for radioactivity logging as well as for sonic logging. In addition to strip chart recorders, these trucks have facilities for recording on magnetic tape, which is more suitable for later computer processing, and some carry small computers for on-site analysis.

Not all electric logging is done with large trucks designed to operate several different

Figure 14–29
Typical field trucks and winch assembly used for well logging. (Courtesy of Schlumberger, Limited.)

sondes. There are small, portable units that can be backpacked into remote areas for obtaining SP and normal logs and natural radioactivity logs.

RADIOACTIVITY LOGGING

The process by which particles of mass or energy are spontaneously emitted from an atom is called radioactivity. These emissions consist of protons, neutrons, electrons, and photons of electromagnetic energy that are called gamma rays. In nature they come from the unstable nuclei or radioactive elements such as uranium, thorium, rubidium, and potassium 40. Similar emissions from the stable nuclei of other elements can be stimulated by bombarding them with gamma rays or neutrons. Radioactivity logging methods make use of both natural and stimulated emissions. We will look at three methods of radioactivity logging. The first uses a detector mounted on a sonde to measure the gamma rays produced by radioactive elements in a formation. In the second method gamma radiation is introduced into

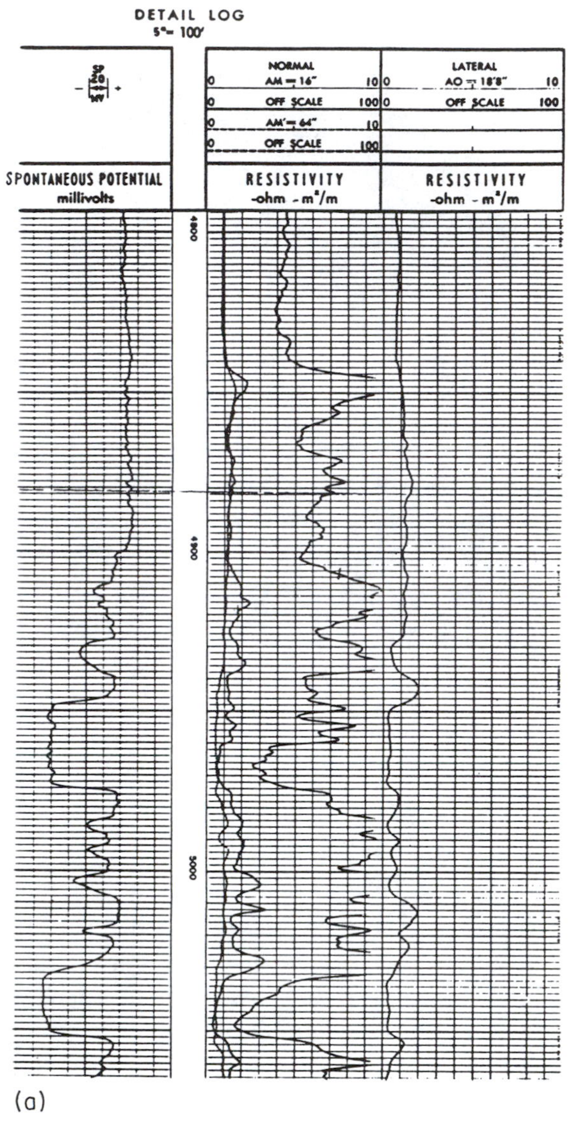

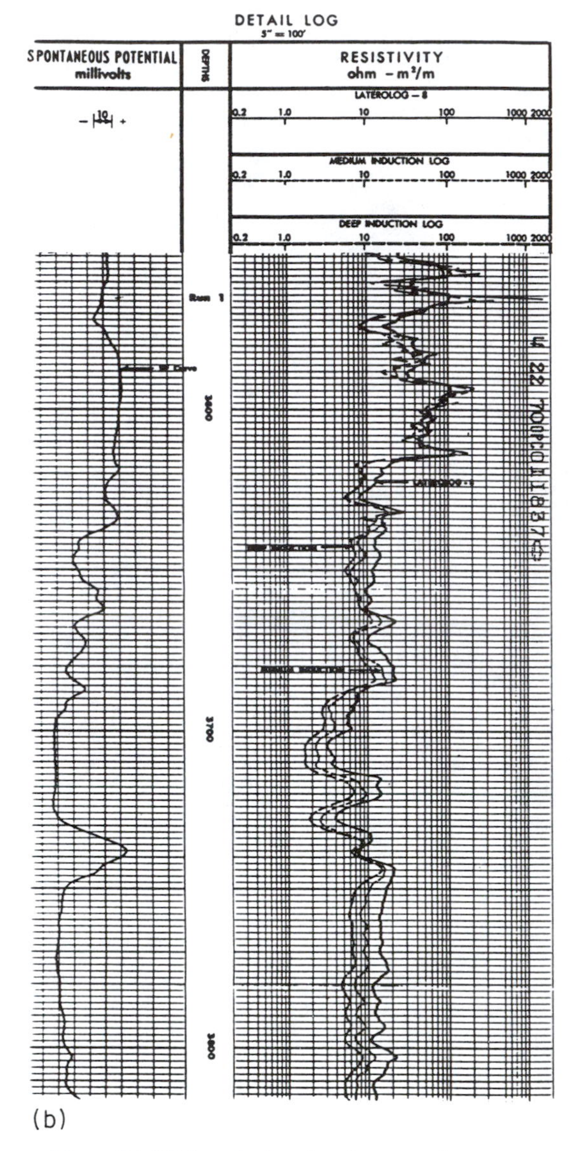

Figure 14–30

Typical electric log combinations: (a) SP log, 16-inch and 64-inch normal logs, and 18-foot, 8-inch lateral log; (b) SP log, induction logs, and laterolog.

the formation from a source mounted at one point on the sonde, and the intensity of this radiation is measured by a detector mounted at another point a fixed distance from the source. In the third method neutron bombardment of the formation stimulates gamma radiation. The neutron source is placed at one point on the sonde, and a gamma-ray detector is mounted at another point.

Common to these three kinds of radioactivity logging is the detection of gamma rays. A device called a *scintillation counter* is the usual instrument; it consists of a special crystal and a photoelectric tube. Certain crystals, such as sodium iodide crystals, emit flashes of light as they absorb gamma-ray photons. A flash of light produced in this way can be converted into a pulse of electric current by a photoelectric tube. Radioactivity logs are obtained by plotting the scintillation counter output on a strip chart.

Natural Gamma Radiation Logging

Small concentrations of radioactive elements exist in most shale beds. Potassium in the micas and the clay minerals produced by decomposition of micas and alkali feldspars includes small proportions of the radioactive isotope potassium-40. Trace amounts of uranium and thorium also occur in shale. These unstable elements produce measurable levels of gamma radiation. Quartz sandstones and carbonate beds have much lower concentrations of these radioactive impurities. Therefore, the principal use of natural gamma-ray logging is to distinguish shale beds and to estimate the clay content in impure sandstones and carbonates.

The scintillation counter used for natural gamma-ray logging counts the number of light flashes that occur in the crystal during a fixed interval of time. This is the value that is recorded on the log. Because of the irregular and sporadic nature of radioactive decay, gamma-ray emissions can vary widely from one moment to the next. Therefore, if the instrument counting time is brief, say, one second, the log can register considerable irregularity as the sonde passes through a uniform shale bed. Over larger periods of time, however, the number of emissions becomes more regular. With the instrument counting time set at, say, five seconds, a much more constant level of radiation would be indicated for the same bed.

The sensitivity of natural gamma-ray logging to boundaries between shale and other less radioactive beds depends on the instrument counting time and the speed at which the sonde is moving. Each value read from a log represents the average radioactivity of all rock through which the sonde has passed during the interval of counting time. Let us consider the three logs in Figure 14–31; log a was obtained with a short counting time and fast sonde speed, log b with a long counting time and fast sonde speed, and log c with a long counting time and slow sonde speed. Observe that if the counting time is too short, the natural irregularity of emissions may obscure the boundary between shale and another bed. This irregularity can be reduced by increasing the counting time, but if the sonde is moving too fast, the log will indicate a gradational rather than a sharp change in radioactivity near the boundary. This gradational change can be sharpened by slowing the sonde speed, but for practical reasons, we cannot operate the sonde too slowly. Therefore, a judgment must be made in choosing the counting time and the sonde speed that will yield a usable log. Typically, good results can be obtained

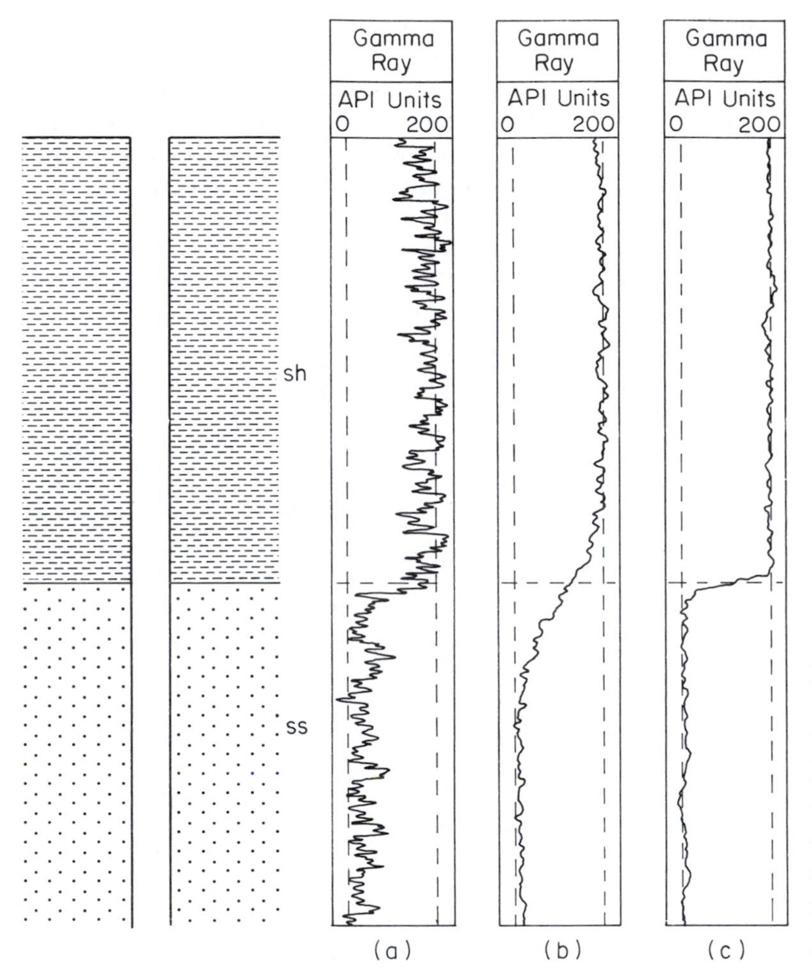

Figure 14–31
Idealized diagram showing the variation of natural gamma radiation near the boundary between sandstone and shale beds obtained by (a) a short counting time and fast sonde speed, (b) a long counting time and fast sonde speed, and (c) a long counting time and slow sonde speed.

with a counting time of 2 seconds and a sonde speed of about 30 feet per minute.

An important advantage of natural gamma radiation logging is that it can be done in cased wells. Electric logging cannot be done in cased wells because most of the current flows in the highly conductive casing pipe rather than through the formation. Gamma radiation detected at the sonde comes from the formation within a few feet from the well.

Gamma-Ray Density Logging

Gamma radiation is emitted in discrete portions of electromagnetic energy called photons. We can think of these photons as particles of energy. While traveling through a substance, they frequently collide with electrons. At each collision, some energy is transferred to the electron, thus reducing the energy of the photon. This process is called

Compton scattering. The number of collisions occurring during a fixed interval of time depends on the abundance of electrons, which is proportional to the density of the substance.

Gamma-ray density logging is the procedure for estimating formation density ρ_0 by measuring gamma radiation that has undergone Compton scattering. A source element mounted on the sonde emits gamma radiation into the formation. A scintillation counter mounted at another point on the sonde detects the portion of this gamma radiation that has been returned to the sonde by Compton scattering. This level of gamma radiation is proportional to the electron density index ρ_e of the formation. The gamma rays produced by the source have a different range of wavelengths than that of natural gamma radiation emitted within the formation. The scintillation counter output is plotted on a strip chart to obtain a log of the electron density index.

The relation between formation density ρ_0 and the electron density index ρ_e depends on the chemical elements present. The number of electrons possessed by each atom of an element is specified by the atomic number N of that element. By summing the atomic numbers of all elements in the formation, we can calculate ρ_0,

$$\rho_0 = \rho_e w / 2\Sigma N \qquad (14\text{–}16)$$

where w is the molecular weight of the combined constituents of the formation, and ρ_e is read from a log. The value of ρ_0 obtained in this way can be used in Equation 14–7, together with estimated values of pore fluid density and solid matrix density, to calculate formation porosity.

Good results can be obtained only if the source element and the detector are pressed firmly against the side of the well. The sonde must be equipped with a blade for cutting through any mud cake to ensure contact with the rock surface. To maintain this contact, the sonde must move quite slowly through the well at speeds of less than 30 feet per minute. Therefore, in deep wells, gamma-ray density logging is usually restricted to zones of particular interest.

Neutron–Gamma-Ray Logging

Nonradioactive elements in a formation can be stimulated to emit gamma radiation if they are bombarded by neutrons. This process yields neutron–gamma-ray logs that provide information about porosity. The sonde contains a neutron source and a scintillation counter mounted a fixed distance apart. The source consists of a small amount of radioactive material such as plutonium–beryllium, which emits neutrons during the process of radioactive decay.

Neutrons emitted from the source travel through the formation, colliding with atomic nuclei. The nuclei of most elements in the formation, such as silicon, oxygen, calcium, and potassium, are much more massive than a neutron. Therefore, a colliding neutron tends to bounce away with almost no loss of kinetic energy. Of all the elements in the formation, only hydrogen possesses approximately the same mass. For this reason, upon collision a neutron can transfer a significant part of its kinetic energy to a hydrogen nucleus, which is a proton. Such collisions reduce the speed of a neutron until its kinetic energy becomes low enough for it to be absorbed into one of the larger nuclei. Absorption, or capture, of a neutron stimulates the emission of capture gamma radiation. A portion of this capture gamma radiation reaches the sonde where it is detected by the scintillation counter.

The intensity of capture gamma radiation detected at the sonde depends on its distance from points of neutron capture. If neutrons travel a considerable distance before capture, only a small portion of the capture gamma rays manages to reach the sonde. But when the neutrons are quickly absorbed close to the well, a high level of capture gamma radiation is recorded.

The concentration of hydrogen in a formation is, by far, the most important factor affecting the distance that a neutron travels before it is captured. Where hydrogen content is high, neutron capture occurs close to the well, and a high level of capture gamma radiation is detected. The neutron–gamma-ray log, then, indicates variations in hydrogen concentration.

What influences hydrogen concentration? Hydrogen exists in molecules of water and petroleum and in crystals of hydrated minerals such as the silicate clays, micas, amphiboles, and gypsum. Therefore, in quartz sandstones and carbonate rocks, hydrogen occurs almost entirely in the pore water or petroleum. Its concentration depends on the formation porosity. In shale, however, micas and clay minerals as well as pore water contribute to the content of hydrogen.

It can be difficult to distinguish a shale bed from porous sandstone or carbonate beds by means of a neutron–gamma-ray log. These different beds may all contain the same concentration of hydrogen. Other kinds of logs, such as the natural gamma-ray log, must be used to make these lithologic distinctions. The principal use of the neutron–gamma-ray log is to detect porosity variations within sandstone beds or carbonate beds, as illustrated in Figure 14–32. We can identify the sandstone and shale beds by means of the natural gamma-ray log. Then we can look at the neutron–gamma-ray log variation within an individual sandstone bed. The log in this example indicates the highest hydrogen content, and therefore the highest porosity, near the center of the bed.

The neutron–gamma-ray log can be run in cased or uncased wells. The best results are obtained with a skid mounting that holds the source and detector against the side of the well. Capture gamma radiation comes mostly from neutron capture at distances of less than 2 feet from the well. Typically, the sonde speed is about 30 feet per minute, and the counting time of the scintillation counter is set at 2 seconds. The scintillation counter is constructed to detect wavelengths of capture gamma radiation, which are different from the wavelengths of natural gamma rays.

SONIC LOGGING

The final logging procedure we will discuss involves a small-scale seismic refraction experiment. This experiment can be conducted with a source and two receivers mounted on a sonde (Figure 14–33a). In the standard design, the receivers are placed one foot apart with the source three feet from the nearest receiver. Sound pulses are emitted from the source at 0.1-second intervals. Because *P*-wave speed is faster in the formation than in the fluid filling the well, the wave refracted in the rock at the side of the well is the first arrival at the receivers. Each pulse activates a timing circuit that records the difference in the travel times to the two receivers. This value is called the *interval transit time* T_0. The interval transit times are plotted on a strip chart to obtain a sonic log.

The procedure we have described is not without problems. If the sonde is tilted in the

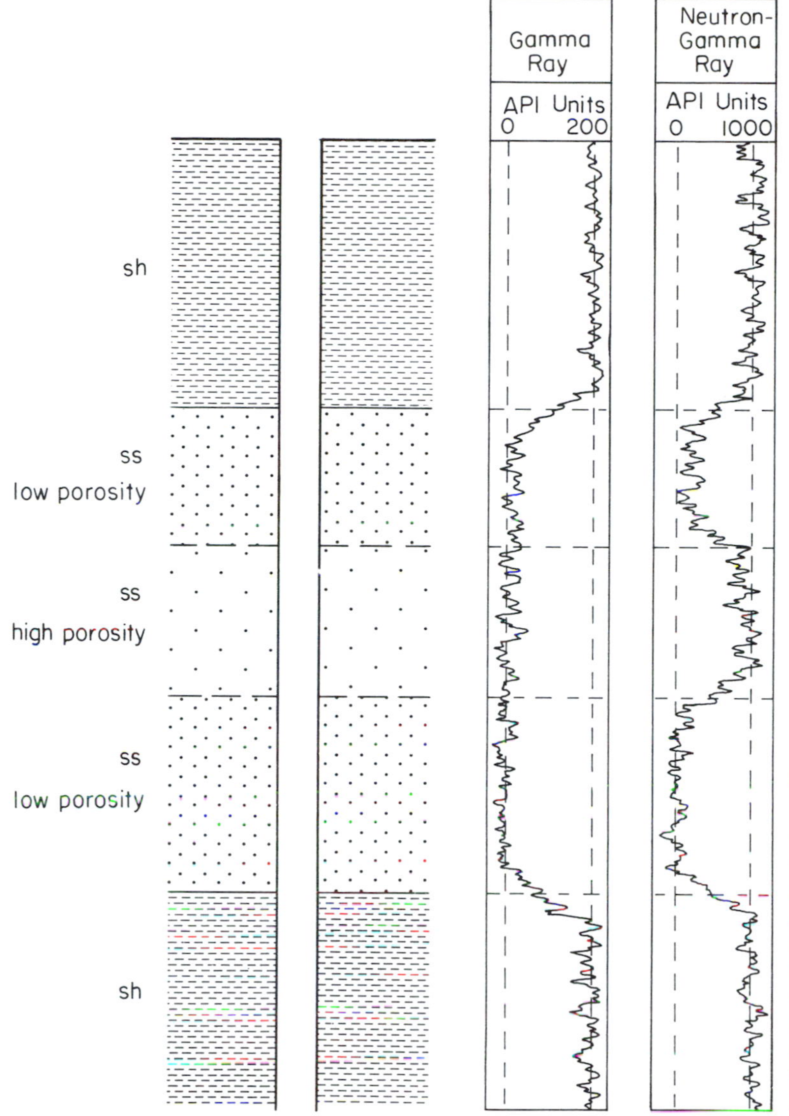

Figure 14–32
Idealized diagram showing the effect of lithology (shale and sandstone) and porosity on natural gamma radiation and neutron–gamma radiation logs. Natural gamma radiation depends on the concentration of radioactive elements in a rock unit, and the neutron–gamma radiation depends principally on the proportion of water-filled pore space in a rock unit.

well, or if the well diameter varies, travel paths through the fluid filling the well to the two receivers will have different lengths (Figure 14–33b). This problem is similar to that with dipping layer refraction (Figure 3-10), which means that an apparent interval transit time is being measured rather than the true value of T_0 in the formation. To overcome this difficulty, we can mount another sound pulse source on the sonde in a position opposite to

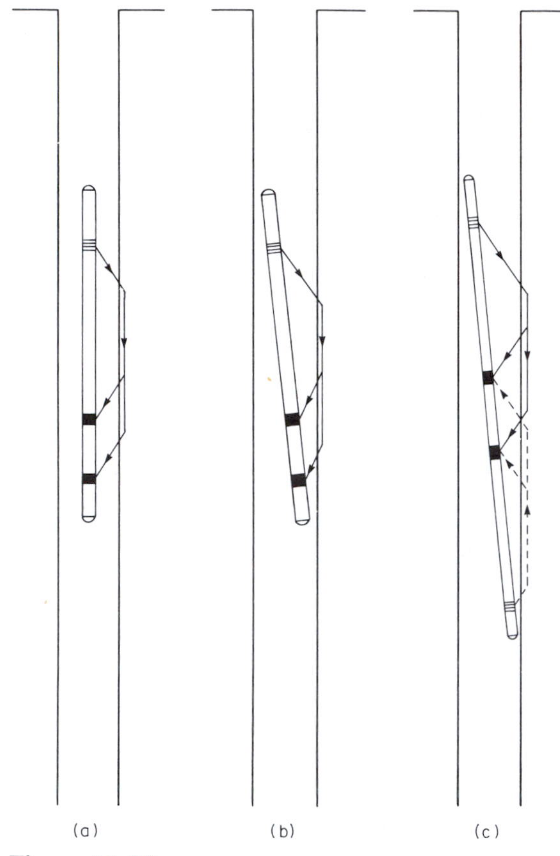

Figure 14–33
Schematic diagram of source–receiver configurations for (a) a nonborehole-compensated single-transmitter device aligned vertically and (b) tilted in the well, and (c) a two-transmitter borehole-compensated device which plots the average of the transit times measured from the two sources.

the original source (Figure 14–33c). Now we have a *borehole-compensated sonde.* Sound pulses are emitted alternately from the sources, and the apparent interval transit times from the two oppositely traveling refracted waves are averaged electronically to obtain the true value of T_0. This practice is similar to that of recording along reversed refraction profiles in order to obtain the apparent velocities needed to solve the dipping layer problem. The averaged interval transit times are plotted on a strip chart to obtain a borehole-compensated (BHC) sonic log.

The *P*-waves reaching the receivers are critically refracted at the side of the well. Therefore, the sonic log tests only the zone situated very close to the well. Interval transit times (T_0) read from a log depend on the proportions of solid matrix and fluid within this zone. These readings together with estimates of the transit times in the solid (T_m) and fluid (T_w) parts are used in Equation 14–4 to obtain values of formation porosity. The estimate of T_m is based on information about lithology obtained from drill cuttings and other kinds of logs. The values given in Table 14–1 are typical for the kinds of rock commonly encountered in oil and gas wells. The estimate of T_w requires some independent knowledge about the formation fluid, which might be obtained

TABLE 14–1 Sonic Transit Time and *P*-wave Speeds in Different Materials

MATERIAL	TRANSIT TIME (μs/foot)	*P*-WAVE SPEED (feet/s)
Sandstone	51–55.5	18,000–19,500
Limestone	43.5–47.6	21,000–23,000
Dolomite	43.5	23,000
Anhydrite	50.0	20,000
Salt	66.7	15,000
Shale	62.5–167	6,000–16,000
Water (pure)	218	4,600
Water (10% NaCl)	208	4,800
Oil	238	4,200
Methane	626	1,600

from electric logs. Typical values of T_w are also given in Table 14–1.

Sonic logs have proved very useful for correlation between nearby wells. Beds as thin as the spacing of the two receivers, commonly one foot, can be detected. Sonic logs are also very useful in the interpretation of seismic surveys of nearby areas. The variations in interval transit time indicate boundaries from which reflections might be expected. By summing the transit time through a continuous succession of intervals reaching to such a boundary, we can calculate the probable reflection travel time.

SUMMARY STATEMENT

Well logs are essential for discovering the nature of rock formations penetrated by a drill. The drill itself does not provide unambiguous information about these formations. Rock cuttings tell us what lithologies are present but are unclear about exactly where they occur. Even core drilling, which can be prohibitively expensive, yields incomplete information about formation fluids, and 100 percent core recovery is seldom possible. Therefore, we need an assortment of well logs for more complete evaluation of the formations.

Electric logs have proved the most useful for evaluating formation fluid properties. But the results from any individual log are distorted by the electrode configuration and by effects of invasion. Only in combination with other logs do they yield reliable information for estimating the salinity of pore water and the proportions of water and petroleum or natural gas. Because both petroleum and natural gas are nonconductors, they are difficult to distinguish from each other by means of electric logs. Their elastic properties, however, are very different, so that sonic logs are useful for making this distinction.

We cannot unambiguously identify different kinds of rock by any single property such as resistivity, density, radioactivity, or transit time. A much clearer picture emerges when these properties are considered altogether. Electric logs, gamma density logs, natural gamma logs, and sonic logs are analyzed together for this purpose.

Formation porosity can be estimated independently from electric logs, a gamma density log, a neutron–gamma log, and a sonic log. Nonetheless, each separate result is subject to error introduced by invasion, effects of the mud cake, and irregularity in the well diameter. To the extent that results from these different logs confirm one another, we can obtain reliable values.

For more than half a century, geophysical well logging has played a central role in the discovery and development of petroleum and natural gas resources. These methods have not been so fully used for mining exploration, in the search and development of freshwater resources, and as a part of engineering test drilling operations. Today modern well logging techniques are being used increasingly for these purposes.

STUDY EXERCISES

1. The cross section in Figure 14–34 shows a borehole that crosses the boundary between two rock layers. The dashed line is an equipotential surface related to the current source shown in the well. From the shape of this surface, determine which resistivity value is largest and which is smallest, where R_m is the resistivity of the drilling fluid filling the well, and R_1 and R_2 are resistivities in the upper and lower layers.

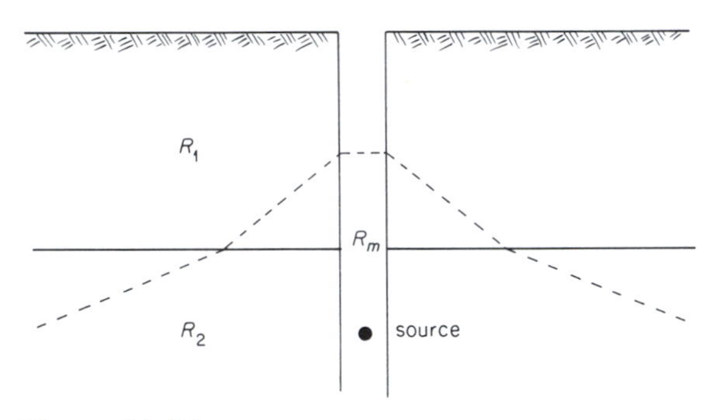

Figure 14–34
Cross section showing a borehole that penetrates two formations, a source of electric current, and an equipotential surface (dashed line).

2. Describe the conditions that cause an annulus of invasion to form in the rock surrounding a borehole. Sketch a graph showing the approximate variation in resistivity with distance away from the center of the hole, assuming that the resistivity of the drilling fluid is higher than the resistivity of the natural fluid filling the pores of the formation.

3. In a sandstone layer, suppose that the resistivity is 30 ohm-m, and that the resistivity of the fluid filling the pores is 6 ohm-m. What is the highest value for porosity that you would expect to calculate for this rock by means of the Archie formula?

4. In a layer of quartz sandstone, assume that the sonic transit time is 82 μs/foot. Using the value of 20,000 feet/s as the *P*-wave velocity in quartz, calculate the porosity of the sandstone.

5. A layer of relatively impermeable limestone 1 foot thick lies between two layers of permeable sandstone, each of which is 50 feet thick. In a well penetrating these layers, a 16-inch normal log, an 18-foot, 8-inch lateral log, and a laterolog were obtained. The central electrode of the laterologging sonde was 10 inches long. Resistivity in the limestone is higher than resistivity in the sandstone.

   a. Which log or logs indicate a higher value of resistivity in the limestone? Explain your answer.

   b. Which log will produce the most accurate value of limestone resistivity? Explain your answer.

   c. Which log or logs will produce the most accurate value of the sandstone resistivity? Explain your answer.

6. Suppose that a layer of sandstone is contaminated by clay mineral grains similar to those ordinarily found in shale. Both a gamma-ray log and a neutron–gamma-ray log from a well penetrating this layer show an increase in radiation with depth through the layer. Does this imply that porosity changes with depth or that clay contamination changes with depth? Does the property that is changing increase or decrease with depth? Explain your answers.

SELECTED READING

Archie, G. E., The electrical resistivity log as an aid in determining some reservoir characteristics, *Transactions of the AIME,* v. 146, pp. 54–62, 1942.

Asquith, George B., and Charles R. Gibson, *Basic Well Log Analysis for Geologists.* Tulsa Okla., American Association of Petroleum Geologists, 1982.

Labo, J., *A Practical Introduction to Borehole Geophysics.* Tulsa, Okla., Society of Exploration Geophysicists, 1986.

Lynch, Edward J., *Formation Evaluation.* New York, Harper and Row, 1962.

Snyder, Donald D., and David B. Fleming, Well logging—A 25 year perspective, *Geophysics,* v. 50, n. 12, pp. 2504–2529, December 1985.

INDEX